D1778107

Terrestrial and Shallow Marine Geology of the Bahamas and Bermuda

Edited by

H. Allen Curran
and
Brian White

Department of Geology
Smith College
Northampton, Massachusetts 01063

THE GEOLOGICAL SOCIETY OF AMERICA · 1888 ·

SPECIAL PAPER

300

1995

Published by The Geological Society of America, Inc.
3300 Penrose Place, P.O. Box 9140, Boulder, Colorado 80301

Printed in U.S.A.

GSA Books Science Editor Richard A. Hoppin

Library of Congress Cataloging-in-Publication Data

Terrestrial and shallow marine geology of the Bahamas and Bermuda /
edited by H. Allen Curran and Brian White.
p. cm. — (Special paper / Geological Society of America ;
300)
Includes bibliographical references (p. -) and index.
ISBN 0-8137-2300-0
1. Geology—Bahamas. 2. Geology—Bermuda Islands. 3. Geology, Stratigraphic—Quaternary. I. Curran, H. Allen (Harold Allen), 1940- . II. White, Brian, 1936- . III. Series: Special papers (Geological Society of America) ; 300.
QE221.T47 1995
557.296—dc20 95-25123
CIP

Cover: Holocene carbonate eolianites of the North Point Member, Rice Bay Formation (see Carew and Mylroie, chapter 1, this volume) crop out along North Point, San Salvador Island, Bahamas. Photograph by Allen Curran.

10 9 8 7 6 5 4 3 2 1

Contents

Holocene Geology

Modern Geologic Settings and Processes

PART II. BERMUDA

Quaternary Geologic History Overview

Pleistocene Sea-Level Changes, Formation of Soils, and Structure

Preface

OVERVIEW

The islands of the Bahama Archipelago and Bermuda long have been recognized as unique geologic sites for the study of modern carbonates. Indeed, these areas have spawned many of the classic geologic studies on the processes and structure of carbonate platforms. Somewhat surprisingly, however, the terrestrial geology of the islands largely has been neglected until recently. Nonetheless, these islands present a varied and well-exposed rock record of all carbonate sequences of shallow subtidal, beach, and eolian dune deposits.

Because the platforms of the Bahamas and Bermuda are essentially tectonically stable, their stratigraphic sequences reveal much about the history of global sea-level change during Quaternary time. Sea-level high-stands are represented by shallow subtidal (commonly reefal) facies and lowstands by terra rossa paleosols, caliche crusts, and karst features. The eolianite beds of different ages that largely cap the islands can be deposited under conditions of both sea regression and transgression. Interpretation of the different characteristics of the Bahamas and Bermuda rock units, their fossil faunas and floras, and the karst surface features of the islands can reveal much about Quaternary climates and rates of global climatic change.

In addition to presenting a near-pristine modern setting (less true of Bermuda), one of the great advantages of these islands is that studies of the modern carbonates system easily can be used to supplement and reinforce rock record investigations and vice versa. Analogues for the environments interpreted for the rocks and sediments of Pleistocene and Holocene age typically can be found not too far away from their outcrops, and studies of modern environments can benefit from concurrent examination of the ancient record. In this regard, the studies of modern shallow marine environments included in this volume can be used to gain a better understanding of the Quaternary rock record of the islands.

This volume reviews current knowledge of the geologic history of the present-day land areas of the Bahamas and Bermuda. In this regard, the lead article for each of the volume's two parts presents an up-to-date summary of the geologic history and models of stratigraphic development for the Bahamas and Bermuda, respectively. Although the following articles of each section cover a broad spectrum of subdisciplines, from paleontology to carbonate systems geochemistry, each article can be viewed as a case-book example for the study of an important aspect of carbonate island geology. In general, the studies of modern shallow marine and karst systems and processes included in the Bahamas section can be viewed as modern analogues for what we might expect to find preserved in the ancient record.

This volume is an outgrowth of "Terrestrial and Shallow Marine Geology, Bahamas and Bermuda," a symposium held in March 1991, at the joint GSA meeting of the Northeastern-Southeastern sections in Baltimore, Maryland, and sponsored by the Eastern Section of SEPM (Society for Sedimentary Geology). Sixteen of the 26 papers presented at the symposium are included in this volume. Several papers not presented at the symposium were added later to fill critical topical gaps.

DEDICATION

We dedicate this volume to Dr. Donald T. Gerace and Kathy Gerace, founders of the Bahamian Field Station on San Salvador Island, Bahamas. In 1971 the Geraces established a small field station at the site of an abandoned U.S. naval base on what was then a remote island. Their vision and pioneering efforts, with the support of the Bahamian government, have enabled many geologists to visit and conduct research in the Bahamas. One result is that San Salvador is now geologically the best known island in the Bahamas and probably one of the most studied carbonate islands in the entire world. In that sense, all scientists interested in carbonates geology owe thanks to the Geraces.

ACKNOWLEDGMENTS

We are most grateful to the many people who helped make this volume possible. First and foremost, we thank the authors and coauthors who contributed papers for the volume and who have cooperated fully, with patience and good humor, throughout the editorial and production process. Many colleagues graciously assisted in the formal reviews of the papers; they are acknowledged individually within the volume. Richard A. Hoppin, GSA books science editor, provided helpful guidance and advice throughout the project. Our special thanks go to Smith College students Susan Timmons for her invaluable assistance with copy editing and Jennifer Christiansen for her excellent drafting work.

H. Allen Curran and Brian White
Northampton, Massachusetts

Geological Society of America
Special Paper 300
1995

Introduction: Bahamas geology

H. Allen Curran and Brian White
Department of Geology, Smith College, Northampton, Massachusetts 01063

The Bahama Archipelago consists of an arcuate chain of carbonate platforms, commonly capped with low islands, separated by inter- or intraplatform basins and troughs that can reach depths greater than 4,000 m. Located to the east and south of the continental margin of North America, the archipelago extends for a distance of about 1,400 km (Fig. 1). Water depths on the platforms commonly are less than 10 m, with the subtidal parts of the banks largely covered by carbonate sands of varied origin.

Research on the modern sediments and sedimentary processes of these platforms and platform margins has generated a large body of geologic literature, as have geophysical and exploration drilling-based studies designed to answer questions about the origin of the Bahamas and the development of carbonate platforms. A review of this voluminous literature is beyond the scope of this short introduction, but much of the most important work about the origin of the Bahamas is cited and discussed by James Carew and John Mylroie (this volume). However, the geology of the islands themselves, the major subject of this volume, has been significantly less studied and reported in the literature. Indeed, many of the Bahamian islands, particularly in the southern part of the archipelago, have hardly been explored from the geologic perspective.

A typical Bahamian island consists of a sequence of late Pleistocene and Holocene carbonate rocks of subtidal, intertidal (beach), and eolian origin. Terra rossa paleosols, caliche crusts, and other karst features mark the disconformity between Pleistocene and Holocene units. Given that the Bahama platforms are tectonically stable (slow rate of subsidence), the major control for the development of sedimentary facies is eustatic change in sea level. However, deciphering the complex pattern of facies present and determining the timing of events that have led to the existence of the islands as they are today have not been easy tasks.

The geologically best-known island in the Bahamas today undoubtedly is San Salvador, owing to the presence of the Bahamian Field Station there and the research that this facility has fostered. In the lead chapter of this section, James Carew and John Mylroie present their summary of more than a decade of work on the stratigraphy of San Salvador and depositional models for the Quaternary development of Bahamian islands. A geologic map of San Salvador is presented (the only Bahamian island that has been mapped geologically). Carew and Mylroie maintain that their stratigraphic scheme and model for development can be extended to other islands in the Bahamas, and they give examples.

The five chapters immediately following all deal with topics on island Pleistocene geology. In an important work on paleosols, Mark Boardman, Richard McCartney, and Matthew Eaton report on the nature of Bahamian paleosols, including analyses of their mineralogy. The authors then discuss in some detail the paleoclimatic and stratigraphic significance of these paleosols. Volumetrically, fossil coral reefs do not form a high percentage of the Pleistocene rocks of Bahamian islands, but they have great significance for determining the sea-level history of the islands. The Sangamonian-age Devil's Point reef on remote Great Inagua Island in the southern Bahamas is the subject of the chapter by Brian White and Allen Curran. They describe in some detail the typical shallow subtidal subfacies associated with a Bahamian reef complex and then demonstrate that the reef was preserved by rapid entombment during glacioeustatic sea-level regression.

The next three chapters address aspects of the Pleistocene geology of San Salvador Island, but the findings clearly can be extended to other Bahamian islands and possibly beyond. Mario Caputo's chapter presents a detailed analysis of the sedimentary architecture of eolian calcarenites that is important owing to the predominance of this facies in the rocks that form the Bahamas and Bermuda. Mollusc-dominated shell beds commonly are a distinctive component of lagoonal and lacustrine facies, and the chapters by Frances Hagey and John Mylroie and by Robinson Noble, Allen Curran, and Mark Wilson investigate the paleoenvironmental significance of such shell beds. Specifically, Hagey and Mylroie show how molluscan assemblages can be used to differentiate Pleistocene lacustrine versus lagoonal facies, and Noble and other analyze differences in lagoonal molluscan assemblages and suggest models for their deposition. Both studies draw on analogies with modern molluscan shell assemblages in their paleoenvironmental interpretations.

Bahamian Holocene geology is the topic of a group of three chapters, beginning with Pascal Kindler's study of Holocene stratigraphy on Lee Stocking Island in the Exuma

Curran, H. A., and White, B., 1995, Introduction: Bahamas geology, *in* Curran, H. A., and White, B., Terrestrial and Shallow Marine Geology of the Bahamas and Bermuda: Boulder, Colorado, Geological Society of America Special Paper 300.

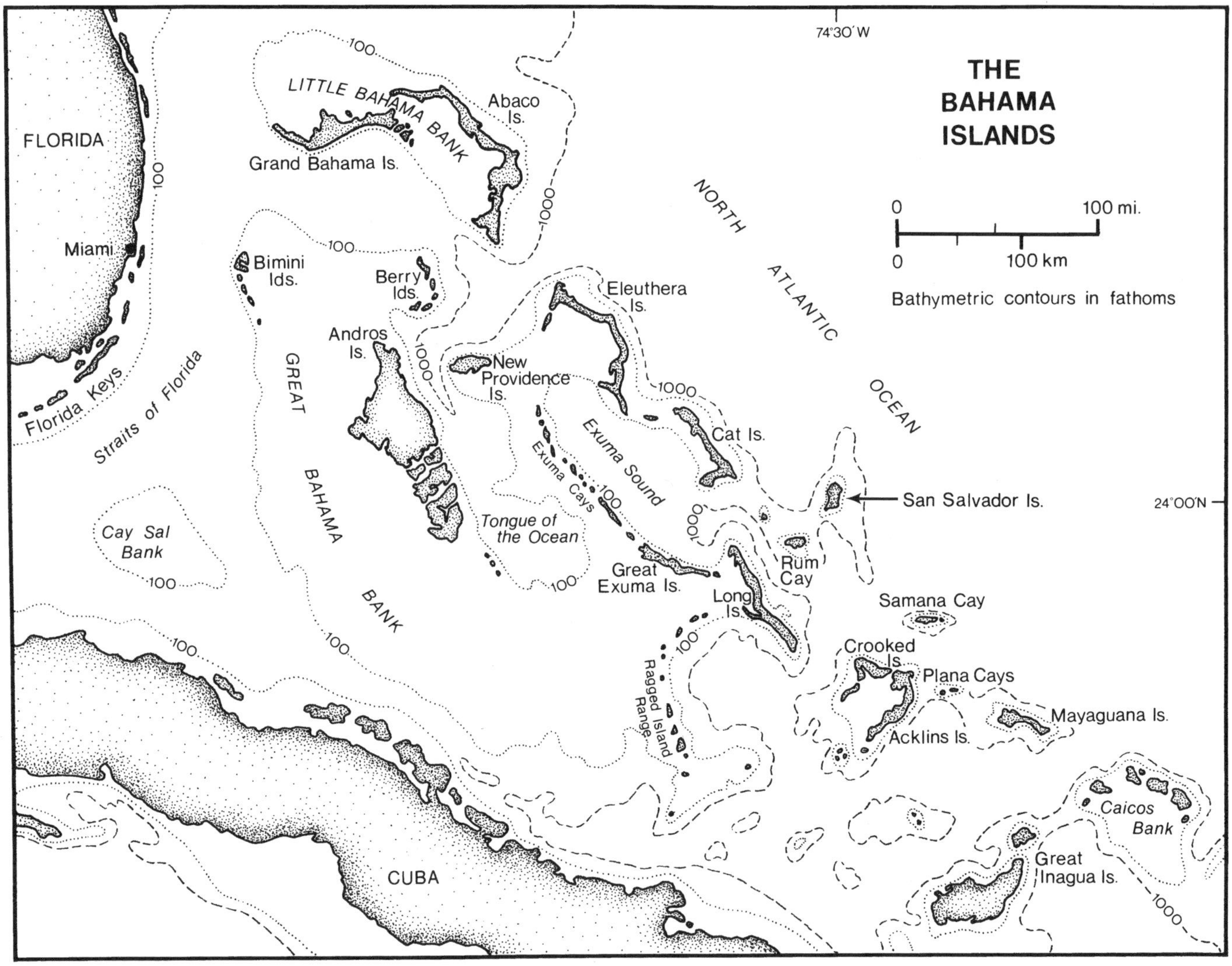

Figure 1. Index map to the islands of the Bahama Archipelago. The Turk Islands and Silver and Navidad Banks lie farther to the southeast.

Cays. Kindler carefully establishes a stratigraphy for the Holocene deposits on this small island and then presents a model to show how sea-level rise, particularly with respect to changes in rate of rise, has acted as a control on sedimentation. Saline lakes and ponds are common features on Bahamian islands, and sedimentation in such lakes can generate distinctive facies, as also discussed in the Hagey and Mylroie chapter. James Teeter's chapter analyzes Holocene history as revealed from cores taken from several saline lakes on San Salvador Island. Ostracode assemblages are defined and analyzed to reveal salinity changes that have occurred in these lakes during mid- to late Holocene time, and Teeter discusses possible causes for the salinity changes. Is mid- to late Holocene sea-level transgression recorded in the cements of the carbonate rocks being transgressed? This is the question addressed by Kathleen White in her field-based petrologic study of the imprint of transgression on Quaternary eolianites on San Salvador. White finds that an imprint of marine aragonite cement is present in the rocks, and she discusses the diagenetic history and broader implications. Again, the models presented in all three of these studies likely are applicable throughout the Bahamas and beyond.

The next group of eight chapters consists of studies of modern geologic phenomena and processes that characterize Bahamian settings and that can be used as analogues for interpretation of the ancient record. One of the most exciting natural history discoveries made in the Bahamas during the last decade was the finding of giant, subtidal stromatolites in the tidal channels of the Lee Stocking Island area by Robert Dill in the mid-1980s. The chapter by Russell Shapiro, Robert Dill, Kenneth Aalto, and Ray Kenny describes in detail the stratigraphic setting of a subtidal stromatolite field in channels near Iguana Cay in the Exumas and presents an interpretation of the late Pleistocene, Holocene, and modern facies sequence. The next two chapters are both studies of sedimentologic processes operative in the large Graham's Harbour lagoon off the north coast of San Salvador Island. Anthony Randazzo and Kathy Baisley report on the sediment facies and patterns of distribution in a part of the lagoon, and they analyze the controls on these patterns, particularly with respect to the effects of the benthic communities present. Benjamin Greenstein's chapter presents a taphonomic analysis of life habit and test microstructure on the preservation potential of echinoid hard parts in the lagoon's sediments. Both studies have clear implications for the interpretation of rocks and fossils from the ancient carbonate record.

A carbonate-geochemistry theme unites the following group of five chapters. In a pair of chapters, Steven Mitchell and Robert Horton investigate dolomitization in modern Bahamian settings. The first describes in detail the varied characteristics of dolomitization occurring in tidal creek sediments on New Providence Island; the second reports on the style of dolomitization found in tidal flat, tidal creek, and lacustrine sediments throughout the Bahamas. Broader implications are obvious and are discussed. As noted previously, paleosols are an integral and significant element of Bahamian (and Bermudan) stratigraphy. The chapter by Annabelle Foos and Roger Bain investigates the mineralogic, chemical, and petrographic characteristics of the modern soils found on San Salvador and Eleuthera Islands and discusses implications for interpretation of paleosols. Susan Gaffey, Victor Zabielski, and Charles Bronnimann have investigated the processes of early alteration of aragonite in coral skeletons. Their work describes in detail how these alteration processes operate and discusses how their findings compare with previous alteration models.

The final chapter of this section forms a link between the geology of the Bahamas and Bermuda. John Mylroie, James Carew, and Len Vacher compare and contrast the development of karst topography in the two areas and discuss reasons for the observed differences. This work provides a good basis from which to consider the characteristics of Bermudan geology.

Manuscript Accepted by the Society January 5, 1995

Printed in U.S.A.

Geological Society of America
Special Paper 300
1995

Depositional model and stratigraphy for the Quaternary geology of the Bahama Islands

James L. Carew
Department of Geology, University of Charleston, Charleston, South Carolina 29424
John E. Mylroie
Department of Geosciences, Mississippi State University, Mississippi State, Mississippi 39762

ABSTRACT

Surficial deposits of the Bahama islands consist of late Quaternary limestones that were deposited during glacioeustatic highstands of sea level. Other than marine deposits formed during the last interglacial highstand (Sangamon) associated with deep-sea oxygen isotope substage 5e (ca. 125,000 yr ago), only the eolian deposits of previous highstands can be seen above present sea level. A complete Bahamian glacioeustatic highstand depositional package consists of transgressive-phase eolianites and beach facies; eolian, beach, and subtidal deposits (including reefs and shoals) largely deposited during the still-stand phase; and regressive-phase deposits (mostly eolianites) that were laid down as sea level dropped and marine water retreated from the bank tops. Because of the patchiness of deposition, these depositional facies cannot often be seen to lie directly atop one another. Successive highstand deposits may be separated from one another by intervening paleosols that were developed largely during lowstands of sea level, but because of the patchy deposition some paleosols represent more than one highstand/lowstand sequence. Based on physical stratigraphic relationships, the rocks exposed on the islands of the Bahamas have been divided into three allostratigraphic units that are currently accorded formation rank: the Owl's Hole Formation and Grotto Beach Formation, which are Pleistocene, and the Rice Bay formation, which is Holocene.

The rocks of the Owl's Hole Formation usually consist of fossiliferous pelsparites (grainstones) that were deposited as eolian dunes during a highstand(s) associated with deep-sea oxygen isotope stage 7 and/or earlier stages. Those deposits are typically capped by a terra-rossa paleosol, or are truncated by a nearly flat wave-cut erosional surface. The younger (sometimes overlying) Grotto Beach Formation comprises a variable suite of subtidal through eolian dune facies formed during the sea-level highstand associated with oxygen isotope substage 5e. At the peak of that highstand, sea level is estimated to have been about 4 to 6 m higher than it is at present. The eolianites of this unit are largely oosparites (grainstones) at elevations above the limit of beach processes at the time of deposition, but regressive-phase eolianites contain markedly fewer ooids than transgressive-phase eolianites. That package of sediment is capped by another terra-rossa paleosol. The youngest stratigraphic unit comprises a suite of Holocene transgressive-phase eolianites, together with eolianites and beach deposits formed in equilibrium with current sea level. The eolianites of this unit are weakly cemented fossiliferous pelsparites (grainstones) that

Carew, J. L., and Mylroie, J. E., 1995, Depositional model and stratigraphy for the Quaternary geology of the Bahama Islands, *in* Curran, H. A., and White, B., Terrestrial and Shallow Marine Geology of the Bahamas and Bermuda: Boulder, Colorado, Geological Society of America Special Paper 300.

may contain a variable amount of superficially coated grains, particularly in the transgressive-phase deposits. Although there are Holocene subtidal deposits, at present, they are not included in the stratigraphy of the existing islands because they are not exposed above sea level. In the future, those deposits, along with regressive-phase deposits, will be part of the Rice Bay Formation. All, or portions, of this tripartite stratigraphy and tripartite intraformational depositional history is recorded by the rocks exposed on the islands of the Bahamas.

INTRODUCTION

The Bahamas are considered to be one of the classic carbonate depositional environments in the world today, and they have often been used as a model for the interpretation of ancient carbonate rocks. However, since the early report by Shattuck and Miller (1905), the surficial geology of the islands of the Bahamas has not been investigated in detail, until recently. Among the vast literature on the Bahamas, only the work of Garrett and Gould (1984) and Carew et al. (1992b) on New Providence; Titus (1980, 1983a, 1984, 1987), Carew and Mylroie (1985, 1987, 1989), and Hearty and Kindler (1993) on San Salvador; Wilber (1987) on Little San Salvador and West Plana Cay; Mitchell (1987) on Rum Cay; Mitchell et al. (1989) on Conception Island; and Carew and Mylroie (1989) on South Andros has dealt with the geology and depositional history of Bahamian islands. This chapter provides a depositional paradigm for the Bahamian islands, presents a workable physical stratigraphy for Bahamian surficial geology, cites specific examples that document aspects of the stratigraphy, and discusses potential pitfalls in stratigraphic and depositional interpretation.

The Bahamian Archipelago comprises a series of carbonate islands and shallow banks, with intervening deep channels and reentrants, that extends 1,400 km from the Florida peninsula in the northwest to the Greater Antilles in the southeast (Fig. 1, Chap. 1). In their discussions, some workers have divided the Bahamas into a northwest section and a southeast section (Uchupi et al., 1971; Mullins and Hine, 1989). The northwest Bahamas consist of islands that occur on the extensive Great Bahama and Little Bahama banks; whereas the southeast Bahamas consist of islands on numerous smaller banks, some of which are not appreciably larger than the islands that occupy them.

Bahamian islands are predominately low lying (<30 m elevation); the topography is dominated by eolianite ridges that extend up to 30 m on most islands, and extend as high as 60 m on Cat Island (Fig. 1, Chapt. 1). All land above 7 m is of eolian origin, but the rocks between present sea level and 7 m in elevation are a facies mosaic of marine, lacustrine, and terrestrial limestones. Pleistocene rocks are covered with an indurated red micritic calcrete, or terra-rossa paleosol, except where it has been removed by erosion. Superimposed on the dune and swale topography of the islands is a well-developed karst that includes flank margin caves, pit caves, banana holes, and blue holes (see Mylroie et al., this volume). Surface streams are absent. Interior lakes and ponds range from fresh to hypersaline, but the majority are of marine salinity or greater. Ground water occurs as fresh or brackish water lenses that are buoyant on underlying

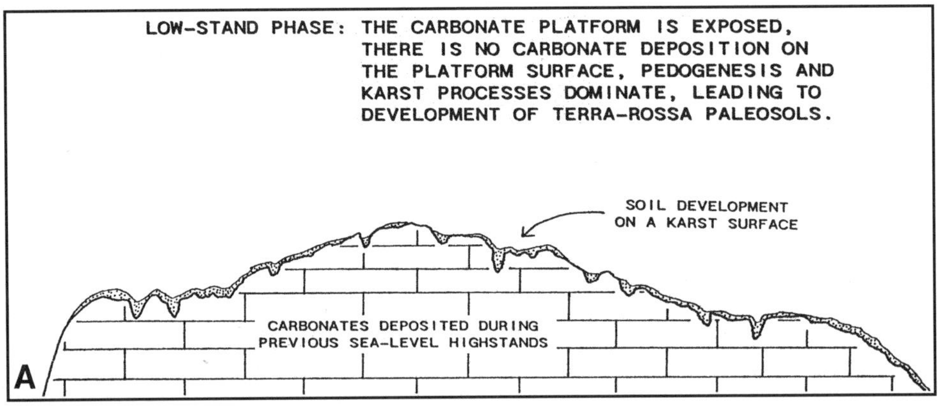

Figure 1. Illustration of the four stages of depositional development of Bahamian islands during each glacial/interglacial sea-level fluctuation (on this and facing page). During highstands, the islands are the highest portions of the steep-walled platforms with quasi-flat tops that are not inundated; during lowstands (below –10 m), the entire platforms are the islands. A, Low-stand phase: sea level >10 m below present sea level. B, Transgressive phase: sea level is above –10 m and the platform tops are being progressively inundated by the sea as it rises to its acme. C, Still-stand phase: sea level hovers around its maximum elevation (usually for 10,000 to 15,000 yr). D, Regressive phase: sea level falls and eventually descends below the platform top.

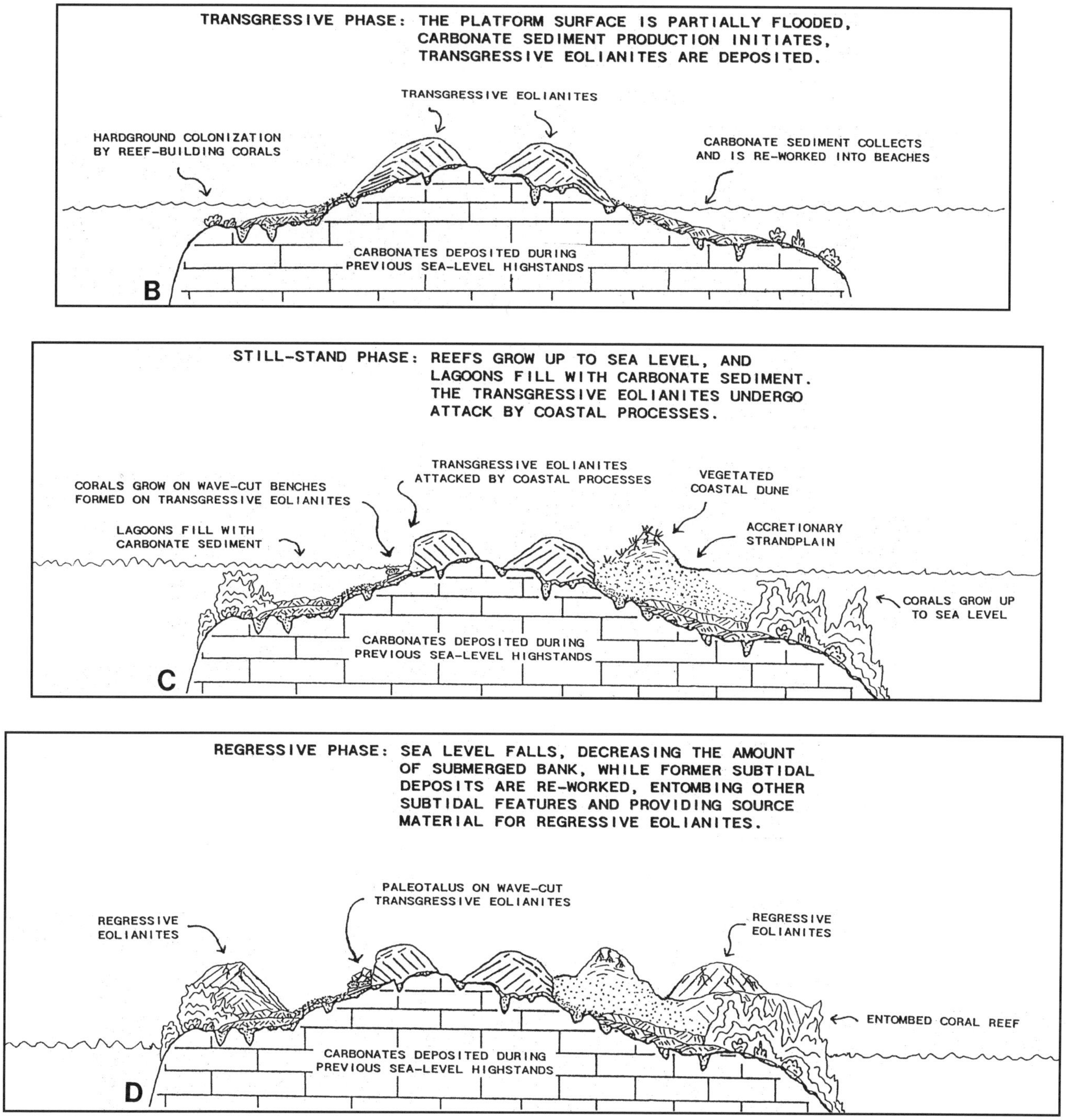

marine water. The position of interior water bodies and groundwater lenses is tied to sea-level position.

GEOLOGIC SETTING OF THE BAHAMAS

Tectonism

The development of the Bahamas began with continental rifting in the Mesozoic that eventually led to the creation of the Atlantic Ocean (Dietz et al., 1970; Meyerhoff and Hatten, 1974; Mullins and Lynts, 1977). Since that time, net carbonate deposition on the banks has kept pace with isostatic subsidence. The result is a series of banks composed almost entirely of shallow-water carbonates, and intervening deep channels or basins (Fig. 1, Chapt. 1). The history of the banks and basins has been the subject of much debate. Mullins and Lynts (1977) proposed that the current topography of shallow banks and linear embayments was inherited from horst and graben structure produced during the early stages of Atlantic rifting. Alternatively, Austin and Schlager (1987) proposed that the current topography reflects the erosional modification of a former single large bank, or megabank. On the other hand, Eberli and Ginsburg (1987) demonstrated that deep channels in the Great Bahama Bank had filled with carbonate sediment, whereas Mullins and Hine (1989) and Mullins et al. (1992) have shown that other bank margins have undergone erosional retreat. Regardless of the mechanism by which the Bahamas have achieved their current configuration, they are generally considered to be tectonically stable, and appear to have been isostatically subsiding at a rate of 1 to 2 m per 100,000 yr (Mullins and Lynts, 1977). Recently, Mullins et al. (1992) have shown that the southeastern Bahamas have been tectonically downwarped as a result of activity along the Caribbean plate margin, but data from the Quaternary geology of the Bahamas indicates that there has not been vertical tectonic motion anywhere in the Bahamas during at least the last several hundred thousand years (Carew and Mylroie, 1995a).

Figure 2. Photo of the north coast of Isla de Mona, Puerto Rico. The 90-m near-vertical cliff of this tectonically uplifted carbonate island illustrates what the Bahama islands looked like during glacioeustatic sea-level lowstands.

Glacioeustasy

Quaternary glacioeustatic sea-level changes have been the major factor in the development of the geology of the existing Bahama islands. The portion of the banks that has been exposed as islands has varied significantly as sea level has changed due to glacioeustasy. During continental glaciation, sea level was low, the Bahamian banks were completely exposed, and the islands comprised entire platforms with near-vertical sides and quasi-flat tops (Figs. 1A, 2). In the northern Bahamas there were two large islands, now the Little and Great Bahama banks. In the southern Bahamas there were some islands that were appreciably larger than the present islands (e.g., Crooked Island and Aklins Island bank), but most experienced only minor increase in size (e.g., San Salvador, Mayaguana, and Great Inagua islands) (Fig. 1, Chapt. 1). During ice minima (interglacials) on the continents, the platforms were flooded, more or less as they are today. During the last interglacial, sea level was about 6 m higher than it is today, and only the preexisting high ridges would have remained exposed as numerous small islands. Throughout the late Quaternary, the glacioeustatic sea-level lows lasted for about 100,000 yr, while the intervening sea-level highs averaged only 10,000 yr in duration (Fig. 3). Considering the $\delta^{18}O$ value for stage 1 (Fig. 3), and that a change of 0.1 ‰ in $\delta^{18}O$ is assumed to equal a 10m change in sea level (Fairbanks and Matthews, 1978), it is clear that during the late Quaternary the Bahamian platforms have been emergent about 10 times longer than they have been submerged.

Subaerial weathering and erosive processes (pedogenesis and karstification) occur at all times on emergent portions of carbonate banks. Carbonate deposition occurs only at times when portions of the banks are submerged. At any time when sea level was >10 m below its present level, the steep-sided platforms were completely exposed, and shallow-marine carbonate sedimentation virtually ceased. During those intervals, the development of the platforms was dominated by weather-

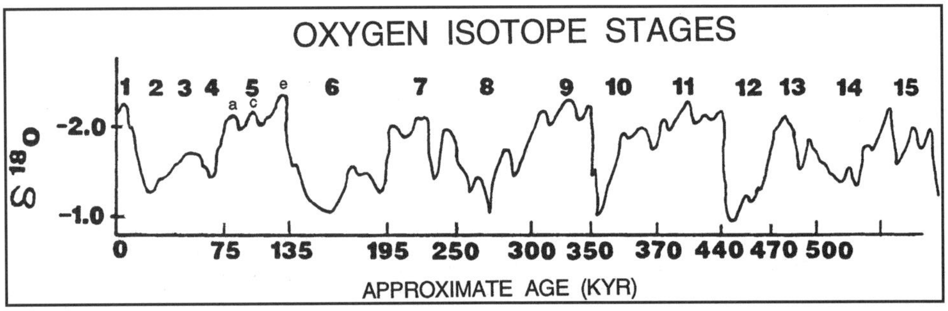

Figure 3. Graph of the deep-sea oxygen isotope curve for the past half million years. Odd-numbered stages are interglacials (sea-level highstands); even-numbered stages are glacials (sea-level lowstands). Stage 1 is the present highstand; stage 2 is the Wisconsin lowstand estimated to be –125 m; the stage 5 (substage 5e) sea-level highstand is estimated to have been at ca. +6 m. (Modified from Shackleton and Opdyke, 1973.) A change of 0.1 ‰ in $\delta^{18}O$ equals 10 m sea-level change (Fairbanks and Matthews, 1978).

ing and erosive processes. At times when sea level rose above the top (–10 m) of the near-vertical walls and flooded the low-lying areas of the relatively flat-topped platforms, abundant shallow-water carbonate sediment production began (Boardman et al., 1987), and those marine sediments were reworked into beaches deposited on the emergent portions of the platforms (Fig. 1B–D). The beaches became the source area for eolian reworking that produced dunes. As sea level high enough to flood the tops of carbonate platforms is a prerequisite for significant marine, coastal, and eolian limestone deposition, stratigraphic units of the Bahamas should be identifiable as sedimentary packages deposited during sea-level highs, that are bounded by terra-rossa paleosols and erosion surfaces produced during sea-level lows.

During the late Quaternary, the Bahamas have been tectonically stable, and isostatically subsiding at a rate of 1 to 2 m per 100,000 yr (Mullins and Lynts, 1977; Garrett and Gould, 1984; Carew and Mylroie, 1985, 1995a). Under those conditions, marine subtidal deposits could not be exposed on Bahamian islands today, unless they were deposited during a sea-level highstand above that of modern sea level. In addition, because of dissolution and isostatic subsidence since the time of deposition, only if those deposits were laid down in the relatively recent past would they remain above sea level today. Considering the generally accepted chronology and elevation of sea level derived from the deep-sea oxygen isotope record (e.g., Shackleton and Opdyke, 1973; Shackleton, 1987) (Fig. 3), only the sea-level highstand of the last interglacial (Sangamon), associated with oxygen isotope substage 5e (ca. 125,000 yr ago), meets those criteria. Therefore, all marine rocks currently exposed on Bahamian islands must have been deposited during the oxygen isotope substage 5e highstand. Geochronologic analyses from the Bahamas substantiate that conclusion (Neumann and Moore, 1975; Carew and Mylroie, 1995a). Like the marine deposits, ancient lacustrine deposits exposed on the islands today must have been deposited during the last interglacial highstand, because the elevation of inland lakes in the Bahamas is governed by the position of sea level (Hagey, 1991; Hagey and Mylroie, this volume).

On the other hand, eolianites escape the constraints mentioned above because they may reach more than 30 m above the sea level at their time of origin. Thus, it is possible that eolianites that formed as long ago as the Middle Pleistocene could still be exposed above sea level today. However, because those eolianites were exposed above all past Quaternary sea-level highstands, they have been exposed to unrelenting erosion since they formed, unless subsequently buried by younger deposits (see Fig. 4). Therefore, it is most likely that the oldest exposed eolianites are less than 500,000 yr old.

DEPOSITIONAL MODEL

It was generally thought that depositional packages in the Quaternary geologic record of carbonate islands were formed largely during regression from a glacioeustatic sea-level highstand (Fairbridge, 1968). That interpretation held that exposed lagoonal sediments were reworked into regressive sequences during sea-level fall (e.g., Titus, 1980, 1983b), or during the still-stand and regressive phases (Garrett and Gould, 1984). Through our extensive field work on San Salvador Island and elsewhere in the Bahamas, we have come to recognize that significant eolianites on San Salvador and other Bahamian islands

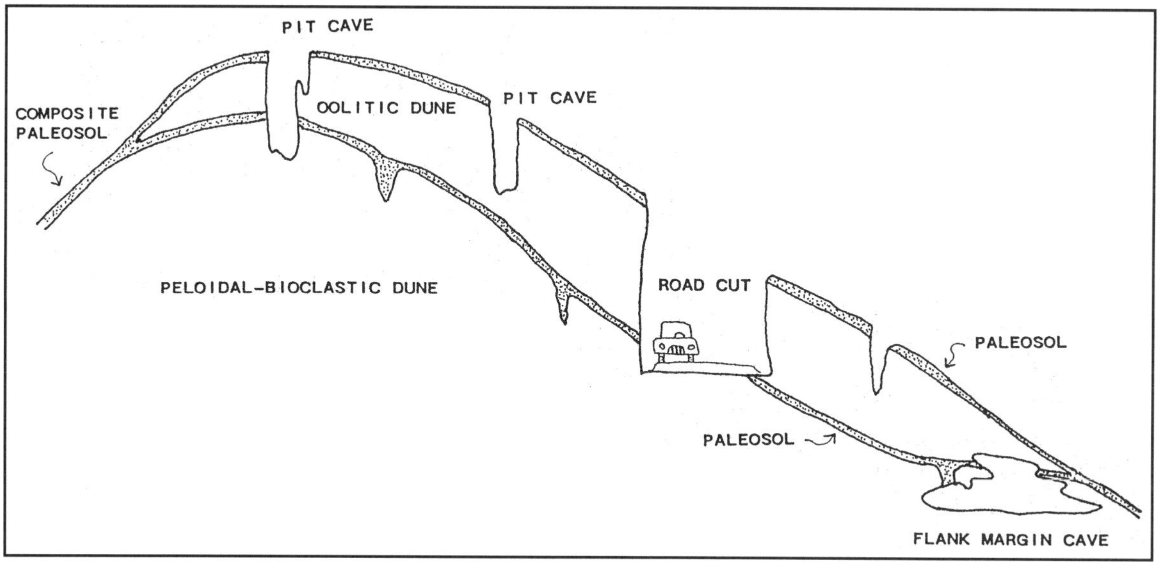

Figure 4. Figure illustrating the potential onlap/overlap conditions between eolianites deposited during separate sea-level highstands (e.g., stages 5e and 7) and terra-rossa paleosols that accumulate largely when sea level is below the platform tops; and the various views of these relationships afforded by flank margin caves, pit caves, and roadcuts or cliff exposures.

are Holocene rather than Pleistocene (Hutto and Carew, 1984; Carew and Mylroie, 1985). Many of those eolianites have foreset beds that extend at least 2 m below present sea level. Given that the Bahamas are tectonically stable, and that the regressive phase of the Holocene highstand has not yet begun, then some of the Holocene eolianites must have been formed during the transgressive phase. Late Pleistocene rocks of the Bahamas also contain a record of transgressive-phase eolianites.

Today, it is generally recognized that a typical Quaternary Bahamian depositional cycle begins with a sea-level rise in response to continental deglaciation. That transgression begins to affect the depositional record when the steep platform margins are overtopped, and the tops of the platforms begin to flood (Fig. 1B). Marine sediments produced early in the platform flooding are not visible above current sea level, but fossil coral reefs and other marine subtidal deposits formed during the acme of oxygen isotope substage 5e are exposed on today's islands. Terrestrial deposits, eolianites, and protosols form during all phases of the highstand, but notable eolianite development occurs during the transgressive phase and again during the regressive phase; and preservation of protosols is most common for those developed during the still-stand and regressive phases. Each depositional cycle ends when sea level retreats from the platform tops. Thereafter, dissolutional erosion and pedogenic processes dominate the geologic development of the islands until the next glacioeustatic highstand that floods the platform tops.

Paleosols

Soil development and maturation occurs during all times that portions of the carbonate banks are exposed, but development of terra-rossa paleosols requires long-term exposure that largely occurs during the long duration of late Quaternary sea-level lowstands. See the section below on lowstand phase and Carew and Mylroie (1991) for further discussion of terra-rossa paleosols.

Calcarenite protosols. Calcarenite protosols are composed of particulate organic detritus and wind-blown sediment captured by the baffling effect of vegetation (Bretz, 1960; Vacher, 1973; Carew and Mylroie, 1991). Protosols can develop during any time when the banktops are at least partially flooded. Because transgressive-phase eolian deposits are continually reworked while sea level continues to rise, protosols are unlikely to be preserved within transgressive-phase eolianites. Protosols commonly accumulate on transgressive-phase eolianites during the still stand, and as such may form the boundary between a transgressive-phase eolianite and ones that accumulate during the still-stand or regressive phase. Protosols are much more common within and between still-stand–phase and regressive-phase eolianites. They represent accumulation during lulls in deposition of the larger depositional packages within which they are contained.

These whitish, unstructured, and poorly indurated buried soils commonly contain abundant fossil pulmonate snails (mostly *Cerion* in the Bahamas). Protosols are generally concordant with the underlying depositional unit, which is usually a backbeach or dune deposit, but occasionally they are seen to lie directly on an underlying marine unit, such as the protosol directly above the reef facies at The Gulf on San Salvador Island (Figs. 5, 6) (Carew and Mylroie, 1985). Extensive protosol development can be seen in roadcuts on Lyford Cay on New Providence Island (Garrett and Gould, 1984; Carew et al., 1992b) (Fig. 7). Recognition of calcarenite protosols, and their differentiation from terra-rossa paleosols, is important for the correct interpretation of stratigraphy and depositional history of Bahamian carbonate sequences (Carew and Mylroie, 1991; Carew et al., 1992b).

Transgressive phase

Transgressive-phase dunes begin to develop as soon as the platform top is flooded by rising sea level (Boardman et al., 1987). Subtidal sediments are transported by waves to beaches along the shoreline, and eolian processes winnow material from the beaches and pile it up into dunes (Fig. 1B). Carbonate dunes do not develop far from, or migrate away from, their beach sources (Bretz, 1960; Carew and Mylroie, 1985). As sea level continues to rise, shoreline processes remobilize beach and dune sediment, and transport it farther landward. As long as sea level continues to rise this activity continues, and the subtidal platform—the ultimate sediment source—increases in area. The beaches and dunes are composed of newly produced allochems plus allochems reworked from deposits (particularly eolianites) formed earlier in that highstand. It is noteworthy, however, that during our study of hundreds of thin sections of Bahamian carbonates (e.g., Hutto and Carew, 1984; Schwabe et al., 1993), it was rare to encounter allochems identifiable as grains form rocks deposited during earlier highstands.

In essence, shoreline processes, driven inland by rising sea level, "bulldoze" large amounts of sediment into high dune ridges that are nucleated on and extend laterally from high grounds remaining from previous highstand deposits (Carew, 1983); this style of development has been termed "catenary" by Garrett and Gould (1984). The transgressive-phase eolian deposits largely develop before the sea-level rise slows down as it approaches a maximum, and before reefs become well established and reduce the amount of wave energy that reaches the shoreline (Boardman et al., 1987).

Bedding features that are well represented in demonstrably transgressive-phase Holocene eolianites and their Pleistocene counterparts include sand flow laminae, grain-fall laminae, and climbing wind-ripple lamination (White and Curran, 1985, 1988; Caputo, 1989, 1993). Such laminae are unlikely to form or be preserved in well-vegetated dunes, because the development of such laminae requires clear windward slopes and lee slip faces. This suggests that the large transgressive-phase eolian ridges of the Bahama islands may have formed rapidly, perhaps largely by storm event(s) that occurred during the later

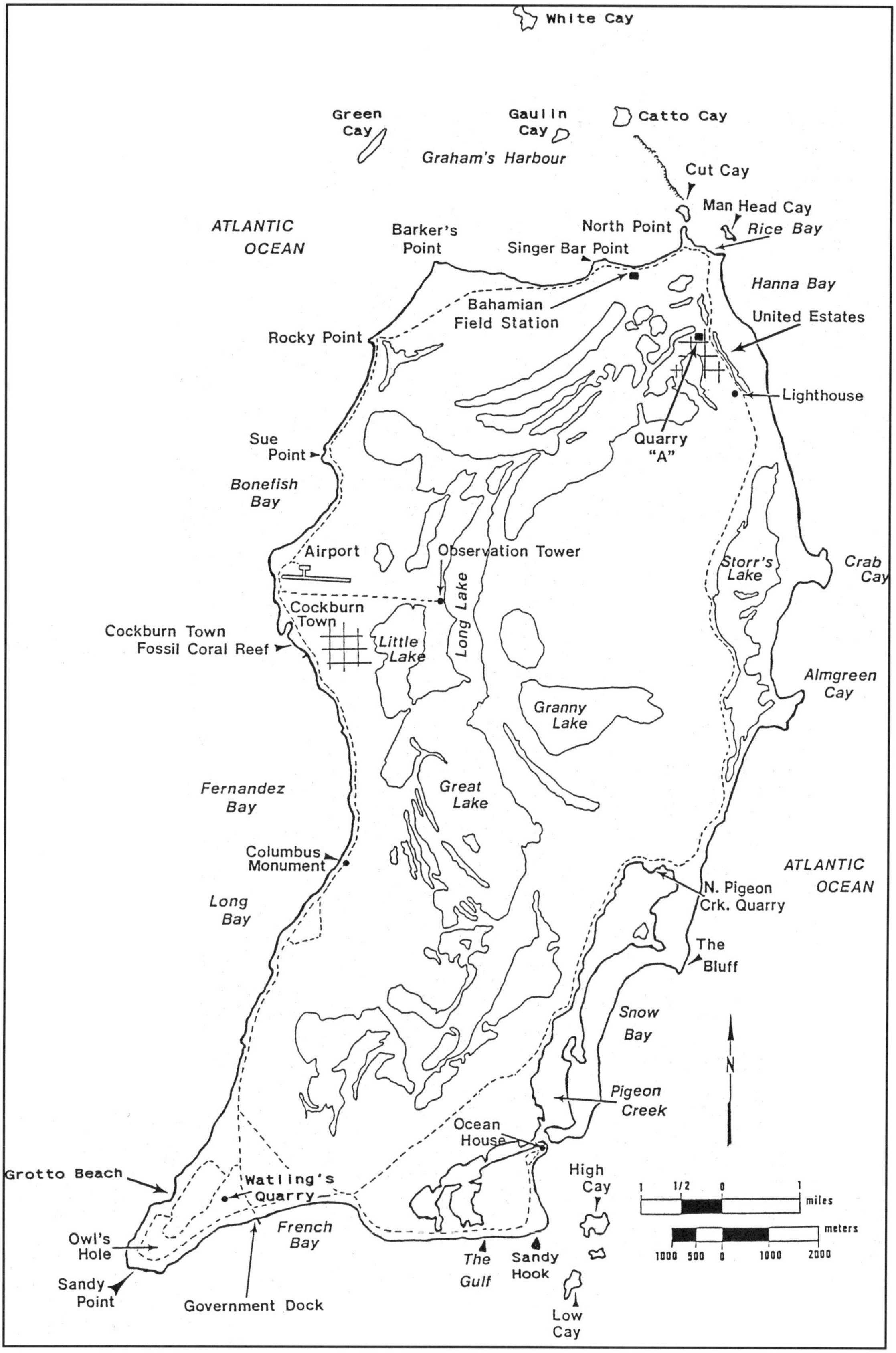

Figure 5. Map of San Salvador Island, Bahamas, showing locations discussed in the text.

stages of the transgression. The presence, in Holocene transgressive-phase eolianites, of cluster burrows attributed to the escape of hatchlings from the brood nests of burrowing wasps attests to the loose nature of at least several meters of sand at the time the insects burrowed their way out (Curran and White, 1991) Among the transgressive-phase dunes, only those formed late in a sea-level rise are likely to survive the destructive effects of marine erosion during that highstand. Because those dunes lie in close proximity to the shoreline for the duration of the highstand, they are subjected to the combined effects of sea

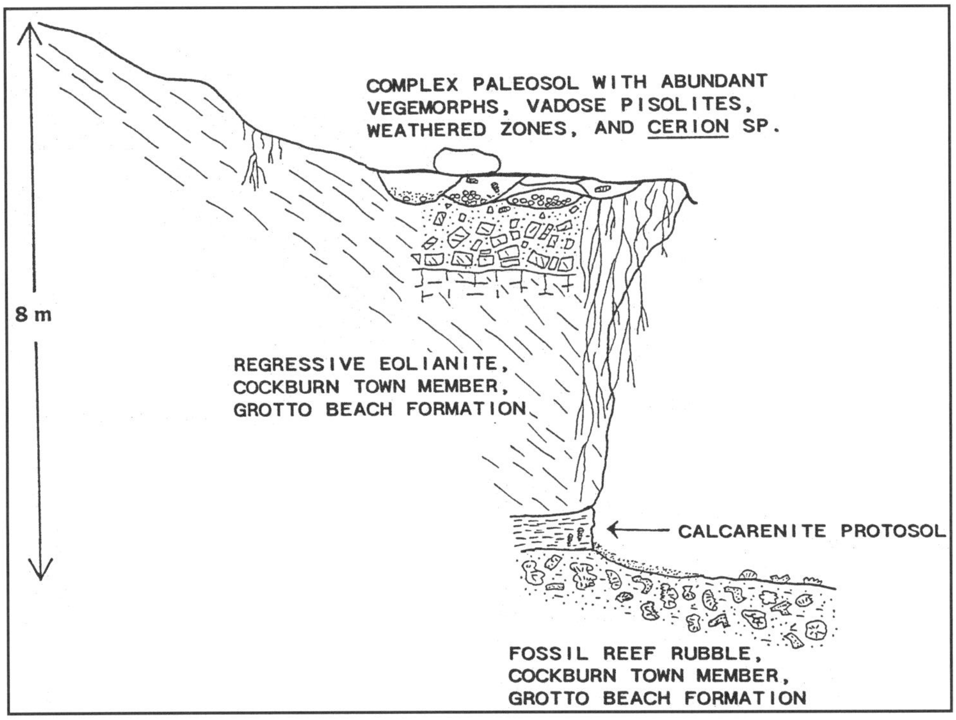

Figure 6. Diagram illustrating the facies relationships seen at The Gulf on San Salvador Island.

Figure 7. Photo of roadcut exposure of Grotto Beach Formation eolianites at Lyford Cay, New Providence Island. This outcrop is capped by a terra-rossa paleosol; the exposure contains several calcarenite protosols that separate individual accumulations of eolianite that were deposited during a single sea-level highstand. A thick, wedge-shaped protosol can be seen immediately above the head of the figure in the left-center foreground. There is a lower terra-rossa paleosol below the road level at this location, which crops out at the western end of these road-cut exposures.

spray and meteoric precipitation, which promote rapid cementation and development of a thin calcrete layer on the dune exteriors. That calcrete inhibits vegetative root development that would otherwise disrupt and obscure the bedding. When sea level falls, these transgressive-phase dunes are already well lithified, so long-term preservation of the fine internal bedding features is likely.

Transgressive-phase dunes are very mobile because of persistent effects of shoreline encroachment, and that probably leads to relatively sparsely vegetated dunes. In addition, plant colonization of transgressive-phase dunes is slow because, during the preceding 100,000+ yr of lowstand conditions, plant taxa adapted to loose and mobile sand bodies would have largely disappeared throughout the Bahamas; hence, colonization would require recruitment from the North American mainland or Caribbean islands that do not have steep bank margins. So, transgressive-phase eolianites have limited development of plant trace fossils, or vegemorphs. We prefer the term vegemorphs for these trace fossils, rather than rhizomorph, rhizolith, or rhizocretion (all of which are common in the carbonate literature), because it more accurately reflects the true character of their origins. They represent not just roots or rhizomes, but also buried vegetation of a variety of types.

With continued rise of sea level during the transgression, transgressive-phase eolianites deposited earlier in the transgression will be subject to wave erosion throughout the remainder of the transgressive and still-stand phases. Those transgressive-phase eolianites that are massive enough, and/or favorably positioned, will survive the wave erosion, but they are likely to develop bioerosion notches, sea caves, wave-cut benches, be substantially eroded to form sea cliffs (particularly on their seaward sides), and may be wholly or partially buried by still-stand and regressive-phase eolianites. Subsequent to wave erosion, corals may become established on submerged portions of wave-cut benches; sea caves may collect rounded boulder rubble deposits, and may be entombed by later still-stand and regressive-phase calcarenites; sea cliffs can become inland scarps by progradation in front of them during the still-stand and regressive phases, and definitely by retreat of sea level below the platform margin. All of these features will be modified by pedogenesis during the ensuing low-stand phase.

Occurrence of the eolianite deposition and the wave erosion on the same highstand may be detected by the lack of paleosol development between the transgressive-phase eolianites and later depositional features (such as corals on a wave-cut bench, boulder rubble on the floor of a sea cave, regressive-phase eolianite) (Fig. 8). Truncated eolianite bedding covered by a terra-rossa paleosol/calcrete indicates either significant wave erosion during the highstand during which the eolianite was formed, thereby indicating a transgressive-phase eolianite; or deposition during one highstand, erosion on a subsequent highstand, and paleosol development during the ensuing lowstand.

Sea cliffs that ultimately become inland scarps are likely to collect talus deposits in reentrants along the cliff base. That talus will be infiltrated by soil and eventually be lithified into a resistant terra-rossa paleosol/paleotalus. Such paleotalus deposits provide evidence for sequences of events as described above. Because eolianites of transgressive, still-stand, and regressive phases can all be subjected to wave erosion that may form benches, sea cliffs, and truncated bedding during subse-

Figure 8. Photos of modern corals growing on the wave-cut bench on the southeast coast of the transgressive-phase (North Point Member, Rice Bay Formation) Holocene eolianites of High Cay, San Salvador Island. A, View from the top of the eroded cliff of the wave-cut bench with corals. B, View of coral-encrusted bench at water level showing truncated eolianite in the background. Circular colonies of *Acropora palmata* in the foreground are nearly 4 m in diameter.

quent highstands of sea level; it may often be that only by proper recognition of paleosols and related exposure surfaces, and their relationships to the deposits in question, that eolianites that are definitely transgressive-phase deposits may be recognized. Many pitfalls exist in the interpretation of paleosols (Carew and Mylroie, 1991).

Without doubt, some Holocene eolianites of the Bahamas are transgressive-phase deposits associated with the oxygen isotope stage 1 (current) highstand of sea level (Fig. 3). Today, at White Cay, Catto Cay, Cut Cay, North Point, High Cay, and along Snow Bay on San Salvador Island (Fig. 5), Holocene eolianites form sea cliffs up to 20 m high. These sea cliffs and their associated sea caves are modern evidence of the features discussed above for transgressive-phase eolianites. For example, at Snow Bay, coastal accretion has isolated a former Holocene sea cliff in an inland setting. At the base of that cliff line, a soil and talus block deposit is now collecting (Fig. 9). Furthermore, at North Point and High Cay, the continued rise of sea level after deposition of those transgressive-phase eolianites has resulted in the erosion of wave-cut benches into the Holocene eolianites, and there are corals growing on the subtidal portions of those eroded surfaces (Fig. 8). Direct analogues to these Holocene relationships can be seen in Pleistocene rocks. See Table 1 for a summary of characteristics associated with deposits of this phase of deposition.

TABLE 1. PHASES OF DEPOSITION AND DIAGNOSTIC CHARACTERS

Transgressive Phase	Still-stand Phase	Regressive Phase
Fine-scale eolian bedding	Disrupted eolian bedding	Disrupted eolian bedding
Few vegemorphs	Abundant vegemorphs	Extensive vegemorphs
Penecontemporary cliffing and boulder paleotalus	Penecontemporary notching of beach and intertidal facies, and beach-face breccia facies	Lack of penecontemporary wave erosion
Penecontemporary sea caves	Rare sea caves	Lack of sea caves
Corals on wave-eroded benches	No corals on eroded benches	No penecontemporary benches
Lack of protosols	Protosols common	Protosols common
On lapped still-stand or regressive-phase deposits	Marine facies abundant	Commonly peleoidal/bioclastic
Predominantly eolianites, marine deposits rare	Ebb-tidal delta, lacustrine, and strand plain deposits	Eolianites overstepping marine deposits

Still-stand phase

During the time when a sea-level highstand remains relatively stable in elevation, carbonate sediment production remains high. Reef growth catches up and lagoons tend to fill because the total wave energy that impinges on the shore of an island decreases compared to that experienced earlier in the transgressive phase, and quieter conditions behind reefs and ridges enhance lagoon filling (Boardman et al., 1987, 1989; Rasmussen and Neumann, 1987; Andersen and Boardman, 1989; Colby and Boardman, 1989). In addition, strandplains and beaches may develop, enlarge, and prograde into the subtidal marine environment, and entomb subtidal deposits (Anderson and Boardman, 1989). Many still-stand-phase progradational deposits may be indistinguishable from regressive-phase deposits, but some (particularly those deposited early in the still

Figure 9. View to the northwest of truncated transgressive-phase eolianite (North Point Member, Rice Bay Formation) with accumulated cliff line talus along Snow Bay, San Salvador Island.

stand) may be recognized because they become semi-lithified, and then cliffed as a result of changes in coastal dynamics. This is happening to some exposures of Holocene still-stand rocks (Hanna Bay Member) at many locations in the Bahamas. Often a shoreface breccia-block deposits that is entombed in nearly contemporary calcarenite will develop at such locales; and similar deposits can be seen in Pleistocene rocks. Both Holocene and Pleistocene examples can be seen at Grotto Beach on San Salvador Island (Fig. 5) (Carew and Mylroie, 1985).

During the still-stand phase, heavily vegetated coastal dunes will develop in some places, and protosols will accumulate on transgressive-phase eolianites and in other locales (see the discussion of protosols, above); ebb-tidal delta deposits may form and prograde at the mouths of tidal creeks, such as the Holocene ebb-tidal delta at the mouth of Pigeon Creek on San Salvador Island (Fig. 5) (see Andersen and Boardman, 1989); and inland water bodies may occupy interdune swales that are of constructional origin, and only slightly modified by subsequent karst processes (Mylroie et al., this volume). Today, on San Salvador Island, the water of such lakes ranges from fresh to hypersaline (commonly >80 ppt), and many of the lakes contain molluscan assemblages (see Hagey, 1991; Hagey and Mylroie, this volume). See Table 1 for a summary of characteristics associated with deposits of this phase of deposition.

Regressive phase

As sea level falls in response to renewed continental glacial advance, beaches and their associated facies retreat toward the platform margin. These regressive-phase beach and dune deposits bury portions of the marine deposits largely formed during the still-stand phase. The shallow subtidal area decreases, which reduces the area of new sediment production; however, sediment (including reefs) that had accumulated in the subtidal environment may be remobilized as the zone of shoreline processes retreats through them. Peloidal and bioclastic allochems would be expected to be important constituents produced where shoreline processes "chew up" reefs and other subtidal deposits. Some of that sediment may be reworked into significant dunes that are likely to accumulate on, or seaward of, the existing vegetated coastal dunes developed late in the still-stand phase. These regressive-phase dunes may bury subtidal deposits that survive the passage through the retreating coastal zone.

Internal bedding of regressive-phase dunes is likely to be disrupted by abundant vegetation, both surface and buried. Several dune ridges are likely to form, probably associated with storms, as sea level falls and the shoreline backs off to the platform margin. Because these dunes are vegetated, it is common for protosols (see the earlier discussion of protosols, above) to develop, often between major dune-building events, as is seen on the headlands of the east coast of San Salvador Island (Fig. 10). Other locales that contain notable exposures with numerous protosols, which indicate deposition during the still-stand or regressive phases, are the walls of the Queen's Staircase in Nassau and the roadcut cliffs in Lyford Cay on New Providence Island (Fig. 7).

Regressive-phase dunes are abandoned by both their sediment source and the coastal processes that enhance cementation. The result is well-vegetated dunes that tend to lack fine-scale bedding features, especially in the upper several meters, where buried vegetation provides preferred paths for descending meteoric water that promotes the development of spectacular, vertically extensive vegemorphs (Fig. 11). Also, because these dunes were vegetated throughout their history, it is common to find abundant populations of fossil pulmonate snails (*Cerion* sp.) associated with these paleosols with extensive vegemorphs (Fig. 11). Regressive-phase eolianites are often seen to overlie or overstep subtidal deposits, commonly reefs, such as at The Gulf (Fig. 6) and on the south side of Crab Cay on San Salvador Island. These regressive-phase eolianites do not undergo wave attack as do transgressive-phase eolianites, but they may be subjected to wave erosion during any succeeding sea-level high-stand. See Table 1 for a summary of characteristics associated with deposits of this phase of deposition.

Low-stand phase

The deep-sea oxygen isotope record is assumed to accurately reflect glacial ice volume and sea-level elevation during the Quaternary. That record indicates that the Bahamian platforms were exposed for most of that time (Fig. 3). During exposure, the primary processes affecting the islands/platforms involve dissolution and pedogenesis, which produce a distinctive geologic signature of karst features (see Mylroie et al., this volume), and terra-rossa paleosols. While sea level is below the

Figure 10. Photo of Cockburn Town Member (Grotto Beach Formation) eolianites showing a nearly horizontal pisolitic calcarenite protosol that forms a slight bench midway up the cliff face at The Bluff, San Salvador Island.

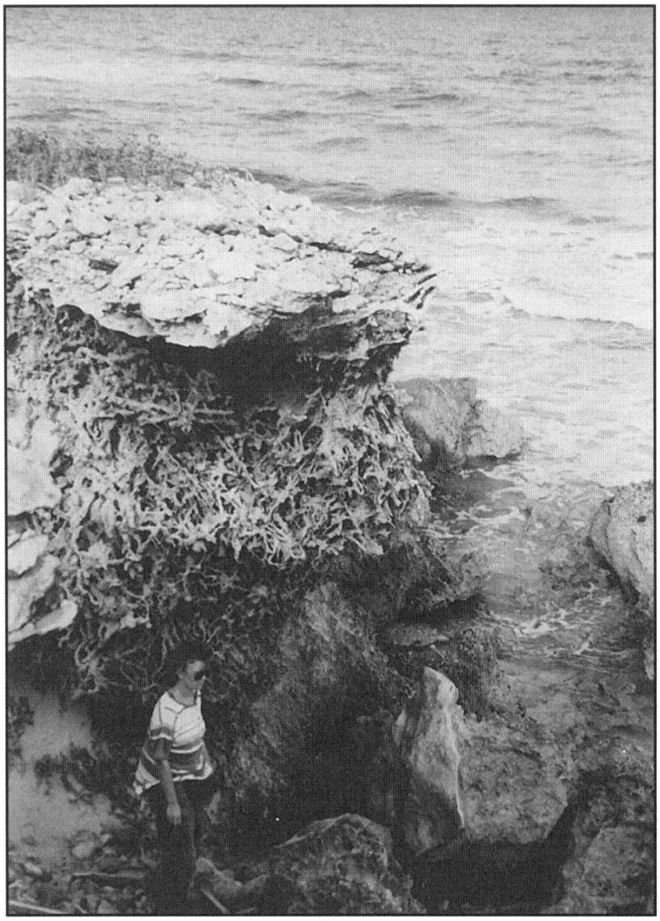

Figure 11. Photo showing massive vegemorphs developed below the terra-rossa paleosol that caps an exposure of Cockburn Town Member (Grotto Beach Formation) eolianites at Almgreen Cay, San Salvador Island. Such spectacular vegemorphs are usually associated with regressive-phase eolianites.

platform margin (–10 m), no significant new sediment is added to the material deposited during the highstand, although atmospheric dust does accumulate. The soils that develop may be represented by simple micritic calcrete layers, or they may be quite complex (Carew and Mylroie, 1991). Those paleosols that developed during prolonged lowstands of sea level mark the boundaries between highstand depositional packages. However, as deposition during highstands, especially of eolianites, is patchy, in some areas eolianites are deposited over a soil that elsewhere remains exposed (Fig. 4). During that highstand and the ensuing sea-level lowstand, pedogenesis continues on both the exposed soil and new eolianites. As a result, paleosols separated by an eolianite at one locality may be merged into one paleosol at another locality. This relationship can be seen on the west end of Watling's Quarry on San Salvador Island (Fig. 5). So, such composite paleosols can represent development during more than one glacial-interglacial cycle (> two oxygen isotope stages).

STRATIGRAPHY

The stratigraphy discussed in this chapter is derived largely from study of the rocks on San Salvador Island, Bahamas, but we have also gathered supporting data from Bimini, Eleuthera, Great Exuma, Great Inagua, Long, New Providence, North Andros, Rum Cay, and South Andros Islands (Fig. 1, Chapt. 1). This physical stratigraphy is applicable throughout the Bahamas.

The physical stratigraphy of the late Quaternary of the Bahamian islands is tripartite. As each of the depositional packages is bounded by an unconformity that represents the long duration (~100,000 yr) of low sea level, they are allostratigraphic units (North American Commission on Stratigraphic Nomenclature, 1983). The oldest exposed surficial rocks in the Bahamas are eolianites capped by a terra-rossa paleosol. The sea-level highstand(s) during which these eolianites were deposited has not been conclusively established, but based on presumed isostatic subsidence rates, and the glacioeustatic sea-level curve for the late Quaternary (Fig. 3), they most likely were deposited during the highstands of oxygen isotope stages 7 (ca. 220,000 yr ago), 9 (ca. 320,000 yr ago), or 11 (ca. 410,000 yr ago). Overlying those oldest eolianites there is a depositional package of transgressive-phase eolian deposits, terrestrial and marine still-stand deposits, and regressive-phase beach and eolian deposits that have been demonstrated to have formed during oxygen isotope substage 5e (ca. 132,000 to 119,000 yr ago) (Chen et al., 1991). The substage5e sedimentary succession is packaged between two terra-rossa paleosols. At many locations throughout the Bahamas, less complete depositional suites have been preserved from that highstand, and in some cases only the eolianites can be seen. Resting on the terra-rossa paleosol that caps the suite deposited during the substage 5e highstand is an uppermost depositional package that is a suite of transgressive-phase eolianites and still-stand-phase marine and eolian deposits, which has been demonstrated to be of Holocene age (Carew and Mylroie, 1985, 1987). As this stratigraphy was initially developed through detailed study of the geology of San Salvador Island, the stratigraphic names are from locales there, and all type locations are on San Salvador. Figure 12 illustrates the physical stratigraphy of the Bahama islands, as modified from Carew and Mylroie (1985).

Owl's Hole Formation

The demonstrably oldest rocks on San Salvador Island are exposed at three well-known locations in the southwest corner of the island: Owl's Hole, Grotto Beach, and Watling's Quarry (Fig. 5). These oldest rocks are assigned to the Owl's Hole Formation. Rocks definitely assignable to the Owl's Hole Formation are eolianites capped by a terra-rossa paleosol that can be shown to be overlain by either an oolitic eolianite that is itself capped by a terra-rossa paleosol, or by subtidal deposits. The type section for this formation is in a large dissolution pit cave,

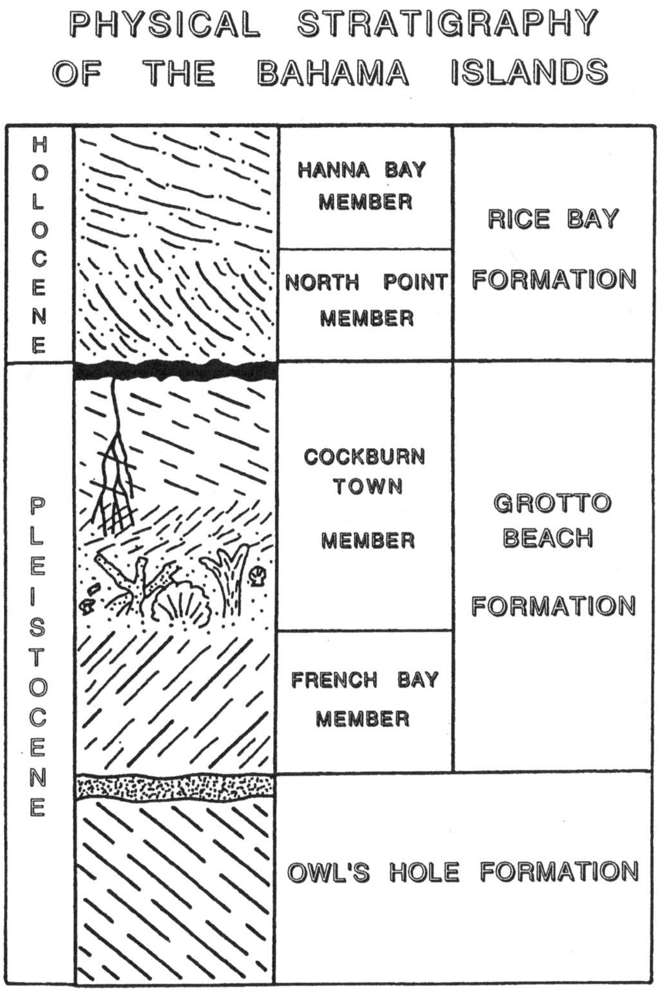

Figure 12. Physical stratigraphic column illustrating the temporal relationship between defined stratigraphic units. In the field, individual units are not necessarily seen stacked atop one another, but are often found lateral to one another. The thin stippled and black layers are terra-rossa paleosols separating deposits formed during separate glacioeustatic sea-level highstands.

named Owl's Hole for its sometimes-resident owl, high on the ridge behind Sandy Point, San Salvador Island (Fig. 5) (Carew and Mylroie, 1985).

From detailed study of the wall rock of many caves in the eolianite ridges of San Salvador, and the outcrop exposures of the ridges themselves, it has been shown that Owl's Hole rocks underlie many of the large Pleistocene eolianite ridges (Carew et al., 1992a; Schwabe et al., 1993). Throughout the Bahamas, nearly all rocks definitely assignable to this unit are peloidal/bioclastic eolianites (fossiliferous pelsparites or fossiliferous peloidal grainstones) (Stowers et al., 1989), but oolitic rocks are also known from this unit on New Providence Island (Schwabe et al., 1993).

At the three outcrops on San Salvador noted above, Owl's Hole rocks exhibit extensive micritization. At the upland localities (Owl's Hole and Watling's Quarry), rocks of the Owl's Hole Formation remain relatively weakly cemented; the top of the unit is a hard, red, micritic terra-rossa paleosol, which is overlain by a younger oolitic eolianite. The Owl's Hole rocks exposed in Owl's Hole exhibit extensive development of vegemorphs beneath the paleosol, which is a characteristic of regressive-phase eolianites. Elsewhere, Owl's Hole rocks do not exhibit features that permit determination of their phase of deposition.

At Grotto Beach, the top of the Owl's Hole Formation is a planar surface on which fossil corals and subtidal deposits of the younger Grotto Beach Formation rest. Rocks assigned to the Owl's Hole Formation at Grotto Beach exhibit alveolar texture in samples taken close to that planar surface, which suggests that a paleosol was present prior to planation. In addition, the underlying eolianite is a peloidal biosparite, whereas even the marine facies of the overlying rocks contain appreciable ooids (Fig. 13A). This petrologic relationship is the same as that seen at the Watling's Quarry and Owl's Hole localities where an intervening terra-rossa paleosol is present (Stowers et al., 1989). The presence of an intervening terra-rossa paleosol indicates an extended period of exposure of the Owl's Hole rocks, such as occurs between glacioeustatic sea-level highstands associated with different oxygen isotope stages. As the overlying eolianites at Watling's Quarry and Owl's Hole are oolitic, and at Grotto Beach the overlying unit is a fossil coral reef facies, the lower unit (Owl's Hole) must be older than the last interglacial (substage 5e).

We have identified rocks definitely assigned to this unit on Eleuthera Island, Long Island, and New Providence Island. Some of the eolianite ridges in the interior of San Salvador, as well as other Bahamian islands, may also belong to this unit, but that usually cannot be demonstrated from outcrop relationships. We believe that rocks assignable to the Owl's Hole Formation exist on most Bahamian islands, but are often mantled by younger deposits, primarily those deposited during the substage 5e highstand. Except for fortuitous coastal exposures and the wall rock of dissolution pit caves and flank margin caves (see Mylroie et al., this volume), only roadcuts and quarries are likely to expose these older rocks. In coastal settings, marine erosion can strip paleosols from older rocks and produce relationships such as that seen at Grotto Beach on San Salvador Island. In those cases, detailed petrologic evidence is needed to avoid misinterpretation.

Paleomagnetic analyses of the paleosols that cap the Owl's Hole rocks at Watling's Quarry and Owl's Hole indicate that they contain an anomalous southeast and shallow magnetic component. The first interpretation of these data was that the signature was that of a magnetic reversal, which suggested that Owl's Hole rocks are at least 780,000 yr old, the age of the Brunhes/Matuyama chron boundary (Mylroie et al., 1985). Subsequent resampling and analyses of these outcrops showed that the paleosol that caps Owl's Hole rocks carries a different paleomagnetic signature than all other paleosols tested from the Bahamas, but

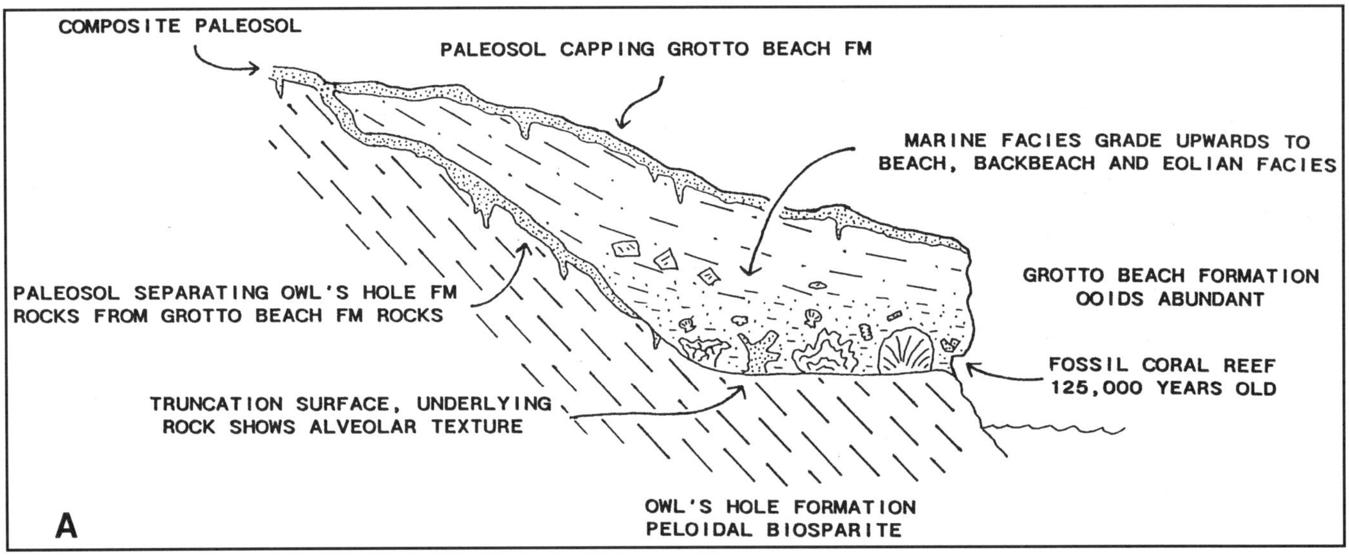

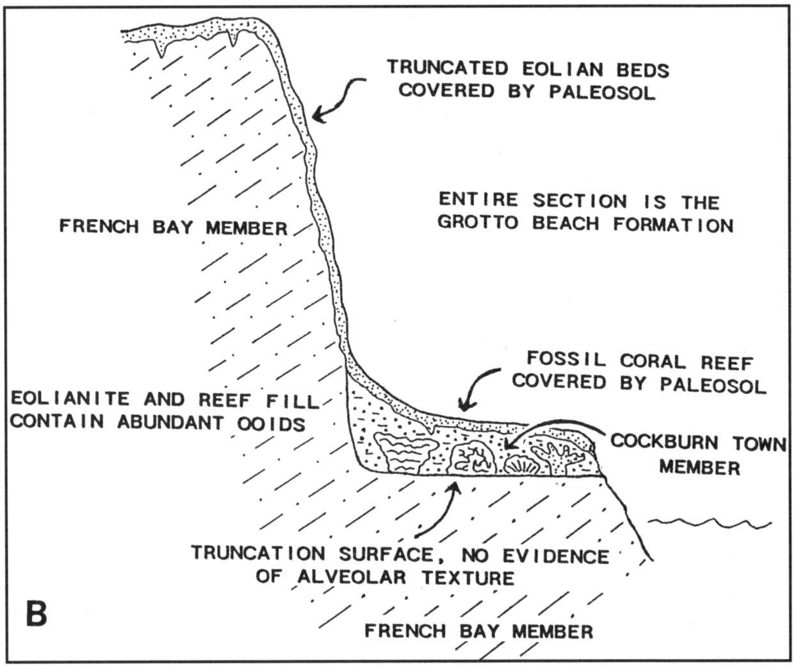

Figure 13. Diagram illustrating the different temporal relationships that may occur where fossil reef deposits are seen to overlie a truncated eolianite. A, Diagram showing composite stratigraphic relationships at Grotto Beach, San Salvador Island. B, Diagram showing stratigraphic relationships at High Cay, South Andros Island.

the data are equivocal, and a true magnetic reversal, while possible, cannot be demonstrated (Panuska et al, 1991). As the Owl's Hole rocks are definitely older than oxygen isotope substage 5e, for reasons discussed earlier it is most likely that they were deposited during stage 7 (ca. 220,000 yr ago), stage 9 (ca. 320,000 yr ago), or stage 11 (ca. 410,000 yr ago).

Grotto Beach Formation

Overlying rocks of the Owl's Hole Formation, and generally separated from them by a terra-rossa paleosol/calcrete, is the Grotto Beach Formation (Fig. 12). The Grotto Beach Formation comprises eolianites and beach-face to subtidal marine limestones that can be subdivided, in places, into members. The formation is capped by a terra-rossa paleosol, except where the paleosol has been stripped by later erosion. Throughout the Bahamas, transgressive-phase and still-stand–phase eolianites of the Grotto Beach Formation contain abundant (up to 80 to 90% of the allochems), relatively large, well-developed ooids similar to those seen at Joulter Cays today (Fig. 14). Most subtidal sand deposits of this formation also contain appreciable ooids, except where they are in close association with a source of bioclastic debris (e.g., reef). Although regressive-phase eolian deposits of the Grotto Beach Formation also usually contain appreciable ooids, they may contain few ooids, perhaps reflecting the emergence of the shallow platforms and less

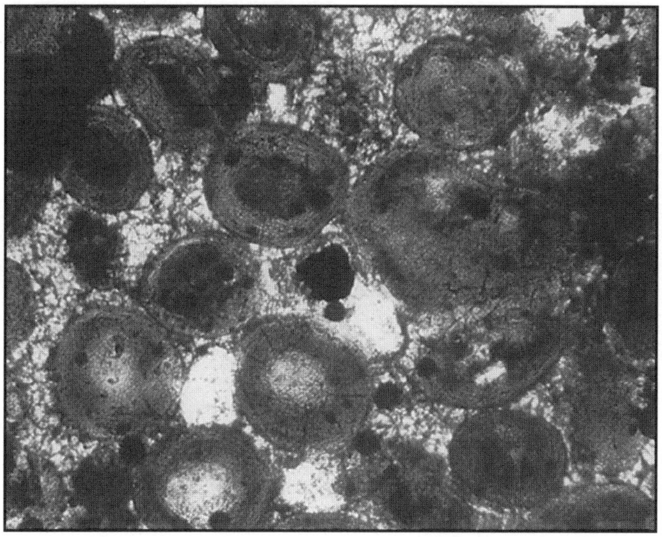

Figure 14. Photo of thin section showing typical ooids from Grotto Beach Formation eolianites. This sample is from South Andros Island. Field of view equals 1.8 mm.

favorable conditions for ooid formation. For reasons explained earlier, because the Grotto Beach Formation contains subtidal facies that are above present sea level., it must have been deposited during the sea-level highstand that is correlated with oxygen isotope substage 5e (ca. 125,000 yr ago).

Large volumes of oolitic eolianites of the Grotto Beach Formation commonly form ridges up to 30 m or more high on all Bahamian islands we have investigated. It is particularly noteworthy that, in a setting as small as San Salvador Island, where the isolated platform is not appreciably larger than the island itself, an enormous volume of ooids was produced during the last interglacial highstand. There are no modern locales in the Bahamas (e.g., Joulter Cays) that approach, even locally, the ooid production seen throughout the Bahamas during the deposition of the Grotto Beach Formation. Such abundant ooid-bearing sediments suggests that, during the Sangamon interglacial highstand (oxygen isotope substage 5e), warmer surface waters and the height of that highstand (+6 m), which must have produced more submerged bank area than is seen today, were conducive to prolific ooid development throughout the Bahamas.

Previously, it was thought that there was little surviving high ground composed of deposits produced during earlier highstands (Carew and Mylroie, 1985), and that the lack of existing land, and consequently greater shallow bank, may have contributed to the profuse ooid development; recent work, however, has revealed that there was much more older high ground, which is now either exposed or mantled by Grotto Beach Formation rock (Bain, 1991; Schwabe et al., 1993). In addition, as was stated earlier, carbonate eolian dunes do not migrate far from their source, which is generally the beach. Today in the Bahamas, only in those few locations where ooids are found in the beach environment are there oolitic eolian deposits (e.g., Joulter Cays). Where there is a lagoon between an ooid shoal and the beach, there are few or no ooids found in the beach sediments (e.g., South Andros Island: Carew and Mylroie, 1989). In comparison, virtually all beach facies rocks of the Grotto Beach Formation are dominated by ooids. Thus, while there is no doubt that there are extensive ooids shoals preserved in Grotto Beach rocks in the Bahamas, it is not the shoals that were the probable source of the eolianites, but rather ooid-rich beaches. Perhaps the beach and shoreface ooid environments reported in the Turks and Caicos Islands (e.g., Lloyd et al., 1987) provide a reasonable model for the abundant oolitic beach and eolianite deposits formed in the Pleistocene. it appears that the reason for the prolific ooid development during the last interglacial was probably not simply the result of a greater amount of flooded platform area, but also may have resulted from warmer surface waters, somewhat greater platform water depths, or other unidentified variables.

French Bay Member. Rocks of the Grotto Beach Formation with the depositional and erosional characters of transgressive-phase eolianites are assigned to the French Bay Member. The type-section for this member is composed of outcrops along the French Bay coast at the southwest tip of San Salvador (Fig. 5) (Carew and Mylroie, 1985). There, the unit is fine to medium oolitic calcarenite (oosparites or oolitic grainstones). Because these eolianites have well-preserved fine-scale internal bedding features that include grain-fall, grain-flow, and climbing wind-ripple laminae, and they have limited vegemorph development, they are interpreted as transgressive-phase eolianites. Additional evidence that these rocks were deposited during the transgressive phase include outcrops containing a fossil sea cave with internal boulder rubble, cliff-line paleotalus deposits, and isolated outcrops of regressive-phase eolian deposits with abundant vegemorphs that lie atop the transgressive-phase deposits (Carew and Mylroie, 1985; Marshall et al., 1984).

French Bay Member rocks can be seen throughout the Bahamas. For example, on High Cay (Fig. 13B), a small island off the eastern coast of South Andros Island (Carew and Mylroie, 1989), on West Plana Cay (Wilber, 1987), and in the Exuma islands (Halley et al., 1991), fossil corals are seen to lie, with no intervening paleosol or evidence of an eroded paleosol, on a wave-cut platform carved into a transgressive-phase eolianite. This relationship is identical to that seen today on Holocene transgressive-phase eolianites at North Point and High Cay on San Salvador Island, discussed earlier in the depositional model section of this chapter.

Cockburn Town Member. The subtidal and still-stand through regressive-phase beach and eolian deposits of the Grotto Beach Formation are assigned to the Cockburn Town Member. This member is characterized by a suite of subtidal marine deposits that extend up to 4 m above current sea level at many places in the Bahamas (Carew and Mylroie, 1995a), and are often entombed by still-stand and regressive-phase beach and dune deposits. All subaerially exposed marine deposits in the Bahamas are assigned to the Cockburn Town Member. The

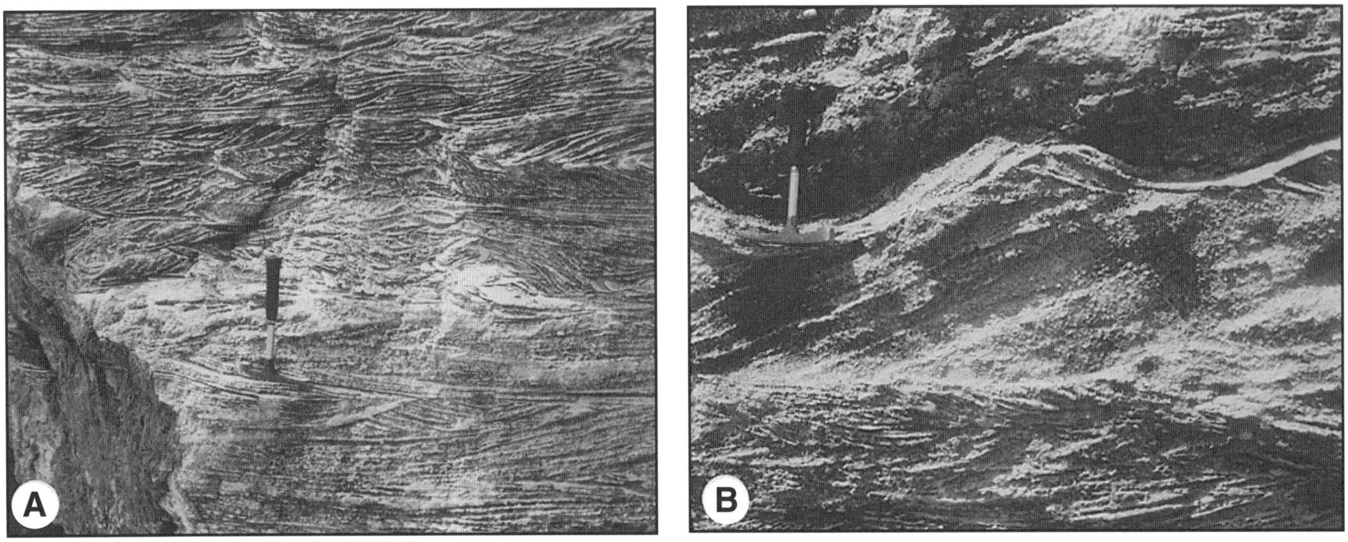

Figure 15. A, Photo showing spectacular cross-bedding and ripples in a Cockburn Town Member (Grotto Beach Formation) subtidal deposit that extends up to 5 m above sea level at Clifton Pier, New Providence Island. B, Close-up of cross-bedding and ripple surface in same outcrop as A.

type-section for this member is seen at exposures behind the Bahamas Electric Company plant in Cockburn Town on San Salvador Island (Fig. 5) (Carew and Mylroie, 1985).

The marine facies of the Cockburn Town Member are recognized in the field by features such as herringbone cross-bedding, asymmetrical ripples (Fig. 15A, B), the presence of abundant macroscopic fossil marine molluscs, corals, and trace fossils (e.g. *Ophiomorpha*), and by fossil coral reefs. At many outcrops on San Salvador, most notably along the coast at Cockburn Town (the type-section: Carew and Mylroie, 1985), Grotto Beach, Sue Point, and Sandy Point (Fig. 5), marine units can be seen to grade upward into beach and back-beach deposits, which in turn grade into eolianites (White et al., 1984; Carew and Mylroie, 1985; Carew et al., 1992b). Curran and White (1985) provided a detailed map and cross section illustrating these facies changes at Cockburn Town fossil reef on San Salvador Island, White and Curran (1987) discussed Devil's Point fossil reef on Great Inagua Island, and White (1989) described and illustrated the Sue Point fossil reef on San Salvador Island. Another impressive example can be seen in the shoreline cliffs at Clifton Point on New Providence Island, where subtidal shoal deposits grade upward to beach facies (Fig. 15A, B) (Garrett and Gould, 1984; Carew et al., 1992b).

The still-stand through regressive-phase beach facies and eolianites are also assigned to the Cockburn Town Member because there is an unbroken gradation from marine to eolian rocks seen at many outcrops; and no terra-rossa paleosol separates the marine and eolian facies. The Cockburn Town Member still-stand through regressive-phase eolian facies present a marked contrast to the transgressive-phase French Bay Member eolianites. The Cockburn Town Member eolianites exhibit some or all of the following, which are characteristic of still-stand through regressive-phase dune deposits: disrupted internal bedding; elaborate paleosols, often with vadose pisolites, complex caliche/calcrete crusts, and abundant fossil pulmonate snails (mostly *Cerion*); calcarenite protosols; exceptional vegemorph development; beach-face breccia facies; and outcrops that show the eolianites overstepping fossil reefs. These eolianites also lack evidence of wave attack during the highstand during which the dunes formed (e.g., fossil sea caves or paleotalus deposits). The sea cliffs observed today in Cockburn Town Member eolianites developed during the Holocene, as can be seen by the truncation of the eolianites *and* the overlying terra-rossa paleosols.

On San Salvador Island, Cockburn Town Member regressive-phase eolianites can be seen to overstep fossil reefs at The Gulf and on the south side of Crab Cay (Fig. 5). Besides those on San Salvador, we have seen Cockburn Town Member eolianites that entomb fossil reefs on North and South Andros Island, Great Inagua Island, New Providence Island, and Great Exuma Island. Such sequences have also been reported on Great Inagua Island (White and Curran, 1987) and on West Plana Cay (Wilber, 1987).

In addition to fossil coral reefs, subtidal shoal, lagoonal, and ebb-tidal delta deposits are represented in the Cockburn Town Member. We have observed subtidal shoal facies of the Cockburn Town Member up to nearly 4 m above present sea level. Excellent examples of such deposits can be seen along the sea cliffs at Sandy Point and along the Altar Cave cliff line on San Salvador Island (Carew and Mylroie, 1985), in the Clifton Pier (Carew et al., 1992b) and Fox Hills areas on New Providence, and along the shorelines of Deep Creek and Little Creek on South Andros Island (Carew and Mylroie, 1989). In fact, the majority of the deposits below an elevation of 6 m on South Andros Island consist of subtidal ooid shoals and associated facies (Carew and Mylroie, 1989).

Cockburn Town Member lagoonal deposits can be seen on San Salvador (e.g., Florentino and Bain, 1984; Sims, 1987; Bain, 1991), and we have observed them on many other Bahamian islands. An excellent example of a fossil ebb-tidal delta facies in the Cockburn Town Member can be seen in North Pigeon Creek Quarry (Fig. 5) on the east side of San Salvador Island (Hinman, 1980; Thalman, 1983; Thalman and Teeter, 1983a, b).

Fossil tidal creek and lacustrine deposits of the Cockburn Town Member are also found on San Salvador (Titus, 1984, 1987; Edwards et al., 1990; Bain, 1991; Hagey, 1991; Noble et al., 1991). According to Hagey (1991) most of these inland fossil deposits comprise the *Chione cancellata* assemblage, which represents normal marine conditions. A few deposits comprise the *Anomalocardia auberiana* assemblage, which is indicative of hypersaline or restricted conditions. Such restricted conditions required the existence of significant preexisting high ground, as suggested earlier in the discussion of ooid formation during deposition of the Grotto Beach Formation. For further discussion, see Hagey and Mylroie (this volume).

Throughout the Bahamas, rocks of the Grotto Beach Formation are capped by a hard, red terra-rossa paleosol, or a beige to red micritic calcrete, except where it has been stripped by later erosion. The crests of eolianite ridges typically have a calcrete layer a few centimeters thick, but in the swales and on regressive-phase eolianites, complex thick paleosols are commonly found. These paleosol/calcrete caps are the youngest Pleistocene rocks on San Salvador Island and elsewhere in the Bahamas.

Suppressed stratigraphic term. Previously, deposits thought to have been deposited during the highstand of sea level associated with oxygen isotope substage 5a (Fig. 3), based on estimated amino acid racemization (AAR) ages of about 85,000 yr, were assigned to the Dixon Hill Member of the Grotto Beach Formation (Carew and Mylroie, 1985, 1987; Carew et al., 1984). Those rocks were correlated with the Southampton Formation of Bermuda, which is also reported to be about 85,000 yr old, based on AAR data (Vacher and Hearty, 1989). The deposits assigned by us and others to the Dixon Hill Member were identified only through AAR age determinations, not by physical stratigraphic relationships. As such, the Dixon Hill Member cannot be a part of the physical stratigraphy, and use of the name should be suppressed. In addition to the improper definition of this unit, the AAR data that supported an 85,000-yr age is open to question (see Mirecki et al., 1993; Carew and Mylroie, 1994, 1995b). We currently consider the rocks of Dixon Hill to be 125,000 yr old, or older, as we suggested in an earlier publication (Carew et al., 1982).

Rice Bay Formation

All rocks that lie above the paleosol at the top of the Grotto Beach Formation are assigned to the Rice Bay Formation (Fig. 12). Throughout the Bahamas, Rice Bay Formation rocks consist of eolianites and beach-facies sediments that have been deposited during the transgressive and still-stand phases of the current sea-level highstand (oxygen isotope stage 1: Fig. 3). Although there is some incipient development of thin calcretes (<1 mm) on some transgressive-phase eolianites of the Rice Bay Formation, hard, reddish-brown terra-rossa paleosols and micritic calcretes are absent on Rice Bay rocks. However, as expected during the still-stand phase, calcarenite protosols are currently forming in coastal areas and in swales between and on transgressive-phase eolianites.

The Rice Bay Formation is divided into two members that can be recognized by differences in bedding character, allochem composition, and their position relative to current sea level. The as-yet unlithified modern eolian dunes, beach ridges, beach sediments, and subtidal sediments, as well as beachrock and reefs, along with any regressive-phase deposits that accumulate, will ultimately be part of the Hanna Bay Member of the Rice Bay Formation (in similar fashion to the Cockburn Town Member of the Grotto Beach Formation).

Unlike the predominantly oolitic beach to dune facies of the Grotto Beach Formation, the Rice Bay Formation rocks are characterized by the following: a generally low abundance of ooids (usually less than 25%, rarely up to 50%), especially high in the section; the superficial nature and small size of those ooids (i.e., only a few laminae); the dominantly bioclastic and peloidal allochem composition of most of the rocks, especially the Hanna Bay Member; limited diagenetic micritization; and generally weak meniscus-style vadose low-magnesium calcite cements (Hutto and Carew, 1984; Carew and Mylroie, 1985; White and White, 1991). Superficial-ooid production occurred during the early phase of Holocene transgression of the San Salvador platform, but that seems to have ceased by about 3,000 B.P., and most Rice Bay rocks, on San Salvador and throughout the Bahamas, are generally ooid-poor or lack ooids (Carney and Boardman, 1991; Boardman et al., 1991). At relatively few locations, such as Joulter Cays, there are abundant well-developed Holocene ooids, but lithified sediments of the Rice Bay Formation throughout the Bahamas generally lack such ooids, even where ooid shoals and bars currently exist off shore, such as along the east coast of South Andros (Carew and Mylroie, 1989). Somewhat better cementation, together with evidence of a developing marine cement, can be seen in the transgressive-phase eolianites of the Rice Bay Formation that are exposed to marine inundation and spray on high-energy coasts (e.g., White, 1991; White and White, 1991).

North Point Member. The North Point Member comprises the transgressive-phase eolianites of the Rice Bay Formation. These eolianites were deposited when sea level was lower than at present, as indicated by steeply dipping foreset beds that continue below current sea level to a depth of at least 2 m. The type-section for this member is located at North Point on San Salvador Island (Fig. 5) (Carew and Mylroie, 1985). North Point Member eolianites exhibit the characteristics of transgressive-phase eolianites, including grain-fall, grain-flow, and climbing wind-ripple laminae, and limited vegemorphs. In addition, as expected from the depositional model, many of these transgressive eolianites have been eroded by wave action that has attacked the deposits during the continued rise of sea level since their deposition. That erosion has produced wave-cut platforms that truncate dune foreset beds, on which living corals can sometimes be seen (e.g., North Point and High Cay, San Salvador) (Fig. 8); formed sea caves (e.g., White Cay, San Salvador); and removed up to half the original dunes to form extensive cliffs that are now sometimes found well inland as a result of beach progradation since the cliffing (e.g., Snow Bay, San Salvador) (Fig. 9). On San Salvador, rocks of the North Point Member are found only on the east coast, where it forms ridges up to 20 m high (e.g., Snow Bay and North Point), and numerous offshore cays such as High Cay, White Cay, Catto Cay, and Cut Cay (Fig. 5). Other than on San Salvador, we have observed North Point rocks on New Providence Island, Great Inagua Island, and on Long Island, which has the most spectacular and extensive deposits of this member that we have seen. Additionally, Wilber (1987) reported North Point rocks on Little San Salvador and West Plana Cay, and Kindler (1991) reported them from Lee Stocking Island. North Point Member rocks may be found on most Bahamian islands.

Hanna Bay Member. The Hanna Bay Member comprises still-stand-phase beach and eolian facies rocks of the Rice Bay Formation. These rocks were deposited in equilibrium with current sea level, as indicated by their concordance with current elevations of similar facies (i.e., intertidal and beach-facies beds of the Hanna Bay member are found at the same elevation as those same facies of the modern beaches). The type-section of this member is located at Hanna Bay, San Salvador Island (Fig. 5) (Carew and Mylroie, 1985). The beach facies are best exposed in areas where changing coastal dynamics have led to erosion that truncates the rocks, such as in Grahams Harbour on San Salvador Island (Fig. 5). In these cases the truncation cuts deposits that are congruent with current sea level. Like the North Point Member, these deposits are most abundant on the east side of San Salvador, but they are present on all coasts. We have identified Hanna Bay Member rocks on the following Bahamian islands: San Salvador, Eleuthera, New Providence, Long, Great Inagua, North Andros, South Bimini, Great Exuma, and South Andros. In addition, Wilber (1987) reported Hanna Bay rocks from Little San Salvador Island and West Plana Cay, and Kindler (1991) and White and Curran (1993) reported them from Lee Stocking Island. While not specifically identified as such, from the information in Mitchell (1987) and Mitchell et al. (1989), Hanna Bay Member rocks are present on Conception Island and Rum Cay. It is probable that there are Hanna Bay Member rocks on all Bahamian islands. As stated earlier, Holocene (oxygen isotope stage 1) subtidal deposits and future regressive-phase deposits are properly assigned to the Hanna Bay Member.

DISCUSSION

Information about the precise timing, and details about the sequence of deposition, of discrete depositional units that are contained within the physical stratigraphy presented in this work may be obtained by careful study of stratigraphic, paleontologic, and petrologic information; a variety of geochronologic analyses have also been utilized with varying success. This discussion provides examples of some of the details of deposition that we have identified, or that have been reported in the literature. We also discuss some difficulties that have arisen from the attempted application of several techniques.

Paleontologic, petrologic, and stratigraphic relationships

Paleontologic and taphonomic information. Hattin and Warren (1989) suggested that the sediments that entomb the reef deposits at Grotto Beach on San Salvador Island prograded during a stillstand, whereas White et al. (1984) and White and Curran (1987) suggested that deposition during actual marine regression entombed the fossil reefs at Cockburn Town on San Salvador Island, and Devil's Point on Great Inagua Island. The excellent preservation of those reef deposits, and others at Sue Point on San Salvador Island, Northwest Point on New Providence Island, and the reef exposed in a cave in the sea cliff at The Bluff on South Andros Island, attest to very rapid burial of the reefs, most likely during storms.

From study of modern settings, such as at Grotto Beach on San Salvador Island where beach progradation is burying nearshore reefs, it is evident that even a few years' delay between death of the corals and their burial in sediment leads to significant degradation in the condition of the corals. So, the superb preservation seen in many of these fossil reefs (Fig. 16) indicates that they were not buried by slow progradation either during the stillstand or during regression, but rather, catastrophically at an undetermined time *before* regression. In order to survive the passage of the coastal zone during regression, these reefs must already have been entombed in calcarenites, and those sediments must have protected the reefs. Those reefs that were not protected from erosion as the coastal zone retreated by their position, are likely to be jumbled and show evidence of truncation. An example can be seen at The Gulf on San Salvador Island (Fig. 5). There, a reef-rubble facies is overlain by a calcarenite protosol that provides evidence for exposure of the reef in the near shore, before being entombed by the overlying eolianite.

Recently, Schellenberg and Hearty (1992) and Hearty et al. (1993) have proposed that, on San Salvador Island, fossil *Cerion* can be utilized to differentiate between deposits formed during different oxygen isotope stages, and even substages (specifically, substage 5e from 5a). Garret and Gould (1984) utilized *Cerion* similarly on New Providence Island. The relationship between *Cerion* morphotypes (whether afforded taxonomic status or not) and inferred depositional history is not the same on these two islands.

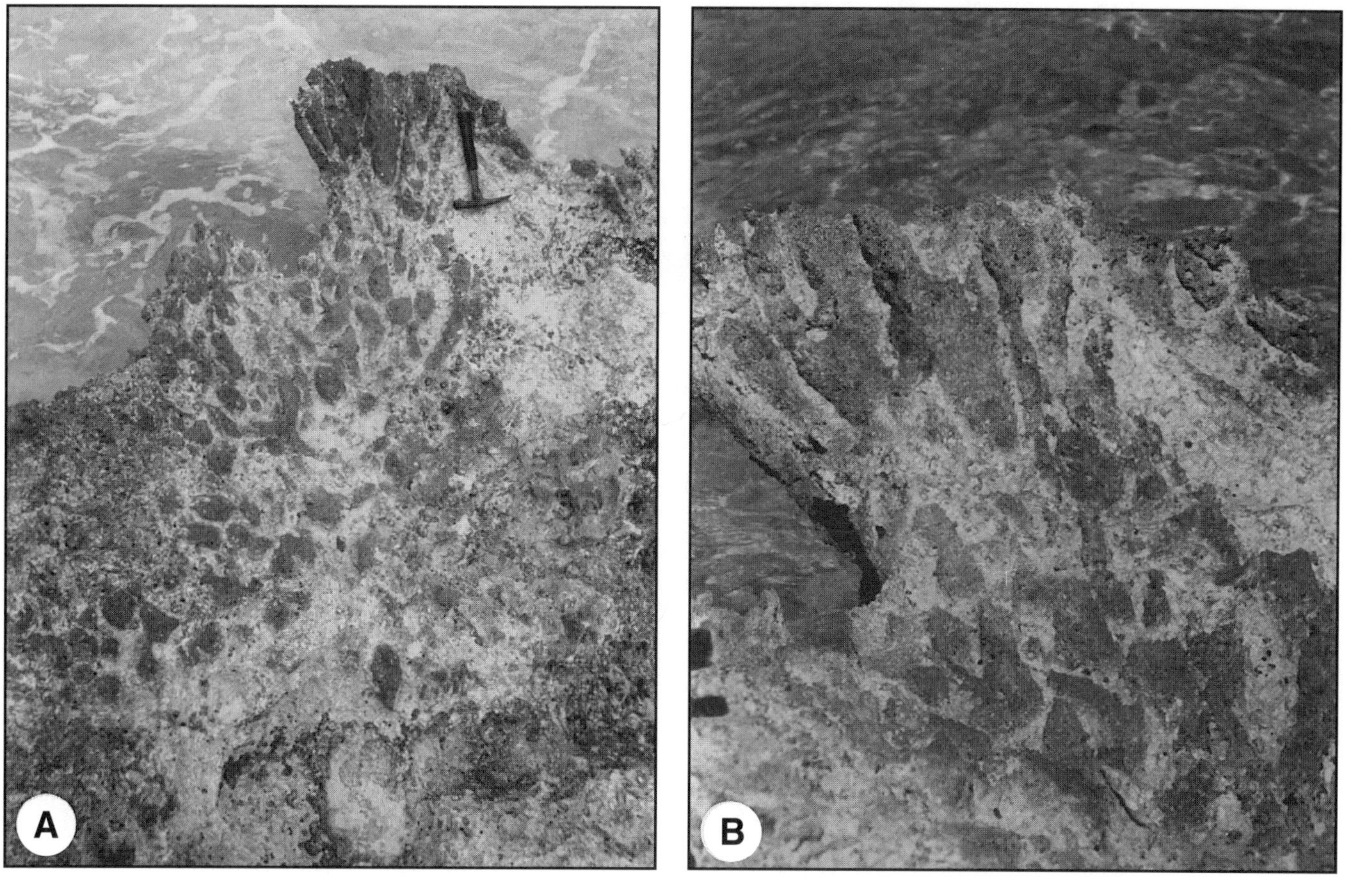

Figure 16. Photos of Pleistocene *Acropora palmata* preserved in current-oriented growth position at Sue Point fossil reef (Cockburn Town Member, Grotto Beach Formation), San Salvador Island. A, Wide-angle; B, Close-up views.

Having done extensive field work on both San Salvador and New Providence Islands, during which we often collected, or made notes about, both fossil and modern *Cerion*, it appeared to us that most, or all, of the morphotypes of *Cerion* assigned to different species (Garrett and Gould, 1984) and times of deposition (Garrett and Gould, 1984; Schellenberg and Hearty, 1992; Hearty et al., 1993) were extant on those islands, and that the distribution of morphotypes seemed to be related more to geographic location than to stratigraphic position. We have begun an investigation of this subject, and to date a study of *Cerion* morphotypes on San Salvador indicates that there is as great, or greater, morphological variability within and between extant populations of *Cerion* than can be identified between any fossil populations (Marcy et al., 1993). Further, this study indicates that *Cerion* from the east side of the island, whether fossil or modern, form a separate statistical cluster from extant *Cerion* from the west side of the island. From the result of our study to date, at least on San Salvador, *Cerion* morphotype does not appear to be a reliable indicator of stratigraphy, or time of deposition. This study is ongoing, and we expect, through analysis of a large number of modern and fossil *Cerion* samples (800 to date), to firmly ascertain the usefulness of *Cerion* morphotypes for identifying deposits of various ages in the Bahamas.

Petrologic data and stratigraphic relationships. Hattin and Warren (1989) concluded that the lower eolianite (discussed earlier in the sections on the Owl's Hole and Grotto Beach formations) seen in outcrops at Grotto Beach on San Salvador Island was produced during the transgressive phase of the same highstand during which the overlying reef was developed. If that assumption were correct, the lower eolianite would belong to the French Bay Member of the Grotto Beach Formation, and the overlying reef and associated deposits would belong to the Cockburn Town Member of the Grotto Beach Formation. However, differences between the petrology of the eolianite that underlies the reef, and the sediments associated with the reef facies, are identical to that seen at localities where a terra-rossa paleosol lies between the two deposits (Stowers et al., 1989). Further, the presence of alveolar texture in the upper part of the lower eolianite at Grotto Beach strongly suggests that the lower eolianite was deposited during a previous highstand, and was pedogenically altered during exposure. It seems likely that the planar surface was cut into an eolianite that had been deposited during an earlier sea-level highstand, and the paleosol that had developed on it was removed by wave action associated with the rise in sea level of the highstand during which the marine deposits were formed. Thus, the subtidal and reef deposits are properly assigned to the Cockburn Town Member of the Grotto Beach Formation, and the underlying eolianite belongs in the Owl's Hole Formation (see Fig. 13A).

Facies relationships, such as that suggested by Hattin and Warren (1989) for the Grotto Beach site, do exist elsewhere in the Bahamas, such as on High Cay, South Andros (Carew and Mylroie, 1989), and in the Exumas (Halley et al., 1991), but at those localities the underlying eolianites and the overlying facies are both ooid-bearing and there is no evidence of a paleosol or former paleosol (Fig. 13B). As previously discussed in the section on the Rice Bay Formation, such relationships can also be recognized in Holocene deposits today, where corals are living on wave-cut benches carved into transgressive-phase eolianites (e.g., North Point and High Cay on San Salvador Island (Fig. 8).

Morphostratigraphy. Titus (1980, 1984, 1987), Hearty and Kindler (1991, 1992, 1993) and Hearty et al. (1993) on San Salvador, and Garrett and Gould (1984) on New Providence, have utilized morphostratigraphy for delineating the sequence of deposition. In general, such studies rely on the geographic position and apparent interrelationships between recognizable deposits (e.g., eolianite ridges) to determine the sequence of deposition. This type of study seems intuitively reasonable; however, because of the different heights to which sea level rose during the highstands of the late Pleistocene and Holocene, and the spatially patchy nature of the deposition, especially that of the eolianites, it is not possible in the Bahamas to rely on this method.

Morphostratigraphic studies generally assume that ridges that occupy positions toward the center of the platform are the earliest deposited in a sequence, but because of the varying height of sea level at times of deposition, such a relationship is not straightforward. For example, today on the northeast coast of San Salvador Island, both the Holocene North Point Member and Hanna Bay Member rocks lie to the interior of Man Head Cay (Fig. 5), which is a regressive-phase deposit of the Pleistocene Grotto Beach Formation. In fact, at that site the inverse of the predicted inland-to-offshore sequence exists. The younger Hanna Bay rocks lie inland of the slightly older North Point Rocks, and both of those lie inland of the Pleistocene Grotto Beach rocks.

Furthermore, recent study of Pleistocene eolianite ridges has indicated that many ridges that appear to be identifiable as individual eolian dunes, are actually a composite of eolianites that were deposited during more than one highstand (Schwabe et al., 1993). We have also seen clear evidence of similar depositional relationships on the west side of Long Island, where Holocene North Point Member eolian deposits wholly, or partially, mantle Pleistocene eolianite ridges. On aerial photos or topographic maps, these deposits appear as single ridges that would be assigned to one depositional event utilizing morphostratigraphy, but, as can be seen in the field, that is not true.

The stratigraphic placement of the rocks that comprise Dixon Hill on San Salvador Island (Titus, 1987; Hearty and Kindler, 1993; Hearty et al., 1993) provides a final example of the questionable usefulness of that methodology in the Bahamas. Using morphostratigraphic principles, Titus (1984, 1987) assigned those rocks to his Dixon Hill Limestone that he suggested (Titus, 1984, p. 215, Fig. 2) was deposited about 80,000 yr ago (oxygen isotope substage 5a), whereas Hearty and Kindler (1993) and Hearty et al. (1993) assigned those same rocks to their Fortune Hill Formation that they reported to be about 200,000 yr old (oxygen isotope stage 7). It is our opinion that use of morphostratigraphy for unraveling the sequence of deposition of units exposed on Bahamian islands is unreliable (see Carew and Mylroie, 1994, 1995b for further comments). Only by detailed field and petrologic study of each ridge can one ascertain its true character; consider the variables depicted in Figures 1A–D and 4.

Geochronology

Uranium/Thorium. In the tectonically stable Bahamas, for reasons discussed earlier, it is evident that all the marine facies exposed in the Bahamas belong in the Cockburn Town Member of the Grotto Beach Formation, and can be correlated with the last interglacial highstand of sea level. According to precise $^{234}U/^{230}Th$ ages from fossil coral reefs on San Salvador Island and Great Inagua Island, that highstand lasted from about 132,000 to 119,000 yr ago (Chen et al., 1991), and is correlated with deep-sea oxygen isotope substage 5e (Shackleton and Opdyke, 1973; Shackleton, 1987). Generally, the data from fossil coral reefs throughout the Bahamas (e.g., Carew and Mylroie, 1985, 1987, 1995a) provide no evidence for a double-peaked substage 5e highstand, as suggested by other workers (e.g., Hollin and Hearty, 1990; Johnson, 1991), but the work of White and Curran (1987) and Chen et al. (1991) does suggest that there may have been a minor regression during the highstand. However, U/Th ages of speleothems and other data from caves throughout the Bahamas may provide evidence for such a double peak during the 5e highstand (Mylroie et al., 1991; Mylroie et al., this volume), but this issue is as yet unresolved.

Radiocarbon dating. Attempts to tease out the details of deposition during the latest Pleistocene and Holocene in the Bahamas have often used whole-rock radiocarbon-age determinations. Those ages actually reflect the collective age of the allochems, modified slightly by the sparse cements, rather than the actual time of deposition. However, these data consistently identify an early stage of Holocene deposition, and a later (more recent) one.

Based solely on stratigraphic relationships, we originally suggested that North Point Member rocks were younger than 10,000 yr old (Carew and Mylroie, 1985). Radiocarbon ages obtained from whole-rock samples of North Point rocks from

San Salvador Island, New Providence Island, Great Inagua Island, and Long Island range between 6,100 and 3,700 B.P., and generally average about 5,000 B.P. (Carew and Mylroie, 1985, 1987; Boardman et al., 1987, 1989; Andersen and Boardman, 1989; Colby and Boardman, 1989). Considering the Bahamian Holocene sea-level curve derived from radiocarbon ages of peats that lie directly on Pleistocene bedrock (Fig. 17), the tops of the Bahamian platforms began to be flooded about 7,000 yr ago (Boardman et al., 1989). Significant sand was produced and incorporated into transgressive-phase eolianites of the North Point Member by about 6,000 B.P. with the time of greatest accumulation at approximately 5,000 B.P. (Carew and Mylroie, 1987; Boardman et al., 1987, 1989). That time coincides with the inflection on the Bahamian sea-level curve that indicates the change from rapid rise to slower rise of sea level that took place after 4,500 to 4,000 B.P. (Boardman et al., 1989) (Fig. 17). Deposition of the transgressive-phase North Point rocks throughout the Bahamas probably took place before reefs became well established, and caught up to sea level; thereafter, they effectively reduced the wave energy that reached the shoreline of the Bahamian Islands (Boardman et al. 1987).

Without radiocarbon ages, and based solely on stratigraphic relationships, it was originally suggested that Hanna Bay Member rocks were deposited in equilibrium with sea level at, or nearly at, its present elevation (Carew and Mylroie, 1985). Subsequent studies (Carew and Mylroie, 1987; Boardman et al., 1987; Andersen and Boardman, 1989), and development of the Bahamian Holocene sea-level curve have reinforced that interpretation (Boardman et al., 1989). Radiocarbon ages of whole-rock samples of Hanna Bay Member rocks from San Salvador Island, New Providence Island, South Andros Island, Great Inagua Island, and Long Island range from about 3,200 B.P. to 300 B.P., and generally are younger than 2,500 B.P. (Carew and Mylroie, 1987; Boardman et al., 1987; Andersen and Boardman, 1989). Hanna Bay Member rocks that yield ages at the older end of that range were probably deposited since 2,000 B.P., but yield older ages because they contain a component of eroded North Point Member rocks (Boardman et al., 1987; Andersen and Boardman, 1989).

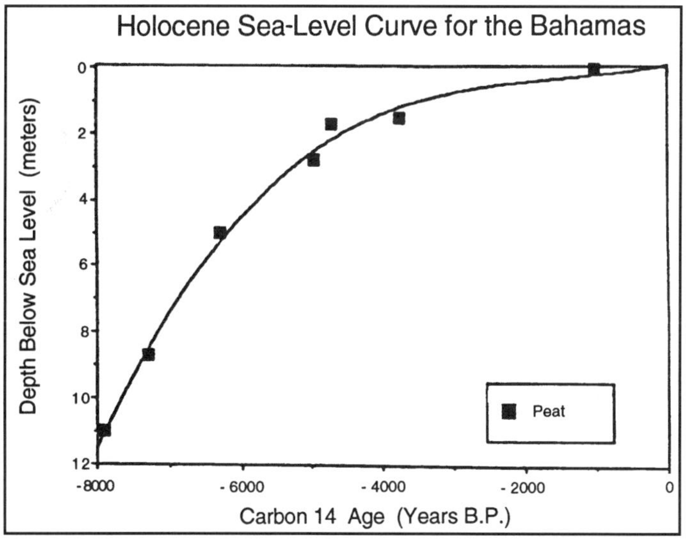

Figure 17. Holocene sea-level curve for the Bahamas based on ^{14}C ages of basal Holocene peat deposits that lie directly on Pleistocene bedrock (modified from Boardman et al., 1989).

Amino acid racemization. Earlier, in attempts to determine which eolianites were deposited during which highstands we utilized amino acid racemization of *Cerion* as a tool (Carew et al., 1984; Carew and Mylroie, 1985, 1987). While some of the data matched well with interpretations based on stratigraphic relationships, some of the data were inconsistent. As mentioned earlier in the stratigraphy section of this chapter, we also erroneously proposed a stratigraphic unit that was thought to be recognizable by the amino acid racemization values of the contained *Cerion*. Because of our concern about the reliability of AAR ratios in *Cerion* from the Bahamas, we recently conducted a blind study of *Cerion* in collaboration with an AAR geochemist. The geochemist was given *Cerion* shells from a variety of geographic and stratigraphic settings, including both Holocene and Pleistocene, but was not told where they came from. The results of that study suggest that AAR ratios of *Cerion* cannot be used to reliably identify deposits that were deposited during different oxygen isotope stages, much less substages (Mirecki et al., 1992, 1993). Further evidence of the lack of reliability of AAR ratios from Bahamian rocks comes from radiocarbon analysis of rocks purported to have been deposited during oxygen isotope substage 5a (ca. 85,000 yr ago) (Hearty and Kindler, 1991, 1992). Those rocks yield Holocene radiocarbon ages (Mirecki et al., 1992, 1993), and regardless of the margin of potential error of those ages, they could not possibly be consistent with deposition 85,000 yr ago (see also Carew and Mylroie, 1994, 1995b). In Hearty and Kindler (1993) and Hearty et al. (1993), those outcrops are shown as Holocene.

Recently, it has been suggested (Hearty and Kinder, 1991, 1992, 1993; Hearty et al., 1993) that, based on AAR whole-rock analyses, some eolianites interpreted by us as regressive-phase deposits of the Cockburn Town Member, were formed during the highstand of sea level associated with oxygen isotope substage 5a, rather than during the regression at the end of substage 5e (Fig. 3). They proposed new stratigraphic names for those deposits. The study of *Cerion* mentioned above suggests that AAR data from a single taxon are unreliable, which makes us even less confident in whole-rock AAR data, so we question whether those deposits can be confidently inferred to have been deposited during oxygen isotope substage 5a. We prefer our interpretation of those outcrops as regressive-phase deposits formed as sea level fell from the substage 5e highstand; however, regardless of whether they were deposited during substage 5e or 5a, the deposits in question belong within the Cockburn Town Member as it is defined in this chapter. According to the North American Commission on Stratigraphic Nomenclature (1983), stratigraphic units cannot be based on geochronologic data alone, so the stratigraphic nomenclature proposed for the alleged substage 5a eolianites should not be a part of the physical stratigraphy of the Quaternary of the Bahama islands. Likewise, a proposed stratigraphic unit composed of eolianites identified by AAR ratios as being older than Grotto Beach Formation eolianites, and younger than Owl's Hole Formation eolianites (Hearty and Kindler, 1991, 1992, 1993; Hearty et al., 1993) is inappropriate (see Carew and Mylroie, 1994, 1995b). AAR ratios from Bahamian carbonates are at best unreliable, and stratigraphic units should not be based on such criteria.

Geologic mapping in the Bahamas. It may at first seem that mapping of the Quaternary rocks of the Bahamas should be relatively simple; after all, they are entirely very young carbonates. Unfortunately, because of the spatial patchiness of the deposition during individual highstands, the varying elevation to which sea level rose at different times, and the variable loss of record during lowstands and from wave erosion during highstands, it is often very difficult to determine the correct stratigraphic position of individual outcrops. It is easiest to obtain a detailed scenario for the depositional history of islands that have the greatest degree of human development (e.g., New Providence Island). There, the abundant roadcuts and quarries expose details of vertical relationships that would be undetectable without them (see Fig. 4). In addition, the presence of the Bahamian Field Station on San Salvador Island has promoted long-term study that has allowed us to uncover the details of deposition there despite limited human development.

Generally, lack of outcrop continuity, highly variable diagenetic alteration (that is, not straightforwardly related to age), complexities of paleosol development and interpretation (see Carew and Mylroie, 1991), onlap and overlap of deposits formed during different highstands, and overall similarity of all these rocks make accurate mapping a complex and painstaking task. Only with very detailed field study, including many measured sections, careful petrologic analysis of many thin sections and hand samples, and a refined understanding of the many vagaries of these deposits can an accurate representation of the geology and stratigraphy of the Bahamian islands be obtained. It turns out, in fact, that the youth of these rocks makes the task more difficult, not easier. Much of the detail that can be seen in these young carbonates is lost through diagenesis in older carbonates, and that ultimately makes it easier to determine an appropriate stratigraphy for the older rocks.

Our geologic map of San Salvador Island (Figs. 18, 19) shows the known distribution of the stratigraphic units of our stratigraphy. One can easily see that the various units are complexly interwoven and commonly do not occur in demonstrable vertical relationship to one another. Our map shows much of the island's interior as undifferentiated Pleistocene because our recent work concerning the full or partial overlap of Grotto Beach Formation rocks and Owl's Hole Formation rocks of some eolianite ridges, and the recognition of the complexity of paleosols that may represent one or more intervals of sea-level lowstands has convinced us that appropriate stratigraphic assignment cannot be reliably determined without detailed work on each ridge. So, we have assigned specific stratigraphic nomenclature only to those locales where we have obtained appropriate identifying criteria for that assignment.

Proposed stratigraphies

Relatively few stratigraphies have been proposed for the surficial geology of Bahamian islands (Titus, 1980, 1984, 1987; Carew and Mylroie, 1985, 1994, 1995b; Carew et al., 1992b; Hearty and Kindler, 1991, 1992, 1993; Hearty et al., 1993). Beach and Ginsburg (1980) assigned all of the late Pliocene and Quaternary rocks in the Bahamas to the Lucayan Limestone. However, as a result of the establishment of the Bahamian Field Station on San Salvador Island, a number of geologists began to investigate the details of the geologic record of that island. Those studies led to the abandonment of the Lucayan Limestone in favor of a more detailed stratigraphy. The first proposed stratigraphy for a Bahamian island was that of Titus (1980). He established the Grotto Beach Limestone and the younger Grahams Harbour Limestone (Table 2). Titus interpreted those units as Pleistocene deposits that were laid down during sea-level regression from a highstand, as the shallow shelf was exposed to subaerial erosion. He made no suggestion concerning when in the Pleistocene he thought they were deposited, and he believed that those two units rested on pre-Pleistocene biomicrite. Because of isostatic subsidence and sea-level highstand elevations, as discussed earlier, it is not possible for pre-Pleistocene deposits to be exposed on any Bahamian islands.

In 1984, Garrett and Gould published on the detailed depositional history of New Providence Island, but they did not propose a physical stratigraphy. Although they proposed phases of deposition, they did not tie them to a precise chronology or stratigraphy. In 1985, Carew and Mylroie proposed a revision to the stratigraphy of San Salvador (Table 2) based on the following: much of the rock assigned to the Grahams Harbour

Figure 18 (on facing page). Geologic map of San Salvador Island, Bahamas. The patterns shown along the coast of the island represent the rock units exposed along that portion of the shore. The width of the pattern is necessary to depict the distribution of the various rock units, but it does not necessarily reflect the actual inland distribution of those rocks. For example, in many places Pleistocene deposits are found immediately inland of Holocene outcrops that form a thin ridge along the shore. This map depicts only that information that we actually have seen and where the unit assignment is supported by field and petrologic relationships. Because of the complexities that exist among paleosols and various deposits (refer to Fig. 4), there is no extension of the data into areas where we have not, or cannot, definitely determine the appropriate stratigraphic unit. Thus, much of the island is mapped as undifferentiated Pleistocene. Those rocks may belong to either the Owl's Hole Formation or Grotto Beach Formation.

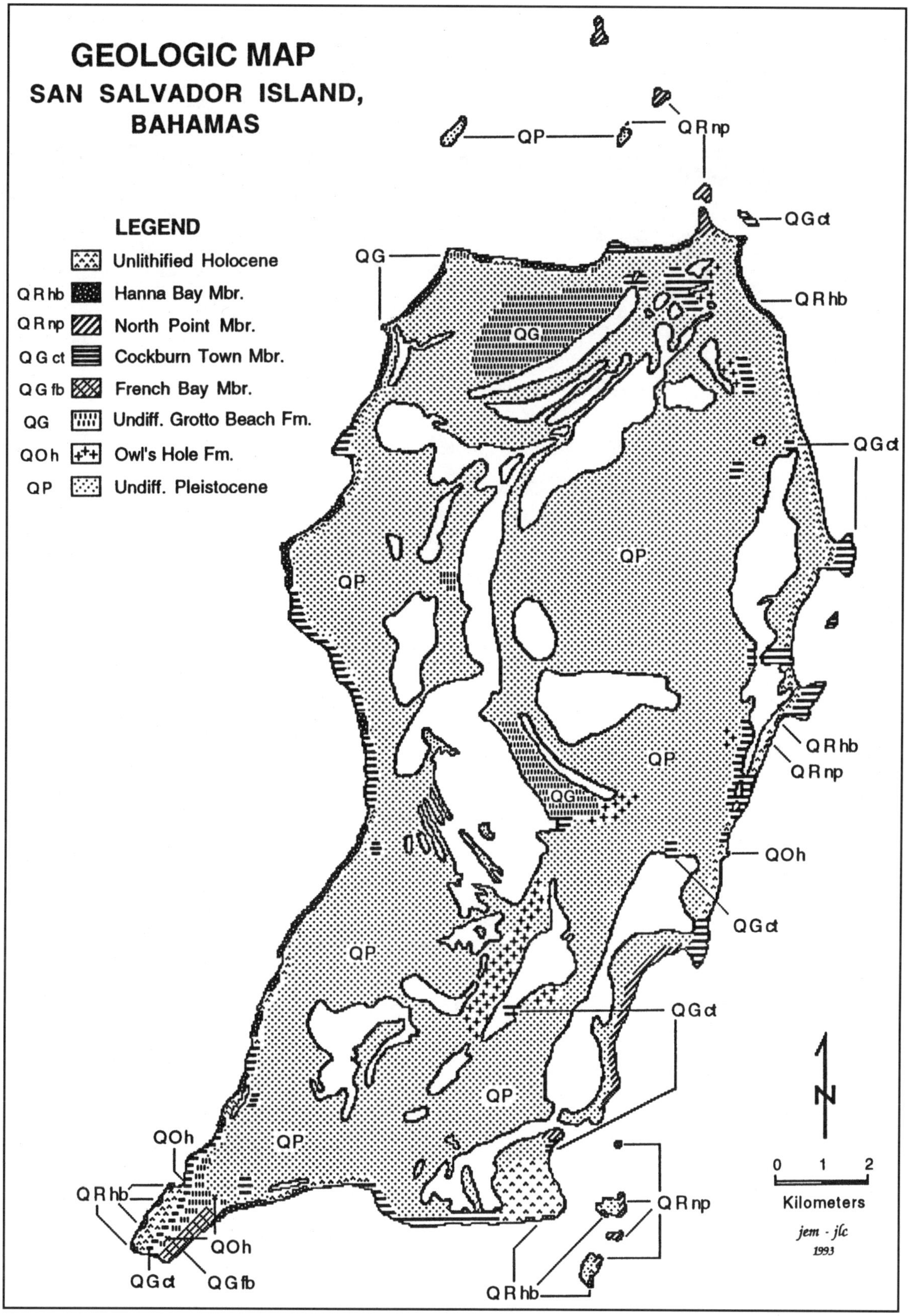
GEOLOGIC MAP
SAN SALVADOR ISLAND, BAHAMAS
LEGEND
Unlithified Holocene
Q R hb Hanna Bay Mbr.
Q R np North Point Mbr.
Q G ct Cockburn Town Mbr.
Q G fb French Bay Mbr.
QG Undiff. Grotto Beach Fm.
QOh Owl's Hole Fm.
QP Undiff. Pleistocene
QP
Q R np
Q G ct
QG
Q R hb
QG
Q G ct
QP
QP
Q R hb
Q R np
QP
QG
QOh
Q G ct
QP
Q G ct
QP
QOh
QP
Q R hb
Q R np
QOh
Q G ct
Q G fb
Q R hb
0 1 2
Kilometers
jem - jlc
1993

Limestone by Titus is Holocene rather than Pleistocene; the rock cited by Titus as the type-section for the Grahams Harbour Limestone is not correlative with the majority of rock assigned to that unit; Titus had failed to recognize an older Pleistocene eolianite beneath the Grotto Beach Limestone at its type locality, and elsewhere; and substantial portions of the rock record on San Salvador were deposited during the transgressive and still-stand phases of sea-level highstands, rather than only during the regression.

As a result of information provided by Carew and Mylroie

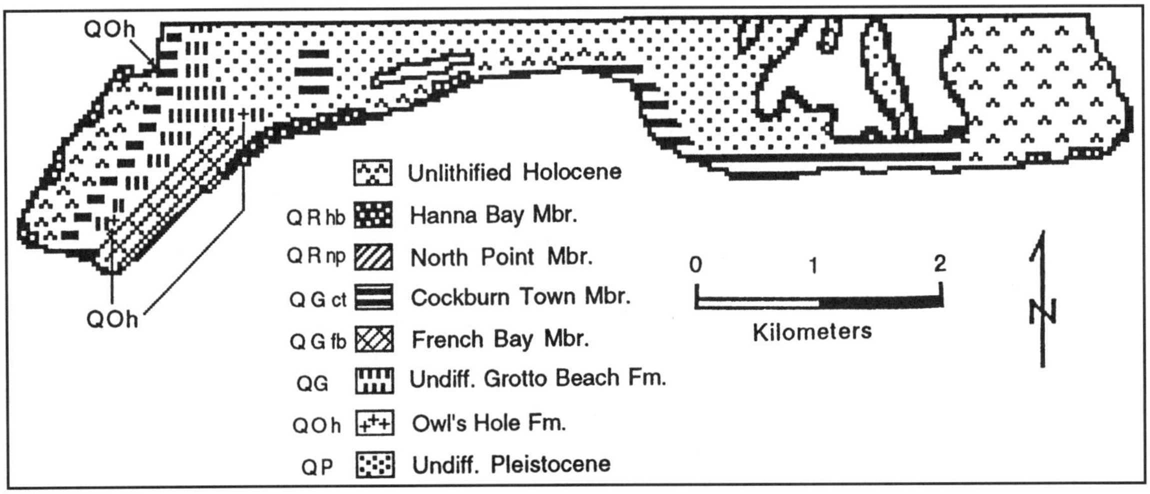

Figure 19. Enlargement of the south coast of San Salvador illustrating the complexity of the geology there, which may be hard to discern in Figure 18.

TABLE 2. PROPOSED STRATIGRAPHIES FOR BAHAMIAN SURFICAL GEOLOGY

Age	Oxygen Isotope Stage	Titus (1980*)	Carew and Myloroie (1985)	Titus (1987)	Hearty and Kindler (1993)	Carew and Mylorie (This Chapter†)
Holocene	1	"Recent sand"	**Rice Bay Formation** Hanna Bay Member North Point Member	Unnamed Holocene	**Rice Bay Formation** East Bay Member§ Hanna Bay Member North Point Member	**Rice Bay Formation** Hanna Bay Member** North Point Member
Pleistocene	3	**Grahams Harbour Limestone**	No deposits recognized	'Granny Lake' Oolite	No deposits recognized	(No positively identifiable deposits of these ages)
	5a		**Grotto Beach Formation** Dixon Hill Member§	**Dixon Hill Limestone**§	**Almgreen Cay Formation**§ Upper Member Lower Member	
	5e		Cockburn Town Member French Bay Member	**Grotto Beach Limestone**	**Grotto Beach Formation** Fernandez Bay Member§ Cockburn Town Member French Bay Member	**Grotto Beach Formation** Cockburn Town Member‡ French Bay Member
	7, 9, 11, or earlier	**Grotto Beach Limestone**	**Owl's Hole Formation**	Unnamed Pre-Sangamonian	**Fortune Hill Formation**§ **Owl's Hole Formation**	**Owl's Hole Formation****

*Titus recognized the Grahams Harbour and Grotto Beach limestones only as Pleistocene, and he considered them to lie on an unnamed pre-Pleistocene biomicrite.
†This physical stratigraphy can be applied throughout the Bahamas.
§Units identifiable only through amino acid racemization data.
**Includes all material assigned to East Bay Member by Hearty and Kindler, 1993.
‡Includes rocks assigned to Almgreen Cay Formation and Fernandez Bay Member by Hearty and Kindler, 1993.
§§Includes rocks assigned to Fortune Hill Formation by Hearty and Kindler, 1993.

in publications and presentations at symposia on the geology of the Bahamas, Titus revised his stratigraphy (1987) to accommodate their then-current information (Table 2). Titus (1987) and Carew and Mylroie (1985) erroneously utilized amino acid racemization data to define parts of their stratigraphies. Beginning in 1988 we began to abandon the Dixon Hill Member because additional amino acid data was raising concerns about its validity, and we recognized that a unit based on AAR data should not be a part of our physical stratigraphy. By 1990, we (Mylroie and Carew, 1990) had ceased to recognize deposits laid down during oxygen isotope substage 5a (Dixon Hill Member), and by 1991 we had entirely eliminated it from our stratigraphy (Mylroie and Carew, 1991; Carew et al., 1992b).

Most recently, Hearty and Kindler (1993) have proposed a refinement of our 1985 stratigraphy (Table 2). Through the use of morphostratigraphy and AAR data, these authors have proposed five additional stratigraphic units. Regardless of the validity of their data, which as discussed earlier is subject to question in our opinion, none of these proposed units can be recognized in the field, and they can be identified only by using AAR data. As such, they are invalid formal stratigraphic units. For more extensive comments on the stratigraphy proposed by Hearty and Kindler (1993), see Carew and Mylroie (1994, 1995b).

The stratigraphy as outlined in this chapter is a valid physical stratigraphy for San Salvador Island. From our work on many other Bahamian islands, as well as that of other workers (e.g., Wilber, 1987; Kindler, 1991; Curran and Dill, 1991), it is evident that this stratigraphy is applicable to surficial rocks throughout the Bahamas. Therefore, the stratigraphy proposed herein may appropriately be seen as the stratigraphy for the Quaternary geology of all the Bahamian islands.

SUMMARY

During the Quaternary, the depositional history of the shallow platforms and islands of the Bahamas was controlled by the glacioeustatic sea-level changes associated with glaciation and deglaciation of the continents. Significant production of carbonate allochems and mud occurred only when highstands of sea level flooded the platform tops (above −10 m). As a result, the sedimentary record consists of packages of transgressive-phase, still-stand-phase, and regressive-phase deposits that were produced during the highest stands of Quaternary sea level. According to the deep-sea oxygen isotope record (Fig. 3), sea level was below the top edge of the platforms for most of the Quaternary. Then, only erosional processes were active on the platforms, and soils that would become paleosols developed on the exposed surfaces.

As a result of the glacioeustatically controlled depositional history, the stratigraphy of the Bahamian islands consists of allostratigraphic units, usually bounded by paleosols. Because of the current high elevation of sea level, and a probable isostatic subsidence rate of 1 to 2 m per 100,000 yr, the only marine subtidal deposits exposed on Bahamian islands are those deposited during oxygen isotope substage 5e (ca. 125,000 yr old). Besides those subtidal rocks, eolianites possibly deposited during oxygen isotope stages 11 (ca. 410,000 yr ago), 9 (ca. 320,000 yr ago), 7 (ca. 220,000 yr ago), and beach facies through eolianites of stages 5 (ca. 125,000 yr ago), and 1 (present) comprise the surficial rocks of the islands of the Bahamas. Based on physical stratigraphy alone, the rocks of the Bahamian islands can be divided into three allostratigraphic units: the older Pleistocene Owl's Hole Formation, the overlying (younger) late Pleistocene Grotto Beach Formation, and the Holocene Rice Bay Formation. The stratigraphy that was developed on San Salvador Island (Fig. 12) (Carew and Mylroie, 1985), is applicable to all other Bahamian islands where the surficial geology has been observed or described in the literature.

ACKNOWLEDGMENTS

We thank Donald T. Gerace, C.E.O., and Daniel Suchy, executive director, Bahamian Field Station, and the staff of the Field Station for their logistical and financial support during the many years that we have worked in the Bahamas. Discussions with many colleagues have added to our understanding of, and led to clarification of our ideas about, the geology of the Bahamas, but only we are responsible for the ideas expressed herein. Over the years, John Goddard, Richard Lively, June Mirecki, Sam Valastro, and John Wehmiller have provided us with U/Th radiometric ages, radiocarbon ages, and amino acid racemization data. We thank all our fellow carbonate enthusiasts with whom we have shared ideas, and we especially thank Mark Boardman, Al Curran, Pascal Kindler, Conrad Neumann, Neil Sealey, Peter Smart, Len Vacher, Brian White, and Jude Wilber. We have also had the benefit of help from the many graduate and undergraduate students who have worked with us in the Bahamas. Walt Manger and Len Vacher reviewed an earlier version of the manuscript, and some of their suggestions greatly improved this work. Additional financial support was provided by the University of Charleston, Mississippi State University, the Southern Regional Educational Board, and the International Blue Holes Research Project. The generous support of the Bahamian government and its officials is greatly appreciated.

REFERENCES CITED

Andersen, C. B., and Boardman, M. R., 1989, The depositional evolution of Snow Bay, San Salvador, *in* Mylroie, J. E., ed., Proceedings, Fourth Symposium on the Geology of the Bahamas: San Salvador, Bahamian Field Station, p. 7–22.

Austin, J. A., Jr., and Schlager, W., 1987, Ocean drilling program leg 101 explores the Bahamas, *in* Curran, H. A., ed., Proceedings, Third Symposium on the Geology of the Bahamas: Ft. Lauderdale, Florida, CCFL Bahamian Field Station, p. 1–33.

Bain, R. J., 1991, Distribution of Pleistocene lithofacies in the interior of San Salvador Island, Bahamas, and possible genetic models, *in* Bain, R. J., ed., Proceedings, Fifth Symposium on the Geology of the Bahamas: San Salvador, Bahamian Field Station, p. 11–21.

Beach, D. K., and Ginsburg, R. N., 1980, Facies succession of Pliocene-Pleistocene carbonates, northwestern Great Bahama Bank: American Association of Petroleum Geologists Bulletin, v. 64, p. 1634–1642.

Boardman, M. R., Carew, J. L., and Mylroie, J. E., 1987, Holocene deposition of transgressive sand on San Salvador, Bahamas: Geological Society of America Abstracts with Programs, v. 19, no. 7, p. 593.

Boardman, M. R., Carney, C., and Kim, N., 1991, Sedimentary compartments of a Holocene carbonate grainstone, San Salvador, Bahamas—Spatial and temporal linkages: Geological Society of America Abstracts with Programs, v. 23, no. 5, p. A225.

Boardman, M. R., Neumann, A. C., and Rasmussen, K. A., 1989, Holocene sea level in the Bahamas, *in* Mylroie, J. E., ed., Proceedings, Fourth Symposium on the Geology of the Bahamas: San Salvador, Bahamian Field Station, p. 45–52.

Bretz, J. H., 1960, Bermuda: A partially drowned, late mature, Pleistocene karst: Geological Society of America Bulletin, v. 71, p. 1729–1754.

Caputo, M. V., 1989, Selective cementation of eolian stratification in Pleistocene calcarenites, San Salvador, Bahamas, *in* Mylroie, J. E., ed., Proceedings, Fourth Symposium on the Geology of the Bahamas: San Salvador, Bahamian Field Station, p. 61–72.

Caputo, M. V., 1993, Eolian structures and textures in oolitic-skeletal calcarenites from the Quaternary of San Salvador Island, Bahamas: A new perspective on eolian limestones, *in* Keith, B. D., and Zuppann, C. W., eds., Mississippian oolites and modern analogs: American Association of Petroleum Geologists Studies in Geology 35, chap. 17, p. 243–259.

Carew, J. L., 1983, The use of amino acid racemization dating for unraveling the chronostratigraphy of San Salvador, Bahamas, *in* Gerace, D. T., ed., Proceedings, First Symposium on the Geology of the Bahamas: San Salvador, CCFL Bahamian Field Station, p. 12–17.

Carew, J. L., and Mylroie, J. E., 1985, The Pleistocene and Holocene stratigraphy of San Salvador Island, Bahamas, with reference to marine and terrestrial lithofacies at French Bay, *in* Curran, H. A., ed., Pleistocene and Holocene carbonate environments on San Salvador Island, Bahamas: Geological Society of America, Orlando, Annual Meeting Field Trip Guidebook: Ft. Lauderdale, Florida, CCFL Bahamian Field Station, p. 11–61.

Carew, J. L., and Mylroie, J. E., 1987, A refined geochronology for San Salvador Island, Bahamas, *in* Curran, H. A., ed., Proceedings, Third Symposium on the Geology of the Bahamas: Ft. Lauderdale, Florida, CCFL Bahamian Field Station, p. 35–44.

Carew, J. L., and Mylroie, J. E., 1989, The geology of eastern South Andros Island, Bahamas: A preliminary report, *in* Mylroie, J. E., ed., Proceedings, Fourth Symposium on the Geology of the Bahamas: San Salvador, Bahamian Field Station, p. 73–81.

Carew, J. L., and Mylroie, J. E., 1991, Some pitfalls in paleosol interpretation in carbonate sequences: Carbonates and Evaporites, v. 6, no. 1, p. 69–74.

Carew, J. L., and Mylroie, J. E., 1994, Comment on: Hearty, P. J., and Kindler, P., 1993, New perspectives on Bahamian geology: San Salvador Island, Bahamas: Journal of Coastal Research, v. 10, p. 1087–1094.

Carew, J. L., and Mylroie, J. E., 1995a, Fossil reefs and flank margin caves: Indicators of Late Quaternary sea level and tectonic stability of the Bahamas: Quaternary Science Reviews, v. 14, p. 145–153.

Carew, J. L., and Mylroie, J. E., 1995b, Rejoinder to Hearty, P. J., and Kindler, P., 1994, Straw men, glass houses, apples and oranges: A response to Carew and Mylroie's Comment on Hearty and Kindler (1993): Journal of Coastal Research, v. 11, p. 256–260.

Carew, J. L., Mylroie, J. E., and Lively, R. S., 1982, Bahamian caves and sea level change: Bahamas Naturalist, v. 6, no. 2, p. 5–13.

Carew, J. L., Mylroie, J. E., Wehmiller, J. F., and Lively, R. S., 1984, Estimates of Late Pleistocene sea level high stands from San Salvador, Bahamas, *in* Teeter, J. W., ed., Proceedings, Second Symposium on the Geology of the Bahamas: San Salvador, CCFL Bahamian Field Station, p. 153–175.

Carew, J. L., Mylroie, J. E., Pace, M., and Schwabe, S. J., 1992a, Development of Late Pleistocene eolianites in The Bahamas: Geological Society of America Abstracts with Programs, v. 24, no. 7, p. A142.

Carew, J. L., Mylroie, J. E., and Sealey, N. E., 1992b, Field guide to sites of geological interest, western New Providence Island, Bahamas; Field Trip Guidebook, Sixth Symposium on the Geology of the Bahamas: Port Charlotte, Florida, Bahamian Field Station, p. 1–23.

Carney, C., and Boardman, M. R., 1991, Oolitic sediments in a modern carbonate lagoon, Graham's Harbour, San Salvador, Bahamas: Geological Society of America Abstracts with Programs, v. 23, no. 5, p. A225.

Chen, J. H., Curran, H. A., White, B., and Wasserburg, G. J., 1991, Precise chronology of the last interglacial period: ^{234}U-^{230}Th data from fossil coral reefs in the Bahamas: Geological Society of America Bulletin, v. 103, p. 82–97.

Colby, N. D., and Boardman, M. R., 1989, Depositional evolution of a windward, high-energy carbonate lagoon, San Salvador, Bahamas, *in* Mylroie, J. E., ed., Proceedings, Fourth Symposium on the Geology of the Bahamas: San Salvador, Bahamian Field Station, p. 95–106.

Curran, H. A., and Dill, R. F., 1991, The stratigraphy and ichnology of a submarine cave in the Exuma Cays, Bahamas, *in* Bain, R. J., ed., Proceedings, Fifth Symposium on the Geology of the Bahamas: San Salvador, Bahamian Field Station, p. 57–64.

Curran, H. A., and White, B., 1985, The Cockburn Town fossil coral reef, *in* Curran, H. A., ed., Pleistocene and Holocene carbonate environments on San Salvador Island, Bahamas: Geological Society of America, Orlando, Annual Meeting Field Trip Guidebook: Ft. Lauderdale, Florida, CCFL Bahamian Field Station, p. 95–120.

Curran, H. A., and White, B., 1991, Trace fossils of shallow subtidal to dunal ichnofacies in Bahamian Quaternary carbonates: Palaios, v. 6, p. 498–510.

Dietz, R. S., Holden, J. C., and Sproll, W. P., 1970, Geotectonic evolution and subsidence of the Bahama platform: Geological Society of America Bulletin, v. 81, p. 1915–1928.

Eberli, G. P., and Ginsburg, R. N., 1987, Segmentation and coalescence of Cenozoic carbonate platforms, northwestern Great Bahama Bank: Geology, v. 15, p. 75–79.

Edwards, D. C., Teeter, J. W., and Hagey, F. M., 1990, Geology and ecology of a complex of inland saline ponds, San Salvador Island, Bahamas, *in* Field Trip Guidebook, Fifth Symposium on the Geology of the Bahamas: San Salvador, Bahamian Field Station, p. 35–45.

Fairbanks, R. G., and Matthews, R. K., 1978, The marine oxygen isotope record in Pleistocene corals, Barbados, West Indies: Quaternary Research, v. 10, p. 181–196.

Fairbridge, R. W., 1968, Periglacial eolian effects, *in* Fairbridge, R. W., ed., Encyclopedia of Geomorphology: Stroudsburg, Pennsylvania, Dowden, Hutchinson, and Ross, p. 825–829.

Florentino, E., and Bain, R. J., 1984, Environment of deposition of the Granny Lake oolite, San Salvador, Bahamas, *in* Teeter, J. W., ed., Proceedings, Second Symposium on the Geology of the Bahamas: San Salvador, CCFL Bahamian Field Station, p. 187–196.

Garrett, P., and Gould, S. J., 1984, Geology of New Providence Island, Bahamas: Geological Society of America Bulletin, v. 95, p. 209–220.

Hagey, D., 1991, Analysis of Pleistocene inland lake and lagoon deposits, San Salvador Island, Bahamas [M.S. thesis]: Mississippi State University, 132 p.

Halley, R. B., Muhs, D. R., Shinn, E. A., Dill, R. F., and Kindinger, J. L., 1991, A +1.5 m reef terrace in the southern Exuma islands, Bahamas: Geological Society of America Abstracts with Programs, v. 23, no. 1, p. 40.

Hattin, D. H., and Warren, V. L., 1989, Stratigraphic analysis of a fossil *Neogoniolithon*-capped patch reef and associated facies, San Salvador, Bahamas: Coral Reefs, v. 8, p. 19–30.

Hearty, P. J., and Kindler, P., 1991, The geological evolution of San Salvador Island, Bahamas: Geological Society of America Abstracts with Programs, v. 23, no. 5, p. A225.

Hearty, P. J., and Kindler, P., 1992, The geological evolution of San Salvador Island, Bahamas: Abstracts and Program, Sixth Symposium on the Geology of the Bahamas, p. 11–12.

Hearty, P. J., and Kindler, P., 1993, New perspectives on Bahamian geology: San

Salvador Island, Bahamas: Journal of Coastal Research, v. 9, p. 577–594.

Hearty, P. J., Kindler, P., and Schellenberg, S. A., 1993, The Late Quaternary evolution of surface rocks on San Salvador Island, Bahamas, *in* White, B., ed., Proceedings, Sixth Symposium on the Geology of the Bahamas: San Salvador, Bahamian Field Station, p. 205–222.

Hinman, E., 1980, Beaches, rocky shores, Pigeon Creek delta, and reefs of San Salvador, *in* Gerace, D. T., ed., Field Guide to the Geology of San Salvador: Miami, Florida, CCFL Bahamian Field Station, p. 106–119.

Hollin, J. T., and Hearty, P. J., 1990, South Carolina interglacial sites, and stage 5 sea levels: Quaternary Research, v. 33, p. 1–17.

Hutto, T., and Carew, J. L., 1984, Petrology of eolian calcarenites, San Salvador Island, Bahamas, *in* Teeter, J. W., ed., Proceedings, Second Symposium on the Geology of the Bahamas: San Salvador, CCFL Bahamian Field Station, p. 197–207.

Johnson, R. G., 1991, Major Northern Hemisphere deglaciation caused by a moisture deficit 140 Ka: Geology, v. 19, p. 686–689.

Kindler, P., 1991, Holocene stratigraphy of Lee Stocking Island, Bahamas—New interpretation with respect to sea level history: Geological Society of America Abstracts with Programs, v. 23, no. 1, p. 53.

Lloyd, R. M., Perkins, R. D., and Kerr, S. D., 1987, Beach and shoreface ooid deposition on shallow interior banks, Turks and Caicos Islands, British West Indies: Journal of Sedimentary Petrology, v. 57, p. 976–982.

Marcy, D. C., Carew, J. L., Colgan, M. W., and Katuna, M. P., 1993, Biometrics of *Cerion* shells using a new computer method (abs.): Bulletin of the South Carolina Academy of Science, v. 55, p. 97.

Marshall, P., Rasor, E., Lawson, D., Dechene, C., Carew, J. L., Schorr, G., Britt, C., and Mylroie, J. E., 1984, Investigation of breccia facies, San Salvador, Bahamas [abs.]: Bulletin of the South Carolina Academy of Science, v. 46, p. 109.

Meyerhoff, A. A., and Hatten, C. W., 1974, Bahamas salient of North America: Tectonic framework, stratigraphy, and petroleum potential: American Association of Petroleum Geologists Bulletin, v. 58, p. 1201–1239.

Mirecki, J. E., Carew, J. L., and Mylroie, J. E., 1992, Precision of amino acid enantiomeric data from fossiliferous late Quaternary units, San Salvador Island, The Bahamas: Abstracts and Program, Sixth Symposium on the Geology of the Bahamas, p. 13–14.

Mirecki, J. E., Carew, J. L., and Mylroie, J. E., 1993, Precision of amino acid enantiomeric data from fossiliferous late Quaternary units, San Salvador Island, The Bahamas, *in* White, B., ed., Proceedings, Sixth Symposium on the Geology of the Bahamas: San Salvador, Bahamian Field Station, p. 95–101.

Mitchell, S. W., 1987, Surficial geology of Rum Cay, Bahama Islands, *in* Curran, H. A., ed., Proceedings, Third Symposium on the Geology of the Bahamas: Ft. Lauderdale, Florida, CCFL Bahamian Field Station, p. 231–242.

Mitchell, S. W., Buening, N., Baldwin, J., Jr., and Westell, B., 1989, Holocene depositional history of Conception Island, Bahamas, *in* Mylroie, J. E., ed., Proceedings, Fourth Symposium on the Geology of the Bahamas: San Salvador, Bahamian Field Station, p. 209–220.

Mullins, H. T., and Hine, A. C., 1989, Scalloped bank margins: Beginning of the end for carbonate platforms?: Geology, v. 17, p. 30–33.

Mullins, H. T., Breen, N., Dolan, J., Wellner, R. W., Petruccione, J. L., Gaylord, M., Anderson, B., Melillo, A. J., Jurgens, A. D., and Orange, D., 1992, Carbonate platforms along the southeast Bahamas–Hispanola collision zone: Marine Geology, v. 105, p. 169–209.

Mylroie, J. E., and Carew, J. L., 1990, Erosional notches in Bahamian carbonates: Bioerosion or ground water dissolution?: Abstracts with Program, Fifth Symposium on the Geology of the Bahamas, p. 15.

Mylroie, J. E., and Carew, J. L., 1991, Erosional notches in Bahamian carbonates: Bioerosion or ground water dissolution?, *in* Bain, R. J., ed., Proceedings, Fifth Symposium on the Geology of the Bahamas: San Salvador, Bahamian Field Station, p. 185–191.

Mylroie, J. E., Carew, J. L., and Barton, C. E., 1985, Paleosols and karst development, San Salvador Island, Bahamas [abs.]: Frankfort, Kentucky, Program, 1985 National Speleological Society Convention, p. 43.

Mylroie, J. E., Carew, J. L., Sealey, N. E., and Mylroie, J. R., 1991, Cave development on New Providence Island and Long Island, Bahamas: Cave Science (Transactions of the British Cave Research Association), v. 18, no. 3, p. 139–151.

Neumann, A. C., and Moore, W. S., 1975, Sea level events and Pleistocene coral ages in the northern Bahamas: Quaternary Research, v. 5, p. 215–224.

Noble, R. S., Curran, H. A., and Wilson, M. A., 1991, Paleoenvironmental and paleoecological analysis of a Pleistocene lagoonal mollusk-rich facies, San Salvador Island, Bahamas: Geological Society of America Abstracts with Programs, v. 23, no. 1, p. 109.

North American Commission on Stratigraphic Nomenclature, 1983, North American Stratigraphic Code: American Association of Petroleum Geologists Bulletin, v. 67, p. 841–875.

Panuska, B. C., Carew, J. L., and Mylroie, J. E., 1991, Paleomagnetic directions of paleosols on San Salvador: Prospects for stratigraphic correlation, *in* Bain, R. J., ed., Proceedings, Fifth Symposium on the Geology of the Bahamas: San Salvador, Bahamian Field Station, p. 193–202.

Rasmussen, K. A., and Neumann, A. C., 1987, Holocene overprints of Pleistocene paleokarst: Bight of Abaco, Bahamas, *in* James, N. P., and Choquette, P. W., eds., Paleokarst: New York, Springer-Verlag, p. 132–148.

Schellenberg, S. A., and Hearty, P. J., 1992, *Cerion* biostratigraphy of San Salvador Island, Bahamas: Program and Abstracts, Sixth Symposium on the Geology of the Bahamas, p. 17.

Schwabe, S. J., Carew, J. L., and Mylroie, J. E., 1993, Petrology of Bahamian Pleistocene eolianites and flank margin caves: Implications for Late Quaternary island development, *in* White, B., ed., Proceedings, Sixth Symposium on the Geology of the Bahamas: San Salvador, Bahamian Field Station, p. 149–164.

Shackleton, N. J., 1987, Oxygen isotopes, ice volume, and sea level: Quaternary Science Reviews, v. 6, p. 183–190.

Shackleton, N. J., and Opdyke, N. D., 1973, Oxygen isotope and paleomagnetic stratigraphy of equatorial Pacific core V28-238: Oxygen isotope temperatures and ice volumes on a 10^5 to 10^6 year scale: Quaternary Research, v. 3, p. 39–55.

Shattuck, G. B., and Miller, B. L., 1905, Physiography and geology of the Bahama islands, *in* Shattuck, G. B., ed., The Bahama Islands: New York, Johns Hopkins Press, Macmillan Co., p. 3–20.

Sims, W. R., 1987, Facies analysis of carbonate rocks on the southern end of San Salvador Island, Bahamas [M.S. thesis]: Akron, Ohio, University of Akron, 135 p.

Stowers, R. E., II, Mylroie, J. E., and Carew, J. L., 1989, Pleistocene stratigraphy and geochronology, southwestern San Salvador Island, Bahamas, *in* Mylroie, J. E., ed., Proceedings, Fourth Symposium on the Geology of the Bahamas: San Salvador, Bahamian Field Station, p. 323–330.

Thalman, K. L., 1983, A Pleistocene lagoon and its modern analog, San Salvador, Bahamas [M.S. thesis]: Akron, Ohio, University of Akron, 166 p.

Thalman, K. L., and Teeter, J. W., 1983a, A Pleistocene estuary and its modern analog, San Salvador, Bahamas, *in* Gerace, D. T., ed., Proceedings, First Symposium on the Geology of the Bahamas: Miami, Florida, CCFL Bahamian Field Station, p. 13–21.

Thalman, K. L., and Teeter, J. W., 1983b, A pleistocene estuary and its modern analog, San Salvador, Bahamas: Geological Society of America Abstracts with Programs, v. 15, no. 2, p. 67.

Titus, R., 1980, Emergent facies patterns on San Salvador Island, Bahamas, *in* Gerace, D. T., ed., Field guide to the geology of San Salvador: Miami, Florida, CCFL Bahamian Field Station, p. 92–105.

Titus, R., 1983a, Quaternary stratigraphy of San Salvador, Bahamas, *in* Gerace, D. T., ed., Proceedings, First Symposium on the Geology of the Bahamas: Miami, Florida, CCFL Bahamian Field Station, p. 1–5.

Titus, R., 1983b, Emergent facies patterns on San Salvador Island, Bahamas, *in* Gerace, D. T., ed., Field guide to the geology of San Salvador (third edition): Miami, Florida, CCFL Bahamian Field Station, p. 97–116.

Titus, R., 1984, Physical stratigraphy of San Salvador Island, Bahamas, *in* Teeter, J. W., ed., Proceedings, Second Symposium on the Geology of the Bahamas: San Salvador, CCFL Bahamian Field Station, p. 209–228.

Titus, R., 1987, Geomorphology, stratigraphy, and the Quaternary history of San Salvador, *in* Curran, H. A., ed., Proceedings, Third Symposium on the Geology of the Bahamas: Ft. Lauderdale, Florida, CCFL Bahamian Field Station, p. 155–164.

Uchupi, E., Milliman, J. D., Luyendyk, B. P., Brown, C. O., and Emery, K. O., 1971, Structure and origin of the southeastern Bahamas: American Association of Petroleum Geologists Bulletin, v. 55, p. 687–704.

Vacher, H. L., 1973, Coastal dunes of Younger Bermuda, *in* Coates, P. R., Coastal Geomorphology; Publications in Geomorphology: Binghamton, State University of New York, p. 355–391.

Vacher, H. L., and Hearty, P. J., 1989, History of Stage 5 sea level in Bermuda: Review with new evidence of a brief rise to present sea level during substage 5a: Quaternary Science Reviews, v. 8, p. 159–168.

White, B., 1989, Field guide to the Sue Point fossil coral reef San Salvador Island, Bahamas, *in* Mylroie, J. E., ed., Proceedings, Fourth Symposium on the Geology of the Bahamas: San Salvador, Bahamian Field Station, p. 353–365.

White, B., and Curran, H. A., 1985, The Holocene carbonate eolianites of North Point and the modern marine environments between North Point and Cut Cay, *in* Curran, H. A., ed., Pleistocene and Holocene carbonate environments on San Salvador Island, Bahamas: Geological Society of America, Orlando, Florida, Annual Meeting Field Trip Guidebook: Ft. Lauderdale, Florida, CCFL Bahamian Field Station, p. 73–93.

White, B., and Curran, H. A., 1987, Coral reef to eolianite transition in the Pleistocene rocks of Great Inagua Island, Bahamas, *in* Curran, H. A., ed., Proceedings, Third Symposium on the Geology of the Bahamas: Ft. Lauderdale, Florida, CCFL Bahamian Field Station, p. 169–179.

White, B., and Curran, H. A., 1988, Mesoscale physical sedimentary structures and trace fossils in Holocene carbonate eolianites from San Salvador Island, Bahamas: Sedimentary Geology, v. 55, p. 163–184.

White, B., and Curran, H. A., 1993, Sedimentology and ichnology of Holocene dune and backshore deposits, Lee Stocking Island, Bahamas, *in* White, B., ed., Proceedings, Sixth Symposium on the Geology of the Bahamas: San Salvador, Bahamian Field Station, p. 181–191.

White, B., Kurkjy, K. A., and Curran, H. A., 1984, A shallowing-upward sequence in a Pleistocene coral reef and associated facies, San Salvador, Bahamas, *in* Teeter, J. W., ed., Proceedings, Second Symposium on the Geology of the Bahamas: San Salvador, CCFL Bahamian Field Station, p. 53–70.

White, K. S., 1991, An imprint for the Holocene transgression found as the marine cement aragonite in Quaternary carbonate eolianites on San Salvador Island, Bahamas: Geological Society of America Abstracts with Programs, v. 23, no. 1, p. 148.

White, K. S., and White, B., 1991, The effects of Holocene sea-level rise on the diagenesis of Quaternary carbonate eolianites, San Salvador Island, Bahamas, *in* Bain, R. J., ed., Proceedings, Fifth Symposium on the Geology of the Bahamas: San Salvador, Bahamian Field Station, p. 235–247.

Wilber, R. J., 1987, Geology of Little San Salvador Island and West Plana Cay: Preliminary findings with implications for Bahamian island stratigraphy, *in* Curran, H. A., ed., Proceedings, Third Symposium on the Geology of the Bahamas: Ft. Lauderdale, Florida, CCFL Bahamian Field Station, p. 181–204.

Manuscript Accepted by the Society January 5, 1995

Printed in U.S.A.

Geological Society of America
Special Paper 300
1995

Bahamian paleosols: Origin, relation to paleoclimate, and stratigraphic significance

Mark R. Boardman, Richard F. McCartney, and Matthew R. Eaton
Geology Department, Miami University, Oxford, Ohio 45056

ABSTRACT

The source and mode of accumulation of insoluble residue (IR) and the potential of paleosols as stratigraphic aids were investigated by megascopic, mineralogic, textural, and scanning electron microscope examination of Bahamian paleosols from San Salvador and Andros Island. Bahamian paleosols are predominantly calcium carbonate and contain minor amounts of mineral insoluble residue (quartz, plagioclase, and clay minerals).

There are at least four megascopically distinct paleosol types: laminated crusts, homogeneous crusts, breccia/conglomerate, and homogeneous matrix. There is no apparent relation between megascopic type of paleosol and underlying carbonate lithology, time of formation, or mineralogic composition of the insoluble residue.

The IR mineralogy and the quartz size distribution of Bahamian paleosols were compared to that of a nearby periplatform core (taken from the Northwest Providence Channel, Bahamas). This comparison suggests that there is a similar source for the insoluble residue, i.e., dust transported from the Sahara Desert. The similarity of quartz size distribution of lowstand (glacial period) sediments to the size distribution of quartz within paleosols suggests that paleosols dominantly form during lowstands of sea level. A comparison of the percentage of IR in the host limestones (<0.01%) with the percentage of IR in paleosols (ca. 6.0%) suggests that, in order for paleosols to have formed by dissolution of host limestones, unacceptably large volumes of limestone would have to have been dissolved in relatively short time periods. However, rates of dust deposition of the deep-sea core suggest that the IR of typical Bahamian paleosols could have formed by dust accretion during the several tens of thousands of years of lowstand conditions. Unfortunately, there is no mineralogic or textural "fingerprint" that can be used to assign specific lowstand (glacial) episodes to specific paleosols.

Paleosols can be classified by IR mineralogy using the relative abundance of two minerals: boehmite and illite. Clay mineralogy in paleosols is primarily a function of diagenesis accompanying climatic variations through the late Quaternary. However, the difference between boehmite-rich paleosols and illite-rich paleosols appears to be due to local environmental conditions (topography) rather than regional climatic changes. Boehmite-rich paleosols result from intense leaching, whereas illite-rich paleosols represent less intense diagenetic alteration and more closely resemble the mineralogy of IR brought to the Bahamas as dust. Interpretation of paleosol IR mineralogy is complicated by erosion and redeposition of soils/paleosols.

All four megascopic types of Bahamian paleosols and all mineralogic varieties of paleosols are found on stratigraphically equivalent rock units. Thus, paleosols in the Bahamas cannot be used as stratigraphic markers to correlate stratigraphic units on a single island or between islands.

Boardman, M. R., McCartney, R. F., and Eaton, M. R., 1995, Bahamian paleosols: Origin, relation to paleoclimate, and stratigraphic significance, *in* Curran, H. A., and White, B., Terrestrial and Shallow Marine Geology of the Bahamas and Bermuda: Boulder, Colorado, Geological Society of America Special Paper 300.

INTRODUCTION

Paleosols associated with limestones are striking visual indicators of subaerial diagenesis that represent unknown, but presumably long, periods of time. Paleosols can form on limestones of marine origin only when relative sea level is lowered, and thus should be an absolute criterion for bounding surfaces and a useful criterion for stratigraphic correlation. This chapter describes Quaternary paleosols from islands in the Bahamas to determine their relationship to fluctuations of Quaternary sea level. In particular, we investigate megascopic aspects of paleosols, the origin of insoluble residue (IR) within the paleosols, the durations of exposure paleosols represent, the mineralogic and textural signatures of IR within paleosols, and the usefulness of paleosols for stratigraphic correlation.

The term paleosol is defined as a "buried soil horizon of the geologic past" and is synonymous with "buried soil" and "fossil soil" (Bates and Jackson, 1980); however, we use this term in a more general sense to embrace other descriptive terms associated with horizons of subaerial exposure such as caliche crusts, calcareous laminated crusts, calcrete, duricrusts, and terra rossa. Because paleosols are indicators of subaerial exposure, they can play a prominent role in stratigraphic analysis (e.g., Bermuda: Vacher et al.., 1989, 1992). The petrography of paleosols has been described for Holocene and Pleistocene limestones (Ruhe et al., 1961; Land et al., 1967; Multer and Hoffmeister, 1968; James, 1972; Read, 1974; Harrison, 1977; Watts, 1980; Braithwaite, 1983; Coniglio and Harrison, 1983), as well as for ancient limestones (Walls et al., 1975; Harrison and Steinen, 1978; Riding and Wright, 1981; Goldhammer and Elmore, 1984; Prather, 1985; Wright, 1986). In addition to petrographic studies, accumulation rates have been estimated (Gile et al., 1966; Robbin and Stipp, 1979), and the origin and diagenesis of insoluble residue in paleosols have been examined (Hose, 1963; Sinclair, 1967).

Paleosols and soils in the Bahamas

In the Bahamas, early descriptions of paleosols (Newell and Rigby, 1957; Kornicker, 1958) have been amplified by recent studies of the late Quaternary paleosols of San Salvador and Eleuthera (Brown, 1984; Hale and Ettensohn, 1984; Bain and Foos, 1993). These studies primarily have been descriptive and have not addressed the source of the insoluble residue (IR), the process of paleosol accumulation/formation, the relation of paleosols to fluctuations of sea level, nor the use of paleosols as stratigraphic markers in the Bahamas.

Modern Bahamian soils have been described based on color and organic content (Little et al., 1977) and insoluble residue mineralogy (Ahmad and Jones, 1969; Foos, 1991). Sealey (1985) recognized four typical modern soils in the Bahamas based on the relative importance of organic matter and clay minerals. He related each soil type to a particular style of vegetation (coppice, coastal grasses, pine forest, cleared pine forest) and substrate (hard limestone, coastal sands, in between ridges).

The dominant source of mineral IR in the Bahamas is dust transported from Africa (Syers et al., 1969; Eaton, 1986; Muhs et al., 1990). The episodic dust "storms" common in Florida today owe their origin to eolian-transported Saharan dust (Glaccum and Prospero, 1980). The supply of dust from Africa and its deposition by "rain-out" has been , during glacial as well as interglacial times. Other regions such as Jamaica, Bermuda, and Barbados may have had significant contributions from volcanic sources (Burns, 1961; Ruhe et al., 1961; Blackburn and Taylor, 1969).

Origins of paleosols

Paleosols typically contain a higher proportion of mineral insoluble residue than the enclosing limestone units, and the mechanism of IR accumulation within paleosols has been debated for decades. There are two widely accepted models that explain the concentration of insoluble residue in limestone paleosols: host limestone dissolution with residual accumulation, and accretion of dust at the limestone exposure surface.

In the limestone dissolution model, selective dissolution and removal of $CaCO_3$ from the carbonate host rock causes the residual accumulation of previously entombed IR. This insoluble residue is mixed with carbonate debris at the exposure surface to form a soil (Ruhe et al., 1961; Ahmad et al., 1966; Sinclair, 1967; Blackburn and Taylor, 1969, 1970). Cementation of the soil mixture by calcite produces a lithified soil that may be preserved. Support for this idea is thought to be the mineralogic similarity between paleosols and the underlying limestone host rock. However, mineralogic similarity would likely exist in either model. Limestone dissolution may generate some paleosols, but, for limestones with low concentrations of IR, the amount of dissolution required to produce a paleosol (typically, 10 to 50 cm thick) is large—on the order of 10 to 20 m (Muhs et al., 1990). Thick soil zones with abundant IR require significant amounts of limestone dissolution.

In the dust accretion model, a rain of IR from the atmosphere accumulates at the limestone surface and is incorporated with limestone debris and reprecipitated carbonate to form one of several types of soils that may be preserved as a paleosol (Multer and Hoffmeister, 1968; Bricker and Mackenzie, 1970; Robbin and Stipp, 1979). In this model, some of the host limestone is dissolved and used to cement the accreting soil; however, the amount of limestone dissolved is presumed to be small. To prove this method of paleosol formation requires, among other things, that the measured dust flux be sufficient, and that there be some characteristic of the IR in the paleosol that was uniquely different from the underlying limestone. To date, both models seem possible, and neither method of formation has been uniquely demonstrated.

Paleosols and stratigraphy

Paleosols have a prominent place in the stratigraphy of many areas. The major use of paleosols is to help resolve correlation problems created by the scarcity of outcrops, the lithologic similarities of the limestone deposits, the distances between limestone units, and the difficulty in dating the rocks. When a paleosol is used as a stratigraphic marker or unit, it is more properly termed a "geosol" (North American Committee on Stratigraphic Nomenclature, 1983). Because paleosols represent hiatuses, they should be important to island stratigraphy and in stratigraphic correlation among islands. In both Bermuda and the Bahamas, paleosols have been used to help differentiate lithologic units (Land et al., 1967; Carew and Mylroie, 1985, 1987; Vacher et al., 1989, 1992). In Bermuda, the significance of four paleosols has been underscored by assigning stratigraphic names to the paleosol units (e.g., Castle Harbor Geosol: Vacher et al., 1989, 1992). In the Bahamas, many Quaternary highstands of sea level have been of short duration (a few thousands of years) and within a few meters of present sea level (Boardman and Neumann, 1984; Boardman et al., 1986). Because of this, relatively thin depositional packages are created, and, around topographically high areas, lateral accretion of carbonate facies as opposed to vertical "layer-cake" accretion has occurred. Thus, an understanding of the formation of paleosols and exposure surfaces should help clarify the complex time and/or stratigraphic relationships among carbonate units on individual platforms and among platforms.

To determine the origin of paleosols, to estimate the duration of exposure required to create a paleosol and to determine the usefulness of paleosols as stratigraphic markers, the insoluble residue of a deep-sea core in the Bahamas was compared with the paleosols of San Salvador. The stratigraphy of San Salvador has been extensively studied (Carew and Mylroie, 1985, 1987; Chen et al., 1991). It is clear that carbonate sediments have been accumulated as laterally and vertically accreted units during several highstand episodes (during times when large portions of the San Salvador platform were flooded). On San Salvador, paleosols might be useful in distinguishing individual stratigraphic units. To test the usefulness of paleosols as stratigraphic markers over longer distances, paleosols from Andros were compared with those from San Salvador.

METHODS

Periplatform core

An 11.7-m-long sediment core was recovered from a water depth of 675 m in the periplatform sediments of Northwest Providence Channel, Bahamas (Fig. 1). The sediment of Northwest Providence Channel is dominated by a calcium carbonate derived from two sources: a continuous rain of planktonic biogenic debris (e.g., calcitic coccoliths and foraminifera), and an episodic

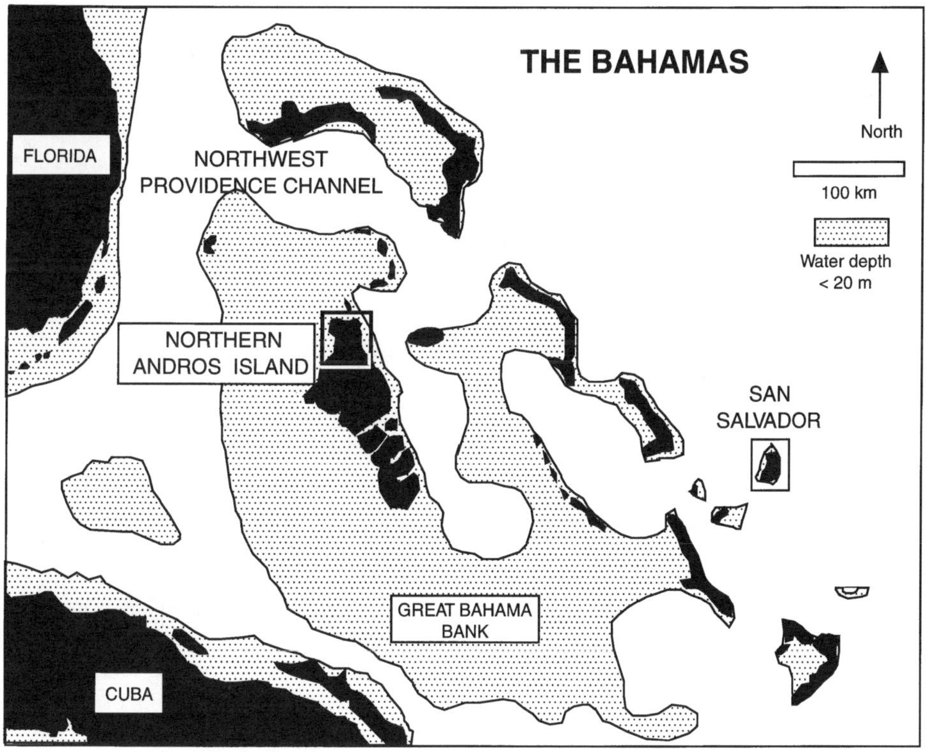

Figure 1. Map of the Bahamas. Paleosol samples came from San Salvador and northern Andros Islands. The deep-sea core is from Northwest Providence Channel located between Great Bahama Bank and Little Bahama Bank.

input of fine-grained material shed from the banktops (e.g., aragonitic material derived from green algae and inorganic precipitation and high-Mg calcite derived from benthic forams and red algae) when the platforms are flooded, i.e., during highstands of sea level. The core from Northwest Providence Channel contains sediments whose mineralogy fluctuates between aragonite-rich carbonate (= highstands of sea level; approximately 80% aragonite) and calcite-rich carbonate (= lowstands of sea level; approximately 20% aragonite; Boardman and Neumann, 1984, 1986; Boardman et al., 1986). Because the aragonite-rich, banktop-shed sediment is added to the continuous rain of calcitic planktonic material, the rate of deposition of these periplatform sediments varies from approximately 2 cm/1,000 yr (calcitic, lowstand sedimentation) to approximately 10 cm/1,000 yr (aragonitic, highstand sedimentation). The percentage of aragonite is, thus, a measure of the rate of sedimentation in addition to an indicator of climate.

Samples were extracted every 10 cm (117 total samples). These samples were sieved to isolate the sand sizes, and the mud fraction was separated into coarse silt (16 to 63 μm), fine silt (4 to 16 μm), and clay (<4 μm) by repeated settling and decanting. The content of IR in each mud fraction was determined by weight loss after dissolution in 10% acetic acid.

Mineralogy of the IR of each size fraction was determined by x-ray diffraction of samples that had been mounted on filters using a vacuum. This minimizes, but does not eliminate, potential errors caused by mineral size segregation (Gibbs, 1965, 1968). Relative proportions of IR were determined by comparing areas of the primary peaks (e.g., Brindley; Biscayne, 1965).

The flux of IR to the sediment was determined using the percentage of IR (Eaton, 1986), sedimentation rate estimated by aragonite content (Boardman and Neumann, 1984), and the sediment density (1.01 g/cm^3; determined by Busby, 1962, for similar periplatform sediment). Details of the procedures and results of the analyses of IR in this core are in Eaton (1986).

Paleosols and host limestones

A total of 116 paleosol and 20 limestone samples were studied by polished sections and thin sections from 31 sites on San Salvador and Andros Island, Bahamas (Figs. 1 and 2). Multiple samples were taken at many sites to examine vertical and lateral continuity of paleosols. Nine modern soil samples were also collected.

Portions of the paleosol and limestone rock samples were powdered, and the percentage of IR was determined by dissolution with 15% acetic acid. Care was taken in sample selection to avoid host limestones contaminated by IR resulting from pedogenic processes. A subsample of the IR of soil and paleosol samples was used to determine the proportion of organic matter by determining the weight loss after combustion at 600°C for several hours (Dean, 1974).

Mineralogy of insoluble residue was determined by x-ray diffraction. Many of the modern soil samples contained significant proportions of organic matter; these samples were treated with 30% hydrogen peroxide to oxidize the organic matter prior to analysis by x-ray diffraction (Tessler et al., 1979). All samples were x-rayed four times: when untreated, after being treated with glycol, after being heated to 400°C, and after being heated to 550°C. Clay mineral identification was determined following the guidelines of Brown and Brindley (1980) and Starkey et al. (1984). Relative proportions of IR were determined by comparing areas of the primary peaks (e.g., Brindley, 1961; Biscaye, 1965).

Aragonite and calcite percentages of bulk limestone rock were determined by x-ray diffraction based on peak area ratios (Chave, 1954; Milliman, 1974) and the calibration curve of Boardman (1978). Porosity of selected limestone and paleosol samples was determined by the water-saturated bulk porosity method.

Quartz was extracted from 16 paleosol samples and 4 samples from the deep-sea sediment core described above using the sodium pyrosulfate–hydrofluorosilicic acid method (Chapman et al., 1969; Dauphin, 1980). Quartz particle size distribution was resolved using a Leeds and Northrup Microtrac small particle analyzer. Scanning electron microscope (SEM) examination was performed on selected samples of the bulk IR and extracted quartz, and an energy dispersive x-ray analyzer (EDAX-9100) was used to aid identification of particles. Details of the methods and results of the examination of paleosols can be found in McCartney (1987).

PERIPLATFORM CORE

Of particular interest in our research is the interpretation of paleosols in terms of sea-level fluctuations and the determination of which characteristics of IR, if any, are related to climate (and thus, sea level). Paleosols are linked to climate through analysis of the periplatform core. A comparison of texture and mineralogy of IR in the deep-sea core, in the trade winds, and in African source areas shows that insoluble residue in the Bahamas originates from two areas in Africa: one area during glacial climatic conditions (lowstands of sea level) and a second area during interglacial climatic conditions (highstands of sea level).

Other sources of insoluble residue in the periplatform core, such as the Western Boundary Undercurrent (WBUC), rivers from southeast North America, and cosmic dust, do not seem to be significant. The core site is located atop an intercanyon, bathymetric high (665 m deep), whereas the WBUC is located at a much greater depth (ca. 4,000 m; Tucholke, 1975). Also, the silt fraction of the WBUC has a different mineralogy than that of the core sediment. Estuaries trap much of the fine sediment from rivers, and the sediment that does bypass the estuaries (both now and during lowstands of sea level) is swept northward and past the Bahamas by the Gulf Stream (Pratt, 1968; Brunner, 1984). Cosmic dust is volumetrically too insignificant to be important in the formation of paleosols in the Bahamas.

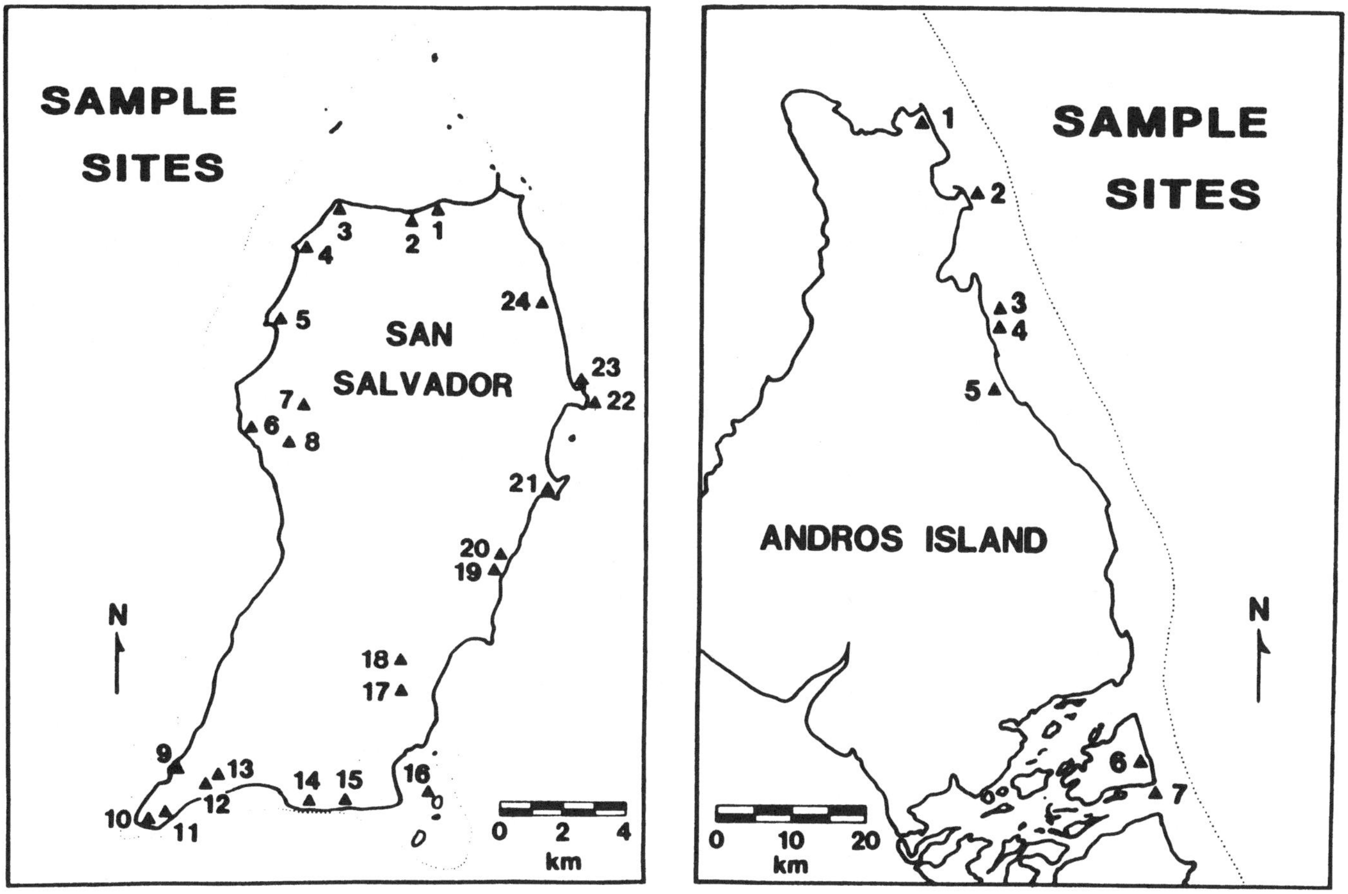

Figure 2. Sample locations. On both San Salvador and Andros, multiple samples were usually taken at each of the numbered sites to provide information on close-spaced lateral and vertical variability of paleosols, as well as to provide a sample of the underlying limestone.

Texture of insoluble residue

The texture of IR of the deep sea core is similar to that of dust collected from today's Trade Wind (Fig. 3). Insoluble residue in the core contains approximately 60% clay-sized material, 30% fine silt-sized material, and 10% coarse silt-sized material. African dust is transported across the Atlantic Ocean today in a layer of the Trade Winds called the "Saharan Air Layer" (1.5 to 7 km high). Air turbulence keeps finer material suspended during the 4 to 6 days required for the journey from Africa to the Bahamas (Prospero and Carlson, 1972; Prospero, 1981a,b). Some of the IR in the cores is coarser than IR collected to date from Trade Wind dust (Fig. 3), and it may be that this coarser IR was transported during periods of increased wind velocity and/or during times when major dust storms were more common.

Concentration of insoluble residue

The concentration of IR is generally between 1 and 2% of the total mud fraction and displays little correlation to climate (and fluctuations of sea level; Fig. 4). There are several regions of the core that have sharply higher concentrations of IR that are not systematically related to the oxygen isotope signal of climatic fluctuations. These five pulses of increased insoluble residue (located in the core at 350, 410, 500, 880, and 1,130 cm) are primarily caused by a significantly higher concentration of coarse and fine silt (Fig. 4). Three of these pulses of coarser IR occur at the transition from a glacial to an interglacial period. Pulses of increased concentration of coarse and fine silt are superposed on the input signal of clay-sized IR, which is systematically related to climate. During glacial periods (measured in the periplatform core by heavy oxygen isotope ratios and low quantities of aragonite), the percentage of IR in the clay-sized fraction is higher than during interglacial periods (Fig. 4). This supports the idea that wind intensities may have been higher during glacial periods when the climatic zones were more compressed (Parkin and Shackleton, 1973; Parkin, 1974). The relatively high proportion of IR in the clay-sized fraction during glacial periods is not necessarily the result of a greater flux of dust, but may result from a lower carbonate deposition rate in this area during glacial episodes (thus causing an increase in the relative percentage of IR; Boardman and Neumann, 1984).

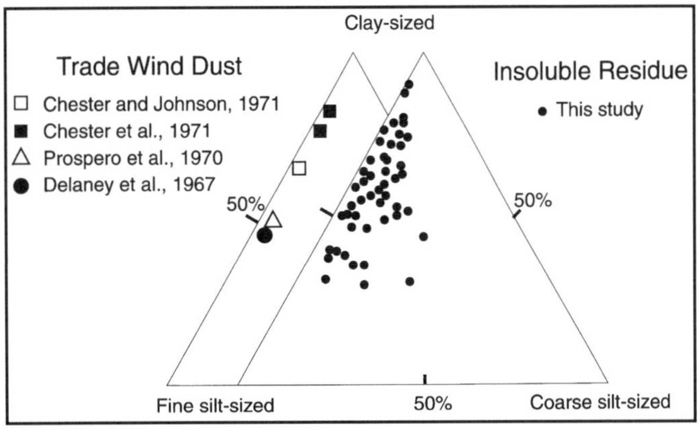

Figure 3. Grain size distribution of dust from Trade Winds today is similar to the texture of IR in a deep-sea core in Northwest Providence Channel.

Mineralogy of insoluble residue

The mineralogy of the fine and coarse silt in the IR from the deep-sea core confirms the importance of climatically influenced dust transport in controlling the input of IR to the Bahamas. Silt-sized dust in the core and Trade Winds predominantly consists of mixtures of quartz, plagioclase, and dolomite.

In the Bahamian deep-sea core, there are two distinct mineralogic suites of silt-sized insoluble residue (Fig. 5). Both suites contain quartz and plagioclase; one suite is dolomite-poor, the other suite contains abundant dolomite. Dolomite is an indicator of dust originating from the southern Sahara region, whereas little dolomite is found in the dust from the northern region of northwest Africa (Fig. 6) (Delaney et al., 1967; Chester and Johnson, 1971; Johnson, 1974, 1979). The two mineralogic suites may result from fluctuations in the size and distribution of deserts in northwest Africa as climate varied from glacial to interglacial periods (Diester-Haas et al., 1973; Parmenter and Folger, 1974; Diester-Haas, 1976). During interglacial periods, deserts (areas likely to generate dust) migrated and/or expanded to the north (dolomite-poor areas). During glacial periods, desert areas were more common in the southern regions (dolomite-rich areas) (Fig. 6). Today (an interglacial period), IR in the top of the periplatform core, as well as dust in Miami, Florida, is distinctly dolomite-poor (Glaccum and Prospero, 1980) and reflects a source from the northern Sahara region.

SEM studies of the dolomite in the periplatform core, in conjunction with other data, suggest that the dolomite is detrital. Most dolomite crystals in the core are corroded and abraded rather than unpitted, euhedral rhombs. If the dolomite had been precipitated in situ, it seems likely that the rhombs would have remained unpitted and euhedral. The lack of evidence of dissolution of associated calcite and aragonite (both of which are more soluble in sea water than dolomite) suggests that dissolution of carbonate phases (especially the more stable dolomite) has not been important in these periplatform sediments. Thus, the surface morphology of the dolomite and the proportion of dolomite relative to other mineral insoluble residue (Figs. 5 and 6) suggest that this dolomite is detrital rather than authigenic (Mullins et al., 1985).

Summary of periplatform insoluble residue

Dust is brought from Africa to the Bahamas by the Trade Winds and is "rained" out. Similarities of texture and mineralogy in a deep-sea sediment core support other studies that indicate African dust as a major source of IR in the Bahamas (Syers et al., 1969; Muhs et al., 1990). The variability in dolomite content and the change in texture is likely a climatic signal that records the changing source area in Africa and changing positions and/or intensities of the Trade Winds. Once at the ocean surface, dust is incorporated into the sediment via the process of settling of fecal pellets formed by filter-feeding plankton. This process is known to be effective in accelerating vertical transport and limiting lateral transport of fine-grained material in the Bahamas (Boardman and Neumann, 1984; Boardman et al., 1986).

INSOLUBLE RESIDUE AND PALEOSOLS

Types of paleosols

Bahamian paleosols are typically micritic deposits that range in color from red to brown and can be laminated to nonlaminated. Paleosols vary from well-indurated crusts that are up to 6 cm thick to nonindurated earthy deposits that can be as thick as 1 m. Four fundamental types are recognized: laminated crust, homogeneous crust, breccia-conglomerate, and homogeneous-matrix (Fig. 7). These megascopically distinct types of paleosols are sometimes found adjacent to each other.

Laminated crusts (Fig. 7A) are the most abundant paleosols found on San Salvador and Andros Islands. Laminae are colored red, tan, and brown, vary from 0.1 to 1 mm, are often regularly spaced, and pinch and swell in association with microtopographic highs and lows. Lamination may be simple or complex. Laminated crusts are rarely associated with breccia-conglomerate and homogeneous-matrix paleosols.

Homogeneous crusts (Fig. 7B) (microcrystalline rind of Multer and Hoffmeister, 1968) vary in thickness from a few millimeters to 1 cm. They are found individually and associated with other paleosols or modern soils.

Breccia-conglomerate paleosols (Fig. 7C) and their modern counterparts have large clasts (>2 mm: Gile et al., 1966) that vary from subangular to rounded. Homogeneous-matrix paleosols (Fig. 7D) lack large clasts. Both may have shell, ooid, or peloid grains engulfed in a micritic matrix with void-filling cement. These two types of paleosols are the thickest, varying from several centimeters to a meter thick.

Characteristics of paleosols

Insoluble residue consists of organic matter and inorganic mineral matter (Fig. 8). Total IR in 70 paleosols averages 6.7%

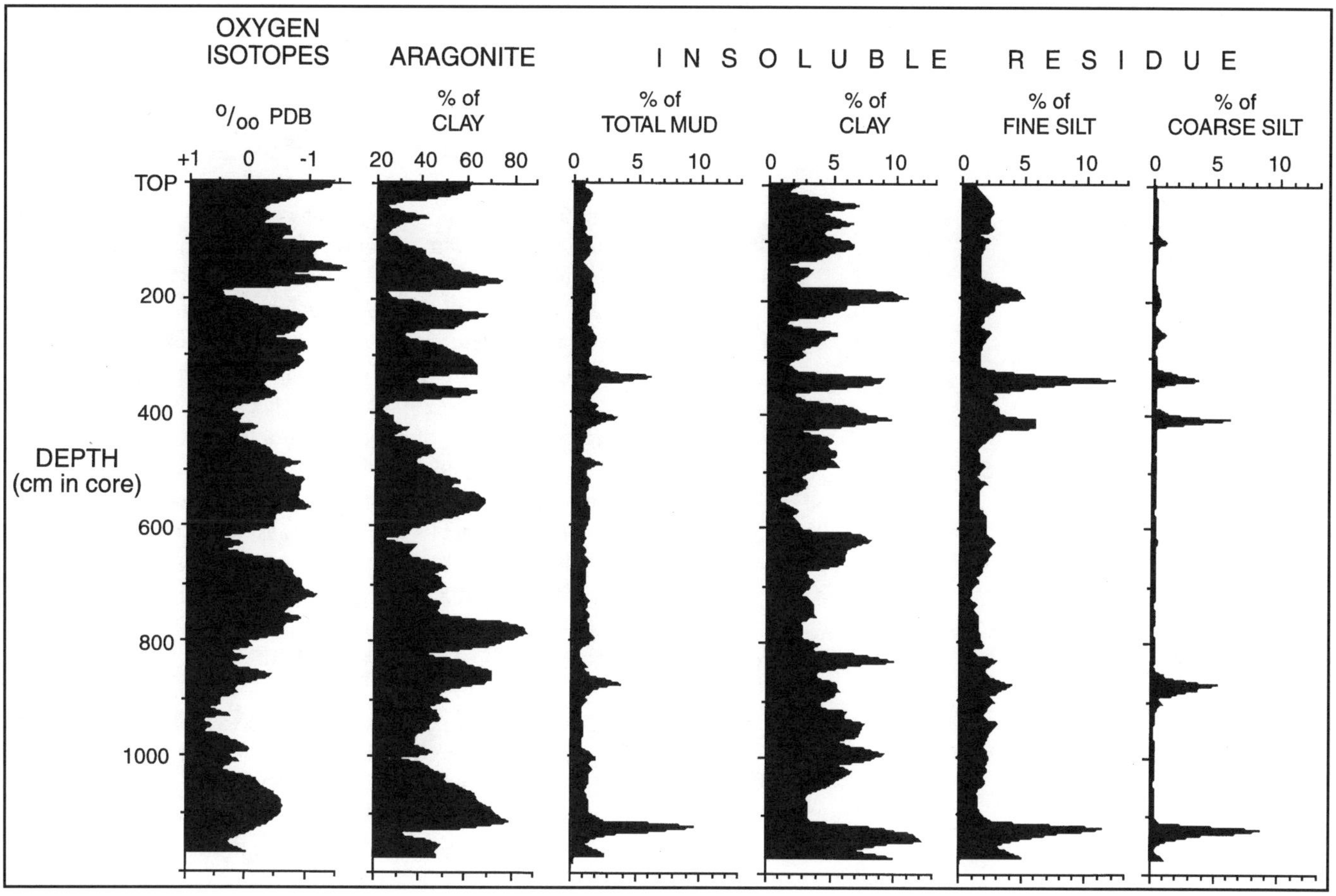

Figure 4. Grain size distribution of insoluble residue in the deep-sea core is compared to climatic indicators. High concentrations of aragonite are caused by fine aragonite shed from the banktops during highstands of sea level (interglacial periods), and low concentrations of aragonite occur during glacial periods (Boardman et al., 1986). Fluctuations of oxygen isotopes of planktonic foraminifera also confirm the record of climate variations.

(range, 0.4 to 34.3%). The average wt% of mineral portion of IR is 5.3% (range, <0.1 to 25%) and the organic component averages 1.4% (range, <0.1 to 6.6%). The mineral and organic components of IR have a positive, linear correlation (r = 0.86).

Quartz and kaolinite are nearly always the most abundant IR minerals in paleosol samples. Other common minerals include boehmite, illite, chlorite, interstratified chlorite-vermiculite, and montmorillonite. Minor minerals in most samples include gibbsite, hematite, K-feldspar, plagioclase, and goethite. This mineral assemblage is similar to that reported for modern soils found on Eleuthera (Foos, 1991). The average carbonate mineralogy in paleosols is 93% calcite (low-Mg calcite) and 7% aragonite.

Characteristics of host limestones

Limestone lithologies underlying the paleosols sampled in this study include bioclastic, oolitic, and peloidal grainstones and packstones. The average carbonate mineralogy of these "host" limestones is 44% calcite and 56% aragonite. Insoluble residue within 11 limestone samples averages 0.3% by weight (range, <0.1 to 0.4%). This value agrees with other studies of Bahamian limestone (0.3 to 0.5%; Kornicker, 1958). Much (if not most) of this IR is not mineral matter, but is organic matter. No mineral IR was detected in the residues of dissolved limestone using x-ray diffraction techniques. However, examination of limestone residues using SEM and EDX (energy dispersive x-ray) analysis clearly shows mineral insoluble residue. Point counting (using SEM) suggests that the mineral IR in limestones is <0.01%.

ORIGIN OF PALEOSOLS IN THE BAHAMAS

A significant distinction between the paleosols and host limestones is the higher concentration of IR in paleosols. The mode of concentration of this IR may be by dissolution of a host limestone. Alternatively, IR may accumulate by direct airborne accretion on a limestone surface without any appreciable dissolution of the host limestone. Evaluation of our data suggests that insoluble residue within paleosols in the Bahamas is predominantly accumulated by dust accretion rather than by dissolution of a host limestone.

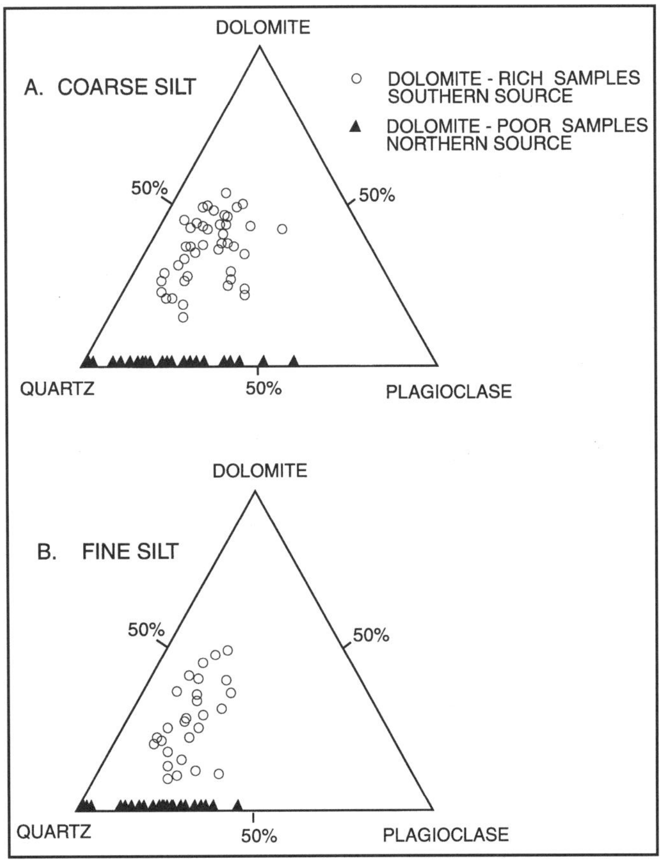

Figure 5. Mineralogy of fine and coarse silt in Bahamian deep-sea core is a mixture of quartz, plagioclase, and dolomite. There are two distinct suites: dolomite-poor and dolomite-rich.

Using an average dust flux of 0.08 g/cm^2/10^3 yr derived from the periplatform core (Fig. 9) (Eaton, 1986), dust accretion could provide the quantity of IR found in 70 cm of typical paleosol every 100,000 yr (Fig. 10). An unknown quantity of dust rained out and initially deposited on the limestone surface will likely be eroded from the limestone surface by wind and or water runoff. Thus the IR accumulated in the paleosol will be less than the potential. However, this calculation demonstrates that dust accretion is a major, if not the dominant, contributor of insoluble residue to Bahamian paleosols.

The case to be made for limestone dissolution as a source of IR is weak. Based on the average amount of IR in paleosols (0.1145 g/cm^3), the amount of mineral IR in limestones (0.01% = 0.000176 g/cm^3), and average porosities in limestones and paleosols, a typical paleosol would require at least 6.5 m of limestone dissolution to produce 1 cm of paleosol (Fig. 10). If there were no intervening highstands that flooded the Bahamas since the last major highstand approximately 125,000 yr ago, the rate of limestone dissolution to produce 1 cm of paleosol would have to be at least 0.052 m/1,000 yr (6.5 m/125,000 yr). Because many paleosols are thicker than 1 cm, the rate of dissolution would have to be several times this rate—on the order of 0.1 to 0.2 m/1,000 yr. This high rate of dissolution is higher than any of the measured rates of limestone dissolution from other nearby areas (Atkinson and Smith, 1976) such as Florida (0.035 m/1,000 yr), Jamaica (0.04 to 0.09 m/1,000 yr), and Puerto Rico (0.045 m/1,000 yr). Perhaps rainfall (and thus dissolution) was higher during lowstands of sea level. The island platforms stood approximately 100 m higher off the ocean surface and would have been more significant loci for vertical convection of moisture-laden air resulting from local heating. Perhaps, the rain that did fall became more acidic because of the thicker, organic-rich soil thought to have existed prior to deforestation by western Europeans.

Although different climatic and soil conditions, such as those mentioned above, probably existed during lowstands of sea level, this magnitude of dissolution (ca 10 m of dissolution in the last 125,000 yr) is not seen in the Bahamas. For example, many islands are comprised of carbonate material deposited in subtidal environments during the previous highstand of sea level (approximately +5 m, 125,000 yr ago; Neumann and Moore, 1975; Garrett and Gould, 1984; Chen et al., 1991), and these subtidal deposits show vertical changes due to regional subsidence of approximately 0.02 m/1,000 yr, but not due to dissolution of this magnitude. In other areas, much of the original, depositional topography is still seen such as Pleistocene linear dunes and paleochannels (Garrett and Gould, 1984; Carew and Mylroie, 1985, 1987; Boardman et al., 1993). This is not to say that dissolution of limestones in the Bahamas is insignificant. There is ample evidence of karst and caves in limestone terrains such as the Bahamas (Mylroie and Carew, 1988; Mylroie et al., 1991; Palmer, 1991) characterized by vertical and lateral shafts, and there is certainly abundant petrographic evidence of moldic porosity production by meteoric dissolution. But surface dissolution that lowers the entire surface and reduces original topography is not common in the Bahamas.

Although it is clear from these calculations that the potential for insoluble residue accumulation by dust accretion is large, the absolute rate of paleosol accumulation probably varies. One assumption made in these calculations is that all the mineral IR is incorporated within a laterally continuous paleosol, and that dissolution and/or accretion occurs uniformly. This assumption is not totally realistic because the IR that forms soils is certainly washed or blown into the ocean during storms and is washed off ridges into valleys and ponds. Robbin and Stipp (1979) found that laminated crust formation is not laterally continuous, and that crusts form rapidly by accretion at rates between 0.25 and 0.5 cm/1,000 yr (25 to 50 cm/100,000 yr). Apparently, when crust formation occurs, it occurs by accretion and at rates comparable to the rates suggested by dust accretion.

Dust accretion is believed to be the dominant method of accumulation of IR in Bahamian paleosols. There is more than an adequate supply of dust transported from Africa to the Bahamas to account for the IR in paleosols. Because there are low quantities of mineral IR within Bahamian limestone and apparently low regional limestone dissolution rates, IR accumulation via limestone dissolution does not appear to be a significant pedogenic

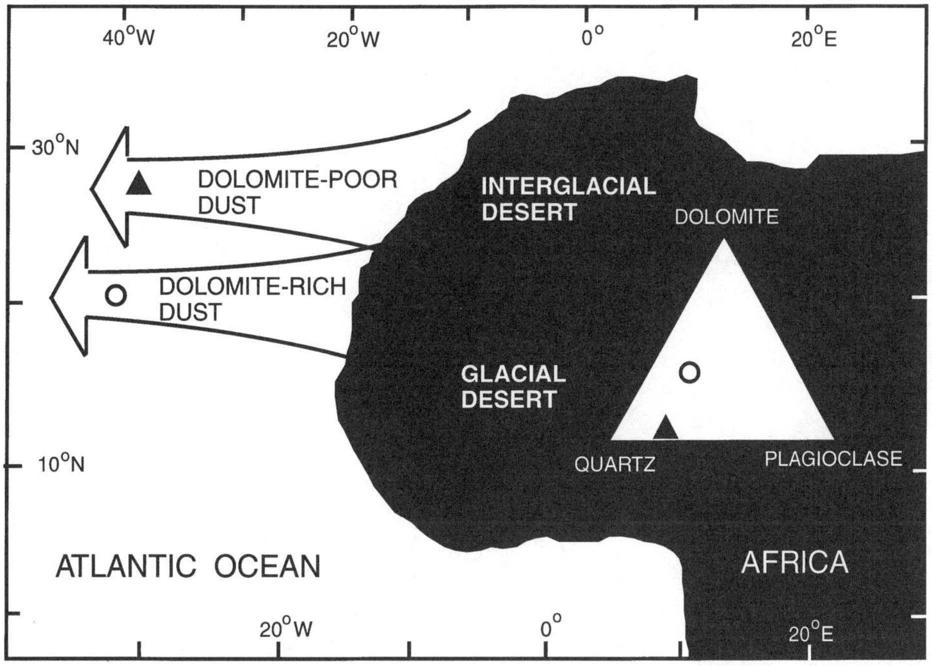

Figure 6. Dust from northern Africa is dolomite-poor, whereas dust originating farther south contains more dolomite. Global climatic variations cause the areas of aridity of northern Africa to expand and contract; this causes a change in the mineralogy of the dust transported by the Trade Winds and accumulated in the Bahamas.

process. Dissolution and reprecipitation of limestone must occur to produce the bulk of the fine-grained, calcite matrix of paleosols, but calculations suggest that unreasonably large amounts of limestone dissolution would be required to equal the IR contributed by dust accretion.

Formation of paleosols in relation to sea level

Mineralogy. Because the bulk of IR distributed to the Bahamas originates in Africa and results from an erosion and transport history that is linked to glaciations (climate) and thus sea level (Figs. 3–6), one would think that mineralogy of the IR of paleosols would be an indicator of when the paleosols formed. However, mineralogic comparisons between the IR of paleosols and the IR found in the periplatform core are inconclusive in assigning the timing of soil accumulation.

Dolomite is deposited in the periplatform core most abundantly during lowstands of sea level (Fig. 6); its absence in paleosols indicates that either paleosol IR is deposited during interglacial (highstand) periods and/or that dolomite is readily lost from the soil profile during diagenesis. Dissolution of dolomite in the soil zone might result from Mg depletion as the aragonite and high-Mg calcite allochems of the host limestone dissolve to provide a solution with a very low Mg/Ca ratio. Even though the solution must become calcite saturated to form the fine-grained calcite matrix of the paleosol and the cement of the host limestone, the Mg/Ca ratio will be low and may cause dissolution of dolomite. Dolomite, if it ever was a part of the IR of a soil, is likely to have been dissolved rather than become part of the paleosol. Thus, timing of paleosol formation based on the absence of dolomite is inconclusive.

Like dolomite, plagioclase is reactive in soil-forming environments, and typically is altered to cation-leached clay minerals. There are several minerals common in paleosols that are not present in the periplatform core: boehmite, gibbsite, hematite, goethite, and interstratified chlorite-vermiculite. These minerals are common pedogenic (diagenetic) minerals and typically form by the alteration of feldspars (Berner, 1975; Valeton, 1972). Thus, the low abundance of plagioclase in paleosols is not a useful criterion for evaluating the origin and timing of insoluble residue in the soil-forming process.

Quartz grain size. Comparisons of quartz size distribution between Bahamian paleosols and periplatform core samples (Fig. 11) suggests that the IR of paleosols is accumulated only during lowstands of sea level by accretion of dust originating from Africa. The size of the quartz accumulated during lowstands of sea level ($\bar{x} = 3.9$ µm) when the platforms are exposed to the subaerial processes and limestone is not being deposited, is similar to the size of quartz in paleosols ($\bar{x} = 4.2$ µm). Quartz that accumulated during highstands of sea level, when carbonate accumulation occurs on platforms, is coarser (8.6 µm). Thus the size distribution of quartz supports the idea that paleosols accumulated by dust accretion only during lowstands of sea level, and that quartz in paleosols is not derived from dust accumulated during highstands (i.e., from dissolution of a preexisting limestone).

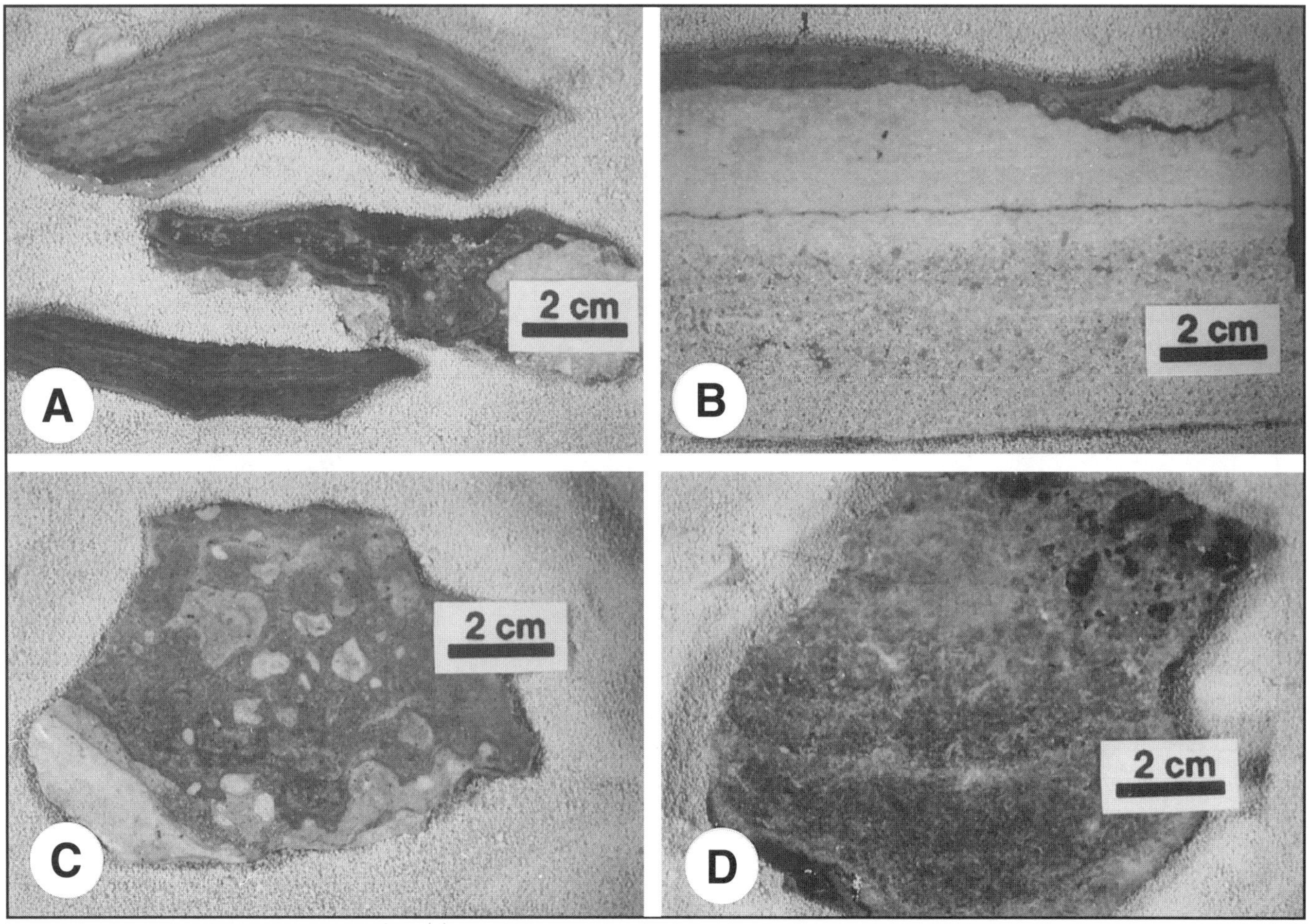

Figure 7. Types of paleosols distinguished megascopically. A, Laminated crusts, B, homogeneous crust, C, breccia/conglomerate, and D, homogeneous matrix.

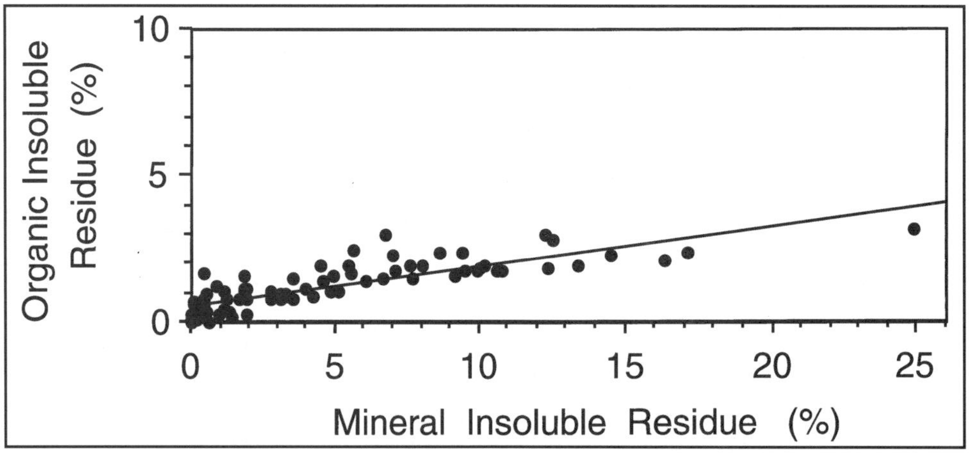

Figure 8. Insoluble residue of paleosols is composed of organic material and inorganic material. There is a positive correlation (r = 0.86) between these two fractions of insoluble residue.

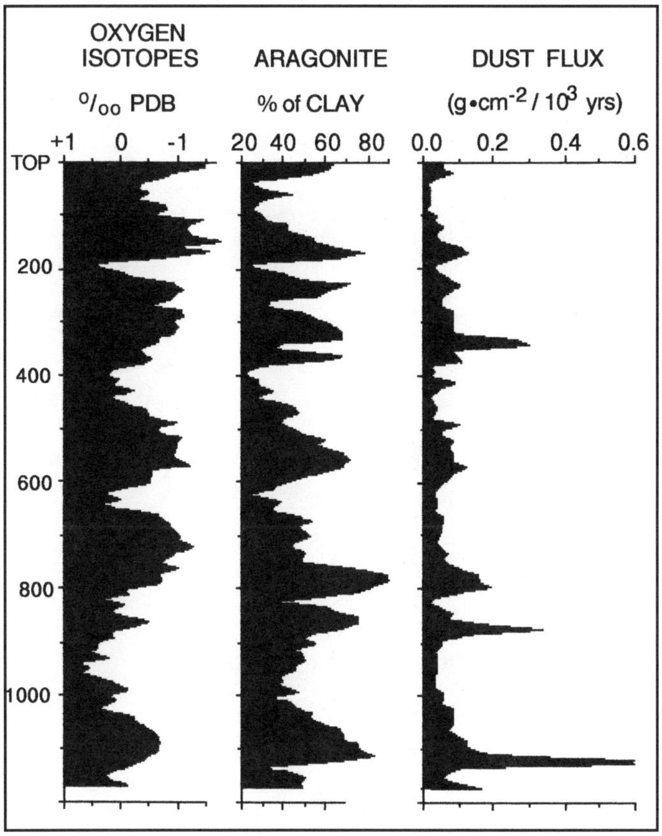

Figure 9. Dust flux to the deep-sea sediment core is compared to climatic indicators. Dust flux is calculated from measurements of the percentage of insoluble residue in the three mud fractions, and the percentage of each of the mud fractions relative to the total sediment, bulk porosity of the sediment, and sedimentation rate. Details and data used to calculate this flux are found in Eaton (1986).

PALEOSOLS AND STRATIGRAPHY

A reasonable assumption is that every sea-level highstand is associated with the development of an additional lithologic "package" that is separated from other "packages" by a subaerial exposure surface produced during the intervening lowstand. Because the Bahama platforms have been slowly subsiding (Mullins and Lynts, 1977; Pierson, 1981; Garrett and Gould, 1984), there have been many episodes of banktop flooding and carbonate production during the many fluctuations of sea level in the late Pleistocene (Shackleton and Opdyke, 1973; Boardman and Neumann, 1984; Boardman et al., 1986; Carew and Mylroie, 1985, 1987). Work on the stratigraphy of San Salvador (Carew and Mylroie, 1985, 1987) indicates that several different episodes of carbonate accumulation are represented there, and the potential use of paleosols to assist stratigraphic correlation and interpretation was investigated. Results of this investigation suggests that the differences in petrology and mineralogy of paleosols in the Bahamas add no new information concerning chronology than that readily gleaned from the fact that paleosols indicate subaerial exposure. Quaternary paleosols are not of assistance in stratigraphic correlation on a single island, and they are not useful for chronostratigraphic correlation between islands within the Bahamas.

Petrography of paleosols and stratigraphy

All four paleosol types (laminated crusts, homogeneous crusts, breccia-conglomerate, and homogeneous matrix) are recognized capping units assigned to the Grotto Beach Formation. Corals from several outcrops of the Grotto Beach Formation, including the type locality and the "alternate" type locality have been isotopically dated at ca. 125,000 B.P. (Carew and Mylroie, 1987; Chen et al., 1991), and these outcrops have different paleosol types capping them. In addition, all four types of paleosols have Holocene analogs in the Bahamian region (Robbin and Stipp, 1979; this study). Therefore, paleosol petrography (type) cannot be used to distinguish paleosols for purposes of stratigraphic correlation.

Clay mineralogy and topography

An examination of the mineralogy of the clay minerals of the paleosols (and soils) suggests that paleosols can be subdivided by the relative abundance of two minerals: illite and boehmite (Fig. 12). A few samples contain both minerals (7%), and 13% of the samples studied did not contain either of these minerals. However, the majority of samples (80%) are dominated by only one of the minerals—either illite or boehmite. Interstratified chlorite-vermiculite is found only in soils that are boehmite-rich.

On San Salvador, boehmite-rich paleosols are primarily found on the northern end of the island, whereas the southern end is dominated by illite-rich paleosols (Fig. 13). This geographic segregation of paleosol mineralogy on San Salvador is not apparent on Andros Island, although sampling on Andros is not as extensive.

The intriguing possibility presented by this observation is that perhaps boehmite, a clay resulting from greater leaching (i.e., more extensive diagenesis), may be more prevalent in the northern portion of San Salvador because that area is older than the southern region. However, stratigraphic analysis and isotopic dating on San Salvador (Carew and Mylroie, 1985, 1987) indicate that the northern portion of San Salvador is not older than the southern portion. In fact, the island appears to have generally grown (prograded) outward in all directions, and the ages of rocks on the periphery are similar. In addition, both illite-rich and boehmite-rich paleosols are found capping several different stratigraphic units assigned to the Grotto Beach Formation.

Rather than simple duration of subaerial exposure, the reason for this geographic distribution of clay minerals is likely a result of diagenetic intensity. The abundance of boehmite in some paleosols (and soils) suggests present or past conditions of higher moisture and extensive leaching (Berner, 1971; Valeton,

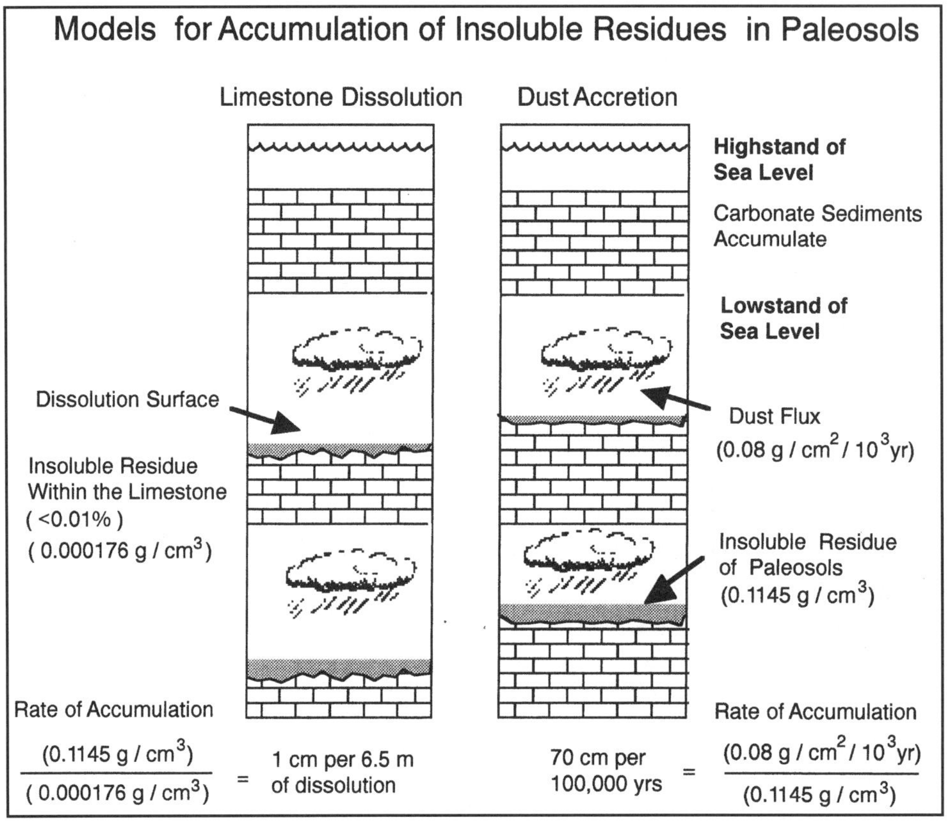

Figure 10. Comparison of mechanisms of incorporating insoluble residue into paleosols (limestone dissolution versus dust accretion) illustrates that the insoluble residue typically found in paleosols can be produced by dust accretion without limestone dissolution. To accumulate the insoluble residue in a paleosol by limestone dissolution requires several meters of dissolution of the surface of the limestone.

1972). The fact that both boehmite-rich and illite-rich paleosols of the same age exist in the Bahamas might indicate the local climate has varied significantly over laterally short distances. However, evaluation of IR mineralogy and sample elevation (Fig. 14) shows that boehmite-rich paleosols are found (or preserved) predominantly at lower elevations (below 2 m above present sea level).

In addition, all peat samples recovered from beneath Holocene sediments contain significant amounts of boehmite. Boehmite-rich soil samples were found in subtidal peat deposits recovered from below Holocene sediments in Grahams Harbor and Pigeon Creek, San Salvador, in Staniard Creek, Andros Island, and beneath the Holocene sediments of Bight of Abaco, Little Bahama Bank (Boardman, 1976; Rasmussen and Neumann, 1988). All contain boehmite and minor amounts of illite except for Grahams Harbor, which has boehmite and no illite. These samples represent soils that have developed in low-lying, wet areas and were buried by the Holocene rise of sea level. Their preservation suggests formation in a "ponded" or protected basin. This topographic control is compatible with the idea that boehmite results from intense diagenetic alteration. Presumably, lower areas are/were areas where ponded water could increase the time for diagenetic reactions to occur, thus

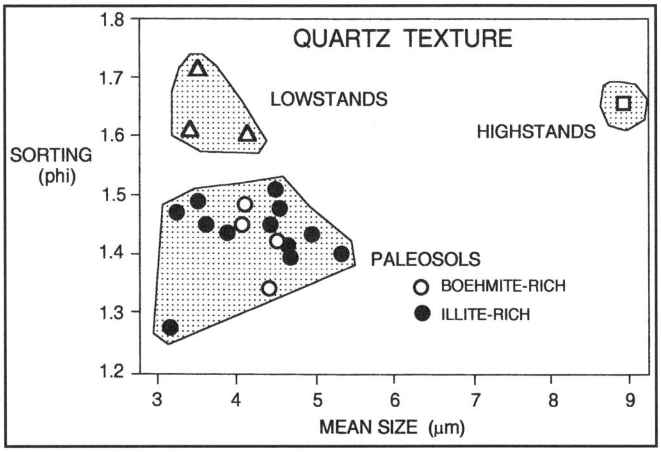

Figure 11. Quartz size distribution versus sorting of paleosols and samples from the periplatform core (highstand quartz, lowstand quartz). Both boehmite-rich and illite-rich paleosols have a quartz size distribution similar to that of lowstand quartz.

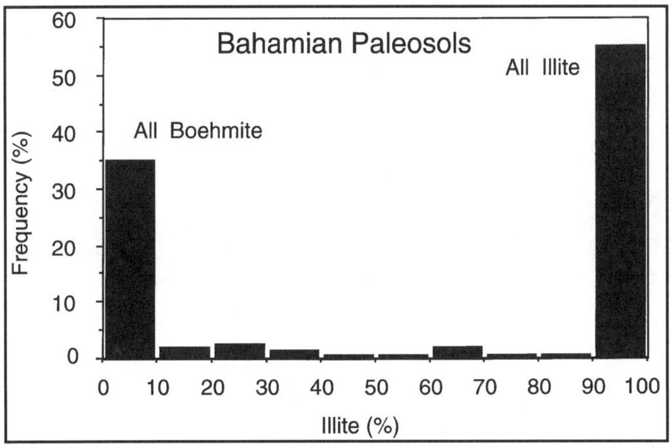

Figure 12. The proportion of boehmite and illite in paleosols indicates that paleosols are either boehmite-rich or illite-rich. Only a few samples have neither mineral or nearly equal amounts of each mineral.

not requiring climatic variation as a mechanism for creating mineralogically distinct paleosols on the same island at the same time.

Boehmite-rich and illite-rich paleosols cannot be differentiated based on the texture of quartz (Fig. 11). Both types of paleosols have a quartz size distribution similar to that of lowstand quartz from the deep-sea core and distinctly different from the highstand quartz.

Studies involving the comparison of Bahamian soils with limestone have inferred a relationship between soil type and its associated underlying limestone (e.g., oolitic grainstone, skeletal grainstone: Little et al., 1977). Our data suggest that soil characteristics may reflect topographic (drainage) variations rather than underlying limestone lithology. To the degree that original limestone lithology is related to topography, the paleosols also will be related, coincidentally, to original lithology.

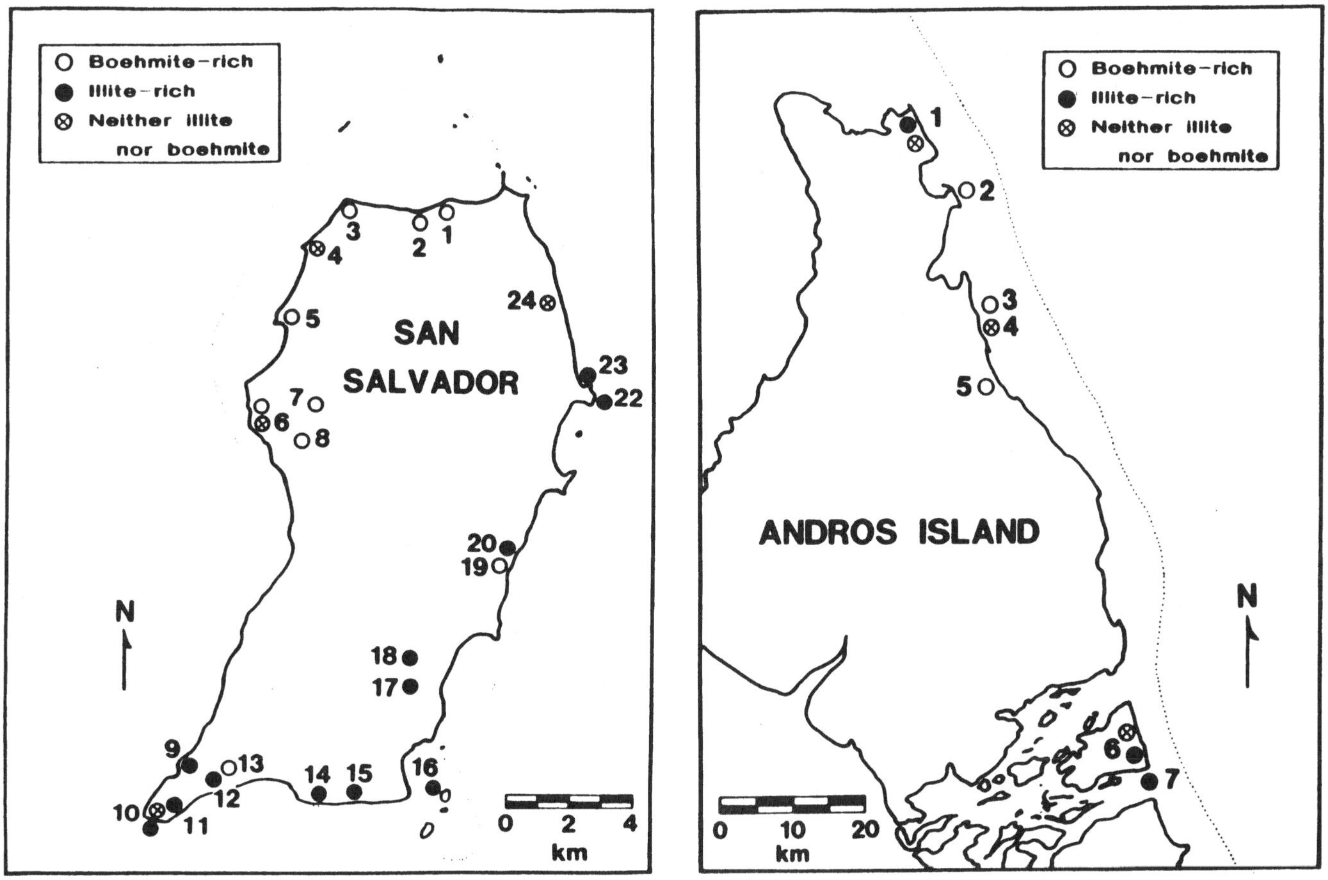

Figure 13. Spatial distribution of boehmite-rich and illite-rich paleosols shows that the northern portion of San Salvador contains paleosols that are boehmite-rich. There is no similar correlation on Andros Island. There is no systematic difference in the ages of the underlying rocks in the northern versus the southern portion of San Salvador. Thus, there is no stratigraphic significance to this portion of the mineralogic signature of paleosols.

Mixing of paleosols

The mixing of soils/paleosols clearly occurs. All nine modern soil samples taken from San Salvador and Andros Island have the same mineralogy as the paleosols found in the immediate vicinity (either illite-rich or boehmite-rich). This suggests that modern soils are influenced by the reworking of preexisting, exposed, paleosols. Perhaps, though, the modern soils merely reflect formation in an environment similar to that of the nearby paleosols. More convincing to the idea of soil reworking and mixing is the occurrence of boehmite-rich clasts within boehmite-rich matrices of paleosols (Fig. 15) found on San Salvador (at sites 2 and 3). The rounding of the boehmite-rich clasts suggests that abrasion of the clasts may have contributed to the boehmite matrix. Whether geochemical conditions for boehmite formation were present during the formation of the second generation paleosol is not clear, but it is clear that some paleosols result from two generations of soil formation. Thus, the usefulness of paleosols as a unique stratigraphic marker is limited.

SUMMARY

The results of this study indicate that paleosols in the Bahamas include four megascopically distinct types. The insoluble residue of these paleosols originates as dust derived from Africa during lowstands of sea level. Based on concentrations of IR in paleosols, host limestones, and fluxes of dust measured in a periplatform core, dust accretion on the exposed limestone surface (during lowstands of sea level) is a more likely method of paleosol formation than limestone dissolution. Highstand (interglacial) and lowstand (glacial) dust can be distinguished from each other on the basis of quartz texture in the periplatform core, and texture of quartz, preserved in the paleosols, also suggests that paleosols formed from dust accumulated during lowstand conditions.

Most of the IR minerals in the periplatform core that are indicative of highstand and lowstand conditions, such as dolomite and plagioclase, are altered in the pedogenic environment. One diagenetic alteration is the preferential formation of boehmite in low-lying areas. In higher areas, with less ponding of water, illite (not boehmite) predominates. Thus, the composition of the paleosol cannot be used to determine whether formation occurred during highstand or lowstand conditions.

Paleosols cannot be the diagnostic criterion to establish stratigraphic correlation within the Bahamas. Paleosols indicate that there was at least one episode of subaerial exposure long enough to accumulate adequate IR from dust to form a resistant, preservable diagenetic record. All four paleosol types and both boehmite-rich and illite-rich paleosols can be found on stratigraphic units that have been isotopically dated and considered contemporaneous. Mixing of paleosols (Pleistocene with Holocene) clearly demonstrates that there is not necessarily one distinct paleosol for every lowstand of sea level (episode of subaerial exposure).

If ancient carbonate sediments were accumulated as vertically stacked units (layer-cake fashion), then paleosols, if they could be distinguished from each other, would be useful aids to chronostratigraphic correlation. This stacking may exist on the interior of platforms, and paleosols may be more useful in such locations. However, the dominance of lateral accretion of Quaternary carbonate sediments seen on islands in the Bahamas obscures the temporal relationships of facies and diminishes the usefulness of paleosols for inter- and intra-island correlations.

ACKNOWLEDGMENTS

Special thanks go to Don Gerace, Kathy Gerace, and the helpful staff of the Bahamian Field Station, San Salvador

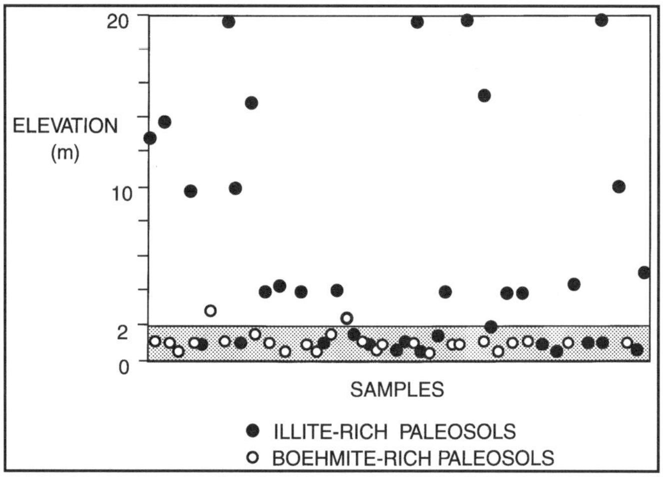

Figure 14. Boehmite-rich paleosols are found at low elevations, whereas illite-rich paleosols occur at many elevations. Both boehmite-rich and illite-rich paleosols are found on limestones of similar age; thus this mineralogic difference is not due to duration of exposure (time, stratigraphy), but is likely due to variation of diagenetic intensity caused by local topographic effects.

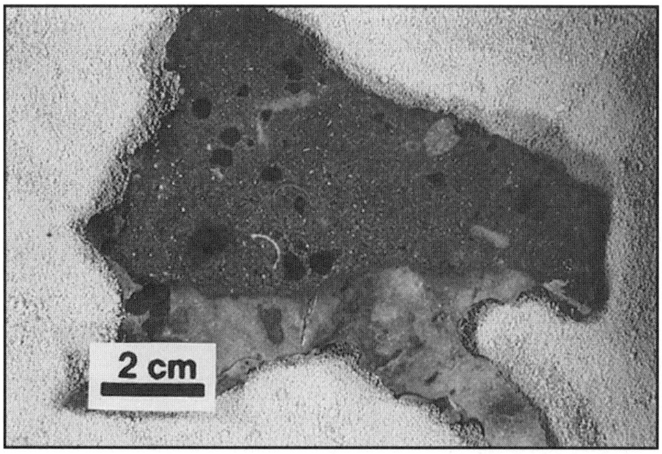

Figure 15. Multigeneration paleosols contain clasts (sometimes rounded) of preexisting paleosols and suggest that the mineralogic characteristics of some paleosols are not created during a single exposure interval.

Island, Bahamas. Ben Bohl and the staff of the Forfar Field Station provided the logistic support on Andros Island, Bahamas. The crew of the *R/V Cape Hatteras* and the scientific participants on this ship helped in the coring operations. Many graduate students helped in collecting samples and performing laboratory work; in particular, we thank Lise Dulin, Cliff Schmitt, Dave Daugherty, Norman Colby, Paul Bergstrand, and Ron Pelle. In addition to the Geology Department of Miami University, this research was supported by National Scientific Foundation Grant OCE 81-17615 (to A. C. Neumann and M. R. B.), by a grant from Sigma Xi (to M. R. E.), and by the Miami University summer workshop on Modern Carbonate Deposition (directed by M. R. B.).

REFERENCES CITED

Ahmad, N., Jones, R. L., and Beavers, A. H., 1966, Genesis, mineralogy and related properties of West Indian soils: 1. Bauxite soils of Jamaica: Proceedings of the Soil Science Society America, v. 30, p. 719–722.

Ahmad, N., and Jones, R. L., 1969, Occurrence of aluminous lateritic soils (bauxites) in the Bahamas and Cayman Islands: Economic Geology, v. 64, p. 804–808.

Atkinson, T. C., and Smith, D. I., 1976, The erosion of limestones, *in* Ford, T. D., and Cullingford, C. H., eds., The Science of Speleology: New York, Academic Press, p. 240.

Bain, R. J., and Foos, A. M., 1993, Carbonate microfabrics related to subaerial exposure and paleosol formation, *in* Rezak, R., and Lavoie, D. L., eds., Carbonate Microfabrics: New York, Springer-Verlag, p. 19–27.

Bates, R. L., and Jackson, J. A., 1980, Glossary of Geology (second edition): Falls Church, Virginia, American Geological Institute, 749 p.

Berner, R. A., 1971, Principles of Chemical Sedimentology: New York, McGraw-Hill, 240 p.

Berner, R. A., 1975, Diagenetic models of dissolved species in the interstitial waters of compacting sediments: American Journal of Science, v.275, p. 88–96.

Biscaye, P. E., 1965, Mineralogy and sedimentation of recent deep-sea clay in the Atlantic Ocean and adjacent seas and oceans: Geological Society of America Bulletin, v. 76, p. 803–832.

Blackburn, G., and Taylor, R. M., 1969, Limestones and red soils of Bermuda: Geological Society of America Bulletin, v. 80, p. 1595–1598.

Blackburn, G., and Taylor, R. M., 1970, Limestones and red soils of Bermuda: Reply: Geological Society of America Bulletin, v. 81, p. 2525–2526.

Boardman, M. R., 1976, Lime mud deposition in a tropical island lagoon, Bight of Abaco, Bahamas [M.S. thesis]: Chapel Hill, University of North Carolina, 121 p.

Boardman, M. R., 1978, Holocene deposition in Northwest Providence Channel: A geochemical approach [Ph.D. thesis]: Chapel Hill, University of North Carolina, 155 p.

Boardman, M. R., and Neumann, A. C., 1984, Sources of periplatform carbonates: Northwest Providence Channel, Bahamas: Journal of Sedimentary Petrology, v. 54, p. 1110–1123.

Boardman, M. R., and Neumann, A. C., 1986, Reply: Banktop responses to Quaternary fluctuations of sea level recorded in periplatform sediments: Geology, v. 14, p. 1040–1041.

Boardman, M. R., Neumann, A. C., Baker, P. A., Dulin, L. A., Kenter, R. J., Hunter, G. E., and Keifer, K. B., 1986, Banktop responses to Quaternary fluctuations of sea level recorded in periplatform sediments: Geology, v. 14, p. 28–31.

Boardman, M. R., Carney, C., and Bergstrand, P. M., 1993, A Quaternary analog for interpretation of Mississippian oolites, *in* Keith, B. D., and Zuppmann, C. W., eds., Mississippian Oolites of North America and Modern analogs: American Association of Petroleum Geologists Studies in Geology 35, p. 285–304.

Braithwaite, C. J. R., 1983, Calcrete and other soils in Quaternary limestones: Structures, processes and applications: Journal of the Geological Society of London, v. 140, p. 351–363.

Bricker, O. P., and Mackenzie, F. T., 1970, Limestones and red soils of Bermuda: Discussion: Geological Society of America Bulletin, v. 81, p. 2523–2524.

Brindley, G. W., 1961, Quantitative analysis of clay mixtures, *in* Brown, G., ed., The X-ray Identification and Crystal Structure of Clay Minerals (second edition): London, Mineralogical Society of London, p. 489–516.

Brown, G., and Brindley, G. W., 1980, X-ray diffraction procedures for clay mineral identification, *in* Brindley, G. W., and Brown, G., eds., Crystal structures of clay minerals and their x-ray identification: Mineralogical Society of London Monograph, v. 5, p. 305–359.

Brown, T. W., 1984, Formation and development of caliche profiles in eolian deposits: San Salvador, Bahamas, *in* Teeter, J. W., ed., Proceedings, Second Symposium on the Geology of the Bahamas: San Salvador, CCFL Bahamian Field Station, p. 245–264.

Brunner, C. A., 1984, Evidence for increased volume transport of the Florida current in the Pliocene and Pleistocene: Marine Geology, v. 54, p. 223–235.

Burns, D. J., 1961, Some chemical aspects of bauxite genesis in Jamaica: Economic Geology, v. 56, p. 1297–1303.

Busby, R. F., 1962, Submarine geology of the Tongue of the Ocean, Bahamas: Washington, D. C., U.S. Naval Oceanographic Office Technical Report, TR-108, 84 p.

Carew, J. L., and Mylroie, J. E., 1985, The Pleistocene and Holocene stratigraphy of San Salvador Island, Bahamas, with reference to marine and terrestrial lithofacies at French Bay, *in* Curran, H. A., ed., Pleistocene and Holocene Carbonate Environments on San Salvador Island, Bahamas, Geological Society of America Field Guidebook, Ft. Lauderdale, Florida: CCFL Bahamian Field Station, p. 11–62.

Carew, J. L., and Mylroie, J. E., 1987, A refined geochronology for San Salvador Island, Bahamas, *in* Curran, H. A., ed., Proceedings, Third Symposium on the Geology of the Bahamas: San Salvador, CCFL Bahamian Field Station, p. 31–60.

Chapman, S. L., Syers, J. K., and Jackson, M. L., 1969, Quantitative determination of quartz in soils, sediments and rocks by pyrosulfate fusion and hydrofluorosilicic acid treatment: Soil Science, v. 107, p. 348–352.

Chave, K. E., 1954, Aspects of the biochemistry of magnesium. 1. Calcareous marine organisms: Journal of Geology, v. 62, p. 266–283.

Chen, J. H., Curran, H. A., White, B., and Wasserburg, G. J., 1991, Precise chronology of the last interglacial period: 234U-230Th data from fossil coral reefs in the Bahamas: Geological Society of America Bulletin, b. 103, p. 82–97.

Chester, R., and Johnson, L., 1971, Atmospheric dust collected off the Atlantic coasts of North Africa and the Iberian Peninsula: Marine Geology, v. 11, p. 251–260.

Coniglio, M., and Harrison, R. S., 1983, Holocene and Pleistocene caliche from Big Pine Key, Florida: Bulletin of Canadian Petroleum Geology, v. 31, p. 3–13.

Dauphin, J. P., 1980, Size distribution of chemically extracted quartz used to characterize fine-grained sediments: Journal of Sedimentary Petrology, v. 50, p. 205–214.

Dean, W. E., 1974, Determination of carbonate and organic matter in calcareous sediments and sedimentary rocks by loss on ignition: Comparison with other methods: Journal of Sedimentary Petrology, v. 44, p. 242–284.

Delaney, A. C., Delaney, A. C., Parkin, D. W., Griffin, J. J., Goldberg, E. D., and Reiman, B. E. F., 1967, Airborne dust collected at Barbados: Geochimica et Cosmochimica Acta, v. 31, p. 885–909.

Diester-Haas, L., 1976, Late Quaternary climatic variations in Northwest Africa deduced from east Atlantic sediment cores: Quaternary Research, v. 6,

p. 299–314.
Diester-Haas, L., Schrader, H. J., and Thiede, J., 1973, Sedimentological and paleoclimatological investigations of two pelagic ooze cores off Cape Barbas, Northwest Africa: Meteor Forchungs-Ergebnisse C, v. 20, p. 1–32.
Eaton, M. R., 1986, Origin of insoluble residue in a deep-sea sediment core from Northwest Providence Channel, Bahamas [M.S. thesis]: Oxford, Ohio, Miami University, 91 p.
Foos, A. M., 1991, Aluminous lateritic soils, Eleuthera, Bahamas: A modern analog to carbonate paleosols: Journal of Sedimentary Petrology, v. 61, p. 340–348.
Garrett, P., and Gould, S. J., 1984, Geology of New Providence Island, Bahamas: Geological Society of America Bulletin, v. 95, p. 741–751.
Gibbs, R. J., 1965, Error due to segregation in quantitative clay mineral x-ray diffraction mounting techniques: American Mineralogist, v. 50, p. 741–751.
Gibbs, R. J., 1968, Clay mineral mounting techniques for x-ray diffraction analysis: A discussion: Journal of Sedimentary Petrology, v. 38, p. 242–243.
Gile, L. H., Peterson, F. F., and Grossman, R. B., 1966, Morphological and genetic sequences of carbonate accumulation in desert soils: Soil Science, v. 101, p. 347–360.
Glaccum, R. A., and Prospero, J., 1980, Saharan aerosols over the North Atlantic—Mineralogy: Marine Geology, v. 37, p. 295–321.
Goldhammer, R. K., and Elmore, R. D., 1984, Paleosols capping regressive carbonate cycles in the Pennsylvanian Black Prince Limestone, Arizona: Journal of Sedimentary Petrology, v. 54, p. 1124–1137.
Hale, A. P., and Ettensohn, F. R., 1984, Micromorphological features observed in pedogenic carbonates, *in* Teeter, J. W., ed., Proceedings, Second Symposium on the Geology of the Bahamas: San Salvador, CCFL Bahamian Field Station, p. 265–278.
Harrison, R. S., 1977, Caliche profiles: Indicators of near-surface subaerial diagenesis, Barbados, West Indies: Bulletin of Canadian Petroleum Geology, v. 25, p. 123–173.
Harrison, R. S., and Steinen, R. P., 1978, Subaerial crusts, caliche profiles, and breccia horizons: Comparison of some Holocene and Mississippian exposure surfaces, Barbados and Kentucky: Geological Society of America Bulletin, v. 89, p. 385–396.
Hose, H. R., 1963, Jamaica-type bauxite developed on limestone: Economic Geology, v. 58, p. 62–69.
James, N. P., 1972, Holocene and Pleistocene calcareous crusts (caliche) profiles: Criteria for subaerial exposure: Journal of Sedimentary Petrology, v. 42, p. 817–836.
Johnson, L. R., 1974, A study of aeolian dust from the lower atmosphere over the Atlantic Ocean [Ph.D. thesis]: Liverpool, United Kingdom, University of Liverpool.
Johnson, L. R., 1979, Mineralogical dispersal patterns of the North Atlantic with particular reference to aeolian dust: Marine Geology, v. 29, p. 335–345.
Kornicker, L. S., 1958, Bahamian limestone crusts: Gulf Coast Association Geological Society, v. VIII, p. 167–170.
Land, L. S., Mackenzie, F. T., and Gould, S. J., 1967, The Pleistocene history of Bermuda: Geological Society of America Bulletin, v. 78, p. 993–1006.
Little, B. G., Buckley, D. K., Cant, R., Henry, P. W. T., Jeffriss, A., Mather, J. D., Stark, J., and Young, R. N., 1977, Land resources of the Bahamas: A summary: Land Resource Study, v. 27, p. 53–57.
McCartney, R. F., 1987, Origin and diagenesis of paleosols in the Bahamas [M.S. thesis]: Oxford, Ohio, Miami University, 88 p.
Milliman, J. D., 1974, Marine Carbonates: New York, Springer-Verlag, 375 p.
Muhs, D. R., Bush, C. A., Stewart, K. C., Rowland, T. R., and Crittenden, R. C., 1990, Geochemical evidence of Saharan dust parent material for soils developed on Quaternary limestones of Caribbean and western Atlantic Islands: Quaternary Research, v. 33, p. 157–177.
Mullins, H. T., and Lynts, G. W., 1977, Origin of the northwestern Bahama Platform: Review and reinterpretation: Geological Society of America Bulletin, v. 88, p. 1447–1461.
Mullins, H. T., Wise, S. W., Land, L. S., Siegel, D. I., Masters, P. M., Hinchey, E. J., and Price, K. R., 1985, Authigenic dolomite in Bahamian periplatform slope sediment: Geology, v. 13, p. 292–295.
Multer, H. G., and Hoffmeister, J. E., 1968, Subaerial laminated crust of the Florida Keys: Geological Society of America Bulletin, v. 79, p. 183–192.
Mylroie, J. E., and Carew, J. L., 1988, Solution conduits as indicators of late Quaternary sea level position: Quaternary Science Reviews, v. 7, p. 55–64.
Mylroie, J. E., Carew, J. L., Sealey, N. E., and Mylroie, J. R., 1991, Cave development on New Providence Island and Long Island, Bahamas: Cave Science, v. 18, p. 139–151.
Neumann, A. C., and Moore, W. S., 1975, Sea level events and Pleistocene coral ages in the northern Bahamas: Quaternary Research, v. 5, p. 215–224.
Newell, N. D., and Rigby, K., 1957, Geologic studies of the Great Bahama Bank, *in* LeBlanc, R. J., and Breeding, J. G., eds., Regional Aspects of Carbonate Deposition: Society of Economic Paleontologists and Mineralogists Special Publication 5, p. 15–22.
North American Committee on Stratigraphic Nomenclature, 1983, American Association of Petroleum Geologists Bulletin, v. 67, p. 841.
Palmer, A. N., 1991, Origin and morphology of limestone caves: Geological Society of America Bulletin, v. 103, p. 1–21.
Parkin, D. W., 1974, Trade winds during glacial cycles: Proceedings of the Royal Society of London, v. A337, p. 73–100.
Parkin, D. W., and Shackleton, N. J., 1973, Trade wind and temperature correlations down a deep-sea core off the Saharan coast: Nature, v. 245, p. 455–457.
Parmenter, C., and Folger, D. W., 1974, Eolian biogenic detritus in deep sea sediments: Possible index of equatorial ice age aridity: Science, v. 185, p. 695–698.
Pierson, B. J., 1981, Late Cenozoic geology of the southeastern Bahama Banks [Ph.D. thesis]: Coral Gables, Florida, University of Miami, 343 p.
Prather, B. E., 1985, An Upper Pennsylvanian desert paleosol in the d-zone of the Lansing–Kansas City Groups, Hitchcock County, Nebraska: Journal of Sedimentary Petrology, v. 55, p. 213–221.
Pratt, R. M., 1968, Atlantic continental shelf and slope of U.S.—Physiography and sediments of the deep-sea basin: U.S. Geological Survey Professional Paper 529B, p. B1–B44.
Prospero, J. M., 1981a, Arid regions as sources of mineral aerosols in the marine atmosphere, *in* Péwé, T. L., ed., Desert dust: Origin, characteristics and effects on man: Geological Society of America Special Paper 186, p. 71–86.
Prospero, J. M., 1981b, Eolian transportation to the world ocean, *in* Emiliani, C., ed., The oceanic lithosphere: The Sea, v. 7: Wiley, New York.
Prospero, J. M., and Carlson, T. N., 1972, The vertical and aerial distribution of Saharan dust over the western equatorial North Atlantic: Journal of Geophysical Research, v. 77, p. 5255–5265.
Rasmussen, K. A., and Neumann, A. C., 1988, Holocene overprints of Pleistocene paleokarst: Bight of Abaco, Bahamas, *in* James, N. P., and Choquette, P. W., eds., Paleokarst: New York, Springer-Verlag, p. 132–148.
Read, J. F., 1974, Calcrete deposits and Quaternary sediments, Edel Province, Shark Bay, *in* Logan, B. W., Read, J. F., Hagan, G. M., Hoffman, P., Brown, R. G., and Gebelein, C. D., eds., Evolution and Diagenesis of Quaternary Carbonate Sequences: American Association of Petroleum Geologists Memoir, v. 22, p. 250–281.
Riding, R., and Wright, V. P., 1981, Paleosols and tidal-flat/lagoon sequences on a Carboniferous carbonate shelf: Sedimentary association of triple disconformities: Journal of Sedimentary Petrology, v. 51, p. 1323–1339.
Robbin, D. M., and Stipp, J. J., 1979, Depositional rate of laminated soilstone crusts, Florida Keys: Journal of Sedimentary Petrology, v. 49, p. 175–180.
Ruhe, R. V., Cady, J. G., and Gomez, R. S., 1961, Paleosols of Bermuda: Geological Society of America Bulletin, v. 72, p. 1121–1142.

Sealey, N. E., 1985, Bahamian Landscapes: London, Collins Caribbean, 96 p.

Shackleton, N. J., and Opdyke, N. D., 1973, Oxygen isotope and paleomagnetic stratigraphy of equatorial Pacific core V28–238: Oxygen isotope temperature and ice volumes on a 10^5 year time scale: Quaternary Research, v. 3, p. 39–55.

Sinclair, I. G. L., 1967, Bauxite genesis in Jamaica: New evidence from trace element distribution: Economic Geology, v. 62, p. 482–486.

Starkey, H. C., Blackmon, P. D., and Hauff, P. L., 1984, The routine mineralogical analysis of clay-bearing samples: U.S. Geological Survey Bulletin 1563, 32 p.

Syers, J. K., Jackson, M. L., Berkheiser, V. E., Clayton, R. N., and Rex, R. W., 1969, Eolian sediment influence on pedogenesis during the Quaternary: Soil Science, v. 107, p. 421–427.

Tessler, A., Campbell, P. G. C., and Bisson, M., 1979, Sequential extraction procedure for the speciation of particulate trace metals: Analytical Chemistry, v. 51, p. 844–851.

Tucholke, B. E., 1975, Sediment distribution and deposition by the Western Boundary Undercurrent: The Greater Antilles Outer Ridge: Journal of Geology, v. 83, p. 177–207.

Vacher, H. L., Rowe, M. P., and Garrett, P., 1989, The geologic map of Bermuda: Oxford, Oxford Cartographers, United Kingdom/Hamilton, Bermuda, scale 1:25,000.

Vacher, H. L., Hearty, P. J., and Mitterer, R. M., 1992, Aminostratigraphy and ages of Pleistocene limestones of Bermuda: Geological Society of America Bulletin, v. 104, p. 471–480.

Valeton, I., 1972, Bauxites: Developments in soil science 1: New York, Elsevier, 226 p.

Walls, R. A., Harris, W. B., and Nunan, W. E., 1975, Calcareous crust (caliche) profile and early subaerial exposure of Carboniferous carbonates, northeastern Kentucky: Sedimentology, v. 22, p. 417–440.

Watts, N. L., 1980, Quaternary pedogenic calcretes from the Kalahari (Southern Africa): Mineralogy, genesis, and diagenesis: Sedimentology, v. 27, p. 661–686.

Wright, V. P., 1986, Paleosols: Their recognition and interpretation: Princeton, New Jersey, Princeton University Press, 315 p.

MANUSCRIPT ACCEPTED BY THE SOCIETY JANUARY 5, 1995

Printed in U.S.A.

Geological Society of America
Special Paper 300
1995

Entombment and preservation of Sangamonian coral reefs during glacioeustatic sea-level fall, Great Inagua Island, Bahamas

Brian White and H. Allen Curran
Department of Geology, Smith College, Northampton, Massachusetts 01063

ABSTRACT

Well-preserved late Pleistocene coral reefs and associated limestones are exposed along the west coast of Great Inagua Island in the southern Bahamas. Lateral facies changes indicate that during the Sangamon Interglacial a lagoon with *Montastrea* and *Diploria* patch reefs was bordered to the north by a bank/barrier reef with *Acropora, Diploria,* and *Montastrea* sp. and to the south by carbonate sand beaches.

Falling sea level resulting from the growth of Wisconsinan ice sheets led to the rapid burial of the coral reefs in a shallowing-upward sequence of subtidal to dunal carbonate sands. Sands that accumulated in the somewhat deeper and lower energy parts of the subtidal environments were extensively burrowed to produce shelly calcarenites with *Ophiomorpha* and *Skolithos*. Shallower environments closer to the beach produced calcarenites with tabular and trough cross bedding, and large boulders of beachrock. Beach-facies calcarenites contain swash-zone laminations, erosion-scarp slumps, and a distinctive rock with a "spongy" texture that probably represents an ancient wrack line of marine grasses and algae (*Sargassum*) near the backshore-foredune interface. The finer grained calcarenites of the capping eolian sediments were deposited in east-west–trending lobate dunes formed by the prevailing easterly trade winds. All facies were affected by the emergence into a terrestrial environment that produced hematitic paleosols, caliche, vadose pisolites, and rhizomorphs.

Rapid entombment of the corals in the overlapping sands led to excellent preservation of corals in growth position. The primary aragonite of many of the corals is preserved despite more than 100 ka of subaerial exposure. This preservation may be due to an arid climate and to development of an impermeable caliche layer that protected the underlying rocks. The preserved aragonite provides ideal material for the determination of coral ages using the technique of mass spectrometric analysis of ^{230}Th. Coral U/Th ages from Great Inagua fall in the range of 130 to 122 ka. Dates from below and above a wave-cut surface suggest a short-lived sea-level lowering at approximately 124 ka.

INTRODUCTION

Great Inagua Island lies in the southern Bahamas northwest of Hispaniola and northeast of Cuba (Fig. 1). Fossil coral reefs and associated facies of late Pleistocene age are well exposed along the west coast of the island (Fig. 2). We first explored the southwest coast of Great Inagua during an oceanographic research cruise in January 1985, when we made preliminary studies of Pleistocene beach and dunes facies exposed between Matthew Town and the lighthouse. More extensive explorations were conducted in January 1986, as part of a scientific expedition supported by the Bahamian Field Station and with the aid of the Bahamas National Trust. During this reconnaissance we made some preliminary observations of the out-

White, B., and Curran, H. A., 1995, Entombment and preservation of Sangamonian coral reefs during glacioeustatic sea-level fall, Great Inagua Island, Bahamas, *in* Curran, H. A., and White, B., Terrestrial and Shallow Marine Geology of the Bahamas and Bermuda: Boulder, Colorado, Geological Society of America Special Paper 300.

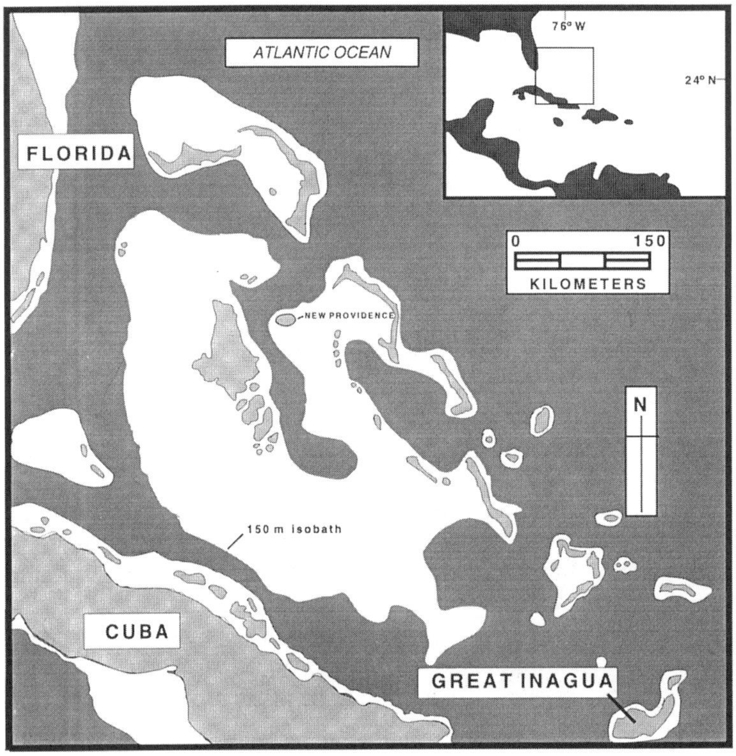

Figure 1. Location of Great Inagua Island, Bahamas.

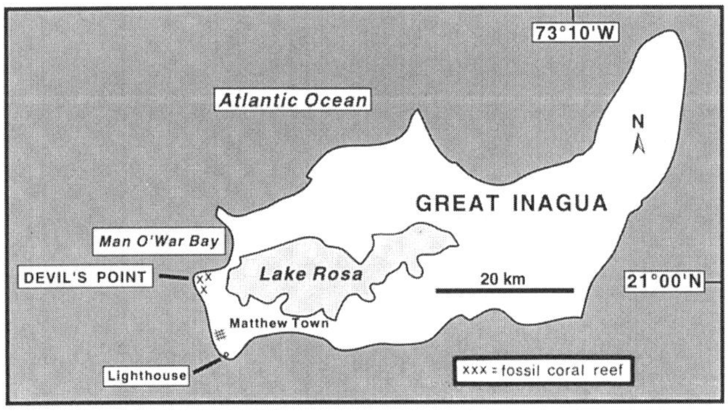

Figure 2. Location of Devil's Point on Great Inagua Island.

standing exposures of fossil coral reefs along the coast north and south of Devil's Point, together with more extensive studies of the associated nonreefal facies in the vicinity of Matthew Town. We returned in June 1986 for more detailed field work on the reefal rocks, and again in June 1987 to collect coral samples from the Devil's Point fossil reef specifically for uranium-thorium age dating (Chen et al., 1991).

Although Great Iguana Island has undoubtedly been visited by other geologists, the only recently published reports on the geology of the island, prior to our field work, are two abstracts by Mitchell (1985a,b). In these, Mitchell briefly mentioned the presence of Pleistocene fossil coral reefs and evidence for shoaling upward to major dune sequences. Based on field work in 1985 and 1986, we described the shallowing-upward sequence from subtidal calcarenites to eolianites and the well-developed fossil coral reef at Devil's Point (White and Curran, 1987; Curran et al., 1989). As a result of relative sea-level fall, all of the above facies were exposed to terrestrial processes, leading to the formation of paleosols, vadose pisolites, caliche, and rhizomorphs.

X-ray diffraction analysis of fossil corals collected in 1986 from the Devil's Point reef indicated that they consisted of 95 to 100% aragonite. More samples were collected in 1987 as part of a project to study the time scale of the last interglacial period using the then newly developed mass spectrometric technique to measure U and Th isotopes. Corals from the Devil's Point reef ranged in age from 130.3 ± 1.3 to 122.1 ± 1.3 ka, with indications of a short-lived sea-level lowering at about 124 ± 1 ka (Chen et al., 1991). In this chapter we describe the principal facies associated with the Devil's Point fossil reef and explore in some detail the history of the reef and its entombment during glacioeustatic sea-level fall. Particular attention is given to the preservation of original aragonite in the corals and associated marine cements of the reef complex.

ROCK FACIES AND DEPOSITIONAL ENVIRONMENTS

Background

We have studied, at least at the reconnaissance level, all of the coastal rock exposures of the southwestern part of Great Inagua Island from the northern limit of exposure in Man O'War Bay to the lighthouse south of Matthew Town (Fig. 2). The rocks, apart from thin hematitic paleosols, are pure limestones. Depositional facies recorded in the limestones include in situ coral reef, coral rubble, subtidal shelly sand bottoms both mobile and stable, beach, and dune. Although not all facies are always present at a given locality, those that are exposed always represent a shallowing-upward sequence. The vertical sequence of facies produced by progradation resulting from falling sea level is shown in Figures 3 and 4, and each facies is described briefly in the following sections.

Coral facies

Introduction. In modern coral reefs, colonial scleractinian corals commonly form massive rocky structures. Strong waves and currents may topple, overturn, or collapse these coral structures without actually moving them any significant distance from their place of growth. Coral debris commonly accumulates near such reefs following breakage of the in-place corals and transport of the fragments a short distance from their source. These two forms of coral occurrence can be distinguished in Pleistocene fossil coral reefs; we use the term coralstone for the former and coral rubblestone for the latter (Curran and White, 1985).

Coralstone. A fossil coral reef is exposed for approximately 7 km along the north- and west-facing shores around Devil's Point (Fig. 2). Species of the genera *Acropora, Diploria, Montastrea,* and *Porites* are extremely well preserved, with many colonies in growth position (Fig. 5A). Farther south along the coast, in situ fossil corals, mainly *Montastrea annularis* and *Diploria strigosa*, become progressively scarcer, and none occur south of Matthew Town. Fossil corals are preserved in growth position at elevations up to 2 m above present sea level, and clearly represent a former higher stand of sea level on this tectonically stable coastline. A well-defined wave-cut surface occurs within the Devil's Point fossil reef, indicating a short-lived fall of sea level (Curran et al., 1989). The age of the sea-level lowering has been dated at 123.8 to 124.9 ka, based on U/Th dating of fossil corals from beneath and above the planed-off surface (Chen et al., 1991).

Coral rubblestone. Fossilized coral debris commonly occurs overlying, underlying, and immediately adjacent to the

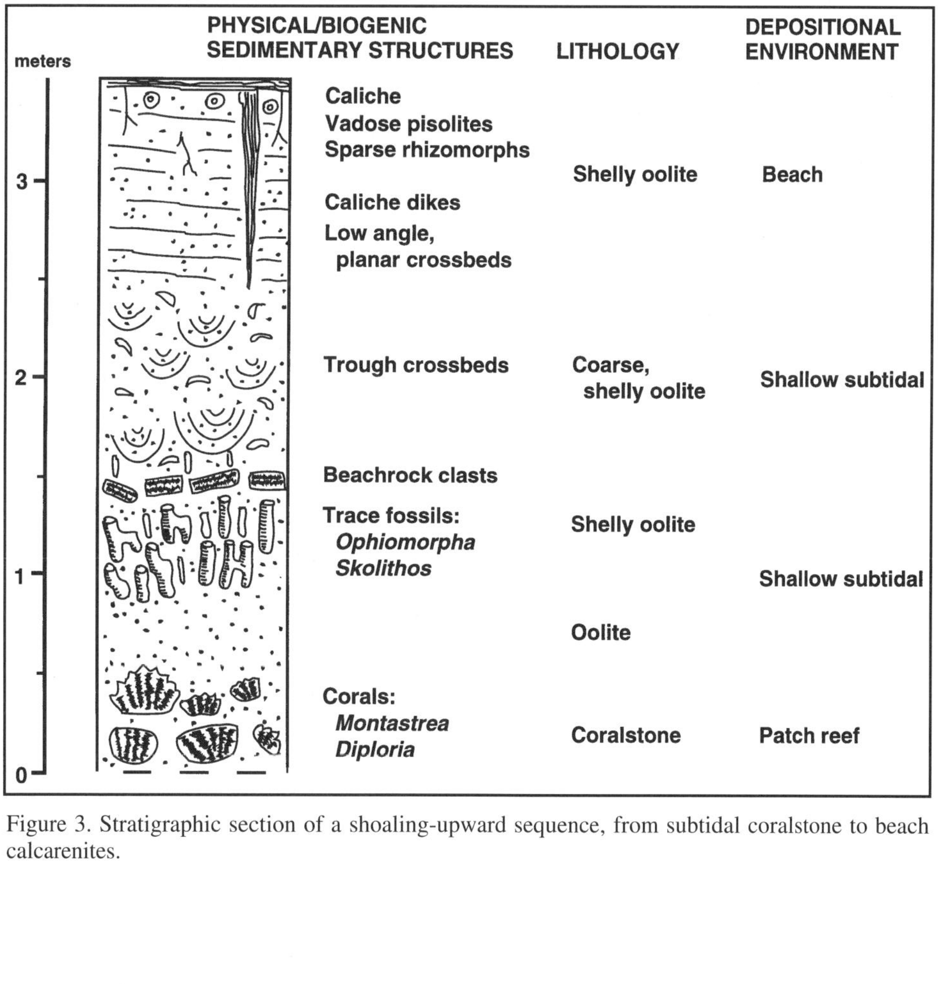

Figure 3. Stratigraphic section of a shoaling-upward sequence, from subtidal coralstone to beach calcarenites.

in situ fossil corals. All coral species found in the fossil reef are represented in the rubblestone, but especially common are small, rounded chunks of *Montastrea annularis* and the long, slender "branches" of *Acropora cervicornis*, the latter commonly unabraded and showing excellent preservation of structural detail (Fig. 5B). Coral rubblestone commonly is overlain by subtidal calcarenites (Fig. 5C).

Subtidal sand facies

Introduction. The subtidal sand facies is composed of medium to coarse, oolitic calcarenites containing a varied and well-preserved shelly fauna. Some of these sands, herein called the burrowed, subtidal sand facies, were sufficiently stable to allow the formation and preservation of numerous trace fossils, whereas others were more mobile and contain many cross beds and no trace fossils. These latter beds are termed the cross-bedded, subtidal sand facies. The subtidal rocks underlie and interfinger with well-preserved fossil corals, including species of *Montastrea, Diploria, Acropora,* and *Porites*.

Burrowed, subtidal sand facies. This facies is characterized by the presence of trace fossils belonging to the ichnogenera *Ophiomorpha* (Fig. 6A,B) and *Skolithos* (Fig. 6B). *Skolithos linearis* specimens are rather short, up to 4.2 cm long, and average 2 mm in diameter. The shafts and tunnels of *Ophiomorpha* sp. are abundant in some strata and commonly form irregular boxwork patterns. The trace fossils were probably formed in water deeper than 1 m and seaward of the swash zone, in a sand bottom that was stable and immobile enough to support lined burrow systems. For comparable trace fossil occurrences on San Salvador Island, Bahamas, the tracemaker organisms were identified as callianassid shrimp for *Ophiomorpha* sp., and polychaetes for *S. linearis* (Curran, 1984). For further information and discussion of the trace fossils that occur in Bahamian Quaternary carbonates, see Curran and White (1991).

Cross-bedded, subtidal sand facies. Other shelly, oolitic, subtidal sands were more mobile; these are represented by calcarenites that contain trough cross beds (Fig. 7A) deposited by currents flowing parallel to the ancient shoreline, and planar tabular cross beds (Fig. 7B) formed by single-event currents flowing perpendicular to the shoreline, probably during storms. Angular clasts of beachrock, up to 80 cm in length, that occur within some of the calcarenites of this facies, indicate the proximity of deposition to a beach (Fig. 7C,D).

Beach facies

The subtidal calcarenites are overlain by somewhat finer, less shelly, oolitic calcarenites, which contain features characteristic of the beach environment. These include long, low-angle cross beds, some with parting lineations on the lamination surfaces that represent the swash zone. The westward dip of these cross beds indicates that the ancient shoreline represented by these deposits was a west-facing beach. Conglomerates of well-

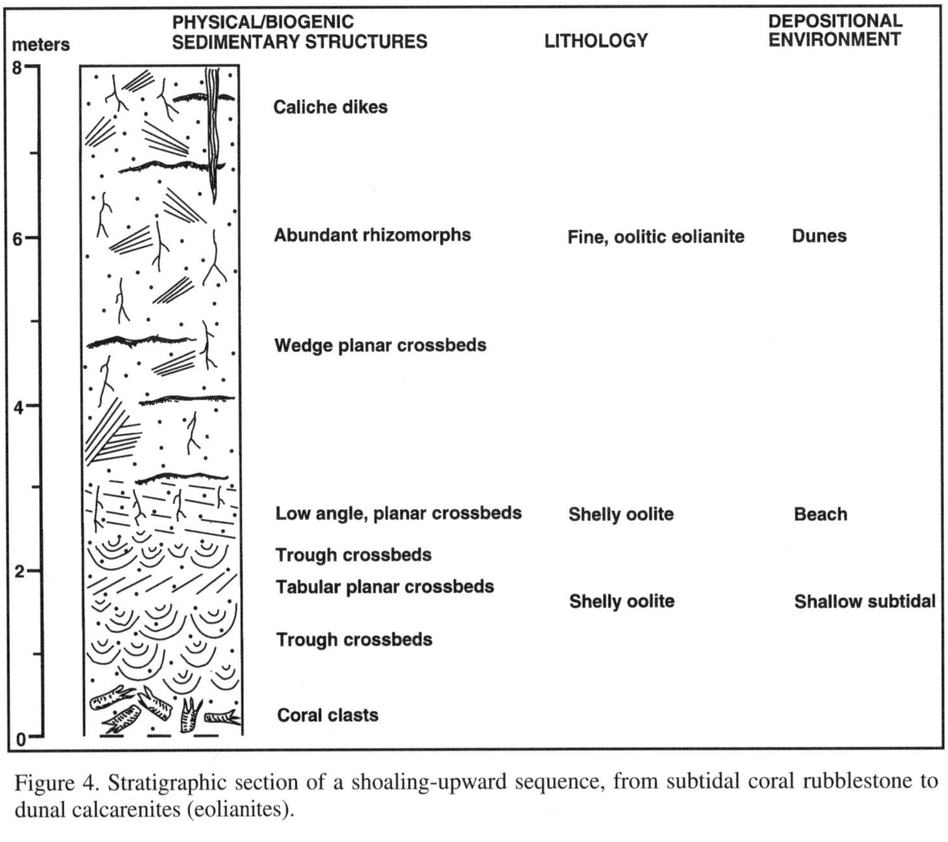

Figure 4. Stratigraphic section of a shoaling-upward sequence, from subtidal coral rubblestone to dunal calcarenites (eolianites).

Figure 5. Photos of coralstone and coral rubblestone facies of the Devil's Point fossil reef. A, Rose coral, *Manicina aveolata*, in growth position on a *Montastrea annularis* head. Scale bar = 2 cm; B, *Acropora cervicornis* coral rubblestone. Note excellent preservation. Scale bar = 3 cm; C, Coral rubblestone overlain by subtidal calcarenites.

rounded coral debris and beachrock clasts are found in some places in the lower beach facies (Fig. 8A), the former indicating the nearby contemporaneous presence of coral reefs. Rare examples of a small, slump-like structure (Fig. 8B) occur in these rocks. These likely resulted from minor slumping on beach erosion scarps. We have also noted such features in modern beach erosion scarps at Sandy Point on San Salvador Island.

An unusual, but widespread, 10- to 20-cm-thick layer, separates the upper beach sediments from the overlying dunal deposits (Fig. 9A). This bed has a porous, sponge-like texture that consists of crowded, rather irregular, rhizomorph-like features (Fig. 9B). It is interpreted as representing a buried wrack line, an accumulation of marine grasses, mainly *Thalassia testudinum* (turtle grass) and *Syringodium filiforme* (manatee grass), and of the floating brown alga *Sargassum*. Such vegetative debris collects at the landward periphery of many modern Bahamian beaches (Fig. 9C), where it commonly is mixed with sand and covered over by incipient dunes.

Eolianite facies

Overlying the fossilized sea grass wrack line bed, there are fine-grained oolitic calcarenites with wedge planar and tabular planar cross beds, climbing wind-ripple laminations, and lee-side sand lens and grain-fall layers. Such structures characterize Holocene dune deposits on San Salvador Island (White and Curran, 1985, 1988), and this facies was deposited as parabolic-like, lobate dunes with generally east-west long axes (Fig. 10A).

Effects of emergence

As a result of prolonged subaerial exposure during Wisconsinan time, the surface exposures of all the facies described above have been subjected to soil development and plant colonization in the terrestrial environment. This exposure has produced a variety of features that are clearly visible in the field. These include draped, laminar caliche (Fig. 10B), vadose pisolites, soil breccias, rhizomorphs, caliche dikes, and hematitic calcretes.

PRESERVATION OF ARAGONITE

Introduction

In modern, tropical and subtropical shallow marine environments, faunal and floral skeletal materials largely consist of aragonite and/or high-Mg calcite. The same two carbonate minerals form the dominant precipitated cements in the sediments and reefs of these marine environments (Tucker and Wright, 1990). Aragonite and high-Mg calcite are metastable, especially in the presence of fresh water. Their long-term preservation is rare, and ancient carbonate rocks consist largely of low-Mg calcite and/or dolomite (Tucker and Wright, 1990). Of particular interest in this study is the survival of scleractinian coral arago-

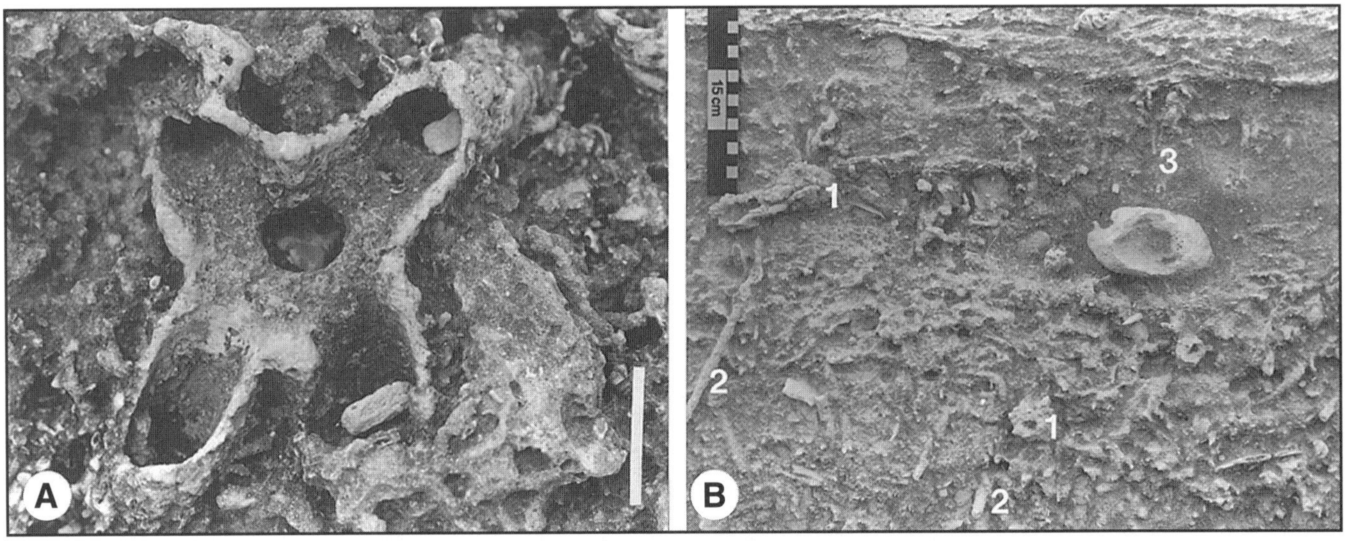

Figure 6. Trace fossils of the burrowed, subtidal sand facies, Devil's Point reef complex. A, *Ophiomorpha* sp. Scale bar = 2 cm; B, Burrow walls (1) and lithified burrow fillings (2) of *Ophiomorpha* sp., *Skolithos linearis* (3), and small beachrock clasts.

nite in a pristine condition, because this is an essential prerequisite for the absolute age-dating techniques that employ the mass spectrometric determination of uranium and thorium isotopes.

Petrographic data

Petrographic analysis of thin sections of fossil corals and associated calcarenites from the Devil's Point fossil reef revealed the preservation of aragonite in *Halimeda* and ooids, in corals fossilized in growth position, and as intergranular (Fig. 11A) and intragranular (Fig. 11B,C) marine cements.

X-ray diffraction data

Samples of *Diploria strigosa* and *Montastrea annularis* from the fossil reef at Devil's Point and from in situ fossil corals between Matthew Town and the airport were collected in 1986 and were analyzed by x-ray powder diffraction (XRD). Methods and results are given in White and Curran (1987). All of the coral samples analyzed contained more than 95% aragonite; many were essentially pure aragonite. These analyses indicated that the rocks were suitable for radiometric age determination and 10 more fossil corals were collected specifically for that purpose from the Devil's Point reef in 1987. These were analyzed by XRD and all were pure aragonite with no detectable calcite found in any of them (Chen et al., 1991).

AGE OF DEVIL'S POINT FOSSIL REEF

The age of the Devil's Point reef and associated rocks was interpreted by White and Curran (1987) as clearly pre-Holocene based on the presence of well-developed paleosols and fossil corals found in growth position up to 2 m above present sea level. Further, a comparison with fossil corals of the same species and at approximately the same elevation with respect to present sea level found on San Salvador Island suggested a Sangamon age of approximately 125 ka for the Inagua sequence. Subsequently, this age was confirmed using mass spectrometric analysis of fossil corals of pristine aragonite composition from Great Inagua Island. Fossil corals from Devil's Point yielded U/Th ages from 122.1 ± 1.3 to 130.3 ± 1.3 ka (Chen et al., 1991, Fig. 3 and Table 1).

INTERPRETATION AND GEOLOGIC HISTORY

The most prominent aspect of the stratigraphic sequence exposed along the southwest coast of Great Inagua Island is the presence, above present sea level, of rocks representing a transition from subtidal marine environments to terrestrial sand dunes. A wide variety of corals, including *Acropora palmata, A. cervicornis, Montastrea annularis, Porites porites,* and several species of *Diploria*, form a large fossil coral reef that is continuously exposed around Devil's Point. Farther south, the corals become restricted to in situ heads of *M. annularis* and *Diploria strigosa*, and these have a patchy distribution. A tentative explanation for this is that the Devil's Point reef represents a bank/barrier reef to the northwest, similar to the Cockburn Town fossil coral reef exposed on San Salvador Island (Curran and White, 1985), and that the in situ fossil corals farther south represent small patch reefs that grew in a lagoon on the landward side of the bank/barrier reef. The coral rubblestones consist of the debris of coral species found in the associated in situ fossil reefs, and clearly they were derived from them, as is commonly seen in modern reefs.

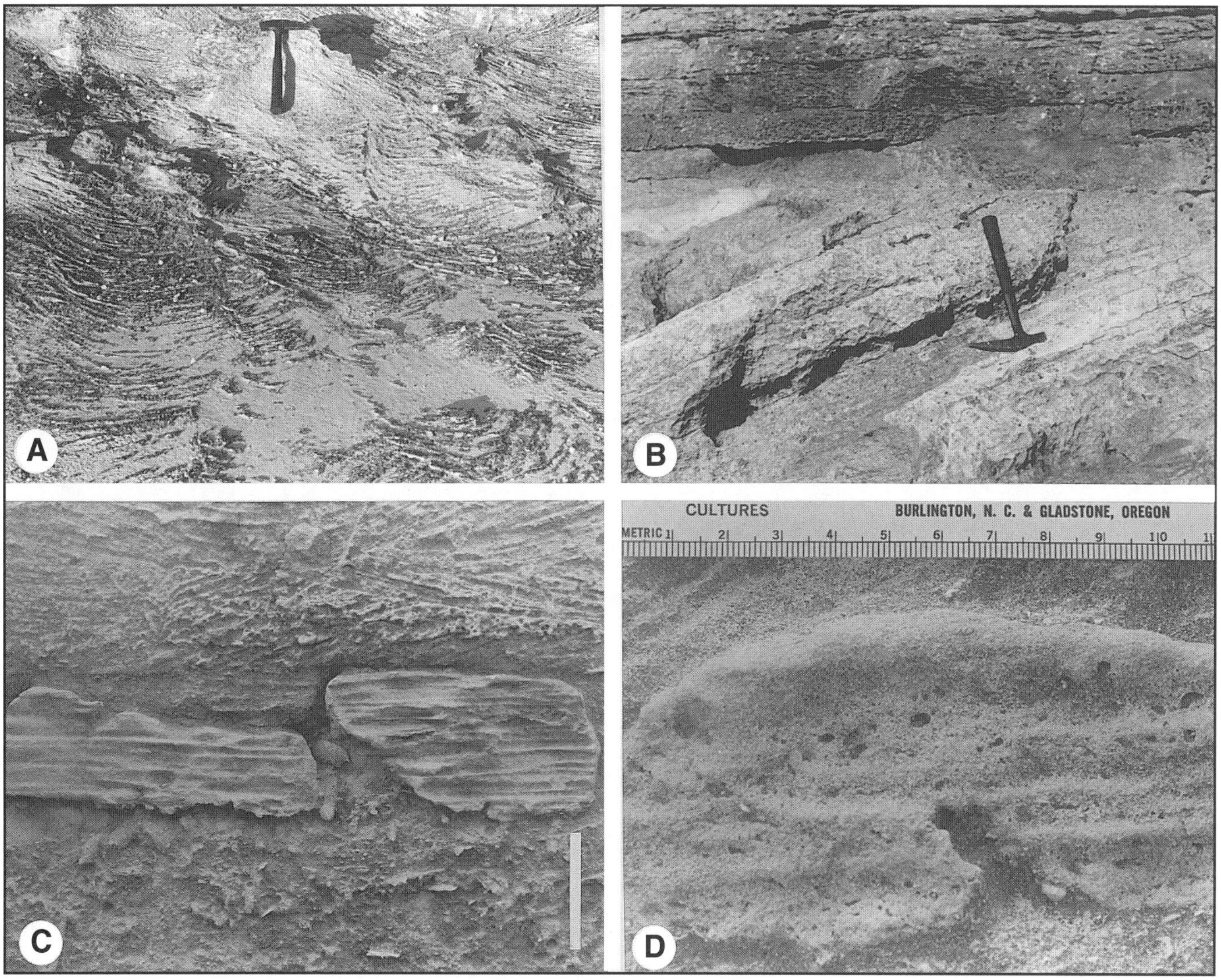

Figure 7. Sedimentary features of the subtidal sand facies, Devil's Point reef complex. A, Bedding surface view of trough cross beds in the cross-bedded, subtidal sand facies; B, Planar, tabular cross beds of the cross-bedded, subtidal sand facies; C, Lithified *Ophiomorpha* sp. burrow fillings and beachrock clasts of the burrowed, subtidal sand facies overlain by trough cross beds of the cross-bedded, subtidal sand facies. Scale bar = 10 cm; D, Beachrock clast with keystone vugs, cross-bedded, subtidal sand facies. Scale in centimeters.

Shelly, subtidal sands accumulated adjacent to, and eventually over, the corals and coral rubble. In places, sand formed a stable substrate where callianassid shrimps, polychaetes, and possibly other animals burrowed extensively, leaving a record now seen as the trace fossils *Ophiomorpha* sp. and *Skolithos linearis*. Formation and preservation of these burrows require that these sands must have been away from the surf zone and protected from significant wave and current activity. Other subtidal sands clearly were more mobile, as they contain planar tabular cross beds and abundant trough cross beds. Angular blocks of beachrock, up to 80 cm, across, found in some of the subtidal sand beds show the proximity of these sands to a former beach.

Shoaling allowed the westward advance of beach sands over the subtidal sands in much of the area. Continued sea regression led eventually to emergence above high tide level and the advance of carbonate sand dunes over the underlying beach and subtidal deposits. The dunes were small and lobate in form, with long axes of the dunes aligned approximately east-west, and roughly parallel to the prevailing easterly trade winds. Similar parabolic-like, lobate eolianite dunes have been described from San Salvador Island (White and Curran, 1985, 1988), and from Bermuda (Mackenzie, 1964a,b).

As a result of sea-level fall, all facies from in situ corals through the shallowing-upward sequence became part of an island, where they were subjected to terrestrial processes. This island was at least in part vegetated, as shown by the numerous

Figure 8. Sedimentary features of the beach facies, Devil's Point reef complex. A, Conglomerate of coral and beachrock clasts, lower beach facies. Scale = 30.5 cm. B, Slump structure thought to have formed on a beach erosion scarp, beach facies. Scale in centimeters.

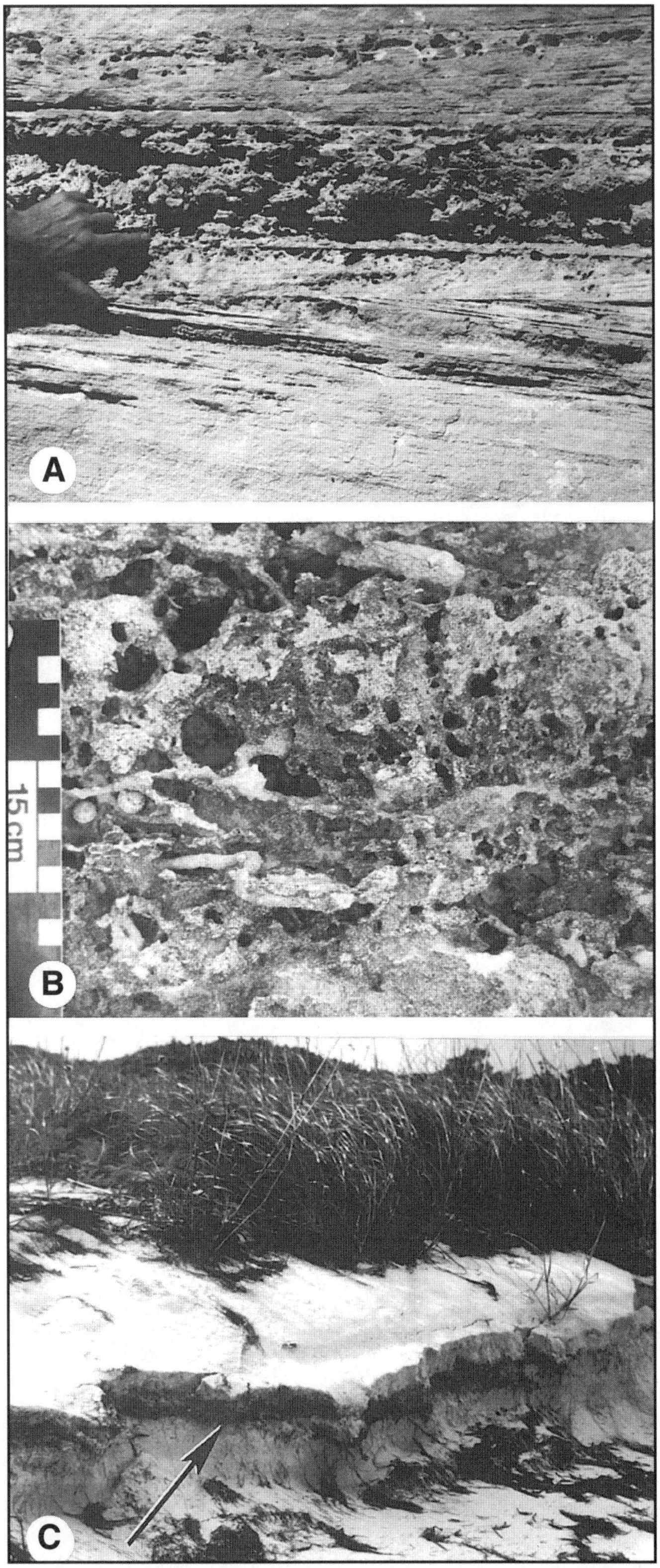

Figure 9. Marine plant accumulations near the backshore-dune contact: A, Bed indicated by the hand believed to represent a fossilized marine plant accumulation near the contact with upper beach and dune calcarenites, Devil's Point reef complex; B, Close-up view of marine plant accumulation layer of (A) showing porous texture and rhizomorph-like structures; C, Modern erosion scarp exposing a layer of accumulated marine plant debris (arrow) near the contact between upper beach and dune sands, Grahams Harbour, San Salvador Island, Bahamas.

Figure 10. A, View along the east-west axis of a parabolic-like, lobate fossil dune, of the eolianite facies, Devil's Point reef complex; B, Draped laminar caliche and vadose pisolites developed on beach facies calcarenites, Devil's Point reef complex.

root traces preserved by the precipitation around individual roots of enclosing, dense micrite. The term rhizomorph was used by Northrop (1890) for plant traces preserved in this fashion in Bahamian carbonate rocks. On Great Inagua Island, rhizomorphs are particularly abundant in the eolianites, but they occur in all facies, including the coralstone facies, thus providing clear evidence of the change from marine to nonmarine conditions. The evidence of a vegetated land surface implies the presence of soil, and the presence of laminar caliche, caliche dikes, vadose pisolites, and soil breccias shows that processes associated with soil formation were active. Many of the paleosols are reddened by hematite, for which there is no obvious local source. A component of these hematite-stained soils probably was formed by weathering of dust derived from the Sahara by long-distance eolian transportation by northeasterly trade winds, as described for Barbados, Jamaica, the Florida Keys, and New Providence Island, Bahamas (Muhs et al, 1990). Thin, laminated caliche can be seen to drape over many of the beds and forms a hard, resistant, micritic layer. Laminated caliche also fills vertical to subvertical fractures and fissures that cut across all facies of the sequence. These are identical to, although narrower than, features found in the Cockburn Town fossil coral reef on San Salvador Island, which were named caliche dikes (Curran and White, 1985).

In an earlier publication, we made a preliminary reconstruction of the paleogeography of the study area, suggesting that an island mantled by wind-blown sand was bordered to the north by a lagoon with patch reefs, and farther offshore, a bank/barrier coral reef (White and Curran, 1987). As sea level fell during the Wisconsinan ice advance, the island expanded and incorporated all facies, including the former coral reefs. The entire shoaling-upward sequence was exposed to terrestrial processes, including the development of vegetated soils on the exposed rocks of all facies types.

The fossil corals and associated rocks with marine cements have retained their original aragonite composition. A similar situation exists in the rocks of the Cockburn Town fossil coral reef on San Salvador Island (White et al., 1984), and in the fossil corals of the Hogsty Reef, Bahamas (Pierson and Shinn, 1985). Preservation of pristine aragonite is a critical prerequisite for the determination of U/Th ages of the fossil corals and thus is of major importance in revealing the geologic history of these rocks. The fossil coral reefs of Great Inagua island have been exposed for almost 120 ka to a terrestrial environment and its associated fresh water since their formation. Such exposure to fresh water conditions is widely believed to lead to the dissolution of aragonite (Scoffin, 1987; Tucker and Wright, 1990). In the case of several Bahamian Pleistocene coral reefs such dissolution has not occurred. The excellent preservation of some in situ corals and coral clasts in coral rubblestone has been attributed to rapid burial in storm-deposited sands in the Cockburn Town fossil reef of San Salvador Island (White et al., 1984; Curran et al, 1989; Carew and Mylroie, this volume), in the Sue Point fossil reef of San Salvador Island (White, 1989), and in the Devil's Point fossil reef of Great Inagua Island (White and Curran, 1987; Curran et al, 1989). Recent taphonomic studies of fossil corals from the Cockburn Town fossil reef suggest essentially instantaneous postmortem burial of the corals by the overlying calcarenites (Greenstein and Moffat, 1995). Coral U/Th ages from the Cockburn Town and Devil's Point fossil reefs (Chen et al., 1991) and the Sue Point fossil reef (White, 1989; Chen et al., 1991) indicate that these burial events occurred during the same time interval. This suggests an external forcing rather than localized progradation caused by the aggradation of the carbonate sediments. The most obvious external cause is global sea-level lowering resulting from the growth of the Wisconsinan ice sheets that terminated the Sangamon Interglacial. Such lowering of sea level made the coral reefs more vulnerable to the effects of major storms, including simultaneous death and burial in rapidly deposited storm-driven carbonate sands.

Although rapid burial in sands explains the excellent preservation of the form of the fossil corals, it does not guarantee the preservation of their primary aragonite mineralogy. This requires the prevention of the dissolution of aragonite by fresh water. Climatic aridity is a possible contributing factor. At the present time there is a northwest to southeast gradient of decreasing rainfall across the Bahamas with Great Inagua Island lying at the drier southeast end of this trend. The average annual rainfall is approximately 60 cm. The potential evapotranspiration for Great Inagua Island is unknown, but with mean monthly temperatures ranging from a low of 24°C in February to a high of 28°C in July, it is likely to exceed the average annual rainfall (weather data from anon., 1985). Thus much of the rainfall may be vaporized back to the atmosphere before the fresh water has time to react with the aragonite. That this cannot be a complete explanation is indicated by the presence of caves, sinkholes, and other karst features in the limestones of the island, clearly demonstrating the localized effects of dissolution by fresh water. Many limestone outcrops in Great Inagua have a surficial layer of hematitic caliche. Once such a layer has formed, it provides an impermeable cover for the underlying rocks, which protects them from further interaction with fresh water. Our studies of cores drilled through similar caliche layers formed on Pleistocene subtidal calcarenites in the interior of San Salvador Island show that aragonitic grains and cements are preserved only millimeters beneath the base of caliche layers. Thus the preservation of corals and their primary aragonite is thought to be due to a combination of rapid burial during storms as they became more vulnerable during glacioeustatic lowering of sea level, postregression exposure to an arid climate, and the progressive development of a protective impermeable cover during calichification.

CONCLUSIONS

1. The sequence of rocks found on the west coast of Great Inagua Island represents a change of depositional environment from shallow subtidal marine to terrestrial. This change was brought about by the lowering of sea level caused by the onset of Wisconsinan glaciation.

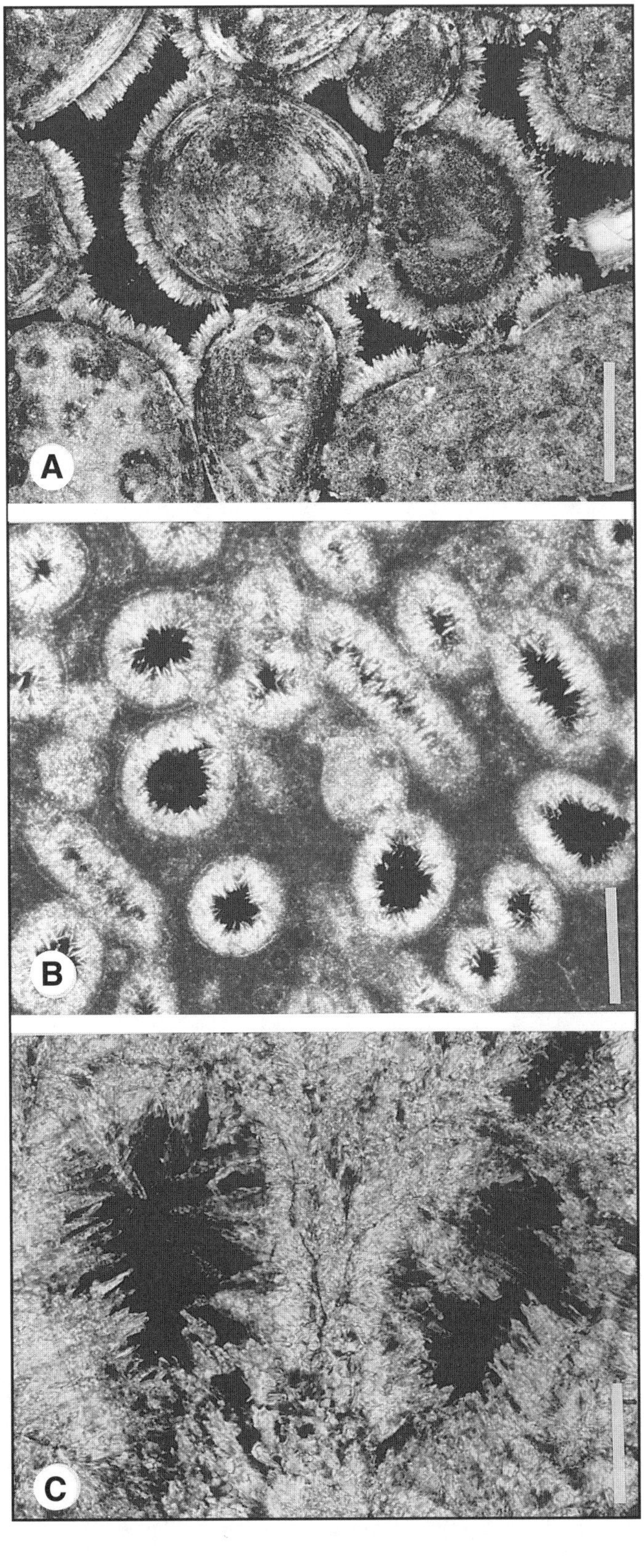

Figure 11. Photomicrographs from Devil's Point reef complex rocks, all with crossed polarizers. A, Intergranular acicular marine aragonite cements and aragonitic ooids in oolitic, peloidal grainstone. Scale bar = 200μ. B, Intragranular acicular marine aragonite cement in utricles of aragonitic *Halimeda* grain. Scale bar = 100μ. C, Intragranular, epitaxial, acicular marine aragonite cement in corallites of aragonitic coral. Scale bar = 100μ.

2. Coralstone, coral rubblestones, burrowed calcarenites, and cross-bedded calcarenites reflect subtle differences in water depth and energy conditions in the subtidal environment. Sediments deposited in beach and backshore environments have distinctive characteristics, including parting lineation on beach face laminations, slumps on erosion scarps, and fossil wrack deposits.

3. Emergence of Great Inagua Island as a result of sea-level lowering permitted the growth of eolian dunes and the formation of fine-grained eolianites and colonization by terrestrial plants. Paleosols developed during the emergent phase with a minor component being contributed by wind-borne African dust.

4. In many examples, the fine details of coral structure have been preserved in exquisite detail. This preservation was facilitated as corals were rapidly buried in sands carried by storm-generated currents. As sea level fell, the corals became more vulnerable to storm damage, and in some instances they appear to have been simultaneously killed and entombed by encroaching sands.

5. Many aragonitic corals, skeletal grains, and cements have retained their original mineralogic composition. Corals with pristine aragonite have been used to determine a U/Th age of 122 to 130 ka for the Devil's Point fossil coral reef. Preservation of unaltered aragonite is thought to be due to a combination of an arid climate and the development of a protective surface layer of impermeable caliche.

ACKNOWLEDGMENTS

We thank Steve Mitchell for alerting us to the presence of the fossil reefs exposed along the west coast of Great Inagua Island, and Jim Carew and John Mylroie for sharing with us their discovery of the reef outcrops early in the 1986 expedition. We express our gratitude to Don Gerace for his support of the 1986 expedition and his long-term encouragement of our research in the Bahamas. We are grateful to Jerry Wasserberg and Jim Chen, California Institute of Technology, for their interest in our fossil reef studies, and for their age determinations of coral samples. We thank Dave White for his careful darkroom work. We are especially grateful to Jimmy Nixon, Henry Nixon, and Carl Farquharson for their generous hospitality and invaluable assistance on Great Inagua Island. This work was much improved through careful reviews by Jim Carew, College of Charleston, and Mark Wilson, College of Wooster, to whom we express our sincere thanks. Grants from the Petroleum Research Fund of the American Chemical Society and from the Committee on Faculty Compensation and Development of Smith College partially supported the field work for this study.

REFERENCES CITED

Anonymous, 1985, Atlas of the Commonwealth of the Bahamas, (second edition): Kingston, Jamaica, Kingston Publishers, 48 p.

Chen, J. H., Curran, H. A., White, B., and Wasserburg, G. J., 1991, Precise chronology of the last interglacial period: ^{234}U-^{230}Th data from fossil coral reefs in the Bahamas: Geological Society of America Bulletin, v. 103, p. 82–97.

Curran, H. A., 1984, Ichnology of Pleistocene carbonates on San Salvador, Bahamas: Journal of Paleontology, v. 58, p. 146–159.

Curran, H. A., and White, B., 1985, The Cockburn Town fossil coral reef, *in* Curran, H.. A., ed., Pleistocene and Holocene carbonate environments on San Salvador Island, Bahamas; Geological Society of America, Orlando Annual Meeting Field Trip Guidebook: Ft. Lauderdale, Florida, CCFL Bahamian Field Station, p. 95–120.

Curran, H. A., and White, B., 1991, Trace fossils of shallow subtidal to dunal ichnofacies in Bahamian Quaternary carbonates: Palaios, v. 6, p. 498–510.

Curran, H. A., White, B., Chen, J. H., and Wasserburg, G. J., 1989, Comparative morphologic analysis and geochronology for the development and decline of two Pleistocene coral reefs, San Salvador and Great Inagua islands, *in* Mylroie, J. E., ed., Proceedings, Fourth Symposium on the Geology of the Bahamas: San Salvador, Bahamian Field Station, p. 107–117.

Greenstein, B. J., and Moffat, H. A., 1995, Comparative taphonomy of Holocene and Pleistocene corals, San Salvador, Bahamas: Palaios, in press.

Mackenzie, F. T., 1964a, Geometry of Bermuda calcareous dune cross-bedding: Science, v. 144, p. 1449–1450.

Mackenzie, F. T., 1964b, Bermuda Pleistocene eolianites and paleowinds: Sedimentology, v. 3, p. 52–64.

Mitchell, S. W., 1985a, Surficial geology of the southernmost Bahama islands: Geological Society of America, Abstracts with Program, v. 17, p. 125.

Mitchell, S. W., 1985b, Quaternary eustatic accretion of southern Bahamas Archipelago: American Association of Petroleum Geologists Bulletin, v. 69, p. 289.

Muhs, D. R., Bush, C. A., Stewart, K. C., Rowland, T. R., and Crittendon, R. C., 1990, Geochemical evidence of Saharan dust parent material for soils developed on Quaternary limestones of Caribbean and western Atlantic islands: Quaternary Research, v. 33, p. 157–177.

Northrop, J. I., 1890, Notes on the geology of the Bahamas: New York Academy of Sciences Transactions, v. 10, p. 4–22.

Pierson, B. J., and Shinn, E. A., 1985, Cement distribution and carbonate mineral stabilization in Pleistocene limestones of Hogsty Reef, Bahamas, *in* Schneidermann, N., and Harris, P. M., eds., Carbonate cements: Society of Economic Paleontologists and Mineralogists Special Publication 36, p. 153–168.

Scoffin, T. P., 1987, An introduction to carbonate sediments and rocks: New York, Chapman and Hall, 274 p.

Tucker, M. E., and Wright, V. P., 1990, Carbonate sedimentology: Oxford, United Kingdom, Blackwell, 482 p.

White, B., 1989, Field guide to the Sue Point fossil coral reef, San Salvador Island, Bahamas, *in* Mylroie, J. E., ed., Proceedings, Fourth Symposium on the Geology of the Bahamas: San Salvador, Bahamian Field Station, p. 353–365.

White, B., and Curran, H. A., 1985, The Holocene carbonate eolianites of North Point and the modern marine environments between North Point and Cut Cay, *in* Curran, H. A., ed., Pleistocene and Holocene carbonate environments on San Salvador Island, Bahamas, Geological Society of America, Orlando Annual Meeting Field Trip, Guidebook: Ft. Lauderdale, Florida, CCFL Bahamian Field Station, p. 73–93.

White, B., and Curran, H. A., 1987, Coral reef to eolianite transition in the Pleistocene rocks of Great Inagua Island, Bahamas, *in* Curran, H. A., ed., Proceedings, Third Symposium on the Geology of the Bahamas: Ft. Lauderdale, Florida, CCFL Bahamian Field Station, p. 165–179.

White, B., and Curran, H. A., 1988, Mesoscale physical sedimentary structures and trace fossils in Holocene carbonate eolianites from San Salvador Island, Bahamas: Sedimentary Geology, v. 55, p. 163–184.

White, B., Kurkjy, K. A, and Curran, H. A., 1984, A shallowing-upward sequence in a Pleistocene coral reef and associated facies, San Salvador, Bahamas, *in* Teeter, J. W., ed., Proceedings, Second Symposium on the Geology of the Bahamas: Ft. Lauderdale, Florida, CCFL Bahamian Field Station, p. 53–70.

MANUSCRIPT ACCEPTED BY THE SOCIETY JANUARY 5, 1995

Printed in U.S.A.

Geological Society of America
Special Paper 300
1995

Sedimentary architecture of Pleistocene eolian calcarenites, San Salvador Island, Bahamas

Mario V. Caputo
Mt. San Antonio College, Department of Earth Sciences, 1100 North Grand Avenue, Walnut, California 91789

ABSTRACT

Exposures of Upper Pleistocene (Sangamon) calcarenites around the southern margin of San Salvador Island, Bahamas, comprise the older French Bay Member and beds thought to be equivalent to the younger Cockburn Town Member, both of the Grotto Beach Formation. A host of lithologic features, including grain and cement textures, stratification, fossils, and related facies that have been described from earlier work, lends support to an eolian origin for most of these calcarenites. This chapter presents a classification of bedding units as components or elements of mesoscopic architecture, the sedimentary framework of carbonate dunes of a Pleistocene dune ridge–swale system of San Salvador Island. Ideas presented herein evolved from a previous examination of eolian microstructure or microscopic architecture of pore, grain, and cement textures. In outcrop, architectural elements are distinguished by scale and bounding surfaces, then later classified by parent bedforms, sedimentary mechanisms, and duration of the sedimentary event. Eolian architectural elements in the Grotto Beach Formation include: first-order sandflow or grainfall strata, second-order wind-ripple strata/cross-laminated sets, third-order wind-ripple cosets, fourth-order cross-bed intrasets, and fifth-order cross-bed sets, given in order of increasing rank and scale.

Collective sedimentary architecture, cross-bed dip patterns, and relict eolian landscape preserved in present-day topography suggest that windblown sand accumulated as sinuous dune ridges with lobate or linguoid projections. Rapid sedimentation and diagenesis account for the excellent preservation of nearly complete internal dune structure. Dune migration was probably brief and restricted to a limited area on a partly submerged isolated carbonate platform.

A prolonged period of sediment starvation and dune deflation accompanied a climate shift from arid to humid. Colonizing plants further stabilized the dunes and enhanced cementation through transpiration. Soil formation on degraded dune topography ultimately prevented the further reworking of sand by wind. These events are recorded in the bounding surface–rhizomorph-paleosol interval that truncates fifth-order cross-bed sets.

INTRODUCTION

Eolian dunes are integral morphosedimentary components of carbonate shorelines both on modern (Tucker and Wright, 1990) and ancient (Hunter, 1993) marine platform settings. With an adequate sediment supply and competent wind, the construction and migration of coastal dunes are generally independent of latitude, sediment composition, and climate (Goldsmith, 1978, 1985; Marzolf, 1988). Large-scale cross-bedding in recent lithified and unlithified oolitic-skeletal deposits is a feature known

Caputo, M. V., 1995, Sedimentary architecture of Pleistocene eolian calcarenites, San Salvador Island, Bahamas, *in* Curran, H. A., and White, B., Terrestrial and Shallow Marine Geology of the Bahamas and Bermuda: Boulder, Colorado, Geological Society of America Special Paper 300.

from marine ooid banks and shoals. However, sedimentologists for decades have recognized the paradoxical eolian nature of some large cross-bed sets composed of marine allochemical grains from the Bahamas, Bermuda, and other carbonate-forming areas (Illing, 1954; Newell et al., 1960; Ball, 1967; Mackenzie, 1964a,b; Vacher, 1973; Ward, 1973, 1975, 1985; Perkins, 1977; Garrett and Gould, 1984). McKee and Ward (1983) clarified this paradox by providing a complete summary of features common among carbonate eolianites.

Only recently, structures and textures of well-studied eolian quartzarenites, mainly from the Colorado Plateau region of the southwestern United States, have been compared with those of eolian calcarenites in Holocene (White and Curran, 1985, 1988, 1993) and Pleistocene (Caputo, 1989, 1993) Bahamian limestones, and older limestones from the midcontinent (Hunter, 1993; Hanford, 1990; Dodd et al., 1993) and western interior of the United States (Loope, 1986; Loope and Haverland, 1988; Boubin and Loope, 1989; Merkley and Dodd, 1991; Rice and Loope, 1991; Loope and Kilibarda, 1992) and the Perth Basin, Australia (Semeniuk and Glassford, 1988).

San Salvador Island, Bahamas

The island of San Salvador lies near the eastern margin of the Bahamas Banks, and owes its origin to cyclic carbonate sedimentation and aggradation that persisted throughout the Quaternary Period (Fig. 1). The emergent terrane consists of moderately lithified, skeletal and nonskeletal limestones deposited on an isolated carbonate platform. A history of fluctuating shorelines and episodic deposition is expressed in the Pleistocene and Holocene bedrock calcarenites, which have been arranged in a well-established lithostratigraphic and geochronologic scheme (Carew et al., 1984; Carew and Mylroie, 1985, 1987, 1989; Carew and Mylroie, this volume). The stratigraphic names are practicable throughout the Bahamas (Carew and Mylroie, this volume) and have been adopted in this chapter (Fig. 2).

In the Carew and Mylroie scheme, the oldest interval of the Pleistocene Series on San Salvador is the partly exposed Owl's Hole Formation. It is overlain by the Upper Pleistocene Grotto Beach Formation, which has been subdivided into the French Bay and Cockburn Town Members. The Holocene Series is represented by the older North Point Member and younger Hanna Bay Member of the Rice Bay Formation. Chen et al. (1991), Foos and Muhs (1991), and Hearty and Kindler (1993) have provided additional information regarding the geochronology of the Quaternary System of San Salvador.

Theme and purpose

The ideas presented in this chapter evolved from field work initiated in 1987 that has been directed toward understanding the origin and architecture of Pleistocene eolian limestones on San Salvador Island, Bahamas. Supported with information on Jurassic eolian quartzarenites from the Colorado Plateau, United States, Caputo (1989, 1993) explained that consistent weathering patterns of eolian calcarenites can be correlated with grain texture and cementation patterns in strata produced by wind. An eolian origin had been assigned tentatively to Pleistocene units on San Salvador based on the presence of terrestrial fossils and pedogenic features (Adams, 1983; Curran and White, 1984, 1985; White et al., 1984; Carew and Mylroie, 1985), and was confirmed by a description of definitive eolian structures (Caputo, 1989, 1993). This chapter continues to focus on outcrop characteristics of Pleistocene limestones on San Salvador Island, Bahamas. The principal aims are to summarize known sedimentary features that indicate an eolian origin, and to propose a preliminary hierarchy of eolian sedimentary architecture and corresponding sedimentary processes.

Methods and location of study

In this chapter, outcrop characteristics are described from stratigraphic units in the Grotto Beach Formation, specifically the French Bay Member and unnamed cross-bedded calcarenites that may be correlated with the type Cockburn Town Member of western San Salvador Island (Carew and Mylroie, 1985). The French Bay Member is known from exposures in the "French Bay cliffs" between Sandy Point and a small, wooden pier called Government Dock. Beds that are equivalent to the Cockburn Town Member crop out in sea cliffs between French Bay and the Gulf (Fig. 1). Some bedding structures in the North Point Member (Holocene Rice Bay Formation) around North Point serve as examples of eolian architecture, where analogous bedding structures in the Grotto Beach Formation are either poorly developed or not exposed.

Stages in the general growth of carbonate island platforms are complex (Garrett and Gould, 1984; Carew and Mylroie, 1985, this volume; Hearty and Kindler, 1993). On San Salvador Island, some of the units, namely, the French Bay and Cockburn Town Members, do not form a continuous vertical succession. Hence, subdividing the Pleistocene Series on San Salvador Island does not lend itself easily to traditional superposition. Instead, lithostratigraphic units are distinguished by geomorphic and petrologic characteristics, and amino acid and radiometric dating. For the purposes of this study, recognizing position in the

Figure 1 (on facing page). Island of San Salvador, Bahamas and location of outcrops examined at Sandy Point, French Bay, the Bluff, and the Thumb. Pleistocene paleocurrent directions for eolian sandflow sets in the French Bay Member, equivalent Cockburn Town Member, and younger strata are indicated by circular histograms for Sandy Point, French Bay, the Bluff, and the Thumb locations (n = number of measurements; $\overline{X}$ = mean azimuth). Wedge-shape histogram segments suggest dunes with multiple slipfaces; mean slip direction (bold arrows) was to the west-southwest and west-northwest. Also shown are Cockburn Town (government-seat) and other type localities for stratigraphic units in Figure 2; lakes (L), Bahamian Field Station (BFS), and Government Dock (GD). Scale of inset map is in nautical miles (N MI) and kilometers (KM) (after Curran, 1989).

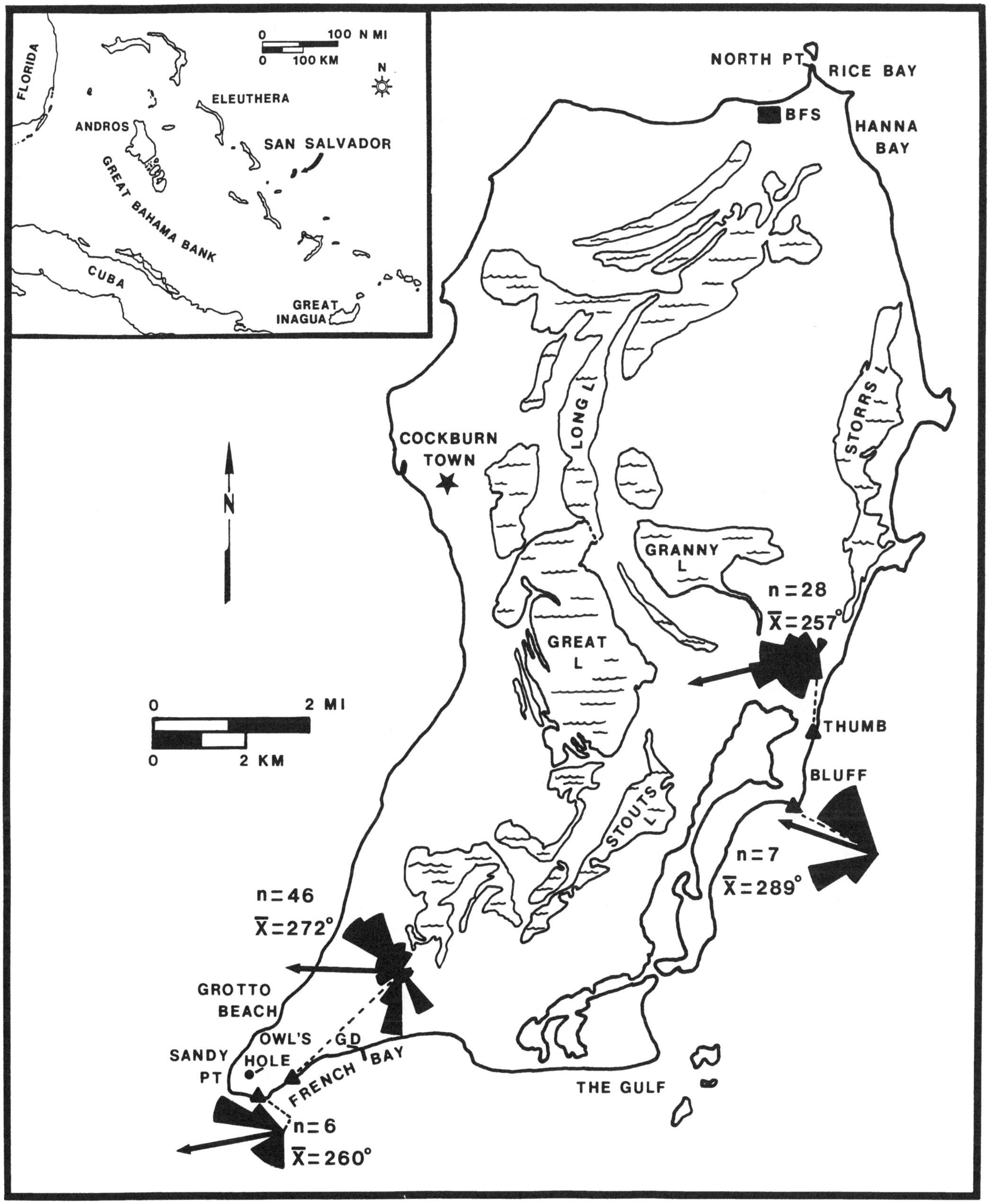

FLORIDA
ANDROS
ELEUTHERA
SAN SALVADOR
GREAT BAHAMA BANK
CUBA
GREAT INAGUA
0 100 N MI
0 100 KM
N
NORTH PT
RICE BAY
BFS
HANNA BAY
LONG L
STORRS L
COCKBURN TOWN
N
GRANNY L
n=28
X̄=257°
GREAT L
0 2 MI
0 2 KM
THUMB
BLUFF
STOUTS L
n=7
X̄=289°
n=46
X̄=272°
GROTTO BEACH
OWL'S HOLE
GD
SANDY PT
FRENCH BAY
THE GULF
n=6
X̄=260°

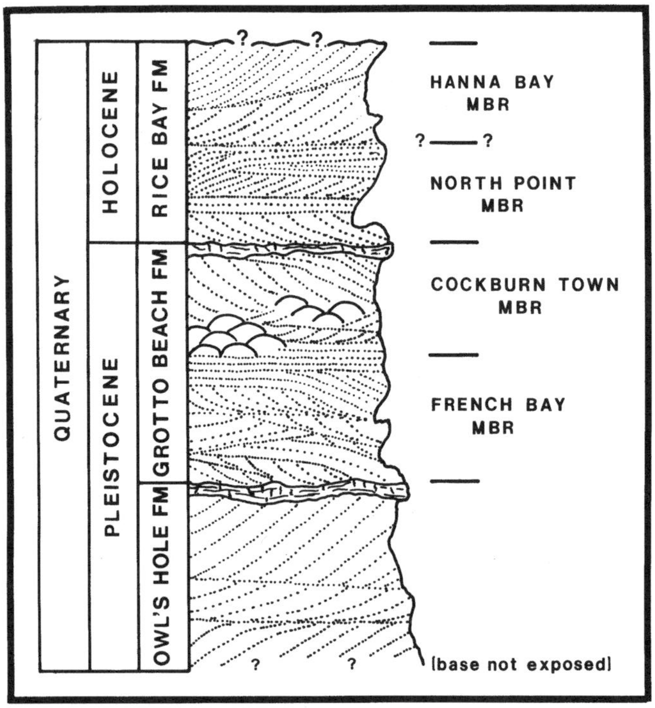

Figure 2. Lithostratigraphic units of the Quarternary System as applied to San Salvador Island and throughout the greater Bahamas (modified from Carew and Mylroie, this volume).

Pleistocene section on San Salvador Island relies on descriptions given in Carew et al. (1984) and Carew and Mylroie (1985, 1987, 1989, this volume).

Specific stratification, produced by eolian mechanisms of ripple-migration, grainfall, and grainflow, are recognized in outcrop by weathering profile, angle of dip, and structural relationships, and in thin section by textural and cementation patterns (Caputo, 1989, 1993). Dip azimuths mainly of sandflow strata and toesets of grainfall and wind-ripple strata were measured in the French Bay Member in the French Bay cliffs near Sandy Point, and in undifferentiated Pleistocene rocks, belonging probably to the Grotto Beach Formation or even younger Pleistocene strata (Hearty and Kindler, 1993) in sea cliff locations known as the Thumb and the Bluff (Fig. 1).

GROTTO BEACH FORMATION: PREVIOUSLY RECOGNIZED EOLIAN FEATURES

In general, most limestones in the stratigraphic record are thought to be marine in origin, or at least the result of some kind of subaqueous deposition, especially those in which the framework consists of ooids and biogenic fragments. Because carbonate grains survive minimal transport and abrasion, their preservation suggests deposition at or close to the sediment source, compared to more durable siliciclastic grains. Therefore, it is reasonable to conclude that the sediment source of the wave-resistant, cliff-forming limestones, marginal to carbonate platforms such as San Salvador and other Bahamian Islands, is the adjacent marine environment, as suggested by mineral, textural, and taxonomic composition of both rock and sediment.

Allochems in the Grotto Beach Formation are of unquestionable marine origin. They comprise skeletal fragments of molluscs, echinoderms, corals, calcareous algae, and foraminifera; and nonskeletal peloids, pellets, aggregates, and ooids. In various proportions, they have been cemented into sorted, rounded, oolitic biopelsparites or grainstones (Fig. 3). However, an assemblage of previously studied lithologic features lends support to a nonmarine eolian origin for much of the Grotto Beach Formation and other Bahamian limestones. The features include grain texture, diagenetic properties, stratification, biogenic properties, and facies associations (Table 1).

Grain texture

Laminations and beds of the French Bay Member and younger strata in the Grotto Beach Formation display consistent trends in grain size and sorting that were controlled by wind-ripple migration, grainfall, and grainflow on eolian dunes (Caputo, 1989, 1993). Clasts range in size from very fine- to slightly medium-grained sand, a texture similar to that of quartzose eolian deposits. As a result of efficient sorting by wind, ripple strata in units of the Grotto Beach Formation are typified by cyclically alternating fine- and medium-grained laminae, both of which are well sorted and together form a complete upward-coarsening ripple stratum. Sandflow beds are thicker than ripple strata but display similar sorting, size, and grading characteristics. Grainfall deposits occur as single, well-sorted laminations among sandflow beds (Fig. 3).

Diagenetic properties

Eolian processes on dunes controlled the degree of grain packing, which affected patterns of cementation in laminations and beds of the Grotto Beach Formation (Caputo, 1989, 1993). Entire grainfall laminae, and basal laminae of wind-ripple and sandflow strata exhibit close grain-packing. Consequently, they are completely cemented and are identified on the outcrop by ledge-forming weathering profiles. Upper parts of wind-ripple and sandflow strata exhibit loose packing, partial cementation, and recess-forming weathering profiles (Figs. 3 and 4).

Figure 3 (on facing page). Photomicrographs of allochems composing selected eolian beds equivalent to the Cockburn Town Member of the Grotto Beach Formation located between French Bay and the Gulf. Bar scale = 0.5 mm (0.0195 in) long. A, Lower fine-grained and upper medium-grained laminae of one complete upward-coarsening wind-ripple stratum (between dashed lines) consisting of peloids and algal fragments. Fine laminae show close packing and complete cementation; medium laminae show loose packing and incomplete cementation. Bright areas around and between grains are cement; dark areas are pores. Bedding surfaces are marked by dashed lines. B, Fine-grained grainfall (GF) (between dashed lines) and medium-grained sandflow (SF) strata composed of peloids, coated grains, and some algal fragments.

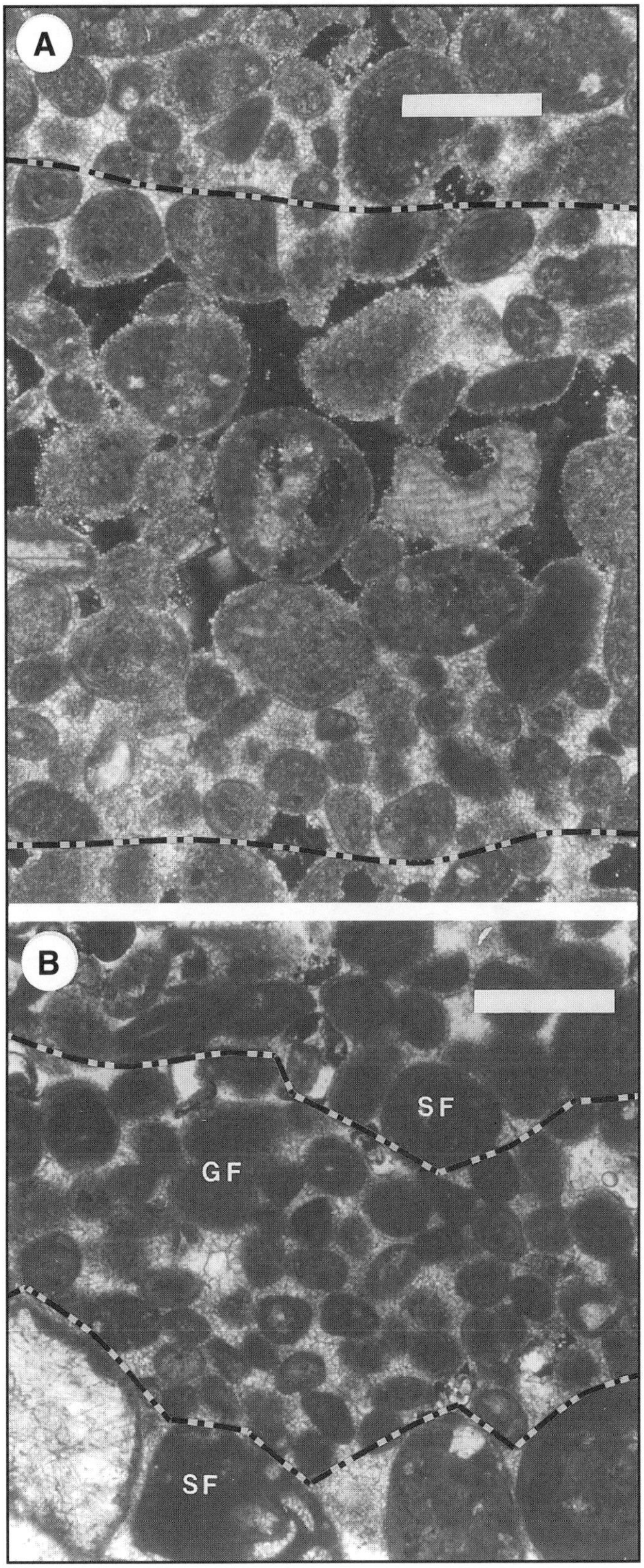

TABLE 1. SUMMARY OF LITHOLOGIC FEATURES THAT SUPPORT AN EOLIAN ORIGIN FOR MUCH OF UPPER PLEISTOCENE GROTTO BEACH FORMATION, SAN SALVADOR ISLAND

Features	Description	References
Grain texture	Good overall sorting; fine to medium grained; sorting differences exist between laminae or beds	Caputo, 1989, 1993
Diagenetic properties	Mixed freshwater vadose and phreatic calcite cement among different strata in overall vadose environment; patterns controlled by eolian processes, grain size, and packing	Caputo, 1993
Stratification	High- to low-angle beds of eolian sandflow, grain-fall, and ripple strata	White and Curran, 1985, 1988; Caputo, 1989, 1993
Biogenic properties	Rhizomorphs; crustacean and insect burrows; terres-trial gastropods	Carew et al., 1984; Curran, 1984, 1992, 1994; Curran and White, 1987, 1991
Associated facies	Upward-shallowing succession of reef-domi-nated shelf, beach, and eolian dune facies	White et al., 1984; Carew and Mylroie, 1985; Curran and White, 1984, 1985; White, 1989

Although certain trends in cement chemistry and texture are not exclusive to eolian strata, they can lend support to an eolian origin when combined with diagnostic eolian features, such as ripple stratification (Hunter, 1977). Quaternary limestones of suspected eolian origin on San Salvador and other tropical carbonate islands have been cemented with low-magnesium calcite (Ward, 1973, 1975; McKee and Ward, 1983; Hutto and Carew, 1984). The cements show meniscus, pendant, needle-fiber, blocky, and drusy geometries originating in freshwater vadose and phreatic diagenetic environments (McKee and Ward, 1983, Caputo, 1993). Conversely, portions of Pleistocene and Holocene eolianites on San Salvador, which have been submerged after the late Holocene rise in sea level, have been re-cemented by aragonite cement (White and White, 1990).

Stratification

Caputo (1989, 1993) recognized definitive eolian structures in Pleistocene calcarenites on San Salvador Island. Bedsets are 0.8 to 6.0 m (2.6 to 19.8 ft) thick and are composed of mainly medium- to high-angle cross-stratification. Foresets are characterized by fine-grained, well-cemented, ledge-forming units, and medium-grained, poorly cemented recessed units of interstratified eolian sandflow and localized eolian grainfall and ripple strata. Wind-ripple strata form low-angle backsets and topsets up

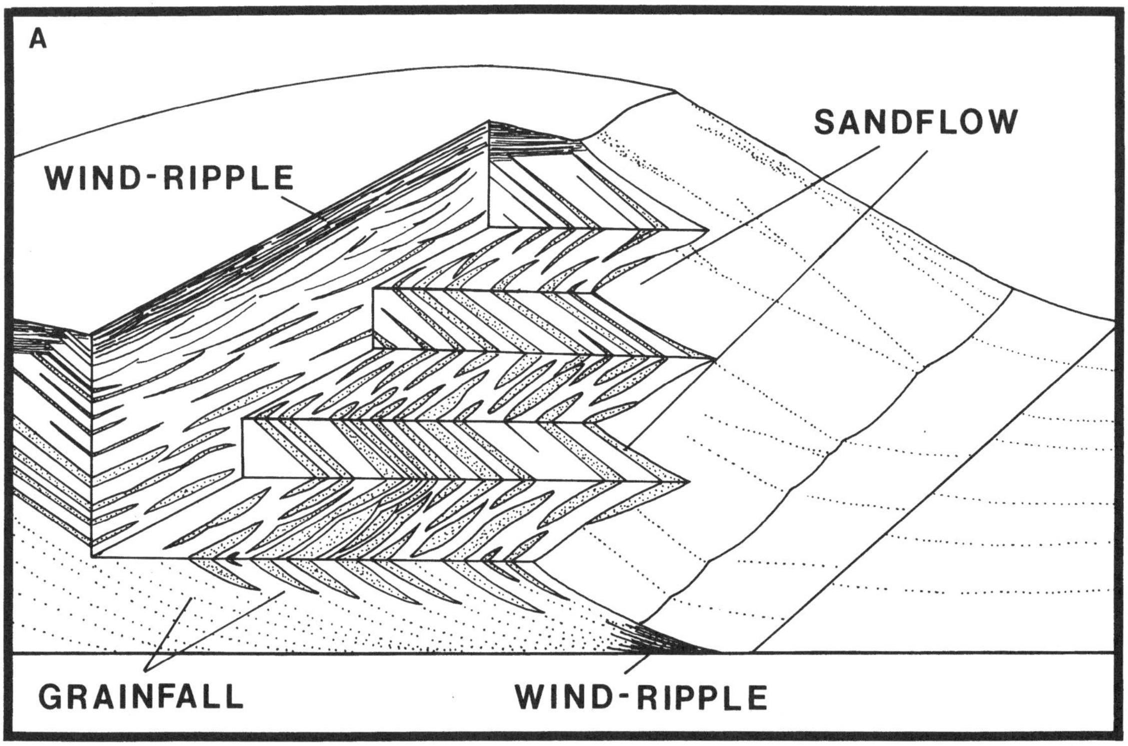

Figure 4 (on this and facing page). Comparative eolian stratification in siliciclastic deposits and in Quaternary limestones on Salvador Island. Scale with arrow is 17 cm (7 in) long. A, Idealized internal architecture of a simple eolian dune (after Hunter, 1977). B, Amalgamated topset wind-ripple strata in eolian beds equivalent to the Pleistocene Cockburn Town Member; note low-angle bounding surfaces (arrows) between cosets. C, Cross-bed intraset consisting of grainfall (GF) and sandflow (SF) strata truncated by reactivation bounding surface (arrows) and overlain by wind-ripple cosets (WR) in Holocene Rice Bay Formation. Note succession from upper to lower foresets: wind-ripple strata (pin-striped laminations), grainfalls (thicker, lighter gray beds), sandflows (thinner, darker gray beds with updip terminations). D, Thick cross-bed set consisting of recessed sandflow and ledge-forming wind-ripple and grainfall strata in eolian beds equivalent to the Pleistocene Cockburn Town Member; cliff is about 6 m (20 ft) high. E, Cross-bed intraset composed of sandflow beds (black arrows) wedging downdip into grainfall toesets. Note upper bounding surface (white arrows) and topset and bottomset wind-ripple strata (WR) in Holocene Rice Bay Formation.

to 1 m (3.3 ft) thick and higher angle brinksets and bottomsets (Fig. 4).

Biogenic properties

Terrestrial trace and body fossils are associated with disrupted bedding in Quaternary eolian limestones on San Salvador Island, particularly on inferred deflated or degraded dune surfaces. Plant structures or rhizomorphs and a variety of crustacean and insect burrows have been described from Quaternary rocks on San Salvador, as well as other Bahamian islands (Curran, 1984, 1992, 1994; White and Curran, 1985, 1988, 1993; Curran and White, 1987, 1991). Terrestrial gastropods of the genus *Cerion* have been collected from the French Bay and Cockburn Town Members of the Grotto Beach Formation (Carew et al., 1984). Fibrous calcite cement, which occurs locally in the Grotto Beach Formation (Fig. 5), is thought to precipitate around the root hairs of plants (Ward, 1975). Rhizomorphs in calcarenites suggest a pause in sedimentation under subaerial conditions, especially when they are associated with paleosols or carbonate crusts, but do not necessarily indicate a terrestrial origin for the host rock (Curran and White, 1987, 1991; White and Curran, 1988). However, in the Grotto Beach Formation, the noticeable disruption of distinct eolian structures by fossil plant structures strengthens a nonmarine eolian origin.

Associated facies

Modern eolian beach-foredune ridges generally accumulate in backshore zones of siliciclastic and carbonate beaches. They form a morphosedimentary facies contemporary with deposits of nearshore, shelf, beach, and back-barrier or coastal plain environments. With time and shoreline progradation, sedimentation produces an upward-shallowing facies succession.

For part of the Pleistocene Epoch, the western margin of San Salvador Island was characterized by reef-dominated nearshore

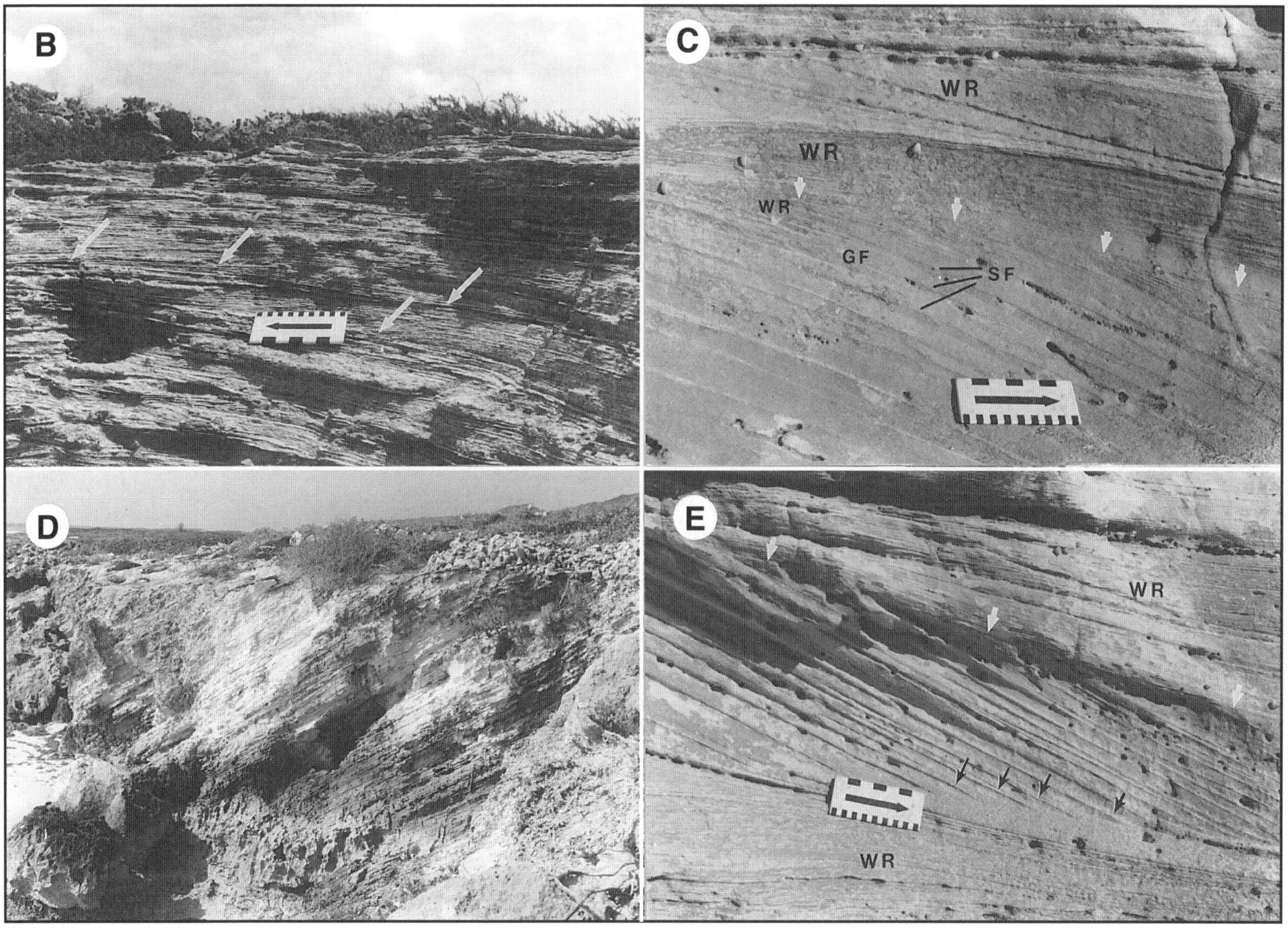

shelf, wave- and tide-influenced foreshore beach, and wind-dominated backshore beach. A drop in mean sea level, related to overall eustatic fluctuations in the Pleistocene, deposited a stratal succession of: (1) a basal coral-reef rudstone and framestone; (2) a medial flat- to cross-bedded beach grainstone; and (3) an upper ripple-laminated and cross-bedded eolian grainstone, which is preserved in the Cockburn Town Member of the Grotto Beach Formation (Carew and Mylroie, 1985; Curran and White, 1984, 1985, 1989; White, 1989; White et al., 1984). Part of a similar facies relationship is visible in basal exposures of the French Bay Member (Fig. 6).

GROTTO BEACH FORMATION: SEDIMENTARY ARCHITECTURE

Sedimentary architecture refers to the three-dimensional internal structure or construction of sedimentary bodies. Defining sedimentary architecture involves: (1) grouping physical sedimentary units into packages, which are genetically related, contemporary, and have environmental significance; and (2) recognizing a scale or hierarchy of sedimentary units as component building blocks for successively larger units, such as beds, bed-

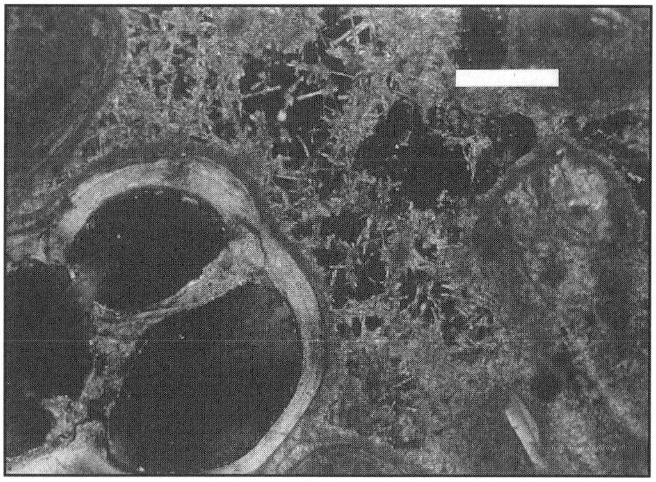

Figure 5. Photomicrograph of coated and skeletal grains bound by whisker or needle-fiber cement related to preexisting plant root hairs; in eolian beds, equivalent to the Cockburn Town Member. Bar = 0.2 mm (0.0078 in) long.

Figure 6. Vertical succession in the French Bay Member near base of sea cliff along beach near Sandy Point: 1 = cross-bedded subtidal-intertidal deposits; 2 = parallel laminated beach deposits; 3 = low-angle eolian deposits. Scale given in centimeters and inches.

TABLE 2. PROPOSED HIERACHY OF ARCHITECTURAL ELEMENTS AND PARENT SEDIMENTARY MECHANISMS FOR PLEISTOCENE GROTTO BEACH FORMATION, SAN SALVADOR ISLAND*

Order	Architectural Element	Parent Mechanism
1°	Grainfall or sandflow stratum	Grainfall or grainflow
2°	Wind-ripple stratum/cross-laminated set	Migration of impact ripples
3°	Wind-ripple coset	Build-up of cross-laminated sets by climbing ripples followed by deflation and erosion
4°	Cross-bed intraset	Dune migration followed by temporary sediment cut-off; pause in dune migration; partial erosion of dune-face and crest; resumed dune migration
5°	Cross-bed set	Dune advance; grainfall, grainflow, and ripple migration on dune lee-slope to produce foresets

*See Figure 7

sets, facies, sequences, and basin-fill complexes (Miall, 1988, 1990, 1991; Walker, 1992). Sedimentary building blocks or architectural elements (Allen, 1983) are the physical results of sedimentary mechanisms, such as, for example, grainfall, grainflow, bedform migration, and bar accretion. Elements are characterized by duration of the sedimentary event and how often it occurs, the magnitude and areal extent of the depositional event, the scale of the bedding or bounding surface that separates one architectural element from another, and the universal occurrence of certain elements in some modern and ancient sedimentary environments (Miall, 1990, 1991; Walker, 1992). Scales or hierarchies of architectural elements, sedimentary mechanisms, and bounding surfaces have been described for eolian (Wilson, 1972; Brookfield, 1977; Kocurek, 1981a,b, 1988) and fluvial (Allen, 1966, 1968, 1983; Miall, 1988) environments, and other environments (for review, see Miall, 1991).

Architectural elements

A classification of sedimentary architecture that is applicable to the eolian calcarenites in the Grotto Beach Formation includes the following elements: grainfall or sandflow stratum, wind-ripple stratum or cross-laminated set, wind-ripple coset, cross-bed intraset, and cross-bed set (Table 2) (Fig. 7). The architectural elements are depositional units and constitute mesoscopic (medium scale) sedimentary architecture or heterogeneity (Tyler and Finley, 1991). Framework grains and pores comprise microscopic architecture and were treated in Caputo (1989, 1993) for the Grotto Beach Formation.

Grainfall or sandflow stratum. A grainfall stratum forms by partially or completely suspended grains settling on the lee slope of a dune; a sandflow stratum forms by grainflow mechanics and the avalanching of sand down the lee slope (Hunter, 1977). Both are produced in a matter of seconds. An individual grainfall or sandflow stratum, comprising foresets, is a first-order building block of cross-bed intrasets in the Grotto Beach Formation. Successive sandflows or interstratified sandflows and grainfalls are separated by diastems.

Wind-ripple stratum/cross-laminated set. A wind-ripple stratum or cross-laminated set is a depositional unit produced by a microform bedform (Jackson, 1975) or an impact ripple, which is thought to be a fourth-order bedform element by Wilson (1972). Wind-ripple strata are herein considered second-order architectural elements that are bounded by parallel planar surfaces and compose either cross-laminated cosets or cross-bed sets. This is consistent with the observation that the time required for their formation is at least an order of magnitude greater than that for first-order grainfall and grainflow elements (Miall, 1991).

Wind-ripple coset. Wind-ripple cosets form amalgamated topset and backset beds (Caputo, 1993) and represent third-order architectural elements in the Grotto Beach Formation (Fig. 4B). A low-angle erosion surface between successive cosets represents a temporary interruption in sedimentation caused by changing conditions along the general depositional surface, such as a shift in either current and transport direction or sediment supply.

Cross-bed intraset. A fourth-order architectural element in the Grotto Beach Formation is the cross-bed intraset composed of sandflow-, grainfall-, and local ripple-strata. Two or more intrasets usually constitute a cross-bed set. Bounding discontinuities are rounded, convex-upward surfaces that occur between topset and foreset beds. They are subparallel to topsets, truncate

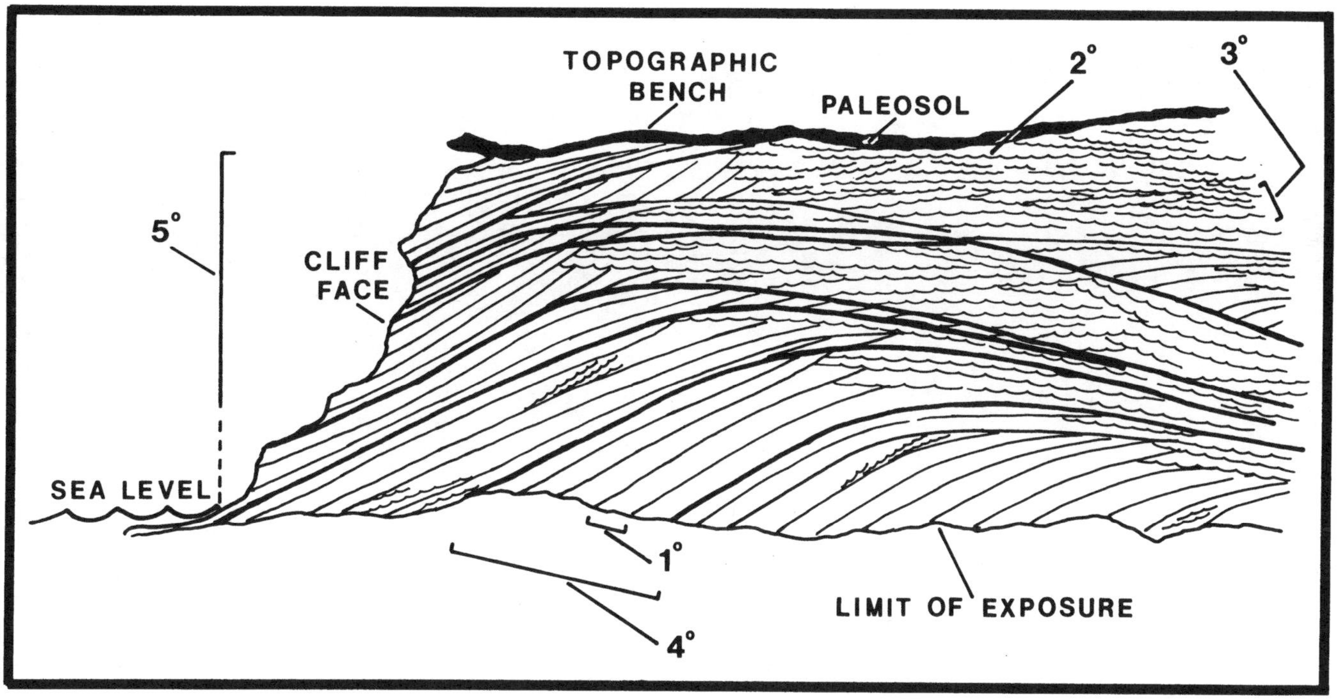

Figure 7. Longitudinal sketch showing general distribution and order of architectural elements (depositional units) in eolian calcarenites of the Grotto Beach Formation. 1° = First-order grainfall or sandflow strata (foresets); 2° = second-order wind-ripple strata (wavy lines); 3° = third-order wind-ripple cosets (separated by angular discordances); 4° = fourth-order cross-bed intrasets bounded by convex-upward reactivation surfaces (heavy curved lines); 5° = fifth-order cross-bed set (partly exposed). Diagram not to scale. Compare with Figure 10.

underlying foresets at high-angles, and extend down-dip concordant with younger foreset strata (Fig. 8). Some intraset bounding surfaces are delineated further by plant-root structures or rhizomorphs (see Fig. 11B).

Intraset bounding surfaces in the Grotto Beach Formation reflect a pause in dune migration and sediment movement, and a reshaping of the dune slipface and crest by shifting or reversing wind. Furthermore, dune migration may have been interrupted briefly by the stabilizing of sediment upwind from the dune either by grain cohesion due to rain or sea spray moistening, or by surficial hardening by salt or calcareous crusts. Consequently, net sediment transport and dune accumulation approached zero.

Cross-bed set. Sea cliff and roadcut outcrops of probable Cockburn Town-age beds near the Gulf on southern San Salvador Island (Fig. 1) are easily accessible and expose longitudinal views of well-developed, large-scale cross-bed sets. These bedding units are fifth-order architectural elements and preserve the development of whole dunes, which comprised a probable dune ridge–swale-shoreline system during Pleistocene time. They are composed of cross-bed intrasets and other lower-order elements and preserve nearly complete dune morphology. Viewed in a down-paleocurrent direction, the internal structure of a cross-bed set is typified by backset and topset strata that

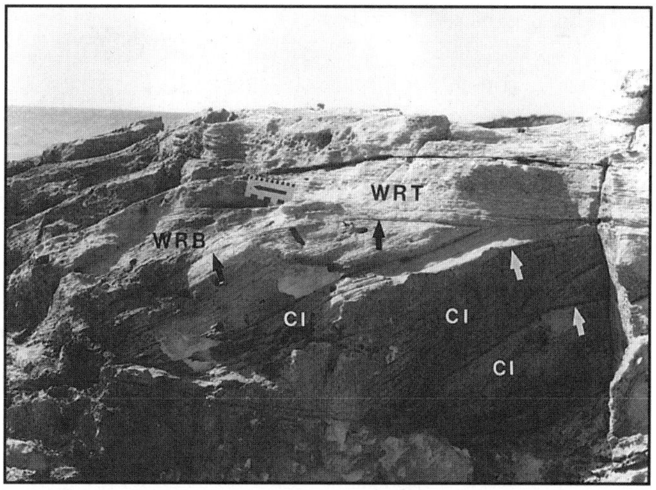

Figure 8. Complete dune-form cross-bed set (medium scale) composed of cross-bed intraset (CI) separated by rounded, convex-upward bounding surfaces (arrows). Note topset (WRT) and brinkset (WRB) wind-ripple strata and foresets composed of mainly sandflows. Sea-cliff exposure of eolian beds equivalent to the Cockburn Town Member near the Gulf. Scale given in centimeters and inches.

Figure 9. Preservation of complete dune-form set in part of a lobate dune ridge. Downwind succession of topset (TS), brinkset (BKS), and foreset (FS) strata. Roadcut exposure of eolian beds equivalent to the Cockburn Town Member near the Gulf. Scale = 17 cm (7 in) long.

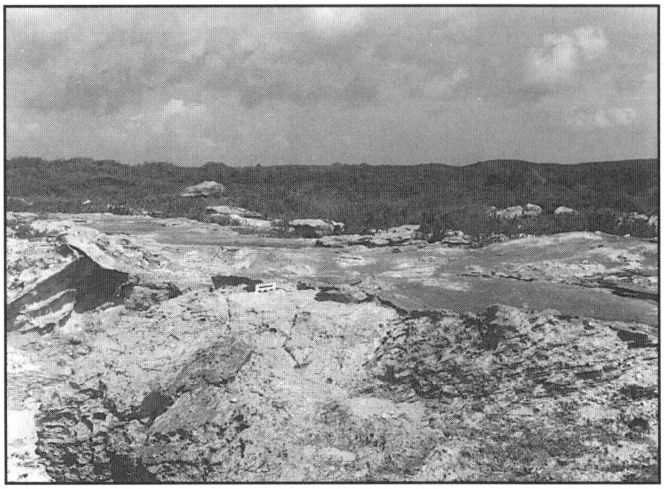

Figure 10. Degraded Pleistocene eolian surface preserved beneath a pedogenic crust, the top of which forms an undulatory bench. Sea-cliff exposure of eolian beds equivalent to the Cockburn Town Member, east of French Bay. Scale = 17 cm (7 in) long.

grade into convex-upward brinkset and foreset strata, and ultimately into toeset strata (Figs. 8 and 9). Such a bedding arrangement suggests stoss, crest, brink, slipface, and toe positions, respectively, of either a parabolic dune, or a sinuous dune ridge having lobate projections along the leeside. The display of paleocurrent data (Fig. 1) supports this conclusion (see the next section on topographic expression and sediment-body geometry). Rapid accumulation and early, rapid cementation were the primary factors favoring the preservation of internal structure of Pleistocene dunes on San Salvador Island. The development of convex-upward cross-strata related to parabolic or lobate dunes has been recognized also in Holocene eolianites of northern San Salvador Island (Adams, 1983; White and Curran, 1988), other Quaternary eolianites (Mackenzie, 1964a; McKee and Ward, 1983), and in recent siliciclastic dunes, both coastal (Bigarella et al., 1969) and interior (McKee, 1966).

Upper boundaries of fifth-order cross-bed sets are typically undulating surfaces with up to 0.5 m (1.6 ft) of relief and truncate topset, brinkset, and foreset strata (Figs. 7 and 10). Lower bounding surfaces overlain by toe- or bottomsets are rarely visible due to estimated set thickness between 6 and 9 m (20 to 30 ft) and submergence below present sea level. Resistant crusts and paleosols further delineate upper bounding surfaces; their distribution serves as a means of correlating and distinguishing stratigraphic units in the Quaternary System throughout the Bahamas and Bermuda (Carew and Mylroie, 1985, this volume; Hearty et al., 1992; Hearty and Kindler, 1993).

A bench-forming paleosol interval, up to 15 cm (6 in) thick, buries an upper bounding surface above cross-bed sets in the French Bay Member and inferred Cockburn Town beds in the sea cliffs between French Bay and the Gulf. Associated structures include fossil terrestrial snails of the genus, *Cerion,* and network of fossil plant roots (rhizomorphs) that locally disrupt eolian bedding (Fig. 11). Other similar surfaces in the Pleistocene of San Salvador Island are identified by karst solution pits and caves, and specific pedogenic features such as vadose pisoids, breccia, terra-rossa paleosols, calcarenite protosols, caliche, calcrete, and laminated micritic crusts (Titus, 1984; Carew et al., 1984; Carew and Mylroie, 1985, 1987, 1989; Foos, 1987, 1990; Foos and Bain, 1991).

The bounding surface-rhizomorph-paleosol association above fifth-order cross-beds in the Grotto Beach Formation indicates the occurrence of several Pleistocene events that were greater in extent and duration than those that produced truncation surfaces separating cross-bed intrasets (described in the previous section). First, wholesale dune growth was interrupted significantly as sand supply diminished in the nearshore shelf and shoreline, and wind deflation modified dune topography. Second, plant colonization, chemical cementation, and soil formation ultimately stabilized dunes following an inferred shift from arid to moist climatic conditions (Carew and Mylroie, 1985; Foos, 1987). These events are common to the forming of super bounding surfaces, particularly those controlled by climate changes (Talbot, 1985; Lancaster, 1993). Super surfaces are widespread unconformities that are recognized in modern and ancient eolian sand bodies; they signify the termination of entire sand seas (ergs) under deflationary, sand-starved conditions (Kocurek, 1988).

The significance of a fifth-order cross-bed set in the Grotto Beach Formation is that it preserves a fairly complete vertical and longitudinal profile of a single dune ridge as part of a dune ridge–swale-shoreline system. The bounding surface associated with fossil plant and soil structures is a scaled-down truncation or deflationary surface analogous to an eolian super surface, in terms of formative processes and effects. However, areal extent and duration were considerably less when considering a small, isolated carbonate platform such as San Salvador Island.

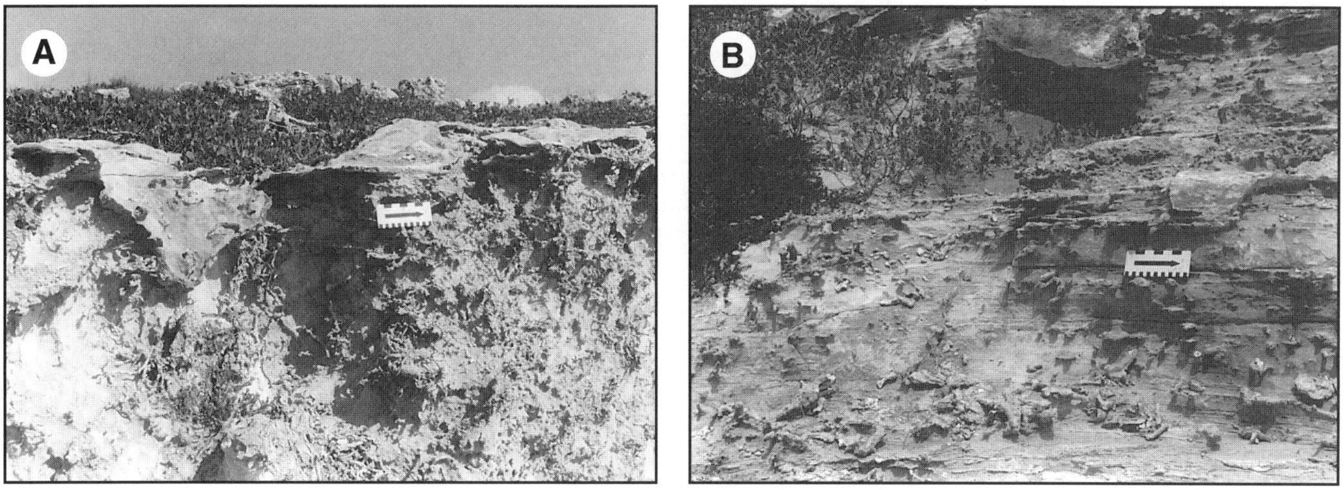

Figure 11. Plant root fossils or rhizomorphs that have penetrated down into eolian dune from paleosol (A) or have grown across dune surfaces and become buried after resumed sediment transport and dune migration (B). French Bay Member, French Bay cliffs. Scale = 17 cm (7 in) long.

Topographic expression and sediment-body geometry

The general occurrence of modern sand dunes, carbonate or noncarbonate, is likely whenever there is sand available, competent wind to transport it, and a low-relief area in which it can accumulate (Hayes and Kana, 1976; Goldsmith, 1985). Coastal dunes may form one or several curvilinear to sinuous ridges, parallel or nonparallel to the existing shoreline, and reflect the rate and direction of shoreline progradation at a given time. Dune ridge growth may be limited by space and some degree of stabilization, either by plants, rainfall, or, in the case of most carbonate dunes, early rapid cementation. Whereas, some siliciclastic dune fields that form on arid coasts tend to extend inland in the absence of vegetation (Goldsmith, 1985). Coastal eolian dunes accumulate above the spring high-tide line; they are commonly described as transverse, barchanoid, lobate, parabolic, and coalesced or compound parabolic dunes that may be semi-permanently or permanently fixed by vegetation (Goldsmith, 1978, 1985; McKee and Ward, 1983; Orme, 1988).

Linear to arcuate ridges and rises contribute to the present-day landscape on San Salvador Island. They are composed mainly of lithified Pleistocene dune sand and are separated by saline lakes, marsh, and dry jungle. Although modified by thousands of years of subaerial processes and vegetal overgrowth, the essential Pleistocene and older Holocene topography has been retained. Relict Pleistocene eolian build-ups follow approximately the outline and trend of the island and partly enclose several large lakes, including Great Lake, Stouts Lake, and Granny Lake, which may have evolved from Pleistocene lagoons at a time when sea level was higher than it is today (Hagey and Mylroie, this volume) (Fig. 1).

Eighty-seven measurements of dip direction of foreset bedding were collected in the French Bay Member of the Grotto Beach Formation in the French Bay cliffs near Sandy Point, and from undifferentiated Pleistocene rocks along cliffed headlands called the Thumb and the Bluff on the east side of San Salvador Island (Fig. 1). Circular histograms show that the trend in Pleistocene cross-bedding on San Salvador Island is bimodal to polymodal. Mean paleowind direction was east-northeast and east-southeast. Total variation in the mean paleowind direction at the four locations analyzed is no more than 32°. The data, coupled with trend in relict topography, suggest shore-parallel, sinuous dune ridges with lobate or parabolic aspects to the slipfaces to account for the bi- and polymodal cross-bed pattern. Assuming that geographic conditions in the Pleistocene were as they are today, an easterly paleowind direction is consistent with the present latitudinal position of San Salvador Island (24°N) in the tropical Northeasterly Trade Wind belt of the Northern Hemisphere.

SUMMARY AND CONCLUSIONS

Deposits of wind-blown carbonate sand or eolianites are composed of distinct marine allochems and are known to have formed in modern and ancient, tropical to subtropical coastal environments worldwide. On San Salvador Island, Bahamas, calcarenites of the French Bay Member and of beds that are probably equivalent to the Cockburn Town Member of the Upper Pleistocene Grotto Beach Formation preserve eolian features that are comparable to those in siliciclastic eolian sands and sandstones. Despite the unequivocal marine origin of the component bioclastic sediment, previous workers have recognized an eolian or nonmarine imprint in grain texture, diagenetic properties, stratification, fossils, and related facies in combination with pedogenic and karst features in limestones of San Salvador and throughout the greater Bahamas.

Framework grains and pores compose microscopic architecture in the Grotto Beach Formation (Caputo, 1989, 1993). They are the essential irreducible elements of all depositional and architectural units and serve as a foundation for understanding mesoscopic sedimentary architecture in outcrop. The classification of sedimentary architecture presented in this chapter is based on the scale of the depositional unit or element, the type of bounding surface, the parent bedform and related processes, and the duration of the sedimentary event that produced both the element and bounding surface. Architectural elements that are applied to the Grotto Beach Formation are: (1) grainfall or sandflow strata, (2) wind-ripple strata/cross-laminated sets, (3) wind-ripple cosets, (4) cross-bed intrasets, and (5) cross-bed sets, in order of ascending rank and scale. These units are the building-blocks of the eolian dune component of a Pleistocene dune ridge–swale-shoreline depositional system of San Salvador Island.

During Grotto Beach time, one or several sinuous-crested dune ridges, and perhaps, scattered independent lobate or parabolic dunes, accumulated on uneven emergent topography of the San Salvador platform. Dune growth and migration was brief; limited to areas elevated above wet interdune swales or lagoons (Hagey and Mylroie, this volume).

Freshwater diagenetic textures and the bounding surface-rhizomorph-paleosol association of fifth-order cross-bed sets indicate a change to moister climatic conditions near the end of Grotto Beach time. Eolian landscape on the isolated carbonate platform of San Salvador was modified during a prolonged period of wind erosion and diminishing sand supply, local-scale conditions that suggest an affinity with the development of super surfaces in siliciclastic eolian sand bodies. Rapid dune growth, short dune migration, climatic shift from arid to humid, and early cementation during late Pleistocene time account for the remarkable preservation of eolian architecture visible today in the Grotto Beach Formation. Dune surfaces were further stabilized by plants, which enhanced cement diagenesis through transpiration. Relict dune topography, evident in the present-day landscape of the island, was ultimately preserved by crusts and paleosols.

The classification of sedimentary architecture proposed here for the Grotto Beach Formation of San Salvador Island is preliminary. It serves as a basis for future work, which may involve a program of subsurface coring, aimed at better comprehending the genesis and evolution of a Pleistocene dune ridge–swale-shoreline system on San Salvador Island and other carbonate platforms, modern and ancient. Microscopic and mesoscopic depositional units as sedimentary building-blocks or architectural elements form the structural framework of progressively larger units such as facies or sequences, are potential pathways of fluid flow in aquifers and petroleum reservoirs, and provide a foundation for basin-scale sequence architecture.

ACKNOWLEDGMENTS

The final draft of this chapter is greatly improved over earlier versions through the time, careful review, and critical comments offered by the following colleagues to whom I am extremely grateful: Ronald Blakey, Northern Arizona University, James Carew, College of Charleston, Ralph Hunter, U.S. Geological Survey, Douglas Jordan, ARCO International, Gary Kocurek, University of Texas, and John Mylroie, Mississippi State University. I also thank Allen Curran and Brian White, Smith College, for their enduring patience and professionalism throughout the course of this project. Material and financial support for this project was provided through the kindness and generosity of Donald Gerace, former executive director of the Bahamian Field Station, San Salvador Island.

REFERENCES CITED

Adams, R. W., 1983, General guide to the geological features of San Salvador, *in* Gerace, D. T., ed., Field guide to the geology of San Salvador: San Salvador, CCFL Bahamian Field Station, p. 1–66.

Allen, J. R. L., 1966, On bedforms and paleocurrents: Sedimentology, v. 6, p. 153–190.

Allen, J. R. L., 1968, The nature and origin of bed-form hierarchies: Sedimentology, v. 10, p. 161–182.

Allen, J. R. L., 1983, Studies in fluviatile sedimentation: bars, bar complexes and sandstone sheets (low sinuosity braided streams) in the Brownstones (L. Devonian), Welsh Borders: Sedimentary Geology, v. 33, p. 237–293.

Ball, M. M., 1967, Carbonate sand bodies of Florida and the Bahamas: Journal of Sedimentary Petrology, v. 37, p. 556–591.

Bigarella, J. J., Becker, R. D., and Duarte, G. M., 1969, Coastal dune structures from Parana (Brazil): Marine Geology, v. 7, p. 5–55.

Boubin, M. A., and Loope, D. B., 1989, Petrology of eolian carbonates, upper Hermosa Formation (Pennsylvanian), southeastern Utah [abs.]: American Association of Petroleum Geologists Bulletin, v. 73, p. 336.

Brookfield, M. E., 1977, The origin of bounding surfaces in ancient aeolian sandstone: Sedimentology, v. 24, p. 303–332.

Caputo, M. V., 1989, Selective cementation of eolian stratification in Pleistocene calcarenites, San Salvador Island, Bahamas, *in* Mylroie, J. E., ed., Proceedings, Fourth Symposium on the Geology of the Bahamas: San Salvador, Bahamian Field Station, p. 61–72.

Caputo, M. V., 1993, Eolian structures and textures in oolitic-skeletal calcarenites from the Quaternary of San Salvador Island, Bahamas: A new perspective on eolian limestones, *in* Keith, B. D., and Zuppann, C. W., eds., Mississippian Oolites and Modern Analogs: American Association of Petroleum Geologists, Studies in Geology 35, p. 243–259.

Carew, J. L., and Mylroie, J. E., 1985, The Pleistocene and Holocene stratigraphy of San Salvador Island, Bahamas, with reference to marine and terrestrial lithofacies at French Bay, *in* Curran, H. A., ed., Pleistocene and Holocene Carbonate Environments on San Salvador Island, Bahamas: San Salvador, CCFL Bahamian Field Station, Field Guide, p. 11–62.

Carew, J. L., and Mylroie, J. E., 1987, A refined geochronology for San Salvador Island, Bahamas, *in* Curran, H. A., ed., Proceedings, Third Symposium on the Geology of the Bahamas: San Salvador, CCFL Bahamian Field Station, p. 35–44.

Carew, J. L., and Mylroie, J. E., 1989, Stratigraphy, depositional history, and karst of San Salvador Island, Bahamas, *in* Curran, H. A., ed., Pleistocene and Holocene Carbonate Environments on San Salvador Island, Bahamas, Field Trip Guidebook T175: Washington, D.C., American Geophysical Union, p. 7–15.

Carew, J. L., Mylroie, J. E., Wehmiller, J. F., and Lively, R. A., 1984, Estimates of late Pleistocene sea level high stands from San Salvador, Bahamas, *in* Teeter, J. W., ed., Proceedings, Second Symposium on the Geology of the Bahamas: San Salvador, CCFL Bahamian Field Station, p. 153–176.

Chen, J. H., Curran, H. A., White, B., and Wasserburg, G. J., 1991, Precise chronology of the last interglacial period: ^{234}U-^{230}Th data from fossil coral reefs in the Bahamas: Geological Society of America Bulletin, v. 103, p. 82–97.

Curran, H. A., 1984, Ichnology of Pleistocene carbonates on San Salvador, Bahamas: Journal of Paleontology, v. 58, p. 312–321.

Curran, H. A., 1989, Introduction to the geology of the Bahamas and San Salvador Island, with an overflight guide, *in* Curran, H. A., ed., Pleistocene and Holocene Carbonate Environments on San Salvador Island, Bahamas, Field Trip Guidebook T175: Washington, D.C., American Geophysical Union, p. 1–5.

Curran, H. A., 1992, Trace fossils in Quaternary, Bahamian-style carbonate environments: The modern to fossil transition, *in* Maples, C. G., and West, R. R., eds., Trace Fossils, Short Course in Paleontology 5: Paleontological Society, p. 105–120.

Curran, H. A., 1994, The palaeobiology of ichnocoenoses in Quaternary, Bahamian-style carbonate environments: The modern to fossil transition, *in* Donovan, S. K., ed., The Palaeobiology of Trace Fossils: New York, Wiley, p. 83–104.

Curran, H. A., and White, B., 1984, Field guide to the Cockburn Town fossil coral reef, San Salvador, Bahamas, *in* Teeter, J. W., ed., Proceedings, Second Symposium of the Geology of the Bahamas: San Salvador, CCFL Bahamian Field Station, p. 71–96.

Curran, H. A., and White, B., 1985, The Cockburn Town fossil coral reef, *in* Curran, H. A., ed., Pleistocene and Holocene carbonate environments on San Salvador Island, Bahamas, Geological Society of America Field Trip Guidebook: San Salvador, CCFL Bahamian Field Station, p. 95–120.

Curran, H. A., and White, B., 1987, Trace fossils in carbonate upper beach rocks and eolianites: Recognition of the backshore to dune transition, *in* Curran, H. A., ed., Proceedings, Third Symposium on the Geology of the Bahamas: San Salvador, CCFL Bahamian Field Station, p. 243–254.

Curran, H. A., and White, B., 1989, The Cockburn Town fossil coral reef of San Salvador Island, Bahamas, *in* Curran, H. A., ed., Pleistocene and Holocene carbonate environments on San Salvador Island, Bahamas, Field Trip Guidebook T175: Washington, D.C., American Geophysical Union, p. 27–34.

Curran, H. A., and White, B., 1991, Trace fossils of shallow subtidal to dunal ichnofacies in Bahamian Quaternary carbonates: Palaios, v. 6, p. 498–510.

Dodd, J. R., Zuppann, C. W., Harris, C. D., Leonard, K. W., and Brown, T. W., 1993, Petrologic method for distinguishing eolian and marine grainstones, Ste. Genevieve Limestone (Mississippian) of Indiana, *in* Keith, B. D., and Zuppann, C. W., eds., Mississippian Oolites and Modern Analogs: American Association of Petroleum Geologists studies in geology 35, p. 49–59.

Foos, A. M., 1987, Paleoclimatic interpretation of paleosols on San Salvador Island, Bahamas, *in* Curran, H. A., ed., Proceedings, Third Symposium on the Geology of the Bahamas: San Salvador, CCFL Bahamian Field Station, p. 67–72.

Foos, A. M., 1990, The mineralogy of Bahamian soils: Fifth Symposium on the Geology of the Bahamas, Abstracts and Programs, p. 11.

Foos, A. M., and Bain, R. J., 1991, Mineralogy and petrography of soils and exposure surfaces from San Salvador Island, Bahamas: Geological Society of America Abstracts with Programs, v. 23, p. 31.

Foos, A. M., and Muhs, D. R., 1991, Uranium-series age of an oolitic-peloidal eolianite, San Salvador Island, Bahamas: New evidence for a high stand of sea at 200-225 ka: Geological Society of America Abstracts with Programs, v. 23, p. 31.

Garrett, P., and Gould, S. J., 1984, Geology of New Providence Island, Bahamas: Geological Society of America Bulletin, v. 95, p. 209–220.

Goldsmith, V., 1978, Coastal dunes, *in* Davis, R. A., Jr., ed., Coastal Sedimentary Environments: New York, Springer-Verlag, p. 171–235.

Goldsmith, V., 1985, Coastal dunes, *in* Davis, R. A., Jr., ed., Coastal Sedimentary Environments (second edition): New York, Springer-Verlag, p. 303–378.

Handford, C. R., 1990, Mississippian carbonate eolianites in southwest Kansas [abs.]: American Association of Petroleum Geologists Bulletin, v. 74, p. 669.

Hayes, M. O., and Kana, T. W., eds., 1976, Terrigenous clastic depositional environments, some modern examples: Coastal Research Division, University of South Carolina, Department of Geology Technical Report No. 11-CRD, 131 p.

Hearty, P. J., and Kindler, P., 1993, New perspectives on Bahamian geology: San Salvador Island, Bahamas: Journal of Coastal Research, v. 9, p. 577–594.

Hearty, P. J., Vacher, H. L., and Mitterer, R. M., 1992, Aminostratigraphy and ages of Pleistocene limestones of Bermuda: Geological Society of America Bulletin, v. 104, p. 471–480.

Hunter, R. E., 1977, Basic types of stratification in small eolian dunes: Sedimentology, v. 24, p. 361–387.

Hunter, R. E., 1993, An eolian facies in the Ste. Genevieve Limestone of southern Indiana, *in* Keith, B. D., and Zuppann, C. W., eds., Mississippian Oolites and Modern Analogs: American Association of Petroleum Geologists Studies in Geology 35, p. 31–48.

Hutto, T., and Carew, J. L., 1984, Petrology of eolian calcarenites, San Salvador Island, Bahamas, *in* Teeter, J. W., ed., Proceedings, Second Symposium on the Geology of the Bahamas: San Salvador, CCFL Bahamian Field Station, p. 197–207.

Illing, L. V., 1954, Bahamian calcareous sands: American Association of Petroleum Geologists Bulletin, v. 38, p. 1–95.

Jackson, R. G., 1975, Hierarchical attributes and a unifying model of bed forms composed of cohesionless material and produced by shearing flow: Geological Society of America Bulletin, v. 86, p. 1523–1533.

Kocurek, G., 1981a, Erg reconstruction: The Entrada Sandstone (Jurassic) of northern Utah and Colorado: Palaeogeography, Palaeoclimatology, Palaeoecology, v. 36, p. 125–153.

Kocurek, G., 1981b, Significance of interdune deposits and bounding surfaces in aeolian dune sands: Sedimentology, v. 28, p. 753–780.

Kocurek, G., 1988, First-order and super bounding surfaces in eolian sequences—Bounding surfaces revisited: Sedimentary Geology, v. 56, p. 193–206.

Lancaster, N., 1993, Origins and sedimentary features of super surfaces, in the northwestern Gran Desierto sand sea, *in* Pye, K., and Lancaster, N., eds., Aeolian Sediments, Modern and Ancient: International Association of Sedimentologists Special Publication no. 16, p. 71–83.

Loope, D. B., 1986, Pennsylvanian eolian limestones, Paradox Basin, U.S.A.: International Association of Sedimentologists 12th International Sedimentological Congress, Canberra, Australia Abstracts, p. 189–190.

Loope, D. B., and Haverland, Z. E., 1988, Giant desiccation fissures filled with calcareous eolian sand, Hermosa Formation (Pennsylvania), southeastern Utah, *in* Kocurek, G., ed., Late Paleozoic and Mesozoic Eolian Deposits of the Western Interior of the United States: Sedimentary Geology, v. 56, p. 403–413.

Loope, D. B., and Kilibarda, Z., 1992, Eolian oolite bodies on the J-3 unconformity, Bighorn Basin, Wyoming: SEPM 1992 Theme Meeting, Mesozoic of the Western Interior, Abstracts, p. 41.

Mackenzie, F. T., 1964a, Geometry of Bermuda calcareous dune cross-bedding: Science, v. 144, p. 1449–1450.

Mackenzie, F. T., 1964b, Bermuda Pleistocene eolianites and paleowinds: Sedimentology, v. 3, p. 52–64.

Marzolf, J. E., 1988, Controls on late Paleozoic and early Mesozoic eolian deposition of the western United States, *in* Kocurek, G., ed., Late Paleozoic and Mesozoic Eolian Deposits of the Western Interior of the United States: Sedimentary Geology, v. 56, p. 167–191.

McKee, E. D., 1966, Structures of dunes at White Sands National Monument, New Mexico (and a comparison with structures of dunes from other selected areas): Sedimentology, v. 7, p. 1–69.

McKee, E. D., and Ward, W. C., 1983, Eolian, *in* Scholle, P. A., Bebout, D. G., and Moore, C. H., eds., Carbonate Depositional Environments: American Association of Petroleum Geologists Memoir 33, p. 131–170.

Merkley, P. A., and Dodd, R. J., 1991, Eolianite-bearing depositional sequences in the Ste. Genevieve Limestone of Indiana and Kentucky: Evidence for Mississippian eustasy?: Geological Society of America Abstracts with Programs, v. 23, p. A226.

Miall, A. D., 1988, Facies architecture in clastic sedimentary basins, *in* Kleinspehn, K. L., and Paola, C., eds., New Perspectives in Basin Analysis: New York, Springer-Verlag, p. 67–81.

Miall, A. D., 1990, Principles of Sedimentary Basin Analysis: New York, Springer-Verlag, 668 p.

Miall, A. D., 1991, Hierarchies of architectural units in terrigenous clastic rocks, and their relationship to sedimentation rate, *in* Miall, A. D., and Tyler, N., eds., The Three-Dimensional Facies Architecture of Terrigenous Clastic Sediments and Its Implications for Hydrocarbon Discovery and Recovery: SEPM Concepts in Sedimentology and Paleontology, v. 3, p. 6–12.

Newell, N. D., Purdy, E. G., and Imbrie, J., 1960, Bahamian oolitic sand: Journal of Geology, v. 68, p. 481–497.

Orme, A. R., 1988, Coastal dunes, changing sea level, and sediment budgets, *in* Psuty, N. P., ed., Dune/Beach Interaction: Journal of Coastal Research, Special Issue 3, p. 127–129.

Perkins, R. D., 1977, Part II, Depositional Framework of Pleistocene Rocks in Southern Florida: Quaternary Sedimentation in South Florida: Geological Society of America Memoir 147, p. 131–198.

Rice, J. A., and Loope, D. B., 1991, Wind-reworked carbonates, Permo-Pennsylvanian of Arizona and Nevada: Geological Society of America Bulletin, v. 103, p. 254–267.

Semeniuk, V., and Glassford, D. K., 1988, Significance of aeolian limestone lenses in quartz sand formations: An interdigitation of coastal and continental facies, Perth Basin, southwestern Australia: Sedimentary Geology, v. 57, p. 199–210.

Talbot, M. R., 1985, Major bounding surfaces in aeolian sandstones—A climatic model: Sedimentology, v. 32, p. 257–266.

Titus, R., 1984, Physical stratigraphy of San Salvador Island, Bahamas, *in* Teeter, J. W., ed., Proceedings, Second Symposium on the Geology of the Bahamas: San Salvador, CCFL Bahamian Field Station, p. 209–228.

Tucker, M. E., and Wright, V. P., 1990, Carbonate Sedimentology: Oxford, United Kingdom, Blackwell, 482 p.

Tyler, N., and Finley, R. J., 1991, Architectural controls on the recovery of hydrocarbons from sandstone reservoirs, *in* Miall, A. D., and Tyler, N., eds., The Three-Dimensional Facies Architecture of Terrigenous Clastic Sediments and Its Implications for Hydrocarbon Discovery and Recovery: SEPM Concepts in Sedimentology and Paleontology, v. 3, p. 1–5.

Vacher, H. L., 1973, Coastal dunes of younger Bermuda, *in* Coates, D. R., ed., Coastal Geomorphology: Binghamton, New York, State University of New York, Publications in Geomorphology, p. 355–391.

Walker, R. G., 1992, Facies, facies models and modern stratigraphic concepts, *in* Walker, R. G., and James, N. P., eds., Facies Models, Response to Sea Level Change: Geological Association of Canada, p. 1–14.

Ward, W. C., 1973, Influence of climate on the early diagenesis of carbonate eolianites: Geology, v. 1, p. 171–174.

Ward, W. C., 1975, Petrology and diagenesis of carbonate eolianites of northwestern Yucatan Peninsula, Mexico, *in* Wantland, K. F., and Pusey, W. C., eds., Belize Shelf—Carbonate Sediments, Clastic Sediments, and Ecology: American Association of Petroleum Geologists Studies in Geology 2, p. 500–571.

Ward, W. C., 1985, Part II, Quaternary geology of northeastern Yucatan Peninsula, *in* Geology and Hydrology of the Yucatan and Quaternary Geology of Northeastern Yucatan Peninsula: New Orleans, Louisiana, New Orleans Geological Society, p. 23–95.

White, B., 1989, Field guide to the Sue Point fossil coral reef, San Salvador Island, Bahamas, *in* Mylroie, J. E., ed., Proceedings, Fourth Symposium on the Geology of the Bahamas: San Salvador, Bahamian Field Station, p. 353–365.

White, B., and Curran, H. A., 1985, The Holocene carbonate eolianites of North Point and the modern marine environments between North Point and Cut Cay, *in* Curran, H. A., ed., Pleistocene and Holocene Carbonate Environments on San Salvador Island, Bahamas, Geological Society of America Field Trip Guidebook: San Salvador, CCFL Bahamian Field Station, p. 73–93.

White, B., and Curran, H. A., 1988, Mesoscale physical sedimentary structures and trace fossils in Holocene carbonate eolianites from San Salvador Island, Bahamas, *in* Hesp, P., and Fryberger, S. G., eds., Eolian Sediments: Sedimentary Geology, v. 55, p. 163–184.

White, B., and Curran, H. A., 1993, Sedimentology and ichnology of Holocene dune and backshore deposits, Lee Stocking Island, Bahamas, *in* White, B., ed., Proceedings, Sixth Symposium on the Geology of the Bahamas: San Salvador, Bahamian Field Station, p. 181–191.

White, B., Kurkjy, K. A., and Curran, H. A., 1984, A shallowing-upward sequence in a Pleistocene coral reef and associated facies, San Salvador, Bahamas, *in* Teeter, J. W., ed., Proceedings, Second Symposium on the Geology of the Bahamas: San Salvador, CCFL Bahamian Field Station, p. 53–70.

White, K. S., and White, B., 1990, The effects of Holocene sea level rise on the diagenesis of Quaternary carbonate eolianites, San Salvador Island, Bahamas: Fifth Symposium on the Geology of the Bahamas, Abstracts and Programs, p. 19.

Wilson, I. G., 1972, Aeolian bedforms—Their development and origins: Sedimentology, v. 19, p. 173–210.

Manuscript Accepted by The Society January 5, 1995

Printed in U.S.A.

Geological Society of America
Special Paper 300
1995

Pleistocene lake and lagoon deposits, San Salvador Island, Bahamas

Frances M. Hagey
H2O Environmental, Inc., Venice, Florida 34293
John E. Mylroie
Department of Geosciences, Mississippi State University, Mississippi State, Mississippi 39762

ABSTRACT

At sea-level highstands during the last interglacial maximum, much of San Salvador Island, Bahamas, was flooded, forming extensive lakes and lagoons. At today's sea level, many of these depressions contain fossil mollusc deposits informally referred to as "lake facies" by some previous workers. Examination of these deposits has revealed that most contain molluscs more typical of a marine environment, and lake deposits are rare.

Two molluscan assemblages have been defined to differentiate between Pleistocene lake and lagoon deposits, based on populations living in Holocene lakes and lagoons. The *Anomalocardia auberiana* Assemblage is characteristic of isolated lake environments that are usually hypersaline to some degree. The *Chione cancellata* Assemblage indicates marine-equivalent salinities; it is characteristic of lagoon and open marine environments. Ideally, because of topography, a Pleistocene transgression-highstand-regression cycle should produce a lake-lagoon-lake depositional sequence; however, deposition or erosion of barriers during the cycle can alter conditions. Additionally, karst conduits allow extension of the marine environment inland to isolated water bodies on San Salvador today and could have in the past. On San Salvador, a depositional sequence representing a transition from lake to lagoon was observed at one location. It may record rising sea level, the breaching of barriers, or both. Petrographic analysis of thin sections from sample sites revealed little useful information bearing on lake or lagoon origins. Molluscan macrofossil assemblages alone were diagnostic of the environment of deposition. Field relationships of all lake and lagoon deposits are consistent with deposition during oxygen isotope substage 5e, approximately 125,000 yr ago.

Hypothetical 40,000-yr-old Lake Cockburn was believed to be the site of deposition of mollusc-rich rocks in the interior of San Salvador. Only two locations in the southern portion of San Salvador were found to contain lake deposits that could potentially be used to support the Lake Cockburn theory. All other deposits found within the boundaries of Lake Cockburn contained marine molluscs. Oxygen isotope stage 3 (ca. 40,000 yr ago) rocks are unknown on San Salvador Island. The mollusc-rich rocks (lagoon or lake in origin) must have been deposited during stage 5. Because of deposition, erosion, and karst processes, caution must be exercised in using fossil assemblages to determine sea level or the configuration of the surrounding topography in the Pleistocene.

Hagey, F. M., and Mylroie, J. E., 1995, Pleistocene lake and lagoon deposits, San Salvador Island, Bahamas, *in* Curran, H. A., and White, B., Terrestrial and Shallow Marine Geology of the Bahamas and Bermuda: Boulder, Colorado, Geological Society of America Special Paper 300.

INTRODUCTION

The Bahama Islands have long been a focal point of research on modern and Pleistocene carbonates. Numerous studies have been made of eolian, beach, shallow marine, fossil reef, and paleosol deposits. Lagoon and lake deposits, formed during past highstands of sea level, have received considerably less attention, although they are known to exist on Andros, Long, New Providence, Great Inagua, Rum Cay, Little San Salvador, and San Salvador Islands.

Pleistocene deposits rich in fossil molluscs have a patchy distribution in both the coastal regions and the interior of San Salvador Island. They have been noted in papers by Titus (1983, 1984, 1987), Florentino and Bain (1984), Vierma et al. (1984), White et al. (1984), Bain (1985, 1989, 1990), Curran and White (1985), Teeter (1985a), Sims (1987), Hagey (1988, 1991a, b), White (1989), and Edwards et al. (1990). Titus (1983, 1984, 1987) and Teeter (1985a) mentioned lacustrine deposits and the probable existence of more extensive lakes on San Salvador during the Pleistocene. Titus's (1983, 1984, 1987) observations of lake deposits were limited to float found in the interior of San Salvador, in agricultural fields and along roads. Likewise, Teeter (1985a) observed similar material in the stone wall surrounding Watlings Blue Hole. The gastropods *Batillaria minima*, *Cerithidea* sp., and *Bulla occidentalis* and the bivalve *Anomalocardia* were observed in this material. Titus (1983, 1984, 1987) believed these lacustrine rocks to have been deposited in a large paleo-lake ("Lake Cockburn"), formed by the coalescence of smaller lakes during a sea-level highstand at about +2.5 m, 40,000 yr ago.

It is necessary to examine present-day and earlier Holocene lake, lagoon, and marine environments to interpret similar Pleistocene environments. The modern lakes and ponds on San Salvador have been well studied with respect to their sedimentology, water chemistry, microfauna, and macrofauna (Bowman and Teeter, 1982; Sanger and Teeter, 1982; Teeter, 1983, 1989; Crotty and Teeter, 1984; Kwolek, 1984; Pacheco and Foradas, 1987; Teeter et al., 1987; Davis and Johnson, 1989; Edwards et al., 1990; Teeter and Quick, 1990; Winter, 1990; Godfrey et al., 1994). Pigeon Creek, San Salvador's only tidal creek, has been studied in detail by Teeter and Thalman (1984), Teeter (1985b), Nutt and Teeter (1986), Mitchell (1987a), and Slone et al. (1990). Open lagoons on San Salvador have been studied by Andersen and Boardman (1989), Colby and Boardman (1989), and Pace et al. (1989). Similar water bodies on other Bahamian islands are documented in recent works by Mitchell (1984, 1985, 1987b), Wilber (1987), Mitchell et al. (1989), and Mitchell and Sigler (1989). A more detailed review of the literature regarding Holocene and Pleistocene lake and lagoon mollusc deposits in the Bahamas can be found in Hagey (1991a).

On San Salvador Island, inland lake and lagoon deposits have been informally referred to collectively as "lake facies" by many workers. Hagey (1991a) undertook to describe in more detail the lake facies on San Salvador. Observations of these deposits, however, revealed that many contained molluscs typically associated with marine depositional environments. It became apparent that differentiation between true Pleistocene lake deposits and lagoon deposits was necessary. Based on lake and lagoon molluscan fauna of the Holocene, two distinct molluscan assemblages were defined that characterize these environments of deposition. The *Anomalocardia auberiana* Assemblage is indicative of lake deposition and the *Chione cancellata* Assemblage is indicative of lagoon deposition. The occurrence of these assemblages reflects the complex interaction between changing sea level and preexisting topographies, barrier formation and destruction, and karst processes. The Pleistocene marine molluscs from the interior of San Salvador indicate that much of what has been called a lake facies actually represents a lagoonal facies. A large inland lake, hypothesized as the site of deposition of 40,000-yr-old lake facies (Titus, 1983, 1984, 1987), is now interpreted as a system of lagoons and lakes deposited during the last interglacial (ca. 125,000 yr ago).

PHYSIOGRAPHIC AND GEOLOGIC SETTING

The Bahamian platform consists of approximately 300,000 km^2 of carbonate islands, shallow banks, and deep intervening troughs, only about 11,406 square kilometers of which lies above sea level (Meyerhoff and Hatten, 1974). San Salvador Island lies approximately 630 kilometers ESE of Miami, Florida, at 24°N, 74° 30′W. San Salvador is located on a small, isolated bank, 25 km long by 12 km wide (Fig. 1). The bank is completely surrounded by ocean waters at least 1,000 m deep. The subaerially exposed portion of San Salvador is 19 km long by 11 km wide and has no more than 40 m of topographic relief. The island is largely constructed of arcuate eolian dune ridges with shallow lakes occupying many of the interdune depressions.

The depositional history and stratigraphic framework for San Salvador Island has been described by Carew and Mylroie (this volume). The rocks of San Salvador can be divided into three distinct depositional packages, usually bounded by paleosols. The Owl's Hole Formation consists of the oldest surficially exposed rocks observed to date. Owl's Hole deposits are exclusively eolian and are thought to have been deposited during oxygen isotope stages 7, 9, and possibly 11—ca. 220,000, 320,000, and 410,000 yr ago, respectively (Carew and Mylroie, this volume). Eolianites, beach facies, and marine subtidal deposits make up the Grotto Beach Formation that accumulated during stage 5 (ca. 125,000 years ago). The most recent deposits are the eolianites and beach facies of the Rice Bay Formation that were deposited during oxygen isotope stage 1 (Holocene). All marine subtidal deposits on San Salvador Island accumulated during oxygen isotope substage 5e. Since the position of lakes and lagoons on San Salvador is directly related to the position of sea level, Pleistocene lake and lagoon deposits found today above present sea level must have also formed during substage 5e. No oxygen isotope stage 3

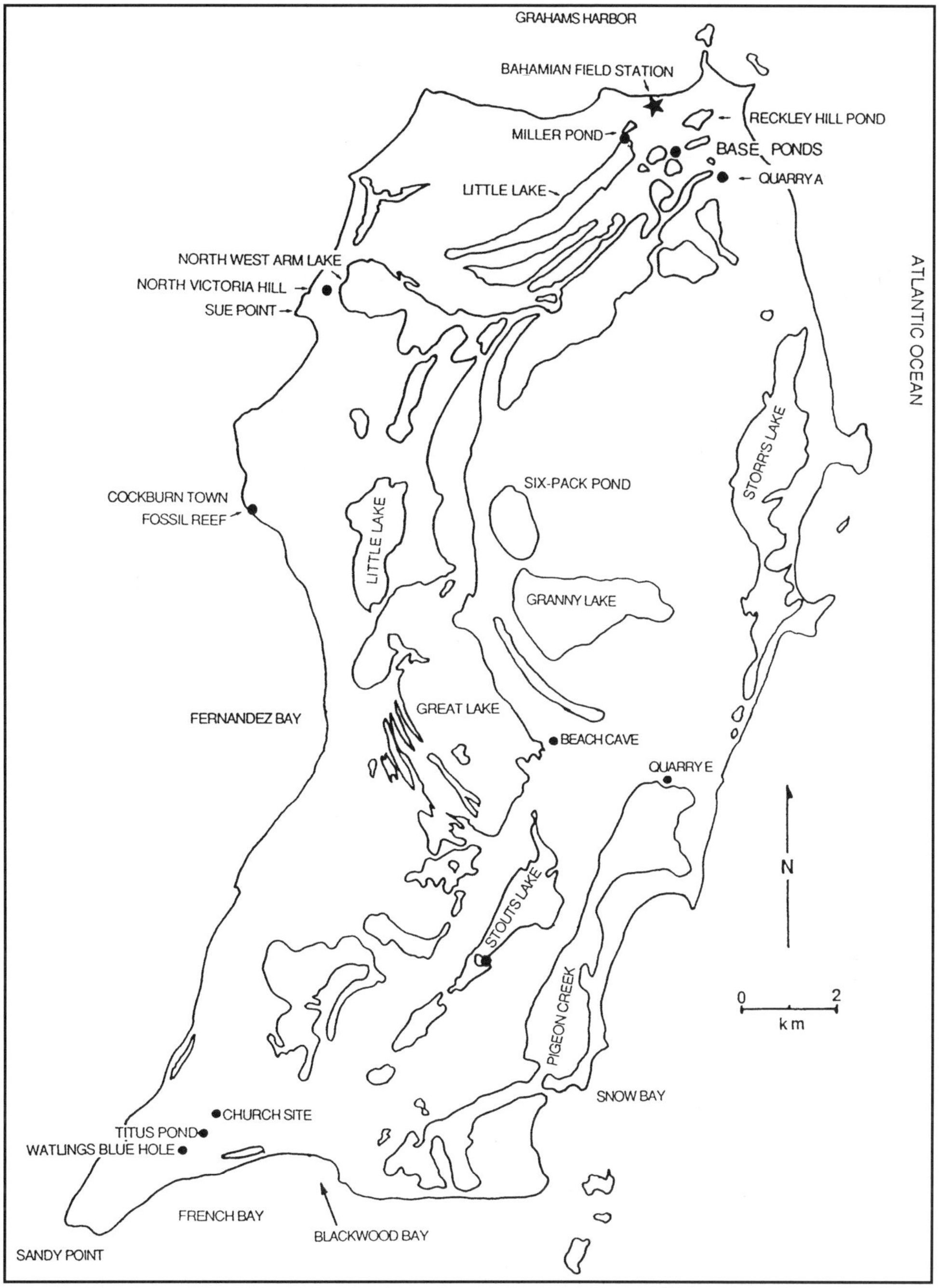

Figure 1. Map of San Salvador Island, Bahamas, showing locations discussed in the text.

rocks, terrestrial or marine, are recognized on San Salvador (Hearty and Kindler, 1993; Carew and Mylroie, this volume).

LAKES AND LAKE DEPOSITS

Lakes and ponds

Lakes on San Salvador are inland bodies of water occupying topographic lows. The water may be fresh, brackish, marine-equivalent, or hypersaline. The lakes do not have direct surface connections to the sea. Seawater may enter a lake, however, by way of porous bedrock, karst conduits, and/or storm flooding. Today, the lakes on San Salvador range from brackish to hypersaline. However, some of these lakes held freshwater earlier in the Holocene (Sanger and Teeter, 1982). Modern lakes include Great Lake, Granny Lake, Storr's Lake, Little Lake, and Stouts Lake (Fig. 1).

Ponds are usually differentiated from lakes only on the

basis of their smaller size. The ponds on San Salvador have salinities ranging from fresh to hypersaline. Reckley Hill Settlement Pond and Six Pack Pond are modern ponds on San Salvador (Fig. 1). Blue holes (i.e., Watlings Blue Hole) are similar to ponds in their surface expression, but they are deep relative to their width ($d/w > 1$). They often connect to submerged cave systems.

On San Salvador, the salinity of a lake or pond is not a direct reflection of its proximity to the ocean. Instead, salinity correlates to the size of the inland water body (Edwards et al., 1990). Smaller lakes and ponds tend to be fresh or brackish, whereas larger lakes tend to be hypersaline. Because San Salvador has a negative water budget in which evapotranspiration exceeds precipitation (Davis and Johnson, 1989), the greater the surface area to volume ratio of a water body, the greater the evaporative withdrawal. Even if a large lake has a karst conduit, active marine exchange may not completely reverse the hypersaline nature of the lake caused by high evaporation rates.

Lake deposits

Pleistocene deposits inferred to have formed in lake environments contain a characteristic assemblage of molluscs here called the *Anomalocardia auberiana* Assemblage. The bivalves *A. auberiana* and *Polymesoda maritima* and the gastropods *Batillaria minima*, *Cerithidea costata*, and *Cerithium lutosum* make up this assemblage (Fig. 2). These five species are typically found in the present day lakes and ponds of San Salvador Island (Teeter, 1985a; Edwards et al., 1990; Godfrey et al., 1994).

Pleistocene rocks sampled at Miller Pond, Watlings Blue Hole, and Church Site 1 are interpreted to have been deposited in lake environments (Fig. 1). Samples from these three sites almost exclusively contain molluscs of the *Anomalocardia auberiana* Assemblage (Table 1). In several samples, a single individual of a species belonging to the *Chione cancellata* Assemblage (i.e., *C. cancellata*, *Americardia media*) was found. These shells are not interpreted to have been living in the same body of water with molluscs of the *A. auberiana* Assemblage. It is possible that they were introduced by birds or storm events.

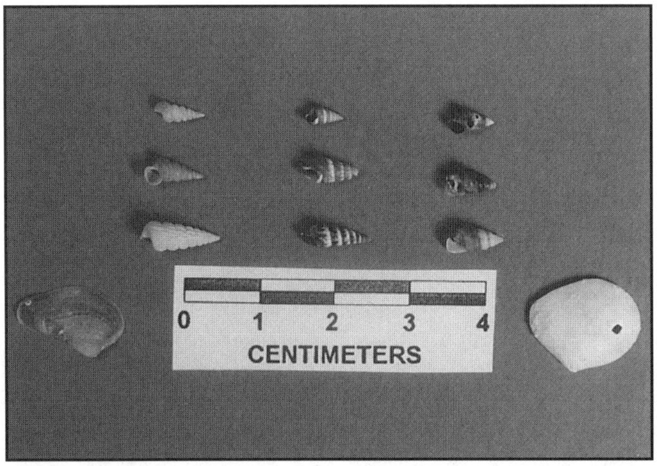

Figure 2. Photograph of the *Anomalocardia auberiana* Assemblage. Molluscs shown are, from left to right: *A. auberiana*, *Cerithidea costata*, *Batillaria minima*, *Cerithium lutosum*, and *Polymesoda maritima*.

Although samples from Miller Pond, Watlings Blue Hole, and Church Site 1 contain the *A. auberiana* Assemblage and are all interpreted to have formed in Pleistocene lakes, they differ from one another in several aspects. For example, in lake deposits from Church Site 1 and Miller Pond, *A. auberiana* is the most abundant species. At Watlings Blue Hole, the most abundant species is *Polymesoda maritima*. Gastropods (*Batillaria minima*, *Cerithidea costata,* and *Cerithium lutosum*), while present at all three sites, are more numerous in samples from Miller Pond and Watlings Blue Hole than from Church Site. Similar molluscan faunal variations in modern ponds in geographically close proximity have been observed by Godfrey et al. (1994).

Bivalves (*P. maritima* and *A. auberiana*) tend to be significantly larger in the Watlings Blue Hole samples than in the Miller Pond or Church Site 1 samples. This may be related to differences in salinity. Abbott (1968) stated that *Anomalocardia cuneimeris* (a synonym of *A. auberiana*) populations from brackish water have smaller shells than those living in normal marine waters. Teeter (1985a), however, observed that on San Salvador, the opposite is true: the size of *A. auberiana* shells is inversely related to salinity.

TABLE 1. MOLLUSCAN FAUNA PRESENT IN SAMPLES FROM PLEISTOCENE DEPOSITS INTERPRETED TO HAVE FORMED IN A LAKE ENVIRONMENT*

Miller Pond	Watlings Blue Hole	Church Site 1
Anodontia alba†	*Anomalocardia auberiana*	*Americardia media*†
Anomalocardia auberiana	*Batillaria minima*	*Anomalocardia auberiana*
Batillaria minima	*Bulla* sp.†	*Batillaria minima*
Bulla sp.	*Bulla striata*†	*Cerithidea costata*
Cerithidea costata	*Cerithidea costata*	*Cerithium lutosum*
Cerithium lutosum	*Cerithium lutosum*	*Marginella apicina*
Chione cancellata†	*Chione cancellata*	*Polymesoda maritima*
Codakia orbicularis†	*Linga pensylvanica*†	
Marginella apicina	*Marginella apicina*	
Polymesoda maritima	*Polymesoda maritima*	

*Lake environment means there was no surface connection to the open ocean. From Hagey, 1991a; taxonomy is that of Abbott, 1974.
†Indicates one individual of a species present in one sample, which is extremely rare.

In an earlier study, Church Site 1 rocks were incorrectly interpreted as having been deposited in an intertidal/beach environment (Sims, 1987). This interpretation was based purely on the abundance of molluscan shell material, not on the actual species present (Sims, 1987; R. J. Bain, personal communication). One of Sims's (1987) cores was taken from a deposit at Church Site containing the *Anomalocardia auberiana* Assemblage. The *A. auberiana* Assemblage indicates a lake environment. The Church Site 1 samples may have formed in a beach environment, but not that of a marine, coastal beach. Along the margins of some modern lakes on San Salvador, there are shell hash beaches composed of molluscs of the *Anomalocardia auberiana* Assemblage. The Church Site 1 deposits are consistent with this type of beach deposit.

It should be noted that there are modern exceptions to the generalization that lakes can be recognized by the occurrence of the *A. auberiana* Assemblage (Edwards et al., 1990; Godfrey et al., 1994). Several small water bodies in the interior of San Salvador Island contain species that may be considered anomalous, such as *Barbatia cancellaria* and *Codakia orbiculata,* which usually live in marine environments. The ponds inhabited by these species are small and connected to the ocean by karst conduits. Normal marine salinities are maintained in these ponds by active exchange with the ocean. There are several possible explanations for the introduction of the faunas into these ponds. Edwards et al. (1990) suggested the following: (1) distribution by shore birds, (2) "flushing" of larvae through conduits, and (3) relics of a higher sea-level stand.

LAGOONS AND LAGOON DEPOSITS

Lagoons and tidal creeks

The term lagoon is used here only in reference to a shallow body of water with a direct surface connection to the ocean. Two types of lagoons are discernable: those with open marine exchange, and those with limited marine interaction (sometimes termed tidal creeks; see below). The degree of connection to the ocean controls the mechanical energy and water chemistry of the lagoon. An open lagoon, or a lagoon with several active tidal inlets, will have conditions similar to the ocean. On San Salvador Island, Grahams Harbor, Blackwood Bay, and Snow Bay (Fig. 1) are open lagoons separated from the ocean by reefs.

A lagoon with limited surface connection to the ocean, via one tidal inlet or several narrow inlets, for example, will have fully marine conditions only in the vicinity of the tidal inlet(s). Water chemistry and mechanical energy will be influenced by other controlling factors in more distal settings: salinity, for one, is dependent on the amount of freshwater input. A large net input of freshwater results in brackish conditions, whereas a low net input results in hypersaline conditions, if evaporation rates are high. On San Salvador, Pigeon Creek is a lagoon almost entirely enclosed by dune ridges where marine influence is limited to a single tidal inlet (Fig. 1). Therefore, Pigeon Creek is marine in nature near the tidal inlet; it becomes increasingly saline farther from the inlet as a result of evaporation.

The term creek (as in Pigeon Creek) needs clarification. In this usage, the word does not imply freshwater or a net outflow from the island (although either may exist, as is true on Andros Island). In British usage, a creek is a lagoon with a tidal inlet (Bates and Jackson, 1987). The Bahamas are a former British colony, and because of this many bodies of water are called creeks that would be called lagoons in American usage.

Lagoon deposits

The assemblage of molluscs associated with marine or marine-equivalent conditions is here called the *Chione cancellata* Assemblage (Fig. 3). The bivalves *Americardia media*, *C. cancellata*, *C. paphia*, *Codakia orbicularis*, and *C. orbiculata,* and the gastropods *Bulla striata*, *Cerithium eburneum*, and *C. litteratum* are characteristic of this assemblage; these are only a few of many species of molluscs that suggest marine conditions or marine-equivalent salinities. Pleistocene molluscs found in known marine deposits, such as the Cockburn Town fossil reef, were used as a basis for characterizing the assemblage. Rocks sampled at the Base Ponds, Beach Cave, Stouts Lake, Titus Pond, Church Site 2, and North Victoria Hill (Fig. 1) probably record lagoon/tidal creek environments; all contain marine faunas of the *Chione cancellata* Assemblage (Table 2). The samples from these locations are interpreted as representing more restricted lagoons/tidal creeks rather than more open marine environments. This is based on the fact that the samples contain lower diversity assemblages of molluscan species than known Pleistocene coastal marine depositional environments, such as the Cockburn Town fossil coral reef (Hagey, 1988)

Figure 3. Photograph of select members of the *Chione cancellata* Assemblage. Molluscs shown are, top row from left to right: *C. cancellata*, *Americardia media*, *Codakia orbicularis*, and *Divaricella quadrisulcata*; middle row from left to right: *Tellina listeri*, *Polinices lacteus*, *Nassarius albus*, and *Modulus modulus*; bottom row from left to right: *Cerithium eburneum* and *C. litteratum*.

TABLE 2. MOLLUSCAN FAUNA PRESENT IN SAMPLES FROM PLEISTOCENE DEPOSITS INTERPRETED TO HAVE FORMED IN LAGOONS, SOME OF WHICH MAY HAVE HAD LIMITED SURFACE CONNECTION TO THE OPEN OCEAN*

Base Ponds	Beach Cave	Stouts Lake	Church Site 2	North Victoria Hill	Titus Pond
Americardia media	*Anomalocardia auberiana*	*Americardia media*	*Americardia media*	*Acteocina canaliculata*	*Americardia media*
Anadara brasiliana	*Batillaria minima*	*Anomalocardia auberiana*	*Anomalocardia auberiana*	*Americardia media*	*Bulla striata*
Anadara notabilis	*Bulla striata*	*Batillaria minima*	*Bulla striata*	*Anomalocardia auberiana*	*Cerithium eburneum*
Anomalocardia auberiana	*Cerithium eburneum*	*Bulla striata*	*Cerithidea costata*	*Barbatia cancellaria*	*Chione cancellata*
Arca imbricata	*Chione cancellata*	*Cerithidea costata*	*Cerithium eburneum*	*Batillaria minima*	*Chione paphia*
Barbatia domingensis	*Codakia orbicularis*	*Cerithium eburneum*	*Cerithium litteratum*	*Brachiodontus modiolus*	*Tellina listeri*
Bulla striata	*Dosinia discus*	*Cerithium litteratum*	*Chione cancellata*	*Bulla* sp.	
Cerithidea costata	*Nassarius albus*	*Cerithium lutosum*	*Chione paphia*	*Cerithidea costata*	
Cerithium algicola	*Polymesoda maritima*	*Chione cancellata*	*Codakia orbicularis*	*Cerithium litteratum*	
Cerithium eburneum	*Tellina tampaensis*	*Codakia orbicularis*	*Codakia orbiculata*	*Cerithium lutosum*	
Cerithium litteratum		*Codakia orbiculata*	*Divaricella quadrisulcata*	*Chione cancellata*	
Cerithium lutosum		*Lima lima*	*Dosinia discus*	*Codakia costata*	
Chione cancellata		*Modulus modulus*	*Modulus modulus*	*Codakia orbiculata*	
Chione paphia		*Nassarius albus*	*Nassarius albus*	*Divaricella quadrisulcata*	
Codakia orbicularis		*Olivella nivea*	*Polinices lacteus*	*Isognomon radiatus*	
Codakia orbiculata		*Polymesoda maritima*	*Pyramidella dolabrata*	*Lithophaga* sp	
Cymatium labiosum		*Tellina radiata*	*Strombus gigas*	*Modiolus americanus*	
Glycymeris pectinata		*Tellina* sp.	*Tellina listeri*	*Modiolus* sp.	
Laevicardium laevigatum			*Tellina* sp.	*Nassarius albus*	
Melampus monile				*Polinices lacteus*	
Modiolus americanus				*Polymesoda maritima*	
Modulus modulus				*Tellina listeri*	
Nassarius albus				*Tellina* sp.	
Periglypta listeri					
Pinctada imbricata					
Polinices lacteus					
Tegula fasciata					
Tellina listeri					
Tellina radiata					
Trachycardium muricatum					

*From Hagey, 1991a; taxonomy is that of Abbott, 1974.

(Table 3). Molluscan fauna present in beach, rocky shore, and subtidal deposits from Quarry A (Hagey, 1988) have been included in Table 3 for comparison.

Pleistocene molluscan fauna of the *C. cancellata* Assemblage were also found at Quarry E. This location is interpreted as having experienced direct communication with the open ocean in the late Pleistocene. The Pleistocene rocks exposed in the quarry walls are analogous to the modern Pigeon Creek tidal delta system located in Snow Bay (Thalman, 1983; Teeter and Thalman, 1984; Teeter, 1985b). The four samples from Quarry E collected for this study contain 28 species of molluscs. Two samples from an earlier study (Hagey, 1988) contain 30 species of molluscs. Overall, 40 species of molluscs have been identified in the Pleistocene deposits of Quarry E (Table 3). The fact that the molluscs belong to the typically marine *C. cancellata* Assemblage supports previous interpretations of direct communication with the open marine environment at this site.

Lake facies was reported to exist at the North Victoria Hill sampling site (Titus, 1983). The samples collected at North Victoria Hill for this study do not support a lake depositional environment. While all species of the *Anomalocardia auberiana* Assemblage are present (including an abundance of *Batillaria minima*), many species of the *C. cancellata* Assemblage are also present (Table 2). The mollusc-rich rocks at the North Victoria Hill site are here interpreted as having been deposited in a lagoon, perhaps one with restricted connection to the open marine environment.

PETROGRAPHY

Rock samples were examined in thin section with the expectation that petrographic distinctions could be made between lake and lagoon deposits. These distinctions could then be used in conjunction with fossil molluscan data to further define the different depositional environments; or, petrographic data could be used to distinguish between types of inland water bodies where no macrofossils are present. Data

TABLE 3. MOLLUSCAN FAUNA PRESENT IN PLEISTOCENE DEPOSITS INTERPRETED TO HAVE FORMED IN A MARINE ENVIRONMENT* OR LAGOON WITH GOOD SURFACE CONNECTION TO THE OPEN OCEAN†

Cockburn Town Fossil Reef	Quarry E	Quarry A
Americardia media	*Americardia media*	*Acmaea lucopleura*
Arca imbricata	*Anodontia alba*	*Americardia media*
Astraea phoebia	*Anomalocardia auberiana*	*Anomalocardia auberiana*
Astraea tecta	*Arca imbricata*	*Astraea tecta*
Barbatia cancellaria	*Barbatia cancellaria*	*Barbatia cancellaria*
Barbatia candida	*Barbatia domingensis*	*Barbatia domingensis*
Barbatia domingensis	*Batillaria minima*	*Batillaria minima*
Batillaria minima	*Bulla striata*	*Brachiodontus modiolus*
Bulla striata	*Cerithidea costata*	*Bulla striata*
Calliostoma jujubinum	*Cerithium eburneum*	*Cerithium eburneum*
Cerithium eburneum	*Cerithium litteratum*	*Chione cancellata*
Cerithium litteratum	*Cerithium lutosum*	*Chione pygmaea*
Chama sinuosa	*Chione cancellata*	*Codakia costata*
Chione cancellata	*Chione paphia*	*Codakia orbicularis*
Chione paphia	*Chione pygmaea*	*Columbella mercatoria*
Chione pygmaea	*Codakia orbiculata*	*Conus jaspideus*
Codakia orbicularis	*Columbella mercatoria*	*Diodora listeri*
Columbella mercatoria	*Conus jaspideus*	*Diodora minuta*
Conus jaspideus	*Diodora listeri*	*Divaricella quadrisulcata*
Cyphoma gibbosum	*Fissurella barbadensis*	*Fissurella barbadensis*
Cypraea cinera	*Glycymeris pectinata*	*Hemitoma emarginata*
Diodora listeri	*Hipponix antiquatas*	*Hipponix antiquatas*
Divaricella quadrisulcata	*Lima scabra*	*Isognomon radiatus*
Fasciolaria tulipa	*Linga pensylvanica*	*Laevicardium laevigatum*
Fissurella barbadensis	*Littorina nebulosa*	*Latirus angulatas*
Glycymeris pectinata	*Lucapina suffusa*	*Linga pensylvanica*
Glycymeris undata	*Marginella* sp.	*Lithophaga* sp.
Hemitoma emarginata	*Modulus modulus*	*Modiolus americanus*
Hipponix antiquatas	*Murex pomum*	*Modulus modulus*
Laevicardium laevigatum	*Nassarius albus*	*Nassarius albus*
Latirus angulatas	*Nerita peloronta*	*Nerita peloronta*
Lima lima	*Olivella nivea*	*Olivella nivea*
Lima scabra	*Polinices lacteus*	*Polinices lacteus*
Linga pensylvanica	*Polymesoda maritima*	*Strigilla mirabilis*
Modiolus americanus	*Pseudochama radiens*	*Tellina radiata*
Modulus modulus	*Tegula fasciata*	*Trivia quadripunctata*
Murex pomum	*Tellina listeri*	
Nassarius albus	*Tellina* sp.	
Nerita peloronta	*Trivia pediculus*	
Olivella nivea	*Trivia quadripunctata*	
Polinices lacteus		
Pseudochama radiens		
Spondylus americanus		
Strigilla mirabilis		
Strombus gigas		
Tegula fasciata		
Tegula lividomaculata		
Tellina listeri		
Tellina radiata		
Terebra glossema		
Trachycardium muricatum		
Trivia pediculus		
Trivia quadripunctata		

*Such as reef, intertidal, subtidal, beach, or rocky shore sites.
†From Hagey, 1988, 1991a; taxonomy is that of Abbott, 1974.

were obtained on grain size, porosity, and allochemical and orthochemical constituents (Table 4). Rock names were assigned based on Folk's (1962) classification. The overall results of this petrographic analysis do not help to define the nature of the water bodies at the time of deposition. Although the data do not conclusively support or negate interpretations based on the fossil molluscan faunas, several general trends can be recognized.

The most abundant allochems are peloids (in 13 samples) and bioclasts (in 11 samples). Overall, peloids are more common in facies otherwise interpreted as lake deposits. Bioclasts are more common in the lagoon deposits. Ooids were the most abundant grain type in three of the samples: two samples containing typical lake molluscs and one sample with no identifiable macrofossils (in two samples ooid abundance was approximately equal to other allochems). The presence of ooids in the lake deposits is not surprising as the substage 5e transgressive dunes on San Salvador are commonly composed of well-developed ooids (Carew and Mylroie, this volume). Micrite tended to be more extensive in the thin sections than sparite. Eighteen sections were predominantly micrite, and 12 sections were mainly sparite: nine micrite and five sparite for lake samples; and nine micrite and seven sparite for lagoon samples.

The average grain sizes and porosities for the lake and lagoon deposits are very similar: 0.28 for the lake deposits and 0.32 for the lagoon deposits. Bioclasts in many of the sections from both interpreted lake and lagoon deposits are much larger than the average, ranging from several millimeters to several centimeters in size. Porosity averages about the same for both environments: lake, 12%, and lagoon, 10%.

Despite the above general trends, there was too much variability among samples from similar environments to draw any substantial conclusions concerning depositional setting based on petrographic data alone. The molluscan faunas present in the samples appear to provide more reliable evidence of the types of Pleistocene environments in which the deposits accumulated.

DISCUSSION

The fossil assemblage of "typical lake molluscs" is helpful in identifying Pleistocene lake deposits. Teeter (1985a) reported that the two bivalves (*Anomalocardia auberiana* and *Polymesoda maritima*) are found living in lakes with salinities ranging from 26 to 65‰. The gastropod *Cerithium lutosum* is also euryhaline (Godfrey et al., 1994). The occurrence of the gastropods *Batillaria minima* and *Cerithidea costata* in the same setting implies that these species, too, are euryhaline over approximately the same range. Being euryhaline, these species cannot be used to determine lake salinities, as has been done in Holocene lakes on San Salvador Island with ostracod assemblages (Bowman and Teeter, 1982; Sanger and Teeter, 1982; Teeter et al., 1987; Teeter, 1989; Teeter and Quick, 1990).

Nonetheless, the euryhaline *A. auberiana* Assemblage in the fossil record on San Salvador Island strongly implicates an

TABLE 4. SUMMARY OF PETROGRAPHIC DATA

Sample	Grain Size (mm)	Porosity (%)	Micrite (%)	Sparite (%)	Ooids (%)	Peloids (%)	Bioclasts (%)	? (%)	Name
SS 89 MP 04	0.25	10	60	9	10	4	6	1	Sparsely fossiliferous oomicrite
SS 89 MP 05	0.22	11	10	21	15	33	10	-	Poorly washed pelsparite
SS 89 MP 06	0.28	9	44	5	15	10	15	2	Oobiomicrite
SS 89 MP 07	0.26	10	20	23	3	31	7	4	Poorly washed pelsparite
SS 89 MP 08	0.34	18	52	7	1	17	5	1	Sparsely fossiliferous pelmicrite
SS 89 MP 11	0.30	6	5	34	1	52	10	3	Fossiliferous pelsparite
SS 89 MP 14	0.26	17	35	25	5	15	6	3	Poorly washed pelsparite
SS 89 MP 15	0.29	31	18	4	15	15	15	1	Oopelbiomicrite
SS 90 MP 02	0.26	4	43	30	3	9	10	-	Poorly washed biosparite
SS 90 MP 06	0.55	8	46	11	<1	7	17	-	Fossiliferous micrite
SS 89 BP 02	0.26	17	32	9	2	20	20	-	Biopelmicrite
SS 89 BP 03	0.20	8	52	<1	<1	6	32	1	Biomicrite
SS 89 BP 04	0.40	15	5	40	10	15	14	1	Biopelsparite
SS 89 BP 05	0.20	6	74	10	1	1	7	-	Fossiliferous micrite
SS 90 BP 02	0.30	5	8	50	<1	15	20	-	Pelbiosparite
SS 89 BC 01	0.34	15	48	21	3	3	10	-	Poorly washed biosparite
SS 89 BC 03	0.20	12	70	1	3	8	6	1	Fossiliferous micrite
SS 89 SL 03	0.20	7	51	1	<1	15	26	-	Pelbiomicrite
SS 89 SL 04	0.45	9	40	25	<1	<1	24	-	Poorly washed biosparite
SS 89 SL 05	0.50	10	61	20	1	2	7	-	Sparsely fossiliferous micrite
SS 89 WB 01	0.20	17	40	30	2	8	3	-	Poorly washed pelsparite
SS 89 WB 03	0.21	10	16	23	35	10	3	2	Poorly washed oosparite
SS 90 WB 01	0.28	10	1	25	43	15	4	2	Oosparite
SS 90 WB 03	0.17	8	10	15	25	33	8	2	Poorly washed oopelsparite
SS 89 TP 01	0.40	5	15	60	3	11	6	-	Pelsparite
SS 89 CS 02	0.44	10	45	15	2	14	16	1	Pelbiomicrite
SS 90 CS 01	0.26	2	55	1	<1	35	7	-	Fossiliferous pelmicrite
SS 90 NV 02	0.35	14	1	24	7	32	19	3	Biopelsparite
SS 90 NV 03	0.19	8	6	22	15	35	10	4	Fossiliferous oopelsparite
SS 90 NV 04	0.27	12	10	20	10	20	27	-	Peloid and ooid-bearing biosparite

inland water body with little or no connection to the open ocean (surface or conduit). Fossil molluscs of the *Chione cancellata* Assemblage, however, do not necessarily indicate a surface connection to the open ocean. Several ponds on San Salvador with active conduits (i.e., Oyster Pond, Pain Pond) currently contain living "marine" molluscs (Edwards et al., 1990), indicating fully marine salinities, and therefore a direct connection to the open marine environment. Through conduit flow, typically marine faunal populations can be maintained in the physiographic setting of a lake.

The different fossil assemblages (*A. auberiana* or *C. cancellata*) may thus be useful as indicators of varying degrees of connection between the open ocean and inland water bodies during the Pleistocene. On San Salvador Island, connection with the open ocean includes (1) surface connections, resulting from the inundation of terrestrial barriers by marine water during a sea-level highstand event or the erosion of a barrier at a given constant sea level; and (2) conduit systems, as discussed in Davis and Johnson (1989) and Edwards et al. (1990). A lake with no conduits would seem to be more sensitive to sea-level fluctuations than one with conduits. In a lake lacking conduits, marine conditions would only exist during (and shortly after) a sea-level highstand at which time a surface connection with the ocean was established. With a subsequent drop in sea level, lake conditions would presumably be reestablished after some period of time (Fig. 4). Given a sufficient length of time or sufficient elevation increase for the higher sea level, a change in faunal assemblages (*A. auberiana* to *C. cancellata*) would be expected. Evidence for this type of change is found for the Pleistocene at Church Site and Watlings Blue Hole (although the sequence of the change is unknown at Watlings Blue Hole).

Conversely, it would appear that lakes connected to the open ocean through an active conduit system would not be as sensitive to sea-level fluctuations. Assuming that a conduit opened into the ocean at a substantial depth below sea level, a small-scale change (drop or rise) in sea level should not affect the ability of the conduit to transmit marine waters to the lake (Fig. 4). If the connection to the marine environment was maintained, the lake waters should remain the same (i.e., marine-equivalent salinities). Thus, even with changing sea level, the faunal population could be maintained. This should result in

Figure 4. Summary diagram showing the complex interaction of topography, sea-level fluctuations, barrier formation and destruction, and karst processes that may result in the formation of lake and lagoon depositional environments on small carbonate platforms.

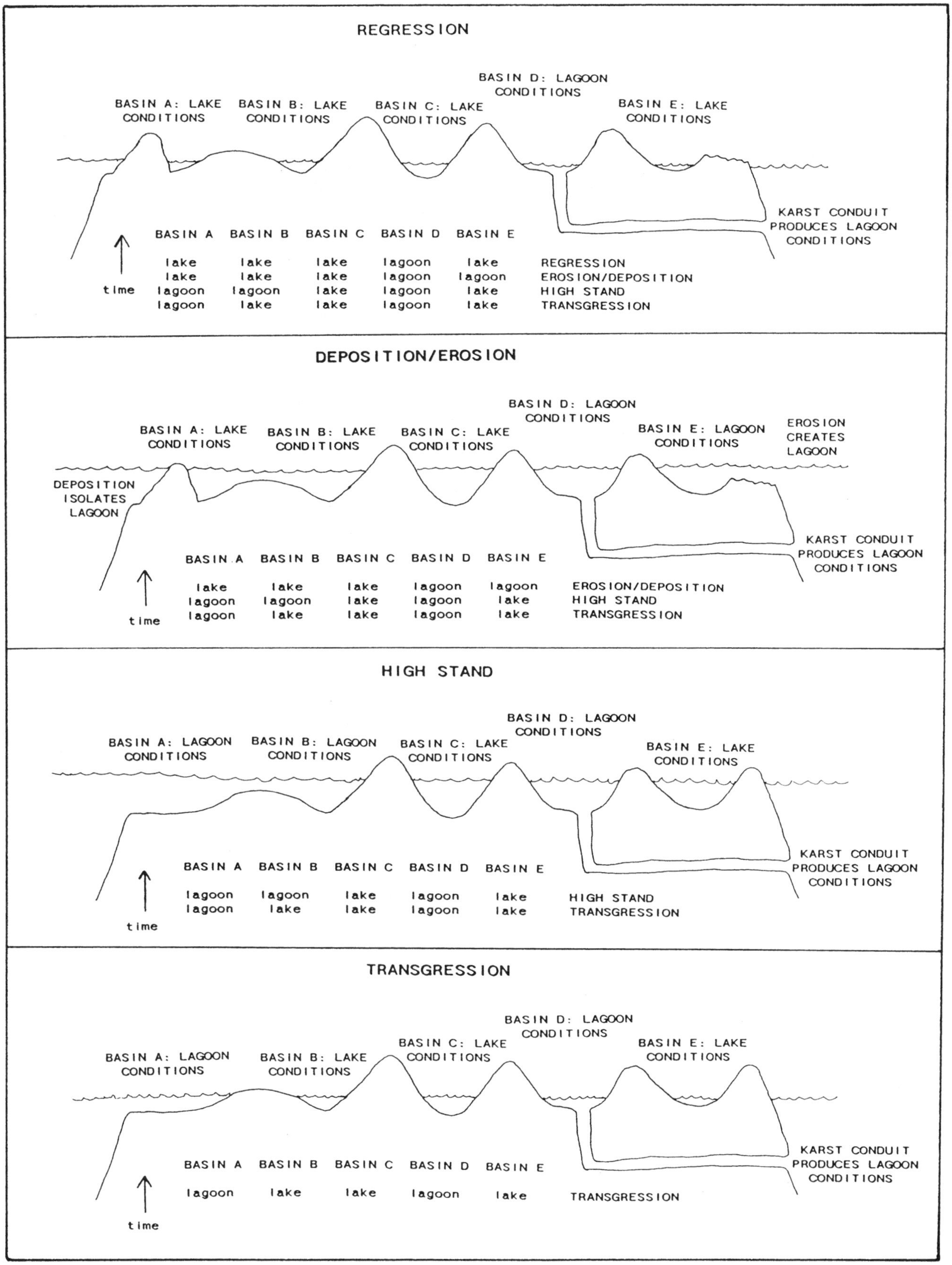
REGRESSION
BASIN A: LAKE CONDITIONS
BASIN B: LAKE CONDITIONS
BASIN C: LAKE CONDITIONS
BASIN D: LAGOON CONDITIONS
BASIN E: LAKE CONDITIONS
KARST CONDUIT PRODUCES LAGOON CONDITIONS
time
BASIN A BASIN B BASIN C BASIN D BASIN E
lake lake lake lagoon lake REGRESSION
lake lake lake lagoon lagoon EROSION/DEPOSITION
lagoon lagoon lake lagoon lake HIGH STAND
lagoon lake lake lagoon lake TRANSGRESSION
DEPOSITION/EROSION
BASIN A: LAKE CONDITIONS
BASIN B: LAKE CONDITIONS
BASIN C: LAKE CONDITIONS
BASIN D: LAGOON CONDITIONS
BASIN E: LAGOON CONDITIONS
EROSION CREATES LAGOON
DEPOSITION ISOLATES LAGOON
KARST CONDUIT PRODUCES LAGOON CONDITIONS
time
BASIN A BASIN B BASIN C BASIN D BASIN E
lake lake lake lagoon lagoon EROSION/DEPOSITION
lagoon lagoon lake lagoon lake HIGH STAND
lagoon lake lake lagoon lake TRANSGRESSION
HIGH STAND
BASIN A: LAGOON CONDITIONS
BASIN B: LAGOON CONDITIONS
BASIN C: LAKE CONDITIONS
BASIN D: LAGOON CONDITIONS
BASIN E: LAKE CONDITIONS
KARST CONDUIT PRODUCES LAGOON CONDITIONS
time
BASIN A BASIN B BASIN C BASIN D BASIN E
lagoon lagoon lake lagoon lake HIGH STAND
lagoon lake lake lagoon lake TRANSGRESSION
TRANSGRESSION
BASIN A: LAGOON CONDITIONS
BASIN B: LAKE CONDITIONS
BASIN C: LAKE CONDITIONS
BASIN D: LAGOON CONDITIONS
BASIN E: LAKE CONDITIONS
KARST CONDUIT PRODUCES LAGOON CONDITIONS
time
BASIN A BASIN B BASIN C BASIN D BASIN E
lagoon lake lake lagoon lake TRANSGRESSION

fossil-rich deposits with an unchanging assemblage reflecting the persistent marine-equivalent waters (no separate assemblages for small-magnitude sea-level fluctuations).

Titus (1983, 1984, 1987) hypothesized that a large Pleistocene lake (Lake Cockburn) formed about 40,000 yr ago (oxygen isotope stage 3 of Shackleton and Opdyke, 1973) during a sea-level rise of +2.5 m (Fig. 5). Lake Cockburn was said to have formed by the coalescence of smaller lakes as a result of the sea-level rise. As described by Titus (1983, 1984, 1987), Lake Cockburn would have been a lake as defined in this study. Fossils at several locations on San Salvador (Miller Pond, Watlings Blue Hole, Church Site 1) indicate that more extensive lakes were present at one or more times in the Pleistocene. Miller Pond is not shown as part of Lake Cockburn, and therefore cannot be used to support Titus's hypothesis (Fig. 5). Lake Cockburn does, however, include several sites sampled in this study: North Victoria Hill, Beach Cave, Stouts Lake, Church Site 2, and Titus Pond. Deposits at the North Victoria Hill site were reported as having formed in Lake Cockburn (Titus, 1983). Deposits examined in this study at all of these sites contain fauna of the *Chione cancellata* Assemblage, indicating marine or lagoonal conditions rather than lake conditions. Recent work on the stratigraphy of San Salvador (Hearty and Kindler, 1993; Carew and Mylroie, this volume) does not include any stage 3 deposits. Based on this stratigraphy, the molluscan fauna of Lake Cockburn must be part of the Grotto Beach Formation deposited during oxygen isotope substage 5e (ca. 125,000 yr ago).

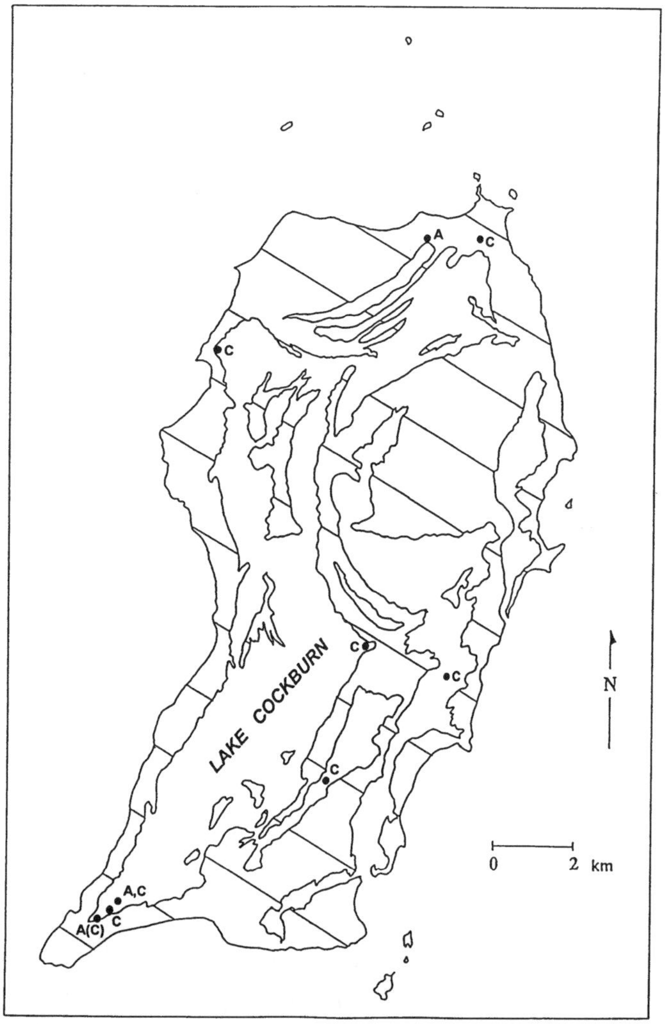

Figure 5. Hypothetical Lake Cockburn. Sampling locations are indicated by the black dots: A = *Anomalocardia auberiana* Assemblage. C = *Chione cancellata* Assemblage. Diagonal lines indicate land.

The lake deposits bearing the *Anomalocardia auberiana* Assemblage could have formed at several possible times during the oxygen isotope substage 5e sea-level event. The deposits could have formed during transgression or regression when the platform was not completely flooded. They could also have formed during the highstand. However, the existing topography of San Salvador Island indicates that a +6-m rise in sea level would produce no isolated inland lakes. Instead, a series of lagoons would have varying degrees of connection to the open ocean (Fig. 6) (Mylroie and Carew, 1990; Vacher and Mylroie, 1991). Some inlets may temporarily have been closed off, however, allowing lakes to exist. Later breaching of barriers could have reopened the inlets, reestablishing lagoons. It was not possible, given the scope of this study, to distinguish among these possibilities.

In an idealized model, a complete sequence of deposits formed during the sea-level event would record the transition from initial transgression (lakes and more restricted lagoons) to highstand (lagoons and open marine) to regression (lakes and more restricted lagoons) (Fig. 4). The rate of sea-level change and duration at any given level would have influenced deposition. Alternatively, during the highstand, spit and bar formation, or breaching of barriers, could alter the inland water body environment, causing a change in the type of shell deposition.

Samples from Watlings Blue Hole and Church Site contain the *Anomalocardia auberiana* Assemblage. Both locations also have deposits that contain the marine *Chione cancellata* Assemblage. The two assemblages are separate and distinct. While no in situ exposures have been found around Watlings Blue Hole, at Church Site there is an in situ exposure of the lacustrine material (covered by paleosol). Close inspection of a large block of float revealed that, at least in places, the lacustrine material is overlain by the marine material, which is in turn covered by a paleosol. This sequence, lake-marine-paleosol, shows that the lacustrine material must have been deposited either during the transgressive stage of a sea-level highstand event, or before breaching of a restrictive barrier during the highstand. The lack of an intervening paleosol between the lake and the marine deposits precludes the possibility that these lake deposits are relicts of an earlier highstand (Carew and Mylroie, 1989).

Some of the more inland deposits contain low-diversity marine faunal assemblages, suggesting restricted conditions. Restrictedness, as indicated by molluscan assemblages, could possibly be used to determine timing of dune formation. For

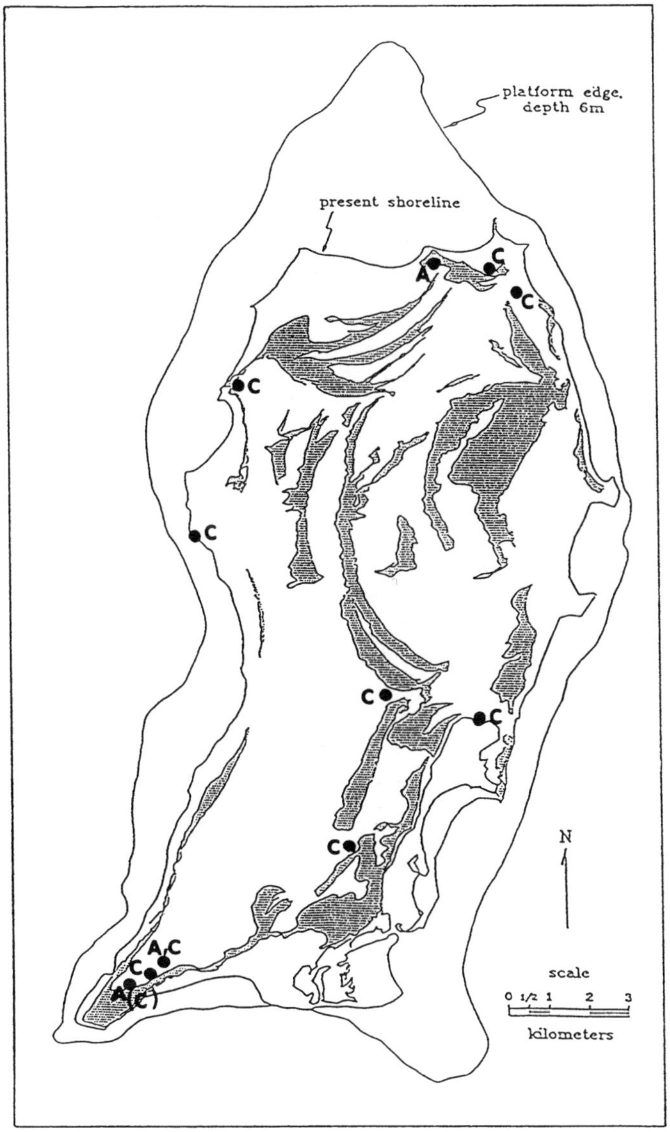

Figure 6. San Salvador Island at a +6-m sea-level highstand. Sampling locations are indicated by the black dots. A = *Anomalocardia auberiana* Assemblage. C = *Chione cancellata* Assemblage. Shaded areas indicate land.

example, based on analysis of Holocene lake deposits and the Holocene transgression (Kwolek, 1984; Teeter, 1985a), an assemblage that suggests more restricted conditions would seem to imply preexisting topography in order to create the restricted environment. Carew and Mylroie (1985; this volume) stated that the large interior dunes of San Salvador Island formed during the oxygen isotope substage 5e transgression or earlier, which has been confirmed by Hearty and Kindler (1993). The presence of restricted faunal assemblages appears to further support this hypothesis (see also Bain, 1990). It is probable that a transgression on substage 5e would have encountered topography remaining from stage 7 (approximately 220,000 yr ago) or earlier. However, the molluscan samples themselves are unlikely to be stage 7 or older, since the elevation of the deposits, coupled with subsidence of the Bahamian platform at 1 to 2 m per 100,000 yr (Mullins and Lynts, 1977), would have brought marine deposits equal or older than stage 7 below present sea level.

SUMMARY AND CONCLUSIONS

In the late Pleistocene, highstands of sea level led to flooding of much of San Salvador Island. Deposits rich in molluscs were formed during these highstands. These deposits have a patchy distribution above present sea level over the island. Several outcrops on the coast and in quarries that have previously been studied were found to represent marine depositional environments. These include the Cockburn Town fossil coral reef, Quarry A, and Quarry E.

Farther inland, mainly along the shores of the present-day lakes and ponds, there are sporadic outcrops of mollusc-rich Pleistocene rocks. Many workers had informally referred to these deposits in the interior as "lake facies." Some of these lake facies rocks were thought to have been deposited during a +2.5-m sea-level rise about 40,000 yr ago (stage 3). In fact, many of the lake facies rocks contain marine molluscan fauna that were not deposited in lake environments.

Samples were collected from the interior of San Salvador Island, the majority of which contained fossil molluscs. The molluscs were identified to define the environments that existed at the time of deposition. Thin sections from 30 of the samples were also examined in an attempt to further characterize the deposits. The following conclusions can be drawn from this study:

1. Two Pleistocene molluscan assemblages have been defined to differentiate between lake deposits and lagoon deposits: (A) The *Anomalocardia auberiana* Assemblage is made up of five species of molluscs that are indicative of a lake depositional environment: *A. auberiana* and *Polymesoda maritima* (bivalves) and *Batillaria minima*, *Cerithium lutosum*, and *Cerithidea costata* (gastropods). This criterion is based on the fact that these molluscs live in the modern lakes and ponds on San Salvador. (B) The *Chione cancellata* Assemblage contains many species of molluscs that suggest marine conditions or marine-equivalent conditions. Some of the most commonly occurring taxa include the bivalves *Americardia media*, *Chione cancellata*, *C. paphia*, *Codakia orbicularis*, and *C. orbiculata*, and the gastropods *Bulla striata*, *Cerithium eburneum*, and *C. litteratum*.

2. Deposits found at Miller Pond, Watlings Blue Hole, and Church Site 1 contain the *Anomalocardia auberiana* Assemblage and are believed to have formed in a lake environment. Deposits at the Base Ponds, Beach Cave, Stouts Lake, Titus Pond, Church Site 2, and North Victoria Hill contain the *Chione cancellata* Assemblage and are interpreted as representing former lagoon or tidal creek environments. Molluscs in samples from Quarry E support previous reports of a marine environment.

3. Lake Cockburn did not exist as it had been conceived.

Rather than a single large lake, it is more probable that a system of lagoons and lakes was responsible for the deposition of the mollusc-rich rocks. Most of the fossiliferous deposits within the borders of the hypothetical lake contain the *Chione cancellata* Assemblage, suggestive of a lagoon or tidal creek environment. Lake deposits containing the *Anomalocardia auberiana* Assemblage are found only at the southernmost end of the hypothetical lake. Current stratigraphic interpretation places the age of these deposits as 125,000 yr old (oxygen isotope substage 5e), not 40,000 yr (stage 3).

4. In an ideal situation, the transgression-stillstand-regression sequence of a sea-level event would be reflected in deposits as a lake-lagoon-lake transition. During a still stand, the formation of a barrier might result in the transition of lagoon to lake deposits; the destruction of a barrier would result in the opposite, a sequence of lake to lagoon deposits. On San Salvador, Pleistocene mollusc-rich deposits reflect only a lake to lagoon transition. It is uncertain whether this transition resulted from marine inundation due to a rising sea level or the breaching of barriers.

5. Marine molluscs found in the interior of San Salvador indicate marine-equivalent conditions of salinity. They may not, however, necessarily indicate a surface connection to the open ocean. Marine-equivalent conditions can be maintained in lakes or ponds with active conduit systems connecting them to the ocean. Lakes and ponds with karst connections to the sea will not be sensitive to minor sea-level fluctuations if the conduit connects to the ocean at depth.

6. Petrographic analysis produced results that were not useful in defining the nature of the inland water bodies at the time of deposition. Thin sections of samples from a single outcrop location were typically variable in both their allochemical and orthochemical constituents.

ACKNOWLEDGMENTS

We thank the Bahamian Field Station, Donald T. Gerace, executive director, for providing logistical support and general encouragement for the research. Roger J. Bain, James L. Carew, Mario V. Caputo, H. Allen Curran, Christopher P. Dewey, D. Craig Edwards, Benjamin J. Greenstein, James W. Teeter, Roger D. K. Thomas, and Robert Titus provided much advice and assistance during the field work, and during the writing of this manuscript.

REFERENCES CITED

Abbott, R. T., 1968, Seashells of North America: New York, Golden Press, 280 p.

Abbott, R. T., 1974, American seashells (second edition): Princeton, New Jersey, Van Nostrand–Reinholt, 663 p.

Andersen, C. B., and Boardman, M. R., 1989, The depositional evolution of Snow Bay, San Salvador, *in* Mylroie, J. E., ed., Proceedings, Fourth Symposium on the Geology of the Bahamas: San Salvador, Bahamian Field Station, p. 7–22.

Bain, R. J., 1985, Subtidal-beach-dune sequence, Quarry A, *in* Curran, H. A., ed., Pleistocene and Holocene carbonate environments on San Salvador Island, Bahamas, Geological Society of America, Orlando Annual Meeting Field Trip, Guidebook: Ft. Lauderdale, Florida, CCFL Bahamian Field Station, p. 63–72.

Bain, R. J., 1989, Pleistocene beach rock in a subtidal-beach-dune sequence, Quarry A, *in* Curran, H.A., ed., Pleistocene and Holocene carbonate environments on San Salvador Island, Bahamas, International Geological Congress, 28th, Field Trip Guidebook T175: Washington, D.C., American Geophysical Union, p. 23–26.

Bain, R. J., 1990, Distribution of Pleistocene lithofacies in the interior of San Salvador Island, Bahamas, and possible genetic models, Fifth Symposium on the Geology of the Bahamas: San Salvador, Bahamian Field Station, Abstracts and programs, p. 7.

Bates, R. L., and Jackson, J. A., eds., 1987, Glossary of Geology (third edition): American Geological Institute, Alexandria, Virginia, 788 p.

Bowman, P. A., and Teeter, J. W., 1982, The distribution of living and fossil foraminifera and their use in the interpretation of the post-Pleistocene history of Little Lake, San Salvador Island, Bahamas: Ft. Lauderdale, Florida, CCFL Bahamian Field Station Occasional Papers, 1983, no. 1, 26 p.

Carew, J. L., and Mylroie, J. E., 1985, The Pleistocene and Holocene stratigraphy of San Salvador Island, Bahamas, with reference to marine and terrestrial lithofacies at French Bay, *in* Curran, H. A., ed., Pleistocene and Holocene carbonate environments on San Salvador Island, Bahamas, Geological Society of America, Orlando Annual Meeting Field Trip Guidebook: Ft. Lauderdale, Florida, CCFL Bahamian Field Station, p. 11–61.

Carew, J. L., and Mylroie, J. E., 1989, The geology of eastern South Andros Island, Bahamas: A preliminary report, *in* Mylroie, J. E., ed., Proceedings, Fourth Symposium on the Geology of the Bahamas: San Salvador, Bahamian Field Station, p. 73–81.

Colby, N. D., and Boardman, M.R., 1989, Depositional evolution of a windward, high-energy carbonate lagoon, San Salvador, Bahamas, *in* Mylroie, J. E., ed., Proceedings, Fourth Symposium on the Geology of the Bahamas: San Salvador, Bahamian Field Station, p. 95–105.

Crotty, K. J., and Teeter, J. W., 1984, Post Pleistocene salinity variations in a blue hole, San Salvador Island, Bahamas, as interpreted from the ostracode fauna, *in* Teeter, J. W., ed., Proceedings, Second Symposium on the Geology of the Bahamas: Ft. Lauderdale, Florida, CCFL Bahamian Field Station, p. 3–16.

Curran, H. A., and White, B., 1985, The Cockburn Town fossil coral reef, *in* Curran, H. A., ed., Pleistocene and Holocene carbonate environments on San Salvador Island, Bahamas, Geological Society of America, Orlando Annual Meeting Field Trip Guidebook: Ft. Lauderdale, Florida, CCFL Bahamian Field Station, p. 95–120.

Davis, R. L., and Johnson, C. R., Jr., 1989, Karst hydrology of San Salvador, *in* Mylroie, J. E., ed., Proceedings, Fourth Symposium on the Geology of the Bahamas: San Salvador, Bahamian Field Station, p. 118–135.

Edwards, D. C., Teeter, J. W., and Hagey, F. M., 1990, Geology and ecology of a complex of inland saline ponds, San Salvador Island, *in* Bain, R. J., ed., Fifth Symposium on the Geology of the Bahamas: San Salvador, Bahamian Field Station Field Guide, p. 35–45.

Florentino, E., and Bain, R. J., 1984, Environment of deposition of the Granny Lake oolite, San Salvador, Bahamas, *in* Teeter, J. W., ed., Proceedings, Second Symposium on the Geology of the Bahamas: Ft. Lauderdale, Florida, CCFL Bahamian Field Station, p. 187–196.

Folk, R. L., 1962, Spectral subdivision of limestone types, *in* Ham, W. E., ed., Classification of carbonate rocks: American Association of Petroleum Geologists Memoir 1, p. 62–84.

Godfrey, P. J., Edwards, D. C., Davis, R. L., Smith, R. R., and Wells, J. A., 1994, Natural history of northeastern San Salvador Island: A "new world" where the New World began: San Salvador, Bahamian Field Station, Bahamian Field Station Trail Guide, 28 p.

Hagey, F. M., 1988, The distribution of Pleistocene molluscan assemblages on San Salvador Island, Bahamas [B.A. honors thesis]: Williamstown, Massachusetts, Williams College, 88 p.

Hagey, F. M., 1991a, Analysis of Pleistocene inland lake and lagoon deposits, San Salvador Island, Bahamas [M.S. thesis]: Mississippi State, Mississippi State University, 132 p.

Hagey, F. M., 1991b, Pleistocene molluscan assemblages on San Salvador Island, Bahamas: Preliminary investigations and interpretations, *in* Bain, R.J., ed., Proceedings, Fifth Symposium on the Geology of the Bahamas: San Salvador, Bahamian Field Station, p. 103–115.

Hearty, P. J., and Kindler, P., 1993, New perspectives on Bahamian geology: San Salvador Island, Bahamas: Journal of Coastal Research, v. 9, p. 577–594.

Kwolek, J. M., 1984, Holocene deposition of a multi-layered carbonate sequence in Reckley Hill Settlement Pond, San Salvador Island, Bahamas, *in* Teeter, J. W., ed., Proceedings, Second Symposium on the Geology of the Bahamas: Ft. Lauderdale, Florida, CCFL Bahamian Field Station, p. 27–39.

Meyerhoff, A. A., and Hatten, C. W., 1974, Bahama salient of North America: Tectonic framework, stratigraphy and petroleum potential: American Association of Petroleum Geologists Bulletin, v. 58, p. 1201–1239.

Mitchell, S. W., 1984, Geology of Great Exuma Island, Second Symposium on the Geology of the Bahamas: Ft. Lauderdale, Florida, CCFL Bahamian Field Station, Field Guide, 45 p.

Mitchell, S. W., 1985, Quaternary lacustrine and tidal creek microbiofacies of the Bahama Archipelago and Florida Keys: Geological Society of America Abstracts with Programs, v. 17, p. 666.

Mitchell, S. W., 1987a, Sedimentology of Pigeon Creek, San Salvador Island, Bahamas, *in* Curran, H. A., ed., Proceedings, Third Symposium on the Geology of the Bahamas: Ft. Lauderdale, Florida, CCFL Bahamian Field Station, pp. 215–230.

Mitchell, S. W., 1987b, Surficial geology of Rum Cay, Bahama Islands, *in* Curran, H. A. ed., Proceedings, Third Symposium on the Geology of the Bahamas: Ft. Lauderdale, Florida, CCFL Bahamian Field Station, p. 231–241.

Mitchell, S. W., and Sigler, M. E., 1989, Twentieth century sedimentological development of Bonefish Pond, New Providence Island, Bahamas, *in* Mylroie, J. E., ed., Proceedings, Fourth Symposium on the Geology of the Bahamas: San Salvador, Bahamian Field Station, p. 221–234.

Mitchell, S. W., Baldwin, J. N., Buening, N., and Westell, B., 1989, Holocene depositional history of Conception Island, Bahamas, *in* Mylroie, J. E., ed., Proceedings, Fourth Symposium on the Geology of the Bahamas: San Salvador, Bahamian Field Station, p. 209–220.

Mullins, H. T., and Lynts, G. W., 1977, Origin of the northeastern Bahama Platform: Review and reinterpretation: Geological Society of America Bulletin, v. 88, p. 1447–1461.

Mylroie, J. E., and Carew, J. L., 1990, The flank margin model for dissolution cave development in carbonate platforms: Earth Surface Processes and Landforms, v. 15, p. 413–424.

Nutt, W. H., and Teeter, J. W., 1986, Holocene developmental history of Pigeon Creek, San Salvador Island, Bahamas: 20th annual meeting of the North-Central Section, Kent, Ohio, Geological Society of America Abstracts with Programs, p. 317.

Pace, W., Mylroie, J. E., and Carew, J. L., 1989, Sedimentology of a Holocene platform-margin carbonate lagoon: Blackwood Bay, San Salvador Island, Bahamas, *in* Mylroie, J. E., ed., Proceedings, Fourth Symposium on the Geology of the Bahamas: San Salvador, Bahamian Field Station, p. 253–265.

Pacheco, P. J., and Foradas, J. G., 1987, Holocene environmental changes in the interior karst region of San Salvador, Bahamas: The Granny Lake pollen record, *in* Curran, H. A., ed., Proceedings, Third Symposium on the Geology of the Bahamas: Ft. Lauderdale, Florida, CCFL Bahamian Field Station, p. 115–122.

Sanger, D. B., and Teeter, J. W., 1982, The distribution of living and fossil ostracoda and their use in the interpretation of the post-Pleistocene history of Little Lake, San Salvador Island, Bahamas: Ft. Lauderdale, Florida, CCFL Bahamian Field Station Occasional Papers, 1982, No. 1, 19 p.

Shackleton, N. J., and Opdyke, N. D., 1973, Oxygen isotope and paleomagnetic stratigraphy of equatorial Pacific core V28-238: Oxygen isotope temperatures and ice volumes on a 10^5 and 10^6 year scale: Quaternary Research, v. 3, p. 39–55.

Sims, W.R., 1987, Facies analysis of carbonate rocks on the southern end of San Salvador Island, Bahamas [M.S. thesis]: Akron, Ohio, University of Akron, 135 p.

Slone, G. B., Boardman, M. R., and Cummins, R. H., 1990, Molluscan skeletal associations as a function of benthic cover and environmental stress: Pigeon Creek, San Salvador, Bahamas, *in* Fifth Symposium on the Geology of the Bahamas: San Salvador, Bahamian Field Station, Abstracts and programs, p. 18.

Teeter, J. W., 1983, The topographic, hydrologic, and sedimentologic setting of Little Lake, San Salvador Island, Bahamas: Ft. Lauderdale, Florida, CCFL Bahamian Field Station Occasional Papers, No. 1, 7 p.

Teeter, J. W., 1985a, Holocene lacustrine depositional history, *in* Curran, H. A., ed., Pleistocene and Holocene carbonate environments on San Salvador Island, Bahamas, Geological Society of America, Orlando Annual Meeting Field Trip Guidebook: Ft. Lauderdale, Florida, CCFL Bahamian Field Station, p. 133–145.

Teeter, J. W., 1985b, Pigeon Creek Lagoon, a modern analogue of the Pleistocene Granny Lake Basin, *in* Curran, H. A., ed., Pleistocene and Holocene carbonate environments on San Salvador Island, Bahamas, Geological Society of America, Orlando Annual Meeting Field Trip Guidebook: Ft. Lauderdale, Florida, CCFL Bahamian Field Station, p. 147–160.

Teeter, J. W., 1989, Refinement and timing of salinity fluctuations in Watling's Blue Hole, San Salvador Island, Bahamas, *in* Mylroie, J. E., ed., Proceedings, Fourth Symposium on the Geology of the Bahamas: San Salvador, Bahamian Field Station, p. 331–336.

Teeter, J. W., and Quick, T. J., 1990, Magnesium-salinity relation in the saline lake ostracode *Cyprideis americana*: Geology, v. 18, p. 220–222.

Teeter, J. W., and Thalman, K. L., 1984, Field trip to Pigeon Creek, *in* Teeter, J. W., ed., Proceedings, Second Symposium on the Geology of the Bahamas: Ft. Lauderdale, Florida, CCFL Bahamian Field Station, p. 177–186.

Teeter, J. W., Beyke, R. J., Bray Jr., T. F., Brocculeri, T. F., Bruno, P. W., Dremann, J. J., and Kendall, R. L., 1987, Holocene depositional history of Salt Pond, San Salvador, Bahamas, *in* Curran, H. A., ed., Proceedings, Third Symposium on the Geology of the Bahamas: Ft. Lauderdale, Florida, CCFL Bahamian Field Station, p. 145–150.

Thalman, K. L., 1983, A Pleistocene lagoon and its modern analogue, San Salvador, Bahamas [M.S. thesis]: Akron, Ohio, University of Akron, 166 p.

Titus, R., 1983, Quaternary stratigraphy of San Salvador, Bahamas, *in* Gerace, D. T., ed., Proceedings, First Symposium on the Geology of the Bahamas: Ft. Lauderdale, Florida, CCFL Bahamian Field Station, p. 1–5.

Titus, R., 1984, Physical stratigraphy of San Salvador Island, Bahamas, *in* Teeter, J. W., ed., Proceedings, Second Symposium on the Geology of the Bahamas: Ft. Lauderdale, Florida, CCFL Bahamian Field Station, p. 187–196.

Titus, R., 1987, Geomorphology, stratigraphy, and the Quaternary history of San Salvador, *in* Curran, H. A., ed., Proceedings, Third Symposium on the Geology of the Bahamas: Ft. Lauderdale, Florida, CCFL Bahamian Field Station, p. 155–164.

Vacher, H. L., and Mylroie, J.E., 1991, Geomorphic evolution of topographic lows in Bermudian and Bahamian islands: Effect of climate, *in* Bain, R. J., ed., Proceedings, Fifth Symposium on the Geology of the Bahamas: San Salvador, Bahamian Field Station, p. 221–234.

Vierma, L., Kwolek, J. M., Heidt, D. A., Hattin, D. E., Hasenmueller, W. A., and Feldman, H. R., 1984, Stratigraphic analysis of a newly discovered Pleistocene reef, San Salvador Island, Bahamas: Compass, v. 62, no. 1, p. 16–30.

White, B., 1989, Field guide to the Sue Point fossil coral reef, San Salvador Island, Bahamas, *in* Mylroie, J. E., ed., Proceedings, Fourth Symposium on the Geology of the Bahamas: San Salvador, Bahamian Field Station,

p. 353–365.

White, B., Kurkjy, K. A., and Curran, H. A., 1984, A shallowing upward sequence in a Pleistocene coral reef and associated facies, San Salvador Island, Bahamas, *in* Teeter, J. W., ed., Proceedings, Second Symposium on the Geology of the Bahamas: Ft. Lauderdale, Florida, CCFL Bahamian Field Station, p. 53–70.

Wilber, R.J., 1987, Geology of Little San Salvador and West Plana Cay: Preliminary findings with implications for Bahamian island stratigraphy, *in* Curran, H. A., ed., Proceedings, Third Symposium on the Geology of the Bahamas: Ft. Lauderdale, Florida, CCFL Bahamian Field Station, p. 181–204.

Winter, J. H., 1990, Recent geological finds on San Salvador, *in* Fifth Symposium on the Geology of the Bahamas: San Salvador, Bahamian Field Station, Abstracts and programs, p. 28.

Manuscript Accepted by the Society January 5, 1995

Printed in U.S.A.

Geological Society of America
Special Paper 300
1995

Paleoenvironmental and paleoecologic analyses of a Pleistocene mollusc-rich lagoonal facies, San Salvador Island, Bahamas

Robinson S. Noble
Department of Geology, The College of Wooster, Wooster, Ohio 44691
H. Allen Curran
Department of Geology, Smith College, Northampton, Massachusetts 01063
Mark A. Wilson
Department of Geology, The College of Wooster, Wooster, Ohio 44691

ABSTRACT

Numerous exposures of a recently discovered late Pleistocene mollusc-rich facies occur just south of the Bahamian Field Station on San Salvador Island. These outcrops surround a network of tidally influenced saline lakes and blueholes. Bedrock exposures of the facies occur as discontinuous lenses of poorly indurated shelly rock. Twenty-four species of fossil molluscs have been identified from this facies; they can be grouped into two distinct molluscan assemblages. The most common assemblage is dominated by two species of the burrowing bivalve *Codakia*. The other assemblage is characterized by the gastropods *Cerithium* and *Bulla* and the bivalve *Trigoniocardia*.

This Pleistocene facies represents a tidally influenced lagoon environment of undetermined size similar in many respects to parts of the modern south arm of Pigeon Creek lagoon on San Salvador, which was studied for comparison. The modern shell assemblages, which occur most commonly as thick deposits in current scour pits and in the main channel of the lagoon, contain many species identical to those identified in the fossil assemblages. The Pleistocene outcrops provide further evidence of a lagoonal paleoenvironment. The *Codakia* spp. assemblage is within a clean sand similar to the subtidal channel sediments in Pigeon Creek. However, the *Cerithium-Bulla* assemblage is present in a burrowed, muddy sand that contains trace fossils preserved as lithified burrow-fill segments attributable to upogebid shrimp, which today burrow in the intertidal flats of Pigeon Creek.

As might be expected, there are significant differences between the modern and Pleistocene molluscan lagoonal facies. Some bivalves, such as arcoids, are common in the fossil deposits yet appear to be absent from the modern Pigeon Creek lagoon; other bivalves are underrepresented or absent from the Pleistocene beds but common to abundant in Pigeon Creek. These differences, along with the similarities between the ancient and modern study areas, provide information on the reliability of paleoenvironmental and paleoecologic reconstructions based on molluscan faunal data.

Noble, R. S., Curran, H. A., and Wilson, M. A., 1995, Paleoenvironmental and paleoecologic analyses of a Pleistocene mollusc-rich lagoonal facies, San Salvador Island, Bahamas, *in* Curran, H. A., and White, B., Terrestrial and Shallow Marine Geology of the Bahamas and Bermuda: Boulder, Colorado, Geological Society of America Special Paper 300.

INTRODUCTION

Molluscan shells are a common and often volumetrically important component of the subtidal marine facies of late Pleistocene carbonate grainstones that cap the islands of the Bahamas. Although fossil molluscs have been mentioned and sometimes used in paleoenvironmental analysis in previous studies of Bahamian carbonates (see Hagey and Mylroie, this volume, for a partial listing of previous work), detailed analyses largely are lacking.

Exposures of a Pleistocene (Sangamon) mollusc-rich facies located in the northeast interior of San Salvador Island, Bahamas (Fig. 1) were discovered by Paul J. Godfrey and D. Craig Edwards of the University of Massachusetts–Amherst in January 1989. These exposures occur in and around a network of tidally influenced saline lakes and ponds just south of the Bahamian Field Station campus.

The primary purpose of this study is to describe the molluscan fossil assemblages preserved in these exposures. A secondary goal is to interpret the paleoenvironmental significance of the assemblages and compare them with modern molluscan assemblages, such as those found within the tidally controlled Pigeon Creek lagoon located on the southeast coast of San Salvador.

STUDY AREA AND ITS GEOLOGIC SETTING

Numerous, richly fossiliferous Pleistocene bedrock exposures are located between the northeast side of Oyster Pond and the west side of Pain Pond just south of the Bahamian Field Station (Fig. 1). The natural history of this area has been well described by Godfrey et al. (1994), and further information on the hydrogeology and ecology of the saline ponds is given in Edwards et al. (1990). Thick vegetation restricts determination of the full extent of the mollusc-rich facies. Bedrock exposures are discontinuous, but all of the main outcrops within the study probably have been discovered. Seven principal exposures (Fig. 1) were evaluated and sampled for this study.

These mollusc-rich exposures are capped by a well-developed karst surface and a discontinuous paleosol. Based on the dominant presence of *Codakia* spp. bivalves, these deposits had a marine origin (see also Hagey, 1991; Hagey and Mylroie, this volume). Although the beds have not been radiometrically dated, their stratigraphic position and the presence of the karst/paleosol surface indicate a late Pleistocene age for these rocks. Following the stratigraphy for San Salvador Island developed by Carew and Mylroie (1985; this volume, and earlier references therein), these rocks are part of the Cockburn Town Member of the Grotto Beach Formation and likely were deposited during the Sangamon sea-level highstand of oxygen isotope substage 5e, about 120,000 to 125,000 B.P. when marine waters reached a maximum of about 6 m above present mean low sea level (Chen et al., 1991). The *Codakia*-dominated beds are different from the reefal shallow subtidal, open shelf deposits that typify the Cockburn Town Member (Carew and Mylroie, 1985; this volume) and thus represent a different facies of the member.

Most of the outcrops in the study area present only hori-

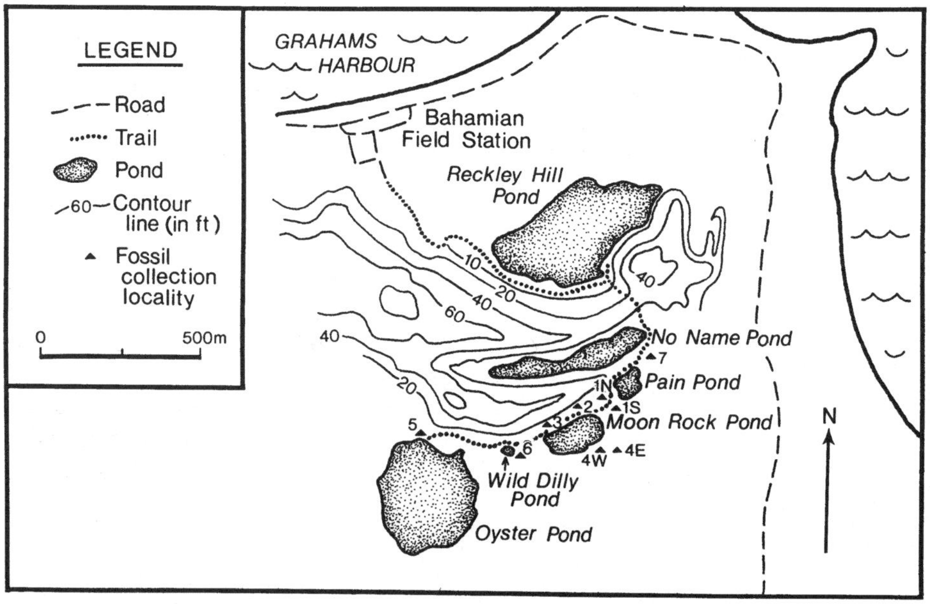

Figure 1. Index map to the study area, northeast corner of San Salvador Island, Bahamas. Sample collection sites are marked by solid triangles. Contours, in feet, are from the topographic map for San Salvador.

zontal exposure and float blocks for study. However, a small cliff (2 m high) on the south side of Wild Dilly Pond (Fig. 1), near the opening of the pond conduit, was measured and described (Fig. 2). The upper 1.5 m of the cliff is characterized by a shelly, fine, massive grainstone. The matrix is well sorted, and small shell fragments are abundant. The rock weathers from a white to a light gray sandy surface to a darker gray karst surface at the top. The lower 0.5 m of the section consists of a highly fossiliferous shell deposit dominated by *Codakia* spp. The contact between the shell bed and the shelly grainstone is gradational, and the matrix of the shell deposit is essentially the same as the grainstone above. Because the base of the shell bed is not exposed, its total thickness is unknown. However, the shell bed can be seen to cover most of the bottom of Wild Dilly Pond (Fig. 3).

A supplementary aspect of this study was the collection and analysis of modern shells from the southern arm of Pigeon Creek lagoon at the southeast corner of San Salvador. Shells were collected for comparison with the Pleistocene molluscan assemblages from the rich accumulations that occur in tidal current scour pits in the main channel of the creek.

GENERAL ATTRIBUTES OF THE FOSSIL SHELL BEDS

Physical characteristics

Nine bedrock exposures of the fossil molluscan facies at seven sample sites were sampled and analyzed. Each exposure is a highly fossiliferous, discontinuous lens of shelly rock, with fairly distinct boundaries (Fig. 4). The areal extent of the exposures ranges from several square meters to several hundred square meters. Thicknesses of the exposures are uncertain; however, the stratigraphic column measured at Wild Dilly Pond (Fig. 2) indicates that the shell beds have the potential to be thicker than 0.5 m.

Elevations of several of the exposures were obtained from Davis (1994; elevations given in feet to conform with the topographic map of San Salvador), and those exposures with no previously calculated elevations were measured with the help of R. L. Davis. The exposures range in elevation from 1.87 ft (0.57 m) below mean low sea level for shell bed 6 to 4.86 ft (1.48 m) above sea level for shell bed 1-S (Table 1). The elevations of the different shell beds have significance for determining the depositional sequence of the mollusc-rich facies.

Associated lithologies

Two distinct carbonate rock subfacies occur within the mollusc-rich beds, which can be readily distinguished in the field. One subfacies consists of shelly, fine-grained, clean grainstone; the other is a burrowed packstone.

Shelly grainstone. The shelly grainstone is composed of a variety of skeletal grains, including abundant gastropod and bivalve fragments, and some fragments of echinoderms and of the red calcareous alga *Neogoniolithon*. The nonskeletal grains include peloids and some ooids. Because a large number of skeletal grains are present, the shelly grainstone is interpreted as subtidal in origin.

Burrowed packstone. This is the dominant rock type at site 4-W only; the muddy sands are characterized by peloids and some ooids and pellets. Skeletal grains are also present, but

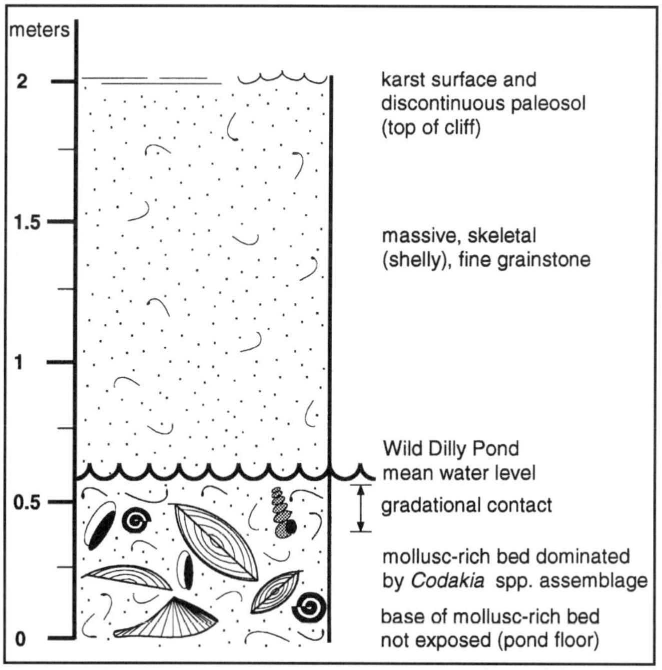

Figure 2. Generalized stratigraphic column at Wild Dilly Pond, near sample collection site 6.

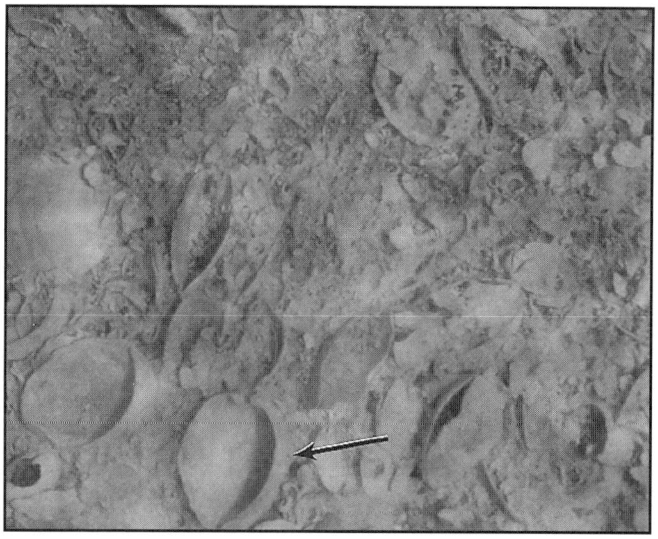

Figure 3. Fossil molluscan assemblage dominated by *Codakia* spp. bivalves exposed on the bottom of Wild Dilly Pond. Arrow marks an articulated *Codakia* that is about 8 cm in length.

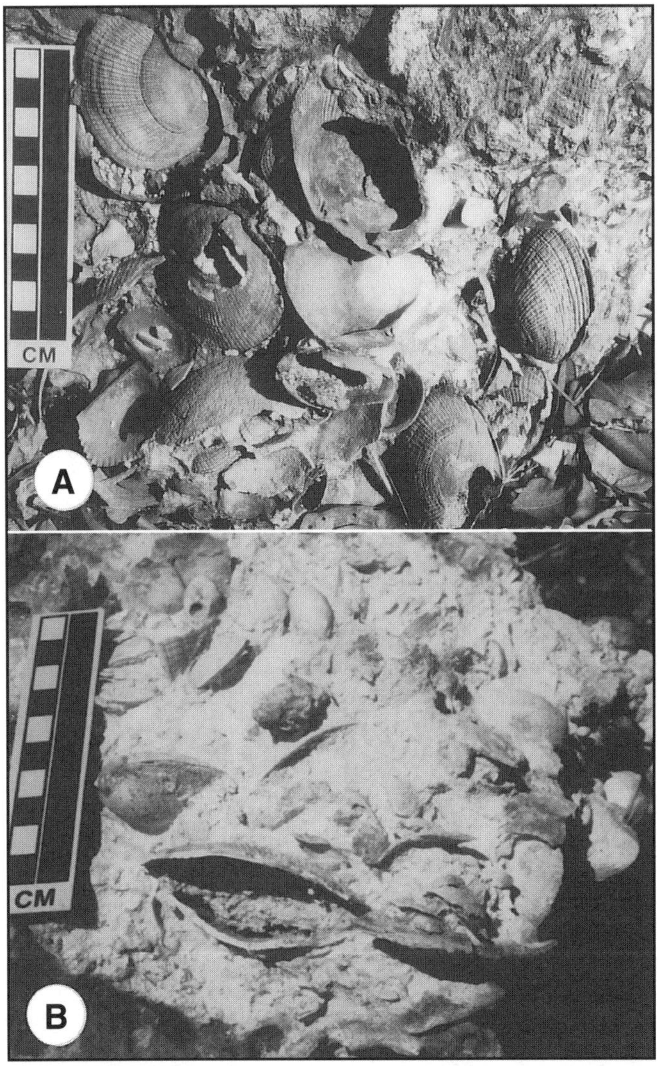

Figure 4. Examples of occurrence and preservation of the Pleistocene mollusc-rich facies. A, Surface exposure dominated by *Codakia* spp. B, Float block showing numerous *Codakia* spp., some articulated.

they are much less abundant than in the shelly grainstone, and fragments of calcareous algae are rare. Also found commonly within this subfacies are trace fossils in the form of segments of well-lithified burrow infills. The characteristics and significance of these trace fossils are discussed later in this chapter. Many of the pellets can be assigned to the ichnogenus *Favreina*, the fossilized pellets formed by callianassid shrimp. These *Favreina* specimens are very similar to those described by White et al. (1984) from the subtidal beds of the shallowing-upward sequence of the Cockburn Town fossil coral reef sequence (Cockburn Town Member) on San Salvador.

Depositional sequence

Field observations indicate that the shell bed exposures are latterally discontinuous, and elevations of the shell beds reveal

TABLE 1. SHELL BED ELEVATIONS*

Shell Bed	Elevation (ft)
1-N	n.a.
1-S	4.86
2	1.41
3	1.47
4-E	4.52
4-W	1.10
5	3.13
6	-1.87
7	4.79

*As shown in Figure 1. Elevations from Davis, 1994, and personal communication with the authors.

that they occur at different stratigraphic levels. The depositional sequence corresponds closely with a model proposed by Wanless (1981) for sedimentation on the modern, carbonate-dominated, littoral sand platforms of south Florida covered by the sea grasses *Thalassia* and *Syringodium*. Molluscan assemblages dominated by *Codakia orbicularis* and *Chione cancellata* accumulate in blowouts or scour pits in this environment. The blowouts then migrate, creating a fining-upward sedimentary sequence consisting of lenses of molluscan shell lag surrounded by shelly sand.

The resulting sedimentary sequence is very similar to the stratigraphic column measured at Wild Dilly Pond (Fig. 2). The rate of deposition and the frequency of blowout reworking will determine the number of molluscan lenses in the facies. Figure 5 illustrates how these sedimentary sequences would look in cross section. Following the Wanless model, we think that the laterally and vertically discontinuous molluscan lenses in the San Salvador Pleistocene facies represent sequences of medium to rapid rates of deposition with medium to low rates of blowout reworking.

Fossil assemblages and census data

Fossil mollusc shells were collected from the weathered exposures at the nine sample sites and identified using standard literature (Morris, 1975; Rehder, 1981; Sterrer, 1986). A total of 24 mollusc species were identified (Table 2). The bivalve *Codakia orbicularis* is the most common species in the fauna and is the only species recovered from all sample sites.

Seven of the nine study sites (1-N through 5) were sampled extensively (i.e., all identifiable shells were collected or identified in the rock from as much of the area of the exposure as possible). Sites 6 and 7 yielded few fossils and are not included in the taxa counts; however, the beds at these sites showed a common occurrence of *Codakia* spp. The fossils of the seven "census" samples then were identified and counted. The census counts included only complete or slightly broken shells. Table 3 lists the species found in each sample and their abundances, and Table 4 shows the percentage frequency of occurrence of each species in the samples.

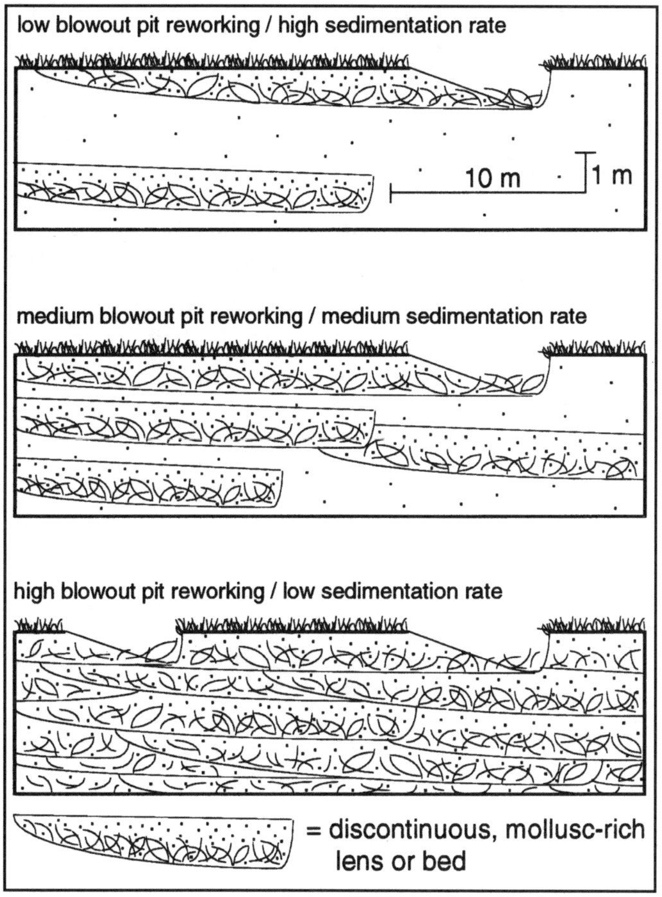

Figure 5. Model for a sedimentary sequence generated by scour pit or blowout migration in a sea grass-dominated, shallow marine environment. (Modified from Wanless, 1981).

DEPOSITIONAL ENVIRONMENTS AND MOLLUSCAN FAUNA OF PIGEON CREEK LAGOON: MODERN ANALOGUE FOR THE FOSSIL DEPOSITS

Subtidal environments

The south arm of Pigeon Creek is a tidally dominated, current-influenced lagoon on the southeast side of San Salvador Island. The lagoon has been the site of several previous studies that have emphasized its usefulness as an analogue to aid with interpretation of Pleistocene rocks on San Salvador and elsewhere in the Bahamas (Teeter, 1989; Boardman and Carney, 1992, and references cited therein). Within a relatively small area, the lagoon displays several significant and distinct depositional environments; Mitchell (1987) recognized 12 lithofacies within the entire Pigeon Creek depositional system. Our study area was confined to the south arm of the creek, near its juncture with the north arm (Mitchell, 1987, Fig. 3). The main channel of the creek is characterized by clean shelly sands with areas of sea grasses and by depressions with shell accumulations (Fig. 6).

TABLE 2. COMPLETE MOLLUSCAN FAUNAL LIST FROM THE FOSSIL BEDS OF THIS STUDY

Bivalves	Gastropods
1. *Anadara lienosa floridana*	16. *Astraea phoebia*
2. *Anodontia alba*	17. *Bulla occidentalis*
3. *Anomalocardia brasiliana*	18. *Cerithium algicola*
4. *Chione cancellata*	19. *Collumbella mercatoria*
5. *Chione paphia*	20. *Modulus carchedonius*
6. *Codakia orbicularis*	21. *Murex recurvirostris*
7. *Codakia orbiculata*	22. *Nerita tesselata*
8. *Lucina floridana*	23. *Strombus gigas*
9. *Macoma hendersoni*	24. *Tegula fasciata*
10. *Noetia ponderosa*	
11. *Periglypta listeri*	
12. *Pitar albida*	
13. *Tellina alternata*	
14. *Tellina angulosa*	
15. *Trigoniocardia antillarum*	

Thick deposits of shells accumulate in both the main channel of the creek and in isolated scour or storm "blowout" pits. The scour pits have a depth range of 0.5 to 1 m, and the main channel has an average depth of 3.5 m. We have not studied the origin of these scour pits or blowouts, but the mechanism proposed by Wanless (1981) of cutting during storms and erosional migration of the scour pits to produce lensoidal deposits (Fig. 5) seems applicable. The walls of the Pigeon Creek channel commonly are abrupt on either side, much like an enlarged scour pit, and allow for the thick accumulation of shells.

Boardman and Carney (1992) reported high rates of sedimentation, shell movement, and shell accumulation in the scour pit areas of Pigeon Creek. We attribute the discontinuous occurrence and different elevations of the Pleistocene shell beds to a paleodepositional environment with medium to high rates of sedimentation and/or medium to low rates of blowout reworking, following the Wanless model scale (Fig. 5).

The subtidal scour pits and the main channel of Pigeon Creek are the only environments in the lagoon with thick accumulation of mollusc shells in death assemblages. Many of the shells in the main channel, particularly in the scour pits, are pristine specimens. In addition, a number of the bivalves are articulated, suggesting that the shells were transported only short distances or were eroded from the walls of the channel or scour pits before being deposited. Slone (1990), Slone et al. (1990), and Cummins and Boardman (1994) confirmed that the molluscan death assemblages closely resemble the living assemblages in the different depositional environments throughout Pigeon Creek. Slone (1990) and Slone et al. (1990) pointed out that, even on the tidal delta at the mouth of the creek, the death assemblage closely resembles the living fauna and is not a mixture of all species in the creek, further indicating that the shells are undergoing little postmortem transportation from their life environment.

TABLE 3. NUMBERS OF FOSSIL MOLLUSCAN ABUNDANCES*

	Assemblages						
	1-N	1-S	2	3	4-E	4-W	5
Bivalves							
Anadara lienosa floridana	29	14	11	12	4	1	11
Anodontia alba	--	2	4	--	--	--	--
Anomalocardia brasiliana	2	--	--	--	--	--	--
Chione cancellata	--	--	2	--	--	--	--
Chione paphia	--	--	2	37	2	1	14
Codukia orbicularis	90	71	80	55	43	1	22
Codakia orbiculata	41	56	31	24	24	2	--
Lucina floridana	1	4	--	--	--	--	--
Macoma hendersoni	1	9	39	10	1	12	--
Noetia ponderosa	6	4	4	--	2	1	2
Periglypta listeri	1	1	--	--	--	--	--
Pitar albida	2	--	--	--	--	--	--
Tellina alternata	--	--	1	3	--	--	5
Tellina angulosa	--	--	--	--	--	--	3
Trigoniocardia antillarum	6	7	29	20	18	31	16
Total	179	168	203	161	94	49	70
Gastropods							
Astraea phoebia	--	1	1	--	--	--	--
Bulla occidentalis	7	26	30	8	11	28	14
Cerithium algicola	3	3	20	10	4	37	--
Collumbella mercatoria	3	1	--	--	--	--	--
Modulus carchedonius	2	2	1	--	2	3	--
Murex recurvirostris	--	1	2	--	--	1	5
Nerita tesselata	--	--	1	--	--	--	--
Strombus gigas	1	--	1	--	--	--	4
Tegula fasciata	14	26	17	7	9	3	16
Total	30	60	73	25	26	72	39
Total shells in each assemblage	209	228	276	186	120	121	109

*Counts include all complete or slightly broken shells.

Intertidal sand flats

Extensive burrowed, intertidal sand flats occur along the flanks of Pigeon Creek (Curran, 1984; 1994, Fig. 3.9). These sand flats are characterized by a hummocky surface produced by the burrowing activity of callianassid shrimp. The callianassid mounds are, in turn, burrowed by upogebid shrimp (identified as *Upogebia vasquezi* by Williams, 1993) and by fiddler crabs, particularly *Uca major*. Casts of *U. vasquezi* burrows are shown in Figure 7. Remarkably, the Pleistocene lithified burrow fills found at site 4-W closely resemble the upogebid burrows (Fig. 7B); therefore, we can attribute the Pleistocene trace fossils to upogebid burrowing. Large bivalve shells (width >3 cm) can be found only as isolated specimens on the sand flats during low tide. Smaller shells, particularly the gastropods *Bulla* spp. and *Cerithium* spp., are more common.

Molluscan fauna

For this study, a faunal list (Table 5) was compiled by sampling the molluscan death assemblages in the main channel and scour pits of Pigeon Creek. Our collections from Pigeon Creek did not equal the species diversity obtained by Slone (1990). We attribute this to the fact that Slone collected 34 samples of mollusc shells from a variety of depositional environments with Pigeon Creek lagoon. Slone's list included all species found with greater than 10% abundance in at least one sample. However, many of the species identified in this study match those of Slone (1990) and represent the most common species found in the main channel and tidal flats of Pigeon Creek. The molluscan faunal list from this study was combined with that of Slone (1990) to create a more complete faunal listing for Pigeon Creek (Table 6).

PALEOECOLOGIC AND PALEOENVIRONMENTAL RECONSTRUCTION OF THE FOSSIL SHELL BEDS

Introduction

Fürsich and Flessa (1987) stated that a thorough knowledge of recent environments and biotas is required for an actualistic approach to paleoenvironmental analysis. Another goal of this study is to provide a realistic paleoecologic and paleoenviron-

TABLE 4. FOSSIL MOLLUSCAN ABUNDANCES BY PERCENT*

	Assemblages						
	1-N	1-S	2	3	4-E	4-W	5
Bivalves							
Anadara lienosa floridana	13.8	6.2	4.4	6.4	3.3	0.8	10
Anodontia alba	0	0.9	1.6	0	0	0	0
Anomalocardia brasiliana	0.9	0	0	0	0	0	0
Chione cancellata	0	0	0.8	0	0	0	0
Chione paphia	0	0	0.8	19.9	1.6	0.8	12.7
Codukia orbicularis	43.1	31.6	32.1	29.5	36.4	0.8	20
Codakia orbiculata	19.6	25	12.4	12.9	20.3	1.6	0
Lucina floridana	0.5	1.7	0	0	0	0	0
Macoma hendersoni	0.5	4	15.6	5.3	0.8	9.8	0
Noetia ponderosa	2.8	1.7	1.6	0	0	0.8	1.8
Periglypta listeri	0.5	0.4	0	0	0	0	0
Pitar albida	0.9	0	0	0	0	0	0
Tellina alternata	0	0	0.4	1.6	0	0	4.5
Tellina angulosa	0	0	0	0	0	0.8	2.7
Trigoniocardia antillarum	2.8	3.1	11.6	10.7	15.2	23.3	14.5
Gastropods							
Astraea phoebia	0	0.4	0.4	0	0	0	0
Bulla occidentalis	3.3	11.6	1.2	4.3	9.3	22.9	12.7
Cerithium algicola	1.4	1.3	8.0	5.3	3.3	30.3	0
Collumbella mercatoria	1.4	0.4	0	0	0	0	0
Modulus carchedonius	0.9	0.9	0.4	0	1.6	2.4	0
Murex recurvirostris	0	0.4	0.8	0	0	0.8	4.5
Nerita tesselata	0	0	0.4	0	0	0	0
Strombus gigas	0.5	0	0.4	0	0	0	3.6
Tegula fasciata	6.6	9.8	6.8	6.8	7.6	2.4	14.5

*Counts include all complete or slightly broken shells.

mental reconstruction of the Pleistocene mollusc-rich facies through comparisons with the modern ecology and environments of Pigeon Creek lagoon. However, the literature documenting this method of paleontologic reconstruction is limited.

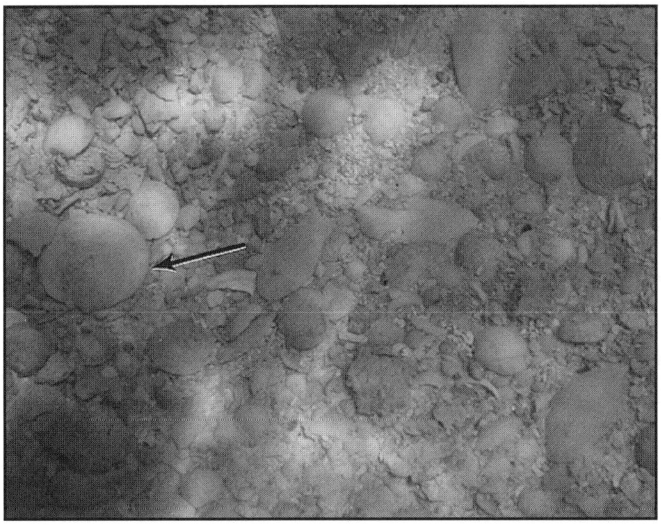

Figure 6. Modern molluscan assemblage dominated by *Codakia* spp. at bottom of a scour pit in Pigeon Creek lagoon. Arrow marks a *Codakia* valve that is about 8 cm in length.

Most previous studies have concentrated on comparisons of "live:dead" ratios of faunal distributions to provide criteria for the recognition of taphonomic bias in fossil communities, but these studies have not applied their criteria from a modern environment to an ancient environment.

Some workers are skeptical as to whether paleoecologic data can provide information on the structure and function of ancient communities (see Hoffman, 1979; Jarvinen et al., 1986). However, many others have concluded that the taphonomic, ecologic, and taxonomic data preserved in fossil assemblages are very useful as paleoenvironmental and paleoecologic indicators (Peterson, 1976; Scott, 1978; Miller, 1988; Dodd and Stanton, 1990). We think that the paleontologic and sedimentologic data preserved in the Pleistocene molluscan beds described in this study are significant indicators that can be directly related to modern carbonate environments in order to develop paleoenvironmental and paleoecologic reconstructions.

Molluscan faunal comparisons

Complete molluscan faunal listings (Tables 2 and 6) were compiled to compare the species present in the Pleistocene beds with the modern death assemblages in Pigeon Creek to evaluate how accurately the fossil assemblages reflect the modern shell distribution pattern. The tables illustrate that the two study

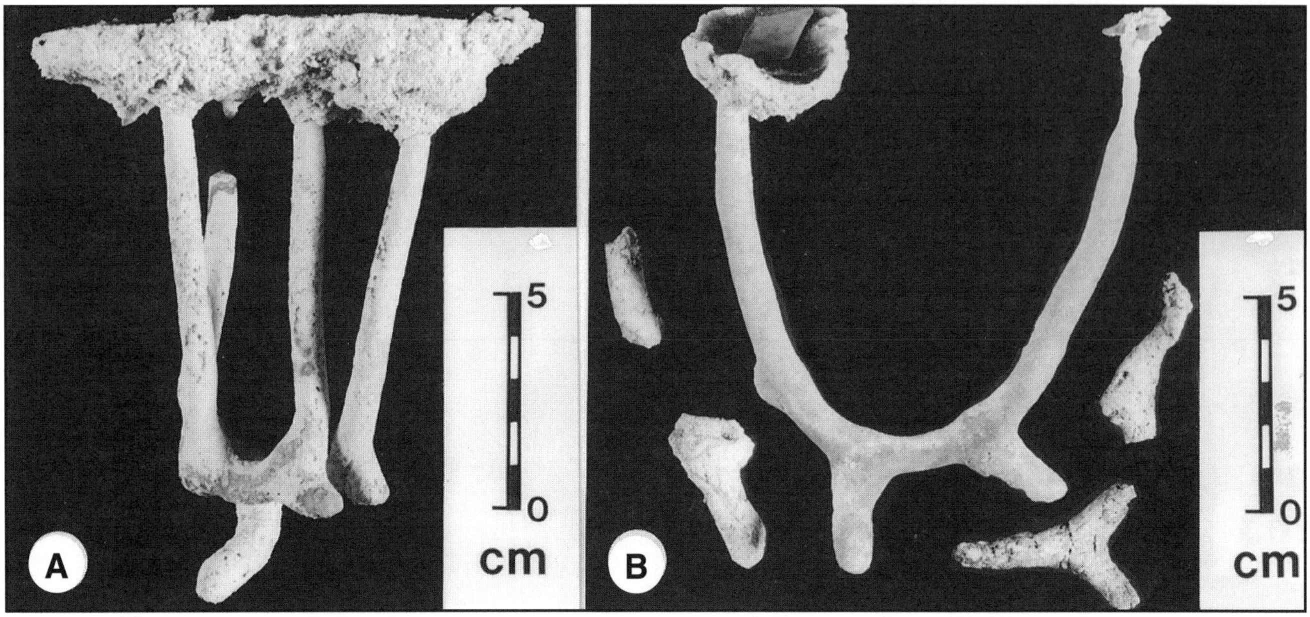

Figure 7. Upogebid shrimp burrows. A, Plastic cast of a modern mud shrimp burrow made by *Upogebia vasquezi* on the intertidal, burrowed, muddy sand flats of the south arm of Pigeon Creek lagoon. Two burrows commonly are intertwined, as in this example. B, Plastic cast of a single modern upogebid burrow surrounded by similar Pleistocene lithified burrow-fill segments from collection site 4-W.

areas possess relatively large numbers of identical species. This initial evidence suggests that the Pleistocene mollusc-rich facies is similar to the modern environment of Pigeon Creek.

The tables also show a degree of difference between the species present in each study area. For instance, the bivalve *Periglypta listeri* is underrepresented in the fossil deposits, and the bivalves *Divaricella quadrisulcata* and *Lucina pensylvanica* are entirely absent from the fossil beds, whereas they are common to abundant in Pigeon Creek deposits. In addition, several of the arcoids, such as *Noetia ponderosa* and *Anadara lienosa floridana*, are common in the fossil remains yet appear to be absent from modern Pigeon Creek deposits.

The greater diversity of species in the modern accumulations can be accounted for in several ways. For one, the Pigeon Creek faunal list (Table 6) is a composite of this study and that of Slone (1990), who sampled a number of environments within Pigeon Creek. The fossil assemblage list is subject to a greater sampling bias (whole and only slightly broken shells) and may be restricted to fewer environmental transitions than the full extent of Pigeon Creek. The differences also may be the result of the life habits of the species that are found in one study area and not found in the other (Tevesz and McCall, 1983; Miller, 1988).

Habitat and life mode implications

Habitat and life mode descriptions also were compiled for comparison purposes for species in the Pleistocene facies and in the modern deposits. Most bivalves in the Pleistocene facies are characterized as slow burrowers in clean to slightly muddy sands of protected waters in the intertidal to shallow subtidal zones. In addition, these clams typically are associated with sea grass-covered bottoms (Stanley, 1970). Such grass-covered bottoms occur in close association with the scour pits and main channel of Pigeon Creek and likely were associated with the clean shelly sand environment of the Pleistocene facies in which the *Codakia* spp. assemblage occurs. Gastropods in the Pleistocene facies commonly are associated with quiet, shallow

TABLE 5. FAUNAL LIST OF MOLLUSCAN DEATH ASSEMBLAGE, PIGEON CREEK LAGOON

Bivalves	**Gastropods**
Chione cancellata	*Astraea phobia*
Codakia costata	*Bulla occidentalis*/striata*
Codakia orbicularis	*Cerithium algicola/eburneum*
Codakia orbiculata	*Columbella mercatoria**
Divaricella quadrisculacata	*Conus* sp.*
*Glycymeris pectinata**	*Modulus carchedonius*
Lucina pensylvanica	*Polinices lacteus*
*Periglypta listeri**	*Strombus gigas**
Tellina sp.	*Tegula fasciata*
Trigoniocardia antillarum	
	Scaphopods
Oysters	*Dentalium occidentale**
Pinctade radiata	

*Species found in Pigeon Creek during this study but not reported by Slone, 1990.

TABLE 6. COMPLETE MOLLUSCAN FAUNAL LIST, PIGEON CREEK LAGOON*

Bivalves	Gastropods
Anomalocardia brasilia	*Acmaea testudinalis*
Chione cancellata	*Anachis sparsa*
Codakia orbicularis	*Astraea phoebia*
Codakia orbiculata	*Bailya parva*
Divaricella quadrisulcata	*Bulla occidentalis/striata*
Glycymeris pectinata	*Cerithidea scarariformis*
Lucina pensylvanica	*Cerithium algicola/eburneum*
Periglypta listeri	*Cerithium literatum*
Semele nuculoides	*Cerithium variable*
Tellina sp.	*Columbella mercatoria*
Trigoniocardia antillarum	*Conus* sp.
	Diodora listeri
Oysters	*Modulus carchedonius*
Pinctade radiata	*Nassarius albus*
	Nerita tessellata
Scaphopods	*Olivella floralia*
Dentalium occidentale	*Polinices lacteus*
	Smaragdia viridemaris
	Strombus gigas
	Tegula fasciata
	Tricolia affinis
	Turbo castaneus
	Vermicularia spirata

*From this study and Slone, 1990.

water in or near the intertidal zone and lived as either carnivores or epifaunal algae grazers. *Bulla* spp. are characterized as burrowing in muddy substrates near the low tide line; this substrate type is characteristic of the burrowed beds of the Pleistocene facies and the intertidal burrowed sand flat of Pigeon Creek.

The species found in the modern environments of Pigeon Creek that are not found in the Pleistocene facies exhibit habitat preferences and life habits similar to those of species occurring in the fossil deposits. Therefore, based on habitat descriptions alone, it is difficult to determine why these modern species are not found in the Pleistocene fossil deposits. Since there are several species (including *Glycymeris pectina* and *Lucina* [=*Linga*] *pensylvanica*) that occur commonly in the modern deposits and have been observed in other Bahamian Pleistocene assemblages (Hagey, 1991; Hagey and Mylroie, this volume) preservational bias in the fossil record alone cannot fully explain their absence here. The Pleistocene beds of this study may reflect more restricted environmental conditions than the comparison deposits, or some of these species may have been naturally or possibly artificially introduced to modern San Salvador Island lagoonal environments since the deposition of this particular Pleistocene facies. These differences in the faunal lists do not affect the paleoenvironmental reconstruction. Rather, they are a significant feature of this study, suggesting that a certain degree of variation should be expected and allowed for when comparing a modern environment to a Pleistocene paleoenvironment (see also Noble et al., 1991).

Faunal distribution, frequencies, and multidimensional scaling

Faunal distributions and frequencies were calculated for the mollusc species present in seven of the Pleistocene samples (1-N through 5, Fig. 8) in an attempt to recognize distinct zonations that may indicate environmental change within the mollusc-rich facies. The histograms show that shell assemblage 4-W, which is characterized by an abundance of *Bulla, Cerithium,* and *Trigoniocardia*, is quite different from the other fossil assemblages, which are dominated by the two species of *Codakia*. The faunal data also were analyzed by Q-mode nonparametric multidimensional scaling (Systat) to delineate further the biofacies. Briefly, multidimensional scaling (MDS) is a data-reduction technique that graphically portrays structure and similarity within multivariate data sets. The algorithm operates on a symmetric matrix of similarity or distance coefficients among cases or variables. MDS is particularly useful for a data set that contains overlap, because it graphically displays gradational relationships among cases or variables (for other examples, see Bjerstedt, 1988; Meldahl and Flessa, 1990). MDS plots cases or variables in an n-dimensional configuration based on similarity and then reduces the dimensionality (usually to two or three dimensions) so relationships can be visualized. Scaled data are displayed on scatter plots where the degree of similarity is reflected by distance; closer cases or variables are more similar.

Euclidean distance values were computed for the percentage of transformed data (Table 4) and scaled using MDS. The results are shown in Figure 9. The distances between sample sites on the MDS are a measure of the degree of their similarity based on faunal content. Samples that are largely composed of the same taxa with similar relative abundances plot close together; samples that are distinctive in the faunal content plot farther apart. Not that sample 4-W is particularly distinctive, suggesting that it contains a different faunal assemblage from the other six samples (as is the case).

The extent to which a fossil assemblage reflects the living assemblages from which it originated remains an important issue in paleontologic reconstruction (Peterson, 1976; Miller, 1988, 1989; Parsons, 1989). Peterson (1976) suggested that three sources can account for discrepancies in relative abundances between living and dead assemblages: (1) mixing of separate communities by postmortem transportation; (2) dissolution and fragmentation of fossil remains at different rates; and (3) the cumulative nature of the fossil record, which commonly may reflect only the pooled production of all component species and not an instantaneous community composition. The latter is sometimes referred to as "time-averaging." Therefore, by determining the degree to which these processes affect the relative abundances between living and dead assemblages, it can be determined how accurately the fossil assemblages reflect the living fauna and environment from which they are derived.

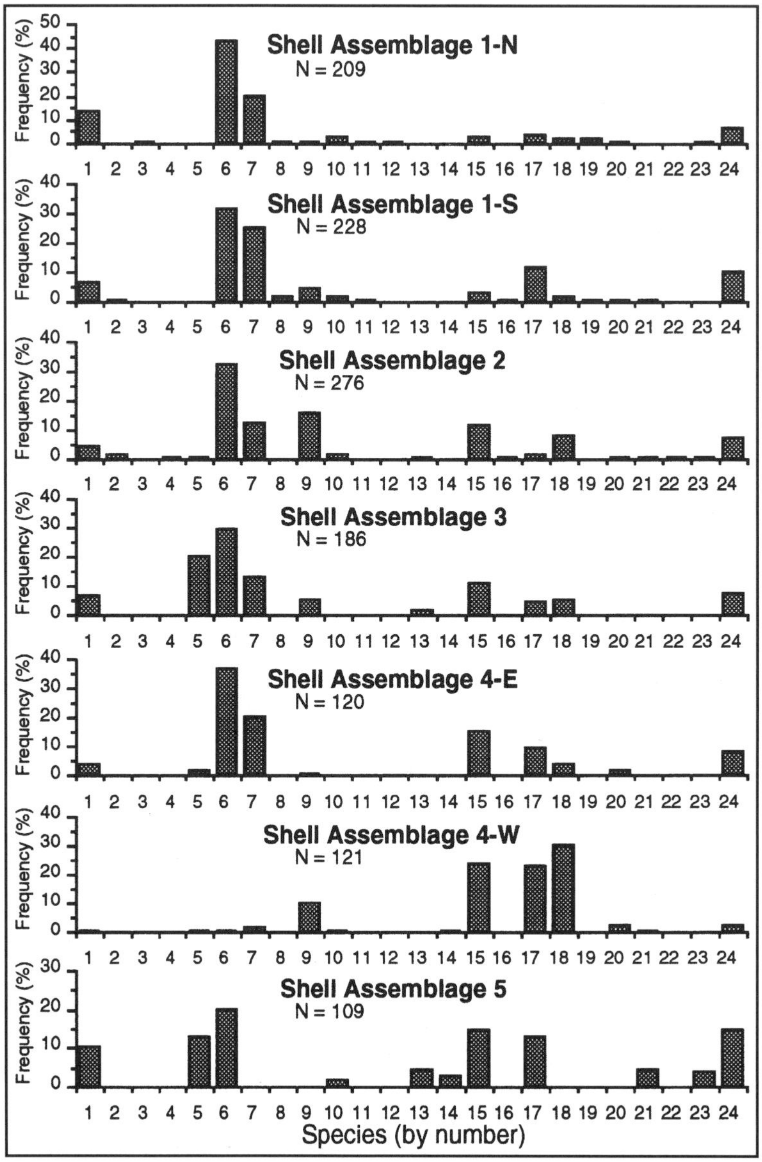

Figure 8. Histograms illustrating the frequency of occurrence of each identified fossil mollusc species from sample collection sites 1-N through 5. Species are listed by numbers assigned in Table 2. Note that sample 4-W has a very low occurrence of *Codakia* spp.

Effects of postmortem transportation

The amount of postmortem transportation is an important factor considered when determining whether a fossil assemblage reasonably reflects the living faunas and environment from which it originated (Miller and Cummins, 1990). Because most of the bivalves in the Pleistocene fossil beds are preserved as whole valves, and since many articulated *Codakia* specimens were found, the evidence from this study suggests that postmortem transport was negligible. Peterson (1976) and Parsons (1989) confirmed that, despite current and wave activity, little postmortem mixing occurs in lagoon environments. Therefore, this parameter largely can be ignored in lagoonal paleoecology. However, Fraser and Greenstein (1993) showed that mixing of modern molluscan death assemblages can be significant in the bays along the San Salvador coast.

If postmortem mixing is negligible, death assemblages will accurately represent distribution patterns of live individuals (Fürsich and Flessa, 1987). The significance for this study is that the death assemblages in the Pleistocene facies most likely do adequately represent the living fauna from which they were derived. Certainly the clear parallels between the modern Pigeon Creek molluscan communities and those in the Pleistocene facies further support this conclusion.

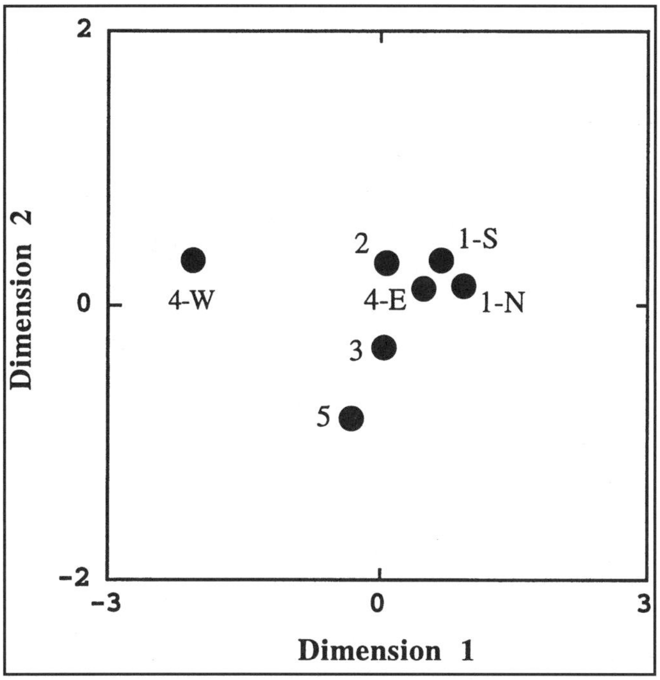

Figure 9. A Q-mode multidimensional scaling plot demonstrating the contrasting faunal assemblage present in the Pleistocene facies represented by sample 4-W. Points are labeled with sample numbers. See text for further discussion.

Zonation

Given that death assemblages in lagoonal settings closely resemble the life assemblages from which they were derived, and since many live species exhibit distinct zonation, the death assemblages of the individuals of such species in a lagoonal environment should exhibit distinct zonations associated with the habitat and life habits of the species. Miller (1988) concluded that, indeed, spatial faunal transitions associated with environmental change are not only recognizable but also preservable in the rock record. Accordingly, we think that the two very different assemblages in the Pleistocene mollusc-rich facies reflect a natural faunal transition. These shell assemblages are distinguishable by the frequencies of the species present in each, and represent environmentally controlled facies changes within the Pleistocene study area.

Effects of time-averaging

Fossil assemblages usually are not an instantaneous account of community composition. Rather, assemblages tend to have a cumulative nature, reflecting a combined collection of all species in a system. The effects of this time-averaging are of particular importance when determining the extent to which a fossil molluscan assemblage accurately represents the life assemblages from which it was derived, particularly if postmortem transportation was active (Peterson, 1976). However, since postmortem transport is not normally a significant factor in lagoons, lagoonal molluscan death assemblages are a time-average of local environments, and they reflect the living molluscan assemblages rather well (Peterson, 1976; Fürsich and Flessa, 1987). The results of time-averaging are small variations in species composition and relative abundances within molluscan assemblages that represent the same depositional environment. The histograms (Fig. 8) constructed from this study illustrate a distinct faunal transition associated with environmental zonation; also shown are small variations associated with time-averaging in species abundances and species composition.

Implications for the Pleistocene fossil record

The paleoenvironments of the Pleistocene mollusc-rich facies can be determined by correlating the locations of the fossil molluscan assemblages to the habitat descriptions of the faunal constituents that comprise each assemblage. This is possible because the assemblages in the Pleistocene facies are valid indicators of the living fauna and environment from which they originated, while the individual mollusc species exhibit distinct zonations associated with environmental change. The habitat descriptions of the *Codakia* spp. assemblage suggest that the locations in the Pleistocene facies where this association is found represent a paleoenvironment that was a protected, moderately shallow lagoon with a clean sandy, to slightly muddy, sea grass-covered substrate subject to scouring and development of blowouts by strong tidal currents, or possibly by storm events. The paleoenvironment of the *Cerithium-Bulla* assemblage, however, was apparently a muddier substrate in the intertidal zone. The common occurrence of lithified burrow segments, likely produced by the burrowing activity of upogebid shrimp, is a strong indication of an intertidal, muddy sand flat paleoenvironment.

The modern processes, environments, and faunal associations observed in the south arm of Pigeon Creek lagoon provide significant information about the origin and paleoenvironmental setting of the Pleistocene mollusc-rich facies. The shelly sediments of blowout scour pits located in the main channel of Pigeon Creek lagoon, which closely resemble the fossil deposits in size, shape, and density of material, provide a model for the accumulation of richly fossiliferous molluscan assemblages such as those found in the Pleistocene facies. The faunal constituents, dominated by *Codakia* spp., found within the scour pits and in the main channel of Pigeon Creek are nearly identical to those found in the Pleistocene facies, suggesting that the Pleistocene molluscan assemblages dominated by *Codakia* spp. were deposited in a channel and scour pits of a paleolagoon with a substrate of shelly, clean, subtidal sand with extensive *Thalassia* grass coverage. In the Pleistocene facies, the assemblage of sample 4-W, characterized by the gastropods, *Cerithium* and *Bulla*, and the probable upogebid shrimp burrows, exhibits a spatial faunal change associated with an environment different from that of the *Codakia* spp. assemblage. The intertidal sand flats in the south arm of Pigeon Creek lagoon, characterized by a muddy sand and hummocky surface of callianassid burrows that are in turn burrowed by *Upogebia* shrimp, appear to be nearly identical to the paleoenvironment suggested by the *Cerithium-Bulla* assemblage and its associated trace fossils.

In conclusion, by comparing the interpreted paleoenvironment of a Pleistocene mollusc-rich facies to a modern analogue in Pigeon Creek lagoon, it is evident that the Pleistocene fossil record provides a significant amount of information from which accurate paleoenvironmental and paleoecologic reconstructions can be made. The fossil assemblage dominated by *Codakia* spp. is representative of a lagoonal, subtidal paleoenvironment with a clean sandy substrate, partially covered by *Thalassia* and other sea grasses. The *Cerithium-Bulla* assemblage with burrows attributable to upogebid shrimp represents the intertidal paleoenvironment of a burrowed, muddy sand flat.

CONCLUSIONS

1. The Pleistocene rocks described in this study on San Salvador Island represent a lagoonal facies of undetermined size. The paleoenvironments bear close similarity to the modern setting of Pigeon Creek lagoon on San Salvador.

2. By determining which processes create discrepancies between living and dead molluscan assemblages in modern environments, the degree to which fossil molluscan death assemblages accurately reflect the living assemblages can be approximated. Our study indicates that the molluscan death assemblages in the Pleistocene lagoonal facies fairly accurately reflect the living assemblages from which they were derived.

3. Species within the Pleistocene molluscan assemblages exhibit distinct faunal associations as a result of environmental differences. Our data indicate that the habitats and life habits of the species in each fossil assemblage accurately represent the paleoenvironment of that assemblage.

4. There are two separate faunal assemblages in the Pleistocene mollusc-rich facies, representing two distinct, shallow marine, lagoonal paleoenvironments. The *Codakia* spp. assemblage represents a subtidal lagoon environment, dominated by a clean sandy substrate with *Thalassia* and other sea grasses. This environment is similar to the scour pits and main channel deposits in Pigeon Creek lagoon. The *Cerithium, Bulla, Trigoniocardia* assemblage with apparent upogebid shrimp burrows represents an intertidal, burrowed, muddy sand environment much like the intertidal sand flats observed in the south arm of Pigeon Creek lagoon.

ACKNOWLEDGMENTS

We thank the directors and staff of the Bahamian Field Station for full logistical support of our field work on San Salvador Island. Paul J. Godfrey and D. Craig Edwards (University of Massachusetts–Amherst) and R. Laurence Davis (University of New Haven) and their students opened the trail to the study area and did the initial survey work. Without their pioneering efforts, this study would not have been possible, and we are grateful to them. Godfrey and Edwards also were the first to bring the fossil-bearing exposures to our attention, and Davis also kindly provided elevation data on the sample collection sites. Helpful criticism of our original manuscript was provided in reviews by Ben Greenstein (Smith College) and Carl Mendelson (Beloit College). Greenstein aided with the statistical analysis, and Kathy Bartus (Smith College) performed word processing for the manuscript with skill and patience. Funding for this research was provided by the Keck Geology Consortium through a generous grant from the W. M. Keck Foundation and by the Senior Independent Study Program at The College of Wooster.

REFERENCES CITED

Bjerstedt, T. W., 1988, Multivariate analysis of trace fossil distribution from an Early Mississippian oxygen-deficient basin, central Appalachians: Palaios, v. 3, p. 53–68.

Boardman, M. R., and Carney, C., 1992, The geology of Columbus' landfall: A field guide to the Holocene geology of San Salvador, Bahamas, *in* Field Trip 3, Cincinnati Annual Meeting of the Geological Society of America: Ohio Department of Natural Resources, Division of Geological Survey, Miscellaneous Report 2, 49 p.

Carew, J. L., and Mylroie, J. E., 1985, The Pleistocene and Holocene stratigraphy of San Salvador Island, Bahamas, with reference to marine and terrestrial lithofacies at French Bay, *in* Curran, H. A., ed., Pleistocene and Holocene carbonate environments on San Salvador Island, Bahamas, Geological Society of America, Orlando Annual Meeting Field Trip Guidebook: San Salvador, CCFL Bahamian Field Station, p. 11–61.

Chen, J. H., Curran, H. A., White, B., and Wasserburg, G. J., 1991, Precise chronology of the last interglacial period: ^{234}U-^{230}Th data from fossil coral reefs in the Bahamas: Geological Society of America Bulletin, v. 103, p. 82–97.

Cummins, R. H., and Boardman, M. R., 1994, The recognition of environmental transitions using species composition, biomass estimates and taphonomic signatures in a Holocene carbonate lagoon, Pigeon Creek, San Salvador, Bahamas, *in* Abstracts and Program, Seventh Symposium on the Geology of the Bahamas: San Salvador, Bahamian Field Station, p. 10–11.

Curran, H. A., 1984, Ichnology of Pleistocene carbonates on San Salvador, Bahamas: Journal of Paleontology, v. 58, p. 146–159.

Curran, H. A., 1994, The palaeobiology of ichnocoenoses in Quaternary, Bahamian-style carbonate environments: The modern to fossil transition, *in* Donovan, S. K., ed., The palaeobiology of trace fossils: New York, John Wiley & Sons, p. 83–104.

Davis, R. L., compiler, 1994, Map of area south of Bahamian Field Station, *in* Godfrey, P. J., and others, eds., Natural history of northeastern San Salvador Island: A "new world" where the New World began, *in* Bahamian Field Station Trail Guide: San Salvador, Bahamian Field Station, p. 16.

Dodd, J. R., and Stanton, R. J., 1990, Paleoecology: Concepts and applications, second edition: New York, John Wiley & Sons, 502 p.

Edwards, D. C., Teeter, J. W., and Hagey, F. M., 1990, Geology and ecology of a complex of inland saline ponds, San Salvador Island, Bahamas, *in* Fifth symposium on the Geology of the Bahamas, Field Trip Guidebook: San Salvador, Bahamian Field Station, p. 35–45.

Fraser, N. M., and Greenstein, B. J., 1993, Taphofacies analysis of a modern molluscan facies: Snow Bay and Bonefish Bay, San Salvador, Bahamas: Geological Society of America Abstracts with Programs, v. 25, no. 3, p. 39.

Fürsich, F. T., and Flessa, K. W., 1987, Taphonomy of tidal flat molluscs in the northern Gulf of California: Paleoenvironmental analysis despite the perils of preservation: Palaios, v. 2, p. 543–559.

Godfrey, P. J., Edwards, D. C., Davis, R. L., Smith, R. R., and Wells, J. A., 1994, Natural history of northeastern San Salvador Island: A "new world" where the New World began, *in* Bahamian Field Station Trail Guide: San Salvador, Bahamian Field Station, 28 p.

Hagey, F. M., 1991, Pleistocene molluscan assemblages on San Salvador Island, Bahamas: Preliminary investigations and interpretations, *in* Bain, R. J., ed., Proceedings of the Fifth Symposium on the Geology of the Bahamas: San Salvador, Bahamian Field Station, p. 103–115.

Hoffman, A., 1979, Community paleoecology as an epiphenomenal science: Paleobiology, v. 5, p. 357–379.

Jarvinen, O., and 10 others, 1986, The neontological-paleontological interface of the community evolution: How do the pieces in the kaleidoscopic biosphere move? *in* Raup, D. M., and Jablonski, D., eds., Patterns and processes in the history of life: Berlin, Springer, Dahlem Konferenzen, Life Sciences Research Report 35, p. 331–350.

Meldahl, K. H., and Flessa, K. W., 1990, Taphonomic pathways and comparative biofacies and taphofacies in a Recent intertidal/shallow shelf environment: Lethaia, v. 23, p. 43–60.

Miller, A. I., 1988, Spatial resolution in subfossil molluscan remains: Implications for paleobiological analyses: Paleobiology, v. 14, p. 19–103.

Miller, A. I., 1989, Lateral mixing and spatial resolutions in molluscan assemblages of Smuggler's Cover, St. Croix, U.S. V.I., *in* Hubbard, D. K., ed., Terrestrial and marine geology of St. Croix, U.S. Virgin Islands: West Indies Laboratory, St. Croix, Special Publication 8, p. 129–134.

Miller, A. I., and Cummins, H., 1990, A numerical model for the formation of fossil assemblages: Estimating the amount of post-mortem transportation along environmental gradients: Palaios, v. 5, p. 303–316.

Mitchell, S. W., 1987, Sedimentology of Pigeon Creek, San Salvador Island, Bahamas, *in* Curran, H. A., ed., Proceedings of the Third Symposium on the Geology of the Bahamas: Fort Lauderdale, Florida, CCFL Bahamian Field Station, p. 215–230.

Morris, P. A., 1975, A field guide to shells: Atlantic and Gulf Coast and the West Indies: Boston, Houghton Mifflin Co., 330 p.

Noble, R. S., Curran, H. A., and Wilson, M. A., 1991, Paleoenvironmental and paleoecological analysis of a Pleistocene lagoonal, mollusk-rich facies, San Salvador Island, Bahamas: Geological Society of America Abstracts with Programs, v. 23, no. 1, p. 109.

Parsons, K. M., 1989, Taphonomy as an indicator of environment: Smuggler's Cove, St. Croix, U.S. V.I., *in* Hubbard, D. K., ed., Terrestrial and marine geology of St. Croix, U.S. Virgin Islands: West Indies Laboratory, St. Croix, Special Publication 8, p. 135–143.

Peterson, P. H., 1976, Relative abundances of living and dead molluscs in two California lagoons: Lethaia, v. 9, p. 137–148.

Rehder, H. A., 1981, The Audobon Society field guide to North American seashells: New York, Alfred A. Knopf, 194 p.

Scott, R. W., 1978, Approaches to trophic analysis of paleocommunities: Lethaia, v. 7, p. 315–330.

Slone, G. B., 1990, Sedimentology and taphonomy of a Holocene carbonate lagoon: Pigeon Creek, San Salvador, Bahamas [M.S. thesis]: Oxford, Ohio, Miami University, 143 p.

Slone, G. B., Boardman, M. R., and Cummins, R. H., 1990, Molluscan skeletal associations as a function of benthic cover and environmental stress: Pigeon Creek, San Salvador, Bahamas, Abstracts and Program, Fifth Symposium on the Geology of the Bahamas: San Salvador, Bahamian Field Station, p. 18.

Stanley, S. M., 1970, Relation of shell form to life habits of the Bivalvia (Mollusca): Geological Society of America Memoir 125, 296 p.

Sterrer, W., 1986, Marine fauna and flora of Bermuda: A systematic guide to the identification of marine organisms: New York, John Wiley & Sons, 742 p.

Teeter, J. W., 1989, Pigeon Creek lagoon, a modern analogue of the Pleistocene Granny Lake Basin, *in* Curran, H. A., ed., Pleistocene and Holocene carbonate environments on San Salvador Island, Bahamas, *in* International Geological Congress, 28th, Field Trip Guidebook T175: Washington, D.C., American Geophysical Union, p. 43–46.

Tevesz, M. J. S., and McCall, P. L., 1983, Biotic interaction in recent and fossil benthic communities: New York, Plenum Press, 837 p.

Wanless, H. R., 1981, Fining-upwards sedimentary sequences generated in seagrass beds: Journal of sedimentary Petrology, v. 51, p 445–454.

White, B., Kurkjy, K., and Curran, H. A., 1984, A shallowing-upward sequence in a Pleistocene coral reef and associated facies, San Salvador, Bahamas, *in* Teeter, J. W., ed., Proceedings of the Second Symposium on the Geology of the Bahamas: San Salvador, CCFL Bahamian Field Station, p. 53–70.

Williams, A. B., 1993, Mud shrimps, Upogebiidae, from the western Atlantic (Crustacea: Decapoda: Thalassinidea): Smithsonian Contributions to Zoology, no. 544, 77 p.

MANUSCRIPT ACCEPTED BY THE SOCIETY JANUARY 5, 1995

Printed in U.S.A.

Geological Society of America
Special Paper 300
1995

New data on the Holocene stratigraphy of Lee Stocking Island (Bahamas) and its relation to sea-level history

Pascal Kindler
Département de Géologie et de Paléontologie, University of Geneva, 13 rue des Maraîchers, 1211 Geneva 4, Switzerland

ABSTRACT

The Holocene deposits exposed on Lee Stocking and other Bahamian islands display a record of two distinctive units that contrasts strongly with the gradual rise of sea level expressed on most regional eustatic curves. This chapter reviews the field criteria for separating the Holocene and Pleistocene units in the Bahamas, presents the results of a detailed sedimentologic and petrographic study of the Holocene succession on Lee Stocking, and proposes a five-phase model that reconciles the observed pattern of episodic sedimentation with the continuous trend of sea-level rise.

The Holocene deposits of Lee Stocking include two newly named lithostratigraphic units: the Dune Pass Bay oolite, a 5,000-yr-old eolianite formed when sea level was lower than today, and the Perry Peak limestone, a younger bioclastic calcarenite deposited in beach and eolian settings when sea level was close to its present position. These units are similar to coeval deposits previously described on San Salvador and are also found on Cat and Eleuthera. On Lee Stocking, the two Holocene units appear in vertical succession and are separated by a pronounced discontinuity. However, regional sea-level curves do not show a major fall that could explain this sedimentation break.

The five-phase model presented in this chapter suggests that changes in the *rate* of sea-level rise ultimately controlled sedimentation during the middle and late Holocene in the Bahamas. The early flooding history of the platform was characterized by shallow waters that favored production and accumulation of ooids on the bank margins. Due to the rapid rate of sea-level rise during the mid-Holocene, ooid production ceased and the newly formed coastal deposits were subjected to erosion or pedogenesis. When the transgression slowed down about 3,800 yr ago, reefs began to catch up with sea level and generate bioclasts that created new shoreline features bankward of the first deposits. Today, reefs have caught up with sea level, reducing the energy from open ocean waters. Sedimentation on most Bahamian islands has again decreased, as shown by the absence of high unvegetated dunes.

This example from the Bahamas shows that discontinuities within carbonate sequences must not always be interpreted in terms of relative sea-level fall but can also result from changes in the rate of a transgression.

INTRODUCTION

The aims of this paper are to review the existing field criteria for differentiating the Pleistocene and Holocene rocks exposed on the Bahamas islands; to describe the Holocene stratigraphic record of Lee Stocking (southern Exumas); and to show that, on emerged bank margins, several sedimentation pulses, separated by periods of quiescence, may occur during a short portion of a marine transgression.

GEOLOGIC SETTING

Lee Stocking is a small (5.5 km long, up to 1.5 km wide) island located at the southeastern end of the Exuma chain, on the eastern margin of the Great Bahama Bank (Fig. 1). Modern

Kindler, P., 1995, New data on the Holocene stratigraphy of Lee Stocking Island (Bahamas) and its relation to sea-level history, *in* Curran, H. A., and White, B., Terrestrial and Shallow Marine Geology of the Bahamas and Bermuda: Boulder, Colorado, Geological Society of America Special Paper 300.

carbonate environments surrounding the island have been well studied (e.g., Kendall et al., 1990; Boardman and Carney, 1991; Dill, 1991), in particular the stromatolite fields found at the bottom of several neighboring tidal channels (Dill et al., 1986, 1989; Dill, 1991). In contrast, the surficial geology of the island itself is relatively unknown. Dill et al. (1989) first pointed out that Lee Stocking is formed of both Holocene and Pleistocene eolian dunes, but until very recently, confusion existed regarding the precise extension of the Holocene deposits and their distinction from Pleistocene limestones. Radiocarbon dating on a whole-rock sample I collected from a unit considered as Pleistocene (Kendall et al., 1990, p. 59–60) yielded an uncorrected age of 5,170 ± 60 B.P. This result showed that the surficial geology of Lee Stocking needed further investigation, and that the field criteria for correctly identifying the Holocene and Pleistocene units exposed on the Bahamas islands needed to be reevaluated.

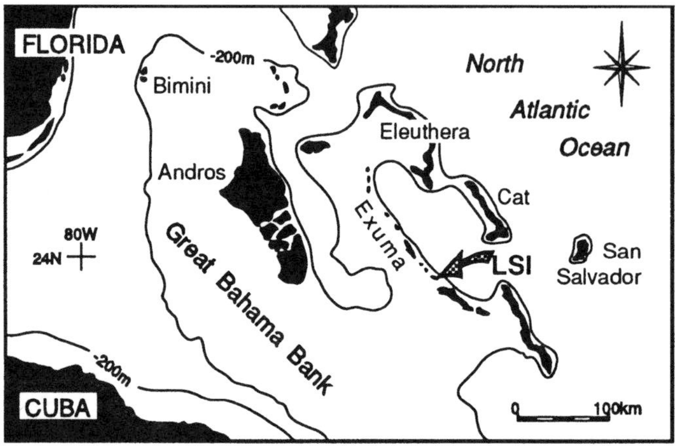

Figure 1. Location of study area. LSI = Lee Stocking Island.

METHODS

Fieldwork

Twenty days of field work were spent constructing a geologic map of the island (Fig. 2). The chart, published by the Bahamas government, scale of 1:25,000, was used as a topographic framework. Locality names follow Kendall et al. (1990). Field determinations of Pleistocene and Holocene deposits were based on six criteria, commonly used by geologists working in the Bahamas: (1) morphostratigraphic relationships, (2) position of the studied unit relative to paleosols, (3) degree of lithification, (4) presence of rhizoliths, (5) petrographic composition, and (6) relation between the depositional environment of the studied unit and the present sea-level position. These criteria are briefly reviewed in the following sections.

Morphostratigraphy. Basic morphostratigraphy has been successfully applied to unravel the succession of Quaternary deposits in a few Bahamas islands (Garrett and Gould, 1984; Hearty and Kindler, 1993). This method is based on the principles of "lateral accretion" and "catenary growth." The former concept (Vacher, 1973) states that, on a prograding shoreline,

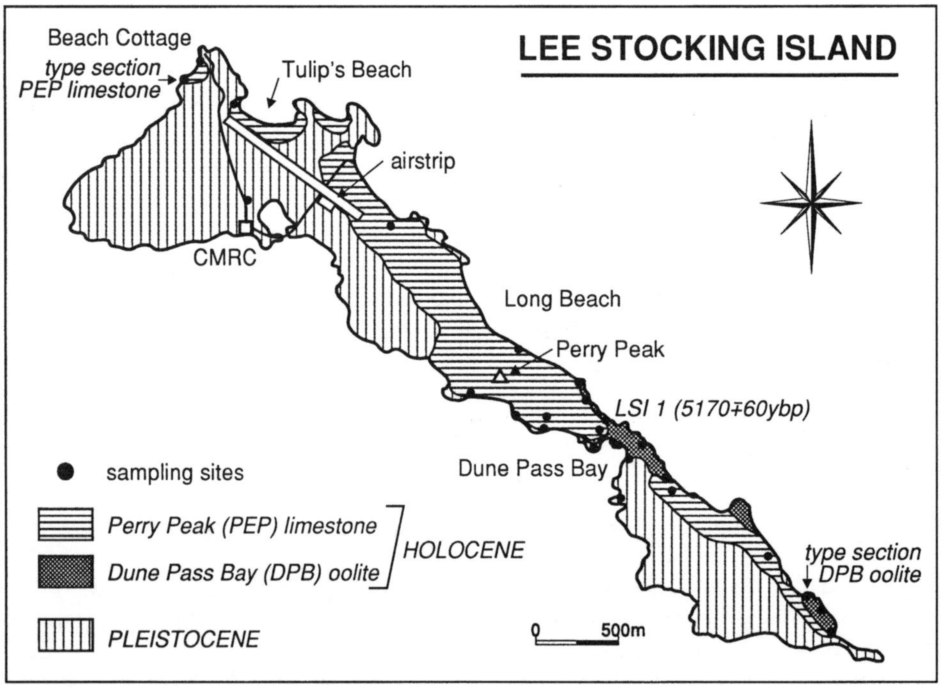

Figure 2. Geologic map of Lee Stocking Island, Bahamas. CMRC = Caribbean Marine Research Center.

deposits get younger seaward; the latter (Garrett and Gould, 1984), which is an exception to the first one, declares that catenary, or suspended, ridges are younger than their anchoring headlands. These principles are useful for giving a relative age to landforms, but they do not always allow identification of Pleistocene and Holocene deposits. In the case of the headland–catenary beach relationship, both deposits can be of Pleistocene or of Holocene age (Fig. 2).

Relative position to paleosols. Previous authors (e.g., Carew and Mylroie, 1985; Wilber, 1987; Mitchell, 1987) have observed that Pleistocene limestones are commonly covered by a laminated calcrete and/or a paleosol, whereas Holocene rocks lack such features. However, in coastal areas, paleosols have usually been stripped by marine erosion from the top of Pleistocene rock bodies. In contrast, pedogenic layers may locally separate Holocene limestone units (Kindler, 1992; Hearty and Kindler, 1993; and herein) and further occur between and within Pleistocene deposits in many places (Carew and Mylroie, 1991; Kindler and Hearty, 1995). This criterion can thus be used in many instances, but it is not infallible.

Degree of lithification. Some workers (Hutto and Carew, 1984; Kendall et al., 1990) refer to Pleistocene rocks as forming the "well-cemented core" of Bahamian islands and to Holocene deposits as being "poorly consolidated." However, at some locations (e.g., The Bluff, San Salvador: Beier, 1987), the former remain uncemented due to protection by an aquifuge. Conversely, the latter, such as the 5,000-yr-old North Point Member (Carew and Mylroie, 1985; White and Curran, 1988; Hearty and Kindler, 1993) are, in many places, more indurated than older Pleistocene limestones. Wilber (1987) rightfully noted that coastal outcrops, whatever their age, are commonly more lithified than coeval inland exposures, due to intergranular precipitation of aragonite cements in the intertidal and spray zones. The degree of lithification is therefore an ambiguous criterion for distinguishing Pleistocene and Holocene units on the Bahamas islands.

Presence of rhizoliths. Ward (1975) and Ward et al. (1985) reported that the Late Pleistocene eolianites of the Yucatan Peninsula contain abundant rhizoliths showing a hard micritic core, whereas Holocene limestones lack these characteristic features. Several weeks of field work, as well as fruitful discussions with many colleagues, convinced me that the nature of rhizocretions can also be used to differentiate Pleistocene from Holocene deposits in the Bahamas Archipelago. Root molds and tubules (Klappa, 1980) can be found within Holocene units, but the occurrence of root casts (sediment or cement-filled root molds) is restricted to Pleistocene limestones.

Petrographic composition. According to several studies (Illing, 1954; Beach and Ginsburg, 1980; Schlager and Ginsburg, 1981; Hutto and Carew, 1984; Carew and Mylroie, 1985; Kendall et al., 1990), ooids and peloids predominate within Pleistocene limestones, whereas bioclasts, particularly pink fragments of the benthic foraminifer *Homotrema rubrum*, characterize Holocene deposits. However, more recent work (White and White, 1991; Hearty and Kindler, 1993; Kindler and Hearty, 1995) shows that Holocene oolites as well as skeletal calcarenites of Pleistocene age are common in the Bahamas. Ongoing research (Kindler and Hearty, 1993) further demonstrates that the composition of the limestones exposed on the Bahamas islands is controlled by the degree of platform flooding, and thus ultimately by regional climatic and eustatic factors. Qualitative petrographic analysis is therefore of little help in differentiating Pleistocene from Holocene deposits in the field.

Depositional environment and sea level. Published Holocene sea-level curves for the Caribbean region (Scholl and Stuiver, 1967; Scholl et al., 1969; Wanless, 1982; Digerfeldt and Hendry, 1987; Boardman et al., 1989; Fairbanks, 1989; Lidz and Shinn, 1991; Pirazzoli, 1991) all indicate that sea level never exceeded the modern datum during this time interval. Raised marine deposits, such as those forming the Clifton Pier section on New Providence (Ball, 1967; Neumann and Moore, 1975; Aurell et al., 1995), can thus be attributed to the Pleistocene, more precisely to the Sangamonian interglacial, when sea level was globally higher than today (e.g., Chappell and Shackleton, 1986; Muhs, 1992; Selivanov, 1992). However, the recent discovery of late Holocene backshore strata at +7 m on the ocean-facing shoreline of Lee Stocking (White and Curran, 1993) shows that this last criterion, as those previously mentioned, should be used with caution.

Laboratory work

Collected samples (see Fig. 2 for location) were impregnated with blue epoxy resin, thin-sectioned, and studied qualitatively and quantitatively under a polarizing microscope. Point-counting was performed according to the method developed by Chayes (1956) and revised by Halley (1978), Harrel (1981), and Flügel (1982). Two 250-grain counts were made on each thin-section to obtain the relative percentages of grains (ooids, peloids, bioclasts, and miscellaneous), cements, and primary and secondary porosity (Table 1). Selected samples were x-rayed in a Philips-Norelco XRG 3000 x-ray diffractometer using Ni-filtered Cu radiation at the University of Miami. Goniometer scans were conducted from 31° to 25° (2Θ) at a speed of 1° per min to identify the main carbonate minerals. Finally, one key sample (LSI 1) was sent to Beta Analytic Laboratories (Miami, Florida) for whole rock ^{14}C dating.

RESULTS

The Holocene deposits of Lee Stocking include two units and form the highest elevations (more than 40 m at Perry Peak) and the broadest ridges on the island. These units overlie a complex Pleistocene substrate that is briefly addressed in the following section.

Pleistocene substrate

Pleistocene rocks form promontories along the ocean-facing shoreline of Lee Stocking (e.g., at both ends of Tulip's

TABLE 1. PETROGRAPHIC DATA BASE USED IN THIS CHAPTER

No. (LSI)	Grains (%)	Cement (%)	Porosity (%)	Bioclasts (%)	Ooids (%)	Peloids (%)	Misc. (%)
			Perry Peak Limestone (Beach Facies)				
3	69.7	4.1	26.2	68.3	55.5	16.7	9.5
6	64.8	4.8	30.4	65.3	14.7	7.8	12.2
7	65.8	7.0	27.2	56.9	9.5	18.1	15.5
12	59.3	9.8	30.9	58.4	4.4	24.8	12.4
14	52.5	9.0	38.5	58.8	6.2	25.7	9.3
19	66.3	6.8	26.9	56.6	9.3	20.2	13.9
Total	63.1	6.9	30.0	60.7	8.3	18.9	21.1
			Perry Peak Limestone (Washover Facies)				
37	63.5	8.8	17.7	46.0	40.2	7.9	5.9
51	84.0	5.0	31.0	48.4	34.5	6.3	10.8
Total	63.8	6.9	29.4	47.2	37.4	7.1	8.4
			Perry Peak Limestone (Eolian Facies)				
2	67.6	0.7	31.7	52.4	4.3	28.0	15.3
4	67.6	1.1	31.3	80.5	3.2	13.1	3.2
8	67.9	0.7	32.4	75.7	2.9	12.0	9.4
13	66.8	8.3	38.5	53.7	4.1	31.5	10.7
21	64.5	4.5	31.0	54.9	10.3	21.0	13.8
30	52.8	1.7	45.5	64.3	20.0	8.5	7.2
35	58.8	3.5	38.7	65.6	16.8	11.0	6.6
39	68.6	2.5	28.9	42.0	26.5	20.4	11.1
45	76.8	16.0	7.2	39.0	36.1	10.3	14.7
48	61.3	2.0	36.7	69.7	2.1	11.5	16.7
Total	65.1	4.1	32.2	59.8	12.6	16.7	10.9
			Dune Pass Bay Oolite				
1	68.0	12.4	19.6	10.0	80.3	5.1	4.6
1B	66.5	9.0	24.5	6.3	68.0	20.2	5.5
23	61.5	19.0	19.5	2.1	77.9	18.5	1.5
29	70.5	11.1	18.4	6.9	66.6	15.5	11.1
31	64.4	22.4	13.2	13.3	69.2	10.2	7.3
33	61.5	20.5	18.0	13.5	70.8	12.2	3.5
34	65.0	12.5	22.5	17.9	51.7	15.4	15.0
36	67.8	11.7	20.5	13.0	55.9	15.9	15.2
38	67.9	14.8	17.3	4.4	73.4	19.2	3.0
41	65.0	16.0	19.0	5.0	80.1	10.0	4.9
46	61.3	15.8	22.9	15.9	59.3	13.6	11.2
48	58.8	20.0	21.2	10.5	67.0	16.1	6.4
50	74.5	16.5	9.0	10.3	62.0	20.3	7.4
52	63.5	16.3	20.2	8.8	73.7	11.3	6.2
Total	65.4	15.6	19.0	9.9	68.3	14.5	7.3
			Pleistocene Oolite				
5	64.7	25.3	10.0	8.3	31.3	42.1	18.3
11	70.3	29.0	0.7	1.9	19.1	54.1	24.9
24	70.3	21.8	7.9	3.6	54.0	16.5	25.9
43	64.3	19.0	16.7	8.6	66.3	16.1	9.0
Total	67.4	23.8	8.8	5.6	42.7	32.2	19.5
			Pleistocene Bioclastic Calcarenite				
22	60.5	0.3	36.5	96.8	1.0	2.2	0.0
42	52.5	2.8	44.7	80.0	4.4	11.9	3.4
Total	56.5	1.6	40.6	88.4	2.7	7.1	1.7

beach, Fig. 2), but they essentially crop out on the leeward side of the island. In this area, these rocks constitute heavily vegetated low-elevation hills that display a karstified surface covered by a patchy laminated calcrete. Preliminary investigation shows that the Pleistocene bedrock of Lee Stocking comprises at least two units. The first one can be observed along the road leading from the Caribbean Marine Research Center (CMRC) to Beach Cottage and also on the headland to the northwest of Dune Pass Bay. It is composed of a well-lithified, well-sorted, oolitic-peloidal grainstone (Fig. 3B). General morphology, rock texture, and the presence of root molds characterize this first unit as an eolianite. The second Pleistocene unit is well exposed along the bank-facing coast of the island, to the southeast of Dune Pass Bay. It is made of friable bioclastic calcarenites (Fig. 3A) that contain abundant rhizoliths and display steep foresets dipping below sea level. These features, as well as widespread freshwater vadose cements (Fig. 3A), show that this second unit accumulated in an eolian environment when sea level was lower than today.

Holocene units

Two lithostratigraphic units of Holocene age are found on Lee Stocking Island. Both present a distinctive morphology, as well as typical sedimentologic and petrologic characteristics. They are informally designated herein as Dune Pass Bay oolite (the older one) and Perry Peak limestone (the younger one).

Dune Pass Bay oolite. The Dune Pass Bay (DPB) oolite derives its name from a marked embayment on the leeward coast of Lee Stocking (Fig. 2). It was hitherto regarded as Pleistocene (Kendall et al., 1990). However, radiometric dating on a whole-rock sample (see Fig. 2 for location) yielded an uncorrected age of 5,170 ± 60 B.P. This new result clearly places the DPB oolite in the Holocene.

This unit can be best observed along the windward shoreline of the island to the south of Dune Pass Bay. At the type section (Figs. 2 and 4A), it overlies a well-lithified, calcrete-capped oolitic limestone of Pleistocene age, but it is not covered by younger deposits. In contrast, to the north of Dune Pass Bay, it underlies the eolian facies of the Perry Peak limestone (Fig. 4D). Two calcarenite protosols (Vacher and Hearty, 1989; Carew and Mylroie, 1991; Fig. 4A) and a brown sandy paleosol (see Kindler, 1992, Fig. 5C) have been observed respectively within and on the top of the DPB oolite. The former are indicative of episodic deposition, whereas the latter reveals a major break in sedimentation following the formation of this unit.

The DPB oolite commonly forms 6- to 8-m-high, lobate rock bodies that are reminiscent of the parabolic dune type of McKee and Ward (1983). These dunes locally merge by overlapping of adjacent flanks to form linear ridges. They are composed of a moderately lithified, well-sorted calcarenite that displays pristine small-scale eolian lamination produced by grainfall, sandflow, and the migration of climbing wind ripples (Hunter, 1977; White and Curran, 1988) (Fig. 4B). These deposits contain

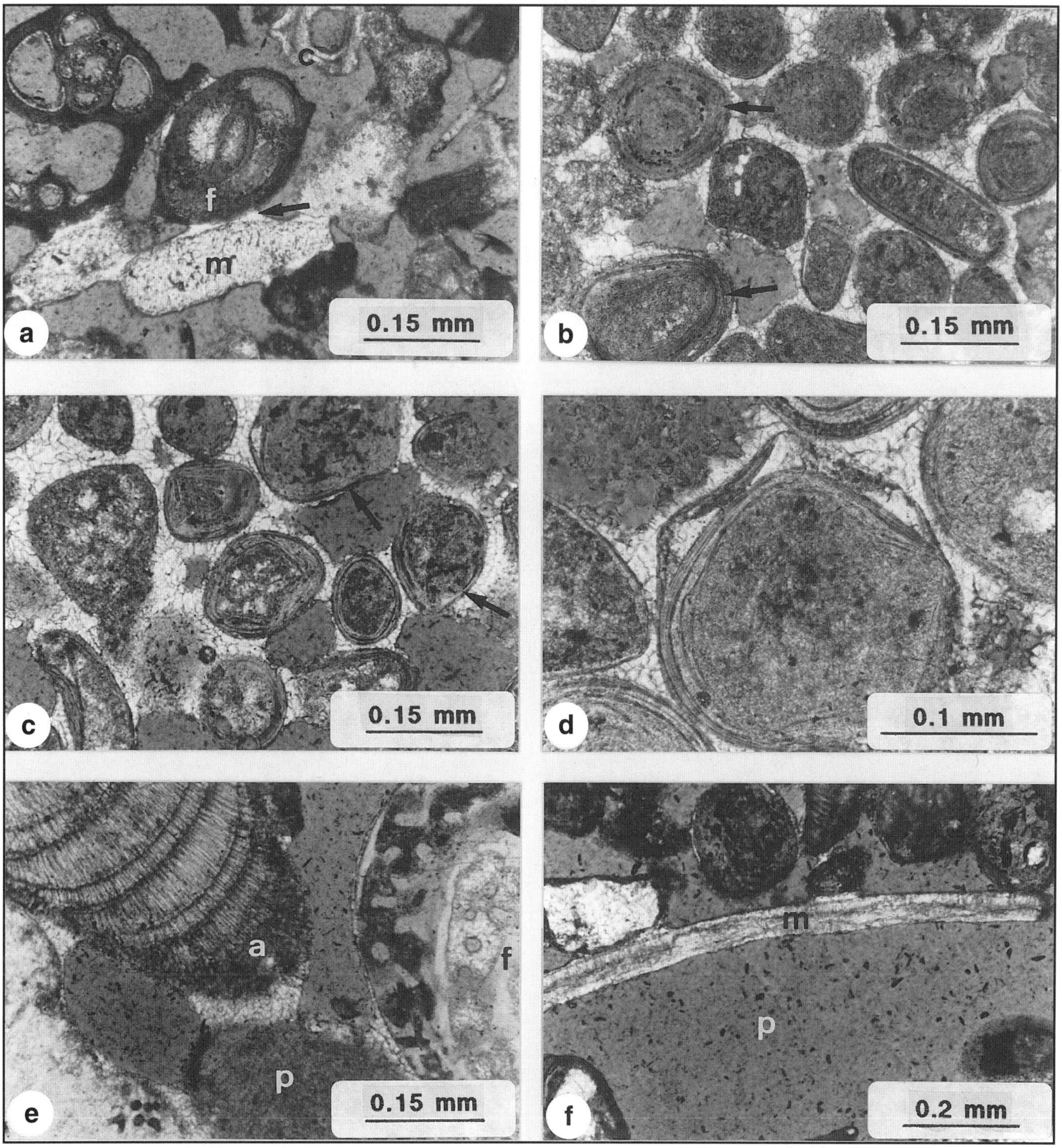

Figure 3**a.** Sample LSI 22, headland located to the southeast of Dune Pass Bay. Poorly cemented calcarenites of Late Pleistocene age essentially contain bioclastic fragments such as mollusk (m) and coral (c) debris and benthic foraminifers (f). Arrow points to LMC meniscus cement. **b,** Sample LSI 43, headland located to the northwest of Dune Pass Bay. Late Pleistocene oolites are characterized by numerous thickly coated ooids (arrows). **c,** Sample LSI 1; see Figure 2 for location. The DPB oolite differs from Late Pleistocene oolites by the predominance of superficial (i.e., thinly coated) ooids (arrows). **d,** Sample LSI 41, Dune Pass Bay. Spiny ooid (Davaud et al., 1990) resulting from precipitation of blocky cement within cortical dissolution voids. **e,** Sample LSI 2, Airport Ridge. PEP limestone contains pristine bioclastic fragments such as red algae (a) and foraminifer debris (f), as well as peloids (p). **f,** Sample LSI 35, Long Beach. Shelter porosity (p) caused by mollusc fragment (m) in PEP limestone. Note scarcity of cement compared to DPB oolite (Fig. 3c).

Figure 4**a.** Type section of the Dune Pass Bay oolite at the southern end of Lee Stocking. Note the calcarenite protosols (p) and the eolian foresets (f) that dip below sea level. Backpack (56 cm long, at arrow) for scale. **b,** Small-scale eolian stratifications within the DPB oolite. s = sandflow lamination; c = climbing wind-ripple lamination. Hammer is 33 cm long. **c,** Type section of the Perry Peak limestone near Beach Cottage. p = Pleistocene substrate; f = foreshore (beach) deposits; b = backshore deposits. Hammer for scale. **d,** Sharp contact (arrow) between the DPB oolite (1) and the PEP limestone (2). Small-scale eolian features are well preserved within the older unit, whereas they have been partly obliterated by weathering and bioerosion within the younger one. **e,** Cliffs of PEP limestone at the southern end of Long Beach. Landward-dipping planar beds are interpreted as washover deposits. Backpack for scale. **f,** Large-scale eolian stratifications within the PEP limestone; cliffs on the bankward shoreline of Lee Stocking to the North of Dune Pass Bay. f = foreset; t = topset. Hammer (at arrow) for scale.

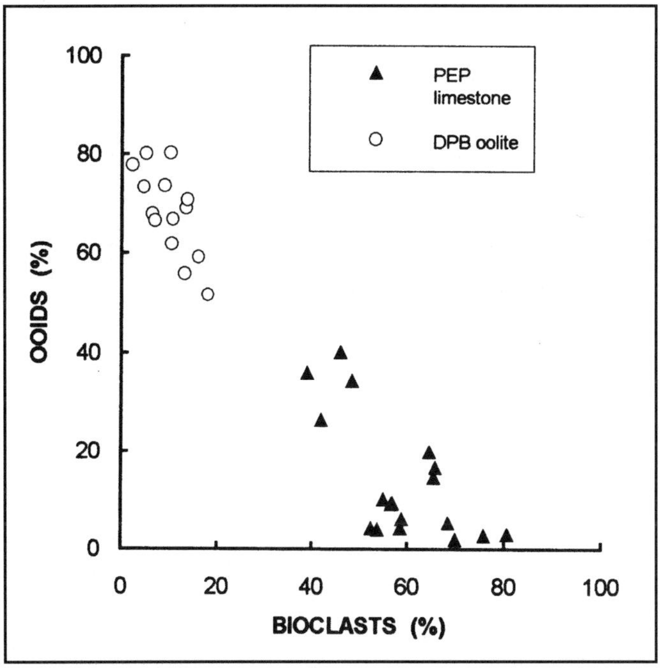

Figure 5. Ooids versus bioclasts ratios of samples collected on Lee Stocking Island. DPB oolite (open circles) and PEP limestone (full triangles) form two distinct petrographic groups. Intermediate position of some PEP samples corresponds to reworking of DPB oolite.

a trace-fossil assemblage including root molds, rare *Skolithos* burrows, and a complex form known as "cluster-burrow" (Curran and White, 1987, 1991; White and Curran, 1988), apparently created by the escape of hatchling digger wasps.

The DPB limestones predominantly contain peloids and superficial ooids (Illing, 1954) (Table 1; Figs. 3C and 5) that present concentric dissolution bands within cortices. Bioclasts make up less than 15% of the rock constituents. The grains are bound by blocky low-Mg calcite (LMC) occurring at grain contacts or filling interstitial pores. The cement textures and mineralogy characterize the freshwater vadose environment (Longman, 1980). These cements probably originate from the selective dissolution of ooid cortices by meteoric waters (Loucks and Kim, 1990). Locally, ooid shape has been altered by precipitation of blocky cement within cortical dissolution voids (Fig. 3D). In these cases, the coated particle shows an irregular shape that evokes "spiny ooids" (Davaud et al., 1990).

No marine facies seems to be associated with the DPB eolianites. Dune foresets always dip below the present sea level (Fig. 4A), indicating that deposition occurred at a lower stand than that of today. The age of 5,170 ± 60 B.P. obtained by radiocarbon dating is coherent with this assumption.

Perry Peak limestone. The name of this unit, Perry Peak (PEP) limestone, originates from the highest (43 m) elevation on Lee Stocking (Fig. 2). These rocks constitute a 1.5-km-long, 40-m-high hill in the central portion of the island, a lower and less extensive ridge along the southeastern, windward shoreline, and modest ridges (up to 2 m high) along the ocean-facing bays of the north coast. They overlie either the calcrete-capped Pleistocene substrate, as at the type-section near Beach Cottage (Fig. 4C), or the DPB oolite. The limit between the two Holocene units usually corresponds to a discrete erosional surface (Fig. 4D); however, at a few places, it can be marked by a brown sandy paleosol (Kindler, 1992, Fig. 5C). The PEP rocks are covered by, or juxtaposed to, modern sand not discussed in this study.

The bulk of the PEP limestone accumulated in an eolian setting, as shown by its overall morphology, large-scale eolian stratifications (Fig. 4F), and the occasional presence of palm tree molds and palm frond imprints (White and Curran, 1993). Exposures near sea level commonly display planar beds dipping slightly seaward (Kindler, 1992, Fig. 5B), coarse shell layers, irregular fenestrae (Tucker and Wright, 1990), and good specimens of ghost crab burrows (*Psilonichnus upsilon*) (Curran and White, 1991; White and Curran, 1993). All these features indicate deposition in a beach environment. At one location, along the 8-m-high sea cliff forming the southeastern end of Long Beach (Fig. 4E), the PEP limestone exhibits moderately steep (25°), landward-dipping foreset beds that can be interpreted as washover deposits.

The PEP limestone consists of yellowish calcarenites that predominantly contain mollusk, coral and echinoid fragments, benthic foraminifers (miliolids, soritids), red and green algae debris (Table 1; Fig. 3E, F, and 5). At the base of the unit, ooids reworked from the DPB oolite may form as much as 30% of the constituent grains. The PEP calcarenites are cemented by LMC equant spar that is confined to grain contacts in the eolianites (Fig. 3E), and that also fills intergranular voids in the beach facies. Cement mineralogy shows that diagenesis of these rocks took place in a meteoric environment (Longman, 1980). Although it is found in the modern intertidal zone, the beach facies of the PEP limestone cannot be considered as "true" beachrock, because it lacks the fibrous and micritic aragonite and high-Mg calcite cements that characterize this type of deposit (Ginsburg, 1953; Halley and Harris, 1979; Tucker and Wright, 1990). Instead, it was most likely cemented in a freshwater setting during a phase of shoreline progradation and later exhumed by marine erosion (Davaud and Strasser, 1984; Strasser and Davaud, 1986).

The PEP limestone was not dated radiometrically; however, geomorphic and stratigraphic evidence clearly indicates that it is younger than the DPB oolite. On the southeastern shoreline of Lee Stocking, the PEP limestone displays several systems of catenary beaches and dunes that are anchored on DPB headlands (Fig. 2), and are hence younger. Along the same coast but farther to the Northwest, between Dune Pass Bay and the southern end of Long Beach, both Holocene units appear in a vertical succession. The younger deposits fill the relict dune and swale topography of the older (Kindler, 1992, Fig. 5C). The elevation of the beach facies of the PEP limestone is conformable with the present stand of sea level (Kindler, 1992, Fig. 4B). Therefore, deposition must have occurred when sea level was at or close to its present position. Sea-level studies in the Caribbean region (Scholl and Stuiver, 1967; Scholl et al., 1969; Wanless, 1982; Digerfeldt and Hendry, 1987; Boardman et al., 1989; Fairbanks, 1989; Lidz and Shinn, 1991; Pirazzoli, 1991) all indicate that sea level reached an elevation close to its modern stand about 3,500 yr ago. This date appears suitable as an age for the PEP limestone until further radiometric dating can be performed.

DISCUSSION

Comparison with other Bahamian islands

Pleistocene substrate. More work is clearly needed to resolve the Pleistocene stratigraphy of Lee Stocking and the aforementioned observations are intended to stimulate further research. The oolitic-peloidal unit visible near the CMRC presents striking petrographic similarities with the Sangamonian oolites found on San Salvador: French Bay Member (Carew and Mylroie, 1985; Hearty and Kindler, 1993) and Fernandez Bay Member (Hearty and Kindler, 1993). However, I did not observe, on Lee Stocking, the marine facies commonly associated with those units, although a Sangamonian reef is exposed on neighboring Norman's Pond Cay (Kendall et al., 1990; Halley et al., 1991). If these marine facies were deposited, they are probably covered by Holocene sediments. On the basis of its petrographic composition, the bioclastic eolianites of the second unit could be considered as an equivalent of the late Sangamonian Almgreen Cay Formation (Hearty and Kindler, 1993) or of the pre-Sangamonian Owl's Hole Formation (Carew and Mylroie, 1985; Hearty and Kindler, 1993). Their position relative to the platform edge supports the second hypothesis.

Holocene units. Holocene units comparable to those observed on Lee Stocking occur on San Salvador Island (Fig. 1). Similar to the DPB oolite, the North Point Member (Carew and Mylroie, 1985; White and Curran, 1989; Kindler, 1992; Hearty and Kindler, 1993) consists of white, oolitic-peloidal limestones, forming small dunes that were built up when sea level was lower than today. Analogous to the PEP limestone, the Hanna Bay Member (Carew and Mylroie, 1985; Kindler 1992; Hearty and Kindler, 1993) is made of a yellowish, bioclastic calcarenite deposited in beach and eolian settings when sea level was close to its present position. Carew and Mylroie (1987) radiometrically dated the North Point Member at about 5,300 B.P., whereas the Hanna Bay Member yielded ages between 3,200 B.P. and 420 yr ago. The former value is comparable to the value measured on the DPB oolite, whereas the latter ones agree with the assumed younger age of the PEP limestone.

I have also observed undated rock units similar to the DPB oolite on Cat Island (Fernandez Bay area; Kindler, 1992) and at the southern tip of Eleuthera Island, and examined rock bodies analogous to the PEP on Cat (Greenwood Bay area; Kindler, 1992), Stocking Island (Exumas), Eleuthera, and Bimini (Davaud and Strasser, 1984). Finally, preliminary results on the Holocene stratigraphy of Little San Salvador and West Plana Cay (Wilber, 1987) are strikingly similar to those obtained in the present study.

In summary, this comparison shows that the Holocene units observed on Lee Stocking extend on other Bahamian islands. Consequently, they are not only related to local morphology or oceanic setting. They could also reflect regional changes of sea level.

Holocene sedimentation and sea level

The following five-phase model (Fig. 6) attempts to reconcile the apparent discrepancy between the punctuated Holocene record observed on the aforementioned islands and the continuous sea-level rise depicted on regional eustatic curves for this time interval (Scholl and Stuiver, 1967; Scholl et al., 1969; Wanless, 1982; Digerfeldt and Hendry, 1987; Boardman et al., 1989; Fairbanks, 1989; Lidz and Shinn, 1991; Pirazzoli, 1991). These curves are steep for the middle Holocene, revealing a rapid rate of sea-level rise. They flatten out towards present time, indicating a slowing down of the transgression.

Phase 1. About 8,000 yr ago, sea level may have been some 10 to 15 m below present MSL. Carbonate production was limited to small reefs scattered on the submerged outer shelf (shelf-edge reefs, Adey and Burke, 1976) and to low-energy environments on the platform. Bioclasts and fine-grained sediments formed during this period were apparently reworked as ooid nuclei in Phase 2. Pedogenesis and karstification were the only geologic processes at work on the future Bahamian islands.

Phase 2. Between 6,000 and 5,000 B.P., larger areas of the platform became flooded by a shallow and energetic layer of water. These hydrodynamic conditions were favorable to ooid production (Newell et al., 1960; Tucker and Wright, 1990) and shoal initiation, particularly on topographic highs (Harris, 1979). Oolitic shoreline features were built on an irregular late-Pleistocene topography at an elevation estimated at about 5 m below sea level. This sedimentation pulse is expressed by the DPB oolite on Lee Stocking and by the North Point Member on San Salvador. Only the eolian facies of these features are emergent today.

Phase 3. Prior to 3,800 B.P., sea level was rising rapidly and the water layer covering the shelf soon became too deep for ooid production to continue. According to several studies (Newell et al., 1960; Flügel, 1982; Tucker and Wright, 1990), oolitic accretion is quantitatively unimportant at depths 2 m below low tide. In some areas, (e.g., Lee Stocking; Kendall et al., 1990), the locus of ooid production shifted from the shelf margin to flood tidal deltas. Along the ocean-facing shorelines, sedimentation decreased and stopped shortly after the formation of the 5-ka deposits. Marine erosion and pedogenesis then became the dominant processes. Phase 2 unconsolidated beaches were washed away by the rising sea, whereas coeval

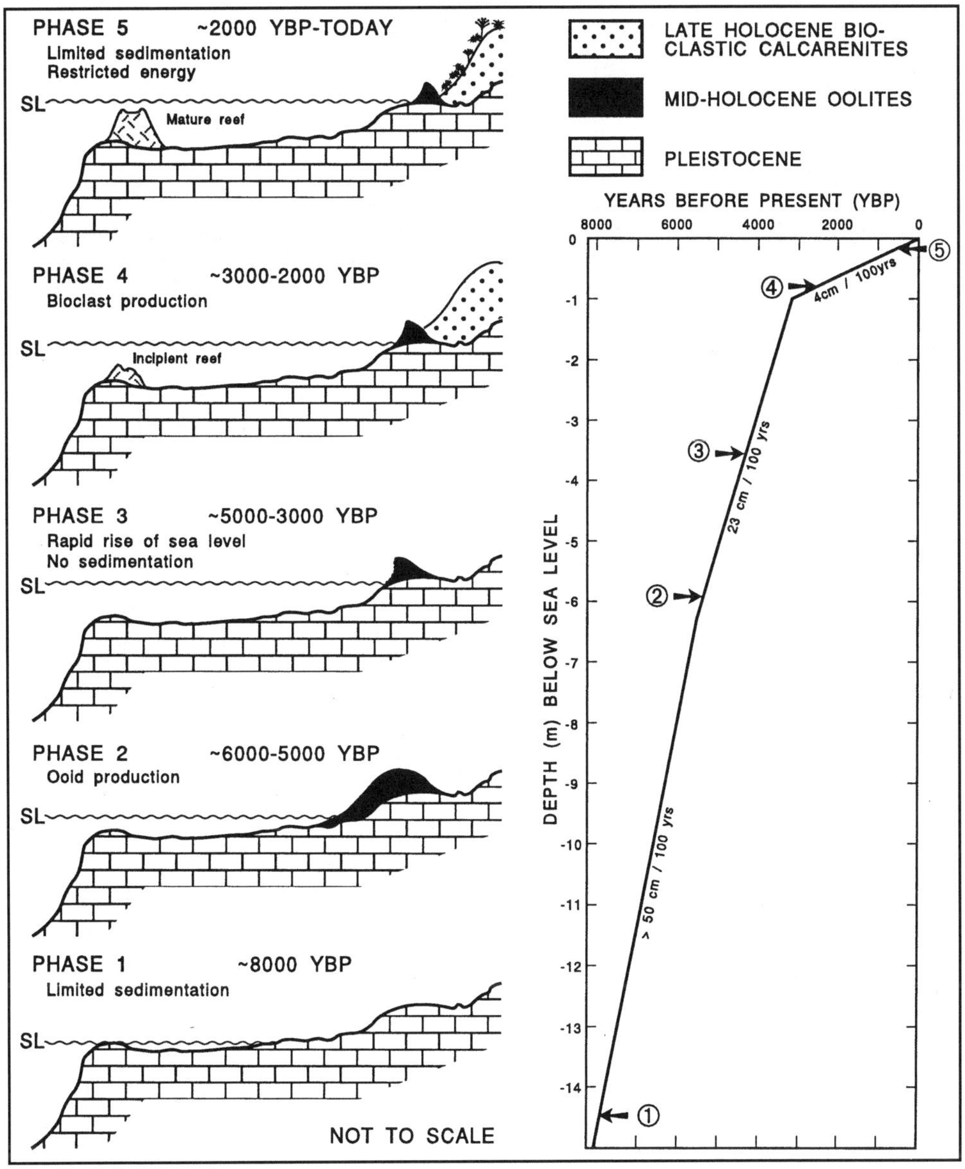

Figure 6. Proposed five-phase model that reconciles continuous sea-level rise and episodic sedimentation in the eastern Bahamas during the middle and late Holocene. Idealized bank margin may not always correspond to existing topography. Refer to text for detailed explanations.

beachrock subsisted as offshore mounds and hardgrounds (R. F. Dill, personal communication, 1992).

Phase 4. Around 3,800 B.P., the rate of sea-level rise abruptly slowed down from 23 cm/100 yr to 4 cm/100 yr (Fig. 6). This change was favorable to reef colonization and subsequent buildup (Neumann and MacIntyre, 1985). A large quantity of bioclastic sands was then produced and transported ashore. This second pulse of sedimentation is marked by the PEP limestone on Lee Stocking and by the Hanna Bay Member on San Salvador. The peculiar occurrence of elevated (+8 m) washover deposits at the southern end of Long Beach and the considerable elevation of the associated eolianites (up to 43 m on Lee Stocking) may be explained by the absence of well-developed offshore energy barriers (sand shoals, reefs) at or near the shelf edge. Interestingly, these elevated deposits are coeval with the numerous high-energy coral-bearing rubble storm benches found on several South Pacific islands (Curray et al., 1970).

Phase 5. During the past 2,000 yr, reefs have started to catch up with sea level, cutting off the available energy from the open ocean. Recently produced bioclasts either accumulate on tidal deltas and pocket beaches (R. F. Dill, personal communication, 1992), or are transported off the platform. Sedimentation on the islands is now limited, as confirmed by the absence of high unvegetated modern dunes. Similar reduction of energy levels in the past 2,000 yr has been documented on many

Caribbean Islands (Adey and Burke, 1976) where decaying algal bioherms, which normally thrive in agitated waters, are now found near shores behind barrier reefs.

Comparison of this model with the well-documented growth history of Joulters Cays (Western Bahamas; Harris, 1979) reveals that development of oolitic rock bodies occurred later in Joulters Cays than in the studied regions, and that it was not followed by extensive bioclastic sedimentation. These differences emphasize the importance of antecedent topography in the nature and repartition of subsequent sediments.

CONCLUSIONS

Two main conclusions can be drawn from the present study: one of local interest, one of a more general application.

At a local scale, this work contributes to a better understanding of the Holocene deposits of Lee Stocking Island. These consist of two rock bodies showing different petrographic and sedimentologic characteristics: (1) the Dune Pass Bay oolite, a 5,000-yr-old, oolitic-peloidal eolianite formed when sea level was lower than today, and (2) the Perry Peak limestone, a younger bioclastic calcarenite deposited in beach, washover, and dune environments when sea-level was at or close to its present elevation.

From a more global standpoint, this example from the Bahamas illustrates that sedimentation is not continuous on the subaerial regions of partly emerged carbonate platforms, but that it occurs in pulses controlled by local (e.g., paleotopography, hydrodynamism) and regional (e.g., rate of sea-level rise) factors. Rock bodies, separated by clear discontinuities, can thus be generated during a short period of the same transgression, without an intervening relative sea-level fall. This should be remembered when interpreting uncomformities within older carbonate sequences.

ACKNOWLEDGMENTS

I am very grateful to R. N. Ginsburg and R. F. Dill, who introduced me to the geology of Lee Stocking; to R. I. Wicklund and G. Wenz, director and assistant director, respectively, of the Caribbean Marine Research Center, for logistical support. Critical reviews by Dill and B. White greatly helped in improving the final draft of this manuscript. Special thanks to P. J. Hearty for profitable discussions; to the entire Lee Stocking crew for their friendly hospitality; to G. Brass for free use of an x-ray diffractometer; to J. Metzger for drafting part of the figures; to B. Ujetz for reviewing the English of the final version; and to Anick for her invaluable presence in the field. This study was undertaken during my tenure at the University of Miami, Florida and was supported by the National Science Foundation of Switzerland (grant 8220-028458/1, project 20-299179.0). Field expenses were partly covered by a grant from the Swiss Academy of Natural Sciences.

REFERENCES CITED

Adey, W. H., and Burke, R., 1976. Holocene bioherms (algal ridges and bank-barrier reefs) of the eastern Caribbean: Geological Society of America Bulletin, v. 87, p. 95–109.

Aurell, M., McNeill, D. F., Guyomard, T., and Kindler, P., Pleistocene shallowing-upward sequences in New Providence, Bahamas: Signature of high-frequency sea-level fluctuations in shallow carbonate platforms: Journal of Sedimentary Research, v. B65, p. 170–182.

Ball, M. M., 1967, Carbonate sand bodies of Florida and the Bahamas: Journal of Sedimentary Petrology, v. 37, p. 556–591.

Beach, D. K., and Ginsburg, R. N., 1980, Facies succession of Pliocene-Pleistocene carbonates, northwestern Great Bahama Bank: American Association of Petroleum Geologists Bulletin, v. 64, p. 1634–1642.

Beier, J. A., 1987, Petrographic and geochemical analysis of caliche profiles in a Bahamian Pleistocene dune: Sedimentology, v. 34, p. 991–998.

Boardman, M. R., and Carney, C., 1991, Origin and accumulation of lime mud in ooid tidal channels, Bahamas: Journal of Sedimentary Petrology, v. 61, p. 661–680.

Boardman, M., Neumann, A. C., and Rasmussen, K. A., 1989, Holocene sea level in the Bahamas, *in* Mylroie, J. E., ed, Proceedings, Fourth Symposium on the Geology of the Bahamas: Fort Lauderdale, Florida, Bahamian Field Station, p. 45–52.

Carew, J. L, and Mylroie, J. E., 1985, The Pleistocene and Holocene stratigraphy of San Salvador Island, Bahamas, with reference to marine and terrestrial lithofacies at French Bay, *in* Curran, H. A., ed., Pleistocene and Holocene Carbonate Environments on San Salvador Island, Bahamas, Geological Society of America, annual Meeting Field Trip Guidebook: Ft. Lauderdale, Florida, CCFL Bahamian Field Station, p. 11–61.

Carew, J. L., and Mylroie, J. E., 1987, A refined chronology for San Salvador Island, Bahamas, *in* Curran, H. A., ed., Proceedings, Third Symposium on the Geology of the Bahamas: Ft. Lauderdale, Florida, CCFL Bahamian Field Station, p. 35–44.

Carew, J. L., and Mylroie, J. E., 1991, Some pitfalls in paleosol interpretation in carbonate sequences: Carbonates and Evaporites, v. 6, p. 69–74.

Chappell, J., and Shackleton, N. J., 1986, Oxygen isotopes and sea level: Nature, v. 324, p. 137–140.

Chayes, F., 1956, Petrographic modal analysis: New York, Wiley, 113 p.

Curran, H. A., and White, B., 1987, Trace fossils in carbonate upper beach rocks and eolianites: recognition of backshore to dune transition, *in* Curran, H. A., ed., Proceedings, Third Symposium on the Geology of the Bahamas: Ft. Lauderdale, Florida, CCFL Bahamian Field Station, p. 243–254.

Curran, H. A., and White, B., 1991, Trace fossils of shallow subtidal to dunal ichnofacies in Bahamian Quaternary carbonates: Palaios, v. 6, p. 498–510.

Curray, J. R., Shepard, F. P., and Veeh, H. H., 1970, Late Quaternary sea-level studies in Micronesia: CARMARSEL expedition: Geological Society of America Bulletin, v. 81, p. 1865–1880.

Davaud, E., and Strasser, A., 1984, Progradation, cimentation, érosion: évolution sédimentaire et diagénétique récente d'un littoral carbonaté (Bimini, Bahamas): Eclogae Geologicae Helvetiae, v. 77, p. 449–468.

Davaud, E., Strasser, A., and Jedoui, Y., 1990, Spiny ooids: Early subaerial deformation as opposed to late burial compaction: Geology, v. 18, p. 816–819.

Digerfeldt, G., and Hendry, M. D., 1987, An 8,000 year Holocene sea-level record from Jamaica: Implications for interpretation of Caribbean reef and coastal history: Coral Reefs, v. 5, p. 165–169.

Dill, R. F., 1991, Subtidal stromatolites, ooids and crusted-lime muds at the Great Bahama Bank margin, *in* Osborne, R. H., ed., From shoreline to abyss: SEPM Special Publication 46, p. 147–171.

Dill, R. F., Shinn, E. A., Jones, A. T., Kelly, K., and Steinen, R. P., 1986, Giant subtidal stromatolites forming in normal salinity waters: Nature, v. 324, p. 55–58.

Dill, R. F., Kendall, C. G. St.C., and Shinn, E. A., 1989, Giant subtidal stromatolites and related sedimentary features, International Geological Congress, 28th, Field Trip Guidebook, T 373: Washington, D.C., American Geophysical Union, 33 p.

Fairbanks, R. G., 1989, A 17,000-year glacio-eustatic sea-level record: influence of glacial melting rates on the Younger Dryas event and deep-ocean circulation: Nature, v. 342, p. 637–642.

Flügel, E., 1982, Microfacies analysis of limestones: New York, Springer-Verlag, 633 p.

Garrett, P., and Gould, S. J., 1984, Geology of New Providence Island, Bahamas: Geological Society of America Bulletin, v. 95, p. 209–220.

Ginsburg, R. N., 1953, Beachrock in south Florida: Journal of Sedimentary Petrology, v. 23, p. 85-92.

Halley, R. B., 1978, Estimating pore and cement volumes in thin section: Journal of Sedimentary Petrology, v. 48, p. 642–650.

Halley, R. B., and Harris, P. M., 1979, Fresh-water cementation of a 1,000-year-old oolite: Journal of Sedimentary Petrology, v. 49, p. 969–988.

Halley, R. B., Muhs, D. R., Shinn, E. A., Dill, R. F., and Kindinger, J. L., 1991, A +1.5-m reef terrace in the southern Exuma Islands, Bahamas: Geological Society of America Abstracts with Programs, v. 23, no. 1, p. 40.

Harrel, J., 1981, Measurement errors in the thin-section analysis of grain packing: Journal of Sedimentary Petrology, v. 51, p. 674–676.

Harris, P. M., 1979, Facies anatomy and diagenesis of a Bahamian ooid shoal: Sedimenta (University of Miami) VII, 163 p.

Hearty, P. J., and Kindler, P., 1993, New perspectives on Bahamian geology: San Salvador Island, Bahamas: Journal of Coastal Research, v. 9, p. 577–594.

Hunter, R. E., 1977, Basic types of stratifications in small eolian dunes: Sedimentology, v. 24, p. 361–387.

Hutto, T., and Carew, J. L., 1984, Petrology of eolian calcarenites, San Salvador, Bahamas, *in* Teeter, J. W., ed., Proceedings, Second Symposium on the Geology of the Bahamas: San Salvador, CCFL Bahamian Field Station, p. 197–207.

Illing, L. V., 1954, Bahaman calcareous sands: American Association of Petroleum Geologists Bulletin, v. 38, p. 1-95.

Kendall, C. G. St.C., Dill, R. F., and Shinn, E. A., 1990, Guidebook to the marine geology and tropical environments of Lee Stocking Island, the southern Exumas, Bahamas: San Diego, California, KenDill Publishers, 82 p.

Kindler, P., 1992, Coastal response to the Holocene transgression in the Bahamas: Episodic sedimentation versus continuous sea-level rise: Sedimentary Geology, v. 80, p. 319–329.

Kindler, P., and Hearty, P. J., 1993, Sea-level control of limestone composition: New data from Quaternary carbonates in the Bahamas: Geological Society of America, Abstracts with Programs, v. 25, no. 4, p. 27.

Kindler, P., and Hearty, P. J., 1995, Pre-Sangamonian eolianites in the Bahamas? New evidence from Eleuthera Island: Marine Geology (in press).

Klappa, C. F., 1980, Rhizoliths in terrestrial carbonates: classification, recognition, genesis and significance: Sedimentology, v. 27, p. 613–629.

Lidz, B. H., and Shinn, E. A., 1991, Paleoshorelines, reefs, and a rising sea: South Florida, USA: Journal of Coastal Research, v. 7, p. 203–229.

Longman, M. W., 1980, Carbonate diagenetic textures from nearsurface diagenetic environments: American Association of Petroleum Geologists Bulletin, v. 64, p. 461–487.

Loucks, R. G., and Kim, P. A, 1990, Closed-system vadose diagenesis in the Holocene Cancun eolianite, Isla Cancun, Yucatan Peninsula, Mexico: American Association of Petroleum Geologists, Annual Meeting, San Francisco, California, Abstracts, p. 128.

McKee, E. D., and Ward, W. C., 1983, Eolian environment, *in* Scholle, P. A., Bebout, D. G., and Moore, C. H., eds., Carbonate depositional environments: American Association of Petroleum Geologists Memoir 33, p. 132–170.

Mitchell, S. W., 1987, Surficial geology of Rum Cay, Bahamas Islands, *in* Curran, H. A., ed., Proceedings, Third Symposium on the Geology of the Bahamas: Ft. Lauderdale, Florida, CCFL Bahamian Field Station, p. 231–241.

Muhs, D. R., 1992, The last interglacial-glacial transition in North America: Evidence from uranium-series dating of coastal deposits, *in* Clark, P. U., and Lea, P. D., eds., The Last Interglacial-Glacial Transition in North America: Boulder, Colorado, Geological Society of America Special Paper 270, p. 31–51.

Neumann, A. C., and MacIntyre, I., 1985, Reef response to sea-level rise: Keep-up, catch-up or give-up, *in* Proceedings, Fifth International Coral Reef Congress: Tahiti, v. 3, p. 105–110.

Neumann, A. C., and Moore, W. S., 1975, Sea level events and Pleistocene coral ages in the Northern Bahamas: Quaternary Research, v. 5, p. 215–224.

Newell, N. D., Purdy, E. G., and Imbrie, J., 1960, Bahamian oolitic sand: Journal of Geology, v. 68, p. 481–497.

Pirazzoli, P. A., 1991, World atlas of Holocene sea-level changes: Amsterdam, Elsevier, 300 p.

Schlager, W., and Ginsburg, R. N., 1981, Bahama carbonate platforms—the deep and the past: Marine Geology, v. 44, p. 1–24.

Scholl, D. W., and Stuiver, M., 1967, Recent submergence of southern Florida: A comparison with adjacent coasts and other eustatic data: Geological Society of America Bulletin, v. 78, p. 437–454.

Scholl, D. W., Craighead, F. C., Sr., and Stuiver, M., 1969, Florida submergence curve revised: Its relation to coastal sedimentation rates: Science, v. 163, p. 562–564.

Selivanov, A. O., 1992, Spatial-temporal analysis of large Pleistocene sea-level fluctuations: Marine terrace data: Journal of Coastal Research, v. 8, p. 408–418.

Strasser, A., and Davaud, E., 1986, Formation of Holocene limestone sequences by progradation, cementation, and erosion: Two examples from the Bahamas: Journal of Sedimentary Petrology, v. 56, p. 422–428.

Tucker, M. E., and Wright, V. P., 1990, Carbonate sedimentology: Oxford, United Kingdom, Blackwell, 482 p.

Vacher, H. L., 1973, Coastal dunes of younger Bermuda, *in* Coates, D. R., ed., Coastal Geomorphology: State University of New York, Binghamton, Publications in Geomorphology, p. 355–391.

Vacher, H. L., and Hearty, P. J., 1989, History of Stage 5 sea level in Bermuda: Review with new evidence of a brief rise to present sea level during Substage 5a: Quaternary Science Reviews, v. 8, p. 159–168.

Wanless, H. R., 1982, Sea level is rising, so what?: Journal of Sedimentary Petrology, v. 52, p. 1051–1054.

Ward, W. C., 1975, Petrology and diagenesis of carbonate eolianites of northeastern Yucatan Peninsula, Mexico, *in* Wantland, K. F., and Pusey, W. C., eds., Belize Shelf-Carbonate Sediments, Clastic Sediments, and Ecology: American Association of Petroleum Geologists Studies in Geology 2, p. 500–571.

Ward, W. C., Weidie, A. E., and Back, W., 1985, Geology and hydrogeology of the Yucatan and Quaternary geology of the northeastern Yucatan Peninsula: New Orleans, Louisiana, New Orleans Geological Society Publications, 160 p.

White, B., and Curran, H. A., 1988, Mesoscale physical sedimentary structures and trace fossils in Holocene carbonate eolianites from San Salvador Island, Bahamas: Sedimentary Geology, v. 55, p. 163–184.

White, B., and Curran, H. A., 1989, The Holocene carbonate eolianites of North Point and the modern environments between North Point and Cut Cay, San Salvador Island, Bahamas, *in* Curran, H. A., ed., Pleistocene and Holocene carbonate environments on San Salvador Island, Bahamas, International Geological Congress, 28th, Field Trip Guidebook, T 175: American Geophysical Union, Washington, D.C., p. 17–22.

White, B., and Curran, H. A., 1993, Sedimentology and ichnology of Holocene dune and backshore deposits, Lee Stocking Island, Bahamas, *in* White, B., ed., Proceedings, Sixth Symposium on the Geology of the Bahamas: San Salvador, Bahamian Field Station, p. 181–191.

White, K. S., and White, B., 1991, The effects of Holocene sea-level rise on the diagenesis of Quaternary carbonate eolianites, San Salvador Island, Bahamas, *in* Bain, R. J., ed., Proceedings, Fifth Symposium on the Geology of the Bahamas: Fort Lauderdale, Florida, Bahamian Field Station, p. 235–247.

Wilber, R. J., 1987, Geology of Little San Salvador and West Plana Cay: preliminary findings with implications for Bahamian stratigraphy, *in* Curran, H. A., ed., Proceedings, Third Symposium on the Geology of the Bahamas: Ft. Lauderdale, Florida, CCFL Bahamian Field Station, p. 181–204.

Manuscript Accepted by the Society January 5, 1995

Printed in U.S.A.

Geological Society of America
Special Paper 300
1995

Holocene saline lake history, San Salvador Island, Bahamas

James W. Teeter
Department of Geology, University of Akron, Akron, Ohio 44325-4101

ABSTRACT

The lakes of San Salvador Island, Bahamas, occupy sinkholes and curvilinear depressions between dune ridges. Most lakes have maximum water depths of 2 m. They lie approximately at sea level and are saline to varying degrees. Sea water from the surrounding ocean infiltrates the lake basins through permeable carbonate bedrock. The average salinity of lake waters is controlled by the degree of development of the conduit system, the size of the basin, its elevation relative to sea level, rainfall, and the presence of local freshwater lenses.

Ostracodes, which are abundant in these lakes, define three salinity-controlled assemblages: freshwater, marine, and euryhaline. The euryhaline assemblage, consisting of four species, predominates, with a salinity range of 10 to almost 100 ppt.

Lake basins typically contain as much as 2 m of unconsolidated Holocene sediment resting on Pleistocene carbonate bedrock. Where different assemblages are present, varying ostracode abundances can be used to subdivide cores into zones. Longer cores exhibit four zones, which are apparently correlative from lake to lake. Consistent ostracode assemblages indicate prevailing or typical salinities in these zones. Different lakes reveal parallel salinity histories. Salinity fluctuations through time are likely to have been caused by changing climate or sea level.

Initial study of salinity-controlled Mg concentrations in the carapace of *Cyprideis americana* gives estimates of salinity similar to those interpreted from assemblages of ostracodes. Salinity minima, which occur at zone boundaries, correlate with Holocene low stands of sea level. In lakes where only the euryhaline assemblage is present, salinity minima may be used for correlation.

INTRODUCTION

The sedimentary sequence in the lakes of San Salvador contains an abundant and diverse ostracode fauna that indicates changing salinity during the Holocene. Since 1977, the salinity history of several of these lakes has been the focus of study (Sanger and Teeter, 1982; Luginbill, 1983; Crotty and Teeter, 1984; Teeter et al., 1987; Zaleha, 1987; Teeter, 1989;. Edwards et al., 1990; Teeter et al., 1991). Pronounced salinity minima appear to correlate well with lowered stands of Holocene sea level and may be the result of lower sea levels or changes in climate.

PRESENT LAKE CONDITIONS

The lakes (Fig. 1) range in shape and size from roughly circular blue holes a few meters across to elongate bodies lying between crescentic Pleistocene dune ridges and extending several kilometers in length. The lakes are floored by Pleistocene carbonate bedrock covered by a veneer of unconsolidated Holocene sediments having a typical maximum thickness of 2 m. Maximum water depths are approximately 2 m, although locally some lakes may be up to 7 m deep. Lake surfaces lie approximately at sea level.

Teeter, J. W., 1995, Holocene saline lake history, San Salvador Island, Bahamas, *in* Curran, H. A., and White, B., Terrestrial and Shallow Marine Geology of the Bahamas and Bermuda: Boulder, Colorado, Geological Society of America Special Paper 300.

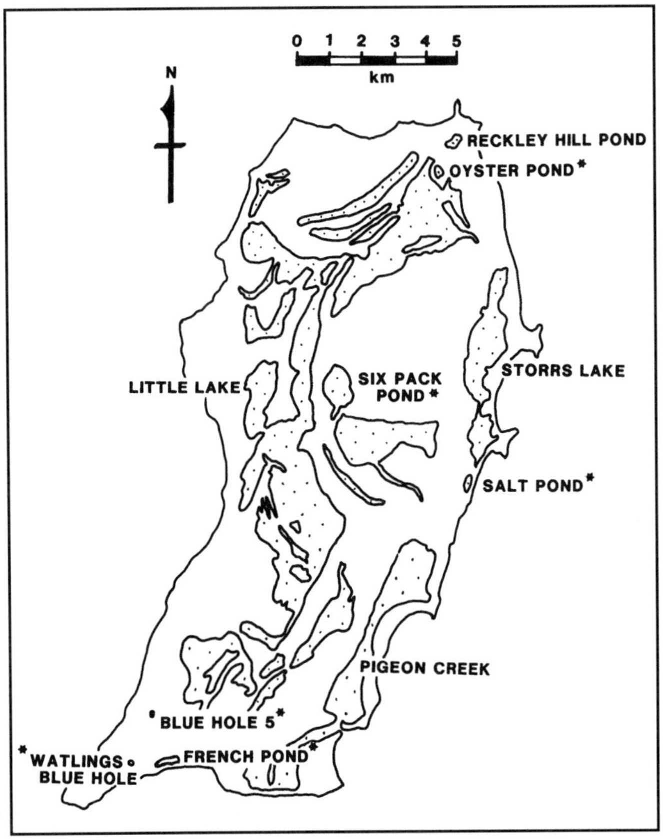

Figure 1. Saline lakes of San Salvador Island, Bahamas. Most lakes identified have been isolated from surface connection with the ocean during the Holocene. Pigeon Creek and Storrs Lake, at least intermittently, have been directly connected to the ocean and are not considered further here. Lakes marked by an asterisk have had paleosalinity histories reconstructed from MgO content of the carapace of the ostracode *Cyprideis americana*.

The lakes are saline to varying degrees. The average salinity of lake water is controlled by the degree of development of the conduit system, the presence of local fresh groundwater lenses, the size of the lake, its elevation relative to sea level, and rainfall. Sea water seeps through permeable carbonate bedrock flooding low-lying depressions. The degree of exchange between lakes and ocean depends on the development of the conduit system. Ponds and blue holes having unimpeded connections to the open ocean, for example, Pain Pond, Moon Rock Pond, and Oyster Pond (Edwards et al., 1990), exhibit normal-marine salinities and tidal change, although the range is diminished and timing lags behind ocean tides. Similar tidal ponds, Watling's Blue Hole and Fortune Hill Pond, are apparently influenced by isolated fresh groundwater lenses, mixing with which produces brackish salinities.

Water in ponds having poor interchange with the ocean has long residence time and very high salinity. Salinity of Salt Pond (Teeter et al., 1987) ranges from 90 to 300 ppt. At times, gypsum and halite are precipitated in Salt Pond. Large lakes, such as Little Lake and Storrs Lake, because of their size and restricted connection with the ocean, have no tide and their waters have long residence times, and are hypersaline. Water level and salinity in hypersaline lakes fluctuate in response to rainfall.

Some lake basins, such as Line Hole Sink, lie slightly above sea level and are floored by relatively impermeable bedrock. Such basins may contain temporary freshwater ponds after heavy rainfall.

LAKE BIOTA

Marshall (1982) identified the phytoplankton from Little Lake and Great Lake. Among the taxa of blue-green algae are several genera known to have increased red pigment at higher salinity. This may account for the turbid, red-brown water that is characteristic of many hypersaline lakes. Green algae commonly present include *Batophora oerstedi* and *Acetabularia crenulata*, the latter of which produces aragonitic skeletal materials including calcispheres, which may be represented by birdseye structures in the geologic record.

Five species of molluscs occur commonly. In some saline lakes all five species are present, but in most at least one or two are lacking. The three common gastropod species include *Batillaria minima, Cerithidea costata*, and *Cerithium lutosum*. Pelecypods are represented by *Polymesoda maritima* and *Anomalocardia auberiana*. The two pelecypod species live at salinities ranging from 20 to 65 ppt and appear to have an inverse salinity-size relationship. Several lakes, which are characterized by normal-marine salinities, have additional pelecypod species, some of which are restricted to marine conditions (Edwards et al., 1990). Windward-facing shores on large lakes often have molluscan shell beaches.

Foraminifera are common constituents of nearly all lakes. Bowman and Teeter (1983) noted the extreme abundance of the miliolid species *Quinqueloculina bosciana* and *Q. costata* in the Holocene sediments of Little Lake.

Ostracodes abound throughout the Holocene of the saline lakes and may be separated into three distinct assemblages characteristic of different salinities (Table 1). The freshwater assemblage is restricted to salinities of a few parts per thousand and is rarely present. The marine assemblage most frequently appears in the range of 30 to 40 ppt, although some species survive slightly beyond these limits. The euryhaline assemblage is most

TABLE 1. SALINE LAKE OSTRACODE SPECIES

Freshwater Assemblage	Marine Assemblage	Euryhaline Assemblage
Candona annae	*Bairdia harpago*	*Perissocytheridea bicelliforma*
Cypridopsis vidua	*Aurila floridana*	*Hemicyprideis setipunctata*
Physocypria denticulata	*Reticulocythereis multicarinata*	*Cyprideis americana*
Limnocythere floridensis	*Loxoconcha purisubrhomboidea*	*Dolerocypria inopinata*
	Xestoleberis curassavica	
	Cytherella arostrata	

TABLE 2. SALINITY RANGES OF THE EURYHALINE ASSEMBLAGE*

Species	Salinity Range (ppt)	Remarks
Perissocytheridea bicelliforma	8.1–27.1	Typically common at 10–20 ppt
Cyprideis americana	10.8–98.5	Least abundant at 30–50 ppt
Hemicyprideis setipunctata	11.3–60.9	Peak abundance at approximately 30–40 ppt
Dolerocypria inopinata	10.0–76.0	

*Based on Crotty and Teeter, 1984; Garbett and Maddocks, 1979; King and Kornicker, 1970; Klie, 1939a,b; Krutak, 1971; and continuing research.

frequently encountered and ranges from brackish to hypersaline conditions (Table 2). Although the euryhaline species *Hemicyprideis setipunctata* has a broad salinity tolerance, it may also be considered a marine species because it prefers salinity of 30 to 40 ppt and lives on the open-marine platform, unlike the other euryhaline species.

Due to the infrequent occurrence of the freshwater and marine assemblages and the broad salinity tolerances of most species of the frequently encountered euryhaline assemblage, it is sometimes difficult to interpret paleosalinity within Holocene lake deposits. An alternate method of paleosalinity reconstruction uses the inverse relationship between salinity and the MgO content of the carapace in the broadly salinity-tolerant ostracode *Cyprideis americana* (Teeter and Quick, 1990).

METHODS

Piston cores though Holocene lake sediments to Pleistocene bedrock have been collected from several lakes (Fig. 1). Cores were subdivided into 2-cm intervals, and selected intervals were split and processed using standard micropaleontologic techniques. Where possible, a minimum of 300 ostracodes was counted per interval. Counts included complete carapaces and right valves of adult through A-2 instars.

For determination of paleosalinity using the MgO content of the carapace of *C. americana*, five well-preserved valves of different adult specimens were selected from each processed interval. Each specimen was mounted in epoxy, sectioned, and microprobed at three different points midway through the valve. The 15 analyses for each interval were averaged to determine the mean MgO value. Ostracodes were analyzed with a WDX electron microprobe on an ETEC SA-3. Instrument conditions were as follows: 15 KV; 0.5×10^{-5} mA; beam diameter of approximately 10 μm. The magnesium standard used was dolomite USNM 10057.

HOLOCENE PALEOSALINITY HISTORY

Where different assemblages are present, changing ostracode abundances can be used to subdivide the Holocene sequence into zones. Four zones can be identified in cores that are sufficiently long or extend far enough below lake level (Figs. 2 and 3). Each zone has a prevailing paleosalinity based on interpretation using ostracode assemblages. Prevailing paleosalinities in Watling's Blue Hole (Fig. 2) are summarized in Figure 4a and Table 3. In Watling's Blue Hole, initial prevailing paleosalinity (Zone 4) was approximately 25 ppt, followed by a decrease to the low to mid-teens in Zone 3. Zone 2 had an approximately normal-marine prevailing paleosalinity, with a return to brackish conditions in Zone 1. In Little Lake (Fig. 3), prevailing paleosalinities (Zone 4) were initially brackish, increasing to marine or slightly hypersaline before declining abruptly at the boundary of Zones 3–4. The beginning of Zone 3 is marked by the rare appearance of the freshwater assemblage. Throughout the rest of Zone 3, salinity increased to brackish to marine conditions. Zone 2 was characterized by marine to hypersaline salinities followed by decreased salinities of Zone 1.

Thus, prevailing paleosalinities based on ostracode assemblages for Watling's Blue Hole and Little Lake are parallel. Other lakes with similar paleosalinity histories are Reckley Hill Pond (Luginbill, 1983) and Six Pack Pond (Zaleha, 1987). Preliminary work reveals similar amino acid ratios in *Polymesoda maritima* and *Cerithidea costata* at the boundaries between Zones 1 and 2 and Zones 2 and 3 in Watling's Blue Hole and Little Lake (Table 4). Similar amino acid ratios and parallel salinity trends suggest that the boundaries of the zones are contemporaneous in the two lakes.

In Salt Pond (Teeter et al., 1987), Blue Hole 5 (Teeter et al., 1991), and French Pond, the euryhaline assemblage dominates the Holocene sequence, making recognition of zones and correlation with other lakes impossible. Furthermore, in Salt Pond (Fig. 5), *Perissocytheridea bicelliforma* predominates, suggesting brackish salinity throughout, a condition difficult to reconcile with the recent salinities ranging from 90 to 300 ppt and the presence of gypsum beds in the sequence. Teeter et al. (1987) suggested that *P. bicelliforma* may be a fecund, opportunistic species repopulating the pond in great numbers during infrequent, lowered salinity events, perhaps caused by storms.

Paleosalinities determined from the MgO content of the carapace of *Cyprideis americana* provide an independent check on prevailing paleosalinities interpreted from assemblages and offer a means of correlation between lakes. The paleosalinity curve of Watling's Blue Hole based on MgO analyses (Fig. 4b) probably gives a more realistic indication of salinity changes through time than does the curve based on data from ostracode assemblages (Fig. 4a). At first glance, the two curves do not appear very similar; however, if the paleosalinities calculated from MgO content are averaged for the intervals within each zone, the results (Table 3) agree well with those originally predicted by Crotty and Teeter (1984).

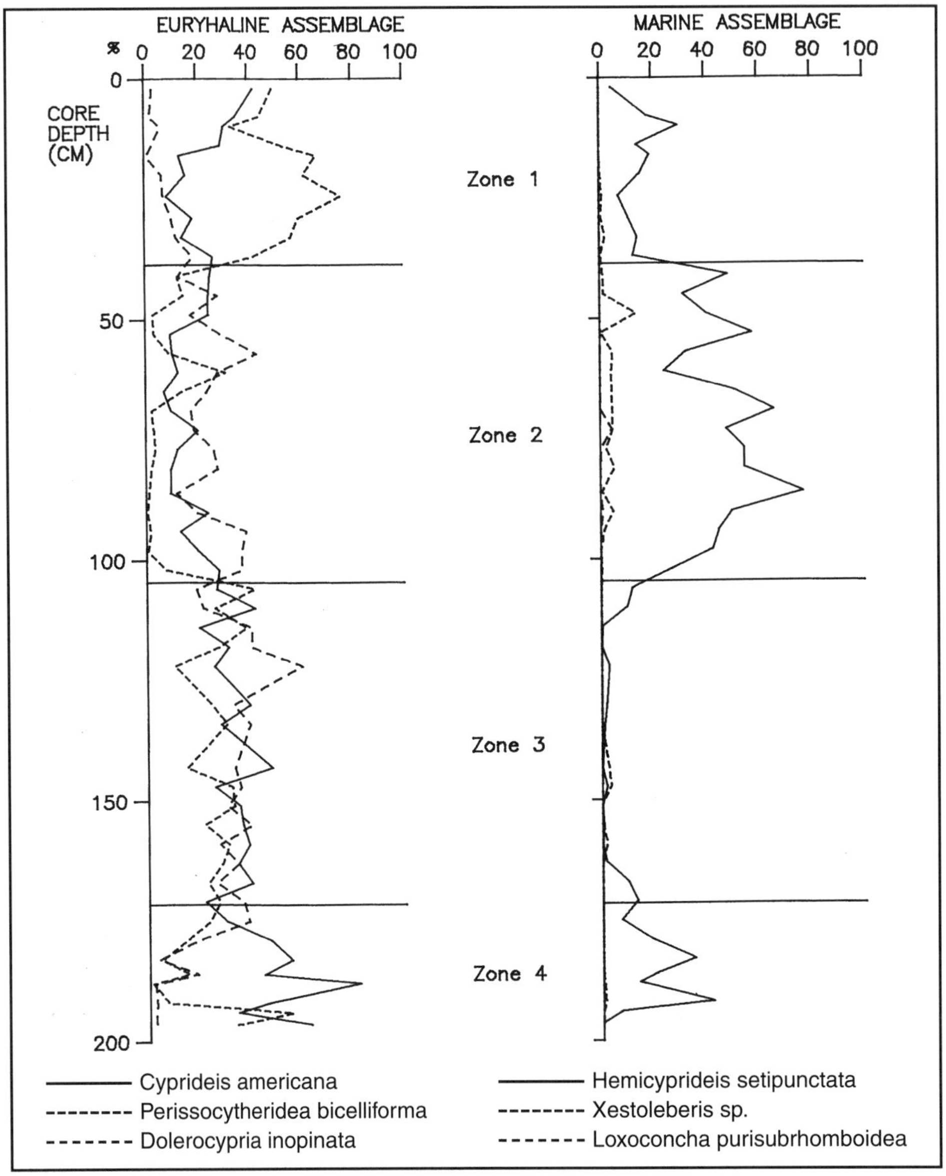

Figure 2. Holocene ostracode distribution, Watling's Blue Hole, core 2 (Crotty and Teeter, 1984). Interval for most samples is 2 cm.

The paleosalinity curve (Fig. 4b) based on MgO analyses reveals pronounced salinity minima at the zone boundaries. Seldom is confirming evidence provided by the assemblages; however, in Little Lake (Fig. 3), a freshwater assemblage occurs at the boundary between the earliest two zones.

Paleosalinity minima occur at approximately the same depth in cores from different lakes (Fig. 4b, 6a-c), suggesting that they represent widespread, contemporaneous events. Lowered salinity can be caused by falling sea level or by such climatic events as increased rainfall or decreased temperature, causing less evaporation. If sea level falls, so does the amount of sea water infiltrating the lake basin. Thus, precipitation and seepage of fresh groundwater will lower salinity. Should sea level fall below the floor of the lake basin and if the lake floor is impermeable, a freshwater lake would result.

As the Wisconsinian glaciers were retreating and sea level rising, temporary glacial advances must have produced oscillations of sea level. Some sea-level curves (Fairbridge, 1961; Morner, 1969; Colquhoun and Brooks, 1987; Ters, 1987) portray similar oscillations. Compared to the curve from Ters (1987), salinity minima occurred at low stands of sea level (Fig. 7).

In Watling's Blue Hole, the evidence of a salinity minimum at 2,680 B.P. lies approximately 3.2 m below present mean sea level. The brackish water ostracode assemblage here indicates that the pond lay at or slightly below sea level at that time. The correlative horizon in Little Lake (Sanger and Teeter, 1982) lies

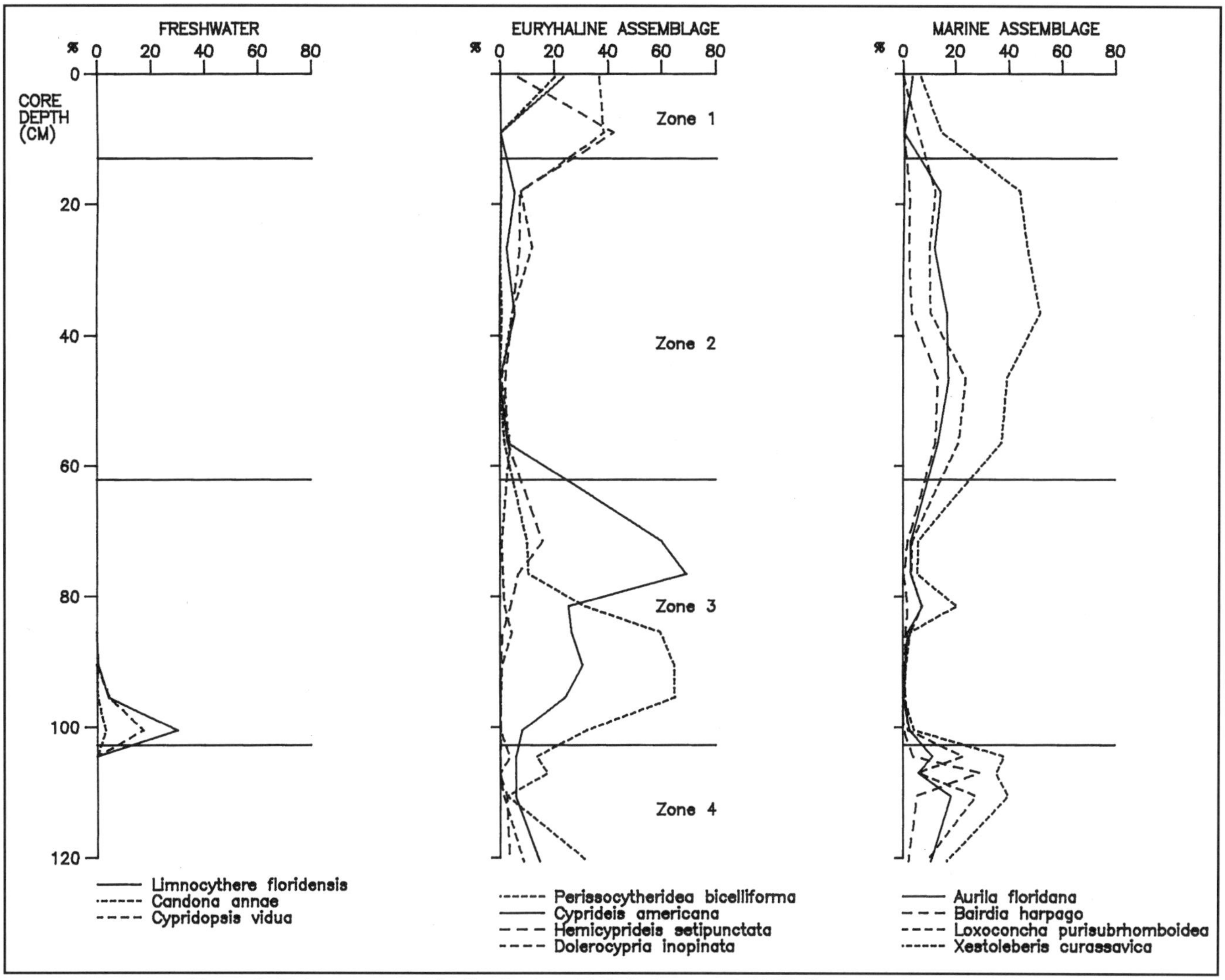

Figure 3. Holocene ostracode distribution, Little Lake, core 9-1977 (Sanger and Teeter, 1982). Interval for most samples is 5 cm.

approximately 2.7 m below present mean sea level and the presence of a freshwater ostracode assemblage indicates that this lake stood above sea level. Thus, approximately 2,700 B.P., sea level on San Salvador lay between 2.7 and 3.2 m below present sea level. This is slightly lower than the sea level determined by Fairbridge (1961) and Colquhoun and Brooks (1987) and slightly higher than the elevation recognized by DePratter and Howard (1981). It is considerably higher than the contemporary sea level (Fig. 7) recognized by Ters (1987) from the French Atlantic coast.

The salinity minima at 1,900 and 1,360 B.P. agree well with lowered stands of sea level recorded by Ters (1987). In Watling's Blue Hole, the presence of marine to brackish ostracodes at these two horizons indicates that sea level was lowered less than 2.5 m 1,900 yr ago and less than 1.9 m 1,360 B.P. Absence of freshwater ostracodes at these horizons in other lakes prevents the determination of minimum sea levels. Lower sea levels recognized by Ters (1987) and other workers during the past 1,000 yr are not obviously reflected by ostracode assemblages or salinity determined from Mg concentrations in *C. americana*.

CONCLUSIONS

1. Longer cores in which different ostracode assemblages are present reveal four faunal zones in lake sediments during approximately the past 3,000 yr.

2. Salinity tolerances of the ostracode species can be used to establish a prevailing salinity for each zone.

3. Different lakes had parallel salinity change during the Holocene.

4. From the inverse MgO-salinity relationship of *Cyprideis americana*, a detailed paleosalinity history can be constructed for Holocene lacustrine deposits. Paleosalinity histories reconstructed by assemblages and the above method agree well.

5. Detailed paleosalinity histories reveal pronounced salin-

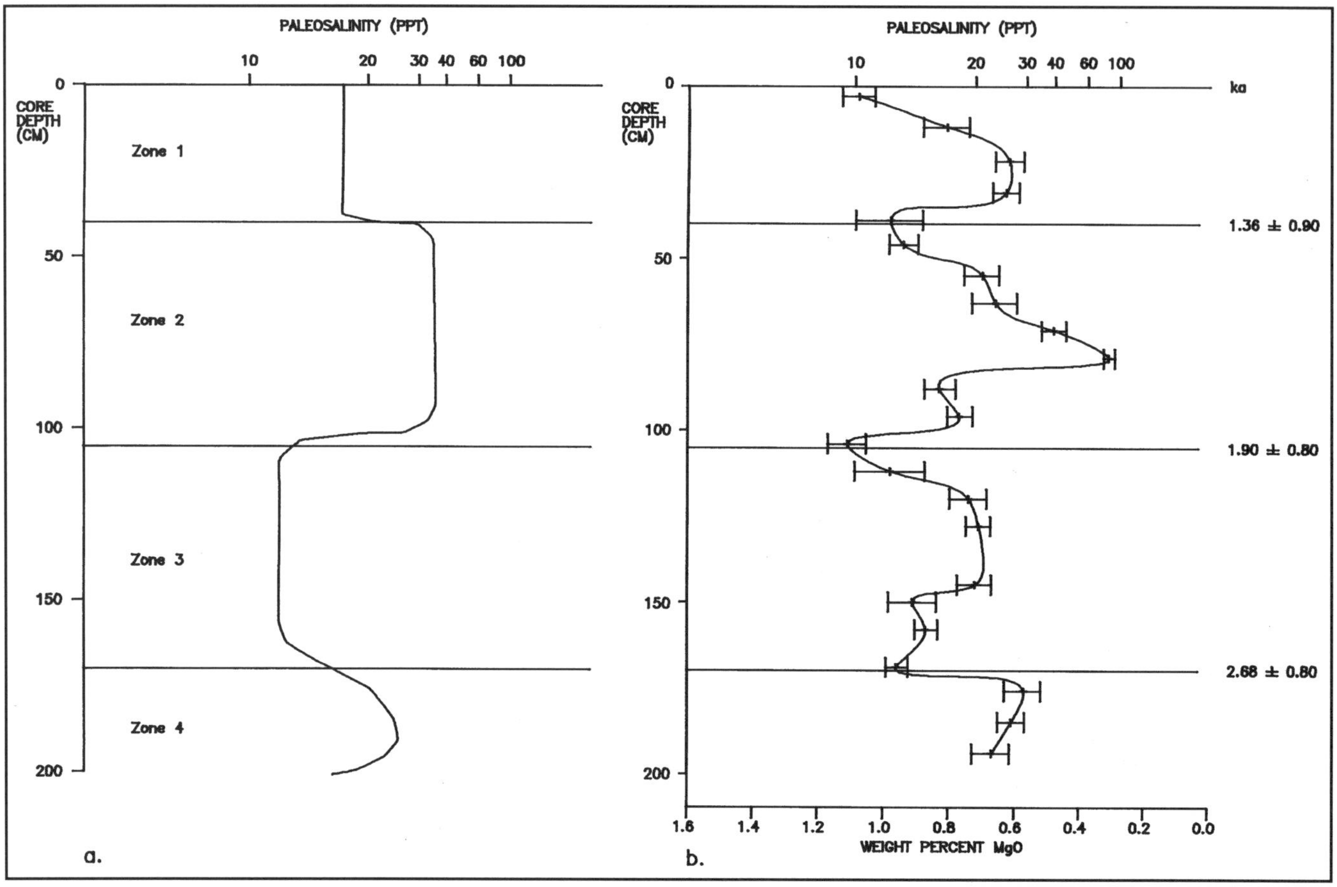

Figure 4. Holocene paleosalinity interpretations for Watling's Blue Hole. a, Curve based on ostracode assemblages (modified from Crotty and Teeter, 1984). b, Spline curve based on means of weight percent MgO in *Cyprideis americana*. Each of the 23 intervals represents 15 analyses. Error bars represent standard error of the mean. Column on right records accelerator mass spectrometry ^{14}C dates of mollusc shells. Dates have been corrected for addition of nonradioactive carbon from Pleistocene bedrock. Radiocarbon dates are recorded in thousands of years B.P.

TABLE 3. COMPARISON OF PALEOSALINITIES BASED ON OSTRACODE ASSEMBLAGES AND MAGNESIUM CONTENT OF *CYPRIDEIS AMERICANA**

Zone	Paleosalinity From Ostracode Assemblages (ppt)	Paleosalinity From Mg Content (ppt)
1	High teens	18
2	25–35	28
3	Low to mid-teens	16
4	25	27

*Data from core 2, Watling's Blue Hole (Crotty and Teeter, 1984; Teeter, 1990).

TABLE 4. AMINO ACID RATIOS IN SELECTED HOLOCENE MOLLUSCS*

Lake (Core No.)	Alloisoleucine:Isoleucine (Zone Boundary) *Polymesoda maritima*	*Cerithidea costata*
Watling's Blue Hole (I)	0.04 (1-2)	0.14 (2-3)
Little Lake (4-1981)	0.04 (1-2)	0.16 (2-3)

*Neither core provided suitable molluscan material at boundary of Zones 3-4.

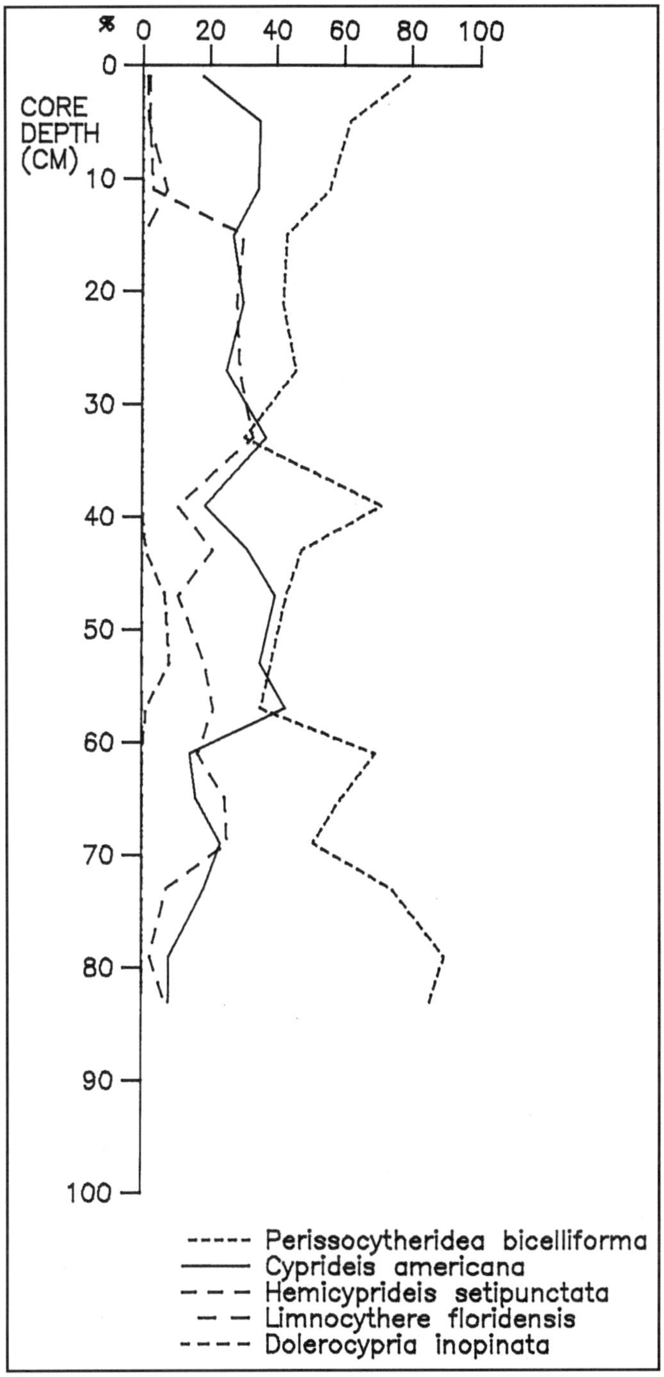

Figure 5. Holocene ostracode distribution, Salt Pond (Teeter et al., 1987). Sampling interval is 2 cm.

ity minima at boundaries of zones established by ostracode assemblages.

6. Where only the euryhaline assemblage is present, salinity minima can be used to correlate from lake to lake.

7. Salinity minima occurred at times of Holocene low stands of sea level and may have resulted from lowered sea level or climatic change.

ACKNOWLEDGMENTS

I am indebted to Richard Mitterer, who provided the amino acid data, and to Tom Quick for the production of computer graphics. This work has benefited considerably from critical review by Roger Kaesler and Laurence Davis.

REFERENCES CITED

Bowman, P. A., and Teeter, J. W., 1983, The distribution of living and fossil Foraminifera and their use in the interpretation of the post-Pleistocene history of Little Lake San Salvador Island, Bahamas: San Salvador, CCFL Bahamian Field Station Occasional Paper 1, 26 p.

Colquhoun, D. J., and Brooks, M. J., 1987, New evidence for eustatic components in late Holocene sea levels, *in* Rampino, M. R., Sanders, J. E., Newman, W. S., and Konigsson, L. K., eds., Climate, History, Periodicity and Predictability: New York, Van Nostrand Reinhold, p. 143–156.

Crotty, K. J., and Teeter, J. W., 1984, Post-Pleistocene salinity variations in a blue hole, San Salvador Island, Bahamas, as interpreted from the ostracode fauna, *in* Teeter, J. W., ed., Proceedings, Second Symposium on the Geology of the Bahamas, June 1984: San Salvador, CCFL Bahamian Field Station, p. 3–16.

DePratter, C. B., and Howard, J. D., 1981, Evidence for a sea level lowstand between 4500 and 2400 years BP on the southeast coast of the United States: Journal of Sedimentary Petrology, v. 51, p. 1287–1295.

Edwards, D. C., Teeter, J. W., and Hagey, F. M., 1990, Geology and ecology of a complex of inland saline ponds, San Salvador Island, Bahamas, *in* Fifth Symposium on the Geology of the Bahamas, Field Trip Guidebook: San Salvador, Bahamian Field Station, p. 35–45.

Fairbridge, R. W., 1961, Eustatic changes in sea level, *in* Ahrens, L. H., Rankama, K., Press, F., and Runcorn, S. K., eds., Physics and Chemistry of the Earth, v. 4: New York, Pergamon Press, p. 99–185.

Garbett, E. C., and Maddocks, R. F., 1979, Zoogeography of Holocene cytheracean ostracodes in the bays of Texas: Journal of Paleontology, v. 53, p. 841–919.

King, C. E., and Kornicker, L. S., 1970, Ostracoda in Texas bays and lagoons: An ecologic study: Smithsonian Contributions to Zoology, v. 24, p. 1–92.

Klie, W., 1939a, Ostracoda aus den marinen salinen von Bonaire, Curacao und Aruba: Capita Zoologica, v. 8, p. 1–19.

Klie, W., 1939b, Brackwasserostracoden von nordostbrasilien: Zoologischer Jahrbucher Abteilung fur Systematik Geographie und Biologie der Tiere, v. 72, p. 359–372.

Krutak, P. R., 1971, The recent Ostracoda of Laguna Mandinga, Veracruz, Mexico: Micropaleontology, v. 16, p. 1–30.

Luginbill, C. P., 1983, Ecology of living Ostracoda from selected lakes and post-Pleistocene history of Reckley Hill Pond, San Salvador Island, Bahamas: Ohio Journal of Science Abstracts with Programs, v. 83, p. 27.

Marshall, H. G., 1982, Phytoplankton composition from two saline lakes in San Salvador, Bahamas: Bulletin of Marine Science, v. 32, p. 351–353.

Morner, N. A., 1969, The late Quaternary history of the Kattegatt Sea and the Swedish West Coast. Deglaciation, shorelevel displacement chronology, isostasy and eustasy: Sveriges Geologiska Undersokning, v. 63, 487 p.

Sanger, D. B., and Teeter, J. W., 1982, The distribution of living and fossil Ostracoda and their use in the interpretation of the post-Pleistocene history of Little Lake, San Salvador, Bahamas: San Salvador, CCFL Bahamian Field Station Occasional Paper 1, 26 p.

Teeter, J. W., 1989, Refinement and timing of salinity fluctuations in Watling's Blue Hole, San Salvador, Bahamas, *in* Mylroie, J., ed., Proceedings, Fourth Symposium on the Geology of the Bahamas: San Salvador, Bahamian Field Station, p. 331–336.

Teeter, J. W., and Quick, T. J., 1990, Magnesium-salinity relation in the saline lake ostracode *Cyprideis americana*: Geology, v. 18, p. 220–222.

Teeter, J. W., Beyke, R. J., Bray, T. F., Jr., Brocculeri, T. F., Bruno, P. W., Dre-

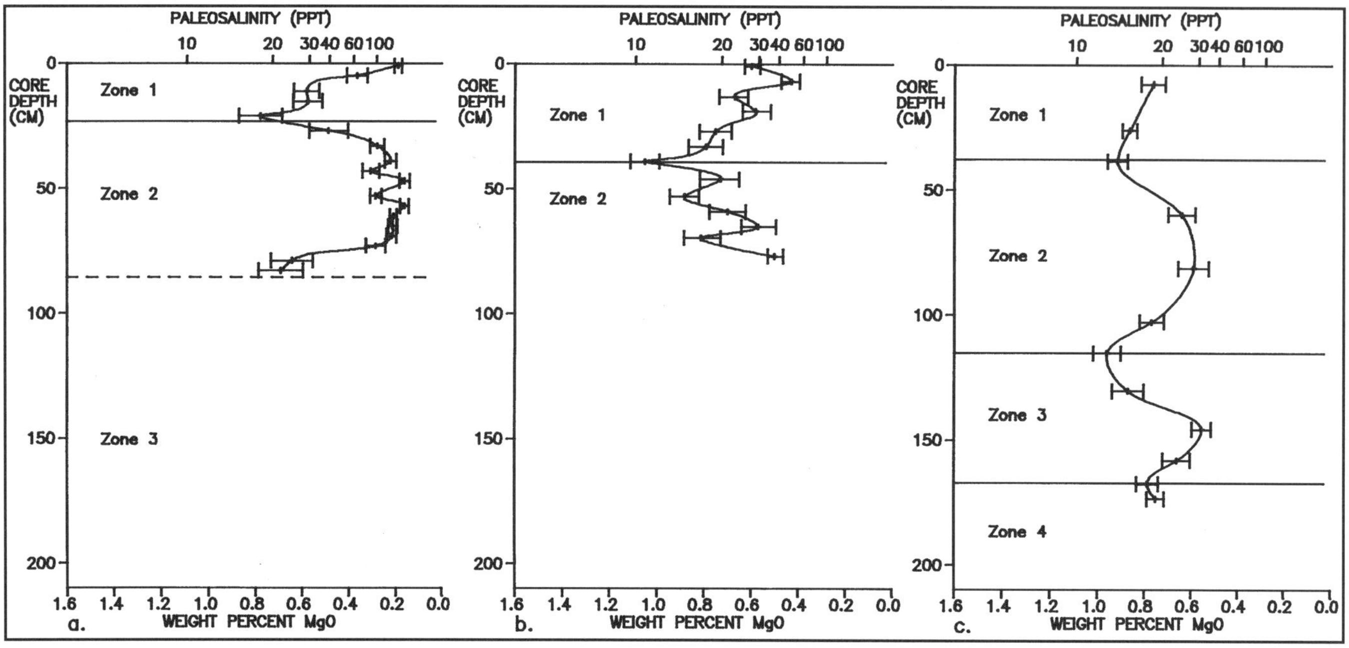

Figure 6. Paleosalinity spline curves based on means of weight percent MgO in *Cyprideis americana*. Mean for each interval based on 15 analyses. Error bars represent standard error of the mean. Zone boundaries inferred on salinity minima. a, Salt Pond (Teeter et al., 1987). b, French Pond. c, Blue Hole 5 (Teeter et al., 1991).

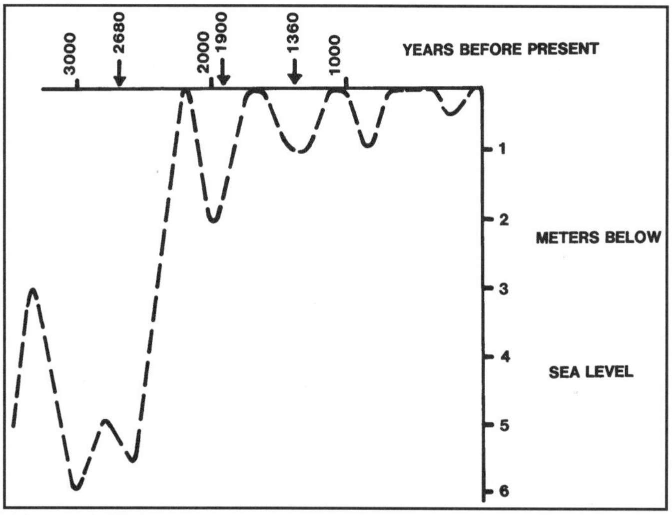

Figure 7. Holocene sea-level curve (after Ters, 1987). Arrows and dates indicate salinity minimum events in Holocene lake deposits.

mann, J. J., and Kendall, R. L., 1987, Holocene depositional history of Salt Pond, San Salvador, Bahamas, *in* Curran, H. A., ed., Proceedings, Third Symposium on the Geology of the Bahamas: Ft. Lauderdale, Florida, CCFL Bahamian Field Station, p. 145–150.

Teeter, J. W., Beltz, J. F., Miller, M. M., Palunas, M. J., and Zurdoky, R. A., 1991, Holocene salinity history of a blue hole, San Salvador Island, Bahamas: Geological Society of America Abstracts with Programs, v. 23, p. 64.

Ters, M., 1987, Variations in Holocene sea level on the French Atlantic coast and their climatic significance, *in* Rampino, M. R., Sanders, J. E., Newman, W. S., and Konigsson, L. K., eds., Climate, History, Periodicity and Predictability: New York, Van Nostrand Reinhold, p. 204–273.

Zaleha, R. D., 1987, Holocene paleoenvironmental history of Six-Pack Pond, San Salvador Island, Bahamas [M.S. thesis]: Akron, Ohio, University of Akron, 100 p.

Manuscript Accepted by the Society January 5, 1995

Printed in U.S.A.

Geological Society of America
Special Paper 300
1995

An imprint of Holocene transgression in Quaternary carbonate eolianites on San Salvador Island, Bahamas

Kathleen S. White
Marine Geology and Geophysics, Rosenstiel School of Marine and Atmospheric Science, University of Miami, 4600 Rickenbacker Causeway, Miami, Florida 33149

ABSTRACT

Through detailed field and laboratory work it has been determined from the presence of marine aragonite cement that marine diagenesis is taking place in the Holocene and Pleistocene eolianites in the northeast corner of San Salvador Island, Bahamas. These eolianites, partially submerged by the Holocene marine transgression, are located in the supratidal, intertidal, and shallow subtidal zones. The characteristics of the marine cement found in these rocks provide criteria for the recognition of earlier marine transgressions in similar rocks.

The aragonite cement is dominantly acicular and forms isopachous rims lining both intergranular and intragranular voids. The acicular aragonite commonly forms an interfingering framework that develops on the grain edges, especially peloids, and grows to fill in pore spaces.

Point count data show that the Holocene rocks all have similar compositions and that they differ greatly from the compositions of the Pleistocene rocks. The Holocene rocks are dominated by peloids, while the Pleistocene rocks are dominated by bioclasts. Therefore it appears that the marine cement is not dependent on the composition of the grains that make up the eolianites. Instead, the source of the aragonite is seawater. The processes that are forming the marine cement in these eolianites may be similar to those that form beachrock along beaches in the tropics, including the beach adjacent to the field area in this study.

INTRODUCTION

San Salvador Island, Bahamas, serves as an excellent place to study the effects of changing sea level on the geology of an island because the island lies on a tectonically stable, or slightly submerging, platform surrounded by deep water. Holocene carbonate sedimentation keeps pace with the subsidence, and so the Holocene and Pleistocene rocks serve as markers for Quaternary sea-level changes (Curran et al., 1989).

The history of the rocks of San Salvador is fairly well understood. Quaternary carbonate environments of San Salvador Island and elsewhere have been subjected to changing sea levels as a consequence of fluctuations in ice volume during the waxing and waning of Pleistocene ice sheets. The depositional and diagenetic history of the Cockburn Town fossil reef on the west coast of San Salvador clearly reflects the Sangamon high sea level at about 125 ka (Chen et al., 1991) and subsequent sea-level fall during the Wisconsinan ice advance. Aragonite and high-magnesian calcite cements that formed in the marine environments predate nonmarine, meteoric low-magnesian calcite cements (White et al., 1984). According to Mylroie and Carew (1988), there have been two subsequent sea-level highstands that would have brought the rocks of the Cockburn Town fossil reef back into the marine or meteoric phreatic diagenetic environment. The rocks of the Cockburn Town fossil reef have been studied extensively (Curran and White, 1985), and the effects of marine regression and subaerial exposure on marine reefal rocks have been documented (White et al., 1984). However, there are

White, K. S., 1995, An imprint of Holocene transgression in Quaternary carbonate eolianites on San Salvador Island, Bahamas, *in* Curran, H. A., and White, B., Terrestrial and Shallow Marine Geology of the Bahamas and Bermuda: Boulder, Colorado, Geological Society of America Special Paper 300.

no marine diagenetic features that postdate the fresh-water vadose cements and calichification textures that developed during the emergent phase. Such features may be absent because the sea-level highstands did not occur, or because such highstands do not leave a recognizable diagenetic imprint.

In northeastern San Salvador Island, Holocene carbonate eolianites that were deposited and lithified in the nonmarine environment (White and Curran, 1988) are being exposed for the first time to marine waters because of the Holocene transgression (Boardman et al., 1989). This provides an ideal opportunity to investigate the diagenetic effect of a change from nonmarine to marine environments on the carbonate rocks. The fact that beachrock is forming on the sandy shores of Rice Bay, just a few hundred meters from the field area, is a strong indication that some types of marine cements may be developing in the terrestrial eolianite rocks exposed along the nearby rocky shores. An important question is whether such submergence leaves a diagenetic imprint, and if so, what are the distinctive mineralogic and textural characteristics. Answers could be useful in better understanding the diagenetic history that is recorded in older carbonate rocks and in relating that history to past changes in sea level.

GEOLOGIC SETTING

Geography

San Salvador Island is located in the southeast corner of the Northwest Bahama Platform at 24°N and 74°34′W and is separated from the other shallow water banks on the Bahama Platform by waters up to 1 km in depth (Fig. 1). It is part of the small San Salvador Bank, approximately 25 km long and up to 12 km wide, made up of low islands and narrow, shallow shelves. San Salvador Bank forms a northerly promontory in the northwest-southeast–oriented Bahama Escarpment that separates the Bahama Platform from the deep Atlantic Ocean to the east.

San Salvador Island is 11 km wide and 19 km long. It is bordered by a narrow shelf bounded by an abrupt shelf-edge break, with a very steep drop-off to oceanic depths. The topography of the island is dominated by arcuate ridges that are thought to represent successive stages of carbonate eolian accretion. The low interdune areas on the island are occupied by extensive shallow lakes. The coastline includes headlands, eolianite cliffs, and beaches made of fine- to medium-grained carbonate sands, many of which contain Holocene beachrock. The best exposures of the rocks of San Salvador Island occur along the coastline, in the few road cuts, or in the quarries found around the island (Curran, 1985).

Geology

As with the other exposed Bahamian islands, the geologic development of San Salvador Island has been strongly influenced by the changes in sea level caused by glacial advance and retreat in the Pleistocene (Carew and Mylroie, 1985). The Bahamas are considered to be tectonically stable, subsiding isostatically at a rate of 1 to 2 m per 100 ka due to carbonate deposition on the

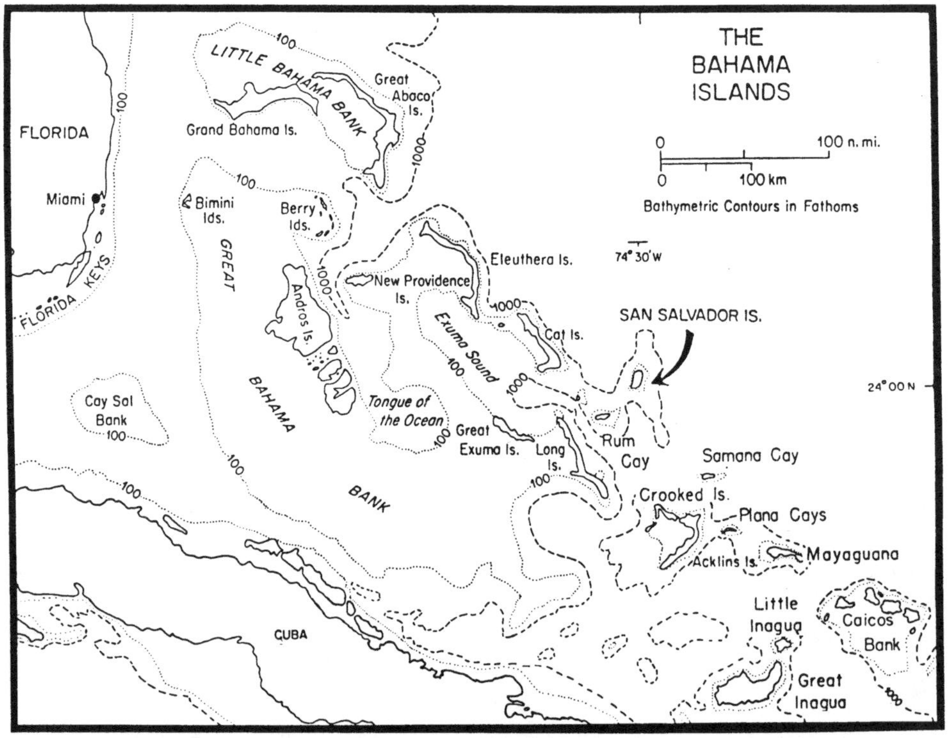

Figure 1. Index map of San Salvador Island, Bahamas (after White and Curran, 1988).

platform surfaces (Mylroie and Carew, 1988). Tectonic stability of the platform makes the Holocene and Pleistocene rocks on the island accurate markers of Quaternary eustatic sea-level change. Vertical facies changes in Pleistocene and Holocene sediments are abrupt, which allows for a precise determination of former positions of sea level (Curran et al., 1989).

During high sea-level stands, San Salvador was partially or wholly submerged and carbonate sediments including shell fragments, pellets, ooids, and grapestone grains were produced. Once sea level fell and emergence occurred, relatively rapid cementation took place, which loosely lithified the grains with low-Mg calcite sparite rim cements (Titus, 1983). This period was followed by the development of a karst surface that includes karst crusts, paleosols, and solution phenomena (Titus, 1983). This karst surface is used as a basis for defining the stratigraphy of San Salvador.

Stratigraphy

An extensive and detailed account of the stratigraphy of San Salvador Island (Fig. 2) has been given by Carew and Mylroie (1985). According to this stratigraphic scheme, the two units that were studied in this project are the Cockburn Town Member of the Grotto Beach Formation and the North Point Member of the Rice Bay Formation. The Cockburn Town Member, composed of eolianites approximately 85,000 yr old, is the youngest member of the Grotto Beach Formation. The top of this member marks the end of the Pleistocene rock record on San Salvador. The eolianites are dominantly oolitic, have significant karst development, and are capped by a paleosol, which marks the top of the Pleistocene on San Salvador. Pleistocene eolianites were studied on Man Head Cay off the northeast corner of San Salvador.

The North Point Member is the oldest member of the Rice Bay Formation, containing eolianites that were deposited close to the bank edge. Eolian cross bedding in this member extends below present sea level, demonstrating that deposition occurred before sea level rose to modern levels (Carew and Mylroie, 1985). These eolianites lack significant karstification or paleosol development, indicating their youth and suggesting that they formed during the present sea-level rise (Hutto and Carew, 1984). The fact that they were lithified well enough to resist simple reworking of the sand as sea level rose suggests that these eolianites were not only deposited but also lithified prior to the present sea-level rise (White and Curran, 1985). Whole-rock radiocarbon ages indicate that these rocks are approximately 5,500 yr old (Carew and Mylroie, 1985). Eolianites of the North Point Member were studied in this project at North Point and Cut Cay along the northeastern coastline of San Salvador.

PREVIOUS WORK

Hutto and Carew (1984) analyzed thin sections from 75 eolianite localities on San Salvador Island, Bahamas. They divided the eolianites into two categories: oomicrosparites, which made up 76% of the sample localities, and biomicrosparites, which made up the remaining 24% of the sample localities. Although they found that the same ridge usually contains deposits that have similar petrographic characteristics, compositional similarity between eolian ridges cannot be used to determine that the dunes were formed at the same time because any similarities or differences seen between the compositions of several ridges can also be related to the sediment source.

Lawlor (1985) examined the Holocene eolianites in the Rice Bay area and found that peloids dominated the grain type, with ooids as the second most abundant grain type. She found that the rocks nearer the shoreline had more cement than the rocks farther from the sea, and suggested that this difference in the abundance of cement indicated a greater volume of pore water nearer the shoreline.

Work done by Mylroie and Carew (1988) has indicated the possibility of a late-stage sea-level rise occurring after the Sangamon highstand and prior to the Holocene transgression.

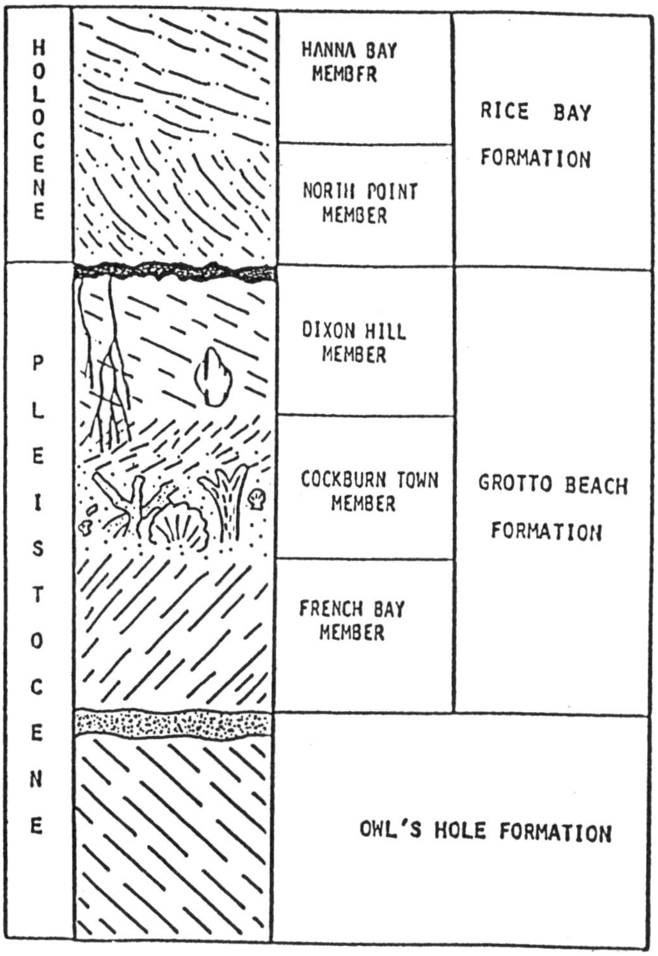

Figure 2. Stratigraphy of San Salvador Island, Bahamas. Note that an updated stratigraphy does not include the Dixon Hill Member of the Grotto Beach Formation (after Carew and Mylroie, 1985; modified by Carew et al., 1992).

They have used solution conduits as sea-level indicators. The size, passage morphology, and sediments in abandoned conduits provide information about the water table that existed when the conduits were active. The morphology of conduits formed in the vadose environment differs from that formed in the phreatic environment. It is assumed that the solution surface of the ceiling of the phreatic conduit indicates the minimum elevation of the top of the fresh-water lens, and so the minimum elevation of sea level, at least to within a few meters (Mylroie and Carew, 1988). Conduits found along the wall of San Salvador Island at depths of –105 and –125 m are difficult to explain without the existence of a low sea-level period (Mylroie and Carew, 1988).

Secondary calcite deposits, especially stalagmites, provide information on the surface conditions occurring after the conduit was abandoned, and can be dated by the uranium/thorium radiometric dating technique. Since stalagmites only grow in air-filled caves, the age at their base reveals the minimum age for when the cave was first drained of water. This also means that reflooding of the cave would stop stalagmite growth. A change in the delivery of drip water also could interrupt the stalagmite growth (Mylroie and Carew, 1988).

Lighthouse Cave, on the northeastern side of San Salvador Island, is formed in the eolianites of the Cockburn Town Member. Amino acid racemization analysis of *Cerion* sp. found in the eolianites indicates that the rocks have an approximate age of 85 ka. The vertical range of the cave is 2 m below sea level to 7 m above sea level, so the cave must have formed when the fresh-water lens was located in that same range. This had to occur after 85 ka, and so suggests that sea level was at a highstand sometime since 85 ka (Mylroie and Carew, 1988). The whole of the bottom part of a stalagmite from this cave was dated at 49 ka. The entire top part of the same stalagmite was dated at 37 ka. A layer of tubes of a marine serpulid, *Filograna* sp., has been found on the boundary between the bottom part and the top part of this stalagmite. Therefore, there must have been a period of marine flooding that was long enough to allow encrustation by the serpulids sometime between 49 and 37 ka. Climatic conditions, as well as a sea-level rise, probably contributed to the lack of growth of the stalagmite during this period (Mylroie and Carew, 1988). This sea-level rise would have occurred after the Sangamon interglacial, about 125 ka, and should have left a marine imprint on coastal rocks such as the Cockburn Town fossil reef rocks of San Salvador Island. As stated above, no such imprint has been detected.

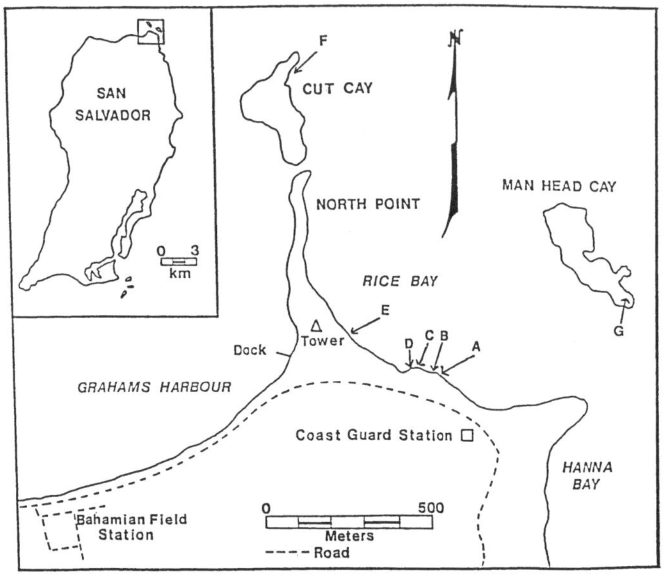

Figure 3. Northeastern San Salvador Island showing the location of the field area and of the measured profiles (modified from White and Curran, 1985).

FIELD WORK

Field studies were conducted mainly at the North Point area in the northeastern corner of San Salvador (Fig. 3). The main field area was selected because it includes differences in topography and morphology of the eolianites. The area contains two promontories, a large embayment area, and a smaller embayment area (see Figs. 4 and 5). The large embayment area has a sloping surface exposing a broad expanse of eolianites, while the promontories have a more vertical section. These differences in morphology and topography result in varying degrees of exposure of the eolianites to seawater. The promontories are subject to heavier wave action than the embayments, which allows the seawater to reach higher elevations on the eolianites. The embayments accumulate seawater in tidal pools, as well as sand, in the lower lying areas. The embayments are set back farther toward the interior of the island, so the seawater reaches only the higher parts of the eolianites in these areas when storm waves affect the coast.

Seven topographic profiles were constructed across the eolianites from the vegetation line down through the color-zoned intertidal zone into the shallow subtidal. Six crossed Holocene-age rocks and one was along rocks of Pleistocene age. Four of the profiles (A through D) were located in the main field area at North Point (Fig. 6). A fifth profile (E) was a cliff section farther northwest along the coast, another was on Cut Cay (F), and the final profile (G) crossed Pleistocene strata on the seaward side of Man Head Cay (Fig. 7).

The cliff section (E) was selected to study the possible effect on marine cementation of waves frequently crashing on a vertical surface. The Cut Cay profile (F) was chosen because it juts off the northeastern corner of San Salvador and is exposed to heavier wave action than the other Holocene profiles. The Man Head Cay profile (G) was chosen to serve as a comparison of Pleistocene eolianites to Holocene eolianites.

Eolianites located in the present shore zone of San Salvador have a surficial color zonation similar to that first described in the Florida Keys by Stephenson and Stephenson (1950). There are four zones on San Salvador: White, Grey, Yellow, and Green.

Figure 4. Field photograph of Profile A, a promontory section of Holocene-age eolianites.

Figure 5. Field photograph of Profile D, an embayment section of Holocene-age eolianites.

The Grey zone on San Salvador corresponds to the combined Black and Grey zones in the Florida Keys described by Stephenson and Stephenson (1950). The White and Grey zones are supratidal, the Yellow zone is intertidal, and the Green Zone is subtidal. Cyanobacteria, at least in part, produce these color zonations (Stephenson and Stephenson, 1950). The position of the color zones on the eolianites along the coast varies with the topography of the coastline. The color zones extend farther inland in the embayment areas than in the promontory areas.

In this part of San Salvador there is no established survey marker, so the elevation of the surveying stations was determined by measuring their elevation above the top of the Yellow zone. Through detailed surveying of the Cockburn Town fossil reef, based on precisely surveyed bench marks related to accurately measured mean sea level (Curran and White, 1985), it has been established that the top of the Yellow zone is 35 cm above mean sea level (B. White, personal communication, 1988). Between 10 and 18 samples along each profile were taken from the surface of the rocks, to preserve the effects of seawater, sea spray, and terrestrial processes. Each sample was taken because of some textural distinction.

LABORATORY TECHNIQUES

The rock chips were impregnated with Petropoxy 154 in a small vacuum impregnator fashioned after Allman and Lawrence (1972, Fig. 44, p. 86) for at least 24 hr. The chips were cured in a drying oven overnight at approximately 60°C. Excess resin was removed from the rock chips using a lapping machine with a 220-mesh polishing wheel. The chips were sonified and dried overnight in the drying oven. They were mounted on glass slides with Norland optical glue and left to cure for approximately 48 hr under an ultraviolet lamp. More glue than usual and a much longer curing period were required to create a bond that was strong enough to survive the cutting and polishing stages of production. This is due to the high porosity and friability of the rocks. The thin sections were not ground down to a uniform thickness; instead, they were ground down to a thickness where the carbonate cements could be seen clearly with transmitted light under crossed nicols. Commonly, if the slides were polished to a uniform thickness, or to a point where the grains could be clearly identified with transmitted light, the cements would be destroyed or plucked from the slide. After the cements were identified, the slides were further polished manually, using 1,000 μ carbo-corundum grit, to permit identification of the grain composition of each sample.

The main focus of the petrographic study centered on thin sections made in the laboratory at Smith College. These approximately 100 thin sections were examined extensively, usually under crossed nicols, for marine cements. More than 50 commercially made thin sections were also examined in the same way to supplement the study. At least one thin section from each rock sample was analyzed for its grain composition, using the standard 300-grain count technique. The percentage of peloids, ooids, and bioclasts of each sample were calculated.

Several of the thin sections were stained to aid in identification of carbonate cements present in the rocks. The presence of high-Mg calcite was tested for using titan-yellow stain, as described by Choquette and Trussel (1978), except for the use of the permanent stain fixer. Other thin sections were stained for aragonite using Fiegl's solution based on Friedman's method as described in Schneidermann and Sandberg (1971). These staining procedures were done to determine if any of the isopachous, needle-like cements were high-Mg calcite and not aragonite.

Samples in which petrographic analysis had determined the presence of marine cements were then chosen for scanning electron microscope (SEM) analysis. The samples were examined to determine the crystal habits of the cements using the JEOL scanning electron microscope in the Department of Biological Sciences at Smith College. The SEM was operated initially at an

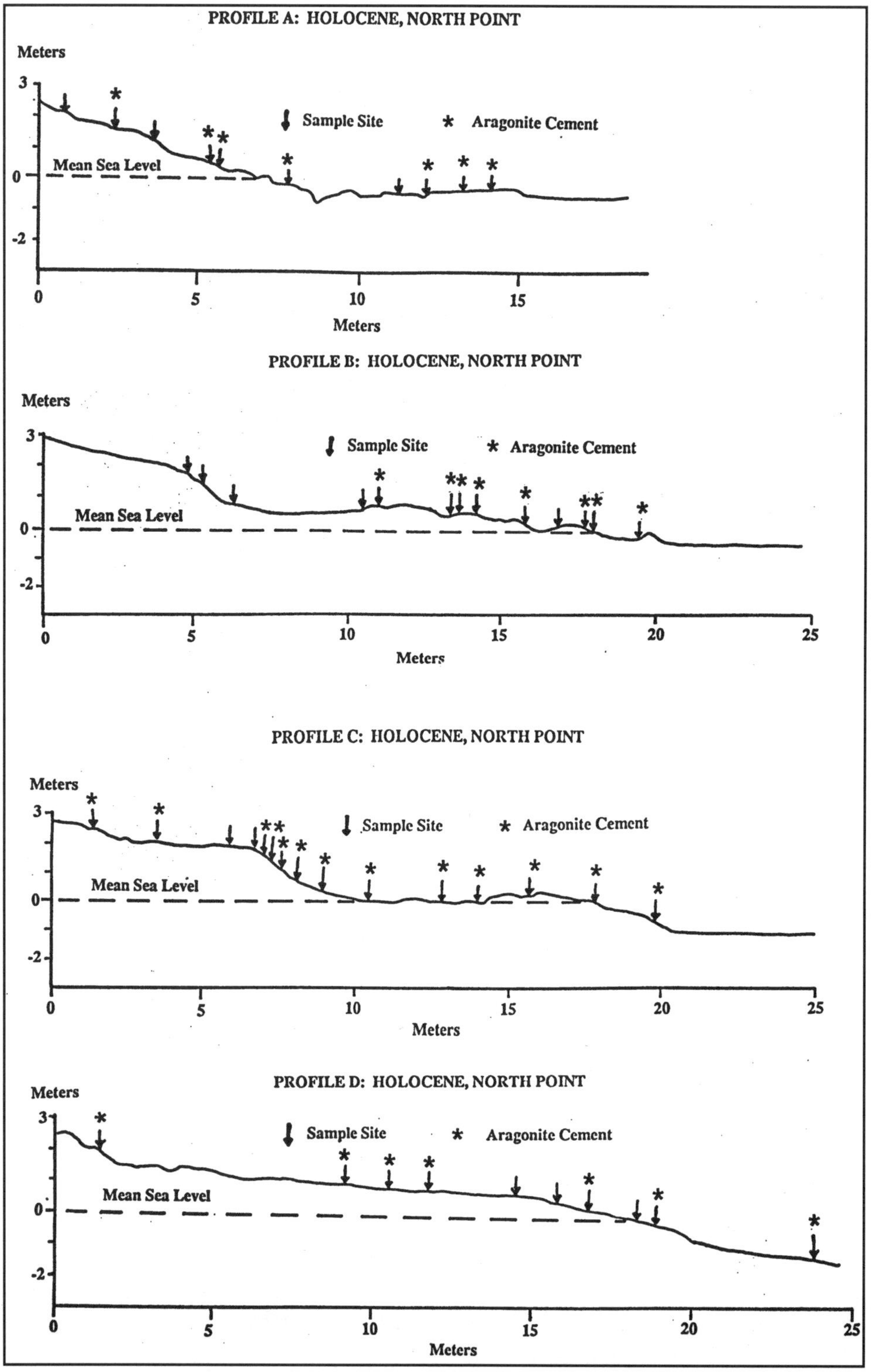

Figure 6. Topographic Profiles A through D in Holocene rocks from North Point, showing sample locations and indicating those samples that contain aragonite marine cement.

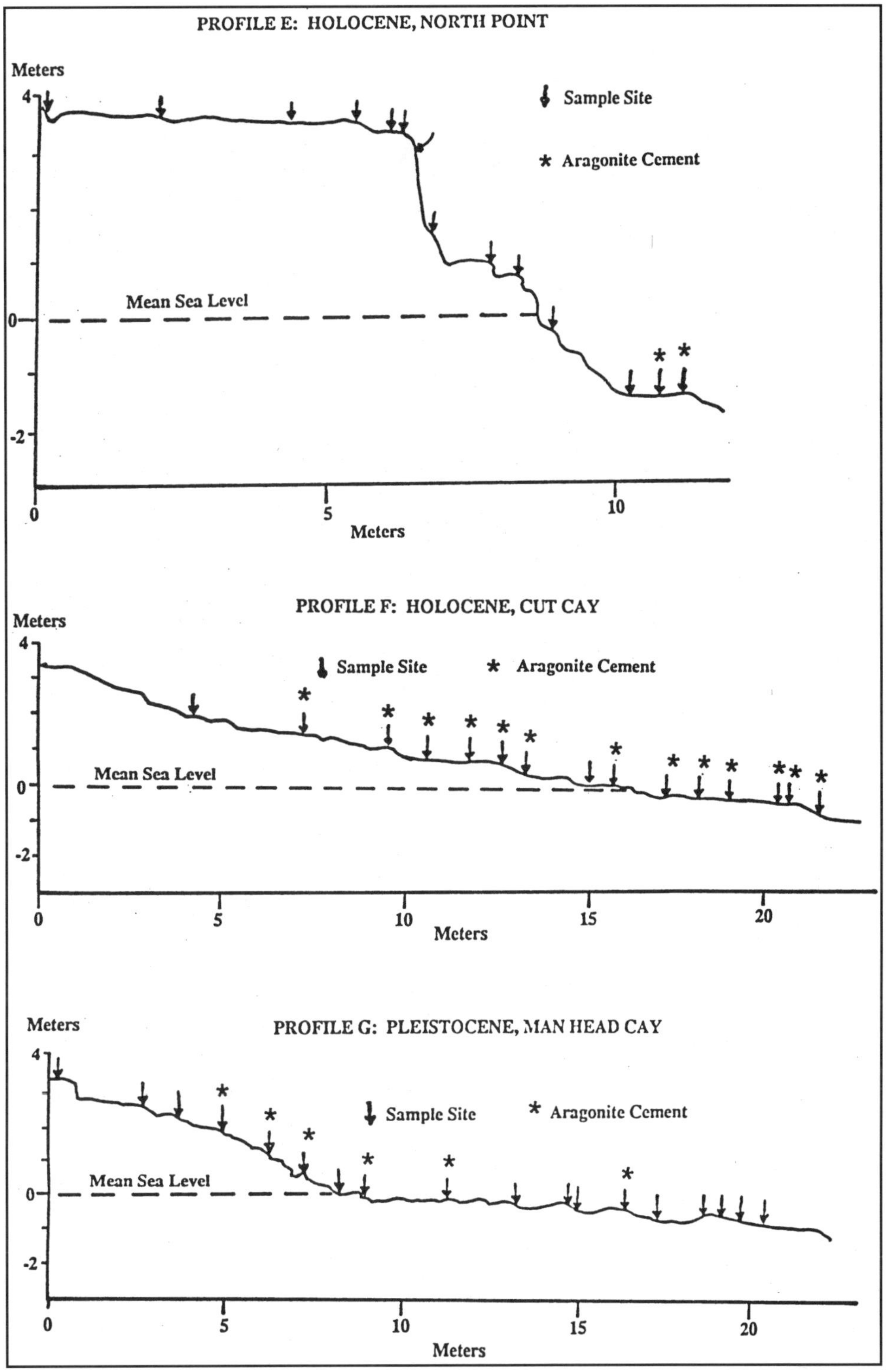

Figure 7. Topographic profiles: Profile E, Holocene rocks from North Point; Profile F, Holocene rocks from Cut Cay; Profile G, Pleistocene rocks from Man Head Cay, showing sample locations and indicating those samples that contain aragonite marine cement.

accelerating voltage of 5,000 volts. Subsequently several samples were analyzed at an accelerating voltage of 2,500 volts. The SEM photographs in this chapter are from this later analysis.

One sample with especially well developed isopachous needle-like cement was analyzed using an energy dispersive x-ray spectrometer, Kevex Microanalyst 8000. The analysis was done to determine the composition and mineralogy of the needle-like cement.

PETROGRAPHY OF THE EOLIANITES OF NORTH POINT AND MAN HEAD CAY

Low-Mg calcite

All of the Holocene eolianite samples from the North Point and Cut Cay localities contain a significant amount of calcite spar cement that binds the grains together (Fig. 8). The calcite spar has a dominant phreatic texture, but in a few places a vadose texture can be seen. The cement crystals are well developed and generally the same size throughout the sample. The spar occurs as both an intergranular and an intragranular cement. Rarely, a calcite rim cement can be found in the Holocene samples.

The Pleistocene eolianites probably were lithified by a calcite spar cement too, but much of this cement has been destroyed. What remains of this original calcite spar is a rim cement found in several of the samples form the seaward end of Profile G. The samples farthest inland contain virtually no cement of any kind and are poorly lithifed. This becomes apparent when walking on these eolianite rocks as they crumble underfoot. This scarcity of cement could be because of the lack of an extensive original cement. In arid climates water is scarce, and its movement is very slow through the sediments (Tucker and Wright, 1990). The percolating water commonly precipitates any excess calcium carbonate it contains by the time it reaches the deeper parts of the freshwater phreatic zone, allowing for very little cementation to occur there. Thus, primary porosity is maintained, and any cement that does occur forms as a small rim of isopachous equant calcite (Longman, 1980). Pleistocene eolianites in the Yucatan have poor induration and a lack of extensive intragranular dissolution. These observations led to the conclusion that the amount of vadose water moving through the dunes during the Pleistocene was small, so rainfall must have been infrequent when most of the Pleistocene rocks were originally lithified (Ward et al., 1985). Similarly, it is possible that the rim cements in the Pleistocene eolianites from Man Head Cay indicate that there was an arid climate at the time that these eolianites were lithified.

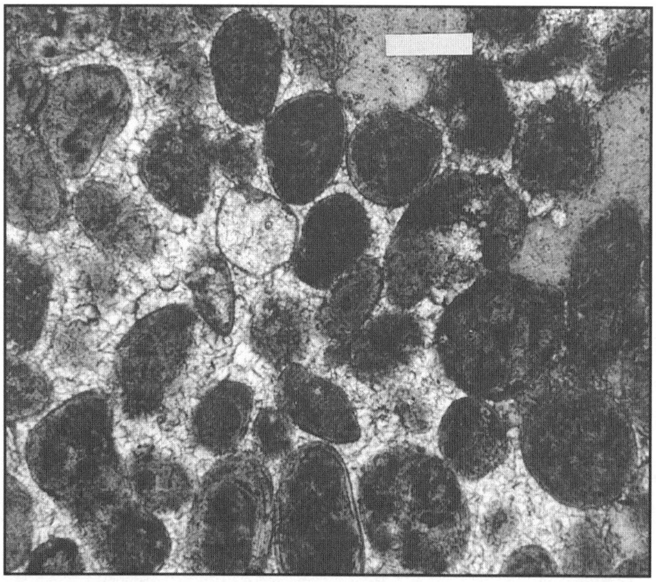

Figure 8. Freshwater calcite spar cement in a Holocene eolianite. Crossed polarizers. Bar scale = 100 μ.

Aragonite cement

Aragonite cement is found in at least one sample from each profile that was measured and sampled in the North Point, Cut Cay, and Man Head Cay localities. Of the 96 samples taken, 55 (57%) contain aragonite. The amount of aragonite cement found ranges from a few needles forming intragranularly in a hole in a grain or a chamber in a bioclast to a well-developed isopachous cement binding the grains together in a large portion of the sample. The aragonite cement most commonly occurs and is most abundant in the intertidal Yellow zone and the overlying Grey zone, but also occurs uncommonly as high as the supratidal White zone and in the only sample collected from the subtidal Green zone.

Aragonite occurs most commonly in the form of needles growing intergranularly and intragranularly, forming isopachous rims both in voids in grains and on grain edges, especially peloids (Fig. 9). Intergranular acicular aragonite is found as an abundant isopachous cement on every grain in some thin sections. In some cases the cement is so well developed that intergrowth of aragonite on neighboring grains forms a meshwork that binds the grains together (Fig. 10). In other samples the intergranular acicular cement is found scattered throughout a thin section on grain edges, but without an isopachous texture. Both the isopachous texture and the acicular habit indicate a marine phreatic cement. Staining with Fiegl's solution and titan-yellow indicated that the needles were aragonite and not high-Mg calcite. This was determined further through SEM analysis, which showed that the crystal morphologies were typical of aragonite (Figs. 11–13) (Scoffin, 1987). Analyses using a Kevex Microanalyst 8000 determined that there is a lack of magnesium in the needle-like cements, which also indicated that the needles are aragonite and not high-Mg calcite. The Fiegl's solution staining revealed that there was some aragonite cement that did not form a needle-like crystal, but was instead micritic, or in rare instances botryoidal, in nature.

The distribution of the aragonite cement is in part dependent on porosity. The cement is best developed and most abun-

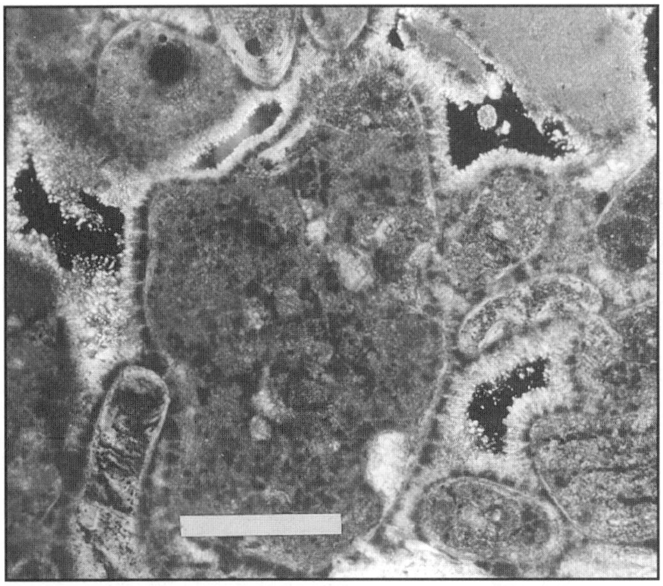

Figure 9. Photomicrograph showing intergranular isopachous aragonite cement. Crossed polarizers. Bar scale = 100 μ.

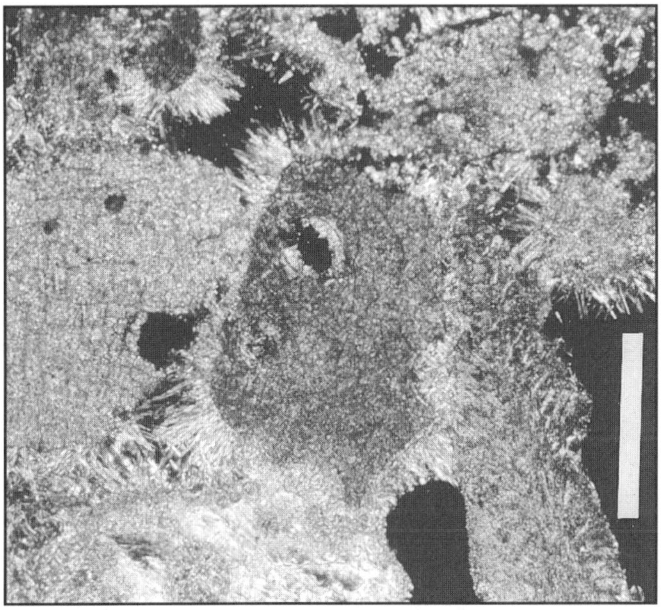

Figure 10. Photomicrograph showing aragonite cement that forms an interlocking meshwork of acicular crystals that in some places occludes much of the intergranular pore space. Crossed polarizers. Bar scale = 100 μ.

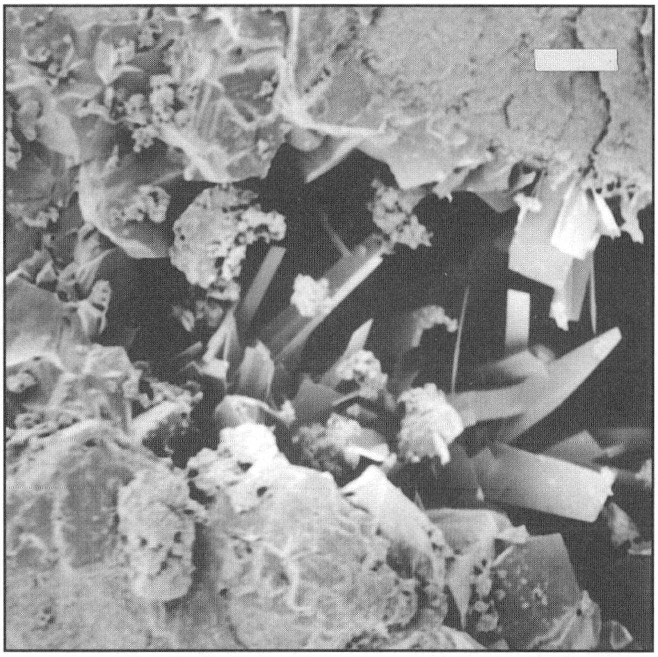

Figure 11. Scanning electron photomicrograph showing an intergranular meshwork of aragonite cement crystals. Bar scale = 20 μ.

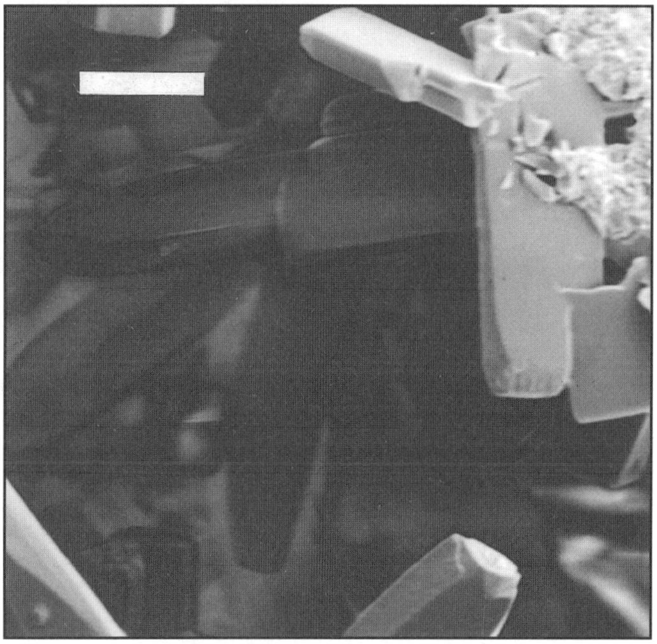

Figure 12. Scanning electron photomicrograph showing a close-up view of the aragonite cement crystals. Bar scale = 10 μ.

dant in areas in the rocks where the fresh-water calcite spar cement is not extensive. This occurs most commonly in coarse-grained layers where the porosity is greater, rather than where the finer grains are more closely packed and the porosity is much less (Fig. 14). These fine- and coarse-grained layers, which are characteristic of eolianites (White and Curran, 1988), are the same as those seen in hand-sample. The aragonite cement also develops in chambers of foraminifera and gastropods (Fig. 15), as well as in dissolution voids in ooids and peloids, and can be found forming on both intergranular and intragranular fresh-water sparite cement (Fig. 16).

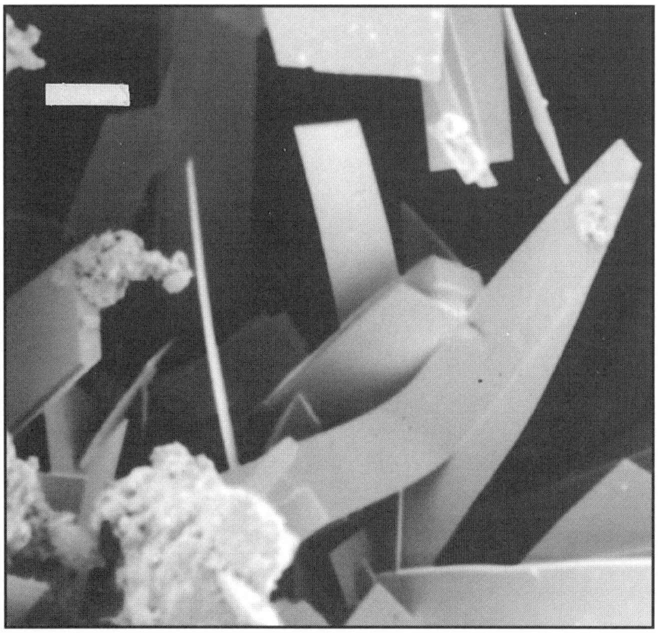

Figure 13. Scanning electron photomicrograph showing a close-up view of the aragonite cement crystals. Note the curvature in one of the needles. This may indicate rapid formation of the aragonite crystals, which may provide an explanation for the presence of marine cements in intertidal and supratidal environments. Bar scale = 10 μ.

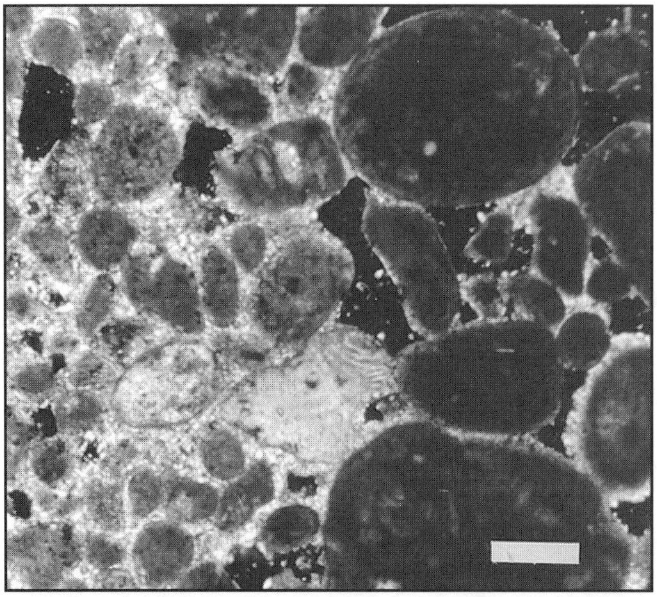

Figure 14. Photomicrograph showing differential freshwater calcite cementation related to differences in grain size within coarser and finer eolianite laminations. Crossed polarizers. Bar scale = 100 μ.

The timing of the formation of aragonite cement is indicated by its relationship to vadose diagenetic features. For example, in Figure 17, an ooid and a peloid are joined, and a dissolution void that cuts through both grains and nearby nonmarine cement is lined with acicular aragonite. The occurrence of acicular aragonite cements on the freshwater calcite spare

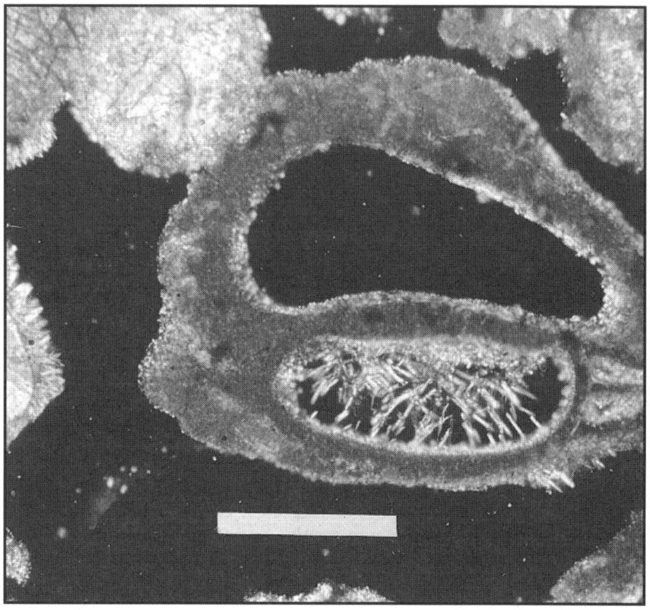

Figure 15. Photomicrograph showing well-developed interlocking needles of aragonite cement within a fossil gastropod. Crossed polarizers. Bar scale = 100 μ.

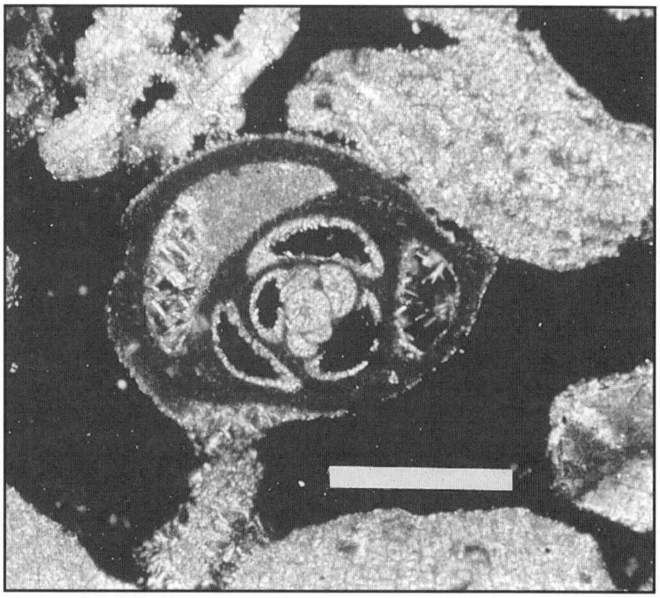

Figure 16. Photomicrograph showing marine aragonite cement postdating nonmarine calcite cement within the chambers of a miliolid foraminiferan. Crossed polarizers. Bar scale = 100 μ.

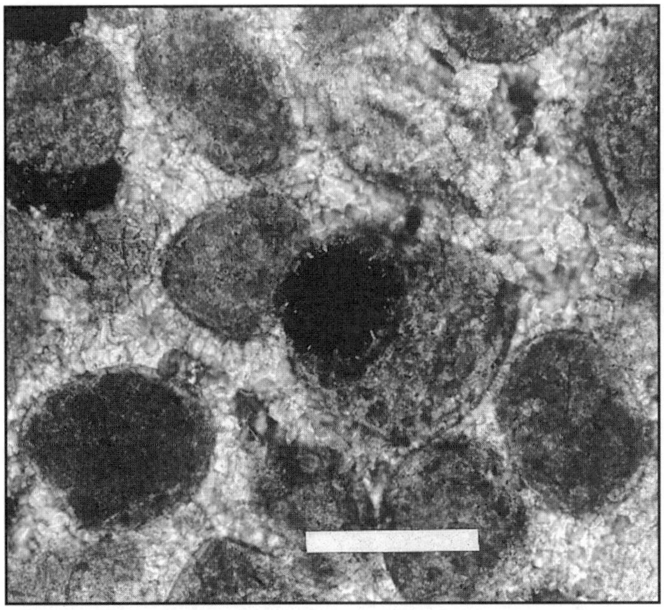

Figure 17. Photomicrograph showing a dissolution pore that cuts across two adjacent grains and some of the nearby intergranular nonmarine calcite cement. The solution pore is lined by needles of marine aragonite cement. Crossed polarizers. Bar scale = 100 μ.

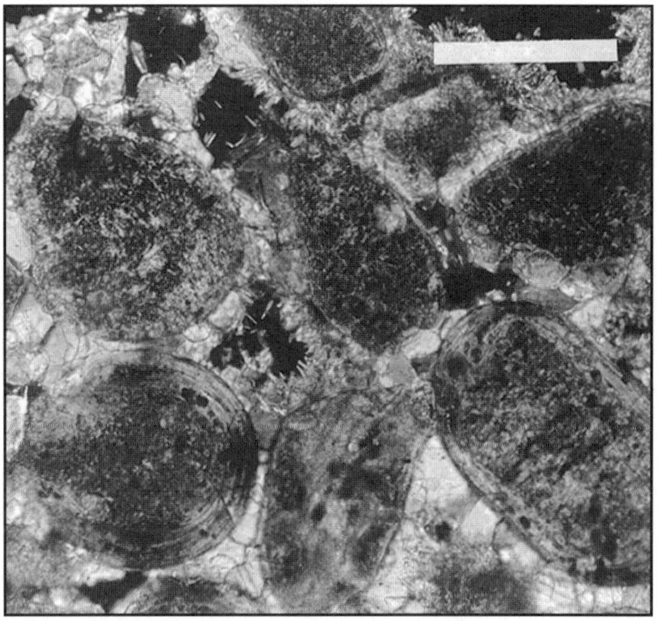

Figure 18. Photomicrograph illustrating the sequence of cementation with an earlier intergranular nonmarine calcite cement on which marine aragonite cement needles grew later. Crossed polarizers. Bar scale = 100 μ.

cement also indicates that the aragonite cement is younger than the spar cement and has formed since the grains were lithified (Fig. 18). The growth of aragonite cement on calcite sparite cement is especially important when it occurs inside chambers of foraminifera and gastropods and in holes in peloid or ooid grains (Figs. 15 and 16) because it indicates that the aragonite cement precipitated after the grains were transported to the subaerial environment, were incorporated into the dunes, and were lithified, rather than precipitating on the sea floor soon after the grains formed.

The abundance and development of the aragonite cement is related to elevation above mean sea level, and therefore to the amount of exposure to seawater, further evidence that the aragonite is a marine cement. The dominant zone for the presence of aragonite is between –1 and +2 m above mean sea level. There are cases in which either limit is exceeded, especially the higher boundary in a promontory section. Waves break with greater force on promontories during high tide, forcing seawater higher than the actual sea level. This does not happen in embayment sections, and only strong storm action can force the seawater up the cliff face. The profiles that are exposed to the waves more regularly are those that are on promontories (Profiles A and C), exposed to more open ocean conditions (Profile F), or located on the windward ocean side of an offshore island (Profile G). The profiles located in embayments, Profile B and Profile D, or the one that is a cliff section, Profile E, contain much less aragonite cement. Although the lower part of Profile E, which is exposed to wave action continuously, does contain very well developed, interwoven needles of aragonite.

ENVIRONMENTS OF MARINE DIAGENESIS

Both high-Mg calcite and aragonite can occur as marine cements (Tucker and Wright, 1990), but aragonite is the dominant marine cement that has been found developing in the eolianite rocks of North Point, Cut Cay, and Man Head Cay. The textures of the aragonite cement are those that are commonly found in the active marine phreatic zone. However Profile B shows aragonite in chambers of bioclasts or in small holes in peloids or ooids, which is somewhat characteristic of the stagnant marine phreatic zone. Aragonite cements are particularly well developed in or near small tidal pools along Profile B. Seawater remains in these pools and is not forcefully moved through the grains. Instead, ion diffusion combined with bacterial action and microboring algae contribute to the development of this aragonite cement (Longman, 1980).

Aragonite cements that formed mainly in the intertidal and supratidal zones commonly are isopachous, indicating precipitation in fluid-filled pores during high tides, beneath semi-permanent tide pools, or where water was held within the rocks by local areas of lower permeability.

There is no indication in the eolianites that there is a mixing zone environment in any of these localities, although their position along the coast would suggest the possibility of one. The absence of a mixing zone may be because of a shortage of fresh water.

The characteristics of beachrock help to explain the sole presence of aragonite and lack of high-Mg calcite as the marine cement developing in these rocks. Beachrock forms in the tropics between the high- and low-tide marks. Cementation of grains on the beach occurs in situ. These cements are normally fibrous aragonite, which may be isopachous, or occur at grain contacts with a meniscus texture. Vadose textures may develop at low tide. The cements are less commonly peloidal, micritic, or fibrous high-Mg calcite (Scoffin, 1987). Thus the most common type of cement found in beachrock is very similar to the type of cement that has been found in the eolianites of this study. This occurrence is especially interesting considering the fact that beachrock is forming today on sandy beaches of Rice Bay, just a few hundreds of meters away from the study area.

GRAIN COMPOSITION

The point count data show that the Holocene rocks all have very similar compositions. Peloids are dominant, occurring usually as 70 to 80% of the grains, but ranging from 50 to 100%. Ooids are the second most abundant grain type. They usually make up 20 to 30% of the grains, but in some cases may be 40 to 50% of the grains. Bioclasts are not present in all of the Holocene rocks. Where present, they commonly make up 1 to 5% of the grains. Bioclasts account for 70 to 90% of the grains in the Pleistocene rocks with the remaining grains being peloids, except for a few samples that have up to 1% ooids (Fig. 19).

DISCUSSION AND CONCLUSIONS

The marine aragonite cement is more abundant, better developed, and extends farther landward in the rocks exposed in the promontories than in the embayments because waves and spray reach farther up the promontory sections. This high-energy environment causes the seawater to be forced through the carbonate grains and to precipitate aragonite cement. High-energy shorelines are areas that are commonly being cemented. Cementation occurs more rapidly along shorelines because evaporation promotes precipitation (Tucker and Wright, 1990).

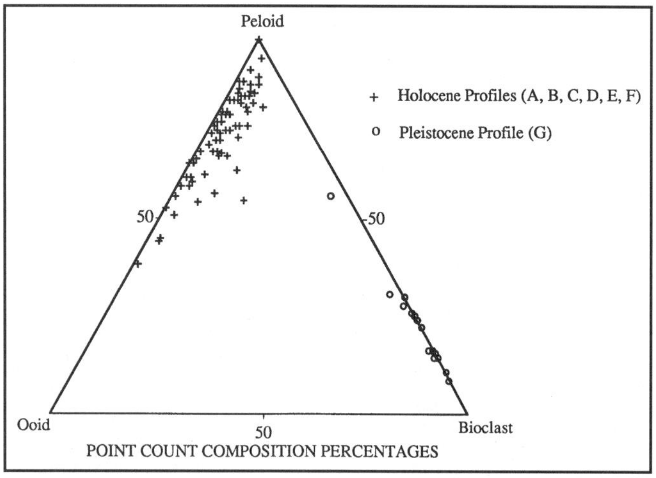

Figure 19. Triangular diagrams showing grain composition of the Holocene and Pleistocene eolianites examined in this study.

The availability of CO_3^{2-} ions is the rate-limiting step in crystal growth. This can account for the acicular nature of aragonite in shallow-marine sediments. The precipitation is mostly driven by CO_2 degassing from agitated seawater, which generates CO_3^{2-} ions from the dissociation of HCO_3^- (Tucker and Wright, 1990).

Although the samples mostly were taken from the vadose environment, the abundance of isopachous aragonite indicates that much of the aragonite formed in fluid-filled pores. Thus the isopachous texture indicates localized pore saturation in the vadose zone, occurring perhaps in tide pools, during high tide, and in pores surrounded by low permeability material.

The fact that aragonite is present in these eolianite rocks, and not high-Mg calcite, suggests that the diagenetic processes occurring in the eolianites of this study are very similar to the majority of those that produce Holocene beachrock along tropical beaches. Cementation takes place preferentially where there is a pumping mechanism to force large quantities of water through a sediment body (Tucker and Wright, 1990), such as the environment of the eolianites in this study and of areas of developing beachrock. Since the mineralogy and the physical environment of the eolianites are similar to those of beachrock, it is possible to conclude that the processes producing marine cement in the Holocene and Pleistocene eolianites examined in this study are similar to those in the beachrock developing along the nearby sandy shores of Rice Bay.

Two things can be determined from the grain composition data. First, the source areas for the Holocene and Pleistocene dunes appear to have been different. The Holocene source was dominated by peloids and the Pleistocene source by bioclasts. The data from the Pleistocene is not extensive enough to make sweeping conclusions, but it does indicate a possibility of different shelf environments during the formation of the Pleistocene and Holocene dunes. Second, the type and amount of marine cement is not dependent on the nature of the grains that make up the rocks. The occurrence and characteristics of the aragonite cement are no different in the Pleistocene samples than in the Holocene samples although the compositions of the rocks are very different (Fig. 20). Thus it is likely that the aragonite is being precipitated directly from the seawater and does not have its source in the grains that make up the rock. This provides supporting evidence that the formation of the aragonite cement is dependent on the exposure of the rocks to seawater.

Because the history of these eolianites is fairly well known, it has been determined that the Holocene eolianites are undergoing their first marine transgression, and therefore their first period of marine diagenesis. The texture, nature of growth, and composition of the marine cement aragonite have been studied to determine the characteristic diagenetic imprints resulting from a marine transgression. The fact that the Pleistocene eolianites have very similar diagenetic history to the Holocene eolianites suggests that they are also undergoing their first period of marine diagenesis.

These characteristics of marine diagenesis are what would be expected in older rocks on San Salvador, for example, the

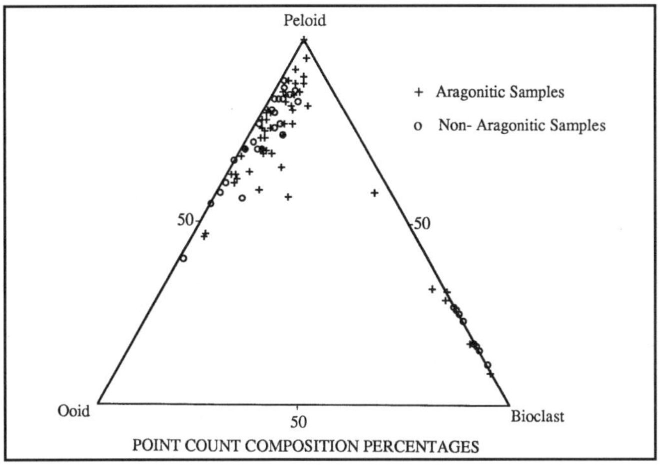

Figure 20. Triangular diagrams showing the distribution of marine aragonite cements in the Holocene and Pleistocene eolianites examined in this study.

rocks of the Cockburn Town fossil reef, if they had undergone a period of marine transgression. In the case of the rocks at the Cockburn Town fossil reef, there are no indications of marine cements that postdate the fresh-water vadose cements or calichification that have developed during the emergent phase (White et al., 1984). Thus the fact that Holocene marine diagenesis is occurring in rocks that are not completely submerged by the transgressing sea indicates that if there had been a period of late Pleistocene sea-level rise to a level equal to, or greater than, present sea level, such as proposed by Mylroie and Carew (1988), there would be marine cements that postdate the fresh-water cements that developed in the reef rocks following emergence caused by post-Sangamon sea regression. It is possible to suggest that the marine cements would have changed to more stable forms of carbonate cement because they are metastable in the fresh-water environment, and so would be unrecognizable as marine cements. However, they could be identified by their acicular habit and isopachous fabric. In addition, older metastable aragonite and high-Mg calcite cements that formed while the reef was still in the marine environment have been preserved (Japy, 1986). Therefore, the evidence of marine diagenesis occurring in the eolianites of the North Point area combined with the detailed work done on the rocks of the Cockburn Town fossil reef indicate that it is unlikely that there was a late Pleistocene sea-level rise to a level similar to present sea level.

More work needs to be done on the Pleistocene eolianites of San Salvador Island to determine if the composition of the Pleistocene eolianites differs as drastically from the composition of the Holocene eolianites, as this study suggests. If the compositions differ enough to indicate a difference in source areas, it may be possible to establish the exact nature of the San Salvador Bank during the formation of the Pleistocene and Holocene eolianites. This may give insight on the possibility, extent, and nature of a late Pleistocene sea-level rise.

Investigation of the older rocks of San Salvador in search of the types of marine cements found in this study to be characteristic of a marine transgression would also provide more information on the possibility of a late Pleistocene sea-level rise. A different sampling method may be necessary to actually find the marine signature. In past studies of the rocks of the Cockburn Town fossil reef, for example, nonweathered samples have been collected in order to determine the older periods of cementation. However, in this study, only rocks from the very surface of the eolianites were collected, and thin sections from the top of these samples were used in the petrographic analysis. Therefore it is possible that evidence for a late Pleistocene sea-level rise may be discovered by sampling the older rocks differently.

ACKNOWLEDGMENTS

I thank the Bahamian Field Station, especially Donald T. Gerace and Kathy Gerace, for full logistical support of the field work. I thank Susan Gaffey and Randolph Steinen for their careful reviews of earlier drafts of this manuscript. I also express my appreciation to Brian White for his help with the field, laboratory, and writing stages of this study. And I thank Virginia White, for her time and help in the production of the photographs, and Dick Briggs, for his help with the scanning electron microscope analysis. Finally, I am indebted to the Keck Foundation for its financial support through grants to the Keck Geology Consortium and to H. Allen Curran for organizing the Keck research project on San Salvador Island during summer 1988.

REFERENCES CITED

Allman, M., and Lawrence, D. F., 1972, Geological Laboratory Techniques: New York, Arco Publishing, 335 p.

Boardman, M. R., Neumann, A. C., and Rasmussen, K. A., 1989, Holocene sea level in the Bahamas, *in* Mylroie, J. E., ed., Proceedings, Fourth Symposium on the Geology of the Bahamas: San Salvador, Bahamian Field Station, p. 45–52.

Carew, J. L., and Mylroie, J. E., 1985, The Pleistocene and Holocene stratigraphy of San Salvador Island, Bahamas, with reference to marine and terrestrial lithofacies at French Bay, *in* Curran, H. A., ed., Pleistocene and Holocene Carbonate Environments on San Salvador Island, Bahamas, Geological Society of America, Orlando Annual Meeting Field Trip Guidebook: Ft. Lauderdale, Florida, CCFL Bahamian Field Station, p. 11–61.

Carew, J. L., Mylroie, J. E., and Sealy, N. E., 1992, Field guide to sites of geological interest, western New Providence Island, Bahamas: Field Trip Guidebook, Sixth Symposium on the Geology of the Bahamas: Port Charlotte, Florida, Bahamian Field Station, p. 1–23.

Chen, J. H., Curran, H. A., White, B., and Wasserburg, G. J., 1991, Precise chronology of the last interglacial period: ^{234}U-^{230}Th data from fossil coral reefs in the Bahamas: Geological Society of America Bulletin, v. 103, p. 82–97.

Choquette, P. W., and Trussel, F. C., 1978, A procedure for making the titan-yellow stain for Mg-calcite permanent: Journal of Sedimentary Petrology, v. 48, p. 639–641.

Curran, H. A., 1985, Introduction to the geology of the Bahamas and San Salvador Island with an overflight guide, *in* Curran, H. A., ed., Pleistocene and Holocene Carbonate Environments on San Salvador Island,

Bahamas, Geological Society of America, Orlando Annual Meeting Field Trip Guidebook: Ft. Lauderdale, Florida, CCFL Bahamian Field Station, p. 1–10.

Curran, H. A., and White, B., 1985, The Cockburn Town fossil reef, *in* Curran, H. A., ed., Pleistocene and Holocene Carbonate Environments on San Salvador Island, Bahamas, Geological Society of America, Orlando Annual Meeting Field Trip Guidebook: Ft. Lauderdale, Florida, CCFL Bahamian Field Station, p. 95–120.

Curran, H. A., White, B., and Thomas, R. D. K., 1989, San Salvador Island Bahamas: A natural laboratory for the study of carbonate sediments and rocks, Pt II, *in* Woodard, H. H., ed., Second Keck Research Symposium in Geology, Colorado Springs, Colorado, Abstracts, p. 18–22.

Hutto, T., and Carew, J. L., 1984, Petrology of eolian calcarenites, San Salvador Island, Bahamas, *in* Teeter, J. W., ed., Proceedings, Second Symposium on the Geology of the Bahamas: San Salvador, CCFL Bahamian Field Station, p. 197–207.

Japy, K. E., 1986, The diagenesis in a Pleistocene coral reef complex, San Salvador, Bahamas [Senior honors thesis]: Northampton, Massachusetts, Smith College, 81 p.

Lawlor, J. F., 1985, A petrographic study of carbonate eolianites at Rice Bay, San Salvador, Bahamas [Senior special studies report, Department of Geology]: Northampton, Massachusetts, Smith College, 58 p.

Longman, M. W., 1980, Carbonate diagenetic textures from nearsurface diagenetic environments: American Association of Petroleum Geologists Bulletin, v. 64, p. 461–487.

Mylroie, J. E., and Carew, J. L., 1988, Solution conduits as indicators of Late Quaternary sea level position: Quaternary Science Reviews, v. 7, p. 55–64.

Schneidermann, N., and Sandberg, P. A., 1971, Calcite-aragonite differentiation by selective staining and scanning electron microscopy: Gulf Coast Geological Transcript, v. 21, p. 349–352.

Scoffin, T. P., 1987, An Introduction to Carbonate Sediments and Rocks: New York, Chapman and Hall, 274 p.

Stephenson, T. A., and Stephenson, A., 1950, Life between the tide-marks in North America: Journal of Ecology, v. 38, p. 354–402.

Titus, R., 1983, Quaternary emergent facies patterns on San Salvador Island, Bahamas, *in* Gerace, D. T., ed., Field Guide to the Geology of San Salvador: San Salvador, CCFL Bahamian Field Station, p. 97–117.

Tucker, M. E., and Wright, V. P., 1990, Carbonate Sedimentology: Oxford, United Kingdom, Blackwell, 482 p.

Ward, W. C., Weidie, A. E., and Back, W., 1985, Geology and Hydrogeology of the Yucatan and Quaternary Geology of Northeastern Yucatan Peninsula: New Orleans, Louisiana, New Orleans Geological Society, 160 p.

White, B., and Curran, H. A., 1985, The Holocene carbonate eolianites of North Point and the modern marine environments between North Point and Cut Cay, *in* Curran, H. A., ed., Pleistocene and Holocene carbonate environments on San Salvador Island, Bahamas: Geological Society of America, Orlando Annual Meeting Field Trip Guidebook: Ft. Lauderdale, Florida, CCFL Bahamian Field Station, p. 73–93.

White, B., and Curran, H. A., 1988, Mesoscale physical sedimentary structures and trace fossils in Holocene carbonate eolianites from San Salvador Island, Bahamas: Sedimentary Geology, v. 55, p. 163—184.

White, B., Kurkjy, K. A., and Curran, H. A., 1984, A shallowing-upward sequence in a Pleistocene coral reef and associated facies, San Salvador, Bahamas, *in* Teeter, J. W., ed., Proceedings, Second Symposium on the geology of the Bahamas: San Salvador, CCFL Bahamian Field Station, p. 53–70.

Manuscript Accepted by the Society January 5, 1995

Printed in U.S.A.

Geological Society of America
Special Paper 300
1995

Stratigraphic setting of a subtidal stromatolite field, Iguana Cay, Exumas, Bahamas

Russell S. Shapiro,* Ken R. Aalto, Robert F. Dill
Humbolt State University, Arcata, California 95521 and Caribbean Marine Research Center, Lee Stocking Island, Bahamas
Ray Kenny*
Institute for Arctic and Alpine Research, University of Colorado, Boulder, Colorado 80309

ABSTRACT

This chapter describes stratigraphic relationships among late Pleistocene through Holocene sediments that floor an interisland stromatolite channel between Iguana and Ralph Cays at the southern end of the Exuma Cays, Bahamas. In addition, descriptions and genetic interpretations are presented for normal marine, giant subtidal stromatolites that formed in conjunction with the growth of an ooid sand shoal near the end of the Holocene marine transgression. This study develops a sequence stratigraphy for the Exumas that could be applied in making regional correlations throughout the Bahama Bank and provides a modern model for stromatolite genesis that has use in interpreting ancient counterparts.

INTRODUCTION

Extensive research has been completed on Pleistocene-Holocene stratigraphy and carbonate sedimentology in both the western and eastern portions o the Great Bahama Bank, located in the vicinity of Andros and on Salvador Islands (see summaries of previous work in Beach, 1982; Sheridan et al., 1988; Dill et al., 1989; Carew and Mylroie, 1990; and other chapters in this volume). However, little attention has been given to the geology of the Exuma Islands (see summaries in Kendall et al, 1989; Dill, 1991). Focus on the stratigraphy and sedimentation in this part of the Bahamas has increased since the discovery of giant, subtidal stromatolites near Lee Stocking Island (Dill et al., 1986) (Fig. 1). While most of the effort has been given to investigating the stromatolites (e.g., Griffin, 1987; Shapiro, 1989; Dill, 1991; Riding et al., 1991), preliminary work has been carried out on the seawater chemistry (Kendall et al., 1990; Falls et al., 1990), petrology (Aalto and Shapiro, 1990), carbonate geochemistry (Kenny et al., 1991) and mud beds (Dill et al., 1988; Shinn et al., 1993) as well as investigations of stromatolite locations outside of Lee Stocking Island (Reid and Browne, 1990, 1991; Shapiro et al., 1990, 1992; Browne and Ginsburg, 1991; Dill et al., 1991).

METHODS

Rocks were hand-sampled from depths 6 m below mean high water to island crests (approximately +5 m in elevation). In the summer of 1992 we drilled four cores, 10 cm in diameter, through the submarine surface and recovered material ranging from 58 to 107 cm in length (Fig. 1, Table 1). Thin-sections were made of samples representative of rock units encountered in the field. Half of each section was stained using alizarin red S and potassium ferricyanide. Point counts of 300 points per thin-section were made to determine relative percentages of grains, cement, and unfilled pore space. Relative percentages of grain constituents were also determined. In order to compare grain assemblages within the stromatolite and those in the surrounding dunes, 300-point grain counts were completed on loose sediment samples collected in the stromatolite field. X-ray diffraction analysis was performed on whole-rock samples of the varied stratigraphic units. Particle-size determination was

*Present address: Shapiro, Department of Geological Sciences, Preston Cloud Research Laboratory, University of California, Santa Barbara, California 93106-9630; Kenny, Geology Department, New Mexico Highlands University, Las Vegas, New Mexico 87701.

Shapiro, R. S., Aalto, K. R., Dill, R. F., and Kenny, R., 1995, Stratigraphic setting of a subtidal stromatolite field, Iguana Cay, Exumas, Bahamas, *in* Curran, H. A., and White, B., Terrestrial and Shallow Marine Geology of the Bahamas and Bermuda: Boulder, Colorado, Geological Society of America Special Paper 300.

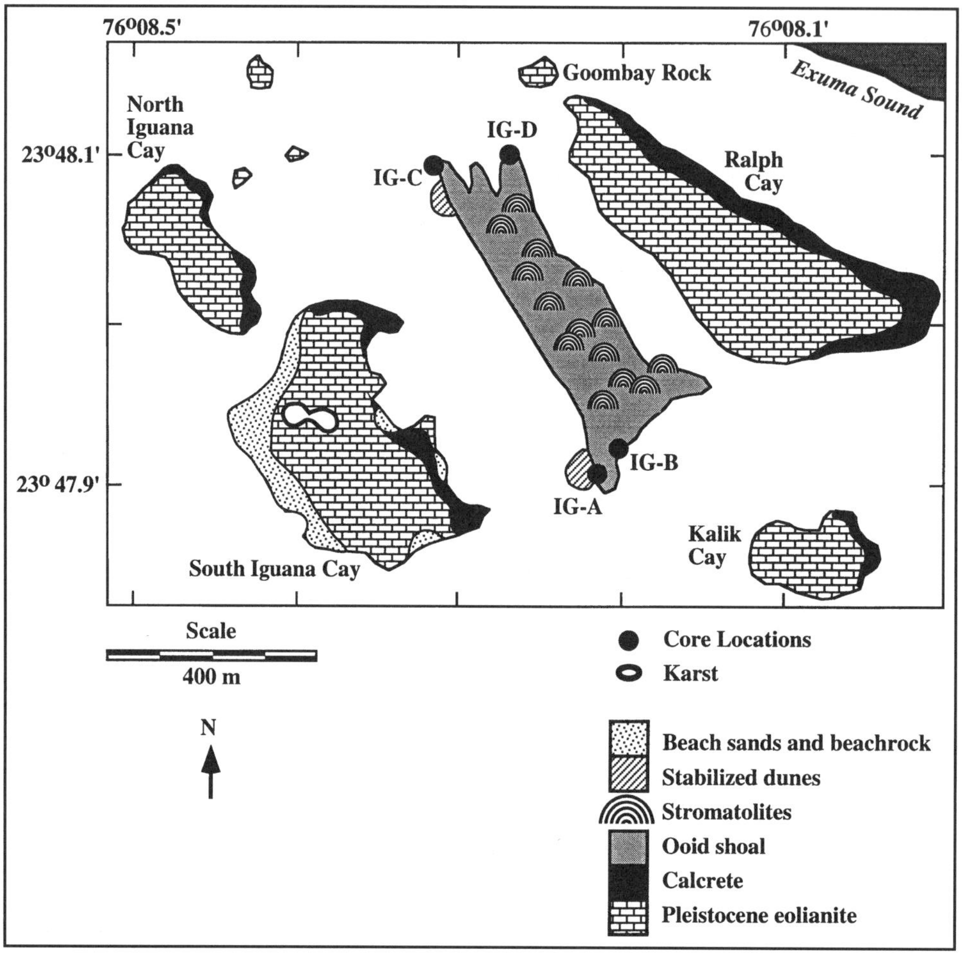

Figure 1. Simplified geologic map of the field area in the southern Exuma Islands showing core locations. Shaded area of Exuma Sound represents depths greater than 40 m below mean sea level.

TABLE 1. DRILLCORE DESCRIPTIONS

IG-A	IG-B	IG-C	IG-D
107-cm eolianite with foreset laminae and rhizomorphs	8- to 12-cm grainstone with conchs	5-cm dead coral head	4-cm stromatolite
	- - - *Sharp contact* - - -	- - - - - -	18-cm dead coral head
	1- to 5-cm gray, bored calcrete*	12-cm brown calcrete with vugs	- - - Bored contact - - -
	- - - - - -	- - - *Gradational contact* - - -	36-cm eolianite with foreset laminae and rhizomorphs
	9- to 13-cm brown calcrete with vugs	27-cm eolianite breccia with rhizomorphs	
	- - - - - -	- - - - - -	
	60-cm eolianite with foreset laminae	30-cm eolianite with foreset laminae and rhizomorphs	
Total thickness: 107 cm	Total thickness: 90 cm	Total thickness: 81 cm	Total thickness: 58 cm

*^{14}C age-dated at 13,230 ± 201 B.P.

carried out on seven loose sediment cores recovered from the stromatolite channel. Carbonate samples were prepared from chips that were quarried from the interior of hand samples and crushed to <74 mm (200 mesh) or subsampled with a 1.0 or 0.5-mm diameter drill. CO_2 was extracted from carbonates by reaction with 100% H_3PO_4 at 25.2°C (McCrea, 1950).

REGIONAL SETTING

The Exuma Cays are a chain of low-lying islands forming the boundary between the eastern margin of the Great Bahama Bank and the deep Exuma Sound (Fig. 1). The sound is one of several deep, steep-sided submarine valleys and oceanic re-entrants that dissect the bank. It lies 2 km to the east of these islands and furnishes oceanic waters to the bank every high tide. The bank margin setting is characterized by a zone of intense mixing of oceanic and platform waters, resulting in active sea-floor cementation and the proliferation of microbial mats on hardgrounds (Dill, 1991). Field reconnaissance and radiometric age-dating of samples collected in the Exumas led to recognition of several lithostratigraphic units within the islands. These include a cemented core of Pleistocene oolitic dunes capped by an extensive paleosol, which in turn is overlain by Holocene eolianite and beach rock (Figs. 1 and 2; Dill et al., 1989; Kendall et al., 1989; Curran and Dill, 1990; Halley et al., 1991; Kindler, 1991). Some islands have elevated Pleistocene reef rock (framestone) formed on their western perimeters (Halley et al., 1991). The lack of framestone of Sangamon age on the eastern, or windward, margin is probably the result of erosion, and not contemporaneous environmental factors (Dill et al., 1991).

Stromatolites are found in some of the interisland channels and windward margins of islands associated with ooid shoals (Dill at al., 1986; Shapiro et al., 1990, 1992; Reid and Browne, 1991). Although all of the stromatolite locales are in the shallow subtidal to intertidal range, there is much diversity in relative physical stresses, sediment budget, meiofauna and microbes, and especially internal construction.

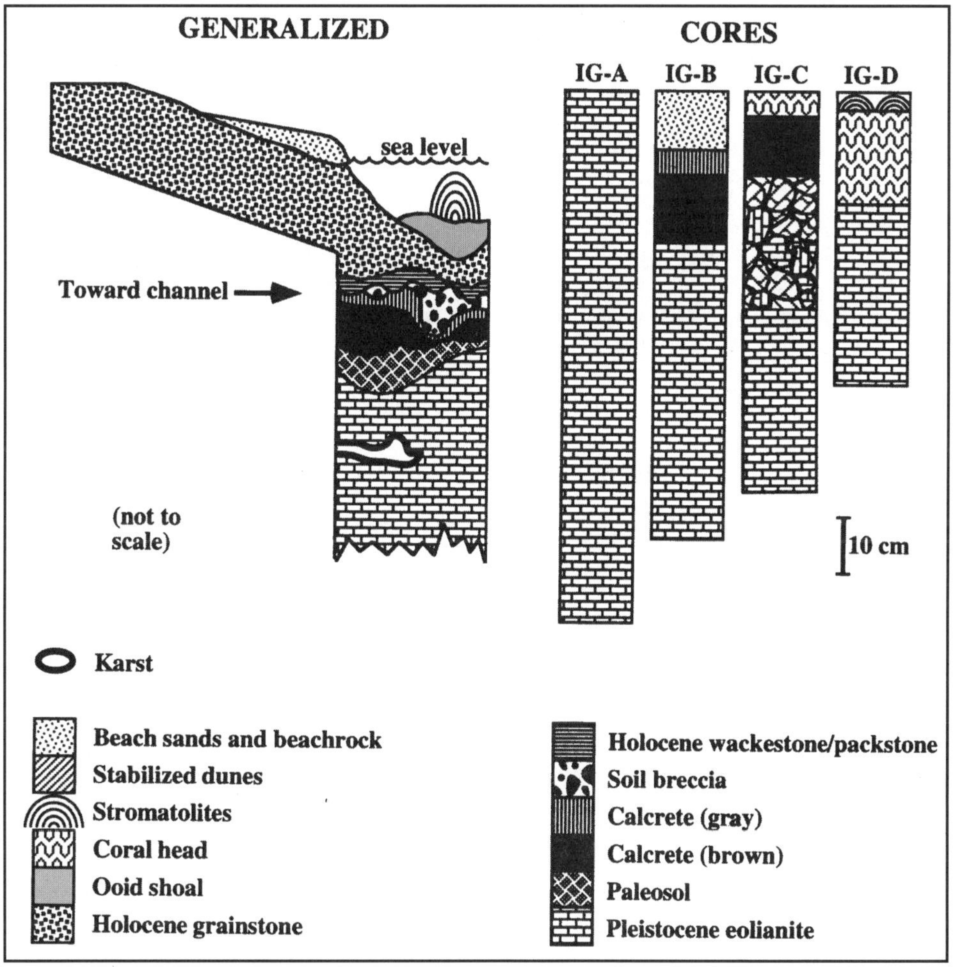

Figure 2. Representative stratigraphic columns through Iguana Cay and the stromatolite channel compiled from field and core data. Core descriptions are given in Table 1.

STRATIGRAPHY AND SEDIMENTOLOGY

Pleistocene rock units

Eolianite. Pleistocene eolianites include grainstones that have bedform structures and *Cerion* fossils. These are exposed in cores taken from the sea floor and are found on the island crests. Carbonate sands are organized into multiple sets of cross-bedded dunes with tabular- to trough-shaped set contacts (Fig. 3). Individual sets commonly range in thickness from 0.5 to 2 m. Large foresets are common on the present-day leeward (western) sides of the islands.

One can trace individual steeply dipping foresets into the Iguana Cay Channel; their presence at depths of up to 5 m is suggested by the alignment of ridges of rock outcrops capped by patches of *Sargassum*, which parallel the trend of foresets exposed on the islands.

The Pleistocene eolianite consists of superficial ooids with intraclast cores and intraclasts cemented by blocky spar (Table 2, Fig. 4). The second most abundant ooid population consists of ooids with pelloidal nuclei, with many concentric laminae. On nearby islands to the southwest these ooids are much more abundant in Pleistocene than Holocene eolianites (Aalto et al, 1992). Some intraclasts have undergone coalescive neomorphism. Skeletal fragments, grapestone, and some single and composite ooids are partially micritized. Foreset laminations are defined by differences in grain size, with finer laminae consisting of well-sorted fine sand and coarser laminae of well-sorted medium sand.

Eolianite from submarine exposures contains an average of 81% grains and 19% cement (n = 5; range, 14 to 23 percent) with no unfilled pore spaces. Bulk x-ray diffraction of one sample indicates the presence of 80 to 90% aragonite and 10 to 20% of both high- and low-Mg calcite. Samples from subaerially exposed outcrops are chiefly calcite and contain an average of 77% grains and 23% blocky spar or empty pore space (n = 3; range, 17 to 33%). Only one sample has unfilled pores (9% by area), which in places are vuggy in appearance.

Paleosol. The windward stoss sides of these fossil dune complexes are the sites of well-exposed and extensive paleosol exposures, rhizomorphs, and relict pre-Holocene solution pits that contain reddish brown soil breccias. Both the durability of these crusts on the exposed high wave energy sides of the windward parts of islands and the tendency for large foreset-bound blocks to calve off into the ocean as island margins are undercut by bioerosion nips on the leeward sides of islands; this has resulted in evolution of a present-day island morphology that mimics the geometry of the fossil dunes. The small sea cliffs afford excellent exposures of the anatomy of soil zones, dune morphology, and the extensive past colonization of the unconsolidated dunes by a dense growth of plants that formed extensive networks of rhizomorphs. A soil cover developed on the cemented Pleistocene carbonates of the Exuma Islands. The remnants of these paleosols occur on most of the islands and

Figure 3. Pleistocene eolianite on Ralph Cay capped by calcrete crust. The foresets dip toward the bank interior. Calcrete coats a fracture surface at the bottom center left.

appear as smooth, reddish brown crusted and brecciated surfaces. They often form the boundary between Pleistocene and Holocene transgressive deposits. Unlithified portions of paleosols are readily eroded; however, the associated calcretes and soil breccias derived from these soils are densely cemented and form resistant bodies that impede physical erosion and chemical dissolution (no permeability). These paleosols commonly are underlain by well-developed pisolths and *Cerion* spp. shells. these surfaces are also found exposed as a resistant feature underwater, down to 8 m below present mean sea level. On islands that surround the Iguana Cay Channel, exposed paleosols cap Pleistocene eolianite, especially on the windward sides of the islands where vegetation is sparse due to sea spray (Aalto et al., 1992).

Calcrete crust. Calcrete crusts on Pleistocene sediments in the study area are commonly laminated and range in thickness from 1 mm to 10 cm. Crusts range in color from medium gray to pink to light brown. At most sites a single crust is developed in profile; however, crusts can bifurcate along fractures and follow several foreset laminations in a single cross-bed set, or be developed along several set contacts. Rhizoconcretions are common beneath the crusts and in some instances are lined by calcrete (Fig. 5). In most places the calcrete is red-stained, especially along joints. In sea-floor core IG-B (Fig. 1, Table 1), 1 to 5 cm of extensively bored, dense, medium gray calcrete overlies an abrupt contact with 9 to 12 cm of brown, crudely laminated, rhizomorph-bearing calcrete. Both gray and brown calcretes have formed as a result of micritization of oosparite. Aragonite-lined borings into the gray calcrete are common, and in the core, unlithified pockets of shells are present in borings. The brown calcrete grades downward into eolianite.

The calcrete is micritic, finely laminated and contains minor filed or partially filled, subhorizontally oriented fenestrae, as well as irregular iron oxide–rich patches and veins of

TABLE 2. POINT COUNT DATA OF SAMPLE CONTENTS, BY PERCENT*

	Pelloids	Pelloids w/ skeletal fragments	Intraclasts	Grapestone	Ooids w/ intraclasts	Ooids w/ skeletal fragments	Ooids w/ pelloids	Composite ooids
				PLEISTOCENE EOLIANITE				
Whg	1	tr	18	2	69	-	8	2
I1a	1	tr	68	-	26	-	5	-
I3a	4	-	47	6	33	tr	10	-
I3b	8	1	31	-	36	1	23	tr
I3c	10	2	26	-	36	-	26	-
I3d	7	tr	47	-	33	tr	13	-
IC1†	1	tr	16	-	53	-	30	tr
RC1†	15	4	11	-	44	-	15	-
RC3†	9	11	49	5	23	-	3	-
Mean	6	2	35	1	41	tr	15	tr
				HOLOCENE SUBTIDAL WACKESTONE/PACKSTONE				
I4a	14	25	48	4	8	-	tr	tr
				HOLOCENE SUBTIDAL GRAINSTONE				
I4b	9	19	43	8	16	-	4	1
In1	17	10	29	-	42	1	1	tr
I1c	5	11	30	10	38	2	2	3
I1d	6	16	23	6	32	7	8	2
Mean	9	14	31	6	32	2	4	2
				HOLOCENE BEACHROCK				
IC5	2	1	21	-	65	1	8	2
				WITHIN A STROMATOLITE				
S1	11	2	35	1	48	-	3	-
				IGUANA CHANNEL OOID SHOAL SANDS				
O1	2	4	26	14	51	1	2	-
O2	4	tr	32	6	54	2	1	1
O3	tr	3	28	11	57	-	1	-
O4	1	5	48	7	38	-	-	1
O5	1	2	29	4	60	2	-	2
O6	1	2	28	6	60	1	2	-
O7	1	2	32	4	58	3	-	-
Mean	2	3	32	7	54	1	1	tr

*300 grains per sample.
†From subaerial outcrops; otherwise subtidal.

microspar (Fig. 6). In places there are floating grains, spherulites, and irregular cracks that may cut through grains (Fig. 6B). Some calcrete zones are extensively cut by subvertically oriented anastomosing sparry calcite veins. Where the lowermost calcrete layers overlie unaltered Pleistocene eolianite, the layers appear to truncate grains sharply along irregular to wavy solution surfaces (Fig. 6C). Elsewhere, the eolianite-calcrete boundary is gradational and defined by increasing micritization of grains and/or sparmicritization of intergranular cement (cf., Kahle, 1977) (Fig. 6A). Within the calcrete zone, remnant grains "float" in micrite.

Laminations are wavy to highly contorted and are defined by several criteria: (1) sizes of crystals within the micritic cement, (2) concentration of insoluble residues, (3) concentration of filled and unfilled fenestrae, and/or (4) grain destruction (Fig. 6B). Gray calcrete contains irregular patches of iron oxide, and has a poorly sorted muddy texture, clotted micrite ("structure grumeleuse"; Esteban and Klappa, 1983), and irregular, anastomosing microspar-filled fenestrae ("alveolar texture"; Esteban and Klappa, 1983). Sediment- and cement-filled veins and vugs are abundant, locally defining solution breccias (Fig. 16). The laminated calcrete–Pleistocene sediment contact roughly approximates the paleohorizontal when viewed over several islands, and veins are largely oriented subvertically with respect to crust.

A sample of calcrete, exposed on submerged karst marginal to the south end of the ooid shoal at a water depth of 7 m, yielded a whole-rock ^{14}C age of 13,203 ± 201 (see core IG-B in

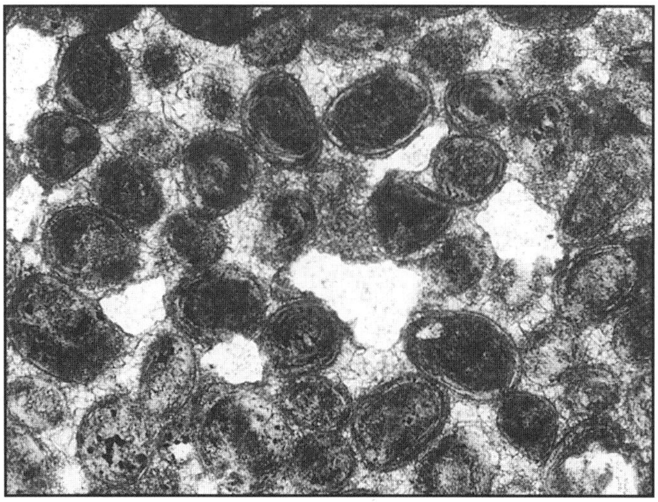

Figure 4. Photomicrograph of Pleistocene eolianite (plane light; field of view = 1.3 mm wide). The superficial ooids and intraclasts are cemented by blocky calcite spar, with crystal size increasing into the center of filled pores. Most of the unfilled interparticle pore spaces were filled with epoxy during thin-section preparation.

Figure 5. Rhizoliths developed in eolianite on Ralph Cay. The resistant layer is a calcrete that caps small karst towers. Hammer handle = 30 cm long.

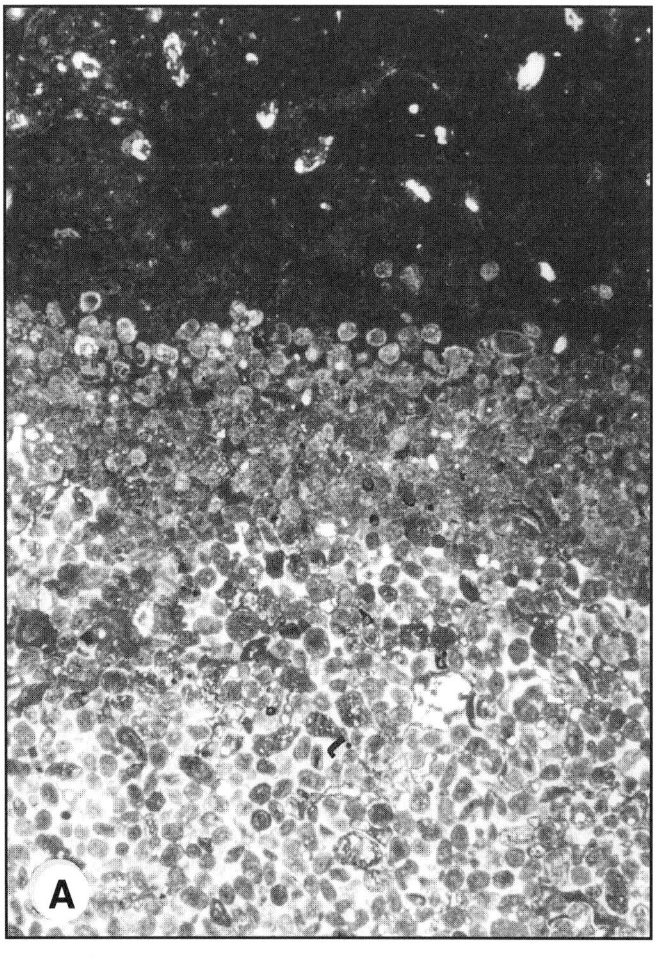

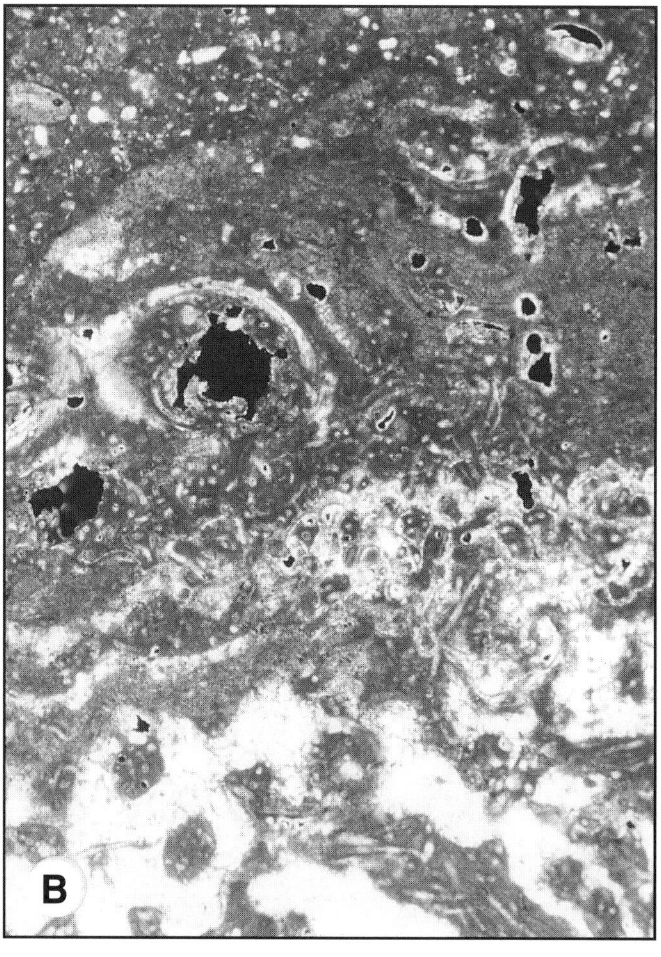

Fig. 1 and Table 1). Calcrete and paleosol from several cays in the Exuma chain are depleted in ^{13}C and ^{18}O relative to modern marine ooids. The $\delta^{13}C$ values range from –3.2 to –2.0‰ (PDB). The $\delta^{18}O$ values range from –6.8 to –9.5‰ (PDB) and form the subaerial (meteoric water) isotope domain in Figure 7. X-ray data show high-Mg calcite in one calcrete sample taken from a submerged paleosol overlying muddy substrate, but show low-Mg calcite plus trace amounts of aragonite for two other samples from submerged areas. Subaerially exposed calcrete contains dominantly low-Mg calcite. Many borings in submerged calcrete are lined with aragonite, which could easily contaminate bulk samples of calcrete.

Soil breccia. Brick red to brown "terra rosa" soil breccias are preserved in solution pipes and pits on many karsted surfaces on the stoss side of Iguana Cay and the southeastern corner of Ralph Cay. They are commonly separated from the Pleistocene eolianite by calcrete crust and appear to grade mesoscopically and microscopically into this crust. These breccias are more resistant to solution than the surrounding limestone. In some instances breccia masses that originally filled depressions stand above the present erosion surface as remnants, a reversed topography to that of deposition. Soil pisoliths may be present in the vicinity of the breccia-calcrete interface. Breccias contain poorly sorted, angular to subangular clasts of light gray to tan eolianite dispersed in the red matrix (Fig. 8). Clasts are commonly granule- to cobble-sized and lack preferred orientation. The breccia fabrics are both matrix- and last-supported. Most solution pits are cylindrical and subvertical in orientation, although in places the breccias are preserved in solution crevices developed along joints. At some sites, red breccia is present in subhorizontal patches.

The boundary between gray calcrete and red soil breccia is marked by an increase in the convolution of laminae, an increase in abundance of filled and unfilled fenestrae and vugs, and the appearance of breccia clasts of less altered light gray eolianite (Fig. 9). Eolianite clast margins commonly display micritization and sparmicritization. In some of the solution pipes, the rims of the clasts have been blackened to a dark gray. Some vugs are infilled with acicular isopachous rim cement and overlain by calcite flower spar cement, which radiates into and fills pore spaces. The opacity of thin sections of breccia attests to high iron content; however, iron hydroxides are too finely disseminated to appear as crystals in thin-section.

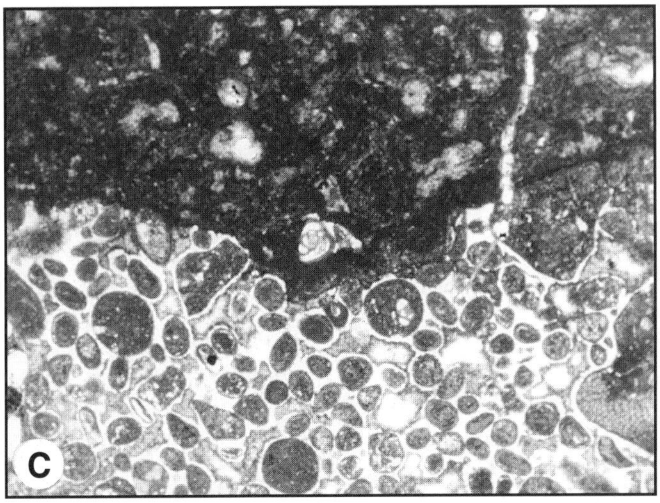

Figure 6 (on this and facing page). Photomicrographs of calcrete overlying Pleistocene eolianite (plane light; field of view = 3.25 mm wide). A, There are three gradational zones visible, top to bottom, in this photograph: unaltered eolianite; eolianite with partially micritized grains and cement; and calcrete with "floating" grains, clotted texture, and vesicles. B, The wavy nature of the laminations is defined by variations in the concentration of iron hydroxides, the proportions of less altered grains, and the presence of microspar veins that fill circumgranular fractures. Wispy microspar veins may be infilled anastomosing root pores (bottom and center left). Larger subvertical, calcite-filled anastomosing veins crosscut the previously formed wispy veins at the bottom. C, Note the fairly abrupt contact between the clotted textured calcrete and unaltered eolianite (top). The outline of some grains can be traced across this contact, while others are sharply truncated.

Paleokarstic surface

The effects of Pleistocene sea-level lowering and consequent genesis of solution topography are evident in both island and sea-floor outcrops of the Bahamas (Benjamin,1970; Dill, 1977). Shoreline shallow caves and sinkholes are extensively developed along fractures, especially at the intersections of joints in the study area (Aby et al., 1992). Several ponds exist in a large group of joint-controlled sinkholes in the interior of South Iguana Cay. A cave formed within the Ralph Cay nip exhibits an undulating floor, ceiling patterns, and speleothems. Many of the cays and islands of the region also have similar shallow shoreline caves. Most of the original oolitic grains forming the limestones in which these caves developed have been altered or have gone into solution, forming a deposit composed of calcite whisker cements.

Holocene rock units

Holocene marine wackestone/packstone. Holocene marine wackestone and packstone, sampled 5 m below mean high water at the southern margin of the sand shoal (Fig. 1), consists of well-indurated, tan aphanitic limestone containing dispersed shell fragments. This rock type has been found only in submerged karst solution pits and scour pockets on the sea floor near where core IG-B was taken. Grain size is too diverse to permit statistically valid point counting for all but one sample. This rock grades from a poorly sorted, muddy wackestone to packstone. The rock lacks unfilled pore space and consists of abundant micrite with a poorly sorted mix of skeletal fragments, ooids, peloids, intraclasts, and grapestone in patchy distribution (Table 1). Acicular aragonite is absent and cements consist of both microspar and spar. Bulk x-ray data show some aragonite, but the composition is chiefly high-Mg calcite. A queen conch shell

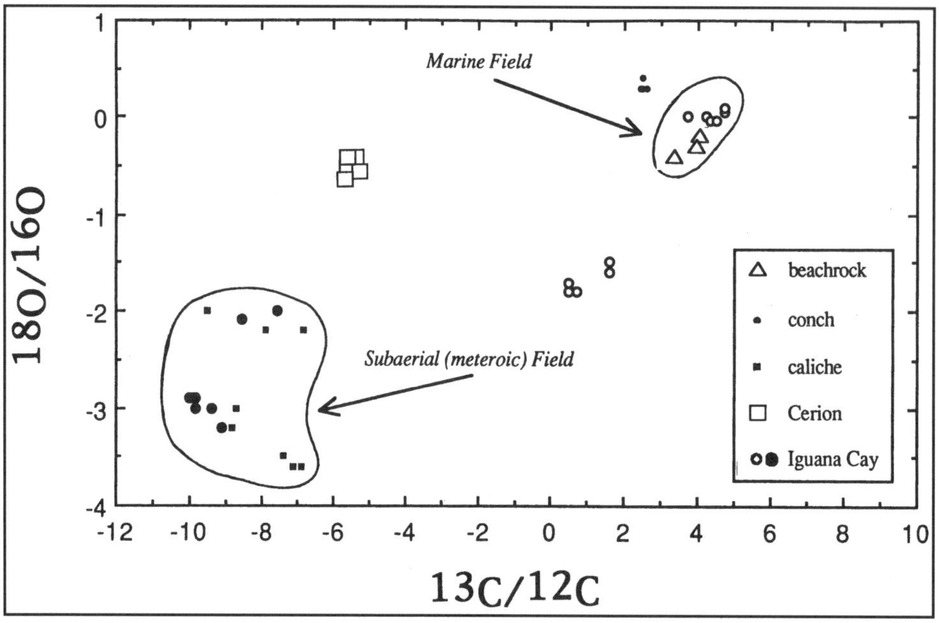

Figure 7. $^{18}O/^{16}O$ and $^{13}C/^{12}C$ ratios of carbonates, southern Exumas, Bahamas. Envelopes indicate general limits of isotopic variation observed in Holocene subaerial and marine carbonate. Data for all Iguana Cay Pleistocene substrate and calcrete samples (large dots) plot in the meteoric water field, as do caliches from other islands (filled squares). Data for Holocene beachrock and Iguana Cay Holocene samples (circles) plot within the marine field, or between the two defined fields. Data for *Cerion* sp. shells (open squares) and conch shells cemented in sea floor Holocene sparites (small dots) lie outside the fields.

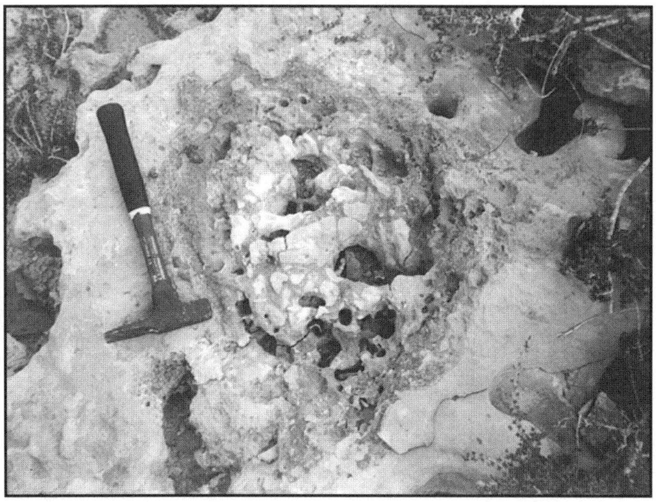

Figure 8. Soil breccia preserved in a solution pit on Ralph Cay. Laminated calcrete surrounds the pit and lines its walls. Length of hammer = 30 cm.

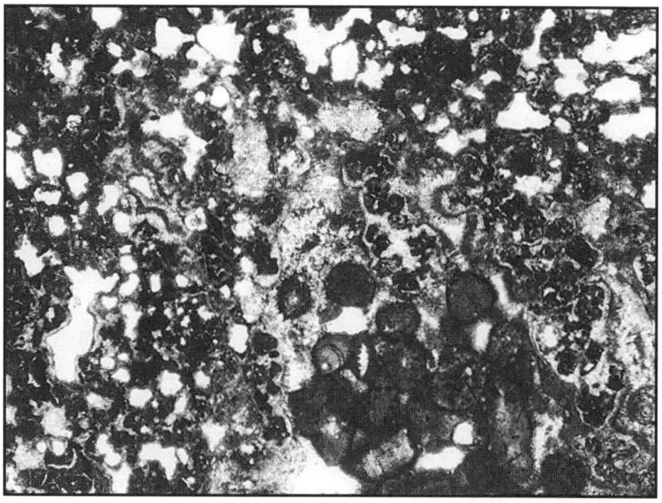

Figure 9. Photomicrograph of terra rossa soil breccia (plane light; field of view = 3.25 mm wide). Note abundant vesicles and vugs, and breccia clasts of less altered light gray eolianite (bottom center left). Some vugs are infilled with acicular isopachous rim cement, overlain by calcite flower spar cement (center and left).

extracted from the submerged wackestone from the northern bank margin side, at a water depth of 7.5 m, yielded a ^{14}C age of 1,500 ± 80 B.P. A conch from the wackestone at the southern bank interior side, at a water depth of 5 m, yielded a ^{14}C age of 1,050 ± 70 B.P. Preliminary stable isotopic analyses of Holocene Iguana Cay carbonate show a maximum depletion of 2% in $\delta^{18}O$ and 4% in $\delta^{13}C$ relative to modern marine values. Stratigraphic relationships between muddy and well-washed Holocene marine rocks were not revealed in any of the cores taken.

Holocene marine grainstone. Grainstones were sampled

Figure 10. Photomicrograph of Holocene fossiliferous packstone (plane light; field of view = 3.25 mm wide) from the south margin of the ooid shoal (bank interior side; depth, 5 m). Shell fragments exhibit sediment-filled and sediment-reduced intraparticle and boring porosity. Note the calcified algal filament, micrite, pelmicrite intraclasts, and abundant superficial ooids. Some of the ooids have interiors aggraded to microspar. Rim cements consist of micrite, blocky microspar and patches of isopachous acicular aragonite. Some of the intergranular micrite appears detrital.

in close proximity to the matrix-rich rock in the southern margin of the ooid shoal. Similar sparry rock directly overlies calcrete-capped Pleistocene eolianite in a swale at the northern margin of the shoal (Fig. 2, Table 1). At other sea-floor exposures at the margins of the shoal, it appears to have been deposited directly on calcrete. In places the grainstone grades into a rudstone that contains large conchs, bivalves, and coral reef debris. The grainstone displays varied grains, including abundant grapestone, ooids, micritic intraclasts, and shell fragments, with an average 3% unfilled interparticle pore space (n = 5; range, 1 to 5%) and 24% cement and/or micritic mud matrix (range, 14 to 30%) (Table 2, Fig. 10). Most samples would be classified as poorly washed and fossiliferous, although textures are highly variable and grainstones grade laterally into packstones over short distances.

Many shell fragments exhibit sediment-filled and sediment-reduced intraparticle, shelter, and boring porosity (Fig. 10). Calcified algal filaments, micrite, and pelmicrite intraclasts are common, and superficial ooids abundant. Some ooids have interiors aggraded to microspar. Rim cements consist of micrite, blocky microspar and patches of isopachous acicular aragonite. Blocky microspar is most abundant and some of the intergranular micrite appears to be detrital. Based on x-ray diffraction study of four bulk samples, these rocks contain 85 to 94% aragonite and 6 to 15% high-Mg calcite. Holocene grainstone samples from Iguana Cay are slightly depleted in ^{13}C and ^{18}O relative to modern marine carbonate, and plot along a covariant alteration trajectory between the marine and subaerial isotopic fields (Fig. 7).

Holocene beachrock. Beachrock is exposed beneath, shoreward, and seaward of modern beach sands, and overlies Pleistocene eolianite and calcrete paleosol in pocket beaches on South Iguana Cay (Figs 1 and 2). At most sites the beachrock exists in tabular, laminated beds that dip gently seaward at an angle similar to modern foreshore sands. Beds are less than 40 cm thick and contain coarse to very coarse skeletal sand and rare conch and reefal debris. The pink coloration of the sand is attributed to abundant fragments of colonial foraminifera (*Homotrema* sp.).

On the eastern (bankward) margin of South Iguana Cay, beachrock is exposed inboard of modern beach sands in an older berm crest that ranges from 0.8 to 2 m above present high tide. Landward of the crest, beds dip at low angles toward the island interior. The highest sands may have been reworked from the berm crest and redeposited in eolian dunes, as is suggested by the presence of tabular cross-bed sets. In the vicinity of the berm crest is a conch midden that has become cemented in beachrock some 80 cm above mean high tide. These shells all have round drillholes near their apices, attesting to their having been harvested by Carib Indians who did not have iron tools for cutting the shell to slice the muscle of the conch (A. Stoner, personal communication, 1991). Based on radiometric age dating and cultural history, such deposits are thought to be pre-Columbian in origin (A. Stoner, personal communication, 1991). The beachrock matrix is fractured in a polygonal pattern with individual polygons commonly subequilateral and 20 to 50 cm across (Fig. 11A). Subvertical burrows are occasionally found in the sediments. Above and seaward of this midden is another midden composed of uncemented fresh shells piled close to the mean high water line.

The highest beachrock exposures lack coarse skeletal debris and consist of fine- to medium-grained oosparite composed of 79% grains (chiefly superficial ooids with intraclast nuclei), 8% acicular, isopachous, aragonitic rim cement and 13% unfilled pore space (Table 2, Fig. 11B). Laminations are defined by slight variations in grain size between fine and medium sand, and sorting is poorer compared to Pleistocene eolianites. Some intraclasts have aggraded to microspar.

Recent sedimentary environments

Beach and storm berm deposits. Beach sands on Iguana Cay are coarser and somewhat richer in shell debris than the ooid shoal containing stromatolites described below. On the eastern windward margins of islands, storm waves have piled up a ridge of large blocks of reefal debris, beachrock, and trash. In areas where surge channels are exposed to incoming seas, meter-high storm berms form 1.4 m above mean high tide. Two main storm berms were recognized on Ralph Cay: a lower one composed of well-rounded cobbles and a higher one of probable hurricane origin containing sharp, subangular to angular fragments. Both berms are composed (~90%) of cobble-sized reefal debris and man-made flotsam and jetsam.

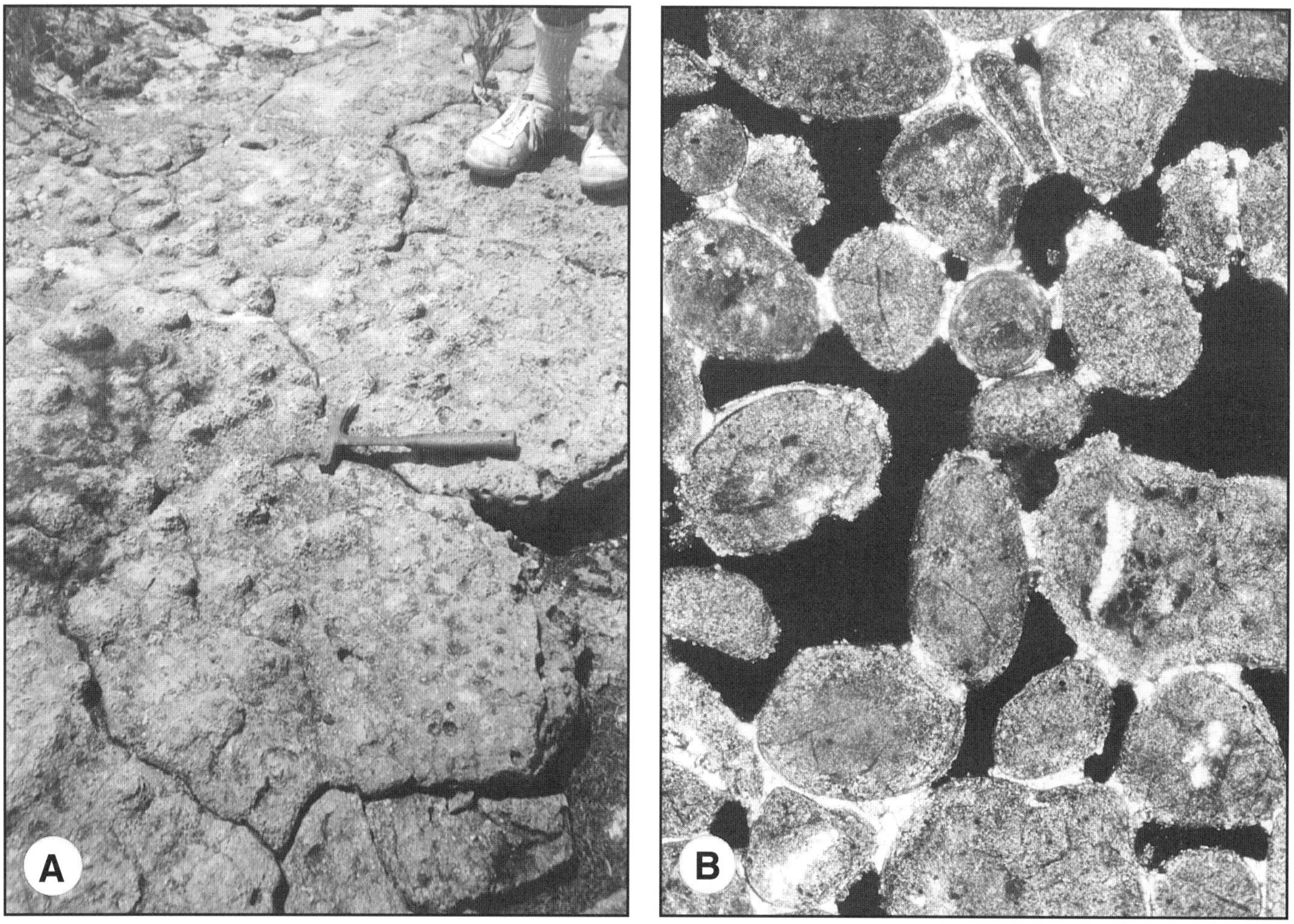

Figure 11. Holocene beachrock from Iguana Cay. A, Polygonal dessication cracks and lithified burrows. The cracks are confined to a 10-cm-thick lithified, massive carbonate sand bed. B, Photomicrograph showing abundant superficial ooids and rim cement (polarized light; field of view = 1.3 mm wide).

Subtidal ooidal shoal. The ooid shoal separating Iguana and Ralph Cay is capped by migrating sand waves and has an elongate shape 275 × 50 m. It is maintained by tidal flow through three interisland channels between South Iguana, Ralph, and Kalik Cays (Fig. 1). Ripple-covered sand waves have sinuous crests, with an average wave length of 15 m (range, 12 to 17 m) and a height of 1 m. The depth to bottom at the northern margin of the shoal is 4.5 m, 5.5 m at the southern margin, 7 m on the edges, and shallows to slightly over 3 m at the crest midshoal.

The bedforms on the shoal shift bidirectionally to the north-northwest and south-southwest as dominant currents change. Between August and May, 1989–1990, a sand wave advanced 10 m to the north, burying 1-m-high stromatolites. Sand waves are reactivated by the changing tides and shift their leeward slip-face orientation while maintaining overall positions. Single sand waves occur with slip faces oriented in opposite directions along the trend of the bedform. Although the majority of the dunes vary their slip faces each tidal cycle, there are several dunes along the Iguana Cay margin that have slip faces that remain oriented to the south, reflecting a flood tide dominance. Because of the dominance of tidal currents over wave-induced bottom surge, the normal wind-driven waves have only a temporal effect on the bottom bedform morphology and migration. Waves produced by large tropical storms, especially hurricanes, probably cause major rearrangement of the field (Dill et al., 1989).

The position of the shoal has been static over 6 yr of observation. Sea-floor scouring by sand in swales that extend both northward and southward from the shoal suggests that at times the shoal has migrated some tens of meters in both these directions. At the southeastern margin of the shoal, a 10- × 20-m portion of the dune field has been stabilized by a dense bed of *Thalassia.* Hardgrounds supporting sponges and soft corals surround the shoal, and there appears to be no net transfer of sand either into or out of the Iguana Cay Channel. Unlike most of the other ooid shoals along the bank margin, the sand does not migrate away from Exuma Sound in flood tidal deltas. Rather, the currents are piling up the sand into an elongated pyramidal mound deposit. This pattern of stationary aggradation has been observed in another shoal 2 km to the south, off the northern tip of Norman's Pond Cay.

The currents in this area are strong, with velocities up to 100 cm/sec (Dill et al., 1989). As the tide shifts, greenish, saline bank water flows under and mixes with blue, cool oceanic water. "Shimmer laminae" can be viewed at the contact between the two water masses due to different indices of refraction.

Based on thin-sections and sieve analysis of seven sand samples taken along the length of the shoal, no significant lateral variations occur in texture or composition. The sands are well sorted, fine to medium grained, and are composed of superficial ooids with intraclast nuclei and uncoated intraclasts (Tables 2 and 3; Fig. 12). Shell fragments and a lag rubble exist in some troughs of dunes and in scour pockets developed around stromatolites. The largest rubble zone is located in the trough between the southernmost south-facing dune, a north-facing dune, and a dune stabilized by *Thalassia*. This is an area of lower current-induced energy. The sand waves here are starved for sediment and are not fully developed. The relationship of a lag deposit surrounding a migrating dune system is common in most of the interisland tidal channels of the southern Exumas that have a source of either ooid or skeletal sands.

Iguana Cay stromatolites

Substrate. The microbial mats that build the stromatolites in the ooid shoal between Ralph Cay and Iguana Cay need a stable surface to initiate growth. Stromatolite-building microbial mats have been found on tin cans, submerged driftwood, exposed mud beds, and even a termite nest that had been trapped at the base of large stromatolites. In core IG-D, taken at the northern margin of the sand shoal, a 4-cm-thick stromatolite capped an 18-cm-high *Diploria* coral head anchored on bored, rhizomorph-rich Pleistocene eolianite (Table 1). In the field this mat-colonized coral head exhibited the morphology of a domal stromatolite. Other dead corals incorporated into stromatolite substrate yielded ^{14}C ages of 630 ± 60 and 950 ± 60 B.P. (5 m below mean high water).

TABLE 3. SIEVE ANALYSIS OF IGUANA CHANNEL SHOAL SANDS

Size*	Sample Site (wt%)						
	Buoy 1	Buoy 2	Buoy 3	Buoy 4	Buoy 5	Buoy 6	Buoy 7
VCS	0.05	0.90	0.42	0.94	0.13	0.04	tr
CS	4.46	1.17	7.42	28.12	2.23	2.63	0.27
MS	91.62	80.66	82.97	70.07	70.93	89.20	73.88
FS	3.81	18.09	9.16	0.87	26.68	8.10	25.83
VFS	0.06	0.07	0.02	0.01	0.04	0.04	0.01

*VCS = very coarse sand; CS = coarse sand; MS = medium sand; FS = fine sand; VFS = very fine sand; tr = trace.

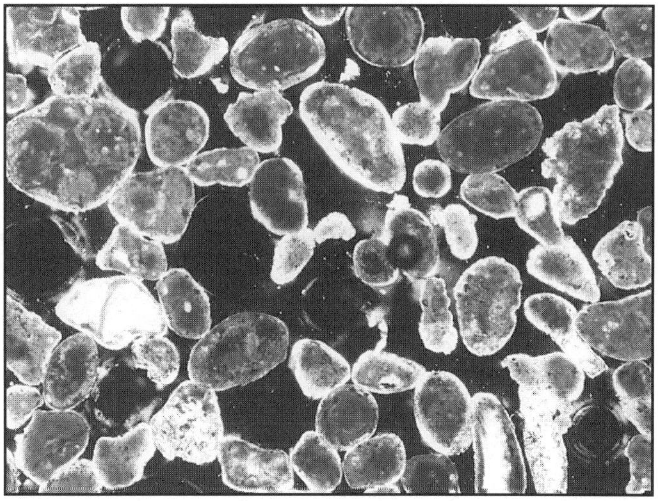

Figure 12. Photomicrograph of modern shoal sand (polarized light; field of view = 1.3 mm wide). Note abundant superficial ooids and the composite ooid in the upper left.

Analysis of a substrate sample 50 × 50 cm recovered from this area revealed a high concentration of large bivalve and gastropod fragments, coral (*Siderastrea*), *Homotrema*, calcareous worm tubes, and one 4-mm calcite rhomb (echinoid plate?), all in a densely cemented oolitic matrix. Bioerosion was thorough and was produced by endolithic algae and sponges, boring clams and gastropods, and several unidentified endolithic invertebrates. On the northern and southern margins of the shoal, the distribution pattern is the result of preferential growth on the high spots of the underlying hummocky Pleistocene substrate. Only a few of the patches are oriented into biohermal walls, perpendicular to the current direction, as are many of the Lee Stocking Island stromatolites (Shapiro, 1990).

Orientation and morphological variance. At Iguana Cay, the stromatolites range in size from small (20 cm high) at the edges of the field, to quite large (2 m high) near the center of the shoal. Stromatolites have never been observed that grow higher than the crests of the dunes that supply the structures with sediment and that migrate to bury and protect the structures from bioerosion (Dill et al., 1986). Although some stromatolites in the Iguana Cay Channel also have increased growth perpendicular to current flow leading to coalescence, most exist as solitary clubs, having diameters 50 to 80% of their height. A patch of small stromatolites having small bases (<10 cm in diameter), that expand outward into large club shapes (15 to 20 cm in diameter) occurs at the northwestern margin of the shoal. At the northern margin of the shoal, a broad, 0.5-m-high stromatolite bench is developed on a coralline algal ridge.

Smaller stromatolites, less than 50 cm high, appear to be concentrated in groups of 10 to 30 clubs, with some areas showing more coalescence than others (Fig. 13). The stromatolites in the central part of the shoal exhibit a two-tiered growth pattern (Fig. 14). The first level extends to small pustules (1 to 2 cm in

Figure 13. Plan view of coalesced stromatolites. Most coalescing in the study area begins as small bridges and grows out. Length of longest head = ~1 m.

Figure 14. Side view of a tiered stromatolite. The lower tier is somewhat uniform, shingled, low, and broad-headed, whereas the upper tier is composed of four coalesced clubs. Note the correlation between the maximum height of the stromatolites and the height of the sand wave in the background. The lower tier (a prevalent feature throughout the channel) may have formed at a time when the sediment budget was smaller, *Siderastrea radians* (~5 cm in diameter) near the top of the side facing the camera. Spacings on the scale bar = 25 cm.

diameter). Although some stromatolites continue upward from this level, most of them abruptly stop, resulting in a planed-off appearance. The second level extends up another 50 cm and is covered by a mucilaginous gel, probably of microbial origin. This height extends to the maximum height of the sand dunes.

With the exception of the bench-type stromatolite at the northern margin, the stromatolites found in the Iguana Cay Channel have club- and molar-form external morphologies. The sediment bounded in the upper portions of stromatolites is usually more fine than the sediment in the surrounding shoal. The largest clasts within stromatolites consist of mollusk shell and coral fragments, mud clasts, and oncolites. These are incorporated into the basal portions of the structures and are derived from interdune rubble zones and lag deposits of coarse material that accumulate in the scour depressions around the base of large stromatolites. As the stromatolites coalesce, this material is incorporated in their basal areas. Coarse material is also incorporated in the early stages of formation, when the tops of the stromatolites lay within the zone of bedload transport (Aalto and Shapiro, 1990). Drilling and underwater observations have shown that coarse shell debris and mud chips 3 to 4 cm in diameter can be incorporated within the central loose sediment-filled depressions of molar-form stromatolites.

Biota. The ecosystem that exists on and within the stromatolites can be divided into three main components: builders, destroyers, and meiofauna. The builders include filamentous green algae, cyanobacteria, diatoms, and other microscopic organisms. They trap and cement ooids and skeletal sands transported by them by fluctuating bottom currents. The dominant cyanobacterium is a large filamentous oscillatoriacian, *Schizothrix* spp., with a well-defined sheath and trichomes up to 7 μm in diameter (K. Browne and S. Golubic, personal communication, 1992; for more information on composition of the microbial mat, see Dill, 1991, and Riding et al., 1991).

The microbially accreted layers are being actively destroyed by endolithic algae, sponges, boring mollusks, and grazing fish. On the surface and within the stromatolites are annelid worms, nematodes, copepods, the coral *Siderastrea radians*, and countless other microscopic invertebrates.

S. radians and the stromatolites occur together in all of the stromatolite locations in this part of the Exumas (Fig. 14), although it is rare to find other stony corals actively growing on the stromatolites. Many macroalgae exist on the stromatolites including *Udotea spinulosa*, *Halimeda* spp., *Rhipocephalus phoenix*, *Penicillus capitatus*, *Sargassum* spp., and *Lobophora variegata*. There appears to be a temporal succession in the composition of the macroalgae as the migrating dunes expose the stromatolites. On first exposure, the surface of the stromatolite becomes colonized by an as-yet unidentified branching filamentous green macroalgae. This is replaced within the first several months by *Batophora oerstedii* and *Acetabularia crenulata*. In certain parts of the channel, *Cladophora prolifera* can also be seen growing on the sides of the stromatolites. After several months, *Sargassum* spp. can completely cover the upper surface. Sponges and hydroid polyps are also found on the upper surfaces. If the stromatolites are not buried under the migrating sand waves (e.g., because of stabilization of the sediment by *Thalassia*), they will be completely covered by soft and hard corals and sponges, and will be bored to an almost unrecognizable form.

Within the upper surfaces of the stromatolites there is a meiofaunal community, dominated by annelid worms and small crustaceans. Although no in-depth study of the meiofauna has been undertaken, a preliminary survey by Shapiro and B. Maguire (University of Texas, Austin) revealed abundant copepods, isopods, free-swimming annelid worms, and small bivalves.

DEPOSITIONAL HISTORY

Based on field observations and laboratory analysis of the rock units, a depositional history of the Pleistocene-Recent can be developed for the Iguana Cay region. Prior to the Sangamon highstand (stage 5, 125 Ka), the Exuma region was an area of migrating dune fields. Bedforms recorded in the eolianiates reflect that prevailing winds were from the east to southeast, much the same as today. The presence of submerged and exposed talus cones that are traceable between several islands suggests that, locally, older dunes may have been considerably larger. During the Sangamon, when relative seal level was 2 m above its present position, reefs developed on the leeward margins of the islands. Because of subsequent erosion, we are not certain if reefs developed on the windward margins of the cays.

During the late Pleistocene regression, the island cores and much of the bank top became exposed. This led to vadose cementation of the dunes and the development of karst topography. A prominent overhanging sea cliff present on Ralph Cay, 5 m above present sea level, very likely formed during this time while modern nips appear to develop chiefly from bioerosion of island margins (Dill et al., 1989). The cave described from Ralph Cay may have originated by dissolution at the margin of a freshwater lens a short distance inland of the Sangamonian shoreline. This hypothesis was suggested for similar features on San Salvador by Mylroie and Carew (1990).

At the peak of the late Pleistocene glacial (~13Ka), a calcrete soil cover developed on the cemented eolianites. Unlike calcretes associated with rhizomorphs, which can occur at any depth in the eolianites, calcretes of this sort are thought to have formed pedogenically in association with root mats (Aalto and Burke, 1992; Rossinsky and Wanless, 1992). Some of the veins and fenestrae within the calcrete may be infilled anastomosing root pores, for they are similar in appearance to the filled root tubule–generated fabrics discussed in depth by Kahle (1977) and Wright et al. (1988).

A sample of calcrete from Ralph Cay yielded a whole-rock ^{14}C age of 13,230 ± 201 B.P. While cautious of the single-age determination, it is noteworthy that it is in agreement with well-dated meltwater pulse, (^{14}C and U-Th), ~13.5-Ka ago, recognized from corals in Barbados by Fairbanks (1989) and Bard et al. (1990). It should also be noted that this age is similar to that of paleosols on San Salvador Island on the eastern margin of the Bahama Platform (Carew and Mylroie, 1990) and those reported for the Florida Keys (Robbin and Stipp, 1979; Robbin, 1981, 1984; Shinn et al., 1989). Thus an occurrence at this level might reflect a platform-wide stage of paleosol formation. The period of paleosol formation at the Iguana Cay Channel site took place when relative sea level was 30 m below its present position.

The gradational eolianite-calcrete boundaries defined by micritization are similar to ones described by James (1972), who attributed them to "vadose micritization." James (1972) noted that calcrete subjected to salt spray characteristically has a higher magnesium content than calcrete precipitated farther inland. the presence of high-Mg calcite in one submerged calcrete sample site may reflect chemical interaction with sea water after submergence.

Soon after and possibly contemporaneous with the formation of the paleosol, soil breccias developed on the calcrete crusts within solution pipes. In some cases, similar structures have been misinterpreted in the past to be tree trunk casts due to their association with rhizomorphs (J. Mylroie, personal communication, 1991). The plants in the pipes may be a secondary feature (as can be seen in modern sinkholes) and are not required for this type of pipe formation. Sediment- and cement-filled veins and vugs defining the solution breccias reflect subaerial exposure and partial dissolution of the substrate (James, 1972; Walls et al., 1975; Knox, 1977; Klappa, 1979; Esteban and Klappa, 1983; Aalto and Burke, 1992). The presence of dark gray crusts on eolianite clasts in submerged soil breccias on Ralph Cay may reflect the presence of trapped organic matter within calcite crystals. Blackening has been attributed to infusion of organic matter into carbonates, subaerial exposure near hypersaline waters (Ward et al., 1970), or precipitation of metallic oxides in conjunction with fungal growth (Esteban and Klappa, 1983), all of which reflect pedogenesis. Blackened carbonate pebbles and clasts at Pleistocene and Holocene unconformities have also been attributed to heating by forest fires. Experiments have shown that the reddish calcretes found in association with organic-rich paleosols convert black breccias on heating (Shinn and Lidz, 1988). Pebble- to boulder-sized clasts are found in many of the paleosols near Iguana Cay.

Flooding of the bank margin occurred approximately 3.5 Ka ago. Reef systems became established, as well as production of skeletal and ooid sediments. Because ooids with many concentric laminae (as opposed to ooids with a superficial coating) are more prevalent in Pleistocene dunes, there may have been a change in the ooid provenance or ooid-forming conditions at this time. It is our contention that the superficial ooids with intraclast nuclei that constitute the bulk of sand-sized material in the ooid shoal are intermittently introduced into the channel—perhaps during strong ebb tides following storms—and remain trapped in this area by balanced tidal currents. Similar movement and trapping of mud chips in ooid shoals was recorded during the August 23–24, 1992, passing of Hurricane Andrew (Shinn et al., 1993).

As sea level rose, muddy and well-washed skeletal wackestones and grainstones accumulated in karst depressions and swales. Based on the data on ages obtained from material incorporated into the stromatolite substrate, the rising sea level began to diminish approximately 4 to 5 Ka. This is in contrast to many eustatic curves that place this inflection point in the 5- to 6-Ka range (e.g., Moore and Curray, 1974). Approximately

3 Ka, sea level stabilized near its present stand, and microbial mats and coralline algae colonized the sea floor.

Because some of the Holocene Iguana Cay carbonate samples plot between the end member geochemical scenarios (marine versus meteoric water fields), several potential interpretations are possible, including carbonate precipitation from isotopically depleted bank waters, subaerial exposure, and/or mixing of Pleistocene grains with the Holocene. One possibility is that the Holocene marine Iguana Cay carbonate precipitated in a depositional environment that did not readily exchange with open ocean water or had a relatively long residence time.

It is also possible that the depleted $\delta^{13}C$ and $\delta^{18}O$ values may have resulted from an influx of depleted meteoric water during early diagenesis and stabilization rather than from a sustained variation in bank water geochemistry (Dunham, 1969). The presence of calcite-filled vugs and veins is not inconsistent with an interval of subaerial exposure. Decay of organic matter and an influx of meteoric water could also account for the marginally depleted $\delta^{13}C$ and $\delta^{18}O$ values.

Beachrock formation occurred in the high tidal zone along pocket beach shorelines. Because mud is absent in the beachrock of Iguana Cay, the polygonal cracking seen today probably resulted from desiccation of pore water and salt crystallization, although precipitation of aragonite cement may have contributed to this deformation.

Stability and moving water needed by the microbes in mats to obtain nutrients and sunlight permitted some to grow up and form stromatolites. The ooid sands, generated in high-energy channels, and biogenic debris from developing reefs provided source material for the sediment-trapping organisms. Two criteria were needed for a stromatolite-building community to be successful: the substrate had to be stable, and the substrate surface had to be situated within the region where sediment was supplied and above permanent sediment cover. Most of the stromatolites at Iguana Cay are growing on high points on the Pleistocene-karsted eolianite or paleosol. Other stromatolites have developed on coralline algal ridges and stony corals. It is interesting to note that, although some "club-shaped stromatolites" recovered from this field were actually mainly coral heads with a thin stromatolitic covering, other coral heads in this area did not have stromatolites forming on them. This is an area of strong current activity and sediment scour; it is possible that the corals could have been covered by stromatolites in the past and then eroded by strong sediment-charged currents. Often, other organisms utilize the stromatolites as growth platforms. *Siderastrea radians* is one of the few corals that exist on the stromatolites because most corals cannot withstand the strong currents and periodic sediment coverage needed for stromatolite accretion. However, *S. radians* can tolerate extreme conditions of physical stress (Colin, 1988). Throughout the Caribbean, *S. radians* is one of the first pioneer corals on newly exposed substrate.

From studies undertaken at other stromatolite fields in the Exumas, it has been shown that the gross outer morphology is largely dictated by community response to environmental factors (current velocity and direction, waves, sedimentary budget, burial time under the migrating ooid sand waves, and water chemistry), not specifically by the make-up of the microbial mat (Dravis, 1983; Dill et al., 1986; Aalto and Shapiro, 1990; Shapiro, 1990; Shapiro et al., 1990; Dill, 1991). The fact that large, streamlined, unbranched, subtidal, columnar stromatolites are found in the fossil record is evidence in favor of the constancy of this form in the light of the evolution of the microorganisms. Remarkable different morphologies can exist within the Iguana Cay channel, which we attribute in part to the highly variable current regimes. Linear, or bench-type stromatolites, such as are found at the northern margin of the field, are affected by both a strong tidal current and by strong swell-induced bottom surges in the same direction. Rounded clubs and molar forms are in the center of the shoal where there is variation in direction and magnitude of tidal currents and wind-driven currents. Bioerosion also affects external morphology but not to the extent of other physical processes. The stromatolites are forming in areas where there are abundant grazers, but they are not in sufficient numbers to overgraze the mats or destroy the internal fabric of the stromatolites. The variation in the internal layering pattern of the stromatolite is a record of the response of the microbial mat to past environmental changes and successions of microbial organisms that create the mat.

It has been proposed recently that the Exuma stromatolites are not strictly comparable to a large number of fossil forms because of differences in the flora (notably the high percentage of diatoms) and the coarseness of the sediment component (e.g., Riding et al., 1991). Many deposits of this age throughout the Great Basin region of the United States host large (~1 m), domical stromatolites and thrombolites, which are streamlined and may display "molar" cavities. The Upper Cambrian stromatolites are often bioturbated and may contain skeletal fragments of potential grazers, such as the polyplacophoran *Matthevia* and the gastropod *Matherella* (Yochelson and Taylor, 1974; Runnegar et al., 1979). Often these stromatolites and thrombolites are associated with cross-bedded oolitic shoals and tidal channels. It is possible that the streamlined, tilted nature of the stromatolites may have been toward the flood current, and therefore may help in reconstructing ancient bank margins (i.e., the direction toward the incoming flood current).

CONCLUSIONS

Nine rock units deposited from the Late Pleistocene to the Recent are exposed at the surface and in shallow cores in the vicinity of Iguana Cay, Exumas, Bahamas. From oldest to youngest, they are eolianite, paleosols, calcrete crusts, soil breccia (Pleistocene), marine wackestone and packstone, marine grainstone, beachrock (Holocene), beach and storm berm deposits, and a subtidal ooid shoal (Recent).

Meter-high chlorophyte-diatom-cyanobacterial stromatolites originated and grew as the ooid shoal developed between the islands. The morphology and distribution of the stromatolites reflect the microbial community response to various environmental factors. In the center of the shoal, the dominant stromatolite forms are solitary and coalesced columns. At the margins of the shoal are predominantly bench-type stromatolites that developed off crustose coralline algal ridges. Ecologic interactions at the macroscopic and microscopic levels lead to both construction and destruction of the stromatolitic buildups.

These various rock units and stromatolites record the carbonate bank's response to eustatic sea-level variation. The Pleistocene eolianite was deposited before the Sangamon highstand (125 Ka) when the eastern rim of the bank hosted extensive dune fields. The Late Pleistocene regression exposed the platform, which then underwent extensive karstification. It was during the peak of the Late Pleistocene glacial period (13 Ka) that calcrete paleosols and calcrete crusts developed over the bank margin. Pockets in the calcrete crusts were the sites of soil breccia formation. As sea level rose and flooded the bank (approximately 3.5 Ka), the buildup and destruction of reef systems and ooid production led to production of skeletal wackestones, packstones, and grainstones. Once sea level stabilized, beachrock began to form around the island margins and storm deposits accumulated above the high water mark. Finally, ooid sands accumulated in the interisland tidal channel along with stromatolites.

These data and interpretations will be useful in creating models for understanding ancient stromatolite development and carbonate bank margin systems. Many stromatolites described from throughout the geologic column, notably the Phanerozoic, have striking similarities in gross morphology and associated facies to the Bahamian forms (e.g., Griffin and Awramik, 1989). There are so many facets of stromatolite biology and sedimentology, however, that the validity of each analogue must be scrutinized closely. The more we understand the various biologic and geologic factors that dictate the distribution, genesis, and growth of the Bahamian stromatolites and their position in overall stratigraphic patterns, the better our analyses will be of the ancient systems.

ACKNOWLEDGMENTS

We acknowledge Robert Wicklund, director, and Geri Wenz, assistant director, of the Caribbean Marine Research Center (CMRC), who have continually given financial and logistical support to our geologic efforts in the Exumas and made our research possible. We extend our thanks to John Perry, Jr., who over the past 11 years has allowed us to use his island as a research base. Field and laboratory assistance were provided by Scott Aby, Bud Burke, Randy Kehl (Humboldt State University), and Bassett Maguire (University of Texas, Austin). ^{14}C dates were provided by Lynton Land (University of Texas, Austin). Harold and Elisabeth Hudson provided the drilling apparatus, instruction on its use, and assistance in the 1992 field core drilling program. Manuscript reviews by S. M. Awramik, H. A. Curran, Jean Hegland, William Miller III, J. Fred Read, and an anonymous reviewer were greatly appreciated. Partial funding for this research was obtained through grants from Humboldt State University, Texaco, and the National Oceanic and Atmospheric Administration's National Undersea Research Program funds to CMRC. Arizona State University provided funds for chemical studies.

REFERENCES CITED

Aalto, K. R., and Burke, R. M., 1992, Pleistocene talus cones on Low Cay, Exumas, Bahamas: Geological Society of America Abstracts with Programs, v. 24, no. 5, p. 1.

Aalto, K. R., and Shapiro, R. S., 1990, The petrology of a modern subtidal stromatolite, Lee Stocking Island, Bahamas, *in* Bain, R. J., ed., Proceedings, Fifth Symposium on the Geology of the Bahamas: San Salvador, Bahamian Field Station, Field Guide, p. 1–10.

Aalto, K. R., Aby, S., Dill, R. F., Shapiro, R. S., and Burke, R. M., 1992, Pleistocene-Holocene reconnaissance stratigraphy of Leaf and Pond Cays, Exumas, Bahamas: Geological Society of America Abstracts with Programs, v. 24, no. 5, p. 1.

Aby, S. B., Aalto, K. R., and Dill, R. F., 1992, Origin and significance of filled fractures along a carbonate bank margin, Lee Stocking Island, Exumas, Bahamas: Geological Society of America Abstracts with Programs, v. 24, no. 5, p. 1.

Bard, E., Hamelin, B., and Fairbanks, R. G., 1990, U-Th ages obtained by mass spectrometry in corals from Barbados: Sea level during the past 130,000 years: Nature, v. 346, p. 456–458.

Beach, D. K., 1982, Depositional and diagenetic history of Pliocene-Pleistocene carbonates of Northwestern Great Bahama Bank; Evolution of a carbonate platform [Ph.D. thesis]: Miami, Florida, University of Miami, 425 p.

Benjamin, G., 1970, Diving into the blue holes of the Bahamas: National Geographic Magazine, v. 138, no. 3, p. 347–363.

Browne, K., and Ginsburg, R. S., 1991, Successions of lamination types in Bahamian Holocene stromatolites: Products of interactions between physical, biological and chemical processes: Geological Society of America Abstracts with Programs, v. 23, p. A439.

Carew, J. L., and Mylroie, J. E., 1990, Potential difficulties in the use of paleosols for stratigraphic interpretation of carbonate sequences: Geological Society of America Abstracts with Programs, v. 22, no. 7, p. A334.

Colin, P. L., 1988, Marine Invertebrates and Plants of the Living Reef: Neptune City, New Jersey, T.F.H. Publications, 512 p.

Curran, H. A., and Dill, R. F., 1990, Stratigraphy and ichnology of a submarine cave, Exuma Cays, Bahamas: Geological Society of America Abstracts with Programs, v. 22, p. A269.

Dill, R. F., 1977, Blue holes: Geologically significant submerged sink holes and caves off British Honduras and Andros, Bahama Islands, *in* Proceedings, Third International Coral Reef Symposium, v. 2: Geology: Miami, Florida, University of Miami, Rosenstiel School of Marine Science, p. 237–242.

Dill, R. F., 1991, Subtidal stromatolites, ooids and crustal-lime mud at the Great Bahama Bank margin, *in* Osbourne, R., ed., From Shoreline by Abyss: SEPM Special Publication 46, p. 147–171.

Dill, R. F., Shinn, E. A., Jones, A. T., Kelly, K., and Steinen, R. P., 1986, Giant subtidal stromatolites forming in normal salinity waters: Nature, v. 324, p. 55–58.

Dill, R. F., Kendall, C. G. St. C., and Steinen, R. P., 1988, Deposition of carbonate mud beds within high-energy, subtidal sand dunes, Bahamas: American Association of Petroleum Geologists Bulletin, v. 72, p. 178–179.

Dill, R. F., Kendall, C. G. St. C., and Shinn, E. A., 1989, Giant subtidal stromatolites and related sedimentary features: International Geological

Congress, 28th, Washington, D.C., American Geophysical Union, 33 p.
Dill, R. F., Grotzinger, J. P., and Read, J., 1991, A comparison of elongated, subtidal-stromatolites in high-energy bays of the Exuma Islands, Bahamas, to Paleozoic and Precambrian forms: Geological Society of America Abstracts with Programs, v. 23, p. A439.
Dravis, J. R., 1983, Hardened subtidal stromatolites, Bahamas: Science, v. 219, p. 385–386.
Dunham, R. J., 1969, Vadose pisolite in the Capitan Reef, (Permian), New Mexico and Texas: Society of Economic Paleontologists and Mineralogists Special Paper 14, p. 182–191.
Esteban, M., and Klappa, C. F., 1983, Subaerial exposure environment, *in* Scholle, P. A., Bebout, D. G., and Moore, C. H., eds., Carbonate Depositional Environments: American Association of Petroleum Geologists Memoir 33, p. 1–54.
Fairbanks, R. G., 1989, A 17,000-year glacio-eustatic sea level record: Influence of glacial melting rates on the Younger Dryas event and deep-ocean circulation: Nature, v. 342, p. 637–642.
Falls, W. F., Williams, D. E., Kendall, C. G. St. C., and Dill, R. F., 1990, Stable oxygen and carbon isotope study of recent sediments and cements, Lee Stocking Island, Bahamas: Organic vs. inorganic precipitation: American Association of Petroleum Geologists Annual Meeting, Abstracts with Programs, v. 74, p. 625.
Griffin, K. M., 1987, A comparison of the depositional environments of Upper Cambrian thrombolites and stromatolites of the Bahamas: Geological Society of America Abstracts with Programs, v. 19, no. 7, p. 446–447.
Griffin, K. M., and Awramik, S. M., 1988, Giant Bahamian stromatolites: A modern analog for what? *in* Mylroie, J., ed., Proceedings, Fourth Symposium on the Geology of the Bahamas: San Salvador, Bahamian Field Station, p. 169–175.
Halley, R. B., Muhs, D. R., Shinn, E. A., Dill, R. F., and Kindinger, J. L., 1991, A +1.5 m reef terrace in the southern Exuma Islands, Bahamas: Geological Society of America Abstracts with Programs, v. 23, no. 1, p. 40.
Hintze, L. F., Taylor, M. E., and Miller, J. F., 1988, Upper Cambrian–Lower Ordovician Notch Peak Formation in western Utah: U.S. Geological Survey Professional Paper 1393, 30 p.
James, N. P., 1972, Holocene and Pleistocene calcareous crust (caliche) profiles: Criteria for subaerial exposure: Journal of Sedimentary Petrology, v. 42, p. 817–836.
Kahle, C. F., 1977, Origin of subaerial Holocene calcareous crust: Role of algae, fungi and sparmicritisation: Sedimentology, v. 24, p. 413–435.
Kendall, C. G. St. C., Dill, R. F., and Shinn, E. A., 1989, Guidebook to the giant subtidal stromatolites and carbonate facies of Lee Stocking Island, Bahamas (4th ed.): San Diego, California, KenDill Publishing, 121 p.
Kendall, C. G. St. C., Dill, R. F., and Shinn, E. A., 1990, Guidebook to the marine geology and tropical environments of Lee Stocking Island, the southern Exumas, Bahamas: Carbonate facies, geologic history, giant stromatolites, oceanography, and biological associations: Covington, Virginia, Caribbean Marine Research Center, and San Diego, California, KenDill Publishing, 82 p.
Kenny, R., Aalto, K. R., and Dill, R. F., 1991, Isotope geochemistry of carbonate from modern subtidal stromatolites, Exumas, Bahamas: Geological Society of America Abstracts with Programs, v. 23, p. A51.
Kindler, P., 1991, Holocene stratigraphy of Lee Stocking Island, Bahamas—New interpretation with respect to sea-level history: Geological Society of America Abstracts with Programs, v. 23, p. A53.
Klappa, C. F., 1979, Calcified filaments in Quaternary calcretes: Organo-mineral interactions in subaerial vadose environment: Journal of Sedimentary Petrology, v. 49, p. 955–968.
Knox, G. J., 1977, Caliche profile formation, Saldanha Bay (South Africa): Sedimentology, v. 24, p. 657–674.
McCrea, J. M., 1950, The isotopic chemistry of carbonates and a paleotemperature scale: Journal of Chemistry and Physics, v. 18, p. 849–857.
Miller, J. F., and Taylor, M. E., 1989, Late Cambrian and Early Ordovician stratigraphy and biostratigraphy, southern House Range ("Ibex Area"), Utah, *in* Taylor, M. E., ed., Cambrian and early Ordovician stratigraphy and paleontology of the Basin and Range province, western United States: International Geological Congress, 28th, Guidebook for Field Trip T125: American Geophysical Union, Washington, D.C., p. 45–58.
Moore, D. G., and Curray, J. R., 1974, Midplate continental margin geosynclines: Growth processes and Quaternary modifications, *in* Dott, R. H., Jr., and Shaver, R. H., eds., Modern and ancient geosynclinal sedimentation: Society of Economic and Paleontologists and Mineralogists Special Publication 19, p. 26–35.
Mylroie, J. E., and Carew, J. L., 1990, Erosional notches in Bahamian carbonates: Bioerosion or groundwater dissolution? *in* Bain, R. J., ed., Proceedings, Fifth Symposium on the Geology of the Bahamas: San Salvador, Bahamian Field Station, p. 185–191.
Reid, P. R., and Browne, K. M., 1990, Intertidal stromatolites: A new discovery in the Bahamas: American Geophysical Union Ocean Science Meeting, New Orleans, v. 71, p. 363.
Reid, P. R., and Browne, K. M., 1991, Intertidal stromatolites in a fringing Holocene reef complex, Bahamas: Geology, v. 19, p. 15–18.
Riding, R., Awramik, S. M., Winsborough, B. M., Griffin, K. M., and Dill, R. F., 1991, Bahamian giant stromatolites: Microbial composition of surface mats: Geological Magazine, v. 128, p. 227–234.
Robbin, D. M., 1981, Subaerial $CaCO_3$ crusts—A tool for defining sea-level changes and timing reef initiation, *in* Proceedings, Fourth International Coral Reef Symposium I, Manila, p. 574–579.
Robbin, D. M., 1984, A new Holocene sea level curve for the upper Florida Keys and Florida reef tract, *in* Gleason, J., ed., Environments of South Florida, Present and Past II: Miami, Florida, Miami Geological Society, p. 437–458.
Robbin, D. M., and Stipp, J. J., 1979, Depositional rate of laminated soilstone crusts, Florida Keys: Journal of Sedimentary Petrology, v. 49, p. 175–180.
Rossinsky, V., Jr., and Wanless, H. R., 1992, Topographic and vegetative controls on calcrete formation, Turks and Caicos Islands, British West Indies: Journal of Sedimentary Petrology, v. 62, p. 84–98.
Runnegar, B., Pojeta, J., Jr., Taylor, M. E., and Collins, D. H., 1979, New species of the Cambrian and Ordovician chitons *Matthevia* and *Chelodes* from Wisconsin and Queensland: Evidence for the early history of polypacophoran mollusks: Journal of Paleontology, v. 53, p. 1374–1394.
Shapiro, R. S., 1989, Morphological variations within a modern stromatolite field, Lee Stocking Island, Exumas, Bahamas [B. S. thesis]: Arcata, California, Humboldt State University, 57 p.
Shapiro, R. S., 1990, Morphological variations within a modern stromatolite field, Lee Stocking Island, Exuma Cays, Bahamas, *in* Bain, R. J., ed., Proceedings, Fifth Symposium on the Geology of the Bahamas: San Salvador, Bahamian Field Station, p. 209–220.
Shapiro, R. S., Aalto, K. R., and Dill, R. F., 1990, Physical control of distribution and morphologies of subtidal stromatolites, Exumas, Bahamas: Geological Society of America Abstracts with Programs, v. 22, p. A93.
Shapiro, R. S., Aalto, K. R., and Dill, R. F., 1992, Zonation in the Bock Cay microbialite field, Bahamas: Geological Society of America Abstracts with Programs, v. 24, no. 5, p. 80.
Sheridan, R. E., Mullins, H. T., Austin Jr., J. A., Ball, M. M., and Land, J. W., 1988, Geology and geophysics of the Bahamas, *in* Sheridan, R. E., and Grow, J. A., eds., The Atlantic continental margin, U.S.: Geological Society of America, The Geology of North America, v. I-2, p. 329–364.
Shinn, E. A., and Lidz, B. H., 1988, Blackened limestone pebbles: Fire at subaerial unconformities, *in* James, N. P., and Choquette, P. W., eds., Paleokarst: New York, Springer-Verlag, p. 117–131.
Shinn, E. A., and Lidz, B. H., Halley, R. B., Hudson, J. H., and Kindinger, J. L., 1989, Reefs of Florida and the Dry Tortugas, International Geological Congress, 28th, Guidebook for Field Trip T1776: Washington, D.C., American Geophysical Union, 53 p.
Shinn, E. A., Steinen, R. P., Dill, R. F., and Major, R., 1993, Lime-mud layers in high energy tidal channels: a record of hurricane deposition: Geology, v. 21, p. 603–606.
Taylor, M. E., Cook, H. E., and Miller, J. F., 1989, Late Cambrian and Early Ordovician biostratigraphy and deposition environments of the Whipple

Cave Formation and House Limestone, central Egan Range, Nevada, *in* Taylor, M. E., ed., Cambrian and early Ordovician stratigraphy and paleontology of the Basin and Range province, western United States, International Geological Congress, 28th, Guidebook for Field Trip T125: Washington, D.C., American Geophysical Union, p. 37–44.

Walls, R. A., Harris, W. B., and Nunan, W. E., 1975, Calcareous crust (caliche) profiles and early subaerial exposure of Carboniferous carbonates, northeastern Kentucky: Sedimentology, v. 22, p. 417–440.

Ward, W. C., Folk, R. L., and Wilson, J. L., 1970, Blackening of eolianite and caliche adjacent to saline lakes, Ilsa Mujeres, Quintana Roo, Mexico: Journal of Sedimentary Petrology, v. 40, p. 548–555.

Wright, V. P., Platt, N. H., and Wimbledon, W. A., 1988, Biogenic laminar calcretes: Evidence of calcified root-mat horizons in paleosols: Sedimentology, v. 35, p. 603–620.

Yochelson, E. L., and Taylor, M. E., 1974, Late Cambrian *Matthevia* (Mollusca, Matthevida) in North America: Geological Society of America Abstracts with Programs, v. 6, no. 1, p. 88.

MANUSCRIPT ACCEPTED BY THE SOCIETY JANUARY 5, 1995

Printed in U.S.A.

Geological Society of America
Special Paper 300
1995

Controls on carbonate facies distribution in a high-energy lagoon, San Salvador Island, Bahamas

Anthony F. Randazzo and Kathy J. Baisley
Department of Geology, University of Florida, Gainesville, Florida 32611

ABSTRACT

Statistical evaluation of fine-scale carbonate facies variability is necessary for an understanding of patterns and processes of carbonate sediment deposition and the influence of benthic communities.

Three distinct ecologic zones are recognized in Graham's Harbor, San Salvador Island, a high-energy lagoon: the *Thalassia* Zone, the *Halimeda* and *Penicillus* Zone, and the Sparsely Vegetated Zone. Three grain-type groups (aggregates, skeletal grains, and nonskeletal grains) and mud mineralogy define the sediments in each zone.

Boundaries between zones are gradational, with much lateral transport of sediments between the Sparsely Vegetated Zone and the *Halimeda* and *Penicillus* Zone. Sediments deposited in the *Thalassia* Zone are very stable and are not resuspended and transported away from this zone, even during storms.

Principal component analysis defines three ecologically related sediment groups based on grain sizes and types. The groups vary from coarse aggregate grain types (Sparsely Vegetated Zone) to fine peloidal and skeletal sediments (*Thalassia* Zone). Statistical analysis further documents the link between fine-fraction mineralogy and benthic community zones. This study illustrates significant differences between large- and fine-scale analyses and verifies the important role of benthic communities in modern sedimentologic processes.

INTRODUCTION

The assessment of fine-scale carbonate facies variability is essential for an understanding of controls and processes of carbonate sedimentation. Many workers have studied large-scale patterns of carbonate deposition. Black (1933), Illing (1954), Imbrie and Purdy (1962), and Purdy (1963) provided important studies that established broad facies relationships over large areas of the Great Bahama Bank. These studies laid the groundwork for subsequent detailed investigations of the interrelated factors that control the distribution of lagoonal carbonate facies.

The benthic community supplies much of the sediment to modern lagoons by the breakdown of skeletal material. Land (1970), Chave et al. (1972), and Neumann and Land (1975), among others, demonstrated the huge carbonate production capabilities of a variety of benthic organisms. Several studies have focused on lateral facies distributions in modern carbonate sediments and on the role of the benthic community in determining these distributions (e.g., Ginsburg and Lowenstam, 1958; Swinchatt, 1965; Scoffin, 1970; Turmel and Swanson, 1976).

Carbonate material produced by the benthos is transported by waves and currents, particularly during storms. Sediments are likely to accumulate in areas other than those in which they were generated (Ball, 1967; Wanless et al., 1988). However, the degree of transport of carbonate sediments across benthic community boundaries has not been well established.

Marine grasses have a baffling effect on currents, thus reducing mechanical abrasion and contributing to the accumu-

Randazzo, A. F., and Baisley, K. J., 1995, Controls on carbonate facies distribution in a high-energy lagoon, San Salvador Island, Bahamas, *in* Curran, H. A., and White, B., Terrestrial and Shallow Marine Geology of the Bahamas and Bermuda: Boulder Colorado, Geological Society of America Special Paper 300.

lation of lime mud. They act as a substrate-stabilizer, especially when present as dense carpets. Grassbeds also serve as habitats for a variety of benthic organisms that contribute their skeletons to the sediment (Ginsburg, 1956; Land, 1970; Heck, 1977). Conversely, burrowing organisms, such as *Callianassa*, act as sediment destabilizers. Biogenic reworking increases the water content and decreases the cohesion of sediments, thus facilitating resuspension of the sediments by currents (Young, 1971; Aller and Dodge, 1974).

Authoritative tenets related to substrate stability and the role of organisms in influencing the type, composition and distribution of carbonate sediments are commonly communicated (cf. Bathurst, 1971; Scoffin, 1987; Tucker and Wright, 1990). This investigation attempts to evaluate the tenets of ecologic controls on substrate stability in terms of the sediment distribution patterns for a modern carbonate lagoon (Graham's Harbor, San Salvador Island, Bahamas). Most previous studies cover a large areal extent and investigate variability on a large scale. The quantitative evaluation of controls of fine-scale facies variability can promote a better understanding of patterns and processes of carbonate deposition and allows for a further refinement of these tenets.

This study examines the environment of deposition to identify the factors contributing to the distribution of fine-scale facies patterns. Such an analysis may facilitate interpretations of ancient carbonate sequences and their associated paleoenvironments.

Colby and Boardman (1989) investigated controls on lateral and vertical facies distribution in their study on the depositional evolution of this same modern carbonate lagoon. Armstrong (1989) and Armstrong and Miller (1990) have applied fine-scale parameters in determining carbonate production rates of calcareous algae and epibionts of parts of this lagoon. The present study incorporates some of their methods and addresses their findings.

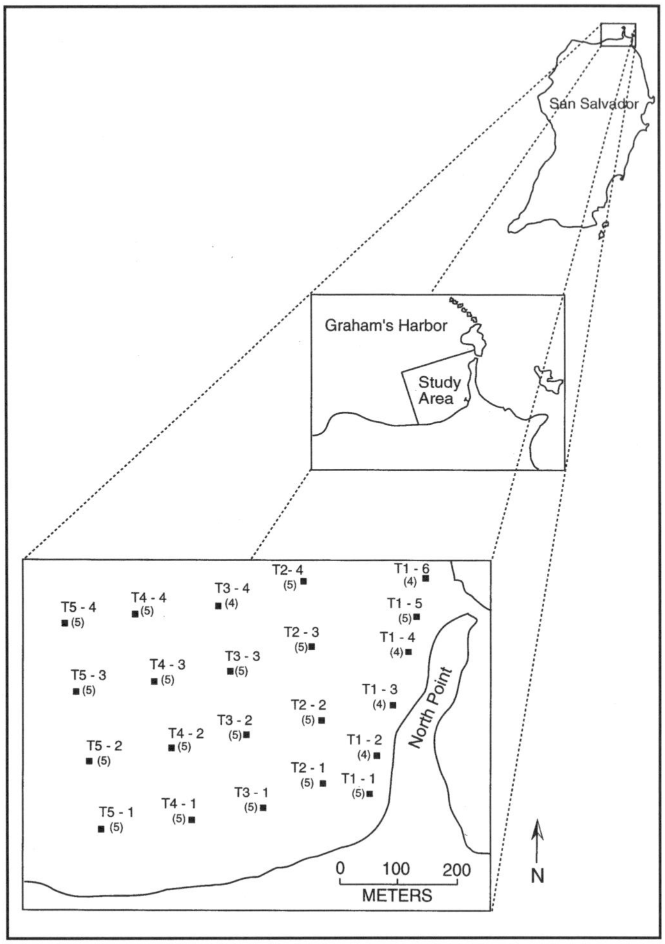

Figure 1. Index map of Graham's Harbor lagoon, showing location of sampling stations. The number of sediment samples obtained at each station is shown in parentheses.

STUDY AREA

Samples were collected from Graham's Harbor, a high-energy, windward lagoon located on the north shore of San Salvador Island, Bahamas, at the eastern edge of the Bahama platform (Fig. 1). The shoreline of the island is characterized by headlands composed of eolianites, with beaches of fine- to medium-grained carbonate sands between the headlands. Holocene beachrock is common. North Point, composed of eolianites, extends northward along the eastern edge of Graham's Harbor. The lagoon is approximately 6 km^2 in area. It is open to the west, allowing free exchange with the Atlantic Ocean. The southern part of the lagoon is more sheltered from the prevailing eastern winds, and provides a relatively lower energy environment than the less sheltered, windward northern area.

Graham's Harbor lagoon deepens toward the center to a maximum depth of 6 m and shallows to the north, east, and south. Sediment thickness increases to at least 4 m toward the center of the lagoon. Evidence suggests Graham's Harbor was a silled basin with a maximum depth of 10 m below present sea level before Holocene sedimentation began (Colby and Boardman, 1989).

This study focuses on the southeastern part of Graham's Harbor lagoon, an area small enough in areal extent (0.5 × 0.5 km) to allow very fine-scale facies variations to be investigated. Colby and Boardman (1989) have described the general bathymetry, sediment thickness distribution, and pre-Holocene history of Graham's Harbor lagoon.

The lagoon supports an extensive benthic flora and fauna. Dense grassbeds, calcareous algae, gastropods, bivalves, echinoderms, bryozoans, crustaceans, corals, and many soft-bodied invertebrates are common. *Callianassa* mounds are frequently encountered. An *Acropora palmata* barrier reef occurs at the northern edge of Graham's Harbor. Armstrong (1989) has reported the annual carbonate production of calcareous green algae and epibionts in Graham's Harbor lagoon to be seasonally variable (higher in June than January).

TABLE 1. SCALE USED FOR ESTIMATING VEGETATION ABUNDANCE IN GRAHAM'S HARBOR*

Value	Description	Substrate Coverage in Quadrat
1	Very sparse	1–2 individuals
2	Sparse	Less than 2%
3	Moderate	2–5%
4	Frequent	5–10%
5	Very frequent	10–15%
6	Abundant	15–25%
7	Very abundant	More than 25%

*Based on relative substrate coverage in 1-m square quadrat.

SAMPLE COLLECTION AND ANALYTICAL METHODS

Field methods

A total of 105 sediment samples were collected by diving using 10-cm-long hand-held core tubes. Several samples were collected at each of 22 stations located approximately every 100 m along each of five transects (Fig. 1).

Vegetation was examined at each station and quantified by methods described by Hiscock (1979) and Russell and Fielding (1981). A $1 m^2$ frame was placed in a representative area at each station. The vegetation within the quadrat was identified and its abundance estimated, using direct counting and visual comparison charts, and quantified on a scale of 1 to 7 (Table 1). Outside the quadrat, changes in vegetation and bedforms were noted when diving to define the boundaries of the ecologic zones.

Laboratory analyses

Sixty-four samples were analyzed for grain size. Standard wet sieve analysis was performed at 0.5Φ intervals, from –0.5Φ (1,400 μ) to 4.0Φ (62 μ). Pipette analysis was performed to determine coarse silt (62 to 16 μ), fine silt (16 to 4 μ), and clay (<4 μ) fractions. Sorting, skewness, and mean diameter were calculated for all samples (Folk, 1966). Modal analyses of 105 thin sections of the coarse fraction (>62 μ) were conducted.

Carbonate mineralogy of the coarse (>62 μ) and fine (<62 μ) fractions was determined for 44 samples by x-ray diffraction analysis (Philips automated powder diffractometer with CuK-alpha radiation) following the methodologies of Chave (1962), Husseini and Matthews (1972), and Fang and Zevin (1985).

Principal component analyses

Two principal component analyses were employed to define and describe sedimentary facies (Joreskog et al., 1976; Fresi et al., 1984; Stevens, 1986).

The first analysis uses compositional data from the thin-section analysis, and includes 12 measurements made on 105 samples. The second uses textural data from the grain-size analysis, consisting of 13 measurements of 64 samples. Principal component analysis can be carried out by two distinct but related processes: R-mode and Q-mode. R-mode analyses compare the relationships between the variables on the basis of all samples. Griffiths and Ondrick (1969) and Saxena and Ekstrom (1970) have provided examples of studies using this type of analysis. Q-mode analyses define the relationships between the samples based on the variables. This technique is useful in determining patterns or groupings among the samples. Imbrie and Van Andel (1964), Solohub and Klovan (1970), and Fresi et al. (1984) have provided examples of the application of Q-mode analyses. The present study utilizes Q-mode principal component analyses.

Several criteria have been proposed to determine the number of principal components to include in the final interpretation of the analysis. Guttman (1954) recommended retaining those components with eigenvalues greater than 1.00; however, this may lead to the retention of some components that account for a very small portion of the variance, and hence may be difficult or impossible to interpret. Joreskog et al. (1976) recommended retaining those components whose cumulative percent variance equals a value considered sufficiently large by the experimenter (e.g., 75, 90, or 95%). In this study, the eigenvalues fall below 1.00 after the fourth component is extracted; however, the first three account for 95% of the variance in both analyses (Table 2). The additional variance accounted for by the fourth component is not considered significant (<2%); therefore, only three components are retained.

TABLE 2. EIGENVALUES AND PROPORTIONS OF VARIANCE EXPLAINED BY EACH COMPONENT*

Principal Component	Eigenvalue	Variance (%)	Cumulative Variance (%)
	I. DATA FROM THIN-SECTION ANALYSIS		
1	86.3411	82.23	82.23
2	8.7952	8.38	90.61
3	4.7786	4.55	95.16
4	1.9530	1.86	97.02
5	0.9665	0.92	97.94
	II. DATA FROM GRAIN-SIZE ANALYSIS		
1	51.7440	80.85	80.85
2	6.0288	9.42	90.27
3	3.0656	4.79	95.06
4	1.1584	1.81	96.87
5	0.6287	0.98	97.85

*Unrotated; from principal component analyses of point count data (I) and grain-size data (II).

The calculation of component loadings provides an important tool to aid in assigning geologic meaning to a principal component. Component loadings may be thought of as the correlation coefficient between a sample (Q-mode) or variable (R-mode) and a principal component. Component loadings are simply the coefficients of the linear equation defined by their respective eigenvector. It is of interest to determine the amount of influence of a component in each sample; the component loadings provide this information. Stevens (1986) suggested a method to determine which component loadings are statistically significant. He recommended doubling the standard error required for significance for an ordinary correlation, at the 0.01 level for a two-tailed test, and using this doubled value to determine whether a loading is significant. For this study, the doubled value was 0.508 for the analysis of the point count data, and 0.650 for the analysis of the grain-size data. Thus, in the analysis of the point count data, all samples with loadings of 0.508 or more on the first principal component, for example, can be thought to be under the same influence. Those samples that load high on the same component must be evaluated in order to determine what they have in common, so that geologic meaning may be assigned to that component.

The component loadings also provide another useful piece of information. The sum of squared loadings for a specific sample is referred to as its communality, and reflects the degree to which that sample vector has been explained by a particular set of component axes. A communality of 1.00 indicates a perfect explanation. In the two analyses performed in this study, all samples except for three have very high communalities (>0.85), indicating that a good description of each sample has been obtained by the use of only three components.

Often the principal components located by extracting the eigenvectors from the covariance or correlation matrix are not in the most meaningful positions with respect to the location of the sample vectors. In order to facilitate interpretation, rotations are performed on the component loading matrix. Two types of rotations can be performed, oblique and orthogonal. Oblique rotations do not require the components to remain orthogonal and hence they become correlated, which may greatly complicate interpretation. In an orthogonal transformation, the components remain uncorrelated. The choice of which rotation to use is not based on statistical grounds, but rather on ease of interpretability. The Varimax orthogonal rotation, developed by Kaiser (1958), is used in this study.

The choice of whether the variance-covariance or correlation matrix is used depends on the nature of the data. A covariance matrix gives more weight to the samples (Q-mode) or variables (R-mode) with greater variances, so would be used if certain samples or variables were felt to be more important to the final interpretation of the data. A correlation matrix scales all samples or variables to unit variance; hence all are assumed to be of equal importance. The use of the correlation matrix also allows the comparison of variables measured in different units. This study uses the correlation matrix in all analyses.

The PROC FACTOR procedure in the SAS (1985) software system was used to accomplish all analyses.

ECOLOGY OF GRAHAM'S HARBOR LAGOON

Vegetation analysis

All potential floral contributors to sediments were identified on the genus level wherever possible (Table 3). Three ecologic zones (Fig. 2) are recognized in Graham's Harbor, named for their dominant elements: *Thalassia* Zone, *Halimeda* and *Penicillus* Zone, and Sparsely Vegetated Zone.

Thalassia Zone. Dense meadows of *Thalassia* occur where the total vegetation cover is greater than 70%. The *Thalassia* blades are often heavily encrusted with epibionts, discoloring the grass blades. *Syringodium* is present in moderate amounts at almost all of the stations. Calcareous algae in this zone are of sparse to moderate abundance and include *Penicillus*, *Halimeda*, *Udotea*, *Rhipocephalus*, *Acetabularia*, and *Ulva*. Very sparse occurrences of coralline algae are encountered. The invertebrate fauna include bivalves, gastropods, and numerous soft-bodied taxa. Bedforms, such as sand waves or ripples, are not encountered in this zone, as the *Thalassia* beds are so dense as to prevent much movement of the sediment. *Callianassa* mounds are present only at one station, which has the lowest total percentage of vegetation coverage in the zone (70%).

Halimeda and Penicillus Zone. The *Halimeda* and *Penicillus* Zone is characterized by intermediate values of vegetation coverage (approximately 50%). *Thalassia*, *Syringodium*, and *Halodule* are encountered in nearly equal amounts. The grass plants occur in small groups or patches rather than a solid, dense carpet. Calcareous algae are more abundant than in the *Thalassia* Zone (*Halimeda* and *Penicillus* predominate). In general, our census found that species diversity is not as great as in the *Thalassia* Zone, although more individuals per species are usually present. Coralline algae rarely occur. The invertebrate fauna includes large gastropods, echinoderms, and crustaceans. *Callianassa* mounds are very common. Large- and small-scale sand ripples are observed trending northeast-southwest.

Sparsely Vegetated Zone. The Sparsely Vegetated Zone has vegetation coverage of less than 30% and at least half the zone has only 10 to 15% vegetation coverage. Marine grasses grow as sparsely distributed single plants or very small clumps. *Penicillus* and *Halimeda* are the most common of the calcareous algae but occur in sparse to moderate amounts. Coralline algae are present in low abundance. Floral species diversity is lowest in this zone. The invertebrate fauna are similar to that encountered in the *Penicillus* and *Halimeda* Zone, *Callianassa* mounds are present at stations along Transect 1, but are rare or absent in the rest of this zone. Large- and small-scale, northeast-southwest–trending sand ripples are a dominant feature of the Sparsely Vegetated Zone.

TABLE 3. VEGETATION ABUNDANCE DATA FOR EACH SAMPLING STATION*

Station	ACE	AVR	CAU	CHA	CA1	CA2	GA1	HAL	PEN	RHI	HAD	SYR	THA	UDO	ULV
T1-1	0	2	2	1	1	0	1	2	3	2	0	0	7	2	1
T1-2	0	2	1	0	1	1	1	2	3	1	0	3	7	1	1
T1-3	3	1	0	0	1	2	1	3	3	0	2	2	3	0	1
T1-4	3	0	0	0	0	0	1	3	3	0	2	2	3	0	1
T1-5	2	0	0	0	1	1	0	2	3	0	0	3	3	1	0
T1-6	2	1	0	0	1	2	1	3	2	0	0	3	7	1	1
T2-1	0	1	2	1	0	0	1	2	3	2	2	3	7	1	1
T2-2	3	0	0	0	2	2	1	5	5	2	3	2	3	1	0
T2-3	3	0	0	0	1	1	0	4	5	2	3	3	3	1	0
T2-4	2	0	0	0	0	1	0	3	3	1	0	3	3	2	0
T3-1	2	1	2	0	0	0	1	2	3	1	2	3	7	1	1
T3-2	3	1	0	0	0	2	1	5	5	2	3	3	3	0	0
T3-3	3	0	0	0	0	2	1	4	5	2	0	3	3	1	0
T3-4	2	0	0	0	0	1	0	3	3	1	0	2	3	0	0
T4-1	3	0	1	0	0	1	1	3	3	2	2	3	3	2	0
T4-2	3	0	0	0	1	2	0	3	3	1	2	3	3	0	0
T4-3	3	0	0	0	1	2	0	4	4	2	0	3	3	2	0
T4-4	2	0	0	0	0	0	0	2	3	2	0	2	2	0	0
T5-1	2	2	1	0	0	1	1	2	3	1	0	4	7	1	0
T5-2	3	2	0	0	1	2	0	3	3	2	0	2	3	1	0
T5-3	2	0	0	0	0	1	0	3	3	2	0	2	3	1	0
T5-4	2	1	0	0	0	1	0	3	3	0	0	2	2	0	0

*Abundance is estimated on a scale of 1 to 7. Values of scale are given in Table 1. Key to genera: ACE = *Acetabularia;* AVR = *Avrainvillea;* CAU = *Caulerpa;* CHA = *Chaetomorpha;* CA1 = Coralline algae 1; CA2 = Coralline algae 2; GA1 = Green algae 1; HAL = *Halimeda;* HAD = *Halodule;* PEN = *Penicillus;* RHI = *Rhipocephalus;* SYR = *Syringodium;* THA = *Thalassia;* UDO = *Udotea;* ULV = *Ulva.*

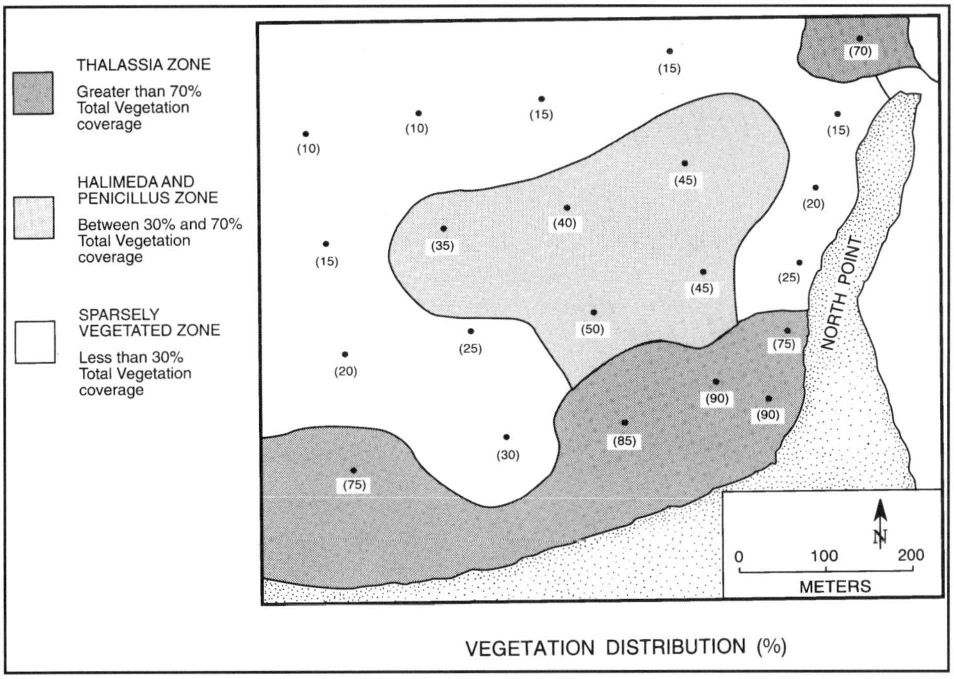

Figure 2. Map of vegetation distribution, based on vegetation abundance data. The total percentage of vegetation coverage is shown in parentheses. Zones are named for dominant elements.

TABLE 4. ARAGONITE PERCENTAGES IN THE COARSE (>63 μ) AND FINE (<63 μ) FRACTIONS OF SEDIMENT SAMPLES FROM GRAHAM'S HARBOR LAGOON*

Station	Coarse Fraction (%)	Fine Fraction (%)
T1-1	64	39
T1-2	70	40
T1-3	73	48
T1-4	80	48
T1-5	70	47
T1-6	78	50
T2-1	73	41
T2-2	69	46
T2-3	69	51
T2-4	65	47
T3-1	60	42
T3-2	82	48
T3-3	70	53
T3-4	68	49
T4-1	70	45
T4-2	71	51
T4-3	64	64
T4-4	62	52
T5-1	72	42
T5-2	69	48
T5-3	73	46
T5-4	67	49

*Two samples per station were analyzed for mineralogy and their values were averaged.

TABLE 5. RATIO OF HIGH- TO LOW-MAGNESIUM CALCITE IN THE COARSE (>63 μ) AND FINE (<63 μ) FRACTIONS OF SEDIMENT SAMPLES FROM GRAHAM'S HARBOR LAGOON*

Associated Ecologic Zone†	Station	High-/Low-Mg Calcite (Coarse)	High-/Low-Mg Calcite (Fine)
T	T1-1	4.2	5.2
T	T1-2	2.5	2.9
SV	T1-3	2.8	2.3
SV	T1-4	2.5	2.1
SV	T1-5	3.0	2.2
T	T1-6	3.6	2.5
T	T2-1	3.0	3.7
H and P	T2-2	3.0	3.0
H and P	T2-3	3.5	3.4
SV	T2-4	3.1	2.8
T	T3-1	3.8	4.3
H and P	T3-2	2.7	2.6
H and P	T3-3	3.8	4.1
SV	T3-4	3.2	2.9
SV	T4-1	2.8	3.0
SV	T4-2	3.5	2.9
H and P	T4-3	4.3	4.0
SV	T4-4	2.9	2.3
T	T5-1	3.6	3.8
SV	T5-2	3.3	2.8
SV	T5-3	3.0	2.5
SV	T5-4	2.8	2.0

*Two samples per station were analyzed for mineralogy and the values were averaged.
†T = *Thalassia* Zone; H and P = *Halimeda* and *Penicillus* Zone; SV = Sparsely Vegetated Zone.

TABLE 6. AVERAGE PERCENTAGE ABUNDANCE OF GRAIN TYPES IN THE THREE SEDIMENTARY GROUPS BASED ON POINT COUNT DATA FROM 105 SAMPLES*

Ecologic Zone	Group	Bioclasts	Coral	Coralline Algae	Forams	*Halimeda*	Molluscs	NSCG	Ooids	Peloids	SCG	CCG
SV	1	6.93	3.56	2.30	10.06	9.08	4.00	13.96	6.24	11.16	26.90	5.81
H and P	2	6.10	1.04	3.86	9.55	12.39	5.30	11.01	5.14	14.78	23.77	7.06
T	3	7.02	0.31	1.76	11.23	9.75	6.38	9.14	2.29	23.56	18.32	10.24

*The groups are recognized with the aid of principal component analysis. Key: NSCG = nonskeletal composite grains; SCG = skeletal composite grains; CCG = cryptocrystalline carbonate grains; SV = Sparsely Vegetated Zone; H and P = *Halimeda* and *Penicillus* Zone; T = *Thalassia* Zone.

SEDIMENTS

Mineralogy and grain types

All of the samples except T4-3 have coarse fractions richer in aragonite than their associated fine fractions (Table 4). The ratio of high- to low-magnesium calcite is approximately equal for the coarse and fine fractions (Table 5).

Sedimentary groups or facies are defined with the aid of a Q-mode principal component analysis of grain-type abundance data (Tables 6 and 7). Table 7 shows a list of communalities and a matrix of component loading on the first three principal components (rotated). Samples that load high on the same principal component are considered to be under the same influences, and thus are members of the same group. Petrographically, three groups are defined by variations in the relative proportions of nonskeletal, skeletal, and aggregate grain types (Table 8). Nonskeletal grains include ooids, peloids, and cryptocrystalline carbonate grains, Skeletal grains include bioclasts, coral, coralline algae, foraminifera, *Halimeda*, and molluscs. Aggregates include nonskeletal and skeletal composite grains.

TABLE 7. RESULTS OF PRINCIPAL COMPONENT ANALYSIS OF GRAIN-TYPE ABUNDANCE DATA FROM 105 SEDIMENT SAMPLES*

Sample†	Communality	Components			Group§
		I	II	III	
T1-1A	0.9194	0.3453	0.3470	0.8245	3
T1-1B	0.9715	0.5003	0.1441	0.8321	3
T1-1C	0.9781	0.2658	0.4920	0.8096	3
T1-1D	0.9523	0.2872	0.0894	0.9283	3
T1-1E	0.9635	0.2429	0.3738	0.8745	3
T1-2A	0.9136	0.2011	0.4093	0.8399	3
T1-2B	0.9867	0.5009	0.2232	0.8221	3
T1-2C	0.9759	0.4936	0.3539	0.7727	3
T1-2D	0.9629	0.2981	0.4901	0.7830	3
T1-3A	0.9089	0.7483	0.4984	0.2369	1
T1-3B	0.9675	0.6848	0.4932	0.5023	1
T1-3C	0.9189	0.7015	0.5860	0.2887	1/2
T1-3D	0.9158	0.7546	0.4723	0.3512	1
T1-4A	0.9605	0.8146	0.4076	0.3616	1
T1-4B	0.9388	0.7169	0.5867	0.2839	1/2
T1-4C	0.9803	0.7276	0.6189	0.2604	1/2
T1-4D	0.9396	0.8032	0.4470	0.3078	1
T1-5A	0.9380	0.7962	0.4238	0.3528	1
T1-5B	0.9175	0.7970	0.3434	0.4035	1
T1-5C	0.9668	0.7752	0.5729	0.1938	1/2
T1-5D	0.9823	0.7619	0.4911	0.4007	1
T1-5E	0.8726	0.7658	0.4359	0.3102	1
T1-6A	0.9059	0.7811	0.4236	0.3411	1
T1-6B	0.9730	0.8974	0.2872	0.2917	1
T1-6C	0.8966	0.8617	0.3159	0.2329	1
T1-6D	0.8998	0.8322	0.2187	0.3992	1
T2-1A	0.9091	0.5018	0.4967	0.5788	3
T2-1B	0.9682	0.4982	0.4828	0.6658	3
T2-1C	0.9462	0.1939	0.6041	0.7374	2/3
T2-1D	0.9376	0.3136	0.6341	0.6605	2/3
T2-1E	0.9748	0.5581	0.4166	0.6999	1/3
T2-2A	0.9724	0.4694	0.7612	0.4154	2
T2-2B	0.8733	0.2586	0.6830	0.5831	2/3
T2-2C	0.9694	0.5008	0.7457	0.4031	2
T2-2D	0.8589	0.3347	0.7596	0.4122	2
T2-2E	0.9442	0.5014	0.7835	0.2681	2
T2-3A	0.9354	0.4976	0.6413	0.5011	2
T2-3B	0.9522	0.4788	0.6548	0.4994	2
T2-3C	0.9416	0.3567	0.7818	0.4507	2
T2-3D	0.9020	0.5613	0.5637	0.5019	1/2
T2-3E	0.9514	0.5779	0.5883	0.4989	1/2
T2-4A	0.9075	0.8712	0.1556	0.3525	1
T2-4B	0.9667	0.7993	0.2492	0.5015	1
T2-4C	0.8894	0.8627	0.0700	0.3744	1
T2-4D	0.9462	0.8288	0.3176	0.3979	1
T2-4E	0.8665	0.8553	0.2185	0.2954	1
T3-1A	0.9348	0.4209	0.4890	0.7200	3
T3-1B	0.9170	0.2442	0.2698	0.8858	3
T3-1C	0.9576	0.3622	0.1426	0.8978	3
T3-1D	0.9651	0.3930	0.4601	0.7739	3
T3-1E	0.9721	0.2910	0.1278	0.9089	3
T3-2A	0.9273	0.3089	0.8078	0.4234	2
T3-2B	0.9311	0.4558	0.7656	0.3704	2
T3-2C	0.9715	0.5728	0.7286	0.3356	1/2
T3-2D	0.9178	0.4106	0.7629	0.4089	2
T3-2E	0.9281	0.3851	0.7592	0.4509	2
T3-3A	0.9370	0.4083	0.8207	0.3111	2
T3-3B	0.9372	0.5024	0.7069	0.4158	2
T3-3C	0.9023	0.5008	0.7145	0.3645	2
T3-3D	0.9744	0.4429	0.7138	0.4983	2
T3-3E	0.8788	0.3771	0.7744	0.3700	2
T3-4A	0.9612	0.7935	0.5022	0.1994	1
T3-4B	0.9174	0.8751	0.1118	0.3729	1
T3-4C	0.9469	0.8378	0.2854	0.4045	1
T3-4D	0.9608	0.8931	0.1843	0.3593	1
T4-1A	0.9380	0.6956	0.4153	0.4936	1
T4-1B	0.8728	0.6502	0.5907	0.3181	1/2
T4-1C	0.8804	0.5739	0.7150	0.1997	1/2
T4-1D	0.8914	0.6658	0.5028	0.4039	1
T4-1E	0.9146	0.6985	0.5873	0.2860	1/2
T4-2A	0.9566	0.7550	0.4570	0.4216	1
T4-2B	0.9224	0.6748	0.6090	0.3101	1/2
T4-2C	0.9271	0.6693	0.5866	0.3675	1/2
T4-2D	0.8027	0.6167	0.5509	0.3449	1/2
T4-2E	0.9573	0.6936	0.6122	0.3187	1/2
T4-3A	0.9348	0.3565	0.7256	0.5003	2
T4-3B	0.9622	0.4583	0.7689	0.4013	2
T4-3C	0.9164	0.5772	0.7097	0.2822	1/2
T4-3D	0.9804	0.4983	0.7418	0.3747	2
T4-3E	0.9796	0.5486	0.7373	0.3676	1/2
T4-4A	0.9687	0.8163	0.4945	0.2405	1
T4-4B	0.9612	0.7861	0.5499	0.2024	1/2
T4-4C	0.9671	0.8554	0.4749	0.0991	1
T4-4D	0.9528	0.7881	0.5628	0.1225	1/2
T4-4E	0.9302	0.7200	0.5679	0.2988	1/2
T5-1A	0.9239	0.1560	0.7145	0.6237	2/3
T5-1B	0.9614	0.1833	0.4978	0.8059	3
T5-1C	0.8565	0.3519	0.6138	0.5966	2/3
T5-1D	0.8815	0.2919	0.4899	0.7179	3
T5-1E	0.8968	0.3594	0.4749	0.7363	3
T5-2A	0.9108	0.6629	0.4887	0.4592	1
T5-2B	0.9727	0.7211	0.5908	0.3220	1/2
T5-2C	0.9951	0.7346	0.5908	0.3261	1/2
T5-2D	0.9015	0.6809	0.4985	0.3979	1
T5-2E	0.8510	0.5539	0.6188	0.3990	1/2
T5-3A	0.9893	0.7213	0.6298	0.2690	1/2
T5-3B	0.8796	0.7498	0.4836	0.1891	1
T5-3C	0.9644	0.8243	0.4944	0.1424	1
T5-3D	0.9856	0.7432	0.4785	0.3697	1
T5-3E	0.9456	0.5877	0.6915	0.3494	1/2
T5-4A	0.9071	0.7729	0.3072	0.4641	1
T5-4B	0.9315	0.7880	0.4954	0.2552	1
T5-4C	0.9215	0.7119	0.5715	0.2970	1/2
T5-4D	0.8840	0.7659	0.4761	0.2658	1
T5-4E	0.9321	0.7248	0.5972	0.2241	1/2

*This table provides a list of communalities and the matrix of component loadings on the first three principal components, rotated by the Varimax method (Kaiser, 1958).

†Samples are labeled according to the site from which they were obtained, e.g., T2-3A is the first sample taken from Station 3 along Transect 2.

§Three sedimentary groups are recognized. Group 1 samples load >0.508 on Component I, Group 2 samples load >0.508 on Component II, and Group 3 samples load >0.508 on Component III.

TABLE 8. AVERAGE PERCENTAGE ABUNDANCE OF THREE GRAIN-TYPE CATEGORIES IN THE THREE SEDIMENTARY GROUPS, BASED ON GRAIN-TYPE ABUNDANCE DATA FROM 105 SAMPLES

Group	Nonskeletal*	Skeletal†	Aggregates§
1	23.21	35.93	40.86
2	26.98	38.24	34.78
3	36.09	36.45	27.46

*Nonskeletal: ooids, peloids, cryptocrystalline carbonate grains.
†Skeletal: bioclasts, coral, coralline algae, foraminifera, molluscs, *Halimeda*.
§Aggregates: nonskeletal composite grains, skeletal composite grains.

Group 1 samples contain the highest percentage of rounded, micritic aggregates and the lowest percentage of nonskeletal grains. Group 2 samples contain nearly equal amounts of aggregates and nonskeletal grains. Skeletal material is slightly more abundant. Group 3 samples have the lowest percentage of aggregates, with the majority of these grains containing skeletal material. They also have the highest amount of nonskeletal grains consisting mainly of elongate and spherical peloids.

Lateral distribution of grain types. Lateral distribution of samples based on a principal component analysis of the grain-type abundance data (Fig. 3) demonstrates that samples taken from the same station may be defined by different principal components. Several stations have combinations of equal amounts of Group 1 and Group 2 samples (T4-1, T4-2, T4-4, and T5-2).

A plot of converted loadings (Klovan, 1966) illustrates which samples are most representative of the influence of a particular principal component (Fig. 4). Mixing trends between Components I and II and Components II and III are apparent, as indicated by the scatter of samples along the sides of the triangle.

Grain size

Three groups of sediment types, or facies, can also be recognized with the aid of principal component analysis of grain-size data (Kaiser, 1958). Table 9 shows a list of communalities and a matrix of component loading on the first three principal components (rotated). Samples that load high on the same component are considered to be members of the same group, and under the same influence of that component.

Group 1 samples are characterized by the largest mean diameter (average, 1.27Φ), the best sorting (2.03Φ), and the least negative skewness (–0.92Φ). The samples from Group 1 have the lowest mud content, averaging 10.8%.

Group 2 samples exhibit slightly smaller mean diameter (1.42Φ), and are more poorly sorted (2.14Φ) and more negatively skewed (–1.04Φ) than Group 1 samples. Their mud content averages 12.2%. The medium sand-size range accounts for approximately 50% of the total sediment.

Group 3 samples are characterized by the smallest mean diameter (1.73Φ), and are the most poorly sorted (2.90Φ) and

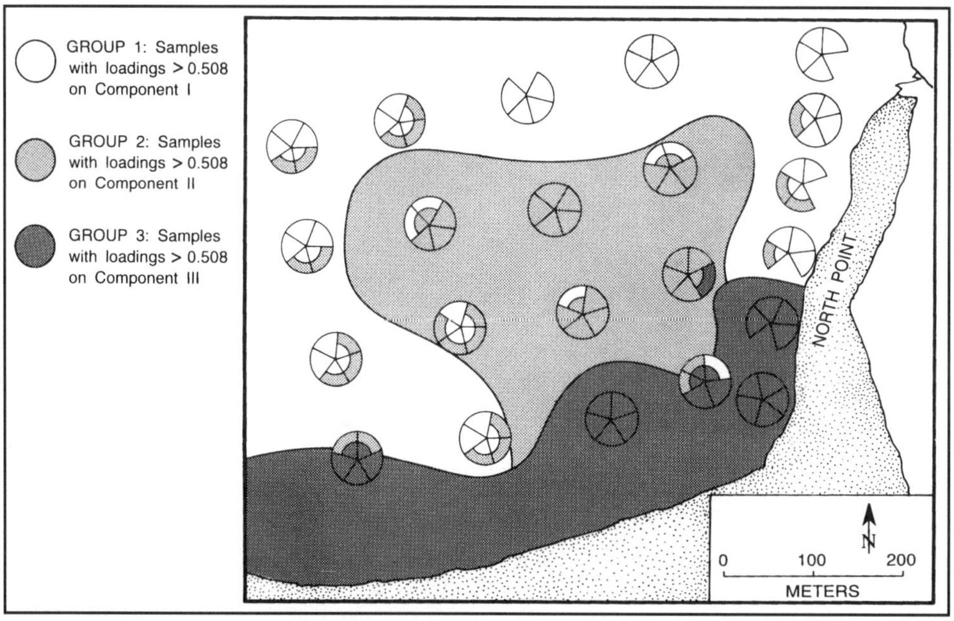

Figure 3. Lateral distribution of sediment groups based on component loadings, using grain-type abundance data. Each wedge represents one sample. Note that some samples are influenced by more than one component, e.g., one sample from site T1-3 loads high on both Components I and II.

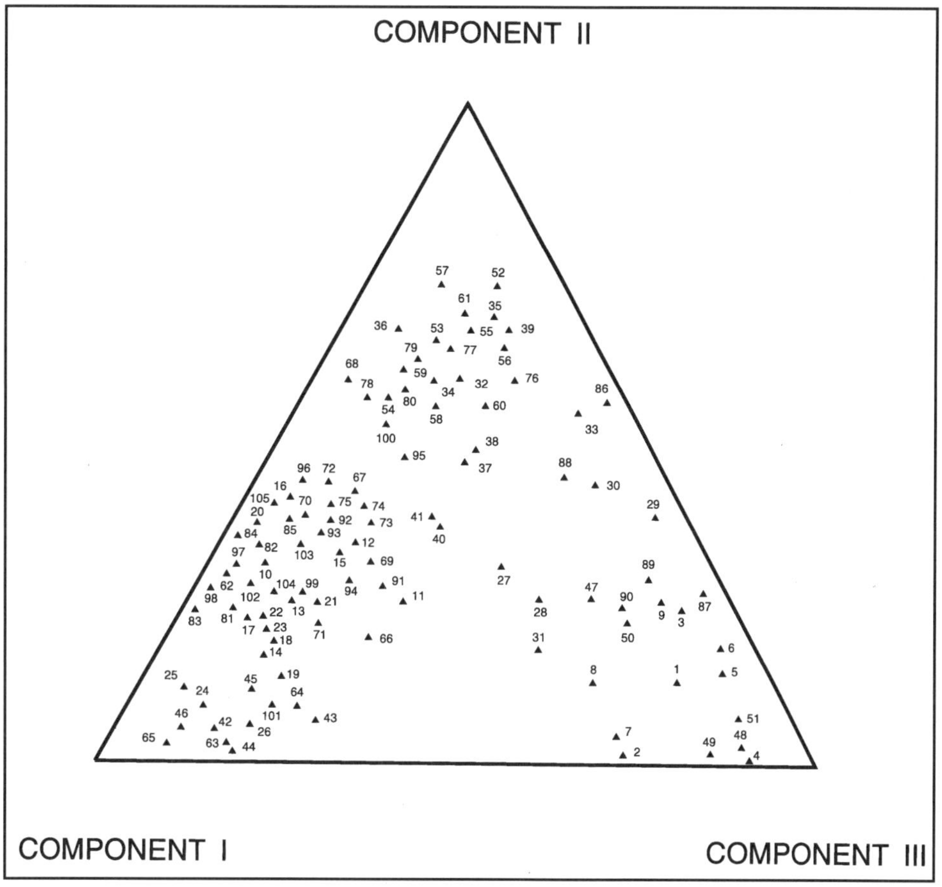

Figure 4. Plot of converted component loadings from principal component analysis of grain-type abundance data prepared according to Klovan (1966). Samples that plot nearest the corners of the diagram can be thought of as end members, as they are most representative of that particular component. Note the scatter of samples along the sides of the triangle, indicating mixing between Components I and II and Components II and III. Samples are numbered from 1 to 105 following the pattern in Figure 1 of sampling stations and number of samples taken (e.g., sample 1 = sample T1-1A; sample 105 = sample T5-4E).

negatively skewed (–1.24Φ) of the three groups. These sediments have the highest mud content, averaging 18.6%. The fine-sand range accounts for the largest portion of the sediment, 45.5%.

Size frequency curves were constructed for representative samples from the three groups (Fig. 5). Group 1 samples are bimodal, with modes at 1.0 and 2.5Φ. Group 2 samples are nearly all unimodal, with a mode at 2.0Φ. Group 2 samples, which also load high on the first- or third-frequency component, show a tendency toward bimodal or trimodal distribution, with modes near 1.0, 2.0, and/or 3.0Φ. The majority of Group 3 samples exhibit a bimodal distribution, although some are trimodal. The bimodal samples commonly have modes at 1.0 to 1.5 and 2.5 to 3.0Φ. The trimodal samples typically have modes at 1.5, 2.5, and 4.0Φ. These samples all tend toward higher representation of the fine-grained size classes.

Lateral distribution of grain sizes. Lateral distribution of samples based on a principal component analysis of the grain-size data (Fig. 6) shows that samples taken from the same station may be defined by different or several components (e.g., stations T1-2, T1-4, and T4-1). A ternary plot of converted loadings for the grain-size data (Fig. 7) indicates a strong mixing trend between Components I and II by the concentrated scatter of samples along the left side of the triangle. The mixing trend between Components I and III is not as pronounced. Mixing between Components II and III does not appear to be very common, as evidenced by the lack of samples along the right side of the triangle.

INFLUENCE OF THE BENTHIC COMMUNITY ON SEDIMENT DISTRIBUTION

Control on sediment mineralogy

Sources of aragonite in the coarse fraction include *Halimeda*, molluscs, corals, and bryozoans. The major source of aragonite fines is thought to be *Penicillus* and other calcareous

TABLE 9. RESULTS OF PRINCIPAL COMPONENT ANALYSIS OF GRAIN-SIZE DATA FROM 64 SAMPLES*

Sample†	Communality	Components			Group§
		I	II	III	
T1-1A	0.9636	0.4105	0.3258	0.8299	3
T1-1B	0.8500	0.3482	0.2459	0.8175	3
T1-2A	0.9722	0.4831	0.7684	0.3853	2
T1-2B	0.9523	0.3473	0.4127	0.8132	3
T1-2C	0.9821	0.2130	0.7502	0.6529	2/3
T1-2D	0.9419	0.4429	0.3759	0.7774	3
T1-2E	0.9349	0.3584	0.8112	0.3851	2
T1-3A	0.9566	0.6561	0.6572	0.3493	1/2
T1-3B	0.9627	0.6618	0.7407	0.1529	1/2
T1-4A	0.8971	0.6737	0.7017	0.1048	1/2
T1-4B	0.9826	0.7884	0.4982	0.3359	1
T1-4C	0.9665	0.8951	0.3481	0.2104	1
T1-4D	0.9462	0.6630	0.6943	0.1274	1/2
T1-5A	0.9529	0.7769	0.4878	0.3339	1
T1-5B	0.9683	0.7891	0.4776	0.3428	1
T1-6A	0.9852	0.7830	0.5221	0.3156	1
T1-6B	0.9710	0.7718	0.4991	0.3553	1
T2-1A	0.9619	0.4717	0.2840	0.8116	3
T2-1B	0.9785	0.4131	0.7864	0.4352	2
T2-1C	0.9660	0.2203	0.3213	0.9022	3
T2-1D	0.9715	0.5284	0.3774	0.7416	3
T2-1E	0.9114	0.2158	0.7107	0.6536	2/3
T2-2A	0.9776	0.5157	0.8077	0.2435	2
T2-2B	0.9391	0.5012	0.7413	0.3720	2
T2-3A	0.9452	0.6639	0.7366	-0.0314	1/2
T2-3B	0.9531	0.8779	0.3875	0.1793	1
T2-3C	0.9169	0.8446	0.4039	0.2009	1
T2-3D	0.9526	0.6510	0.7360	-0.0102	1/2
T2-4A	0.9584	0.8611	0.4210	0.1993	1
T2-4B	0.9298	0.7903	0.4979	0.2397	1
T3-1A	0.4464	0.1269	-0.0796	0.6591	3
T3-1B	0.3813	0.0299	-0.1598	0.6780	3
T3-2A	0.9790	0.5049	0.7707	0.3607	2
T3-2B	0.9714	0.3907	0.7637	0.4853	2
T3-2C	0.9401	0.5425	0.7331	0.3293	2
T3-2D	0.9924	0.6521	0.7184	0.2407	1/2
T3-2E	0.9592	0.7915	0.0205	0.6641	1/3
T3-3A	0.9673	0.7340	0.4713	0.4542	1
T3-3B	0.9674	0.7016	0.4505	0.5217	1
T3-4A	0.8809	0.7790	0.4110	0.3244	1
T3-4B	0.9217	0.6566	0.6721	0.2713	1/2
T4-1A	0.8976	-0.0217	0.6517	0.7238	2/3
T4-1B	0.8626	0.4636	0.6933	0.4088	2
T4-1C	0.9167	0.3463	0.3599	0.8168	3
T4-1D	0.9290	0.5272	0.7717	0.2357	2
T4-2A	0.9788	0.5416	0.8140	0.1517	2
T4-2B	0.9949	0.5184	0.8181	0.2385	2
T4-3A	0.9766	0.6960	0.4685	0.5223	1
T4-3B	0.9690	0.8423	0.4441	0.2497	1
T4-4A	0.9583	0.6763	0.5364	0.4618	1
T4-4B	0.9860	0.8435	0.4481	0.2716	1
T4-4C	0.9755	0.8167	0.4944	0.2531	1
T4-4D	0.9919	0.8507	0.4092	0.3175	1
T4-4E	0.9883	0.6607	0.7426	0.0164	1/2
T5-1A	0.9486	0.1247	0.2843	0.9232	3
T5-1B	0.9549	0.1732	0.3148	0.9088	3
T5-2A	0.9609	0.6572	0.3569	0.6659	1/3
T5-2B	0.9651	0.6744	0.2881	0.6997	1/3
T5-3A	0.9378	0.6516	0.6775	0.3324	1/2
T5-3B	0.8540	0.7870	0.2650	0.4055	1
T5-3C	0.9281	0.6545	0.5048	0.4949	1
T5-3D	0.9368	0.8042	0.4080	0.3516	1
T5-4A	0.8308	0.7329	0.1641	0.5149	1
T5-4B	0.9136	0.8549	0.3139	0.2901	1

*This table provides a list of communalities and the matrix of component loadings on the first three principal components, rotated by the Varimax method (Kaiser, 1958).

†Samples are labeled according to the site from which they were obtained, e.g., T2-3A is the first sample from Station 3 along Transect 2.

§Three sediment groups are recognized. Group 1 samples load >0.650 on Component I, Group 2 samples load >0.650 on Component II, and Group 3 samples load >0.650 on Component III.

algae (Stockman et al., 1967; Neumann and Land, 1975; Armstrong, 1989). The sediments of the *Halimeda* and *Penicillus* Zone have more aragonite fines than the other zones, probably owing to the abundance of calcareous algae.

Halimeda and *Penicillus* are capable of very high production rates, often causing them to be overrepresented in the sediment relative to their abundance in the biota (Swinchatt, 1965; Chave et al., 1972; Neumann and Land, 1975; Armstrong, 1989). Aragonite needles, released on the disintegration of the algae, are often transported great distances from their sites of production (Neumann and Land, 1975; Hine et al., 1981). These factors may account for the similar aragonite percentages in the fine fraction of the Sparsely Vegetated Zone and *Halimeda* and *Penicillus* Zone sediments. Lower aragonite content in the fines of the *Thalassia* Zone is probably due to the high contribution of calcitic epibionts to the mud.

Benthic foraminifera (including Soritidae), coralline algae, and echinoids provide sources of high-magnesium calcite in the coarse fraction. The three facies contain nearly equal total amounts of these combined skeletal elements, thus accounting for the lack of a consistent trend in distribution.

The ratios of high- to low-magnesium calcite in the fine fraction of the sediments appear to parallel the benthic community boundaries quite closely (Table 5). The highest ratios are found in the sediments of the *Thalassia* Zone (average, 4.0). The *Halimeda* and *Penicillus* Zone sediments exhibit intermediate ratios, averaging 3.2. The Sparsely Vegetated Zone sediments have the lowest ratios of high- to low-magnesium calcite in the fine fraction (average, 2.5).

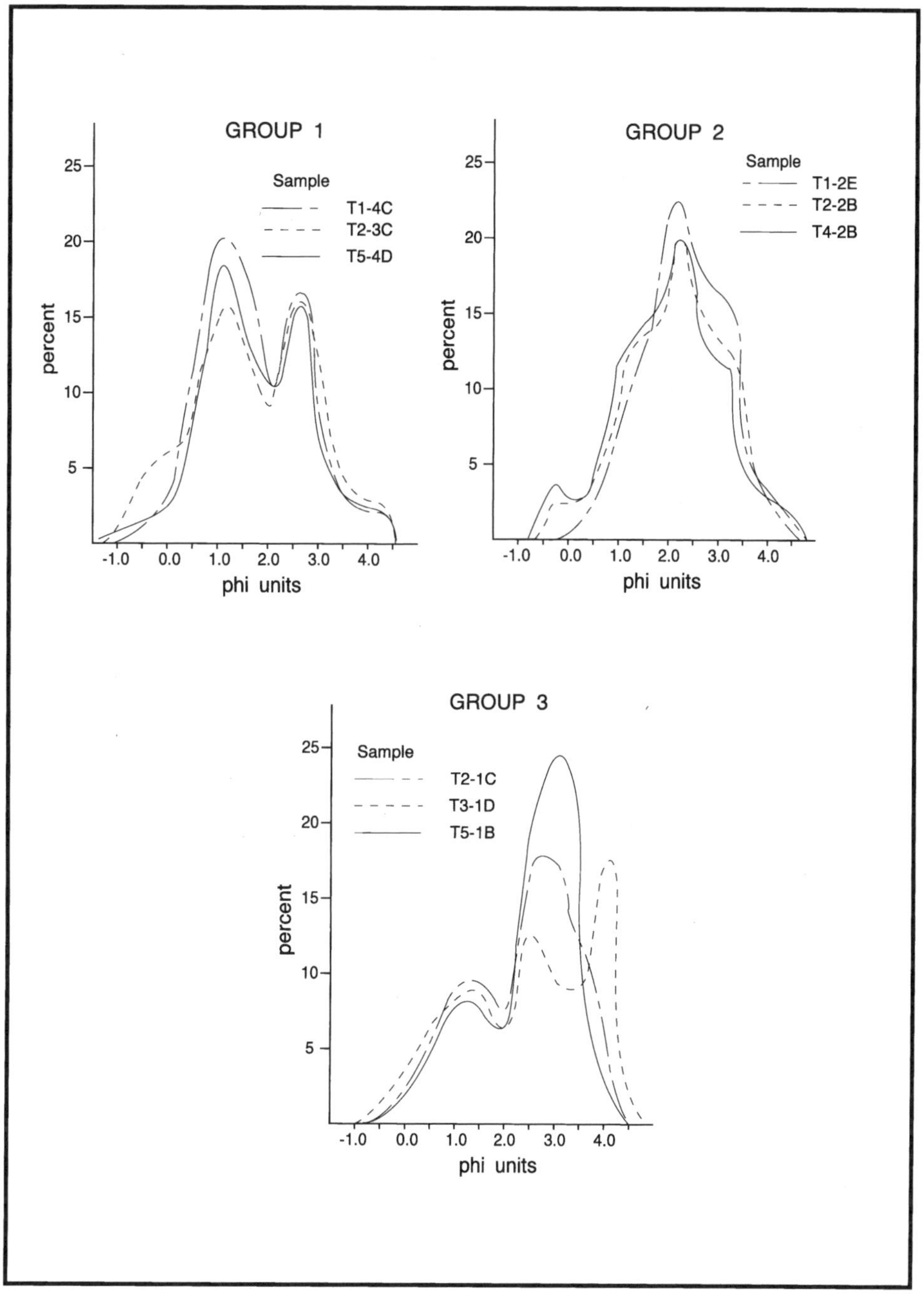

Figure 5. Grain-size frequency curves of representative samples from each sediment group defined by principal component analysis.

Major sources of high-magnesium calcite in carbonate fines include coralline algae and epibionts of *Thalassia* (Land, 1970; Patriquin, 1972). The distribution of high-magnesium calcite in the fine versus the coarse fractions of the sediments may result from the significant contribution of epiphytes to the mud (Chave, 1962; Neumann, 1965; Husseini and Matthews, 1972). This would explain the highest ratios occurring in the *Thalassia* Zone. The grass blades at station T1-6 in the *Thalassia* Zone lack the dense epiphytic growth of other areas of heavy *Thalassia* colonization. The sediments here have high- to low-magnesium calcite ratios similar to stations outside the grassbeds. The contribution from epibionts would also account for the lower aragonite content in the fines of the *Thalassia* Zone, as the production from calcitic epiphytes may overshadow that of aragonitic algae.

The higher proportions of high-magnesium calcite in the fine fraction of the *Thalassia* zone facies suggest that much of the fine sediment is locally derived and is not transported out of

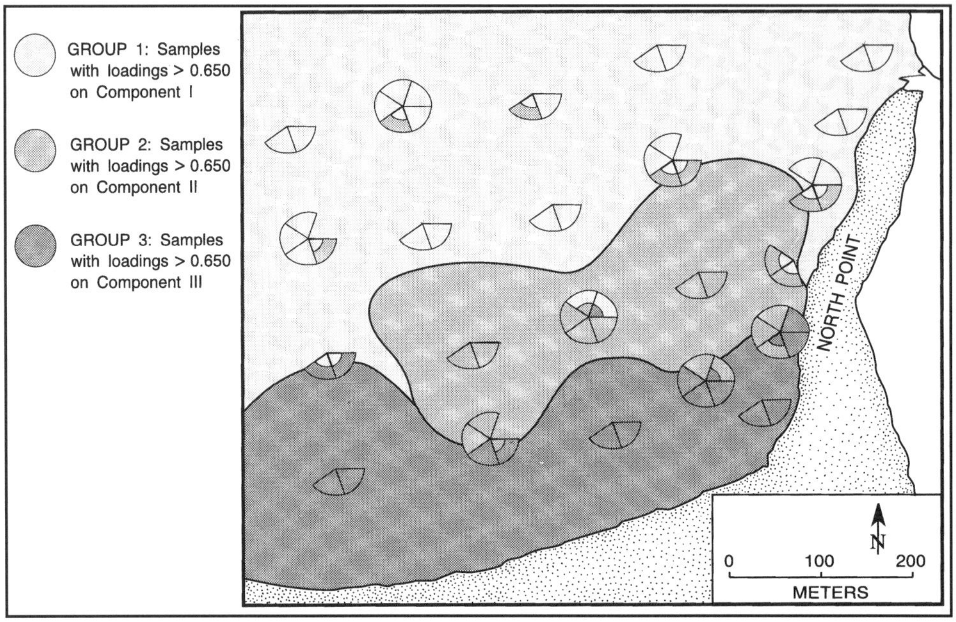

Figure 6. Lateral distribution of sediment groups based on component loadings, using grain-size data. Each wedge represents one sample. Note that some samples are influenced by more than one component, e.g., both samples from site T1-3 load high on Components I and II.

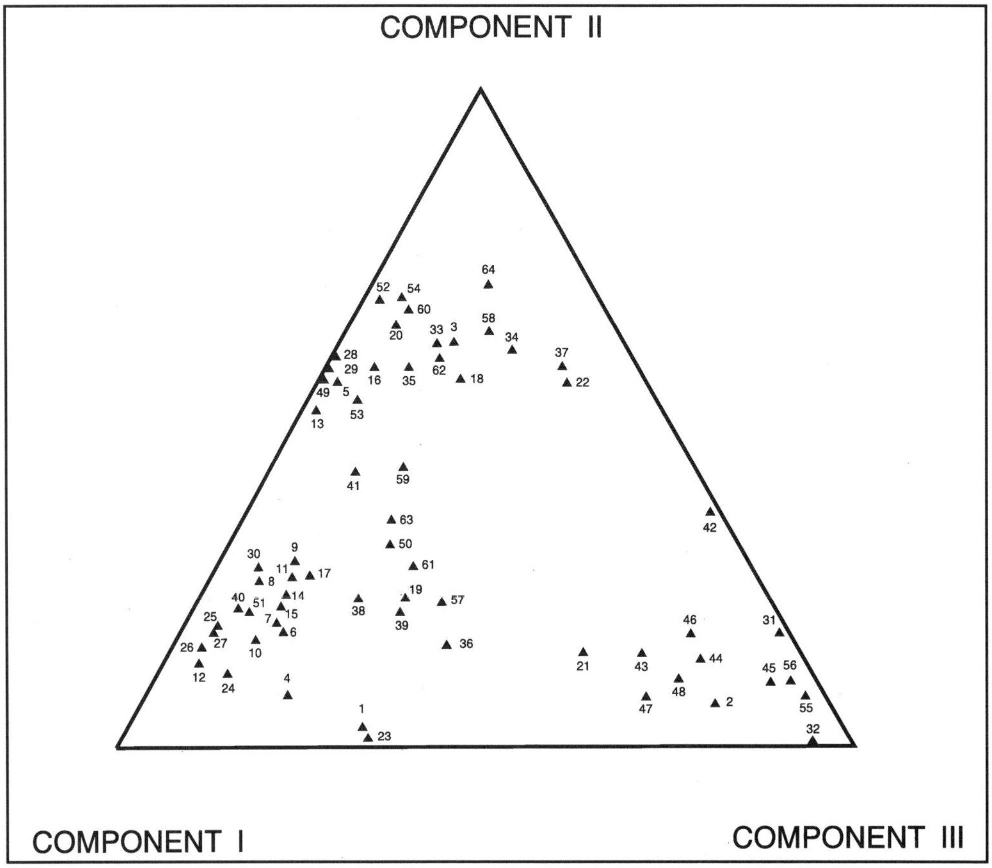

Figure 7. Plot of converted component loadings from principal component analysis of grain-size data, prepared according to Klovan (1966). Samples that plot nearest the corners of the diagram can be thought of as end members, as they are most representative of that particular component. Note the concentration of points along the left side of the triangle, indicating mixing between Components I and II. Samples are numbered 1 through 64 following the number of samples analyzed at each of the sampling stations (see Fig. 6).

the grassbeds once deposited. Principal component analysis of the allochemical constituents data, and to a lesser extent, the grain-size data, support this observation of stability of the sediment deposited in the *Thalassia* Zone.

These data differ from the findings of Colby and Boardman (1989) in Graham's Harbor, who noted approximately equal ratios of high- to low-magnesium calcite in the fine fractions of all samples. They concluded that mud is transported from *Thalassia* communities into other areas of the lagoon. Colby and Boardman (1989) did not provide a detailed assessment of benthic community boundaries, nor were samples collected on as fine a scale as in the present study. These factors may explain the discrepancies in results.

Control on grain types

The lateral distribution of grain types in Graham's Harbor appears to be closely tied to benthic community boundaries. A clear relationship can be seen between the three ecologic zones and the lateral distribution of the sedimentary groups (Figs. 2 and 3). Group 1 samples occur predominantly in the Sparsely Vegetated Zone, where sediment mobility is high. Group 2 samples occur primarily in the *Halimeda* and *Penicillus* Zone, an area of intermediate vegetation abundance where these calcareous algae are dominant. Group 3 samples are confined to the *Thalassia* Zone, which supports dense carpets of marine grasses. Several stations contain samples that load high on different principal components, indicating that group boundaries are gradational or locally influenced by other factors (Fig. 3).

Group 1, defined by Component I, is characteristic of the Sparsely Vegetated Zone (Tables 6 and 7). The sediments have a high percentage of aggregate grains and fragmented skeletal material and represent a high-energy, winnowed environment and a periodically disrupted substrate.

Aggregate grains tend to form in areas where there has been appreciable water circulation, winnowing of fines, and mobilization of the sediment, followed by periods of stabilization and cementation (Boyer, 1972; Winland and Matthews, 1974; Hine, 1977; Wanless, 1981). Aggregates would not be expected to form in areas of dense vegetation. This study demonstrates a clear negative relationship between the abundance of aggregates and vegetation abundance in Graham's Harbor lagoon (Fig. 8). Colby and Boardman (1989), however,

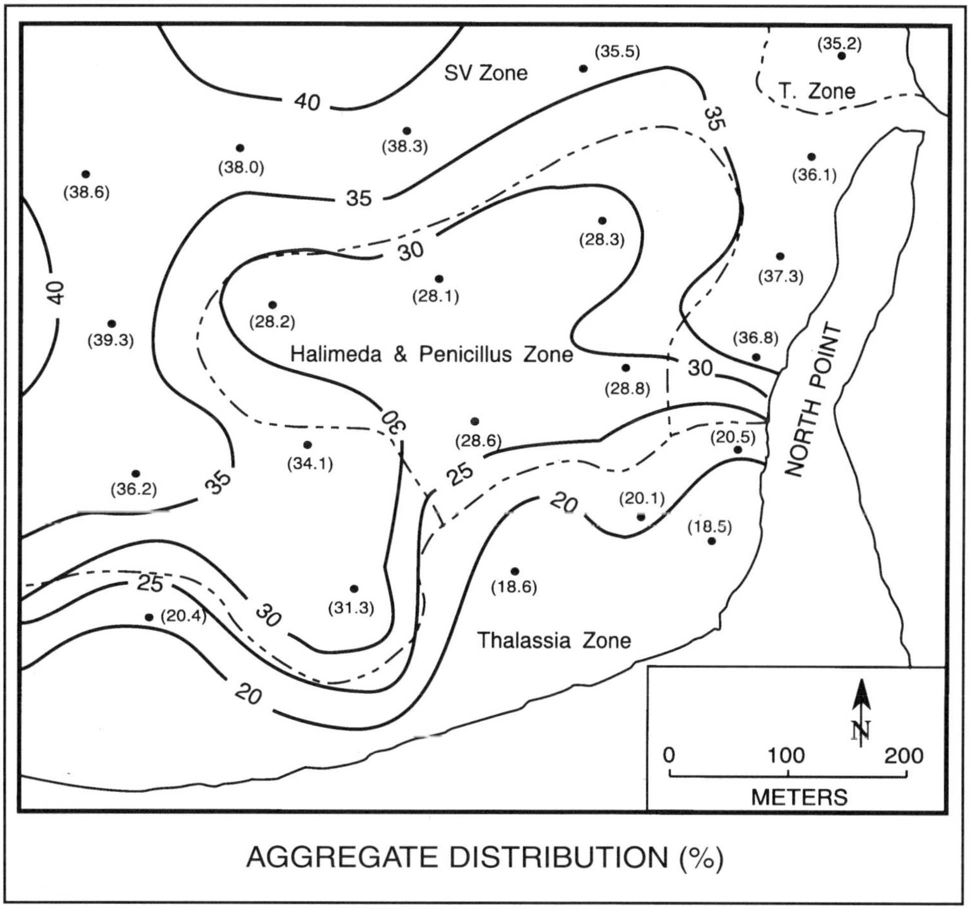

Figure 8. Map showing lateral distribution of aggregates in the study area of Graham's Harbor lagoon. The average percentage of aggregate abundance at each sampling site is shown in parentheses. Note the low percentages of aggregates in areas of high benthic vegetation. The dashed lines mark the boundaries of the three principal ecologic zones.

noted high percentages of aggregates in the areas of dense *Thalassia*. Different scales of the two studies can account for these contradictions. As stated by Armstrong and Miller (1990), significant variability in carbonate sediment production and algal types can occur over a short distance. Fine-scale surveys minimize the uncertainties of extrapolations.

Component II defines the sedimentary group (2) of the *Halimeda* and *Penicillus* Zone (Tables 6 and 7). Aggregates and ooids are less abundant than in the Sparsely Vegetated Zone, suggesting a decrease in energy, although most grains still show evidence of abrasion. Abundant *Callianassa* mounds indicate that bioturbation is an important reworking process. The preservation of these mounds reflects less vigorous current energy than in the Sparsely Vegetated Zone.

Group 3, defined by Component III, occurs in the *Thalassia* Zone (Tables 6 and 7). Aggregates and ooids are least abundant here, as the mud content increases and the mobility of the sediment is greatly reduced. Peloids and cryptocrystalline carbonate grains are very abundant in this facies.

The more common skeletal biota of the grassbeds include foraminifera, principally Miliolidae (a family that often inhabits areas with poor circulation) and Soritidae (a family that frequently lives on the grass blades), and molluscs. The sediments in the *Thalassia* Zone contain fresh, unbroken *Halimeda* segments—evidence of a calmer, lower energy environment.

Station T1-6 is the only station in the *Thalassia* Zone that does not display any influence of Component III (Figs. 2 and 3). This location has the lowest vegetation abundance in the zone (70%), is the farthest offshore, and is in a higher energy environment. The samples at this station all exhibit the influence of Component I, which relates to a higher energy environment with a less stable substrate.

Gradational boundaries between groups result from transport of sediment across benthic community boundaries (Fig. 3). It is also possible that these gradational boundaries reflect a fluctuating migration of community boundaries that produced a time-averaged, mixed accumulation (A. I. Miller, personal communication). Stations located in the parts of the Sparsely Vegetated Zone with the greatest vegetation (e.g., T1-3, T1-4, T4-1) display the influence of Component II, that component relating to the *Halimeda* and *Penicillus* Zone. Conversely, the influence of Component I is seen in many of the stations located on the periphery of the *Halimeda* and *Penicillus* Zone.

The influence of Component III, which relates to the *Thalassia* Zone, is not seen outside this ecologic zone except in one instance, at Station T2-1. Sediments in the *Thalassia* Zone also exhibit the influence of Components I and II. It appears that sediments may be transported from the Sparsely Vegetated Zone or the *Halimeda* and *Penicillus* Zone into the *Thalassia* Zone. Armstrong (1989) reported the presence of ephermal sandy patches in the *Thalassia* Zone, which suggests that some sediments may be derived from within this zone itself. However, once sediments are deposited in the grassbeds, they are not resuspended and transported out, as demonstrated by the negligible influence of Component III outside the *Thalassia* Zone. Transport of sediment over large areas frequently occurs during storms, especially in areas where vegetation is low. With the exception of blowouts, the sediments in the grassbeds may remain stable, even during storms of hurricane force (Ball, 1967; Hine, 1977; Hine and Neumann, 1977; Wanless et al., 1988). The facies distributions in Graham's Harbor confirm this observation.

Control on grain size distributions

Sediment-size distributions in Graham's Harbor show a much more poorly defined relationship with benthic community boundaries. Group boundaries are more gradational when defined on the basis of grain size than on grain type. Many stations contain samples that have high component loadings on different principal components. This is especially true in the case of Component II, which relates only roughly to the *Halimeda* and *Penicillus* Zone (Fig. 6). It appears that an overprint of some other factor may be obscuring the relationships between the benthos and sediment sizes.

Group I samples, those associated with the Sparsely Vegetated Zone, have bimodal frequency distributions (Fig. 5) and the best sorting values, lowest mud contents, and largest mean diameters. These size characteristics are consistent with a higher energy, winnowed environment. Because nearly 80% of the sediment is composed of skeletal grains or skeletal-dominated aggregates, its size distribution is probably primarily related to the mechanical breakdown of skeletal materials through vigorous current action.

Group II samples have slightly poorer sorting, smaller mean diameters, and higher mud contents than Group I samples. Group II samples are roughly associated with the *Halimeda* and *Penicillus* Zone, although their influence can be seen throughout the study area. Size distributions are probably partly related to the mechanical breakdown of skeletal material by current action and reworking of sediments through bioturbation. *Callianassa* mounds occur frequently in the *Halimeda* and *Penicillus* Zone; they are absent or less common in other areas. Construction of these mounds also may have resulted in the ejection of fine-sized sediments into the water column and redistribution to other zones (Shinn, 1968; Tudhope and Scoffin, 1984).

Group III samples, those associated with the *Thalassia* Zone, are much more poorly sorted and have the lowest mean diameters and highest mud contents of the three sediment groups. Their size frequency distributions are trimodal or bimodal, with emphasis on the fine size classes (Fig. 5). This group contains material that, for the most part, was produced in place, as well as sediment that was imported from other areas of the lagoon. The size distribution of sediments in the areas of dense *Thalassia* colonization represents a balance between the rate of skeletal production and the rate of its breakdown (Swinchatt, 1965). The polymodal character of the Group III sedi-

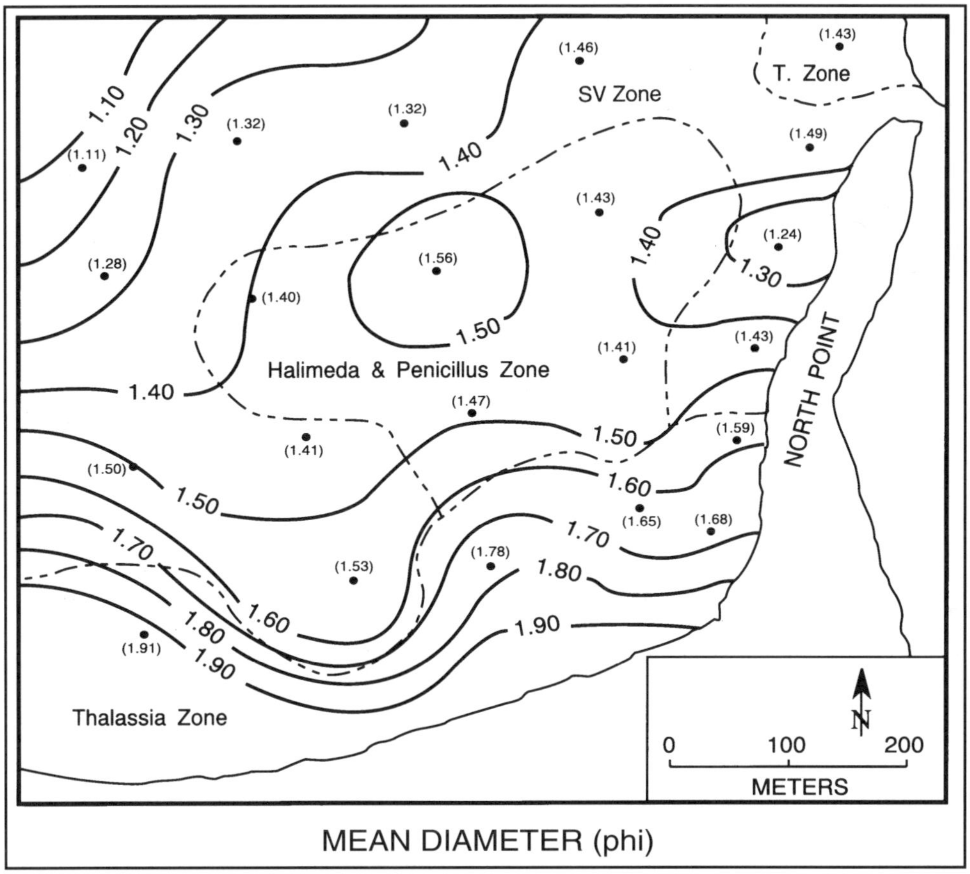

Figure 9. Map showing lateral variability in mean diameter (Φ units) of the sediments in the study area of Graham's Harbor lagoon. The average mean diameter for each sampling site is shown in parentheses. Contour intervals are in Φ units. The dashed lines mark the boundaries of the three principal ecologic zones.

ments reflects the supply of carbonate from a variety of skeletal sources and their differential rates of biologic breakdown.

Mean diameter and sorting show a general increase offshore and toward the northwest portion of the study area (Figs. 9 and 10). The changes in these parameters follow broad bands paralleling the shore, obscuring any obvious relationships with the benthos. Lime mud content decreases offshore and to the northwest, and exhibits only vague ties to benthic community boundaries (Fig. 11).

The most clearly defined relationships between sediment size and vegetation are seen in the *Thalassia* Zone. These sediments exhibit the lowest mean diameters, lowest sorting values, and high-lime mud contents. Colby and Boardman (1989) reported similar results, noting an increase in mud content and a decrease in sorting occurring in areas of dense seagrass cover. They concluded that benthic flora exert an important control on sediment size in Graham's Harbor. Results of the present study do not fully support that conclusion, however. Although dense *Thalassia* growth appears to influence sediment sizes, links between sediment size and the benthic community are difficult to decipher outside the grassbeds.

In Graham's Harbor, sediment size clearly is not as sensitive to changes in the benthic communities as is grain type. Three reasons are proposed to account for this discrepancy:

1. Vegetation has little influence on such size features as sorting, mean diameter, or percentage of fines. Purdy (1963) and Bathurst (1971) noted percentage of fines to be an unreliable indicator of benthic community composition.

2. Both lateral transport by currents or storms and reworking by bioturbation quickly obscure any initial ties between grain size and benthic communities. Grain types generally retain their identity longer.

3. Standard grain-size analysis techniques are not well suited to carbonate sediments, thus making it difficult to uncover relationships between size and the benthic community (e.g., Poole, 1957; Folk and Robles, 1964; Jindrich, 1969).

SYNOPSIS

Wave and current energies influence carbonate deposition by controlling processes of skeletal breakdown and grain winnowing and transport. These energies also have a direct control

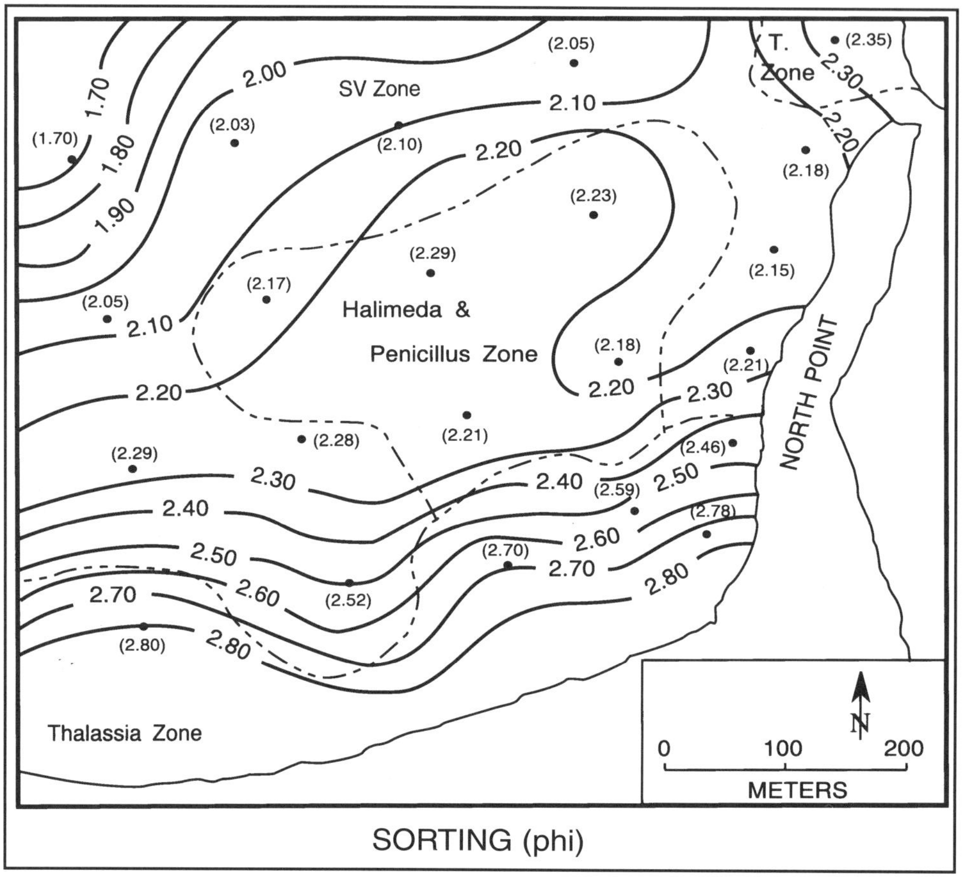

Figure 10. Map showing lateral variability in sorting (Φ units) of the sediments in the study area of Graham's Harbor lagoon. The average sorting value for each sampling site is shown in parentheses. Contour intervals are in Φ units. The dashed lines mark the boundaries of the three principal ecologic zones.

on the distribution of the variety of benthic organisms, relating to their ecology.

In the part of Graham's Harbor studied, three distinct ecologic zones are recognized: the *Thalassia* Zone, the *Halimeda* and *Penicillus* Zone, and the relatively barren Sparsely Vegetated Zone. A partial barrier to the east (North Point) and north (barrier reef) and the prevailing northeasterly wind direction permit the development of relatively strong waves and currents. Sediment mineralogy in Graham's Harbor reflects the influence of the benthic community and current energy on sediment facies distribution. The ratio of high- to low-magnesium calcite in the fine fraction is highest in the sediments of the *Thalassia* Zone (Table 5). Epibionts on grass blades are a major source of high-magnesium calcite in the fine fraction and probably account for the greater proportion of this mineral phase in the *Thalassia* Zone.

The sediment facies distribution in Graham's Harbor based on grain type appears to be closely linked to the benthic community. The three groups recognized by Q-mode principal component analysis can be defined by the abundance of three key grain-type categories: aggregates, nonskeletal grains, and skeletal grains. The abundance of these grain types is related to water circulation, substrate stability, and mud content of the sediment. These factors are in turn tied to the composition and abundance of the benthic community.

Waves and currents also control the transport of grains across benthic community boundaries. The results of the principal component analysis of grain-type abundance data suggest a high degree of lateral mixing between the sediments of the Sparsely Vegetated Zone and the *Halimeda* and *Penicillus* Zone. *Halimeda* fragments are common in the Sparsely Vegetated Zone, despite the low abundance of the living algae in this habitat. The similar appearance and the percentages of aggregate grains and ooids in these two zones suggest a high degree of transport across the benthic community boundaries.

Mixing of sediments between the *Thalassia* Zone and areas outside this zone occurs less frequently than mixing between the Sparsely Vegetated Zone and the *Halimeda* and *Penicillus* Zone. Principal component analysis of both the grain-type and grain-size data indicates the *Thalassia* Zone receives sediments that have been transported from outside this zone. (Figs. 3 and 6).

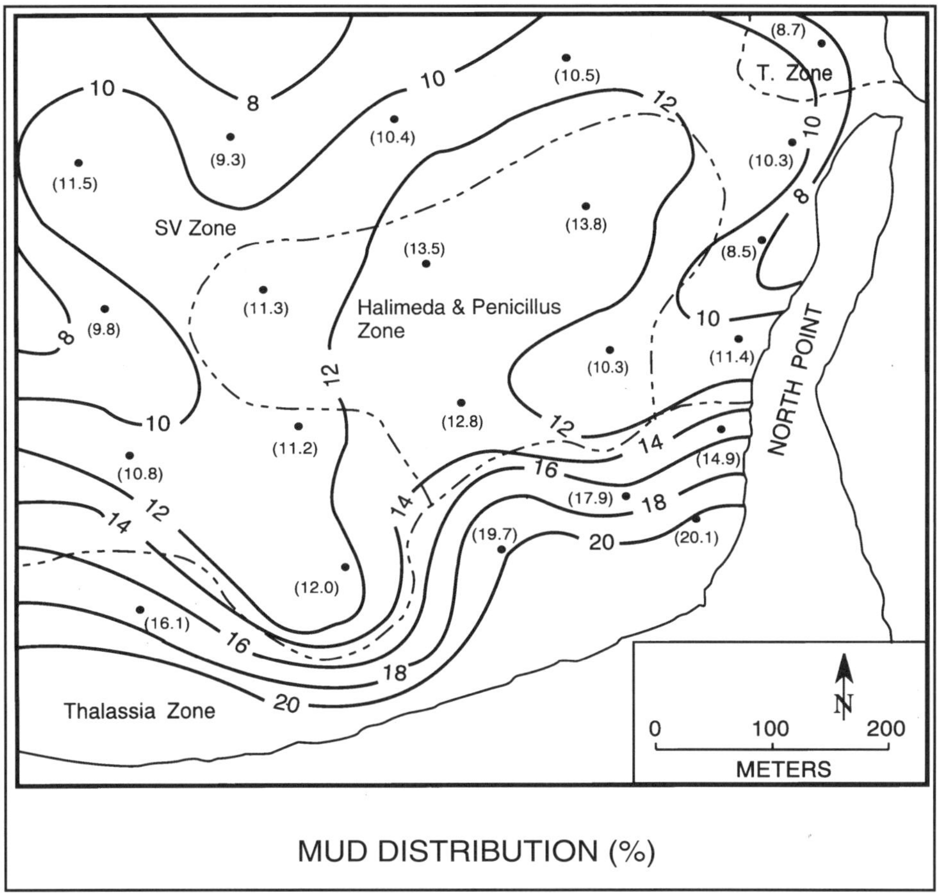

Figure 11. Map showing percentage of lateral variability in lime mud content of the sediments in the study area of Graham's Harbor lagoon. The average percentage of lime mud at each sampling site is shown in parentheses. The dashed lines mark the boundaries of the three principal ecologic zones.

The results of the principal component analysis of the grain-size data also indicate strong lateral transport of sediments between the *Halimeda* and *Penicillus* Zone and the Sparsely Vegetated Zone. Links between the benthic community and size parameters are difficult to decipher. The effect of wave and current energy clearly overprints the control the benthic community may have on the distribution of sediment facies based on size parameters.

The findings of this study differ from those of Colby and Boardman (1989) on sediment distribution in Graham's Harbor. They observed that sediment facies boundaries, based on 10 measured variables, are not tied to benthic community boundaries. They noted that, while size parameters such as mud content and sorting are linked to the benthos, relationships between vegetation and the distribution of variables such as aggregates and fine-fraction mineralogy, are not apparent. The Colby and Boardman (1989) study was conducted on a larger scale and did not include a detailed assessment of the distribution of the benthic communities.

This study found that grain type and fine-fraction mineralogy in Graham's Harbor closely parallel benthic community boundaries. Transport of sediment across these boundaries is evident but does not obscure the relationship between the benthos and sediment distribution. Grain-size parameters are not as clearly linked to the vegetation, and group boundaries based on grain-size data are more gradational than those based on grain type. Wave and current energy have a major influence on grain-size distribution.

The tenets previously expressed (Ginsburg and Lowenstam, 1958; Swinchatt, 1965; Bathurst, 1971; Turmel and Swanson, 1976; Bosence et al., 1985; Tucker and Wright, 1990) regarding the control of the benthic community on sediment facies distributions in modern carbonate lagoons is verified. Benthic communities play a significant role in controlling sediment grain type and fine-fraction mineralogy. Further, the seagrass community exerts a stabilizing influence on sediments imported or formed in this ecozone. Grain-size distributions are not indicative of specific benthic communities because of bioturbation and the vagaries of transport mechanisms and grain-size analysis techniques.

This study demonstrates the patterns and controls of fine-scale carbonate sediment variability, and it quantifies the degree to which sediments may be transported across benthic community boundaries. Such information contributes to the understanding of processes and patterns in modern carbonate deposition and can provide insight about ancient carbonate sequences and their associated paleoenvironments.

ACKNOWLEDGMENTS

We express our gratitude to Donald T. Gerace and Kathy Gerace and staff of the Bahamian Field Station for their support and courtesies extended. They provided valuable assistance during the sample collection phase of this study. We are grateful for the advice and suggestions offered by H. A. Curran, M. R. Boardman, A. I. Miller, and B. White.

REFERENCES CITED

Aller, R. C., and Dodge, R. E., 1974, Animal-sediment relations in a tropical lagoon, Discovery Bay, Jamaica: Journal of Marine Research, v. 32, p. 209–232.

Armstrong, M. E., 1989, Modern carbonate sediment production and its relation to bottom variability, Graham's Harbor, San Salvador, Bahamas [M.S. thesis]: Cincinnati, Ohio, University of Cincinnati, 100 p.

Armstrong, M. E., and Miller, A. I., 1990, Modern carbonate sediment production and its relation to bottom variability, Graham's Harbor, San Salvador, Bahamas, *in* Mylroie, J., ed., Proceedings, Fourth Symposium on the Geology of the Bahamas: San Salvador, Bahamian Field Station, p. 23–32.

Ball, M. M., 1967, Carbonate sand bodies of Florida and the Bahamas: Journal of Sedimentary Petrology, v. 37, p. 556–591.

Bathurst, R. G. C., 1971, Carbonate Sediments and Their Diagenesis; Developments in Sedimentology Ser. 12: New York, Elsevier, 658 p.

Black, M., 1933, The precipitation of calcium carbonate on the Great Bahama Bank: Geological Magazine, v. 70, p. 455–466.

Bosence, D. W. J., Rowlands, R. J., and Quine, M. L., 1985, Sedimentology and budget of a Recent carbonate mound, Florida Keys: Sedimentology, v. 32, p. 317–343.

Boyer, B. W., 1972, Unconsolidated surface sediments from the Florida reef tract: Journal of Sedimentary Petrology, v. 42, p. 205–210.

Chave, K. E., 1962, Factors influencing the mineralogy of carbonate sediments: Limnology and Oceanography, v. 7, p. 218–223.

Chave, K. E., Smith, S. V., and Roy, K. J., 1972, Carbonate production by coral reefs: Marine Geology, v. 12, p. 123–140.

Colby, N. D., and Boardman, M. R., 1989, Depositional evolution of a windward, high-energy lagoon, Graham's Harbor, San Salvador, Bahamas: Journal of Sedimentary Petrology, v. 59, p. 819–834.

Fang, J. H., and Zevin, L., 1985, Quantitative x-ray diffractometry of carbonate rocks: Journal of Sedimentary Petrology, v. 55, p. 611–613.

Folk, R. L., 1966, A review of grain size parameters: Sedimentology, v. 6, p. 73–93.

Folk, R. L., and Robles, R., 1964, Carbonate sands of Isla Perez, Alacran Reef complex, Yucatan: Journal of Geology, v. 72, p. 255–292.

Fresi, E., Gambi, M. C., Focardi, S., Bargagli, R., Baldi, F., and Falciai, L., 1984, Benthic community and sediment types, A structural analysis: Marine Ecology, v. 4, p. 101–121.

Ginsburg, R. N., 1956, Environmental relationships of grain size and constituent particles in some south Florida carbonate sediments: American Association of Petroleum Geologists Bulletin, v. 40, p. 2384–2427.

Ginsburg, R. N., and Lowenstam, H. A., 1958, The influence of marine bottom communities on the depositional environment of sediments: Journal of Geology, v. 66, p. 310–318.

Griffiths, J. C., and Ondrick, C. W., 1969, Modeling the petrology of detrital sediments, *in* Merriam, D. F., ed., Computer Applications in the Earth Sciences: New York, Plenum Press, p. 73–97.

Guttman, L., 1954, Some necessary conditions for common factor analysis: Psychometrika, v. 19, p. 149–161.

Heck, K. L., Jr., 1977, Comparative species richness, composition, and abundance of invertebrates in Caribbean seagrass (*Thalassia testudinum*) meadows (Panama): Marine Biology, v. 41, p. 335–348.

Hine, A. C., 1977, Lily Bank, Bahamas: History of an active oolite sand shoal: Journal of Sedimentary Petrology, v. 47, p. 1554–1581.

Hine, A. C., and Neumann, A. C., 1977, Shallow carbonate bank-margin growth and structure, Little Bahama Bank, Bahamas: American Association of Petroleum Geologists Bulletin, v. 61, p. 376–406.

Hine, A. C., Wilber, R. J., Bane, J. M., Neumann, A. C., and Lorenson, K. R., 1981, Offbank transport of carbonate sands along open, leeward bank margins, northern Bahamas: Marine Geology, v. 42, p. 327–348.

Hiscock, K., 1979, Systematic surveys and monitoring in nearshore sublittoral areas using diving, *in* Nichols, D., ed., Monitoring the Marine Environment: New York, Praeger, p. 55–74.

Husseini, S. I., and Matthews, R. K., 1972, Distribution of high-magnesium calcite in lime muds of the Great Bahama Bank; Diagenetic implications: Journal of Sedimentary Petrology, v. 42, p. 179–182.

Illing, L. V., 1954, Bahamian calcareous sand: American Association of Petroleum Geologists Bulletin, v. 38, p. 1–95.

Imbrie, J., and Purdy, E. G., 1962, Classification of modern Bahamian carbonate sediments, *in* Ham, W. E., ed., Classification of Carbonate Rocks: American Association of Petroleum Geologists Memoir 1, p. 252–272.

Imbrie, J., and Van Andel, T. H., 1964, Vector analysis of heavy mineral data: Geological Society of America Bulletin, v. 75, p. 1131–1156.

Jindrich, V., 1969, Recent carbonate sedimentation by tidal channels in the lower Florida Keys: Journal of Sedimentary Petrology, v. 39, p. 531–553.

Joreskog, K. G., Klovan, J. E., and Reyment, R. A., 1976, Geological Factor Analysis: Amsterdam, Netherlands, Elsevier, 178 p.

Kaiser, H. F., 1958, The Varimax criterion for analytic rotation in factor analysis: Psychometrika, v. 23, p. 187–200.

Klovan, J. E., 1966, The use of factor analysis in determining depositional environments from grain-size distributions: Journal of Sedimentary Petrology, v. 36, p. 115–125.

Land, L. S., 1970, Carbonate mud production by epibiont growth on *Thalassia testudinum*: Journal of Sedimentary Petrology, v. 40, p. 1361–1363.

Neumann, A. C., 1965, Processes of Recent carbonate sedimentation in Harrington Sound, Bermuda: Bulletin of Marine Science, v. 15, p. 987–1035.

Neumann, A. C., and Land, L. S., 1975, Lime mud deposition and calcareous algae in the Bight of Abaco, Bahamas: A budget: Journal of sedimentary Petrology, v. 45, p. 763–786.

Patriquin, D. G., 1972, Carbonate mud production by epibionts on *Thalassia*: An estimate based on leaf growth rate data: Journal of Sedimentary Petrology, v. 42, p. 687–689.

Poole, D. M., 1957, Size analysis of sand by a sedimentation technique: Journal of Sedimentary Petrology, v. 27, p. 460–468.

Purdy, E. G., 1963, Recent calcium carbonate facies of the Great Bahama Bank: Journal of Geology, v. 71, p. 334–355.

Russell, G., and Fielding, A. H., 1981, Individuals, populations, and communities, *in* Lobban, C. S., and Wynne, M. J., eds., The Biology of Seaweeds: Berkeley, University of California Press, p. 125–276.

SAS, 1985, User's Guide—Statistics: Cary, North Carolina, SAS Institute, 956 p.

Saxena, S. K., and Ekstrom, T. K., 1970, Statistical chemistry of calcic amphiboles: Contributions to Mineralogy and Petrology, v. 26, p. 276–284.

Scoffin, T. P., 1970, The trapping and binding of subtidal carbonate sediments by marine vegetation in Bimini Lagoon, Bahamas: Journal of Sedimentary Petrology, v. 40, p. 249–273.

Scoffin, T. P., 1987, An Introduction to Carbonate Sediments and Rocks: New York, Chapman and Hall, 274 p.

Shinn, E. A., 1986, Burrowing in Recent lime sediments of Florida and the

Bahamas: Journal of Paleontology, v. 42, p. 879–894.

Solohub, J. T., and Klovan, J. E., 1970, Evaluation of grain-size parameters in lacustrine environments: Journal of Sedimentary Petrology, v. 40, p. 81–101.

Stevens, J., 1986, Applied Multivariate Statistics for the Social Sciences: London, United Kingdom, Lawrence Erlbaum, 515 p.

Stockman, K. W., Ginsburg, R. N., and Shinn, E. A., 1967, The production of lime mud by algae in south Florida: Journal of Sedimentary Petrology, v. 37, p. 633–648.

Swinchatt, J. P., 1965, Significance of constituent composition, texture, and skeletal breakdown in some Recent carbonate sediments: Journal of Sedimentary Petrology, v. 35, p. 71–90.

Tucker, M. E., and Wright, V. P., 1990, Carbonate Sedimentology, Oxford, United Kingdom, Blackwell, 482 p.

Tudhope, A. W., and Scoffin, T. P., 1984, The effects of *Callianassa* bioturbation on the preservation of carbonate grains in the Davies Reef Lagoon, Great Barrier Reef, Australia: Journal of Sedimentary Petrology, v. 54, p. 1091–1096.

Turmel, R. J., and Swanson, R. G., 1976, The development of Rodriquez Bank, a Holocene mudbank in the Florida reef tract: Journal of Sedimentary Petrology, v. 46, p. 467–518.

Wanless, H. R., 1981, Fining-upwards sedimentary sequences generated in sea grass beds: Journal of Sedimentary Petrology, v. 51, p. 445–454.

Wanless, H. R., Tyrrell, K. M., Tedescom, L. P., and Dravis, J. J., 1988, Tidal-flat sedimentation from Hurricane Kate, Caicos Platform, British West Indies: Journal of Sedimentary Petrology, v. 58, p. 724–738.

Winland, H. D., and Matthews, R. K., 1974, Origin and significance of grapestone, Bahama Islands: Journal of Sedimentary Petrology, v. 44, p. 921–927.

Young, D. K., 1971, Effects of infauna on the sediment and seston of a subtidal environment: Vie et Milieu, supp., v. 22, p. 557–571.

Manuscript Accepted by the Society January 5, 1995

Printed in U.S.A.

Geological Society of America
Special Paper 300
1995

The effects of life habit and test microstructure on the preservation potential of echinoids in Graham's Harbour, San Salvador Island, Bahamas

Benjamin J. Greenstein*
Department of Geology, Smith College, Northampton, Massachusetts 01063

ABSTRACT

Live and dead populations of the regular echinoid *Tripneustes ventricosus* were censused along a 700-m transect in a high-energy lagoon adjacent to San Salvador Island, Bahamas, to determine whether the distribution of the death assemblage accurately reflects that of the live population. Although few living individuals were encountered along the transect, several carcasses were present in various states of degradation. The presence of bore holes in many of the echinoid coronas indicates that predation by gastropods is important as a source of mortality that allows *Tripneustes* carcasses to begin the postmortem interval essentially intact. Additionally, many coronas were encrusted by a variety of epibionts, including calcareous green algae. Scanning electron microscope examination of the encrusted material revealed that encrustation is largely a surface phenomenon. The occurrence of abundant subfossil material with a living individual indicates that coronas of *Tripneustes* are sufficiently robust to survive in the lagoon environment long enough for encrustation to occur. This is in marked contrast with previously studied regular echinoid taxa whose habitat preference and/or fragile skeleton combine to prevent the accumulation of subfossil material.

Tumbling experiments performed with bleached carcasses of *Tripneustes* indicated that coronal material dominates the >2-mm size fraction; no significant (α = 0.05) increase in breakage occurs during tumbling periods of 1, 10, and 100 hr. However, significant (α = 0.05) differences do exist between *Tripneustes* and other common regular echinoids in the amount of breakage inflicted by tumbling. Calculation of a breakage coefficient demonstrates that *Tripneustes* suffer less disarticulation than *Diadema* and *Eucidaris* but more than *Echinometra.* With the exception of *Echinometra,* these results correlate differences in skeletal microstructure with resistance to tumbling. The fact that *Echinometra* prefer higher energy environments accounts for the lack of subfossil material associated with living populations. Thus environment is an important factor influencing the preservation potential of both echinoids, positively affecting preservation of *Tripneustes* and negatively affecting preservation of *Echinometra.* Using unpublished generic diversity data, results of this study were then applied to the evolutionary histories of both groups to assess the amount of taphonomic bias affecting them.

*Present address: Department of Geology, Whittier College, 13406 East Philadelphia Street, P.O. Box 634, Whittier, California 90608.

Greenstein, B. J., 1995, The effects of life habit and test microstructure on the preservation potential of echinoids in Graham's Harbour, San Salvador Island, Bahamas, *in* Curran, H. A., and White, B., Terrestrial and Shallow Marine Geology of the Bahamas and Bermuda: Boulder, Colorado, Geological Society of America Special Paper 300.

INTRODUCTION

Shallow water environments of the Bahamas Archipelago house diverse and abundant echinoid populations. Reef and near-reef environments adjacent to San Salvador are particularly suited for the study of echinoids, because rocky coastlines, shallow water reefs, and protected areas contain abundant populations of different taxa. Moreover, these environments are readily accessible for study. Finally, many species of irregular, burrowing echinoids are present, inhabiting large areas of carbonate sand adjacent to the island. Thus San Salvador is an ideal place to study the taphonomy of echinoids. Taphonomy, the systematic study of processes affecting the preservation potential of organisms, has undergone a renaissance in the last decade. Much recent work (see Donovan, 1991, and Wilson, 1988, for reviews) has demonstrated the usefulness of taphonomic studies of extant organisms for interpreting fossil data. The general purpose of the research described herein is to study the taphonomy of a population of echinoids adjacent to San Salvador and to apply the results to an assessment of the evolutionary history of the group.

As suggested above, the fossil record of a related group of organisms is, in part, a function of taphonomic processes that serve to obscure or alter the underlying evolutionary signal(s). Evolutionary histories as deduced from fossil data therefore carry a distinct taphonomic overprint. Among echinoids, this overprint is particularly strong for regular echinoids. Kier (1977) demonstrated that the fossil record of irregular echinoids is better than that of regular echinoids and argued that low preservation potential of regulars (differences in life habit and skeletal construction) rather than low original abundance was primarily responsible. Among regular echinoids, the orders Temnopleuroida and Echinoida have the best fossil records. Members of the order Echinoida possess skeletons relatively resistant to disarticulation (Greenstein, 1991), but many live in high-energy environments of net erosion. Members of the Temnopleuroida also possess skeletons relatively resistant to disarticulation. Additionally, many live in relatively low-energy lagoonal environments of net deposition. This study investigates the importance of skeletal microstructure and habitat on the preservation potential of the temnopleuroid *Tripneustes ventricosus*.

The echinoderm skeleton is unique in the animal kingdom in that it is composed of stereom, a lattice-work constructed from lathes of high-Mg calcite (trabeculae). With the exception of echinothuroids and some diadematoids, skeletons of extant echinoids are constructed from a series of plates that are rigidly held together by penetrating muscle fibers and collagenous connective tissue (Smith, 1984). Studies of the skeletal microstructure of the echinoid test (Régis, 1977; Smith, 1980) have demonstrated that, in addition to organic connective tissues, the plates of some echinoid taxa are held in place by the interlocking of stereom trabeculae across plate suture faces. Moreover, these studies indicate that the degree of interlocking varies between taxa; some groups show extensive interlocking across plates while other groups show very little. Smith (1984) suggested that, once organic connective tissues have decayed, the degree of interlocking of stereom across plates becomes an important factor in determining the preservation potential of echinoids. Although this suggestion is borne out by anecdotal observations (coronas of various echinoid taxa that have survived a great deal of transport intact, for example, in sand dunes, or lack of carcasses of echinoids with high live abundances), studies of the influences of skeletal microstructure on preservation potential are few.

Kier (1977) suggested that differences in skeletal microstructure were partially responsible for the relative fidelities of the fossil records of regular and irregular echinoids, and a variety of actualistic studies and surveys of Pleistocene echinoid material have explored this theme. Actualistic studies include the tumbling experiments of Kidwell and Baumiller (1990, p. 257) using echinoids that had decayed for varying time intervals. The authors concluded that test of microstructure would influence preservation potential only after a "decay threshold" related to decay-weakening of connective tissues had been reached. Greenstein (1991) demonstrated that such a threshold is reached rapidly in reef and near-reef environments, and quantified the effect of skeletal microstructure on test rigidity using bleached echinoid carcasses. Greenstein has also demonstrated (Greenstein, 1990, 1992) that, in reef and near-reef environments, the distribution of echinoid carcasses does not reflect the distribution of living echinoids for taxa that exhibit little or no interlocking of stereom across plate suture faces (cidarids and diadematids, respectively) and taxa with limited interlocking that live in high-energy environments (echinometrids).

Surveys of Pleistocene material include those of Gordon (1991), who suggested that the relatively poor Pleistocene record of Echinometra was the result of its preference for high-energy habitats. Gordon and Donovan (1992) and Donovan and Gordon (1993) have described the taphonomy of a variety of regular echinoids preserved in Pleistocene strata exposed in Jamaica, and demonstrated the utility of disarticulated echinoid ossicles for paleoecologic analysis. This study concerns the Order Temnopleuroida, a group that exhibits the most extensive interlocking across plate suture faces of all regular echinoids (Smith, 1984). Moreover, *Tripneustes ventricosus* (Temnopleuroida: Toxopneustidae) live primarily in, or adjacent to, sea grass beds in generally low-energy lagoonal environments (Clark, 1933; Kier and Grant, 1965; Keller, 1983). Sea grass beds are areas of net sediment deposition (Ginsburg and Lowenstam, 1958). Thus, it might be expected that preservation is less tenuous than in the high-energy environments of net erosion preferred by other regular echinoid taxa. Population censuses reveal that, unlike other regular echinoids I have studied, carcasses of *Tripneustes* remain intact sufficiently long for encrustation to occur. Scanning electron microscopy (SEM) revealed that mineralization by encrusting organisms is a surface phenomenon and consequent strengthening of the corona may therefore be unlikely. Finally, tumbling experiments indicated that bleached *Tripneustes* carcasses are more robust than those of cidarids and diadematids, but weaker than echinometrids. This suggests that habitat preference may be a more important factor than skeletal durability in determining the preservation potential of this group.

METHODS

Population census

A census of a population of *Tripneustes* was conducted in Graham's Harbour, a high-energy lagoon (sensu Colby and Boardman, 1989) adjacent to the north end of San Salvador Island, Bahamas (Fig. 1). Graham's Harbour contains abundant *Tripneustes* inhabiting a substrate of sea-grass beds alternating with sandy areas. A 700-m transect line was established bearing 350° from the Bahamian Field Station boat ramp in Graham's Harbour. Maximum water depth along the transect was 4 m. At stations 100 m apart, a 1-m^2 quadrat was placed adjacent to the transect line; live and dead echinoids within the quadrat were counted and the condition of the dead material recorded. The quadrat was then flipped over and another square meter studied. This process was repeated 10 times at each station resulting in a census of a 1- × 10-m^2 "row" situated perpendicular to the transect line and occurring every 100 m.

Tumbling experiments

The rationale for the tumbling experiments was provided by the results of earlier field experiments in which freshly killed echinoids were observed to be reduced to essentially "bleached" carcasses, devoid of organic material and denuded of spines within 6 to 12 days (Greenstein, 1991). In this condition, the durability of the corona is determined by the amount of interlocking of stereom across plate suture faces. Specimens of *Tripneustes* were preserved in 95% ethanol for transport to the laboratory. Each specimen was placed in a 4:1 solution of water and household bleach to oxidize the organic tissue contained within the skeleton. This process caused the spines, Aristotle's Lantern, apical plates, and peristomial plate to detach from the corona. The denuded skeleton was then dried for several days under a laboratory hood. Dried skeletons were weighed and placed in room temperature (22°C) synthetic seawater (Instant Ocean) in a baffled plastic tumbler attached to a variable speed motor. The baffle ensured that the contents would tumble once with each rotation rather than slide around the wall of the tumbler. Five trials were run at 1, 10, and 100 hr. At the end of each trial, the contents of the tumbler were wet-sieved through a stack of nested sieves, and the stack was dried for several days under a laboratory hood. Once dry, the >2-mm, 1- to 2-mm, 500 μ to 1-mm and 125 to 500 μ size fractions were isolated for analysis. Spines, coronal material, and lantern elements in the three larger size fractions were counted and weighed. Because of the small grain size, the 125- to 500 μ size fraction was weighed without identifying different skeletal elements.

RESULTS AND DISCUSSION

Population census

Tripneustes was the only regular echinoid encountered along the transect in Graham's Harbour. Although only one live individual was counted, eight dead individuals, in various states of degradation, were observed distributed along the transect (Fig. 2). Condition of the carcasses ranged from freshly killed with organic tissue present, to isolated fragments of interambulacra separated along ambulacral sutures; many carcasses were heavily encrusted (Fig. 3). With the exception of the freshly killed individual, spines, apical systems, and elements of the Aristotle's Lantern were not associated with coronal material. Without exception, carcasses observed along the transect (and in Graham's Harbour in general) exhibited bore holes if the corona was sufficiently intact to permit their identification.

The association of *Tripneustes* carcasses with a live individual in Graham's Harbour is in marked contrast to previous live/dead population censuses (Greenstein, 1990) that revealed no correspondence of subfossil material within living populations. This suggests that postmortem degradation is not as rapid for *Tripneustes* in the lagoonal environment as it is for other regular echinoid taxa in higher energy environments. The ubiquitous presence of bore holes in the carcasses implies that predation is an important source of mortality. Cassid gastropods are present in Graham's Harbour and have been shown to prey almost exclusively on echinoids (Hughes and Hughes, 1981). Once preyed upon, *Tripneustes* lose their spines, apical system, and Aristotle's Lantern. Although loss of these elements results primarily from decay of organic tissues, spine loss can also result from the feeding technique of cassids (Hughes and Hughes, 1981).

In addition to being an important source of mortality, gastropod predation in Graham's Harbour has important taphonomic implications: coronas of *Tripneustes* begin the postmortem interval essentially intact. Moreover, the occurrence of many encrusted coronas indicates that once denuded of organic material, coronas of *Tripneustes* are sufficiently robust to remain intact long enough to become encrusted with various organisms including calcareous green algae (Fig. 3C). In the lagoonal environment, source of mortality, lack of physical disturbance, and skeletal durability combine to positively affect preservation potential. Consequently, fossil *Tripneustes* in ancient strata may indicate a lagoonal environment and provide indirect evidence for the presence of sea-grass beds, which are not likely to be represented by direct fossil evidence. SEM study of encrusted coronas was conducted to determine the degree to which stereom was penetrated by mineral material, and tumbling experiments were performed to quantify skeletal durability.

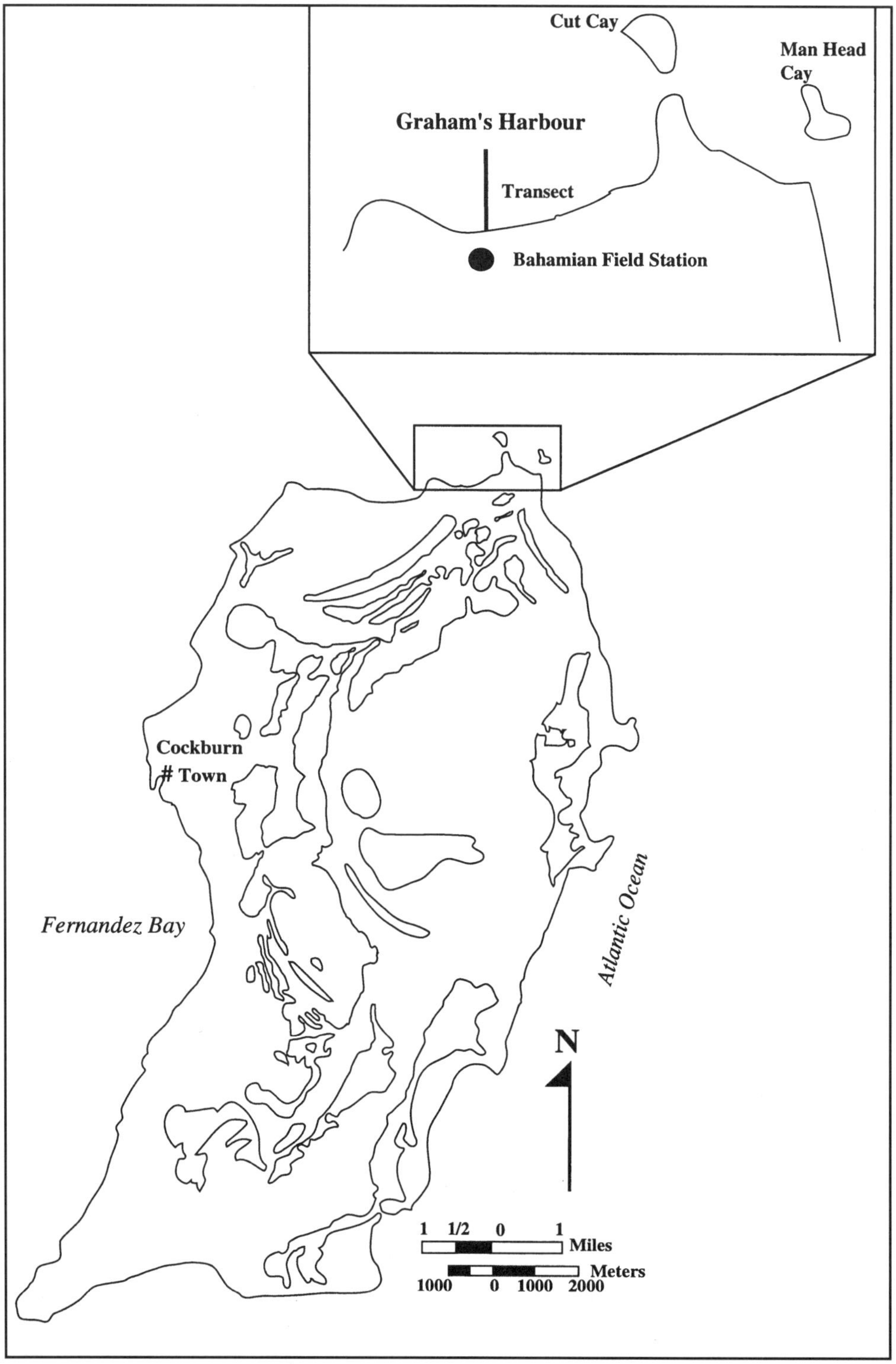

Figure 1. Location of transect in Graham's Harbour, San Salvador, Bahamas.

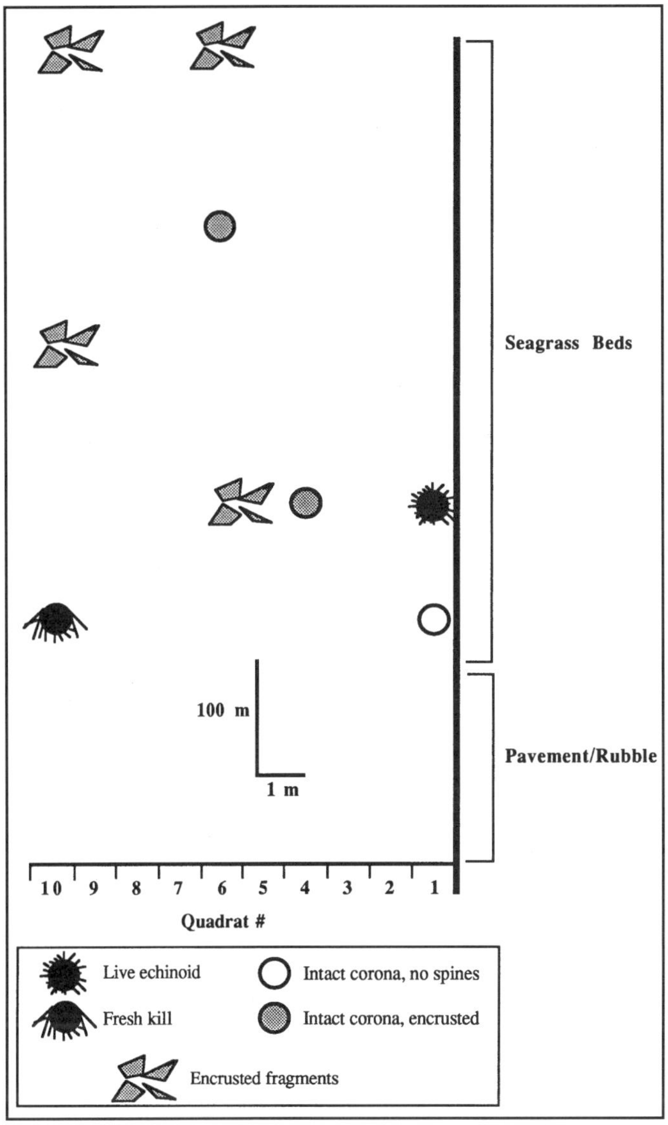

Figure 2. Census results. Population of *Tripneustes ventricosus* in census area consists of one live individual and carcasses of eight others in various states of degradation.

Figure 3. States of degradation of *Tripneustes* carcasses in Graham's Harbour: A, Fresh kill; the spines were still moving on this individual. Note bore hole most likely made by a cassid gastropod. B, Bleached carcass; all organic material has been removed from the corona. Spines, lantern elements, peristomial plates, and apical plates are all missing. Note gastropod bore hole. C, Encrusted corona; note the alga *Acetabularia* sp. (arrow) growing off the corona. D, Large, encrusted coronal fragments that, in the absence of organic connective tissues, have separated along ambulacral plate sutures. E, Encrusted interambulacral fragments.

Scanning electron microscopy

As a result of the unique structure of stereom, the echinoid skeleton is extremely porous. The occurrence of heavily encrusted, but otherwise intact, coronas in Graham's Harbour suggests that encrustation may influence preservation potential by preferentially strengthening the skeleton. Anecdotal observations of heavily encrusted coronas of regular and irregular echinoids out of their natural habitats also suggest that strengthening occurs. Specifically, mineral precipitation within stereom galleries could serve to further strengthen the skeleton.

In order to test this hypothesis, unencrusted and encrusted *Tripneustes* coronas were collected from Graham's Harbour (Fig. 4). Clean and heavily encrusted fragments of corona were coated with a 60 to 40% gold-palladium mix for observation under SEM. Specimens were initially oriented under the electron beam so that surface textures could be studied, and then oriented to assess whether any infilling of the stereom had occurred. The following skeletal fragments were examined: (1) fragment of an ambulacral plate taken from the apical margin of an unencrusted corona (Fig. 4A); (2) an ambulacral fragment taken from the apical margin of a heavily encrusted specimen (Fig. 4B); (3) an interambulacral fragment taken between the ambitus and apical end of the same encrusted specimen (Fig. 4B); and (4) an ambulacral fragment, encrusted by the foram *Homotrema rubrum*, removed from the peristomial margin of an otherwise unencrusted corona.

The unencrusted fragment reveals the open framework of labyrinthic stereom (Fig. 5A) and serves for comparison to the encrusted fragments examined. In contrast to unencrusted stereom, the surface of encrusted stereom is completely filled in by calcite, obscuring any stereom fabric (Fig. 5B). Examination of the edge of the same fragment (Fig. 5C) reveals that the encrustation is largely a surface phenomenon; the stereom fabric is visible within 10 to 20 μ of the surface of the specimen. Encrustation by *Homotrema* is also a surface phenomenon; the test secreted by *Homotrema* does not penetrate into the labyrinthic stereom of the echinoid (Fig. 5D).

Examination of the second encrusted fragment (fragment 3, above) also reveals that encrustation is largely a surface phenomenon (Fig. 6A–C). The lack of penetration by encrusting organisms into the stereom is surprising, considering the qualitative difference in rigidity exhibited by encrusted and unencrusted coronas. The type of encrusting organism may be responsible for the lack of penetration of mineral material. The calcareous green alga, *Acetabularia*, was observed using a *Tripneustes* carcass as a substrate (Fig. 3C). The amount of available light within the stereom structure may limit growth of calcareous green algae to the surface of the corona. Further experimental work is necessary to determine whether surface encrustation alone is sufficient to increase the durability of the echinoid corona.

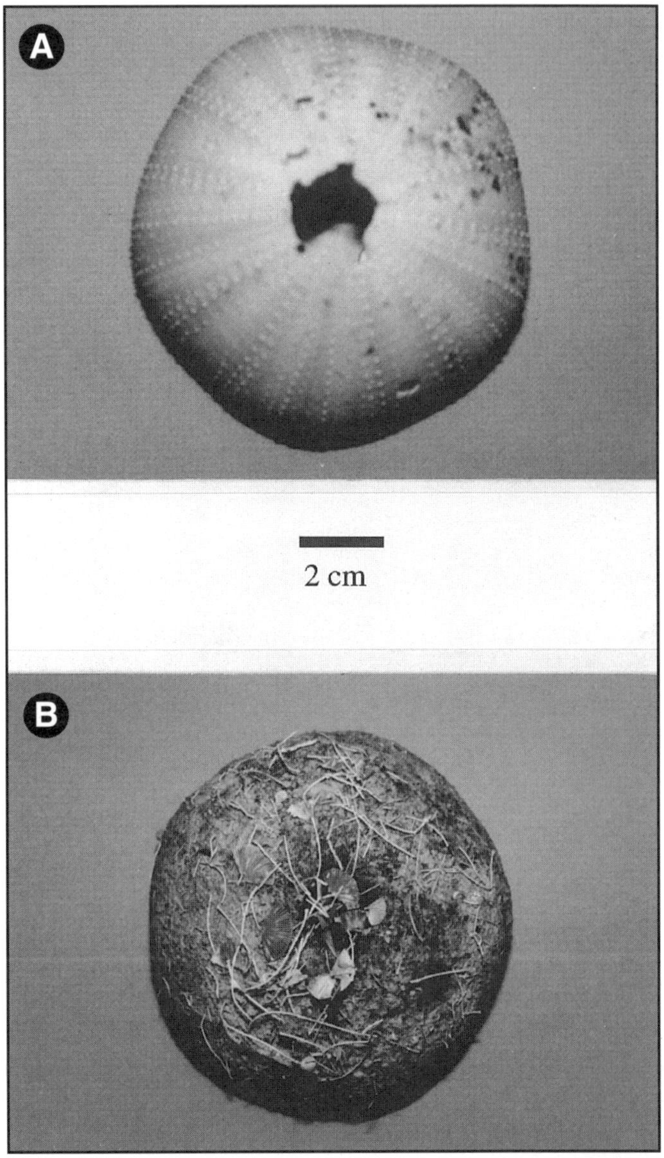

Figure 4. A, Bleached corona selected for SEM examination. B, Encrusted corona (note *Acetabularia*) selected for SEM examination. Two-centimeter-long scale bar applies to both specimens.

Tumbling experiment

Size distribution of skeletal material. When subjected to tumbling, bleached carcasses disarticulated into the five delineated size fractions within 1 hr. The average magnitude of disarticulation of tumbled *Tripneustes* is expressed qualitatively in Figure 7. Eighty percent of the weight of the *Tripneustes* skeletons was rapidly divided between the >2-mm and 1- to 2-mm size fractions; additional tumbling resulted in neither statistically significant (α = 0.05) increases nor corresponding decreases in the size fractions when expressed as weight percent of the entire skeleton (Fig. 8A). These fractions contained approximately 60 and 20% of the weight, respectively. Coronal material predominates in the >2-mm size fraction (Fig. 8B). As might be expected from study of Figure 8, the relative contribution of coronal material, spines, and elements of the Aristotle's Lantern to the >2-mm fraction does not vary significantly (α = 0.05) with tumbling time. In terms of recognizing *Tripneustes* in the fossil record, a size of 2 mm is a logical cut-off value, since fragments smaller than 2 mm would probably go undetected in outcrop and are unlikely to retain features that permit identification at lower taxonomic levels. Further examination of the condition of the coronal material is therefore warranted.

Degree of breakage. To quantify the amount of breakage incurred by tumbling, a coefficient of breakage (CB) was calculated for the coronal material present in the >2-mm size fraction. The coefficient was calculated according to the formula:

$$CB = \frac{N}{W} * \frac{1}{WP}$$

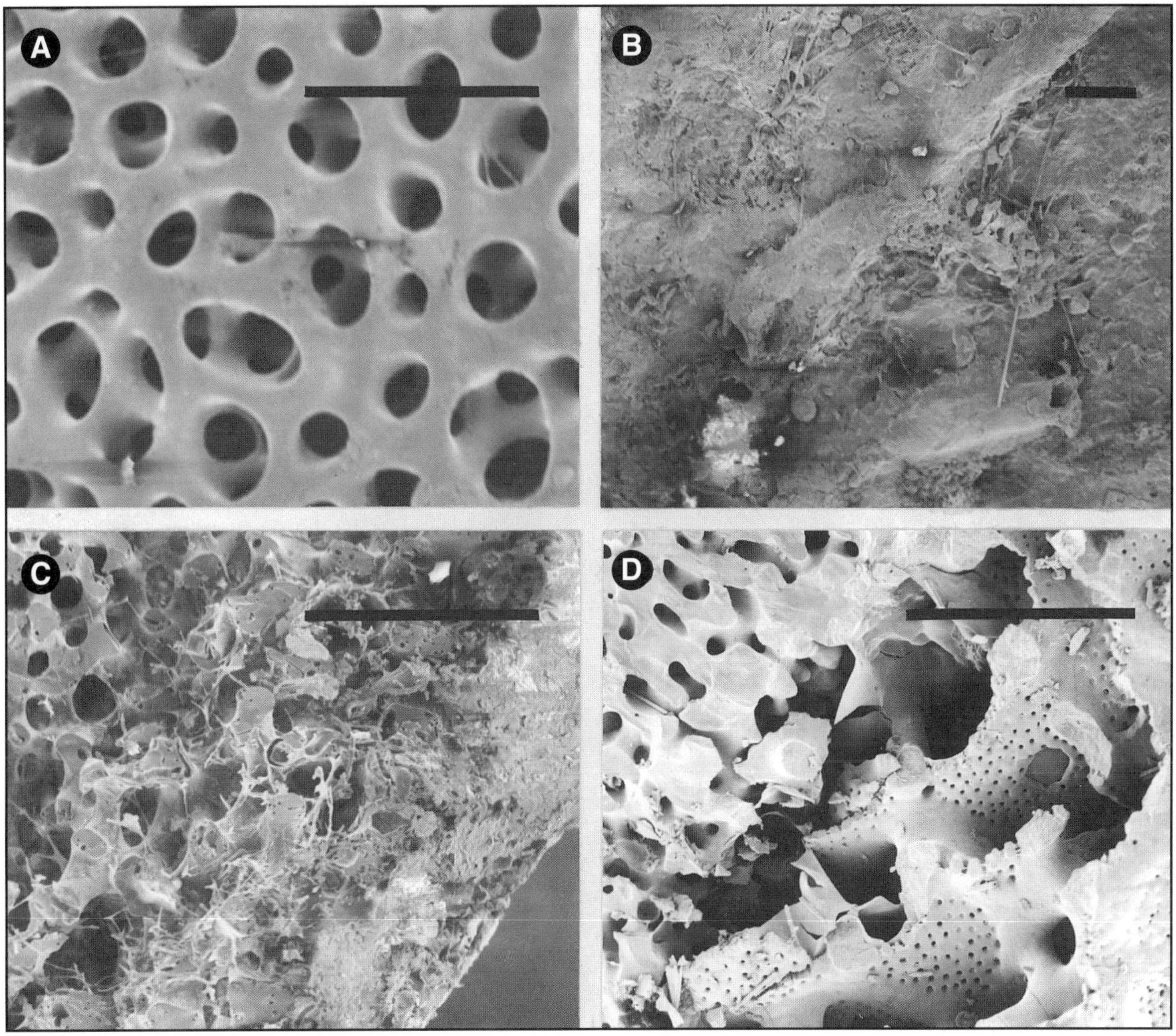

Figure 5. Results of SEM analysis. Scale bar in all photographs is 100 µ. A, Fragment from unencrusted corona exhibiting the porous nature of echinoderm stereom (this particular stereom structure is labyrinthic). B, Surface of encrusted specimen; note the degree of infilling of stereom interstices. C, Edge of same fragment; note that encrustation does not penetrate into stereom galleries. D, View of contact between test of *Homotrema rubrum* (right) and echinoid corona (left); note that the test of *Homotrema* is not secreted into stereom galleries.

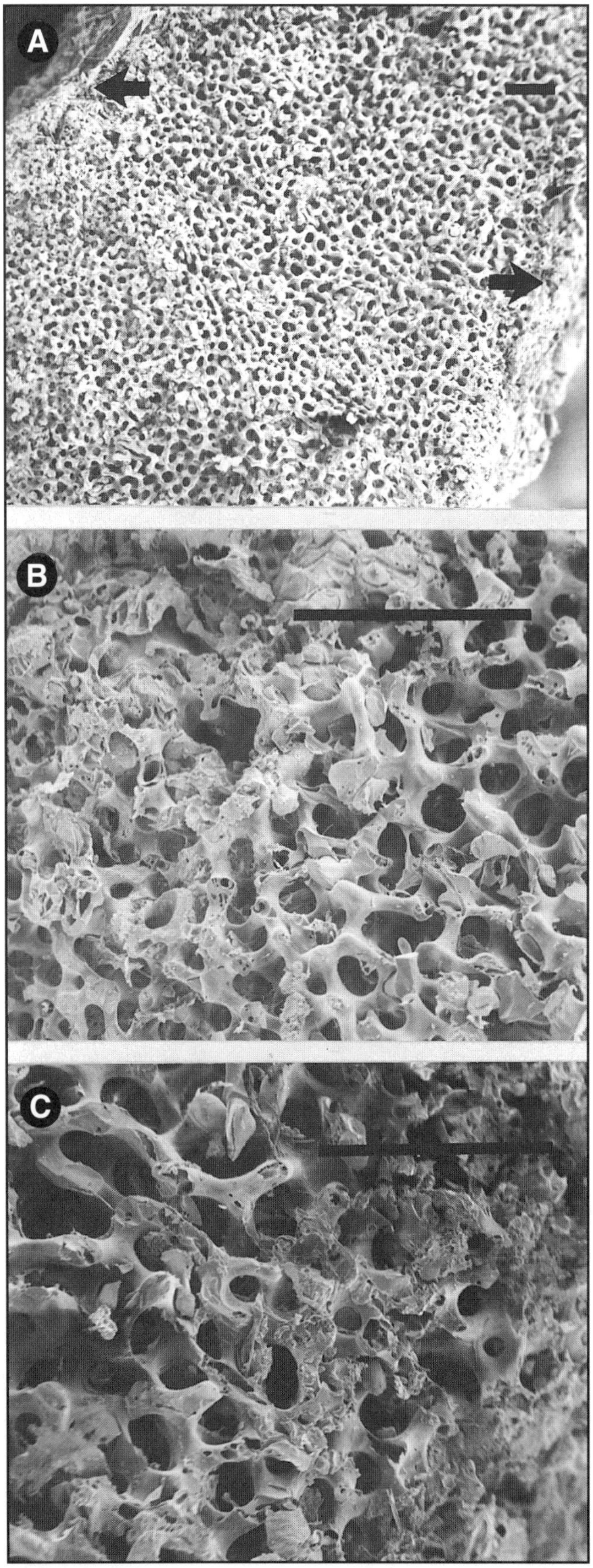

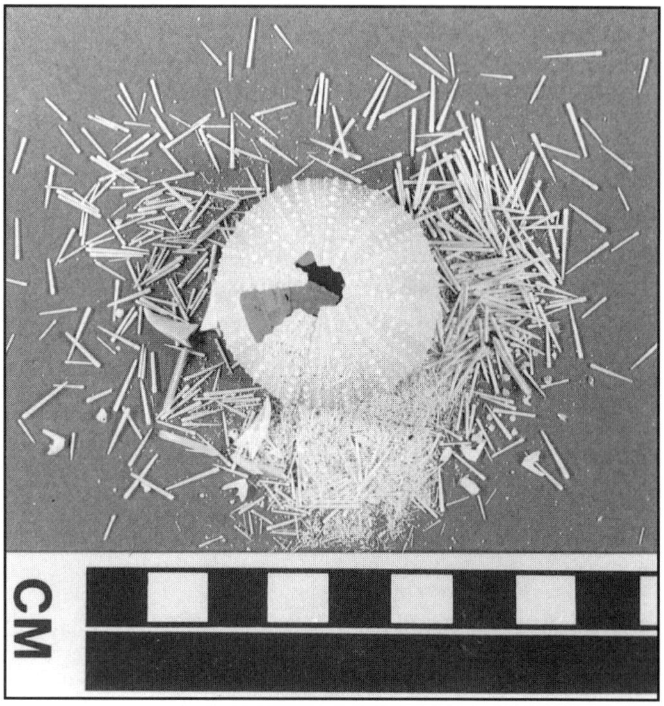

Figure 7. Bleached specimen of *Tripneustes ventricosus* after 100 hr of laboratory tumbling.

where N is the number of coronal fragments in the >2-mm size fraction, W is the weight of the fragments, and WP is the weight percent of the coronal fragments in the >2-mm fraction relative to the entire echinoid skeleton. A high coefficient value represents relatively more breakage. The inclusion of the second term (1/WP) normalizes the coefficient for differences in test size. Further discussion of calculation of the coefficient of breakage is given in Greenstein (1991) and not repeated here.

No significant ($\alpha = 0.05$) increase in the amount of breakage suffered by the corona occurred with increased tumbling times (Fig. 9A). Trials run for 1, 10, and 100 hr were therefore combined to calculate the average coefficient of breakage for *Tripneustes* coronas: 34.31 ± 25.51 (N = 15). Comparison of this value with those obtained under the same experimental regime for other regular echinoid taxa common in shallow water environments in the Caribbean region demonstrate the influence of degree of interlocking of stereom across plate suture faces on skeletal durability (Fig. 9B). Taxa with little or no interlocking of stereom exhibit relatively high coefficient values (*Eucidaris* [Cidaroida] and *Diadema* [Diadematoida], respectively), while those with more extensive interlocking

←

Figure 6. SEM analysis of an additional encrusted fragment (scale bar in each photograph is 100 μ): A, Edge of encrusted fragment; outside margin of corona is to top left. Areas of detail indicated by arrows. B, Detail of edge at outside margin of corona; encrustation is a surface phenomenon. C, Detail of inside margin; note lack of infilling of stereom.

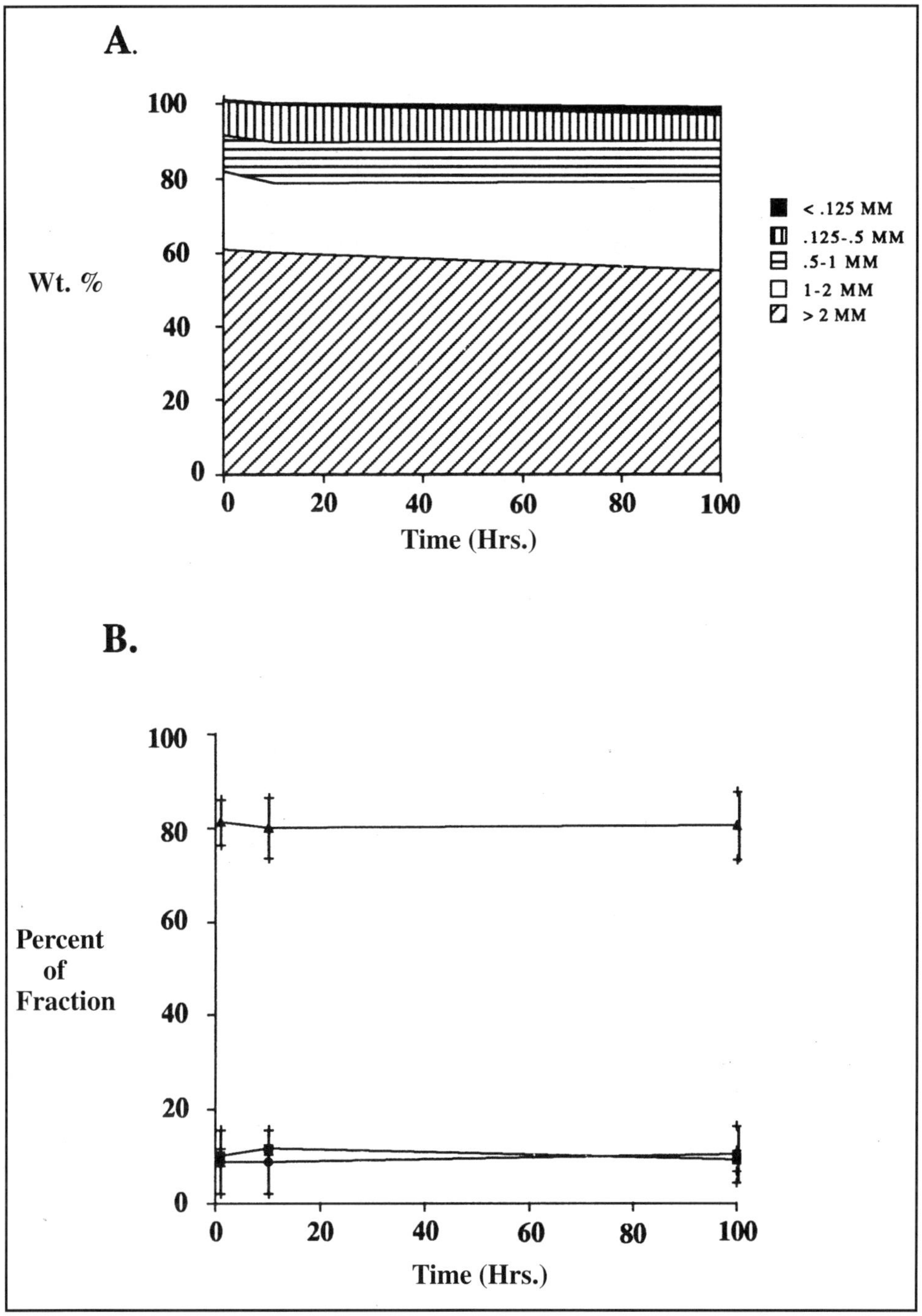

Figure 8. Results of tumbling experiment. A, Weight percent of size fractions discriminated by sieving. B, Composition of the >2-mm size fraction.

exhibit relatively low values (*Tripneustes* [Temnopleuroida] and *Echinometra* [Echinoida]).

As members of the Order Echinoida, *Echinometra* exhibit less interlocking across plate suture faces than do Temnopleuroids (*Tripneustes*) (Smith, 1984). However, when subjected to tumbling, the coronas of *Echinometra* suffered less breakage than those of *Tripneustes*. Coronal plates of *Echinometra* are thicker than those of *Tripneustes*. Moreover, the distribution of stereom fabrics within coronal plates is quite different (see Smith, 1980, for a discussion of coronal plate construction for various echinoid taxa). These differences may account for the rigidity of the echinometrid corona, even though the extent of interlocking across plate suture faces is not as great as that found in temnopleuroids.

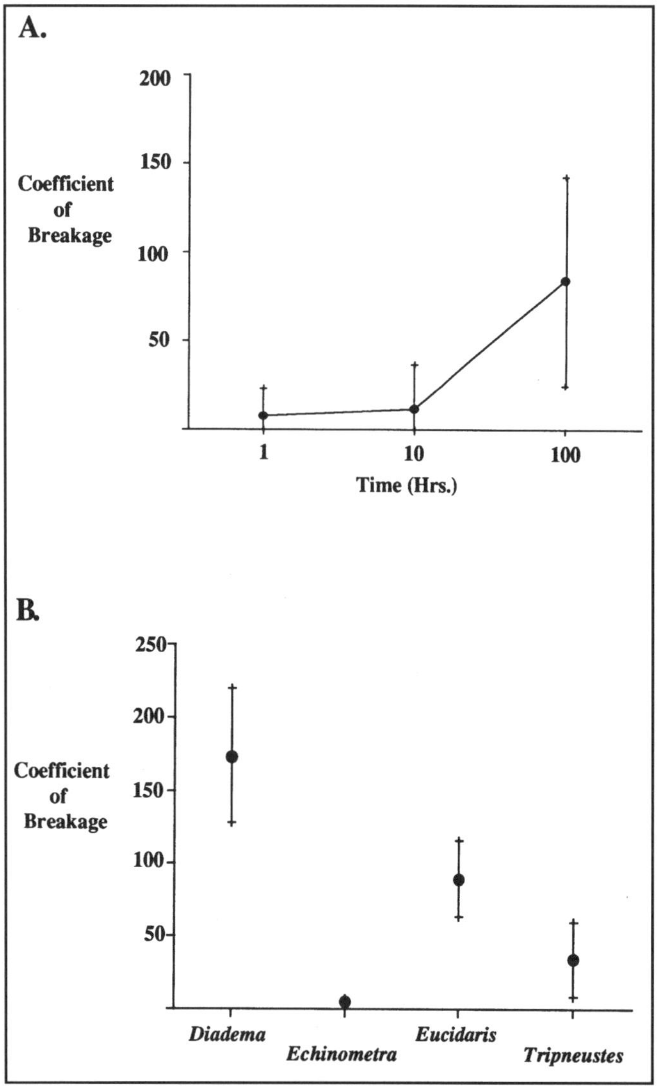

Figure 9. A, Breakage coefficient calculated for each tumbling interval. Error bars represent 95% confidence intervals about the mean. B, Comparison of breakage coefficient calculated for *Tripneustes* to those calculated for additional regular echinoids common in the Caribbean. Error bars represent 95% confidence intervals about the mean (for each echinoid, N = 15).

Although bleached echinometrid coronas survived 100 hr of laboratory tumbling essentially intact, the habitat preference of living echinometrids (shallow, rocky substrates; Kier and Grant, 1965) serves to decrease their chances of preservation. The conclusion that habitat is thus more important than skeletal microstructure in determining preservation potential is corroborated by the occurrence of *Tripneustes* carcasses in various states of decay in Graham's Harbour, while no carcasses of *Echinometra* were counted during a census along a rocky shoreline, where live abundance reached 60 individuals per m^2 (Greenstein, 1990). Moreover, an examination of the fossil record of each family (Echinometridae; Toxopneustidae) indicates differences in preservation potential that appear to be independent of skeletal durability. Study of type material (Greenstein, 1992) revealed that, through 1988, a total of nine species have been placed in six genera on the basis of fossil material since the apparent origination of the Family Echinometridae in Eocene time. In contrast, since the apparent origination of the Family Toxopneustidae in Eocene time, more fossil species (38) have been assigned to more genera (8) (data collected through 1988). A further test of the influence of environment on preservation potential may be accomplished by accumulating sedimentologic data for the intervals from which type material has been collected; this type of analysis is currently underway.

Evolutionary history and taphonomic bias

Clade diagrams for genera within the Families Echinometridae and Toxopneustidae were generated using data kindly supplied by J. J. Sepkoski, Jr., University of Chicago. After their apparent origination in Middle Eocene time, echinometrids are represented by one genus (*Echinometra*) until Pliocene time, when a second genus (*Evechinus*) apparently originated (Fig. 10A). Twelve genera are currently recognized in the Holocene. The resulting clade shape may be described as "top-heavy" (sensu Bambach, 1985). Top-heavy clades have been interpreted to represent large increases in diversity in relatively recent geologic time after a lengthy interval of relatively low diversity (Bambach, 1985). The results of this study suggest that this interpretation must be viewed with some caution, since the discrepancy between fossil and extant diversity that produces the top-heavy clade shape is likely to be an artifact of taphonomic bias. Careful searching for fossil echinometrids in deposits representing high-energy environments should determine whether the group is truly undergoing a rapid diversification in the Holocene or is simply underrepresented in the geologic past.

In contrast, the clade diagram generated for the Toxopneustidae is a relatively smooth and even column (Fig. 10B), representing early diversification followed by maintenance of diversity without much fluctuation. Results of this study reveal that taphonomic bias may be less likely to affect the perceived diversity of this group and thus the evolutionary history suggested by the clade diagram may be more accurate. Moreover, this clade shape resembles that of the Superorder Echinacea, which contains both these and other regular echinoid families (Fig. 11). Gould et al. (1977) have argued that, all else being equal, clade diagrams derived from fossil data should be more irregular at lower taxonomic levels because origination and extinction probabilities are higher, and more regular at higher taxonomic levels because probabilities of origination and extinction are lower. The fact that clade shapes are similar at both low (generic) and high (superorder) taxonomic levels may attest further to the fidelity of the toxopneustid fossil record: the diversity curve for the family accurately reflects that of the superorder.

CONCLUSIONS

Field and laboratory studies of *Tripneustes ventricosus* yielded information useful in assessing the evolutionary history of the Family Toxopneustidae by demonstrating the importance of preferred habitat in determining preservation potential. Research results suggest a caveat against strict interpretation of clade diagrams based on fossil data:

1. The extensive degree of interlocking of stereom across plate suture faces in Temnopleuroids manifested itself in Graham's Harbour, San Salvador, where many carcasses of *T. ventricosus* were counted during a systematic census.

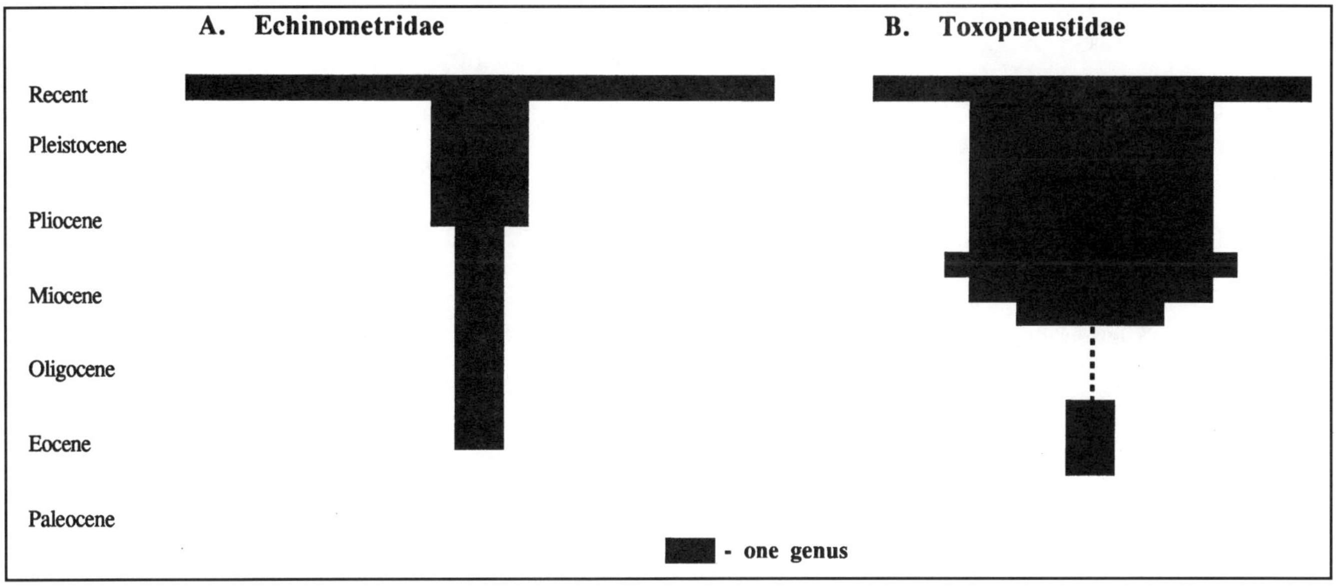

Figure 10. Clade diagrams generated using unpublished generic diversity data supplied by J. J. Spekoski, Jr., University of Chicago. A, Family Echinometridae. B, Family Toxopneustidae.

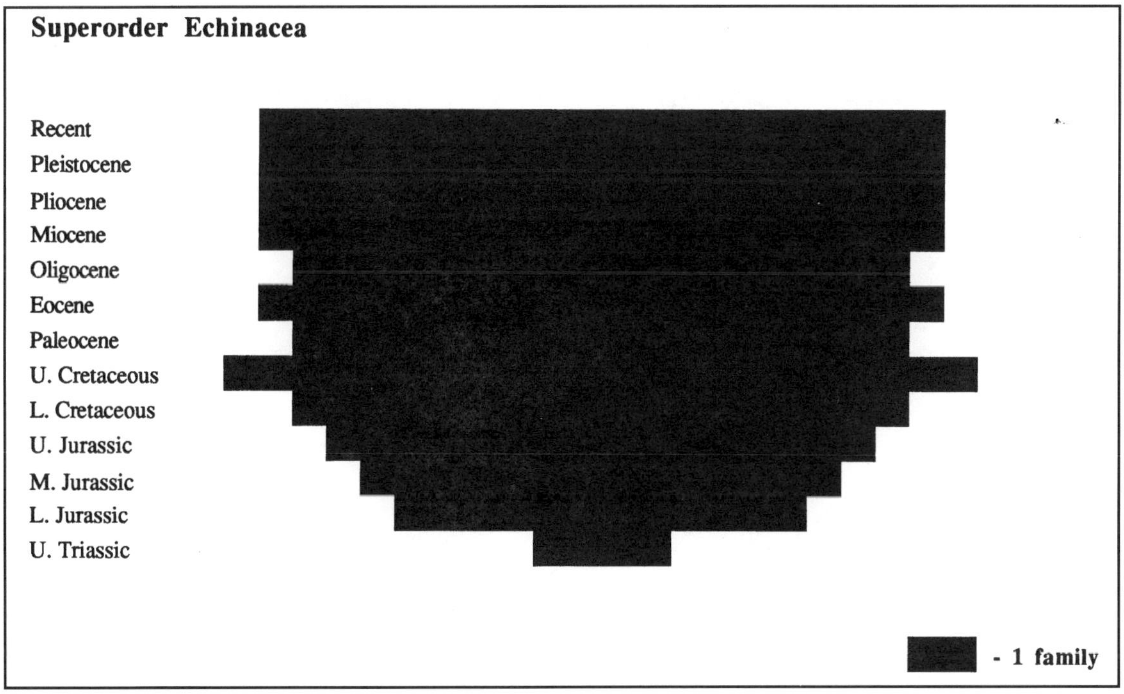

Figure 11. Clade diagram for the Superorder Echinacea. (After Bambach, 1985.) Note change in stratigraphic resolution from Figure 10.

2. Encrustation of much of the dead material suggests that *Tripneustes* coronas are sufficiently sturdy to survive intact for a considerable length of time.

3. SEM study of encrusted skeletal material revealed that mineralization by encrusting organisms is a surface phenomenon. Further study is necessary to quantify the effect of surficial mineralization on skeletal rigidity.

4. When bleached carcasses of *Tripneustes* are tumbled under the experimental regime used in this study, the amount of material contributed to >2-, 1- to 2-mm, 500 μ to 1-mm, 125 to 500 μ, and <125 μ size fractions does not differ significantly between tumbling periods of 1, 10, and 100 hr. Coronal material predominated in the >2-mm size fraction.

5. Calculation of a coefficient of breakage indicates that, under the experimental regime used in this study, the amount of breakage incurred by tumbling does not increase significantly over periods of 1, 10, and 100 hr.

6. With the exception of *Echinometra, Tripneustes* coronas suffered less breakage during tumbling than those of other regular echinoids common in shallow Caribbean environments.

7. Although they are less rigid than those of echinometrids, toxopneustid coronas are more likely to survive the initial postmortem period because live individuals prefer relatively low-energy environments of net deposition.

8. The evolutionary histories of the Families Echinometridae and Toxopneustidae are quite different, exhibiting Holocene diversity expansion and diversity maintenance, respectively. The results of this study suggest that the perceived diversity history of the Family Echinometridae is likely to be more an artifact of taphonomic bias than real; this hypothesis may be tested by careful examination of suitable sedimentary sequences.

9. The perceived diversity history of the Family Toxopneustidae is suggested to be relatively accurate and is similar to the diversity history of the Superorder Echinacea. The similarity of clade shapes at low and high taxonomic levels supports the hypothesis that the diversity curve for the Family Toxopneustidae is real.

ACKNOWLEDGMENTS

The facilities and staff of the Bahamian Field Station, San Salvador, were indispensable to this project; I particularly thank Don and Kathy Gerace, directors of the station while field work was completed. Mike Armstrong served as dive buddy for all work done in Graham's Harbour. Tumbling experiments were performed at the University of Cincinnati. SEM work was performed at Smith College and I thank Dick Briggs (Smith College, Department of Biological Sciences) for instruction on the use of the microscope. I appreciate the work of Brian White and H. Allen Curran in convening the Bahamas symposium and editing this volume. I thank Janet Lauroesch for preparing the final artwork. Financial assistance from Sigma Xi is gratefully acknowledged.

REFERENCES CITED

Bambach, R. K., 1985, Diversity and adaptive variety, *in* Valentine, J. W., ed., Phanerozoic Diversity Patterns: Profiles in Macroevolution: Princeton, New Jersey, Princeton University Press, p. 191–253.

Clark, H. L., 1933, A handbook of the littoral echinoderms of Puerto Rico and the other West Indian islands: New York Academy of Science, Scientific Survey of Puerto Rico and the Virgin Islands, v. 16, 147 p.

Colby, N. D., and Boardman, M. R., 1989, Depositional evolution of a windward high-energy lagoon, Graham's Harbour, San Salvador, Bahamas: Journal of Sedimentary Petrology, v. 59, p. 819–834.

Donovan, S. K., ed., 1991, The Processes of Fossilization: New York, Columbia University Press, 303 p.

Donovan, S. K., and Gordon, C. M., 1993, Echinoid taphonomy and the fossil record: Supporting evidence from the Plio-Pleistocene of the Caribbean: Palaios, v. 8, p. 304–306.

Ginsburg, R. N., and Lowenstam, H. A., 1958, The influence of marine bottom communities on the depositional environment of sediments: Journal of Geology, v. 66, p. 310–318.

Gordon, C. M., 1991, The poor fossil record of *Echinometra* (Echinodermata: Echinoidea) in the Caribbean region: Journal of the Geological Society of Jamaica, v. 28, p. 37–41.

Gordon, C. M., and Donovan, S. K., 1992, Disarticulated echinoid ossicles in paleoecology and taphonomy: The last interglacial Falmouth Formation of Jamaica: Palaios, v. 7, p. 157–166.

Gould, S. J., Raup, D. M., Sepkoski, J. J., Jr., Schopf, T. J. M., and Simberloff, D. S., 1977, The shape of evolution: A comparison of real and random clades: Paleobiology, v. 3, p. 23–40.

Greenstein, B. J., 1990, Taphonomic biasing of subfossil echinoid populations adjacent to St. Croix, U.S. V.I., *in* Larue, D. K., and Draper, G., eds., Transactions, 12th Caribbean Geological Conference (St. Croix, U.S. V.I.): Miami Geological Society, South Miami, Florida, p. 290–301.

Greenstein, B. J., 1991, An integrated study of echinoid taphonomy: Predictions for the fossil record of four echinoid families: Palaios, v. 6, p. 519–540.

Greenstein, B. J., 1992, Taphonomic bias and the evolutionary history of the Family Cidaridae (Echinodermata: Echinoidea): Paleobiology, v. 18, p. 50–79.

Hughes, R. N., and Hughes, H. P., 1981, Morphological and behavioural aspects of feeding in the Cassidae (Tonnacea: Mesogastropoda): Malacologia, v. 20, p. 385–402.

Keller, B. D., 1983, Coexistence of sea urchins in seagrass meadows: An experimental analysis of competition and predation: Ecology, v. 64, p. 1581–1598.

Kidwell, S. M., and Baumiller, T. K., 1990, Experimental disintegration of regular echinoids: Roles of temperatures, oxygen and decay thresholds: Paleobiology, v. 16, p. 247–271.

Kier, P. M., 1977, The poor record of the regular echinoid: Paleobiology, v. 3, p. 168–174.

Kier, P. M., and Grant, R. E., 1965, Echinoid distribution and habits, Key Largo Coral Reef Preserve, Florida: Smithsonian Miscellaneous Collections, v. 149, p. 1–68.

Régis, M. B., 1977, Organization microstructurale du stéreom de l'Echinoide *Paracentrotus lividus* (Lamarck) et ses éventuelles incidences physiologiques: Comptes Rendus de la Academie des Sciences Paris Séries D, v. 285, p. 189–192.

Smith, A. B., 1980, Stereom microstructure of the echinoid test: Special Papers in Palaeontology, v. 25, p. 1–81.

Smith, A. B., 1984, Echinoid palaeobiology: London, United Kingdom, Allen & Unwin, 190 p.

Wilson, M. V. H., 1988, Taphonomic processes: Information loss and information gain: Geoscience Canada, v. 15, p. 1331–148.

MANUSCRIPT ACCEPTED BY THE SOCIETY JANUARY 5, 1995

Printed in U.S.A.

Geological Society of America
Special Paper 300
1995

Dolomitization of modern subtidal sediments, New Providence Island, Bahamas

Steven W. Mitchell and Robert A. Horton, Jr.
Department of Geology, California State University, Bakersfield, California 93311

ABSTRACT

Dolomitization of Recent carbonate sediments is currently occurring in the Bonefish Pond tidal creek system, New Providence Island. Dolomite is forming in supra- and intertidal crusts along the creek margin. The crust is being eroded and fragments in the <4ϕ size fraction are being distributed by wave and current activity. Widespread dolomitization (as 1- to 2-μm crystals) of subsurface sediments apparently has occurred during the past 100 yr. $\delta^{18}O$ and $\delta^{13}C$ values indicate that modified seawater and organic carbon, oxidized during microbially mediated degradation of organic matter, are involved in the dolomitization of aragonitic sediments. The rapidity of dolomitization seems related to the development of ideal conditions for the formation of dolomite: high alkalinity, low level of dissolved sulfate, fine aragonitic sediment, probability of a high Mg^{2+}/Ca^{2+} ratio, and particularly the availability of allogenic dolomite seed crystals. A similar, synsedimentary, micron-sized dolomite characterizes many ancient platform carbonates. The dolomitized sediments of Bonefish Pond provide evidence for the conditions that may be necessary for the syndepositional dolomitization of platform carbonates.

INTRODUCTION

The origin of dolomite continues to be an important, and increasingly complicated, problem in geology (McKenzie, 1991). Most dolomite has formed through geochemical alteration of calcitic and aragonitic sediments and rocks; most workers now agree that seawater or modified seawater is the only geochemical fluid that contains sufficient Mg^{2+} to account for the large volumes of pervasive dolomite found in ancient strata (Land, 1985; Machel and Mountjoy, 1986). Recent studies have shown that dolomite is forming in many modern environments (Zenger et al., 1980; Shukla and Baker, 1988), but the amounts of dolomite described are generally very small and the use of these occurrences as analogues for ancient dolostones is tenuous at best (Machel and Mountjoy, 1986). The discovery that dolomite is widespread in lacustrine and tidal creek environments of the central and southern Bahamas (Mitchell and Horton, 1989, 1991; Horton and Mitchell, 1991) provides an important new analogue for the study of ancient dolomite.

Surficial dolomite crusts occur along the margins of tidal flats of the western portions of Andros and Abaco Islands and the Caicos Banks (Shinn et al., 1965; Shinn et al., 1969; Shinn, 1983, 1986) and western Long Island (Mitchell and Horton, this volume). These crusts are supratidal and are attributed to cement precipitation due to tidal pumping in areas flooded by very high tides and storm events (Shinn et al., 1965; Lasemi et al., 1989). Dolomite also has been reported from surficial lake and tidal creek sediments and in shallow cores from Conception, Long, New Providence, Providenciales, and San Salvador Islands, and from Samana Cay (Kwolek, 1984; Leaver, 1985; Mitchell and Sigler, 1989; Mitchell et al., 1989; Neumann et al., 1991). Overall, these occurrences fall into two general categories: supratidal and intertidal crusts, and subaqueous subsurface zones of concentrated dolomite. They are variably developed in three major depositional environments: tidal flat margins, tidal creeks, and lakes (Mitchell and Horton, this volume).

Intermittent surficial dolomitic crusts are currently forming

Mitchell, S. W., and Horton, R. A., Jr., 1995, Dolomitization of modern subtidal sediments, New Providence Island, Bahamas, in Curran, H. A., and White, B., Terrestrial and Shallow Marine Geology of the Bahamas and Bermuda: Boulder, Colorado, Geological Society of America Special Paper 300.

in supra- and intertidal environments along the upper reaches of tidal creek systems, where salinities typically exceed 45‰. Composition of crusts is up to 90% dolomite. Dolomite is also present in the subtidal sediments of many tidal creek systems. The highest dolomite concentrations (90%) that have been discovered to date are found in Chalk Sound (Providenciales Island) and Bonefish Pond (New Providence Island). Dolomitized tidal creek sediments are also present beneath modern lakes on Conception, Crooked, Great Exuma, and Long Islands (Mitchell and Horton, this volume). This chapter focuses on the formation of dolomite in one modern tidal creek system, Bonefish Pond of south-central New Providence Island (Fig. 1).

DEPOSITIONAL ENVIRONMENT

The creek system currently has three narrow inlet channels constricted by a spit, as well as islands formed since 1881 (Fig. 2). Land areas in 1881, which bordered a shallow bay in the location of the present creek, were composed of Pleistocene oolitic eolianites. The development and physical parameters of Bonefish Pond are summarized by Mitchell and Sigler (1989). Measured salinities range from 40 to 60‰, and maximum observed tidal channel current velocities are about 22 m/min. Surficial sediments are predominantly encrusted macrophyte tubes (macrophyte crusts of Schneider et al., 1983); the tests of foraminifera and fragments of molluscs and calcareous algae are less common. Modern foraminiferal, ostracod, and mollusc biofacies correlate closely with salinity, current velocities, and sediment size.

Two types of dolomitic crusts occur along the margins of Bonefish Pond. A thicker (3.5 cm) supratidal crust is present in the center part of the creek (Fig. 3A). Supratidal crusts contain up to 40% dolomite. A much thinner (0.5 cm or less) crust is formed in the intertidal zone (Fig. 3B–C). Intertidal crusts contain up to 90% dolomite. The crusts are eroded and transported by wave and current activity. Fragments of the crusts are found in the <4ϕ size fraction throughout the tidal creek system, but are most common in subtidal sediments deposited adjacent to the areas of crust formation (Figs. 1 and 4). High-Mg calcite, produced through the degradation of foraminifera and encrusted macrophyte tubes, is the dominant mineral comprising the <4ϕ size fraction in Bonefish Pond.

MINERALOGY

The dolomite is poorly ordered calcic dolomite (Ca:Mg approximately 60:40 by weight) and occurs as 1- to 2-µm rhombs that form in clusters around nucleation sites (Fig. 5A). The carbonate mineralogy of 10 shallow cores from the tidal creek system (Fig. 1) was investigated by x-ray diffraction analysis, which provides a semiquantitative estimate of dolomite

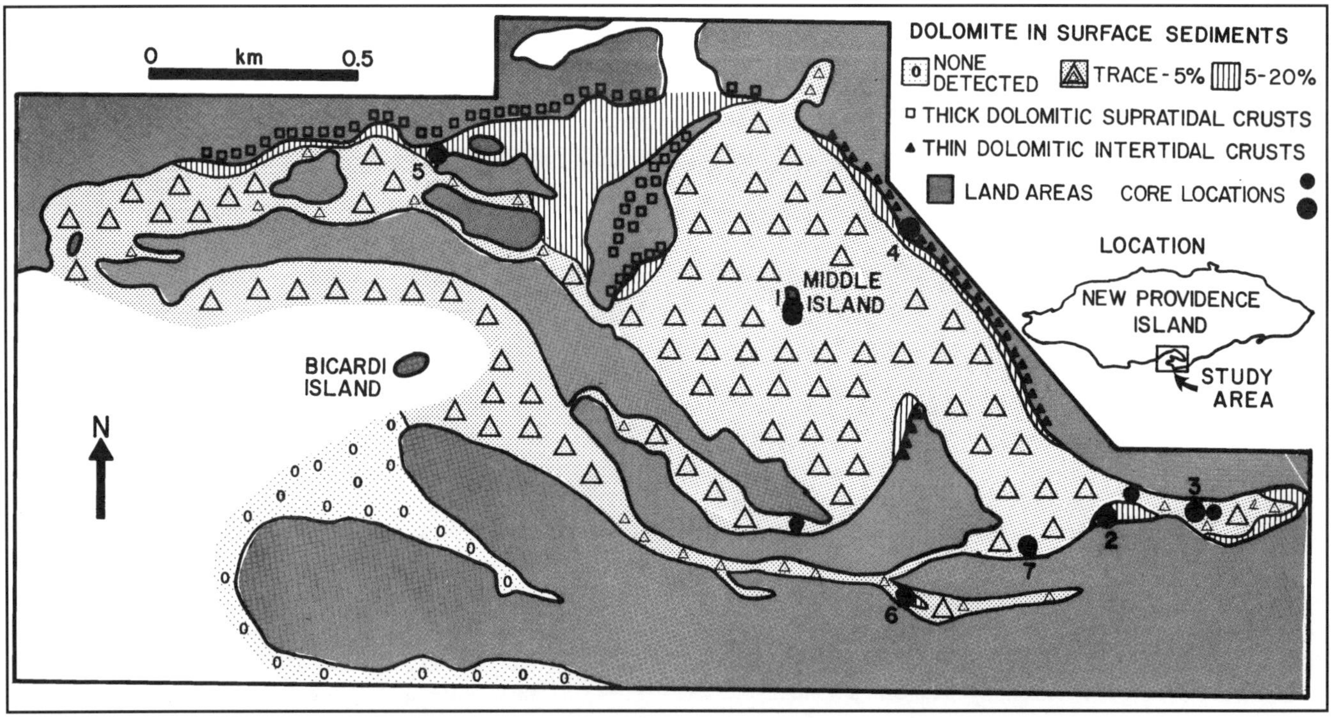

Figure 1. Location map of Bonefish Pond, indicating percentage of dolomite in <4ϕ size fraction of surface subtidal sediments (based on 100 samples), areas of development of dolomitic crusts, and core locations. Larger numbered solid circles indicate locations of cores for data presented in Figures 6, 7, and 8. Smaller solid circles show locations of additional cores not specifically mentioned in text.

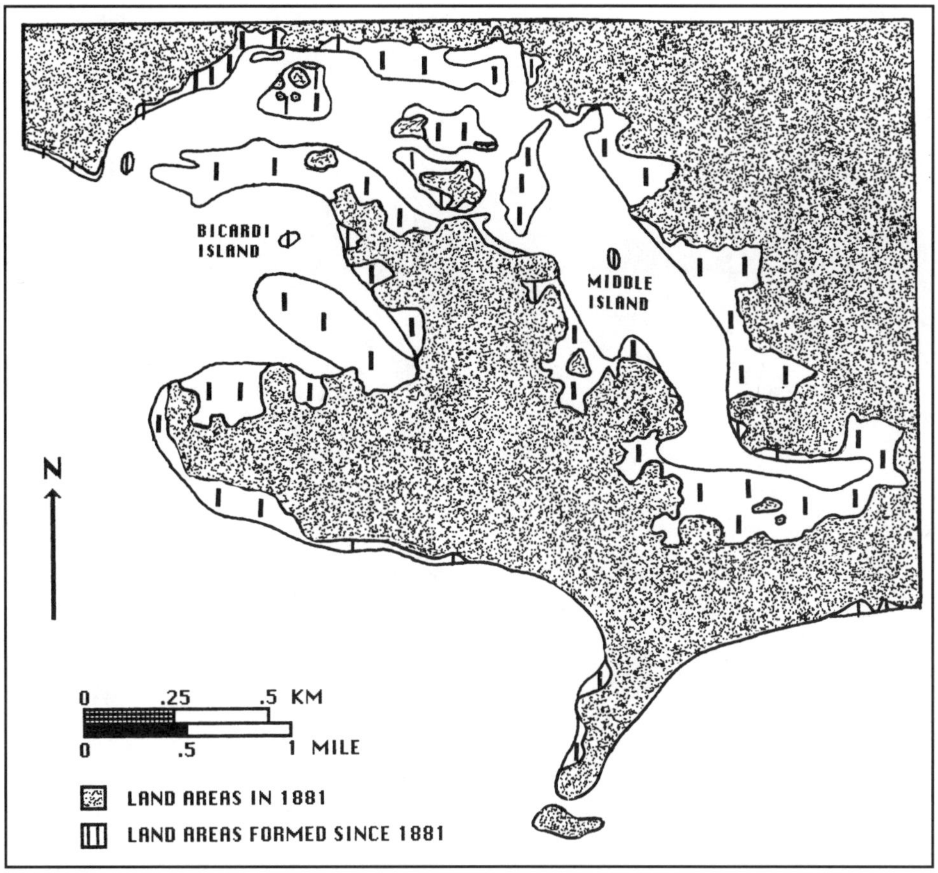

Figure 2. Map of Bonefish Pond showing land areas in 1881 (gray) and land areas formed during the past 115 yr (white). Base map is from U.S. Navy Hydrographic Office (1936).

abundance (Royse et al., 1971). Representative cores are presented in Figure 6. In cores 1 through 3, which were obtained from subtidal locations, dolomite increases at the expense of aragonite; Mg calcite remains fairly constant in cores 2 and 3 and shows a decrease in cores 1 and 4 in only the most dolomite-rich samples. This suggests but does not prove dissolution of the aragonite; aragonite appears to have made up 20 to 40% of the $<4\phi$ size fraction prior to dolomitization. Scattered zones of gypsum disks are also present. A zone of dolomitic cement is present in core 1 (12 to 19 cm). In all subtidal cores, significant dolomitization occurs only when the $<4\phi$ size fraction exceeds 60% of the sediment. Core 4 is from a swale area behind a coastal beach-dune ridge. The beach, dune-ridge, and swale areas are covered with fragments of intertidal dolomitic crusts (Fig. 3D). The crust fragments are also present in the upper 10 cm of core 4.

In all 10 cores a thin transition zone occurs in the upper 8 to 12 cm (Fig. 6). The boundary marks the appearance of encrusted macrophyte tubes and the extensive development of hypersaline foraminiferal, ostracod, and molluscan biofacies. Biofacies of near-marine salinity (36 to 38‰) are present below the transition zone. The onset of encrusted macrophyte tube formation is believed to result from a rapid restriction of the creek system as the spit formed, as well as from the widespread post–World War II use of fertilizers in fields adjacent to Bonefish Pond (Mitchell and Sigler, 1989). The boundary, then, may represent an isochronous surface formed approximately 40 to 50 yr ago.

Isotopes of carbon and oxygen are plotted in Figure 7. Surficial carbonate sediments fall into three distinct groups: (1) subtidal sediments composed primarily of aragonite and Mg calcite, which have $\delta^{18}O$ values between –1.5‰ and +0.8‰ PDB and $\delta^{13}C$ values between +0.8‰ and +3.0‰ PDB; (2) thin dolomitic crusts deposited in intertidal environments, which have $\delta^{18}O$ values between +2.2‰ and +3.7‰ PDB and $\delta^{13}C$ values between –0.8‰ and +2.0‰ PDB; and (3) thick dolomitic crusts deposited in supratidal environments, which have $\delta^{18}O$ values between +0.6‰ and +1.5‰ PDB and $\delta^{13}C$ values between –4.2‰ and –3.3‰ PDB.

The shift in $\delta^{18}O$ between aragonitic-calcitic and dolomitic surficial sediments may indicate evaporative concentration of ^{18}O in interstitial water during exposure of the crusts (McKenzie et al., 1980). The differences between oxygen isotope compositions of these primary sediments is, however, within the range of estimated values for equilibrium isotopic fractionation between calcite and dolomite (Land, 1980) and may simply represent for-

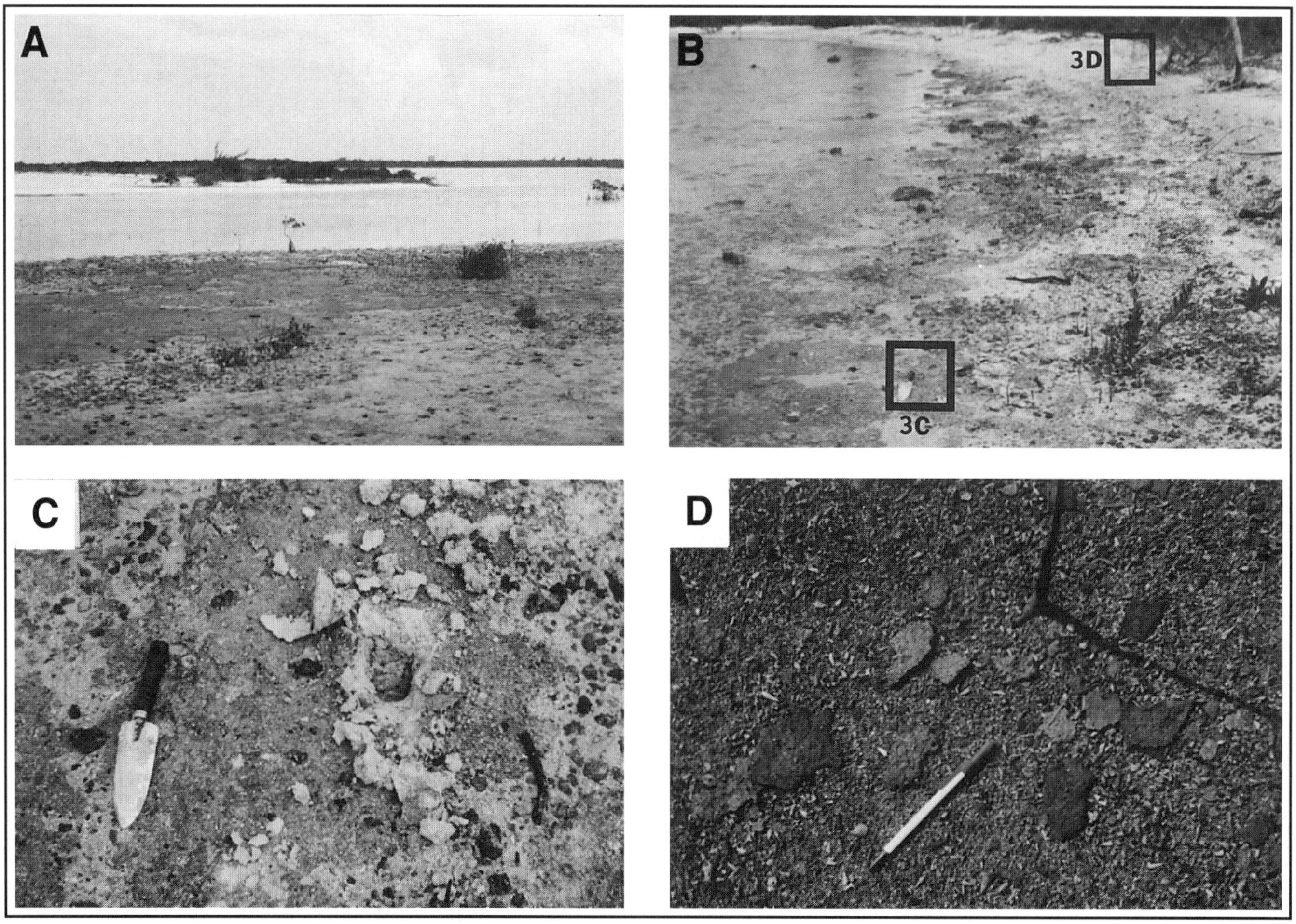

Figure 3. A, Photograph of supratidal crust formed along north-central edge of Bonefish Pond. B, Photograph of intertidal zone along beach near site of core 4 and locations of close-up views shown in C and D. C, Photograph of intertidal crusts in place. D, Photograph of eroded intertidal crusts incorporated in beach sediments.

mation from the same water. The shift in $\delta^{13}C$ in dolomites formed in thin crusts reflects oxidation of organic carbon during microbially mediated degradation of organic material (Burns et al., 1988; Patterson and Walter, 1994). The presence of pyrite within intertidal crusts (Fig. 5B) supports this contention. The very negative $\delta^{13}C$ of the thicker crusts is more difficult to interpret as these crusts do not currently contain abundant organic material. These crusts occur on narrow supratidal flats bordering the central portion of the creek system (Fig. 1) and are not as well cemented as the thin, intertidal variety. The cementing waters may be derived from adjacent low inland areas flooded during the rainy season when decaying cyanobacterial mats are present. Such waters contain CO_2 generated by organic decay, as well as a meteoric component that is reflected in the lighter $\delta^{18}O$ signature of these crusts.

$\delta^{13}C$ values for aragonite/calcite and dolomite are plotted versus depth in Figure 8. The $\delta^{13}C$ content of calcite/aragonite is fairly constant within each core (Fig. 8A), and there does not seem to be any correlation between the shift from normal marine to hypersaline conditions that accompanied the formation of the restricted tidal creek. The lightest carbon occurs in core 1, which is located in the center, and formerly the deepest part, of the basin (Figs. 1 and 4). Cores 2 and 3 are located in the shallow, restricted uppermost part of the tidal creek system, while core 4 is located in a swale area behind a beach ridge on what is now land. Thus, there is some correlation between $\delta^{13}C$ of the calcite/aragonite fraction in these sediments and their location relative to the edge of the tidal creek, possibly due to increased accumulation of organic matter within the deeper, quieter portions of the basin. Some supporting evidence of this comes from cores 5, 6, and 7, all located in narrow, shallow tidal channels. At these locations, strong tidal currents reduce the accumulation

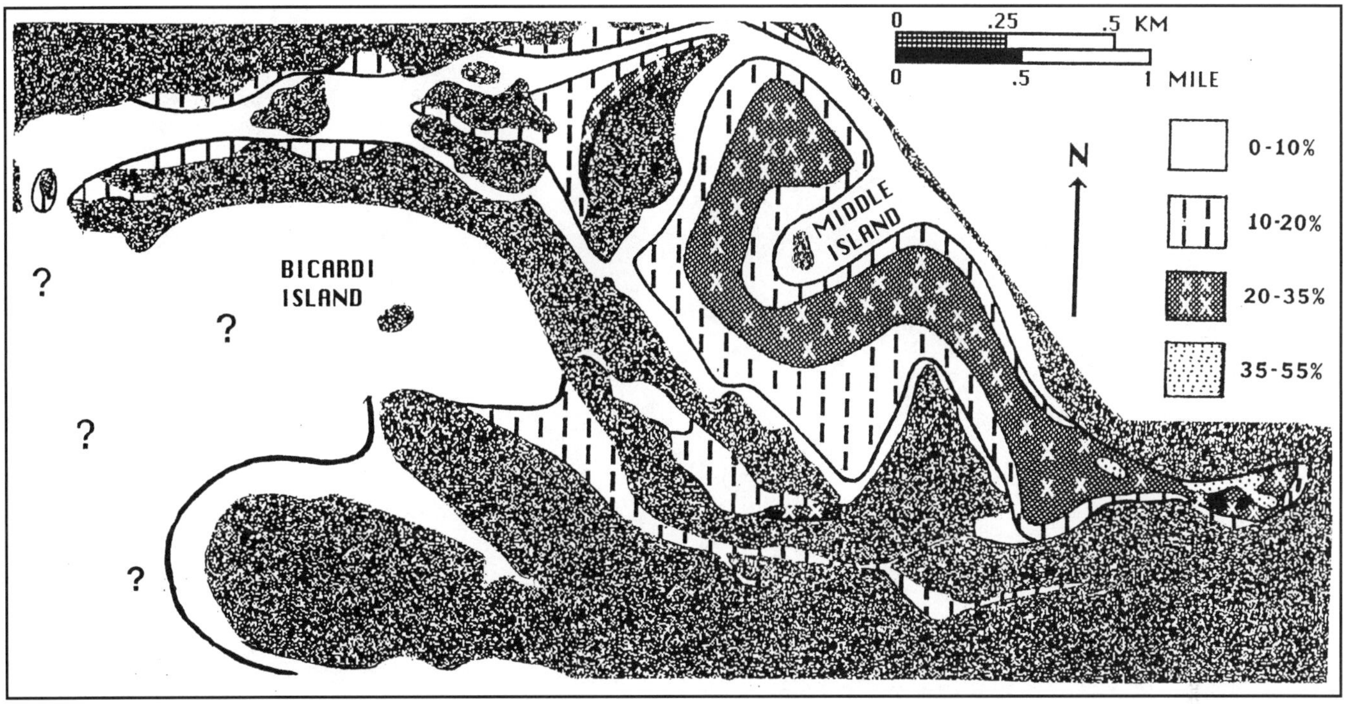

Figure 4. Map of Bonefish Pond indicating percentage of <4ϕ size fraction making up surficial sediment of Bonefish Pond.

of fine-grained organic debris. Sediments from these cores have a high $\delta^{13}C$ similar to that of surficial crusts.

Dolomites in the core samples have $\delta^{13}C$ values between −2.4‰ and +1.2‰ and show a distinct shift toward carbon that is isotopically lighter than that of surficial aragonite and Mg calcite sediments in the same cores (Fig. 8B). This is interpreted to be the result of an input of CO_2 derived from oxidation of organic material within the sediment column. Evidence for oxidation of organic matter is the presence of micron-sized authigenic pyrite, indicating microbially mediated sulfate reduction, within these sediments. The $\delta^{18}O$ values of these subsurface dolomites range from −1.37 to +1.64‰. This range of values exceeds the range of $\delta^{18}O$ values for aragonite and high-Mg calcite in surficial samples. Comparison of $\delta^{18}O$ in mixed Mg calcite/aragonite-dolomite splits shows that $\Delta^{18}O$ ($\delta^{18}O_{dolomite} - \delta^{18}O_{calcite/aragonite}$) ranges from −3.19 to +1.48‰ (Table 1). The equilibrium value of $\Delta^{18}O$ for coexisting calcites and dolomites under normal sedimentary conditions is not known with certainty; most evidence suggests a value of +3 to +4‰ (Land, 1983), although Major et al. (1992) reported values between +1.5 and +3.5‰. Thus, the dolomite contains light oxygen relative to the Mg calcite and aragonite in these samples.

There are several possible explanations for light $\delta^{18}O$ values in dolomite: contamination by calcite during analyses of mixed samples, incorporation of oxygen from meteoric water, dolomite formation at elevated temperature, or dolomite formation during bacterially mediated sulfate reduction. Contamination during analysis of mixed carbonate samples can be ruled out as the source of light oxygen for two reasons. First, dolomite extracts (>95% dolomite) from these samples contain light $\delta^{18}O$. The extracts were prepared by concentrating dolomite-rich samples using dilute acetic acid for 5 to 10 min and then immediately washing and drying the samples. Second, in many cases $\Delta^{18}O$ is negative, indicating that the light oxygen signature of the dolomite could not be due to contamination by the coexisting heavier Mg calcite or aragonite. Exposure to meteoric water can also be ruled out as the source of light oxygen. Except for core 4 (collected on land), all of these samples are subtidal and contain no evidence of subaerial exposure. In addition, the dolomite analyzed from core 4, which is exposed to both meteoric water and tidally pumped marine water, has a positive $\Delta^{18}O$. Nor can elevated temperature be invoked to explain light oxygen in these very shallowly buried, recent sediments.

Recent studies (Aharon et al., 1977; Coleman and Raiswell, 1981; Sass et al., 1991) suggest that low values of $\delta^{18}O$ in carbonate rocks may result from incorporation of oxygen liberated during sulfate reduction. In particular, Sass et al. (1991) presented a theoretical model to explain why this occurs. The high organic content of surficial and core samples in Bonefish Pond coupled with formation of authigenic pyrite provides compelling evidence of sulfate reduction in these sediments. Thus, it seems likely that the low $\delta^{18}O$ of dolomites and the low $\Delta^{18}O$ of

Figure 6 (on facing page). Percentages of dolomite, high-Mg calcite, aragonite, and <4ϕ size fraction in selected shallow cores from Bonefish Pond. Zones that contain gypsum or significant cement are also shown. Heavy dashed line indicates change in biofacies and abrupt appearance of encrusted macrophyte tubes.

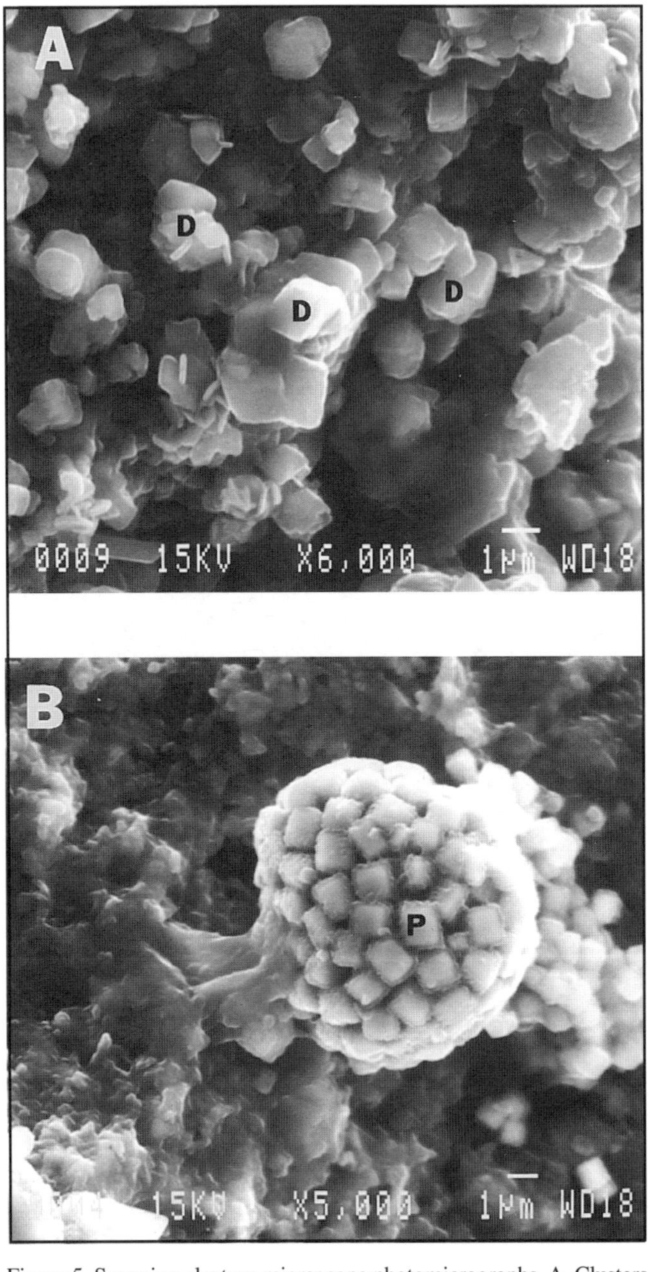

Figure 5. Scanning electron microscope photomicrographs. A, Clusters of dolomite rhombs (D) in eroded, thin intertidal crust collected from subtidal zone. B, Pyrite crystals (P) in thin intertidal crust. Scale bars = 1 µm.

calcite/aragonite-dolomite pairs are due to dolomitization during anaerobic sulfate reduction.

DISCUSSION/CONCLUSIONS

Three types of dolomite are forming in Bonefish Pond. Supratidal crusts develop as a result of the tidal pumping of meteoric-influenced pore waters containing biogenic bicarbonate (Carballo et al., 1987). Intertidal crusts form in areas where intertidal sediments are rich in biogenic carbon. Subsurface sediments are dolomitized, with the concurrent dissolution of aragonite, in a sulfate-reducing environment. In each situation, tidal pumping of Mg-rich pore fluids provides Mg^{2+} for the formation of dolomite. Surficial sediments are enriched with disseminated organics and low levels of <4ϕ allogenic dolomitic grains derived from eroding inter- and supratidal crusts.

Four factors are considered necessary for dolomite to form in modern environments: (1) high Mg^{2+}/Ca^{2+} ratio, (2) high alkalinity, (3) high temperature (>30°C), and (4) very low levels of dissolved sulfate (Folk and Land, 1975; Baker and Kastner, 1981; Gunatilaka et al., 1984; Burns and Swart, 1992). Sulfate in solution seems to retard the rate of calcite dissolution, rather than prevent dolomite precipitation (Morrow and Ricketts, 1988). Sulfate depletion can be produced by gypsum precipitation (Illing and Taylor, 1993) or the activity of sulfate-reducing bacteria (Slaughter and Hill, 1991). Although gypsum disks are present in Bonefish Pond cores, they do not appear to be related to the zones of dolomitization (Fig. 6). Also, the ranges of $\delta^{13}C$ and $\delta^{18}O$ values indicate that biogenic carbon has been incorporated into the dolomite and that evaporative processes have not significantly altered the pore waters.

Sibley et al. (1987) and Sibley (1990) have proposed that there are two stages in the formation of dolomite. The duration of an initial induction stage (in which no detectable products form) is controlled by sediment grain size and mineralogy, as well as pore water chemistry (Ca^{2+}/Mg^{2+} ratio, alkalinity, and dissolved sulfate level). This is followed by a nucleation/growth stage at the expense of calcium carbonate. Dolomitization is enhanced by finer grain size (higher surface area) and aragonitic composition. Subsurface dolomitization of fine-grained aragonitic sediments is taking place in Bonefish Pond in association with bacterially mediated sulfate reduction and pyrite formation. Limited data (two samples) suggest a pore fluid Mg^{2+}/Ca^{2+} of at least 3.5:1. Although no data on alkalinity have been obtained, high alkalinity is produced during sulfate reduction (Hardie, 1987). These are conditions that could promote a very short induction stage.

An additional factor in speeding nucleation may be the allogenic dolomite grains in the <4ϕ size fraction. Allogenic dolomite grains are also common in the sediments of Bonaire

Figure 7 (on facing page). Stable carbon and oxygen isotope data for <4ϕ size fraction of Bonefish Pond sediments. Core locations shown in Figure 1.

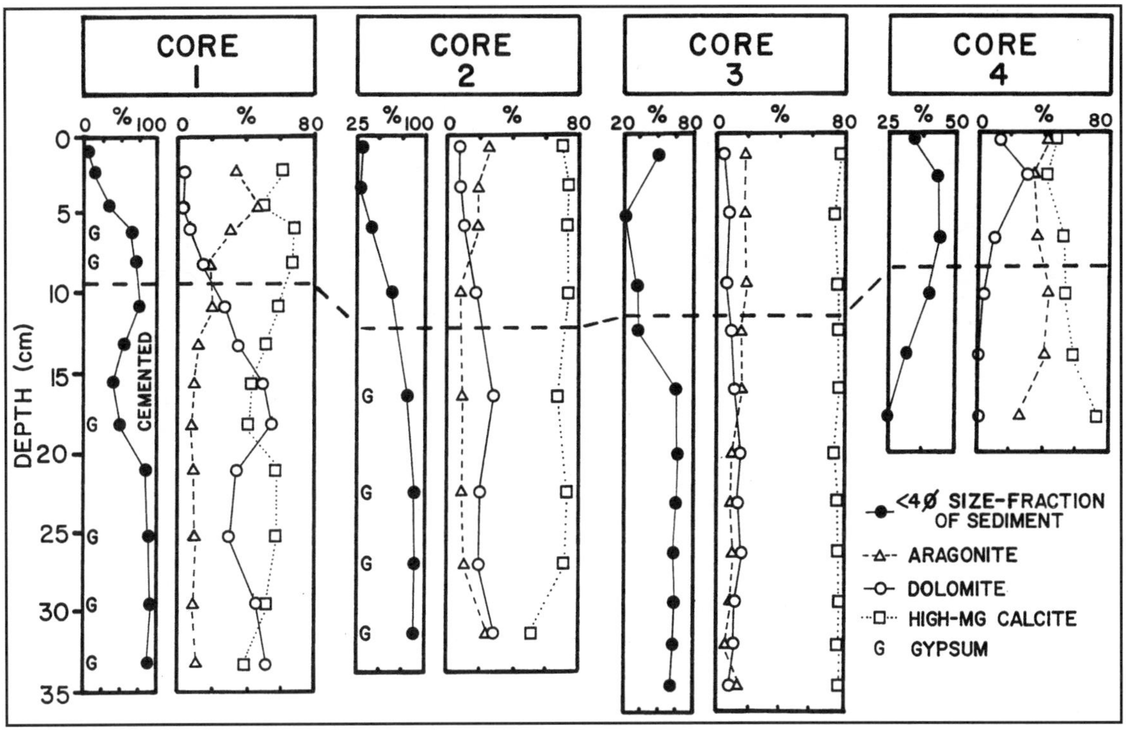
CORE 1
CORE 2
CORE 3
CORE 4
DEPTH (cm)
CEMENTED
<4Ø SIZE-FRACTION OF SEDIMENT
ARAGONITE
DOLOMITE
HIGH-MG CALCITE
G GYPSUM

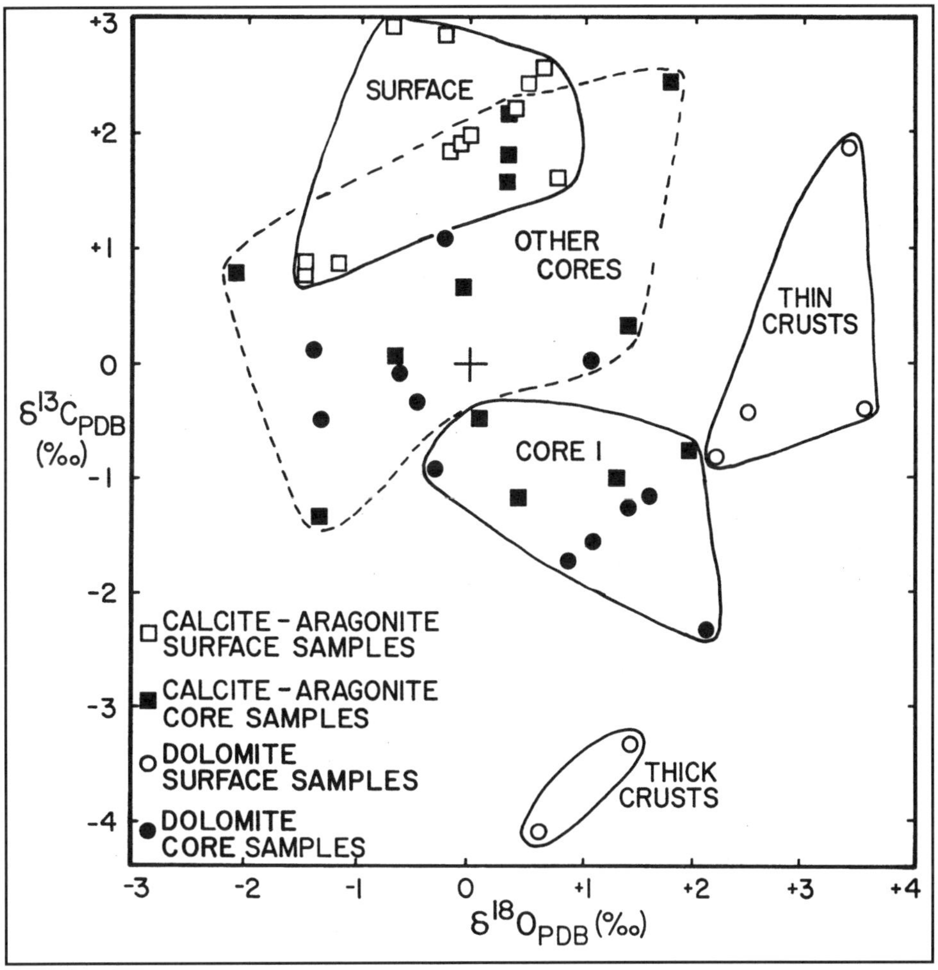
SURFACE
OTHER CORES
THIN CRUSTS
CORE I
THICK CRUSTS
CALCITE-ARAGONITE SURFACE SAMPLES
CALCITE-ARAGONITE CORE SAMPLES
DOLOMITE SURFACE SAMPLES
DOLOMITE CORE SAMPLES
δ13CPDB (‰)
δ18OPDB (‰)

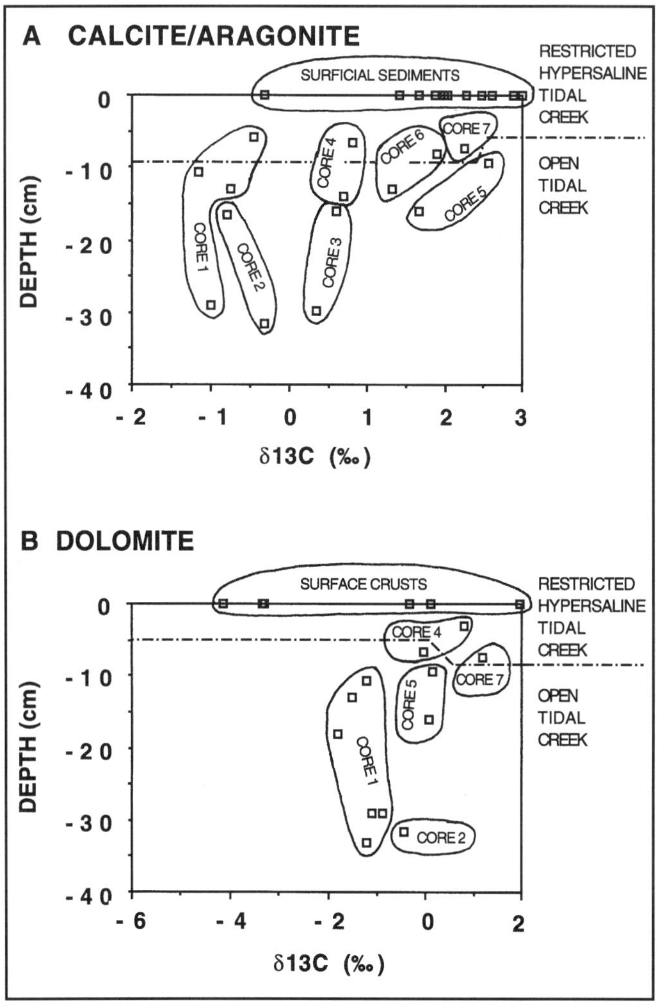

Figure 8. A, Plot of $\delta^{13}C$ versus depth in calcite-aragonite. B, Plot of $\delta^{13}C$ versus depth in dolomite. Core locations shown in Figure 1.

(Deffeyes et al., 1965), Sugarloaf Key, Florida (Carballo et al., 1987), Florida Bay (Swart et al., 1989), and western Andros Island (Hardie, 1977), and allogenic dolomite grains occur over very large areas (hundreds of square kilometers) in the Persian Gulf and the Lacepede Shelf of Australia (Pilkey and Noble, 1966; Bone et al., 1992). Shallow subsurface zones of dolomite enrichment, present along western Andros Island, are similar to those reported here (Gebelein et al., 1980; Lasemi et al., 1989), as are those described from aragonitic muds of a Kuwait bay (Gunatilaka et al., 1984). Allogenic dolomitic grains may serve to reduce further the induction period by providing ideal nucleation points for diagenetic dolomite formation. A similar "seeding" mechanism has been proposed by Lindholm (1969) for the dolomitization of the Devonian Onondaga Limestone of New York and by Bone et al. (1992) for Quaternary sediments of southern Australia. Gregg et al. (1992) observed recrystallization of dolomite and an increase in the dolomite content of buried peritidal crusts in the Holocene carbonates of Belize.

Extensive subsurface dolomitization of the sediments of Bonefish Pond has occurred extremely rapidly, apparently within the past 100 yr. The pore waters of this tidal creek must be providing a near-ideal geochemical environment for the formation of low-temperature dolomite, one in which there is a very short induction period. These ideal factors seem to include an Mg^{2+}/Ca^{2+} ratio of at least 3.5:1, very low levels of sulfate in solution, high alkalinity, tidal pumping, fine-grained aragonitic sediments, and particularly, the presence of allogenic dolomitic seed grains ($<4\phi$) to serve as initial nucleation points.

The sediments of Bonefish Pond and the dolomitized modern tidal creek–lacustrine sediments of other islands are similar to Bahamian Cenozoic dolomitized regressive sequences (Swart et al., 1987; Dawans and Swart, 1988). The preferential dolomitization of these sequences, as well as that of many ancient regressive sequences (Morrow, 1982), could be explained by the development of a similar ideal subsurface geochemical environment for early dolomitization. Many ancient platform dolostones preserve an early micron-sized syndepositional dolomite, along with later, more coarsely crystalline diagenetic dolomite (Morrow, 1982; Machel and Mountjoy, 1986). Such sequences frequently contain evidence of intertidal to supratidal environments (Horton and De Voto, 1990; Rao, 1990; Amthor and Friedman, 1992), including crusts that could have provided a source of allogenic dolomite seed crystals for lagoons and tidal creeks containing organic material. During regressive build-up or eustatic sea-level fluctuations, similar geochemical conditions (high Mg^{2+}/Ca^{2+}, high alkalinity, temperature >30°C, low sulfate) favorable for dolomite formation are likely as organic material decays (Davies et al., 1975; Andrews, 1991; Gunatilaka, 1991; Montañez and Read, 1992a,b). The earliest dolomitization consequently may be attributed to a period of rapid and extensive, but incomplete, dolomitization while the buried sediments were in the zone of bacterially mediated sulfate reduction (between the sediment-water interface and the zone of anoxic methane oxidation) (Raiswell, 1988; Swart et al., 1989; Middelburg et al., 1990). The dolomitization of sediments of Bonefish Pond containing allogenic dolomite seed crystals provides a modern analogue for interpreting the extensive, synsedimentary, micron-sized dolomitization characterizing ancient platform carbonates.

ACKNOWLEDGMENTS

This research was supported by grants from the Center for Field Research, Watertown, Massachusetts; the Foundation for Field Research, Alpine, California; and the California State University Bakersfield Research Council. Isotopic analyses of calcite-aragonite and dolomite extracts were performed by Krueger Enterprises; isotopic analyses of mixed calcite–aragonite–dolomite samples were performed at the Stable Isotope Laboratory, Institute for the Study of Man, Southern Methodist University, Dallas, Texas. The manuscript was improved significantly by the very constructive reviews of Carol B. deWet and Donald H. Zenger.

TABLE 1. STABLE ISOTOPES OF CARBON AND OXYGEN*

Sample Type	Depth (cm)	Calcite $\delta^{13}C$	Calcite $\delta^{18}O$	Dolomite $\delta^{13}C$	Dolomite $\delta^{18}O$	$\Delta^{13}C$	$\Delta^{18}O$	Dolomite (%)
Surface	0	2.62	0.69					
Surface	0	2.47	0.54					
Surface	0	1.4	-1.42					
Surface	0	2.02	0.03					
Surface	0	2.26	0.42					
Surface	0	1.4	1.13					
Surface	0	1.98	-0.02					
Surface	0	2.99	-0.65					
Surface	0	2.89	-0.19					
Surface	0	-0.3	2.27					
Surface	0	1.86	-0.14					
Core 3	14-18	0.6	-0.64					
Core 3	28.5-31	0.36	1.45					
Core 6	6.5-10	1.89	0.38					
Core 6	6.5-19.5	1.3	-1.42					
Core 5	8-11	2.56	1.82	0.15	-1.37	-2.41	-3.19	7
Core 5	14.5-17.5	1.65	0.34	0.07	1.12	-1.58	0.78	8
Core 4	2-5	0.8	0.23					
Core 4	5-8.5	0.81	-2.08	-0.03	-0.6	-0.84	1.48	9
Core 4	12-16	0.69	-0.02					
Core 7	6.5-8.5	2.25	0.4	1.21	-0.19	-1.04	-0.59	20
Core 2	13-20	-0.8	-1.32					
Core 2	29-34	-0.32	-0.42	-0.46	-1.31	-0.14	-0.89	20
Core 1	3-5	-0.75	2	-1.52	1.11	-0.77	-0.89	32
Core 1	5-6.5	-0.44	0.11					
Core 1	9.5-12	-1.16	0.46	-1.23	1.44	-0.07	0.98	22
Core 1	17-19			-1.8	2.17			
Core 1	27-31a	-0.99	1.32	-1.21	1.64	-0.22	0.32	32
Core 1	27-31b			-0.9	-0.32			
Core 1	31-35			-1.2	0.91			
Crust	0	1.66	0.82					
Crust	0			0.1	2.56			
Crust	0			-3.29	1.45			
Crust	0			-3.34	1.47			
Crust	0			1.97	3.49			
Crust	0			-4.15	0.64			
Crust	0			-0.35	3.62			

*As plotted in Figures 7 and 8. Core locations shown in Figure 1.

REFERENCES CITED

Aharon, P., Kolodny, Y., and Sass, E., 1977, Recent hot brine dolomitization in the "Solar Lake," Gulf of Elat: Isotopic, chemical and mineralogical study: Journal of Geology, v. 85, p. 27–48.

Amthor, J. E., and Friedman, G. M., 1992, Early- to late-diagenetic dolomitization of platform carbonates: Lower Ordovician Ellenburger Group, Permian Basin, west Texas: Journal of Sedimentary Petrology, v. 62, p. 131–144.

Andrews, J. E., 1991, Geochemical indicators of depositional and early diagenetic facies in Holocene carbonate muds, and their preservation potential during stabilization: Chemical Geology, v. 93, p. 267–289.

Baker, P. A., and Kastner, M., 1981, Constraints on the formation of sedimentary dolomite: Science, v. 213, p. 214–216.

Bone, Y., James, N. P., and Kyser, T. K., 1992, Synsedimentary detrital dolomite in Quaternary cool-water sediments, Lacepede Shelf, south Australia: Geology, v. 20, p. 109–112.

Burns, S. J., and Swart, P. K., 1992, Diagenetic processes in Holocene carbonate sediments: Florida Bay mudbanks and islands: Sedimentology, v. 39, p. 285–304.

Burns, S. J., Baker, P. A., and Showers, W. J., 1988, The factors controlling the formation and chemistry of dolomite in organic-rich sediments: Miocene Drakes Bay Formation, California, *in* Shukla, V., and Baker, P. A., eds., Sedimentology and Geochemistry of Dolostones: Society of Economic Paleontologists and Mineralogists Special Publication 43, p. 41–52.

Carballo, J. D., Land, L. S., and Miser, D. E., 1987, Holocene dolomitization of supratidal sediments by active tidal pumping, Sugarloaf Key, Florida: Journal of Sedimentary Petrology, v. 57, p. 153–165.

Coleman, M. L., and Raiswell, R., 1981, Carbon, oxygen, and sulfur isotope variations in concretions from the Upper Lias of N. E. England: Geochimica et Cosmochimica Acta, v. 45, p. 329–340.

Davies, P. J., Ferguson, J., and Bubela, B., 1975, Dolomite and organic material: Nature, v. 255, p. 472–474.

Dawans, J. M., and Swart, P. K., 1988, Textural and geochemical alternations in Late Cenozoic Bahamian dolomites: Sedimentology, v. 35, p. 385–403.

Deffeyes, K. S., Lucia, F. J., and Weyl, P. K., 1965, Dolomitization of recent and Plio-Pleistocene sediments by marine evaporite waters on Bonaire, Netherlands Antillies, *in* Pray, L. C., and Murray, R. C., eds., Dolomitization and limestone diagenesis, a symposium: Society of Economic Paleontologists and Mineralogists Special Publication 13, p. 71–88.

Folk, R. L., and Land, L. S., 1975, Mg/Ca ratio and salinity: Two controls over crystallization of dolomite: American Association of Petroleum Geologists Bulletin, v. 59, p. 60–68.

Gebelein, C. D., Steinen, R. P., Garrett, P., Hoffman, E. J., Queen, J. M., and Plummer, L. N., 1980, Subsurface dolomitization beneath the tidal flats of central West Andros Island, Bahamas, *in* Zenger, D. H., Dunham, J. B., and Ethington, R. L., eds., Concepts and Models of Dolomitization: Society of Economic Paleontologists and Mineralogists Special Publication 28, p. 31–49.

Gregg, J. M., Howard, S. A., and Mazzullo, S. J., 1992, Early diagenetic recrystallization of Holocene (<3000 years old) peritidal dolomites, Ambergris Cay, Belize: Sedimentology, v. 39, p. 143–160.

Gunatilaka, A., 1991, Dolomite formation in coastal Al-Khiran, Kuwait, Arabian Gulf—A re-examination of the sabkha model: Sedimentary Geology, v. 72, p. 35–53.

Gunatilaka, A., Saleh, A., Al-Temeemi, A., and Nassar, N., 1984, Occurrence of subtidal dolomite in a hypersaline lagoon, Kuwait: Nature, v. 311, p. 450–452.

Hardie, L. A., 1977, Algal structures in cemented crusts and their environmental significance, *in* Hardie, L. A., ed., Sedimentation on the modern carbonate tidal flats of Northwest Andros Island, Bahamas: Baltimore, Maryland, Johns Hopkins University Press, p. 159–177.

Hardie, L. A., 1987, Dolomitization: A critical review of some current views: Journal of Sedimentary Petrology, v. 57, p. 166–183.

Horton, R. A., Jr., and De Voto, R. H., 1990, Dolomitization and diagenesis of the Leadville Limestone (Mississippian), central Colorado: Economic Geology Monograph 7, p. 86–107.

Horton, R. A., Jr., and Mitchell, S. W., 1991, Modern near-surface dolomitization of Holocene tidal creek sediments, New Providence Island, Bahamas: Geological Society of America Abstracts with Programs, v. 23, p. 47.

Illing, L. V., and Taylor, J. C. M., 1993, Penecontemporaneous dolomitization in Sabkha Faishakh, Qatar: Evidence from changes in the chemistry of the interstitial brines: Journal of Sedimentary Petrology, v. 63, p. 1042–1048.

Kwolek, J. M., 1984, Holocene deposition of a multilayered carbonate sequence in Reckley Hill Settlement Pond, San Salvador Island, Bahamas, *in* Teeter, J. W., ed., Proceedings, Second Symposium on the Geology of the Bahamas: San Salvador, CCFL Bahamian Field Station, p. 27–39.

Land, L. S., 1980, The isotopic and trace element geochemistry of dolomite: The state of the art, *in* Zenger, D. H., Dunham, J. B., and Ethington, R. L., eds., Concepts and Models of Dolomitization: Society of Economic Paleontologists and Mineralogists Special Publication 28, p. 87–110.

Land, L. S., 1983, Dolomitization: American Association of Petroleum Geologists Short Course Notes 24, 20 p.

Land, L. S., 1985, The origin of massive dolomite: Journal of Geological Education, v. 33, p. 112–125.

Lasemi, Z., Boardman, M. R., and Sandberg, P. A., 1989, Cement origin of supratidal dolomite, Andros Island, Bahamas: Journal of Sedimentary Petrology, v. 59, p. 249–257.

Leaver, J., 1985, Sedimentology, mineralogy, and pore water chemistry of schizohaline pond sediments, Turks and Coicos Islands, British West Indies [M.S. thesis]: Durham, North Carolina, Duke University, 76 p.

Lindholm, R. C., 1969, Detrital dolomite in Onondaga Limestone (Middle Devonian) of New York: Its implications to the "dolomite question": American Association of Petroleum Geologists Bulletin, v. 53, p. 1035–1042.

Machel, H.-G., and Mountjoy, E. W., 1986, Chemistry and environments of dolomitization—A reappraisal: Earth-Science Reviews, v. 23, p. 175–222.

Major, R. P., Lloyd, R. M., and Lucia, F. J., 1992, Oxygen isotope composition of Holocene dolomite formed in a humid hypersaline setting: Geology, v. 20, p. 586–588.

McKenzie, J. A., 1991, The dolomite problem: An outstanding controversy, *in* Muller, D. W., McKenzie, J. A., and Weissert, H., eds., Controversies in modern geology: London, United Kingdom, Academic Press, p. 37–54.

McKenzie, J. A., Hsü, K. J., and Schneider, J. F., 1980, Movement of subsurface waters under the sabkha, Abu Dhabi, U.A.E., and its relation to evaporitive dolomite genesis, *in* Zenger, D. H., Dunham, J. B., and Ethington, R. L., eds., Concepts and Models of Dolomitization: Society of Economic Paleontologists and Mineralogists Special Publication 28, p. 11–30.

Middelburg, J. J., de Lange, G. J., and Kreulen, R., 1990, Dolomite formation in anoxic sediments of Kau Bay, Indonesia: Geology, v. 18, p. 399–402.

Mitchell, S. W., and Horton, R. A., Jr., 1989, Dolomitization of Holocene tidal creek–lacustrine transition sediments, Bahama Islands: Geological Society of America Abstracts with Programs, v. 21, p. A77.

Mitchell, S. W., and Horton, R. A., Jr., 1991, Protodolomite precipitation in modern lacustrine environments, Bahamas: Geological Society of America Abstracts with Programs, v. 23, p. 105.

Mitchell, S. W., and Sigler, M. E., 1989, Twentieth century sedimentological development of Bonefish Pond, New Providence Island, Bahamas, *in* Mylroie, J. E., ed., Proceedings, Fourth Symposium on the Geology of the Bahamas: San Salvador, Bahamian Field Station, p. 221–234.

Mitchell, S. W., Buening, N., Baldwin, J. N., Jr., and Westell, B., 1989, Holocene depositional history of Conception Island, Bahamas, *in* Mylroie, J. E., ed., Proceedings, Fourth Symposium on the Geology of the Bahamas: San Salvador, Bahamian Field Station, p. 209–220.

Montañez, I. P., and Read, J. F., 1992a, Fluid-rock interaction history during stabilization of early dolomites, Upper Knox group (Lower Ordovician), U.S. Appalachians: Journal of Sedimentary Petrology, v. 62, p. 753–778.

Montañez, I. P., and Read, J. F., 1992b, Eustatic control on early dolomitization of cyclic peritidal carbonates: Evidence from the Early Ordovician Upper Knox Group, Appalachians: Geological Society of America Bulletin, v. 104, p. 872–886.

Morrow, D. W., 1982, Diagenesis 2. Dolomite. Pt. 2. Dolomitization models and ancient dolostones: Geoscience Canada, v. 9, p. 95–107.

Morrow, D. W., and Ricketts, B. D., 1988, Experimental investigation of sulfate inhibition of dolomite and its mineral analogues, *in* Shukla, V., and Baker, P. A., eds., Sedimentology and Geochemistry of Dolostones: Society of Economic Paleontologists and Mineralogists Special Publication 43, p. 25–38.

Neumann, C. A., Paull, C., Zebielski, V., and Bebout, B., 1991, Modern hypersaline stromatolites of San Salvador, Bahamas, and associated sediments: Geological Society of America Abstracts with Programs, v. 23, p. 108.

Patterson, W. P., and Walter, L. M., 1994, Syndepositional diagenesis of modern platform carbonates: Evidence from isotopic and minor element data: Geology, v. 22, p. 127–130.

Pilkey, O. H., and Noble, D., 1966, Carbonate and clay mineralogy of the Persian Gulf: Deep-Sea Research, v. 13, p. 1–16.

Raiswell, R., 1988, Evidence for surface reaction-controlled growth of carbonate concretions in shales: Sedimentology, v. 35, p. 571–575.

Rao, C. P., 1990, Marine to mixing zone dolomitization in peritidal carbonates: The Gordon Group (Ordovician), Mole Creek, Tasmania, Australia: Carbonates and Evaporites, v. 5, p. 153–178.

Royse, C. F., Jr., Wadell, J. S., and Petersen, L. E., 1971, X-ray determination of calcite-dolomite: An evaluation: Journal of Sedimentary Petrology, v. 41, p. 483–488.

Sass, E., Bein, A., and Almogi-Labin, A., 1991, Oxygen-isotope composition of diagenetic calcite in organic-rich rocks: Evidence for ^{18}O depletion in marine anaerobic pore water: Geology, v. 19, p. 839–842.

Schneider, J., Schroder, H. G., and Le Campion–Alsumard, T., 1983, Algal micro-reefs — Coated grains from freshwater environments, *in* Peryt, T. M., ed., Coated Grains: Berlin, Springer-Verlag, p. 284–298.

Shinn, E. A., 1983, Birdseyes, fenestrae, shrinkage pores, and loferites: A reevaluation: Journal of Sedimentary Petrology, v. 53, p. 619–628.

Shinn, E. A., 1986, Modern carbonate tidal flats: Their diagnostic features: Colorado School of Mines Quarterly, v. 81, p. 7–35.

Shinn, E. A., Ginsburg, R. N., and Lloyd, R. M., 1965, Recent supratidal dolomite from Andros Island, Bahamas, *in* Pray, L. C., and Murray, A. C., eds., Dolomitization and Limestone Diagenesis: A Symposium: Society of Economic Paleontologists and Mineralogists Special Publication 13, p. 112–123.

Shinn, E. A., Lloyd, R. M., and Ginsburg, R. N., 1969, Anatomy of a modern carbonate tidal-flat, Andros Island, Bahamas: Journal of Sedimentary Petrology, v. 39, p. 1202–1228.

Shukla, V., and Baker, P. A., eds., 1988, Sedimentology and Geochemistry of Dolostones: Society of Economic Paleontologists and Mineralogists Special Publication 43, 266 p.

Sibley, D. F., 1990, Unstable to stable transformations during dolomitization: Journal of Geology, v. 98, p. 739–748.

Sibley, D. F., Dedoes, R. E., and Bartlett, T. R., 1987, Kinetics of dolomitization: Geology, v. 15, p. 1112–1114.

Slaughter, M., and Hill, R. J., 1991, The influence of organic matter in organogenic dolomitization: Journal of Sedimentary Petrology, v. 61, p. 296–303.

Swart, P. K., Ruiz, J., and Holmes, C. W., 1987, Use of strontium isotopes to constrain the timing and mode of dolomitization of upper Cenozoic sediments in a core from San Salvador, Bahamas: Geology, v. 15, p. 262–265.

Swart, P. K., Berler, D., McNeill, D., Guzikowski, M., Harrison, S. A., and Dedick, E., 1989, Interstitial water geochemistry and carbonate diagenesis in the sub-surface of a Holocene mud island in Florida Bay: Bulletin of Marine Science, v. 44, p. 490–514.

U.S. Navy Hydrographic Office, 1936, West Indies–Bahama Islands–New Providence Island—Original British surveys to 1881 (21st ed.), H.O. 1377, scale 1:36,350.

Zenger, D. H., Dunham, J. B., and Ethington, R. L., eds., 1980, Concepts and Models of Dolomitization: Society of Economic Paleontologists and Mineralogists Special Publication 28, 320 p.

MANUSCRIPT ACCEPTED BY THE SOCIETY JANUARY 5, 1995

Printed in U.S.A.

Geological Society of America
Special Paper 300
1995

Dolomitization of modern tidal flat, tidal creek, and lacustrine sediments, Bahamas

Steven W. Mitchell and Robert A. Horton, Jr.
Department of Geology, California State University, Bakersfield, California 93311

ABSTRACT

Poorly ordered Ca-rich dolomite is currently precipitating in sediments of modern shallow-water tidal flat, tidal creek, and lacustrine environments of the Bahama Archipelago. Isotopic data and field observations indicate that precipitation occurs in cemented surficial crusts with low organic content (with positive $\delta^{13}C$), in subsurface sediments with intermediate organic content ($\delta^{13}C$ between –1 and –5‰), and in near-surface microbialites with high organic content below cyanobacterial mats ($\delta^{13}C$ between –5 and –10‰). The dolomitized surficial crusts form in tidal flat, tidal creek, and lacustrine environments. Subsurface dolomitized sediments with intermediate organic content occur in tidal creek and lacustrine environments. Dolomitized near-surface microbialites are deposited only in lacustrine environments where cyanobacterial mats are well developed. The dolomite currently seems to be forming in all three environments simultaneously. The dolomitization of carbonate sediments comprising a typical modern regressive sequence provides an important analogue to explain the preferential dolomitization of ancient regressive sequences.

INTRODUCTION

The occurrence of dolomite in modern Bahamian tidal flat sediments has been used as an analogue for the interpretation of dolomitized sections in the geologic record (for example, see Laporte, 1967; Shinn, 1983a; Hardie, 1986; Waters et al., 1989). Bahamian Pleistocene regressive sequences may be dolomitized, as are many similar sequences in the geologic record (for example, see Morrow, 1982; Pierson, 1982; Dawans and Swart, 1988). The dolomitized Pleistocene strata are composed of tidal flat–tidal creek–lacustrine sediments capped by paleosols. The present investigation reviews the previous research on the dolomitization of modern Bahamian tidal flat and tidal creek sediments and presents evidence for the dolomitization of modern Bahamian lacustrine sediments. A synthesis of evidence for dolomitization in three major modern Bahamian depositional environments provides an expanded analogue for the interpretation of dolomitized regressive sequences in the geologic record. This chapter presents new sedimentologic and stratigraphic evidence related to dolomite formation in the three environments. Stable isotope data, pore water chemistry, and geochemical models for the formation of dolomite in the environments are also included.

TECHNIQUES

For this investigation, more than 150 cores were randomly collected from throughout the central and southern Bahamas Archipelago (Fig. 1). Approximately 75% of the cores had at least low levels of dolomite in subsurface sediments (Mitchell and Horton, 1991). Cores with the highest levels of dolomite are presented here. The cores were collected using 3-ft sections of polyvinyl chloride tubing. After extrusion the cores were subdivided into units several centimeters in thickness and then sampled. Wherever possible, units were based on textural differences, which suggested changes in the depositional environment. Each sample was wet-sieved to separate the <4ϕ and sand-size fractions. Relative percentages (by weight) of dolomite, high-Mg calcite, and aragonite in surficial crusts and in the <4ϕ-size fraction of core samples were determined by comparing the areas of

Mitchell, S. W., and Horton, R. A., Jr., 1995, Dolomitization of modern tidal flat, tidal creek, and lacustrine sediments, Bahamas, *in* Curran, H. A., and White, B., Terrestrial and Shallow Marine Geology of the Bahamas and Bermuda: Boulder, Colorado, Geological Society of America Special Paper 300.

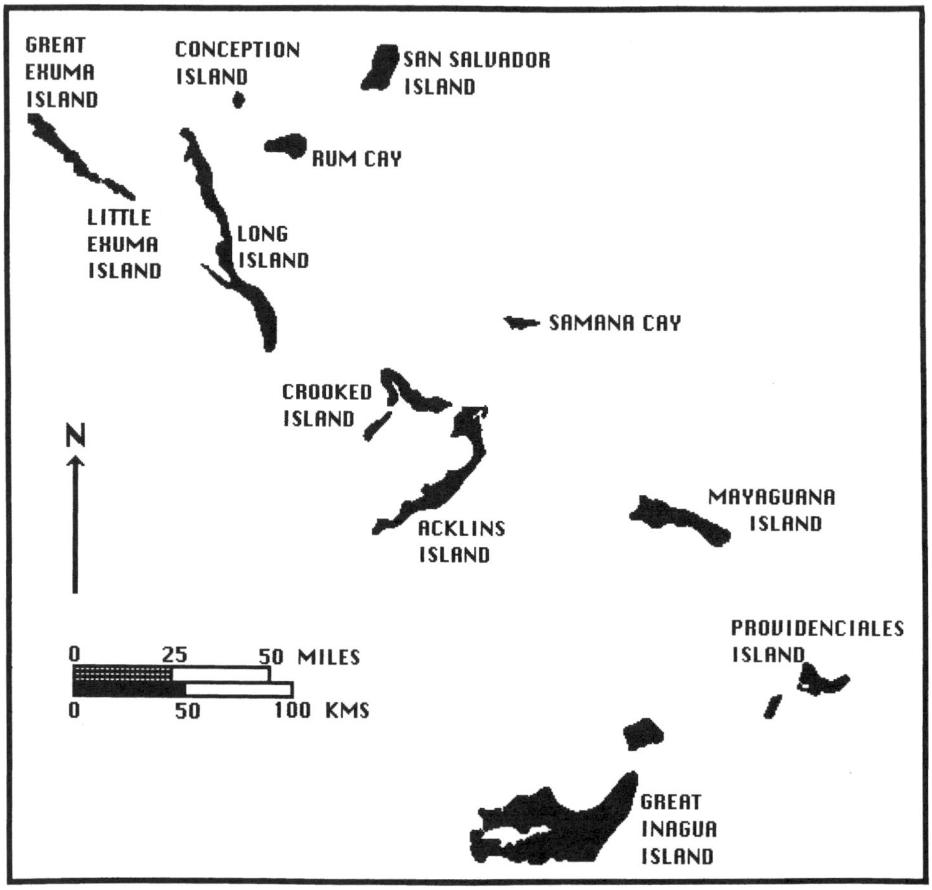

Figure 1. Location map of central and southern Bahamas showing islands included in this study.

major peaks on x-ray diffractograms (Milliman, 1974). This approach provides a semiquantitative estimate of dolomite abundance (Royse et al., 1971). The Mg content of dolomites were determined using the method of Goldsmith and Graf (1958) and Blatt et al. (1972), which is reportedly accurate to ± 0.02 mol % (Lumsden and Chimahusky, 1980), but which we consider to be semiquantitative due to the effects of lattice disorder (Goldsmith et al., 1961). Reconstructions of paleoenvironments in cores were determined by an analysis of ostracod, foraminiferal, and mollusc biofacies and by bulk mineralogy (Mitchell, 1984). Radiocarbon age determinations were performed by Krueger Enterprises of Cambridge, Massachusetts.

Stable isotope analyses of coexisting calcite/aragonite–dolomite samples were performed by the Institute for the Study of Earth and Man at Southern Methodist University, Dallas, Texas, following the procedures of McCrea (1950). Surface-water samples were collected from the surface waters of lakes and at sediment-water interfaces; pore-water samples were collected from shallow pits excavated beneath mats and surficial crusts. However, in situ samples of pore waters from the cores used for mineralogic and isotopic studies were not obtained. Salinity was determined in the field using a Reichert automatic temperature-compensated hand-refractometer. Water samples (without preservatives) were analyzed by the Kern County Water Agency. Cl^- and SO_4^{2-} values were determined by titration and are quantitative; some Ca^{2+} and Mg^{2+} values for samples from Red Pond, Long Island, were determined by atomic absorption spectrometry and are also quantitative. All other Ca^{2+} and Mg^{2+} values were calculated from values obtained from hardness titrations and should be considered semiquantitative.

ENVIRONMENTS OF DOLOMITE OCCURRENCE

Dolomite is currently forming in three major sedimentary environments of the Bahama Archipelago: (1) Extensive fine-grained carbonate tidal flats are present along the west coasts of Andros and Abaco Islands, as well as the northern Caicos Banks (Shinn, 1983a, 1986). The bulk mineralogy of tidal flat sediments is predominantly aragonite. (2) Tidal creeks are present along the coasts of all major islands in the Bahamas. The creeks are enclosed, but with one or more inlets to the open ocean; tidal ranges are considerably lower than those along the open coast (Mitchell, 1987a, b). The bulk mineralogy of tidal creek sediments is predominantly aragonite. (3) Bahamian lakes are completely enclosed bodies of water that are affected, at least very weakly, by ground-water movement due to tidal fluctuations.

Salinities vary seasonally in response to changing rainfall and evapotranspiration levels. The bulk mineralogy of lake sediments is commonly dominated by high-Mg calcite. The locations of each of the tidal flats, tidal creeks, and lakes discussed in this chapter are given in Tables 1 and 2.

TIDAL FLAT DOLOMITIZATION

Supratidal crusts

Modern dolomite formation along the western coast of Andros Island has been studied extensively (Shinn et al., 1965, 1969; Hardie, 1977; Shinn, 1983a, b, 1986). The dolomite is formed as a void-filling cement that nucleates on aragonite needles in a microenvironment of sulfate reduction (Lasemi et al., 1989). Mg^{2+} is provided by the periodic flooding of supratidal areas by seawater during storms (Bourrouilh–Le Jan, 1978). Apparently similar supratidal dolomitic crusts have been reported from western Abaco Island and the Caicos Banks (Shinn et al., 1965; Shinn, 1983a, 1986). However, these occurrences have not been investigated in detail.

Supratidal dolomitic crusts also are associated with a narrow tidal flat along the western coast of Long Island. One of the most extensively developed areas is present along the northern edge of Stella Maris Bay, Long Island (Table 1, Fig. 2A). A core (core 1) from this location indicates that the crust is approximately 2 cm thick with dolomite abundance (approximately 30%) and the amount of cementation decreasing with depth; in the crust, dolomite increases as high-Mg calcite decreases (Fig. 3A). Dolomitic crusts are also present in the supratidal zones of other bays along western Long Island. Samples from Wemyss Bight are about 1 cm thick and contain 10 to 35% dolomite; samples from Greys Bight are up to 2 cm thick and contain 40 to 80% dolomite (Table 1). At both Wemyss Bight and Greys Bight the levels of high-Mg calcite and aragonite decrease relatively as the dolomite content of the crusts increases.

Subsurface dolomitization

Cores from the tidal flats of western Andros Island contain horizons with low levels of dolomite. The zones are most common beneath mangrove hammocks where freshwater lenses may be developed (Steinen, 1980, 1982). The 1-mm dolomite rhombs occur as isolated crystals nucleated around aragonite needles (Gebelein et al., 1980).

A shallow core (core 2) from the supratidal zone along eastern Stella Maris Bay, Long Island (Table 2; Fig. 3B) also provides evidence for subsurface dolomitization. Although there is no surficial crust present at this locality, two partially cemented dolomitic zones are present at depths of 12.5 to 15 and 20 to 22.5 cm, with dolomite levels of the $<4\phi$ sediment-size fraction at approximately 50 and 10%, respectively (Fig. 3B). The lower zone occurs directly above Pleistocene bedrock; as dolomite increases there is a relative decrease in aragonite. Nearby, at the edge of the Stella Maris Bay tidal flat, meteoric water seeps from Pleistocene limestone outcrops into the tidal flat sediments.

TIDAL CREEK DOLOMITIZATION

Supratidal crusts

Supratidal crusts are well developed in the central part of the Bonefish Pond tidal creek system of New Providence Island

TABLE 1. LOCATIONS OF SURFICAL DOLOMITIZED LACUSTRINE, TIDAL FLAT, AND TIDAL CREEK CRUSTS DESCRIBED IN THE TEXT*

Island		UTM Grid Zone	East	North	Latitude (N)	Longitude (W)
	Lake					
Long Island	Middle Fish Pond	18Q VB	478.7	2602.25		
	Red Pond	18Q VA	490.85-491.35	2571.6-2572		
Rum Cay	Carmichael Pond	18Q WB	507.15	2618.35		
	Tidal Flat					
Long Island	Greys Bight	18Q VA	487.6	2575.55		
	Stella Maris Bay	18Q VB	470.8	2608.8		
	Wemyss Bight	18Q VA	478.6	2590.95		
	Tidal Creek					
Long Island	4.5 km N. of Diamond Crystal Salt Company	18Q WA	505.05	2548.15		
Providenciales Island	Chalk Sound				20°45′02″	72°18′17″

*From Bahamas Government Lands and Surveys Department topographic maps, Universal Transverse Mercator Grid 18.

TABLE 2. LOCATIONS OF SURFICAL DOLOMITIZED LACUSTRINE, TIDAL FLAT, AND TIDAL CREEK CORES DESCRIBED IN THE TEXT*

Island		UTM Grid Zone	East	North	Latitude (N)	Longitude (W)
	Lake					
Conception Island	Southeast Lake	18Q VB	489	2635.15		
Crooked Island	Marine Farm Salt Pond	18Q WA	568.65	2525.05		
Great Exuma Island	Isaac Cay Pond	18Q VA	426.3	2595.5		
Little Exuma Island	Pelican Cays Pond	18Q VA	439.4	2591.95		
	Williams Town Salt Pond	18Q VA	442.65	2590.35		
Great Inagua Island	Red Pond	18Q XU	640.15	2332.85		
Long Island	Carmichael Pond	18Q VA	489.5	2577.75		
	North Danes Pond	18Q WA	500.35	2554.05		
	Middle Fish Pond	18Q VB	478.65	2602.25		
	South Fish Pond	18Q VB	478.75	2602.1		
	McKanns Pond	18Q VA	485.25	2586.6		
	West Munroe Pond	18Q WA	501.7	2554.35		
	North Salt Pond	18Q VA	487.65	2582.6		
	South Salt Pond	18Q VA	488.05	2581.45		
	South End Pond	18Q WA	514.3	2527.95		
	Red Pond	18Q VA	490.9	2571.85		
Mayaguana Island	North Pirate Well Pond	18Q XV	698.5	2483.15		
	South Pirate Well Pond	18Q XV	697.75	2481.15		
Rum Cay	Carmichael Pond	18Q WB	507.15	2618.35		
San Salvador Island	Clear Pond	18Q and R WB	546	2651.1		
	Northeast Arm Lake	18R WB	554.05	2666.3		
	Reckley Hill Settlement Pond	18R WB	555	2667		
	Salt Pond	18Q and R WB	555.75	2656.7		
	Storrs Lake	18Q and R WB	556.9	2663.2		
Samana Cay	South-Central Coast Pond	18Q XA	625.8	2551.55		
Providenciales Island	Airport Pond				21°47′16″	72°14′37″
	Tidal Flat					
Long Island	Stella Maris Bay					
	Core 1	18Q VB	470.8	2608.85		
	Core 2	18Q VB	471.65	2608.5		
	Tidal Creek					
Providenciales Island	Chalk Sound				20°45′02″	72°18′17″

(Mitchell and Sigler, 1989; Mitchell and Horton, this volume). The crusts have a maximum thickness of 3.5 cm; dolomite content and cementation decrease with depth. The dolomite levels range up to 40%; dolomite increases as the relative percentages of aragonite and high-Mg calcite decrease. Similar crusts (1 cm thick) are developed in the upper reaches of the Chalk Sound tidal creek system, Providenciales Island (Table 1). The dolomite content is 65 to 90%, decreases with depth, and increases relative to aragonite as well as high-Mg calcite.

Intertidal crusts

Mitchell and Horton (this volume) have described 0.5- to 1.5-cm intertidal crusts from the Bonefish Pond tidal creek system. The crusts have high levels of dolomite (up to 80%) that increase as high-Mg calcite and aragonite decrease. Similar (0.5 cm) well-cemented crusts, forming in the intertidal zone of a small tidal creek along the west coast of Long Island (4.5 km northwest of the Diamond Crystal Salt Company office buildings) (Table 1), contain lower levels of dolomite (5 to 10%).

Subsurface dolomitization

The subsurface sediments of Bonefish Pond also contain significant levels of dolomite. Levels as high as 50% of the $<4\phi$ sediment-size fraction are present in cores from the tidal creek system (Mitchell and Horton, this volume). The dolomite increases as high-Mg calcite decreases. A core from the southern edge of the Chalk Sound tidal creek system (Table 2) also contains evidence of subsurface dolomitization. Up to 15% of the $<4\phi$ sediment-size fraction is composed of dolomite with an equivalent decrease in high-Mg calcite (Fig. 3D).

In shallow cores from sequences where tidal creek sediments underlie lacustrine sediments, dolomite is more common in the lake sediments with abundant high-Mg calcite. However, the tidal creek sediments commonly contain dolomite as well.

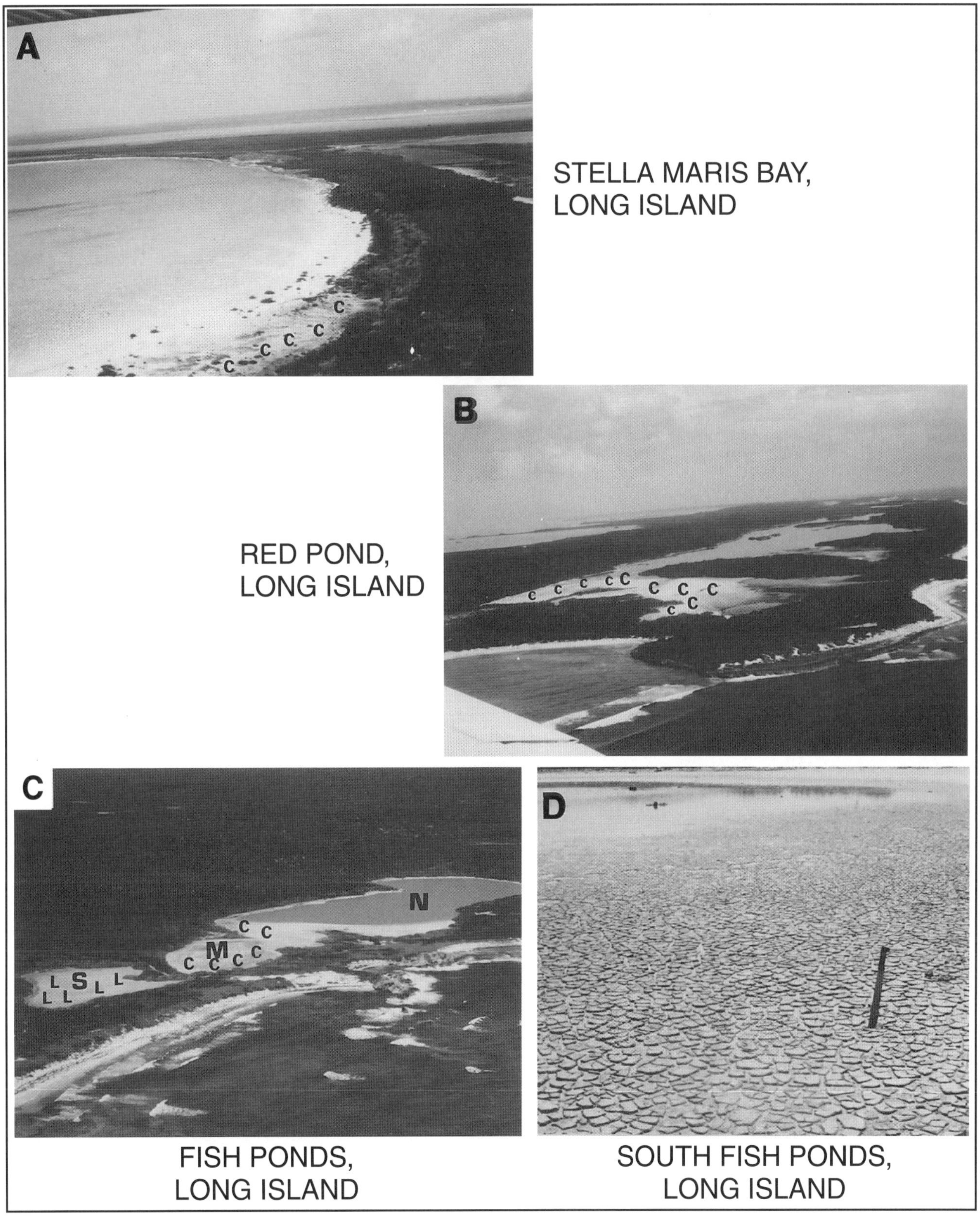

Figure 2. A, Aerial view showing area of supratidal dolomitic crust development along northern Stella Maris Bay, Long Island (looking northwest). B, Aerial view of southern Red Pond, Long Island, showing area of dolomitic crust development (looking northwest). C, Aerial view of North (N), Middle (M), and South (S) Fish Ponds, Long Island, showing areas of dolomitic crusts and laminated sediments (looking west). D, Ground view of mud-cracked, dolomitic, laminated sediments of South Fish Pond, Long Island (looking northwest), with scale indicated by machete (70 cm long, 6 cm wide). C = surficial dolomitic crusts; L = unconsolidated dolomitic laminated sediments.

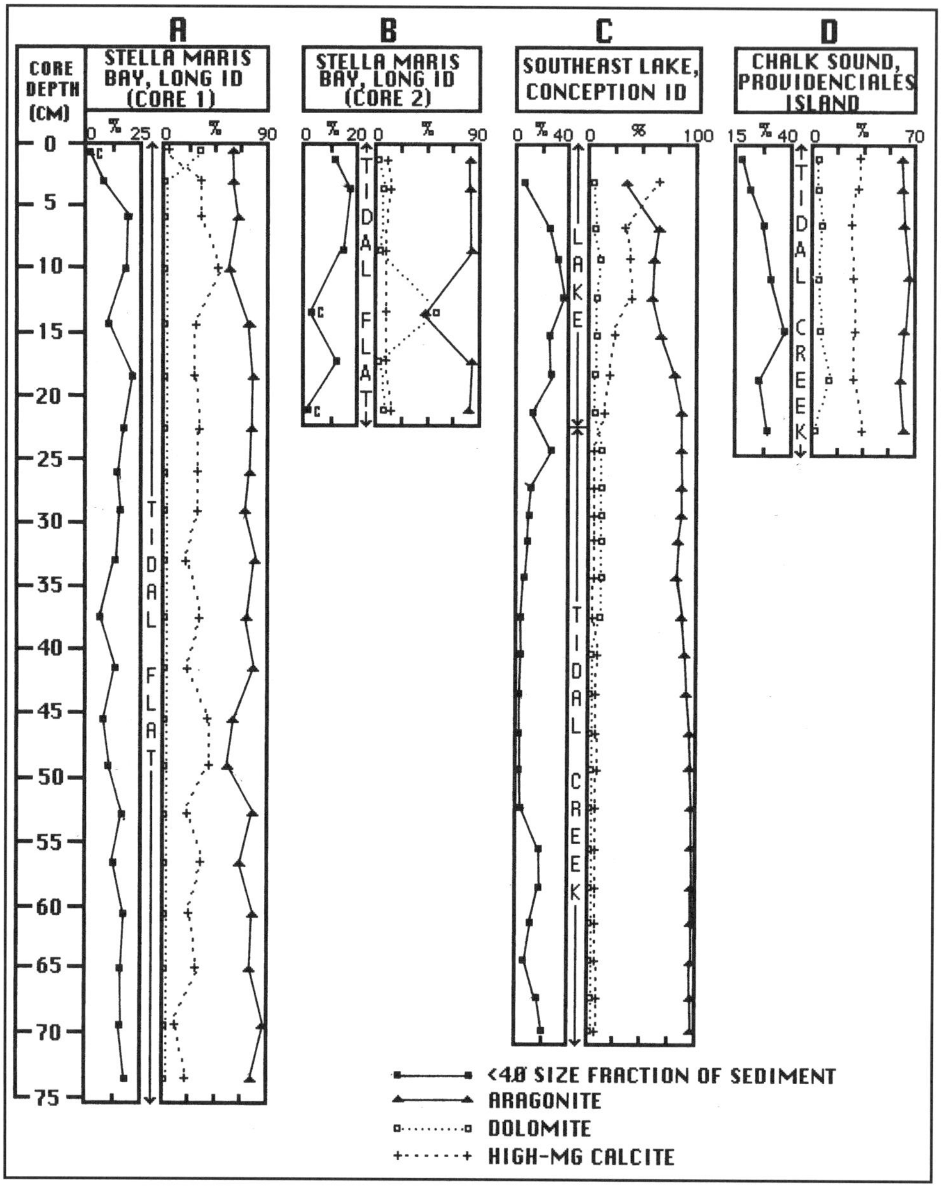

Figure 3. Percentages of dolomite, high-Mg calcite, and aragonite in <4ϕ fraction and weight % of <4ϕ size fraction versus depth in tidal flat and tidal creek cores. Environment of deposition, based on benthic ostracod and foraminifera assemblages, is given in column between <4ϕ size fraction and mineralogy percentages. Core locations are given in Table 2. A, Core from swale behind former beach along north-central part of Stella Maris Bay, northernmost Long Island. B, Core from high intertidal zone along northwest part of Stella Maris Bay, northernmost Long Island. C, Core from lake along southeastern coast of Conception Island due east of inlet to interior tidal creek on west coast (Mitchell et al., 1989). D, Core from west side of southwesternmost part of Chalk Sound tidal creek, southwestern Providenciales Island.

Cores from southern Long and Conception Islands (Table 2) contain low levels of dolomite in the uppermost tidal creek sediments. In the Southeast Lake core from Conception Island, approximately 10% of the <4ϕ size fraction is dolomite in the interval between 25 and 40 cm (Mitchell et al., 1989; Fig. 3C). The South End Pond core from Long Island contains a higher concentration (up to 20%) at a depth of 8 to 25 cm (Fig. 6A).

The occurrence of dolomite in the tidal creek sediments of cores from central Long and Little Exuma Islands (Table 2) is quite different. The entire sequence of cored tidal creek sediments from South Salt Pond, Long Island, is dolomitized, with levels reaching 35% of the <4ϕ size fraction (Fig. 7B). The tidal creek sediments of Pelican Cays Pond, Little Exuma Island, are also dolomitized, with up to 10% dolomite in the lower part of the core (55 to 80 cm) (Fig. 5A). In the aragonite-rich tidal creek sediments, an increase in dolomite commonly coincides with a decrease in aragonite.

LACUSTRINE DOLOMITIZATION

Description of lacustrine environments

Modern Bahamian lakes have formed by one of two mechanisms: (1) by early Holocene flooding of interior depressions as the water table followed rising sea level; or (2) by the closing off of former bays and tidal creeks through sediment deposition by longshore transport, which produces regressive bay–tidal creek–lake transitions in the stratigraphic record (Mitchell, 1984; Mitchell and Keegan, 1987). Depositional rates in Bahamian lakes are influenced primarily by distance from the coast and levels of soil erosion due to agricultural activity. Maximum depositional rates (over 15 cm/100 yr) occur in coastal lakes subject to significant storm deposits (transported over coastal beaches/dunes) (Mitchell and Keegan, 1987). Lowest depositional rates (averaging 1.5 to 4.5 m/100 yr) occur in inland lakes and blue holes in areas of low surface soil erosion (Teeter and Quick, 1990; Paull et al., 1992; Furman et al., 1993). Radiocarbon dates of peat samples from three lakes included in the present study are given in Table 3. In the three lakes on Long and Crooked Islands, average depositional rates are between 1.7 and 2.2 cm/100 yr.

Bahamian lakes are affected significantly by seasonal changes in precipitation and evapotranspiration. Such factors as water depth, ground-water level and movement, water chemistry, and cyanobacterial mat types vary significantly between the dry season (November–May) and the wet season (June–October). Water levels (surficial and ground) are highest during the wet season, salinities are lowest, and flocculose mats are developed in the stagnant anaerobic waters. These are similar to winter cyanobacterial mats, composed of *Lyngbya, Microcoleus, Oscillatoria,* and *Spirulina,* described from a sea-marginal lake along the Gulf of Aqaba (Krumbein and Cohen, 1977). During the dry season, water levels (surficial and ground) are the lowest. Meteoric water seeps, from adjacent higher ground, may flow over cyanobacterial mat surfaces into the lake. Salinities are the highest at this time and the flocculose mats decay. They are replaced by flat, shallow-water mats that often contain dome-shaped polygonal dessication cracks produced by subsurface gas bubbles (Cornee et al., 1992). These mats are similar to the summer mats formed by the diatoms and coccoid blue-green algae *Aphanothece* and *Aphanocapsa* also reported by Krumbein and Cohen (1977).

Dolomite is widespread in the modern sediments of Bahamian lakes (Mitchell and Horton, 1991). It forms in three different situations: in cemented surficial crusts, in organic-rich near-surface laminated sediments, and in subsurface sediments with intermediate amounts of organic carbon. There is a close correlation between dolomite increase and high-Mg calcite decrease in lacustrine sediments.

TABLE 3. RADIOCARBON DATES FOR LACUSTRINE PEAT SAMPLES AND CALCULATED AVERAGE LAKE DEPOSITIONAL RATES

Core	Sample Depth (cm)	Age (^{14}C yr B.P.)	Average Depositional Rate (cm/100 yr)
Marine Farm Salt Pond, Crooked Island (Core 1)	39-44	2,455 ± 125	1.7
Carmichael Pond, Long Island	58.5-61.5	2,880 ± 175	2.1
North End Pond, Long Island	50-53	2,350 ± 130	2.2

Surficial crusts

Surficial crusts occur along the exposed margins of some Bahamian lakes during the dry season. The crusts (up to 2 cm thick) are overlain by thin cyanobacterial mats; the levels of dolomite and cementation decrease with depth. Dolomite ranges up to 30% and increases as aragonite and high-Mg calcite uniformly decrease. Some crusts contain significant levels of gypsum. The mineralogy of shallow cores from four lakes with surficial dolomitic crusts is presented in Figure 4.

Two cores from Long Island lakes contain surficial dolomitic crusts underlain by about 30 cm of partially dolomitized lacustrine sediments. The southernmost part of Red Pond, Long Island, is completely exposed during the dry season. A surficial gypsiferous-dolomitic crust has formed and extends over an area of about 200,000 m^2 of the southern part of the pond (Table 1, Fig. 2B). A shallow core through the lake sediments indicates a gradual reduction in dolomite with depth as high-Mg calcite and aragonite increase (Fig. 4A). The surficial crust contains approximately 45% dolomite and a very low level of aragonite. Undolomitized tidal creek sediments are present at depths below about 32 cm in cores from southern Red Pond (Mitchell, 1984).

Surficial dolomitic crusts are well developed along the margins of Middle Fish Pond (Table 1, Fig. 2C). Cored lake sediments indicate that dolomite increases to about a depth of 15 cm and then decreases rapidly as older tidal creek sediments are encountered (Fig. 4B). Calcification by cyanobacteria in west-central Storrs Lake, San Salvador Island (Table 2), has been documented by Mann and Nelson (1989), Neumann et al. (1989), and Paull et al. (1992). Dolomite has been reported as one of the precipitated carbonates (Neumann et al., 1991). Also, Zabielski and Neumann (1990) and Zabielski (1991) have reported dolomite in the surficial sediments of Storrs Lake.

A thin surficial crust occurs in the cored sediments of Clear

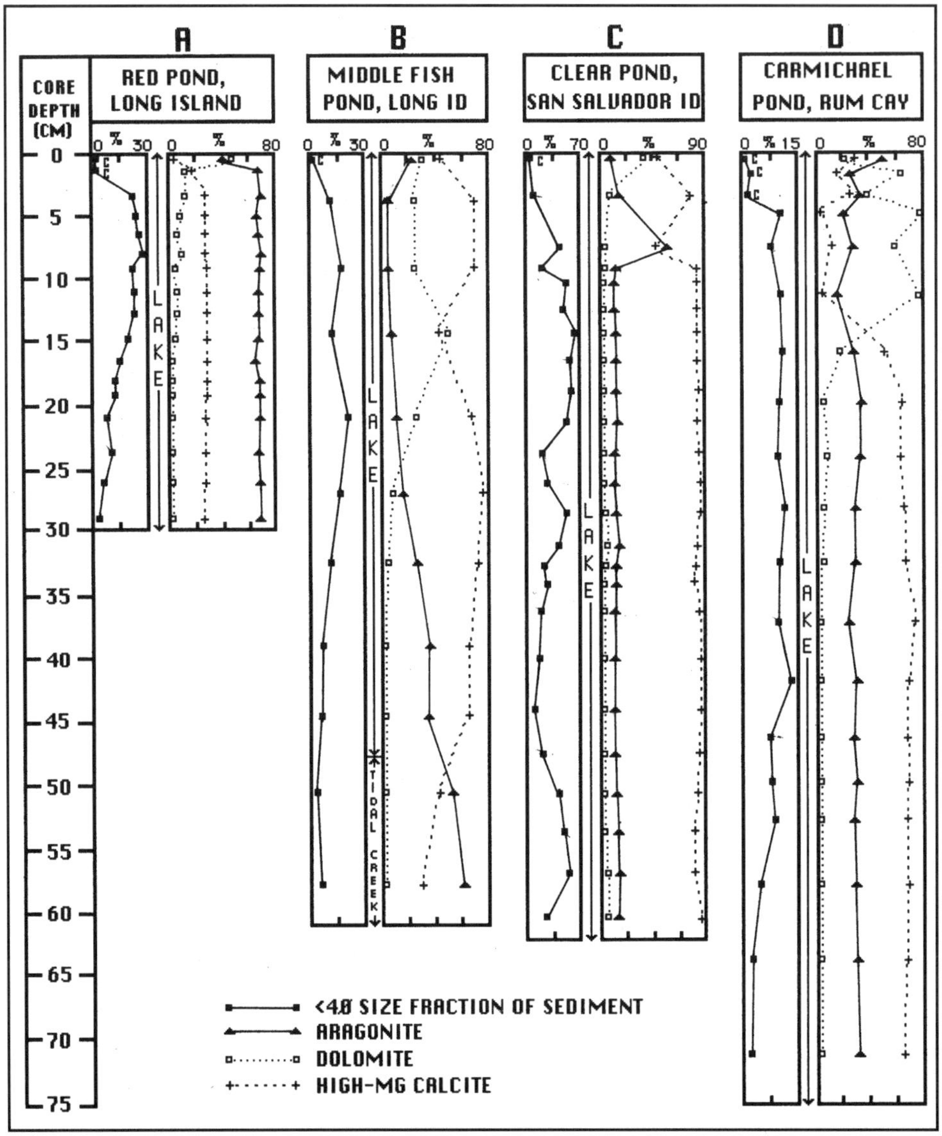

Figure 4. Percentages of dolomite, high-Mg calcite, and aragonite in <4ϕ fraction and weight % of <4ϕ size fraction versus depth in cores from lakes with surficial dolomitic crusts. Environment of deposition, based on benthic ostracod and foraminifera assemblages, is given in column between <4ϕ size fraction and mineralogy percentages. Core locations are given in Table 2. A, Core from southernmost part of Red Pond, east-central coast of Long Island. B, Core from east-central side of Middle Fish Pond, northeast coast of Long Island (4 km south-southeast of Millerton). C, Core from central part of Clear Pond, southwesternmost San Salvador Island. D, Core from northern side of central Carmichael Pond, northwesternmost coast of Rum Cay.

Pond, San Salvador Island (Table 1). Dolomite content (with a comparable decrease in high-Mg calcite) is at a maximum of about 40% of the <4ϕ size fraction in the shallowest sediments; it occurs only as trace levels throughout the remainder of the core (Fig. 4C). The aragonite incursion at a depth of about 7 cm is attributed to an influx of beach sediments during a storm event. A much thicker (up to 4 cm) dolomitic crust is formed along the margins of Carmichael Pond, Rum Cay (Table 1). The underlying sediments, to a depth of about 15 cm, are significantly dolomitized (up to 80% of the <4ϕ size fraction) and contain fragments of former dolomitized surficial crusts. As the dolomite increases, there is an equivalent decrease in the levels of high-Mg calcite (Fig. 4D). Similar dolomitized aggregates of preexisting cemented sediments have been reported in cores from other Bahamian lakes. These are evidently the remnants of older surficial dolomitic crusts that were fragmented and redistributed during periods of wave activity. For example, Kwolek (1984) described the occurrence of a former surficial lacustrine crust (26% dolomite) in cores from Reckley Hill Settlement Pond, San Salvador Island (Table 2).

Dolomite in near-surface microbialites

Some Bahamian lakes contain significant levels of dolomite in near-surface sediments but lack surficial crusts. The laminated organosedimentary deposits of benthic microbial communities (microbialites of Burne and Moore, 1987) grade into an overlying surficial cyanobacterial mat. Unlike the mats associated with surface crusts, these mats are present in areas of lakes containing water even during the dry season. Dolomite formation seems to be associated with a zone of mat degradation directly underlying the living mat; the living mat occurs in the upper several centimeters of cores. The degradation zone extends to a depth of approximately 30 cm (Krumbein and Cohen, 1977; Krumbein et al., 1977). The mineralogy of shallow cores from five lakes with near-surface dolomitic microbialites is presented in Figures 5 and 6.

Three distinct dolomitization signatures in near-surface microbialites can be recognized in cores: (1) dolomite content increases upward to a maximum directly below the living cyanobacterial mat (Figs. 5B and 6A); (2) within microbialites associated with decaying mats, dolomite content declines until it is no longer present 20 to 30 cm below the living cyanobacterial mat (Fig. 5A, D); and (3) dolomite content is highest in lower zones of the microbialite sequence (Fig. 5C). The first two categories are probably identical because analysis of the mineralogy of the <4ϕ size fraction associated with living or decaying mats is dependent on obtaining a large enough sample size for x-ray diffraction analysis. Cores providing large enough samples show an ascending decrease in dolomite in the uppermost 3 to 5 cm of sediment associated with mat degradation, directly underlying the living mat. The third category appears to reflect a former period when processes of mat degradation were more favorable for dolomite formation than they are at present. A core from McKanns Pond, Long Island (Table 2), seems to include dolomite maxima at levels too deep to be associated with the near-surface degradation of modern cyanobacterial mats (Fig. 5C).

Cores from Long and Little Exuma Islands (Fig. 3C, D) best document the apparent relationship between mat degradation and sub-living mat dolomite content. The core from Pelican Cays Pond, Little Exuma Island, contains 15 cm of microbialites with dolomite levels (approximately 35%) in the <4ϕ size fraction underlying a 3-cm-thick living mat sequence (Fig. 5A). Within the decaying mat, dolomite content decreases rapidly upward to the base of the living mat. The core from South Fish Pond, Long Island, exhibits a similar trend (Figs. 2D and 5D). A core from South End Pond, Long Island (Table 2) also suggests an upward decrease in dolomite content in the degrading mat sequence. However, the upper 8 cm of this core contains thin, laminated dolomitic crusts that deformed the sediment when fractured during coring (Fig. 6A). Consequently, the observed slight decrease may have been more marked prior to the deformation of the microbialites. Because the uppermost 4 cm of a core from Carmichael Pond, Long Island (Table 2), have very low levels of <4ϕ size fraction sediment incorporated into the degrading mat sequence, no estimate of the mineralogy could be made. However, dolomite comprises about 40% of this size fraction within the underlying microbialites where complete mat degradation has not yet occurred (Fig. 5B).

The four cores illustrated in Figure 5 exhibit a strong inverse relationship between dolomite and high-Mg calcite levels. Contrarily, the core from South End Pond, Long Island, seems to preserve evidence of aragonite decreasing, and high Mg-calcite increasing, as dolomite increases (Fig. 6A). However, this trend is complicated by a transition from tidal creek to lacustrine sediments at a depth of about 8 cm. As the lake environment became established, the predominant sediment mineralogy typically would shift from aragonite to high-Mg calcite. Consequently, subsurface dolomite formation in the sediments of South End Pond likewise may have resulted in a relative, but difficult to recognize, decrease in high-Mg calcite. Alternatively, the trend could be similar to that observed in surficial crusts where an increase in dolomite results in a relative decrease in aragonite and high-Mg calcite. Overall, it appears that benthic microbial processes dominate the formation and distribution of sediments in many Bahamian lakes, as is the case in Lake Thetis of Western Australia (Grey et al., 1990).

Subsurface dolomitization

The occurrences of dolomite not associated with zones of dolomitic aggregates (probably broken remnants of former surficial crusts), or in microbialites, seem to have a different origin. Dolomitization occurs much more extensively in fine-grained lacustrine sediments than in the coarser tidal creek sediments. Generally, dolomite levels are highest in zones with increased levels of sediment in the <4ϕ size fraction. In the subsurface, an increase in dolomite is mirrored by a reduction in high-Mg calcite. The mineralogy of 12 shallow cores from lakes with evidence of significant subsurface dolomitization is presented in Figures 3, 4, 6, 7, and 8.

The cores can be grouped into two general categories based on the intensity of dolomitization. Low levels of dolomite (<25%) are present in the <4ϕ size fraction of all sampled horizons of lake sediments in the cores. However, some cores contain zones of much higher concentrations of dolomite (35 to 85% of the <4ϕ size fraction). Cores from North and South Salt Ponds, Long Island (Table 2), are strongly dolomitized in the upper 30 cm (Fig. 7A, B). A core from northern Storrs Lake, San Salvador Island (Table 2), also is strongly dolomitized in a zone 25 to 45 cm below the sediment-water interface (Fig. 8A), as is the lower part (25 to 50 cm) of a core (core 1) from Marine Farm Salt Pond, Crooked Island (Table 2, Fig. 6D). In addition, cores from lakes on Samana Cay, Long and Mayaguana Islands (Table 2), preserve thin zones of unusually high levels of dolomite. The Samana Cay core contains 45% dolomite in the <4ϕ size fraction at a depth of 25 cm (Fig. 8D), while the core from North Pirate Well Pond, Mayaguana Island, has about 35% dolomite in the same size range at a depth of about 30 cm (Fig. 6C). In the core

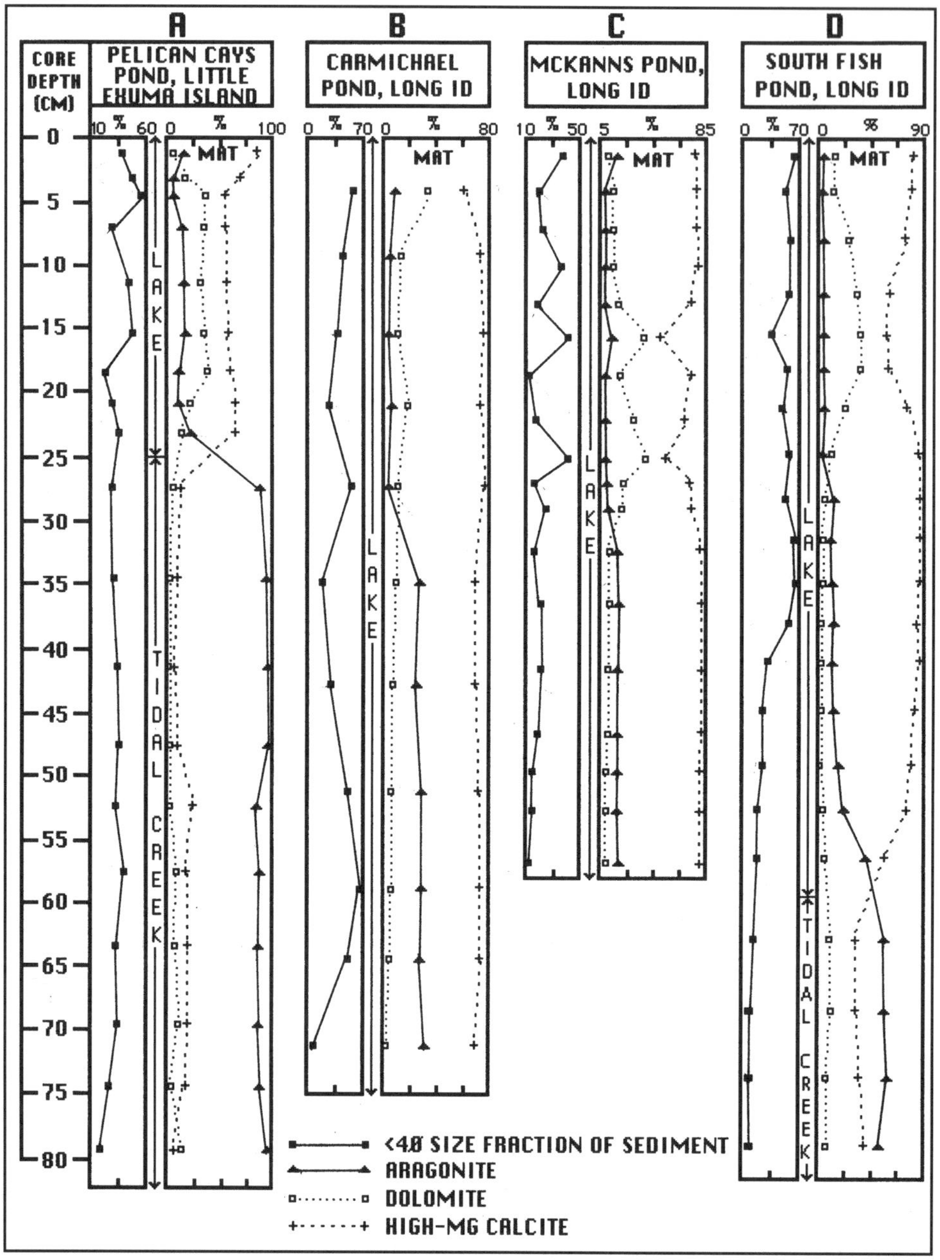

Figure 5. Percentages of dolomite, high-Mg calcite, and aragonite in <4ϕ fraction and weight % of <4ϕ size fraction versus depth in cores from lakes with near-surface dolomitic laminated sediments. Environment of deposition, based on benthic ostracod and foraminifera assemblages, is given in column between <4ϕ size fraction and mineralogy percentages. Core locations are given in Table 2. A, Core from central part of coastal pond south of Pelican Cays, north-central Little Exuma Island. B, Core from eastern side of southernmost part of Carmichael Pond, east-central coast of Long Island. C, Core from west side of central McKanns Pond just north of causeway, east-central coast of Long Island. D, Core from east-central part of South Fish Pond, northeast coast of Long Island (4 km south-southeast of Millerton).

from Middle Fish Pond, Long Island, 50% of this fraction is dolomite at a depth of 15 cm (Fig. 4B).

Cored lake sediments with low (<25% of <4ϕ size fraction), but extensive, dolomitization can be subdivided further by lake environmental history. Cores from Long Island and the Exuma Islands (Table 2) are from lakes without tidal creek–lake transitions. The sediments of North Danes and West Munroe Ponds on Long Island provide evidence that many shallow interior lakes, with apparently limited cyanobacterial mat development, contain extensive dolomite (Fig. 7C, D). Cores from Isaac Cay

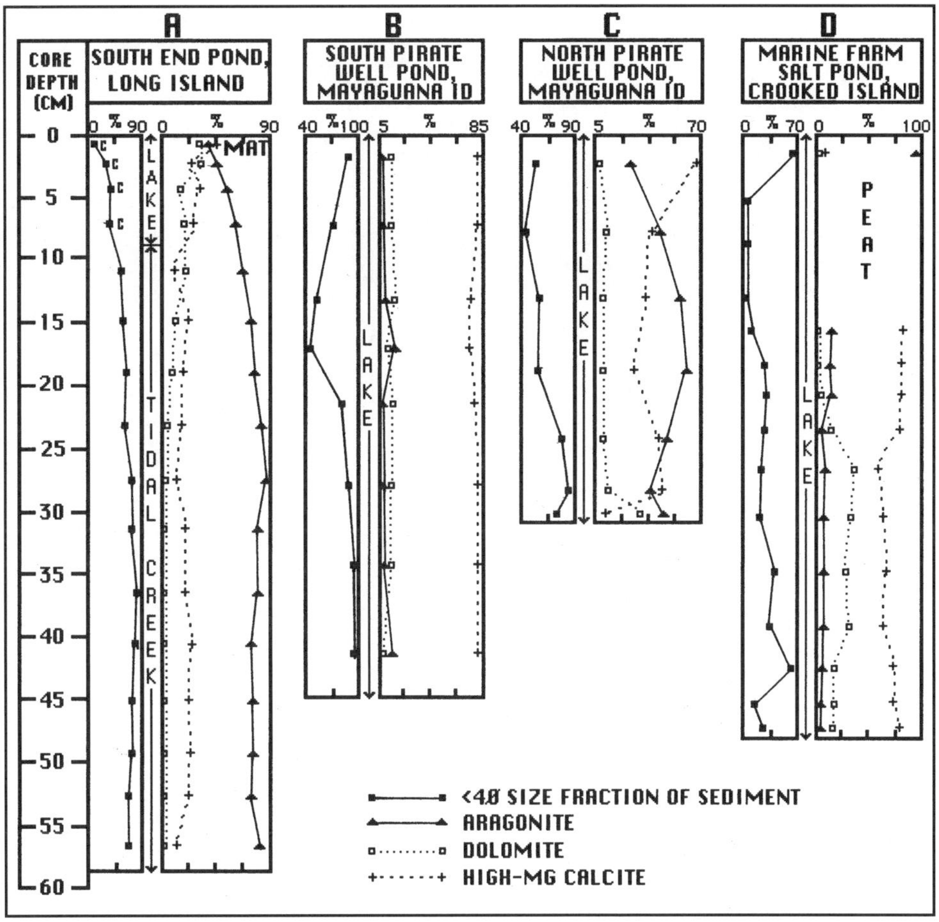

Figure 6. Percentages of dolomite and high-Mg calcite aragonite in <4ϕ fraction and weight % of <4ϕ size fraction versus depth in cores from lakes with near-surface dolomitic laminated sediments or subsurface replacement dolomitization. Environment of deposition, based on benthic ostracod and foraminifera assemblages, is given in column between <4ϕ size fraction and mineralogy percentages. Core locations are given in Table 2. A, Core from west-central side of west-coast pond 1 km north of South End, southernmost Long Island. B, Core from northeasternmost part of southern pond east of Pirate Well, northwesternmost coast of Mayaguana Island. C, Core from northwesternmost coast of Mayaguana Island. D, Core (core 1) from central part of northeast-southwest arm of Marine Farm Salt Pond, northwesternmost Crooked Island.

Pond, Great Exuma Island (Fig. 8C), Williams Town Salt Pond, Little Exuma Island (Fig. 8B), and South Pirate Well Pond, Mayaguana Island (Fig. 6B), preserve similar sequences, as do those from the northernmost part of Northeast Arm Lake, San Salvador Island and Airport Pond, Providenciales Island (Table 2). Identical dolomitized sedimentary sequences from other lakes on Providenciales Island are described by Leaver (1985). Gregg et al. (1992a) and Furman et al. (1993) have described the replacement of high-Mg calcite by submicron dolomite crystals in a core from Salt Pond, San Salvador Island (Table 2). The dolomite occurs below a depth of 25 cm during a period of higher lake salinities. Dolomite formation associated with the shallow interior lakes of Great Inagua Island also has been reported (Miller, 1961; Bubb and Atwood, 1968). A very shallow core from Red Pond, along the northwest coast of Great Inagua Island (Table 2), contains up to 20% dolomite in the <4ϕ size fraction.

GEOCHEMISTRY

Mineralogy

Based on x-ray diffraction patterns of the 015 dolomite reflection, the dolomite is a poorly ordered protodolomite. Samples containing halite show a pronounced shift in the 104 dolomite peak relative to the halite 2.88Å reflection, indicating that these dolomites are Ca-rich. Dolomite stoichiometry calculated from this shift provides molar Ca:Mg ratios of dolomite from surficial crusts averaging 57:43 (seven samples; range % Ca, 51 to 61); these are similar to values reported by Gregg et al.

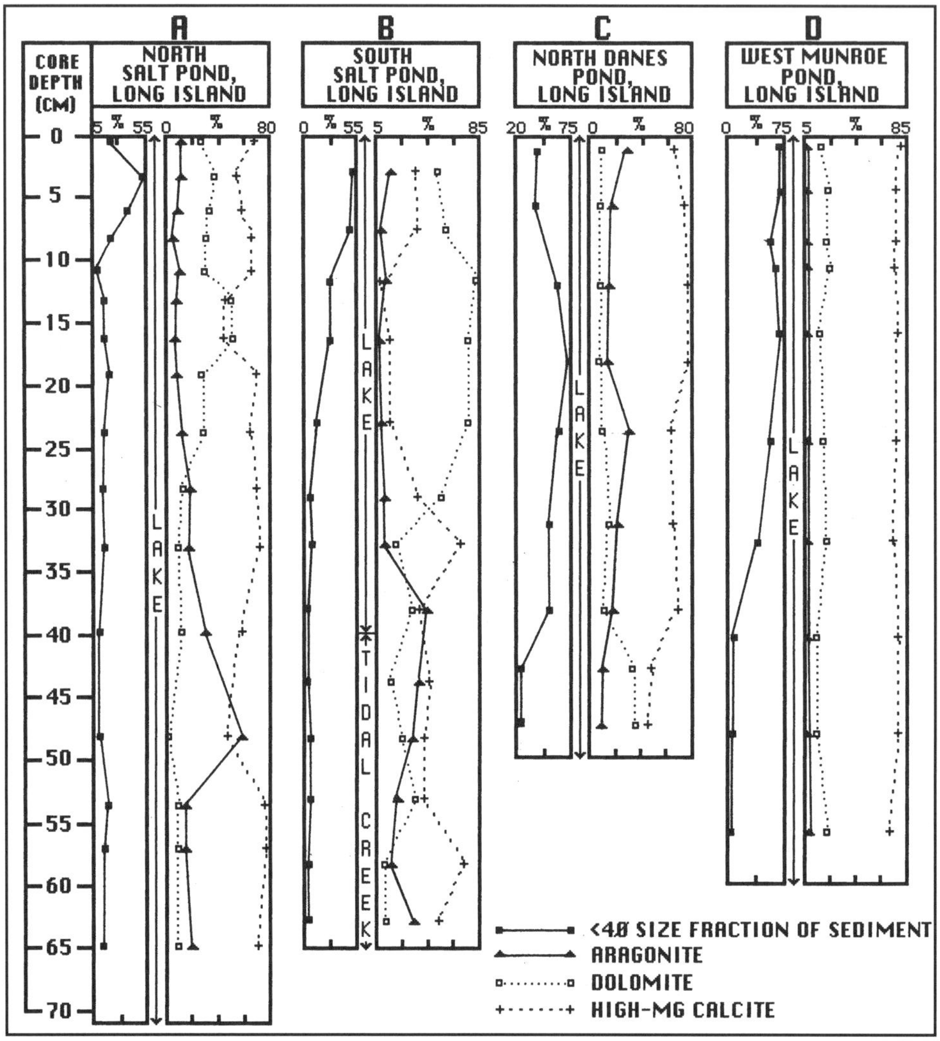

Figure 7. Percentages of dolomite and high-Mg calcite aragonite in <4ϕ fraction and weight % of <4ϕ size fraction versus depth in cores from lakes with subsurface replacement dolomitization. Environment of deposition, based on benthic ostracod and foraminifera assemblages, is given in column between <4ϕ size fraction and mineralogy percentages. Core locations are given in Table 2. A, Core from southeast part of northern pond 2 km north of Salt Pond Settlement, east-central coast of Long Island. B, Core from southeasternmost part of southern pond 1.5 km north-northeast of Salt Pond Settlement, east-central coast of Long Island. C, Core from north-central part of northern pond at Danes, southern Long Island (3.5 km west-southwest of Clarencetown). D, Core from north-central side of pond west of Munroe Pond, southern Long Island (2 km west of Clarencetown).

(1992a) and Mitchell and Horton (this volume). Dolomite molar Ca:Mg ratios of samples from lacustrine cores average 60:40 (seven samples; range % Ca, 59 to 61). Furman et al. (1993) describe submicron-sized poorly ordered Ca-rich dolomite (average molar Ca, 56.3%), which has replaced high-Mg calcite in Salt Pond, San Salvador Island (Table 2). Their published x-ray diffraction patterns are similar to those obtained during this study. The x-ray patterns are also similar to those of micron-sized Ca-rich protodolomite from Bonefish Pond, New Providence Island (Mitchell and Horton, this volume). Oxygen and carbon isotope signatures (see the following section) of surficial dolomitized crusts lacking abundant organics (Fig. 9) are also similar to the values reported by Mitchell and Horton (this volume) for Bonefish Pond. SEM-EDS analysis of dolomites was not available for this study. However, because of the similarities of occurrence, x-ray patterns, and isotope signatures to those of Mitchell and Horton (this volume) and Furman et al. (1993), it is probable that the dolomites described in this study are similar to the submicron- to micron-sized Ca-rich dolomites of Salt and Bonefish Ponds.

Stable isotope data

Fifty-six samples were analyzed for $\delta^{18}O_{PDB}$ and $\delta^{13}C_{PDB}$ in calcite/aragonite–dolomite pairs (Table 4). These data are

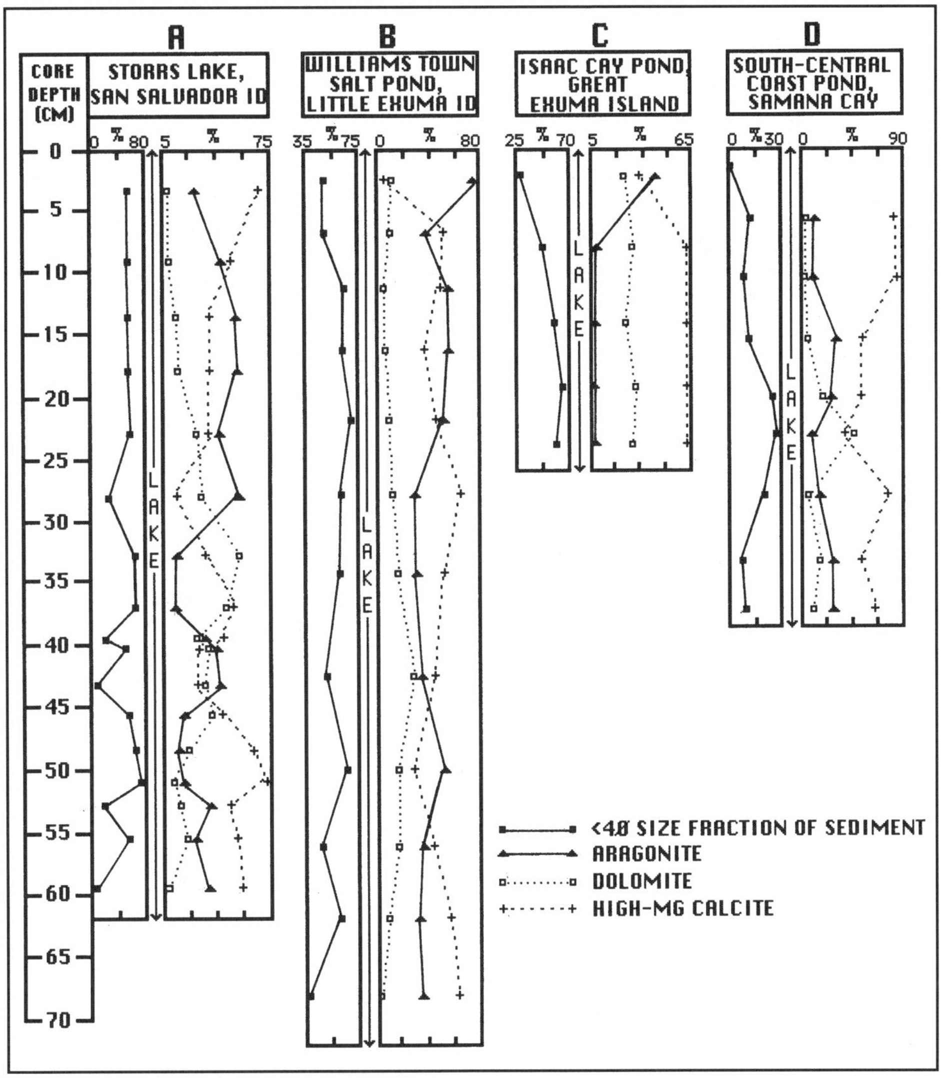

Figure 8. Percentages of dolomite and high-Mg calcite aragonite in <4ϕ fraction and weight % of <4ϕ size fraction versus depth in cores from lakes with subsurface replacement dolomitization. Environment of deposition, based on benthic ostracod and foraminifera assemblages, is given in column between <4ϕ size fraction and mineralogy percentages. Core locations are given in Table 2. A, Core from northernmost Storrs Lake, northeast coast of San Salvador Island. B, Core from east-central part of large salt pond southwest of Williams Town, southeastern Little Exuma Island. C, Core from northeast corner of western pond along coast southeast of Isaac Cay, southeasternmost Great Exuma Island. D, Core from southeastern side of pond along south-central coast of Samana Cay, along east side of pond along south-central coast of Samana Cay, along east side of cape 5.5 km southeast of west end of the island.

plotted in Figure 9. Values for $\delta^{18}O$ in calcite/aragonite range from –0.6 to +3.0‰, with most values falling between –0.5 and +1.5‰. Values for $\delta^{13}C$ range from –9.0 to +4.0‰, but the maximum variation in any single location is less than 3‰. Dolomite $\delta^{18}O$ values fall between –3.2 and +2.9‰, while dolomite $\delta^{13}C$ ranges from –9.6 to +1.7‰. Dolomite in surficial laminates is enriched in ^{18}O by about 1‰ but depleted in ^{13}C by 1 to 2‰ relative to coexisting Mg calcite/aragonite. In core samples $\Delta^{18}O$ ($\delta^{18}O_{dolomite}$ minus $\delta^{18}O_{calcite/aragonite}$) and $\Delta^{13}C$ are highly variable, even within samples from a single location. $\Delta^{18}O$ ranges from –3.8 to +2.6‰ and $\Delta^{13}C$ ranges from –3.3 to +2.2‰.

Although the range of values, especially for $\delta^{13}C$, is large for the entire data set, individual lakes have narrower and somewhat distinctive isotope signatures (Fig. 9). Supratidal crusts (Greys Bight, Red Pond, Wemyss Bight, and Stella Maris Bay, Long Island; Chalk Sound, Providenciales Island) contain carbon with positive $\delta^{13}C$. One intertidal crust sample from Stella Maris Bay, Long Island, contains carbon that is slightly negative. Surficial crusts associated with organic-rich sediments (Middle Fish

TABLE 4. ISOTOPES OF CARBON AND OXYGEN*

Location	Lake	Environment	Depth (cm)	Calcite/Aragonite $\delta^{18}O$	Calcite/Aragonite $\delta^{13}C$	Dolomite $\delta^{18}O$	Dolomite $\delta^{13}C$	$\Delta^{18}O$	$\Delta^{13}C$
Long Island	Carmichael Pond	LM	2-6 a	0.66	-7.24	1.59	-6.30	0.93	0.94
		LM	2-6 b	0.21	-7.19	-0.57	-7.03	-0.78	0.16
		LM	2-6 c			-1.49	-7.20		
		LM	6-13 a	0.19	-5.59	2.24	-4.20	2.05	1.39
		LM	6-13 b			0.07	-5.28		
		LM	13-18	-0.40	-5.26	-1.31	-5.28	-0.91	-0.02
		LM	18-24	0.24	-5.65	0.81	-5.43	0.57	0.22
		LM	24-31 a	0.19	-6.00	-0.59	-5.64	-0.78	0.36
		LM	24-31 b			-1.72	-5.90		
		LM	31-39 a	0.27	-0.02	-0.09	-0.62	-0.36	-0.60
		LM	31-39 b	0.28	0.13				
	McKanns Pond	LM	0-3	2.82	-7.65	2.63	-9.55	-0.19	-1.90
		LM	6-9	-0.14	-9.00	2.48	-8.69	2.62	0.31
		LM	12-15	0.59	-8.05	0.32	-8.22	-0.27	-0.17
		LM	15-17 a	1.09	-7.34	0.47	-7.53	-0.62	-0.19
		LM	15-17 b	1.11	-7.32	0.20	-7.49	-0.91	-0.17
		LM	17-21	0.20	-6.50	1.39	-7.86	1.19	-1.36
		LM	21-24	0.51	-8.43	1.10	-8.35	0.59	0.08
		LM	24-27 a	1.64	-8.82	-0.06	-9.38	-1.70	-0.56
		LM	24-27 b	1.25	-8.91				
		LM	28-31 a	0.16	-7.43	0.03	-7.57	-0.13	-0.14
		LM	28-31 b	0.25	-7.41				
	South End Pond	LM	1-3	0.07	-1.33	-0.20	-4.64	-0.27	-3.31
	Red Pond	SC	Surface	0.46	3.46	1.20	1.49	0.74	-1.97
	Red Pond	LN	3-4	-0.29	3.98	-0.52	1.66	-0.23	-2.32
	South Fish Pond	LM	0-2 a	-0.28	-7.63	-3.20	-7.51	-2.92	0.12
		LM	0-2 b	-0.46	-7.67	-1.18	-7.53	-0.72	0.14
		LM	2-5	-0.32	-7.70	-2.36	-7.35	-2.04	0.35
		LM	5-10	0.02	-7.20	-0.32	-6.43	-0.34	0.77
		LM	14-17	0.68	-6.18	-0.36	-6.96	-1.04	-0.78
		LM	23-26	-0.21	-6.77	-0.83	-7.24	-0.62	-0.47
	Wemyss Bight	SC	Surface	1.93	2.03	2.89	2.01	0.96	-0.02
	Grey Bight	SC	Surface	2.41	3.80	1.64	2.78	-0.77	-1.02
	Stella Maris Bay	SC	Surface	2.82	2.75	1.73	2.97	-1.09	0.22
		TF	0-3	0.59	2.11	0.57	1.12	-0.02	-0.99
	Middle Fish Pond	SC	Surface	-0.08	-3.46	0.59	-4.46	0.67	-1.00
San Salvador Island	Storrs Lake	LN	0-7	0.67	-3.45	-0.29	-3.00	-0.96	0.45
		LN	12-16	1.20	-2.59	-1.69	-3.84	-2.89	-1.25
		LN	20-26	1.06	-3.92	-0.68	-3.75	-1.74	0.17
		LN	26-30	1.85	-4.80	0.63	-3.12	-1.22	1.68
		LN	30-35 a	1.11	-4.79	1.00	-2.59	-0.11	2.20
		LN	30-35 b	1.02	-4.85	1.74	-2.74	0.72	2.11
		LN	40-42	1.14	-4.23	1.01	-2.86	-0.13	1.37
		LN	44-47	1.44	-3.70	0.93	-2.61	-0.51	1.09
		LN	52-54 a	1.63	-2.11	0.02	-2.07	-1.61	0.04
		LN	52-54 b	1.70	-1.98	0.31	-2.28	-1.39	-0.30
		LN	54-57	1.13	-2.83	0.07	-2.62	-1.06	0.21
	Clear Pond	LN	0-3 a	2.42	-4.37	2.07	-5.22	-0.35	-0.85
		LN	0-3 b	2.37	-4.44	2.15	-4.29	-0.22	0.15
Little Exuma Island	Pelican Cays Pond	LM	4-5	0.84	-5.94	-0.23	-6.77	-1.07	-0.83
		LM	17-20	2.98	-5.52	0.91	-6.78	-2.07	-1.26
Rum Cay	Carmichael Pond	LN	0.-2 a	-0.12	-2.41	0.82	-3.12	0.94	-0.71
		LN	0.-2 b	-0.62	-2.52	0.86	-3.15	1.48	-0.63
		LN	4-6	0.51	-2.32	1.49	-2.59	0.98	-0.27
		LN	9-13	0.46	-2.95	1.79	-2.98	1.33	-0.01
Providenciales Island	Chalk Sound	SC	Surface	2.75	1.56	3.03	1.73	0.28	0.17

*All values in per mil relative to PDB. $\Delta^{18}O = \delta^{18}O_{dolomite} - \delta^{18}O_{calcite}$; $\Delta^{13}C = \delta^{13}C_{dolomite} - \delta^{13}C_{calcite}$.

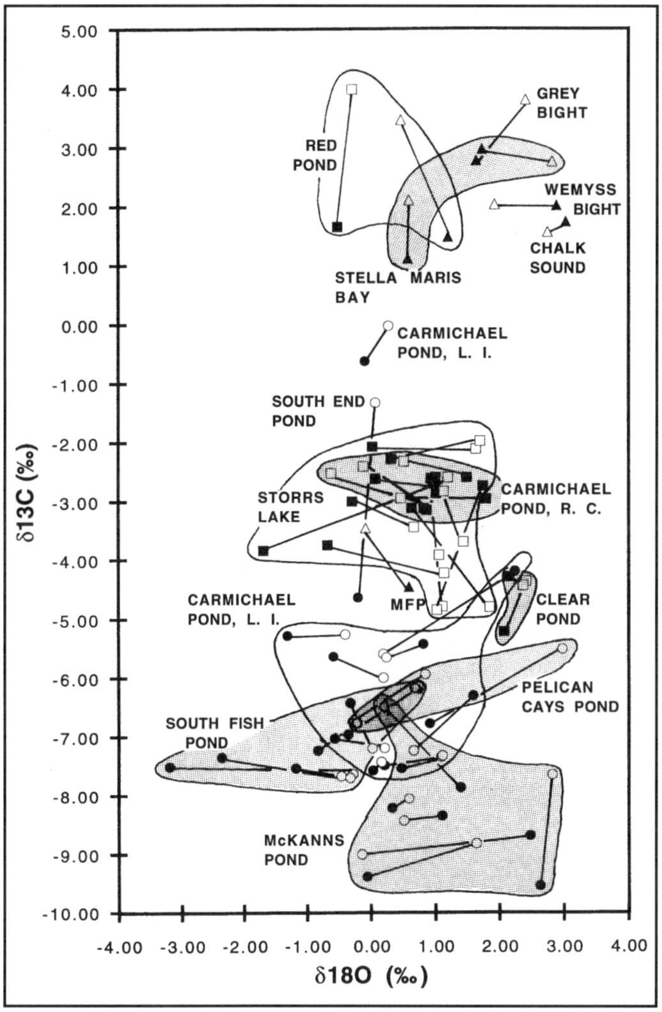

Figure 9. Stable isotopes of carbon and oxygen for coexisting calcite/aragonite–dolomite from Table 4 shown in per mil relative to PDB. Open symbols = calcite/aragonite; solid symbols = dolomite; tie lines connect coexisting calcite/aragonite–dolomite pairs; triangle = surficial crusts and tidal flat sediments (one value); circles = lacustrine sediments in cores taken below well-developed cyanobacterial mats; squares = lacustrine sediment in cores from sequences without well-developed cyanobacterial mats. MFP = Middle Fish Pond. Stipple pattern highlights areas of overlap. See Tables 1 and 4 for locations of lakes.

Pond, Long Island; Carmichael Pond, Rum Cay) contain carbon with $\delta^{13}C$ values in the –2.4 to –4.5‰ range. A sample from Red Pond, Long Island, containing little organic material, has positive carbon and a $\Delta^{13}C$ value of –2.32‰. Lacustrine sediments containing abundant organics, but without well-developed cyanobacterial mats (Storrs Lake and Clear Pond, San Salvador Island; Carmichael Pond, Rum Cay) contain carbon with $\delta^{13}C$ values between –2.07 and –5.22‰. A core from the center of Storrs Lake contains carbon with very similar $\delta^{13}C$ values (–1.3 to –5.1) (Paull et al., 1992). Sediments from lakes with well-developed cyanobacterial mats (Carmichael Pond, McKanns Pond and South Fish Pond, Long Island; Pelican Cays Pond, Little Exuma Island) contain carbon with $\delta^{13}C$ ranging from –4.64 to –9.55‰, with two exceptions. One sample (from sediment immediately below the living cyanobacterial mat) from South End Pond, Long Island, which contains abundant sediment derived from beach erosion, has calcite/aragonite with a $\delta^{13}C$ –1.33‰ and a $\Delta^{13}C$ of –3.31‰, which is the largest value for $\Delta^{13}C$ observed during this study. A single sample (the lowest sample analyzed from the core) from Carmichael Pond, Long Island, has a $\delta^{13}C$ value that is only slightly negative; this sample contains very high levels of aragonite that appear to have been derived from the reworking of underlying tidal creek sediments.

Water chemistry

Eighty-four samples were submitted for analysis of Ca^{2+}, Mg^{2+}, Cl^-, and SO_4^{2-}. Table 5 shows quantitative anion and semiquantitative cation data for a number of lacustrine and coastal samples, while Table 6 shows quantitative data from Red Pond, Long Island. Values for Cl^- and SO_4^{2-} range from about one-half to 10 times that of local normal sea water. Quantitative values for Ca^{2+} range from slightly greater to approximately three times normal seawater concentrations, while Mg^{2+} concentrations range from normal seawater levels to about six times normal concentrations. No trends versus depth could be established for any species due to the lack of a sufficient number of pore-water samples from any single locality. Halite was not normally observed, so Cl^- is considered conservative; Ca^{2+}, Mg^{2+}, and SO_4^{2-} all increase as Cl^- increases. However, when Mg^{2+}, Ca^{2+}, and SO_4^{2-} are normalized against Cl^-, several trends can be seen in both pore and surface waters (Fig. 10). Ratios of Mg^{2+}/Cl^- and Ca^{2+}/Cl^- both decrease, while Mg^{2+}/Ca^{2+} increases as SO_4^{2-}/Cl^- decreases.

INTERPRETATIONS

The geochemical environment for the formation of dolomite can be inferred from the isotopic signature of the sediments. Two significantly different environments of formation are indicated by the isotopic data.

1. Surficial crusts contain carbon with positive $\delta^{13}C$ (Fig. 9) except for an organically rich sample from Middle Fish Pond, Long Island. Oxygen in these sediments is slightly enriched in ^{18}O. These values are similar to carbon and oxygen isotopes found in sediments of the Bonefish Pond tidal creek system on New Providence Island (Mitchell and Horton, this volume). These values are consistent with precipitation from seawater HCO_3^{2-}, which has been slightly enriched in ^{18}O by evaporation. Such a scenario fits with the known hydrologic regime, in which both tidal creek and lake waters have salinities elevated above that of local coastal seawater. $\Delta^{18}O$ for these samples ranges from –1.09 to +0.96‰, indicating that dolomite is not in equilibrium with calcite/aragonite (Land, 1980), but has incorporated isotopically light oxygen. Positive $\delta^{13}C$ values are also consistent with an origin from marine HCO_3^{2-} (Land, 1980). The

TABLE 5. Ca^{2+}, Mg^{2+}, Cl^{-}, AND SO_4^{2-} VALUES FOR WATER SAMPLES*

Island	Location	Sample	Type	Depth (cm)	Ca^{2+} (ppm)	Mg^{2+} (ppm)	Cl^{-} (ppm)	SO_4^{2-} (ppm)	Salinity
Long Island	South Fish Pond	LISFP210	P	17.8	2,800	3,419	30,800	4,462	30
		LISFP420	P	12.7	2,800	4,639	31,191	5,652	55
		LISFP434	S	0	4,000	5,616	42,231	7,000	74
	Middle Fish Pond	LIMFP465	M	3.8	2,000	4,884			
		LIMFP470	P	15.2	1,200	1,465	17,255	4,015	30
		LIMFP475	M	0	2,800	4,884	86,617	8,633	153
		LIMFP75	S	0	1,200	1,953	23,697	3,007	42
		LIMFP413	P	5.1	2,000	2,198	38,239	5,070	67
		LIMFP431	P	10.2	1,200	2,930	23,682	3,255	42
		LIMFP448	P	8.9	2,400	4,395	64,138	7,890	113
		LIMFP449	S	0	3,600	10,500	75,751	8,819	133
		LIMFP454	P	10.2	1,200	2,198	29,067	4,479	51
		LIMFP459	P	8.9	1,600	3,174	48,522	5,666	85
		LIMFP464	S	0	18,400	3,174	140,555	17,270	247
	North Fish Pond	LINFP303	S	0	3,200	6,105	35,486	5,152	63
		LINFP408	P	7.6	1,600	1,953	19,233	2,101	34
		LINFP416	S	0	1,200	1,456	11,221	1,714	20
		LINFP422	S	0	1,200	977	11,199	3,052	20
		LINFP437	P	7.6	2,800	5,128	42,145	8,896	74
		LINFP441	M	0			31,808	4,661	56
		LINFP450	S	0	3,200	5,616	80,336	11,156	141
		LINFP468	S	0	1,600	2,686	33,943	5,658	60
	Red Pond	LIRP176	P	19.5	3,200	4,883	51,005	6,305	90
		LIRP187	P	23			22,849	3,834	40
		LIRP252	P	28	1,600	2,442	23,761	3,681	42
		LIRP282	P	15	2,800	5,372	49,422	4,927	87
		LIRP287	P	28	2,400	3,907	30,700	4,674	54
		LIRP405	P	28	2,000	4,395	55,479	7,793	98
		LIRP410	P	35	2,000	4,639	56,357	8,282	99
		LIRP418	S	0	3,200	8,058	104,065	14,490	183
		LIRP429	M	0	2,800	3,419	38,089	7,280	7,194
		LIRP430	S	0	3,200	7,081	Too large	11,194	12,214
		LIRP443	S ocean	0	15,600	13,186	141,946	28,408	250
		LIRP446	S	0	800	1,953	17,988	2,392	32
	McKanns Pond	LIMP73	S	0	1,200	1,465	18,105	2,427	32
	South End Pond	LISEP412	S	0	3,200	5,616	76,022	9,317	134
		LISEP439	P	7.6	3,200	6,349	80,921	10,496	142
	North End Pond	CINEP423	S	0	2,800	3,663	55,413	7,985	98
	Cape Verde Pond	LICVP258	M	0	2,000	2,686	28,877	4,837	51
		LICVP	S	0	2,400	4,639	58,141	7,345	102
	Millerton	LIMSP215	S	0	2,000	4,151	43,667	7,361	77
	Settlement	LIMSP253	S	0	2,800	2,189	31,269	5,884	55
	Ponds	LIMSP300	S	0	1,200	1,221	15,442	2,328	27
	Gunther's Pond	LISP433	S	0			25,777	3,949	45
	Salt Pond	LICSP26	S	0	12,400	8,791	185,966	20,995	327
		LICSP41	P	1	2,400	6,593	94,737	9,055	167
		LICSP474	S	0	11,600	6,104	129,389	13,524	228
	Carmichael Pond	LICP44	S	0	3,200	9,279	152,947	14,140	269
		LICP74	P	5.1	3,200	7,325	102,527	11,302	181
	Columbus	LICHP411	P	7.6	2,400	3,174	43,464	7,782	77
	Harbor Pond	LICHP444	S	0	2,800	4,151	30,633	8,028	54
		LICHP453	P	12.7	2,000	3,419	42,124	7,438	74
		LICHP456	S	0	2,800	3,419	43,563	9,020	77
		LICHP487	S	0	2,800	2,930	34,461	7,222	61
	Stella Maris Bay	LISM36	P	2.2	1,600	3,419	35,601	4,146	63
		LISM43	M	0	2,000	2,930	42,943	5,521	76
		LISM45	P	17.8	1,600	2,686	39,077	4,444	69
		LISM52	S ocean	0	1,200	2,198	26,761	3,390	47
		LISM55	S	0	2,000	3,419	44,428	5,676	78

TABLE 5. Ca^{2+}, Mg^{2+}, Cl^-, AND SO_4^{2-} VALUES FOR WATER SAMPLES* (continued)

Island	Location	Sample	Type	Depth (cm)	Ca^{2+} (ppm)	Mg^{2+} (ppm)	Cl^- (ppm)	SO_4^{2-} (ppm)	Salinity
		LISM57	S	0	800	1,709	18,170	2,366	32
		LISM58	P	11.4	2,000	2,930	35,471	3,926	62
	Burnt Ground TC	LIBGC263	S	0	3,600	5,182	36,875	5,677	65
	Newton Cay	LINCC190	S	0			17,475	3,061	31
	Tidal Creek	LINCC421	P	0	1,600	1,709	23,618	4,384	42
		LINCC440	S	0	1,200	1,953	23,417	4,350	41
Rum Cay	Carmichael Pond	RKCP2	S	0	2,400	10,500	155,697	14,308	274
		RKCP234	P	10.2	2,400	3,419	28,990	3,587	51
		RKCP250	P	1			51,566	6,958	91
		RKCP296	M	0	14,000	7,081	96,017	13,927	169
		RKCP406	P	11.4	2,000	2,198	32,460	3,212	57
		RKCP445	P	5.1	1,200	2,930	25,969	3,347	46
		RKCP447	P	15.2	1,600	2,442	32,602	5,148	57
Conception Island	Booby Cay Pond	CIBCP227	S	0	1,040	1,416	17,812	2,112	31

*S = surface water sample; P = pore water sample; M = water from cyanobacterial mat (Mg^{2+} and Ca^{2+} determined by semiquantitative methods; SO_4^{2-} and Cl^- determined by quantitative methods).

TABLE 6. QUANTITATIVE WATER CHEMISTRY FROM RED POND, LONG ISLAND*

Island	Location	Sample	Type	Depth (cm)	Ca^{2+} (ppm)	Mg^{2+} (ppm)	Cl^- (ppm)	SO_4^{2-} (ppm)	Salinity
Long Island	Red Pond	LP1	S	0	450	1,220	26,656	3,880	
		LP4	P	35	740	3,831	35,341	6,468	
		LP7	P	32	1,081	3,977		5,973	
		LP0	S ocean	0	480	1,293	16,652	2,355	
		LP5	P	45	1,050	4,240	18,078	8,513	
		LP3a	Swi	0	460	1,574	16,212	2,317	
		LP2	Swi	0	1,180	7,771	56,851	13,699	
		LP3	S	0	500	1,476	18,086	2,534	
		LP11	P	30	580	2,220		4,231	
		LP1a	Swi	0	480	1,452	20,756	3,055	
		LP10	S	0	820	9,504		16,183	

*Symbols same as in Table 5. M = water from cyanobacterial mats; P = pore water; S = lake water; Swi = sediment-water interface.

dolomitized laminites appear to be somewhat analogous to the supratidal crusts described from Andros Island (Hardie, 1977; Gebelein et al., 1980; Lasemi et al., 1989), which also contain negative carbon (Lyons et al., 1989), in that they are forming as a cement without preferential replacement of either high-Mg calcite or aragonite.

2. In contrast, cored microbialites have lighter carbon and generally lighter oxygen (Fig. 10). Each lake has a distinctive isotope signature, reflected primarily in the $\delta^{13}C$ values, which correlates with the amount of organic material present versus that which is inferred to have been oxidized. Calcite/aragonite and dolomite have similar carbon signatures in each lake, with $\Delta^{13}C$ generally between –1 and +1‰. Among the few samples where $\Delta^{13}C$ falls outside these limits, the dolomite is lighter in all but one instance. Negative values for $\delta^{13}C$ indicate that a significant portion of the carbonate was derived from decay of organic material in the sediments (Behrens and Frishman, 1971; Burns et al., 1988; Andrews, 1991; Sass et al., 1991). This is supported by field observations. Samples from microbialites below living cyanobacterial mats, in which the organic material is largely oxidized, contain the lightest carbon ($\delta^{13}C$, –5 to –10‰). Sediments in lakes lacking well-developed cyanobacterial mats contain less negative carbon ($\delta^{13}C$, –1 to –5‰). This is in agreement with the data of Behrens and Frishman (1971), who determined that cyanobacterial mats had lighter carbon (average $\delta^{13}C$, –16‰) than nonmat organic carbon in sediments. The values for $\Delta^{18}O$ in these samples is variable and may be both positive or negative within a given lake.

The values, however, are generally less than estimates for calcite-dolomite equilibrium values in the +2 to +4‰ range (Land, 1980) and suggest an oxygen source in addition to the seawater. A likely possibility is the influx of meteoric water. South Fish Pond, Long Island, and Storrs Lake, San Salvador Island, have predominantly negative values for $\Delta^{18}O$. This could

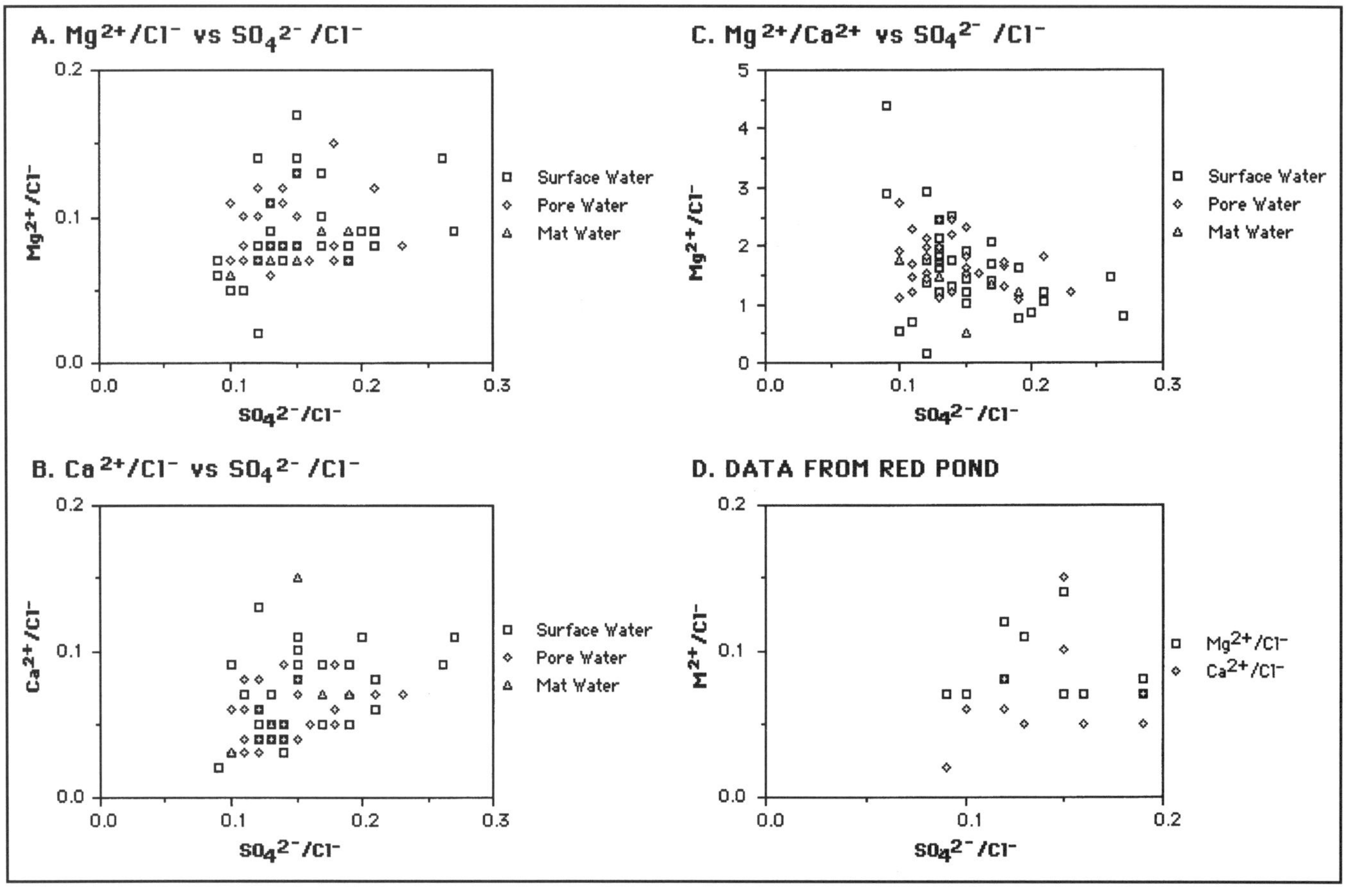

Figure 10. A, Variation of Mg^{2+}/Cl^- versus SO_4^{2-}/Cl^-. B, Variation of Ca^{2+}/Cl^- versus SO_4^{2-}/Cl^-. C, Variation of Mg^{2+}/Ca^{2+} versus SO_4^{2-}/Cl^- (Mg^{2+} and Ca^{2+} determined by semiquantitative methods; SO_4^{2-} and Cl^- by quantitative methods). D, Quantitatively determined Mg^{2+}/Cl^- and Ca^{2+}/Cl^- versus SO_4^{2-}/Cl^-.

be caused by incorporation of CO_2 liberated during sulfate reduction and organic decay (Shinn et al., 1965; Andrews, 1991; Paull et al., 1992).

Lyons et al. (1984) and Chafetz and Buczynski (1992) have shown that carbonates in mats precipitate as cements due to liberation of CO_2 during bacterially mediated organic oxidation accompanied by sulfate reduction. Chelated Mg^{2+} is released by this process and incorporated into the sediments (Andrews, 1991; Canfield and Des Marais, 1991; Burns and Swart, 1992). The water chemistry data generated during this study must be used with caution due to the semiquantitative nature of much of the data and the lack of pore water samples corresponding to samples analyzed mineralogically and isotopically. However, several trends support dolomite formation during sulfate reduction and organic decay. Pore waters show a correlation between the relative amounts of Mg^{2+}, Ca^{2+}, and SO_4^{2-} (Fig. 10). Both Ca^{2+} and Mg^{2+} decrease as SO_4^{2-} decreases, but the decrease in Ca^{2+} is more pronounced and the Mg^{2+}/Ca^{2+} ratio increases with decreasing SO_4^{2-}. There are several possible explanations for this. Mg^{2+} is more abundant than Ca^{2+}, but they are incorporated into dolomite in equal amounts so that a larger percentage of the Ca^{2+} is being incorporated into the cement. Also, Mg^{2+} released by the decay of cyanobacterial mats (Carballo et al., 1987) raises the Mg^{2+}/Ca^{2+} ratio, thus providing additional thermodynamic drive for dolomite formation. However, the same geochemical trends are present in samples from lake waters, so it is possible that other processes are acting here as well.

TIMING OF DOLOMITIZATION

Tidal flat dolomitization

Dolomitized supratidal crusts previously have been reported from the tidal flats of Abaco and Andros Islands. Radiocarbon dates are oldest for buried crusts (2200 B.P.); surficial crust dolomite samples provide dates from 0 to 160 B.P. (Shinn et al., 1965). The period of formation of dolomitized supratidal crusts from Sugarloaf Key, Florida has been estimated to have been between 160 and 1420 B.P. (Carballo et al., 1987). Supratidal dolomitized crusts from Ambergris Cay, Belize, have been dated at 845 to 2925 B.P., but are believed to be still precipitating (Mazzullo et al., 1987; Gregg et al., 1992b). The supratidal dolomitized crusts and shallow subsurface dolomitized zones along Stella Maris Bay reported here have not be radiocarbon-dated. However, the surficial crusts are present along a former beach face situated at the edge of a large Native American (Taino) site containing Spanish artifacts (Keegan and Mitchell, 1984). The site was therefore occupied until the late 1490s to early 1500s. In the past 500 yr, longshore transport has rapidly filled in the bay and extended the tidal flats seaward. The supratidal crusts no longer appear to be forming, but probably were as recently as 500 yr ago. The cored subsurface dolomitized zones of the Stella Maris Bay tidal flat (Fig. 3B) have formed in sediments deposited during the past 500 yr.

Tidal creek dolomitization

Mitchell and Horton (this volume) describe the extensive dolomitization of surficial crusts and subsurface sediments of the Bonefish Pond tidal creek system of New Providence Island, Bahamas. Reconstructions of coastline change and the depositional environments of the tidal creek indicate that the dolomitized crusts and sediments were formed during the past 100 yr. The surficial sediments of the Chalk Sound tidal creek system, Providenciales Island, also contain dolomite (Fig. 3D). This dolomite could have been very recently precipitated in the sediment and/or be allogenic (derived from the break-up of dolomitized supratidal crusts).

Lacustrine dolomitization

Radiocarbon dating of Bahamian lacustrine sediments indicates a typical average depositional rate of 1.5 to 4.5 cm/100 yr. If a very conservative estimate of 1.5 cm/100 yr is used to interpret the ages of the dolomite present in the cores shown in Figures 4 through 8, it is apparent that very significant levels of dolomite have formed within the past several hundred years. This is especially true if the living cyanobacterial mats in the upper few centimeters of the cores in Figure 5 and 6 are ignored due to their rapid decomposition rate. Since a period of only about 1 yr is required for the complete decomposition of cyanobacterial mats, the underlying microbialitic sediments can be considered to be very recent (Doemel and Brock, 1977). Some sediments directly underlying the living mats, as well as those present at the sediment-water interface of lakes without well-developed cyanobacterial mats, already contain high levels of dolomite. For example, the surficial sediments in Carmichael Pond, Long Island, contain 35% dolomite in the $<4\phi$ size fraction (Fig. 5B), while over 40% of the same size fraction is dolomite in the uppermost sediments of South Salt Pond, Long Island (Fig. 7B).

Archaeologic and historic evidence indicates that Red Pond, Long Island, was a tidal creek system that has been closed off to form a lake during the past 200 yr (Mitchell and Keegan, 1987). Cores encounter tidal creek sediments at depths of about 30 cm below the present sediment surface. This suggests a very high depositional rate of approximately 15 cm/100 yr for the cored lake sediments at the southern part of Red Pond. However, this can be explained by evidence that the low coastal dunes separating the pond and the open ocean have been breached during major storms. Significant amounts of beach and dune sand and large pieces of abraded coral have been transported into the pond system producing the high depositional rate. The upper half of the lacustrine sequence contains dolomite and is capped by an extensive surficial gypsiferous-dolomitic crust in the southern part of the pond (Fig. 2B). Thus, dolomitization has occurred since the tidal creek inlet was closed, 200 yr ago, to form Red Pond.

CONCLUSIONS

1. The investigation of over 150 shallow cores and numerous surficial crusts from tidal flat, tidal creek, and lacustrine environments of the central and southern Bahama Archipelago indicates that modern dolomite formation is widespread.
2. The dolomite is a poorly ordered Ca-rich protodolomite.
3. Dolomite precipitation is occurring in surficial crusts with low organic content, in subsurface sediments with intermediate organic content, and in near-surface microbialites with high organic content.
4. The dolomite appears to be forming currently in all three depositional environments simultaneously.
5. Dolomite has formed extensively in sediments and crusts formed during the past 200 yr.
6. Dolomites from each environment have distinctive carbon isotope signatures that reflect the amount and type of organic carbon in the sediments.

ACKNOWLEDGMENTS

This manuscript was improved significantly by the very constructive reviews of J. E. Andrews, Jay M. Gregg, and Brian White. Steven Harris, Kern County Water Agency, arranged for the analysis of water samples. The field research was supported by grants from the Center for Field Research (Watertown, Massachusetts) and Foundation for Field Research (Alpine, California).

REFERENCES CITED

Andrews, J. E., 1991, Geochemical indicators of depositional and early diagenetic facies in Holocene carbonate muds, and their preservation potential during stabilization: Chemical Geology, v. 93, p. 267–289.

Behrens, E. W., and Frishman, S. A., 1971, Stable carbon isotopes in blue-green algal mats: Journal of Geology, v. 79, no. 1, p. 94–100.

Blatt, H., Middleton, G., and Murray, R., 1972, Origin of sedimentary rocks: Englewood Cliffs, New Jersey, Prentice-Hall, 634 p.

Bourrouilh–Le Jan, F. G., 1978, Rôle des ouragans et des cyclones tropicaux sur la sédimentation carbonatée: La plaine d'estran de l'ouest d'Andros, Bahama. Interférences de la climatologie, de l'hydrologie et de la diagenèse: Comptes Rendus de l'Académie des Sciences, Paris, v. 287, p. 907–910.

Bubb, J. N., and Atwood, D. K., 1968, Recent dolomitization of Pleistocene limestones by hypersaline brines, Great Inagua Island, Bahamas: American Association of Petroleum Geologists Bulletin, v. 52, p. 522.

Burne, R. V., and Moore, L. S., 1987, Microbialites: Organosedimentary deposits of benthic microbial communities: Palaios, v. 2, p. 241–254.

Burns, S. J., and Swart, P. K., 1992, Diagenetic processes in Holocene carbonate sediments: Florida Bay mudbanks and islands: Sedimentology, v. 39, p. 285–304.

Burns, S. J., Baker, P. A., and Showers, W. J., 1988, The factors controlling the formation and chemistry of dolomite in organic-rich sediments: Miocene Drakes Bay Formation, California, *in* Shukla, V. and Baker, P. A., eds., Sedimentology and Geochemistry of Dolostones: Society of Economic Paleontologists and Mineralogists Special Publication 43, p. 41–52.

Canfield, D. E., and Des Marais, D. J., 1991, Aerobic sulfate reduction in microbial mats: Science, v. 251, p. 1471–1473.

Carballo, J. D., Land, L. S., and Miser, D. E., 1987, Holocene dolomitization of supratidal sediments by active tidal pumping, Sugarloaf Key, Florida: Journal of Sedimentary Petrology, v. 57, p. 153–165.

Chafetz, H. S., and Buczynski, C., 1992, Bacterially induced lithification of microbial mats: Palaios, v. 7, p. 277–293.

Cornee, A., Dickman, M., and Busson, G., 1992, Laminated cyanobacterial mats in sediments of solar salt works: Some sedimentological implications: Sedimentology, v. 39, p. 599–612.

Dawans, J. M., and Swart, P. K., 1988, Textural and geochemical alternations in late Cenozoic Bahamian dolomites: Sedimentology, v. 35, p. 385–403.

Doemel, W. N., and Brock, T. D., 1977, Structure, growth, and decomposition of laminated algal-bacterial mats in alkaline hot springs: Applied and Environmental Microbiology, v. 34, p. 433–452.

Furman, F. C., Woody, R. E., Rasberry, M. A., Keller, D. J., and Gregg, J. M., 1993, Carbonate and evaporite mineralogy of Holocene (<1900 RCYBP) sediments at Salt Pond, San Salvador Island, Bahamas: Preliminary study, *in* Proceedings, Sixth Symposium on the Geology of the Bahamas: San Salvador, Bahamian Field Station, p. 47–54.

Gebelein, C. D., Steinen, R. P., Garrett, P., Hoffman, E. J., Queen, J. M., and Plummer, L. N., 1980, Subsurface dolomitization beneath the tidal flats of central West Andros Island, Bahamas, *in* Zenger, D. H., Dunham, J. B., and Ethington, R. L., eds., Concepts and Models of Dolomitization: Society of Economic Paleontologists and Mineralogists Special Publication 28, p. 31–49.

Goldsmith, J. R., and Graf, D. L., 1958, Relations between lattice constants and composition of the Ca-Mg carbonates: American Mineralogist, v. 43, p. 84–101.

Goldsmith, J. R., Graf, D. L., and Heard, H. C., 1961, Lattice constants of the calcium-magnesium carbonates: American Mineralogist, v. 46, p. 453–457.

Gregg, J. M., Furman, F. C., Woody, R. E., Rasberry, M. A., and Keller, D. J., 1992a, Evidence for a step-wise nucleation of calcian dolomite in sediments from an evaporite pond, San Salvador Island, Bahamas: Geological Society of America Abstracts with Programs, v. 24, p. A106.

Gregg, J. M., Howard, S. A., and Mazzullo, S. J., 1992b, Early diagenetic recrystallization of Holocene (<3000 years old) peritidal dolomites, Ambergris Cay, Belize: Sedimentology, v. 39, p. 143–160.

Grey, K., Moore, L. S., Burne, R. V., Pierson, B. K., and Bauld, J., 1990, Lake Thetis, western Australia: An example of saline lake sedimentation dominated by benthic microbial processes: Australian Journal of Marine and Freshwater Research, v. 41, p. 275–300.

Hardie, L. A., 1977, Algal structures in cemented crusts and their environmental significance, *in* Hardie, L. A., ed., Sedimentation on the Modern Carbonate Tidal Flats of Northwest Andros Island, Bahamas: Baltimore, Maryland, Johns Hopkins University Press, p. 166–183.

Hardie, L. A., 1986, Ancient carbonate tidal-flat deposits: Colorado School of Mines Quarterly, v. 81, p. 37–57.

Keegan, W. F., and Mitchell, S. W., 1984, The archaeological survey of Long Island, Bahamas: Final report: PRIDE Foundation, Publications in Caribbean Science, no. 1, 106 p.

Krumbein, W. E., and Cohen, Y., 1977, Primary production, mat formation and lithification: Contribution of oxygenic and facultative anoxygenic cyanobacteria, *in* Flugel, E., ed., Fossil Algae: Recent Results and Developments: Berlin, Germany, Springer-Verlag, p. 37–56.

Krumbein, W. E., Cohen, Y., and Shilo, M., 1977, Solar lake (Sinai). 4. Stromatolitic cyanobacterial mats: Limnology and Oceanography, v. 22, p. 635–656.

Kwolek, J. M., 1984, Holocene deposition of a multilayered carbonate sequence in Reckley Hill Settlement Pond, San Salvador Island, Bahamas, *in* Teeter, J. W., ed., Proceedings, Second Symposium on the Geology of the Bahamas: San Salvador, CCFL Bahamian Field Station, p. 27–39.

Land, L. S., 1980, The isotopic and trace element geochemistry of dolomite: The state of the art, *in* Zenger, D. H., Dunham, J. B., and Ethington, R. L., eds., Concepts and Models of Dolomitization: Society of Economic Paleontologists and Mineralogists Special Publication 28, p. 87–110.

Laporte, L. F., 1967, Carbonate deposition near mean sea-level and resultant facies mosaic: Manlius Formation (Lower Devonian) of New York State: American Association of Petroleum Geologists Bulletin, v. 51, p. 73–101.

Lasemi, Z., Boardman, M. R., and Sandberg, P. A., 1989, Cement origin of supratidal dolomite, Andros Island, Bahamas: Journal of Sedimentary Petrology, v. 59, p. 249–257.

Leaver, J., 1985, Sedimentology, mineralogy, and pore water chemistry of schizohaline pond sediments, Turks and Caicos Islands, British West Indies [M.S. thesis]: Durham, North Carolina, Duke University, 76 p.

Lumsden, D. L., and Chimahusky, J. S., 1980, Relationship between dolomite nonstoichiometry and carbonate facies parameters, *in* Zenger, D. H., Dunham, J. B., and Ethington, R. L., Concepts and Models of Dolomitization: Society of Economic Paleontologists and Mineralogists Special Publication no. 28, p. 123–127.

Lyons, W. B., Long, D. T., Hines, M. E., Gaudette, H. E., and Armstrong, P. B., 1984, Calcification of cyanobacterial mats in Solar Lake, Sinai: Geology, v. 12, p. 623–626.

Mann, C. J., and Nelson, W. M., 1989, Microbialitic structures in Storr's Lake, San Salvador Island, Bahama Islands: Palaios, v. 4, p. 287–293.

Mazzullo, S. J., Reid, A. M., and Gregg, J. M., 1987, Dolomitization of Holocene Mg-calcite supratidal deposits, Ambergris Cay, Belize: Geological Society of America Bulletin, v. 98, p. 224–231.

McCrea, J. M., 1950, On the isotope chemistry of carbonates and a paleotemperature scale: Journal of Chemical Physics, v. 18, p. 849–857.

Miller, D. N., Jr., 1961, Early diagenetic dolomite associated with salt extraction process, Inagua, Bahamas: Journal of Sedimentary Petrology, v. 31, p. 473–476.

Milliman, J. D., 1974, Marine carbonates: New York, Springer-Verlag, 375 p.

Mitchell, S. W., 1984, Late Holocene tidal creek-lake transitions, Long Island, Bahamas, *in* Teeter, J. W., ed., Proceedings Addendum, Second Symposium on the Geology of the Bahamas: San Salvador, CCFL Bahamian Field Station, p. 1–28.

Mitchell, S. W., 1987a, Sedimentology of Pigeon Creek, San Salvador Island, Bahamas, *in* Curran, H. A., ed., Proceedings, Third Symposium on the Geology of the Bahamas: San Salvador, CCFL Bahamian Field Station, p. 215–230.

Mitchell, S. W., 1987b, Surficial geology of Rum Cay, Bahama Islands, *in* Curran, H. A., ed., Proceedings, Third Symposium on the Geology of the Bahamas: San Salvador, CCFL Bahamian Field Station, p. 231–241.

Mitchell, S. W., and Horton, R. A., Jr., 1991, Protodolomite precipitation in modern lacustrine environments, Bahamas: Geological Society of America Abstracts with Programs, v. 23, p. 105.

Mitchell, S. W., and Keegan, W. F., 1987, Reconstruction of the coastlines of the Bahama Islands in 1492: American Archaeology, v. 6, p. 88–96.

Mitchell, S. W., and Sigler, M. E., 1989, Twentieth century sedimentological development of Bonefish Pond, New Providence Island, Bahamas, *in*

Mylroie, J. E., ed., Proceedings, Fourth Symposium on the Geology of the Bahamas: San Salvador, Bahamian Field Station, p. 221–234.

Mitchell, S. W., Buening, N., Baldwin, J. N., Jr., and Westell, B., 1989, Holocene depositional history of Conception Island, Bahamas, *in* Mylroie, J. E., ed., Proceedings, Fourth Symposium on the Geology of the Bahamas: San Salvador, Bahamian Field Station, p. 209–220.

Morrow, D. W., 1982, Diagenesis 2. Dolomite—Part 2: dolomitization models and ancient dolostones: Geoscience Canada, v. 9, p. 95–107.

Neumann, C. A., Bebout, B. M., McNeese, L. R., Paull, C. K., and Paerl, H. A., 1989, Modern stromatolites and associated mats: San Salvador, Bahamas, *in* Mylroie, J. E., ed., Proceedings, Fourth Symposium on the Geology of the Bahamas: San Salvador, Bahamian Field Station, p. 235–251.

Neumann, C. A., Paull, C., Zebielski, V., and Bebout, B., 1991, Modern hypersaline stromatolites of San Salvador, Bahamas, and associated sediments: Geological Society of America Abstracts with Programs, v. 23, p. 108.

Paull, C. K., Neumann, A. C., Bebout, B., Zabielski, V., and Showers, W., 1992, Growth rate and stable isotope character of modern stromatolites from San Salvador, Bahamas: Palaeogeography, Palaeoclimatology, Paleoecology, v. 95, p. 335–344.

Pierson, B. J., 1982, Cyclic sedimentation, limestone diagenesis and dolomitization in Upper Cenozoic carbonates of the southeastern Bahamas [Ph.D. dissertation]: Coral Gables, Florida, University of Miami, 343 p.

Royse, C. F., Jr., Wadell, J. S., and Petersen, L. E., 1971, X-ray determination of calcite-dolomite: An evaluation: Journal of Sedimentary Petrology, v. 41, p. 483–488.

Sass, E., Bein, A., and Almogi-Labin, A., 1991, Oxygen-isotope composition of diagenetic calcite in organic-rich rocks: Evidence for ^{18}O depletion in marine anaerotic pore water: Geology, v. 19, p. 839–842.

Shinn, E. A., 1983a, Tidal flat environment, *in* Scholle, P. A., Bebout, D. G., and Moore, C. H., eds., Carbonate Depositional Environments: Tulsa, Oklahoma, American Association of Petroleum Geologists, p. 171–210.

Shinn, E. A., 1983b, Birdseyes, fenestrae, shrinkage pores, and loferites: A reevaluation: Journal of Sedimentary Petrology, v. 53, p. 619–628.

Shinn, E. A., 1986, Modern carbonate tidal flats: Their diagnostic features: Colorado School of Mines Quarterly, v. 81, p. 7–35.

Shinn, E. A., Ginsburg, R. N., and Lloyd, R. M., 1965, Recent supratidal dolomite from Andros Island, Bahamas, *in* Pray, L. C., and Murray, A. C., eds., Dolomitization and Limestone Diagenesis: A Symposium: Society of Economic Paleontologists and Mineralogists Special Publication 13, p. 112–123.

Shinn, E. A., Lloyd, R. M., and Ginsburg, R. N., 1969, Anatomy of a modern carbonate tidal-flat, Andors Island, Bahamas: Journal of Sedimentary Petrology, v. 39, p. 1202–1228.

Steinen, R. P., 1980, Cementation of lime-mud and pellet mud beneath tidal flats of southwest Andros Island, Bahamas [abs.]: American Association of Petroleum Geologists Bulletin, v. 64, p. 788.

Steinen, R. P., 1982, SEM observations on the replacement of Bahamian aragonite mud by calcite: Geology, v. 10, p. 471–475.

Teeter, J. W., and Quick, T. J., 1990, Magnesium-salinity relation in the saline lake ostracode *Cyprideis americana:* Geology, v. 18, p. 220–222.

Waters, B. B., Spencer, R. J., and Demicco, R. V., 1989, Three-dimensional architecture of shallowing-upward carbonate cycles: Middle and Upper Cambrian Waterfowl Formation, Canmore, Alberta: Bulletin of Canadian Petroleum Geology, v. 37, p. 198–209.

Zabielski, V. P., 1991, The depositional history of Storr's Lake, San Salvador, Bahamas [M.S. thesis]: Chapel Hill, University of North Carolina, 89 p.

Zabielski, V. P., and Neumann, A. C., 1990, Field guide to Storr's Lake, San Salvador, Bahamas: Fifth Symposium, Geology of the Bahamas, Field Trip Guidebook: San Salvador, Bahamian Field Station, p. 49–55.

MANUSCRIPT ACCEPTED BY THE SOCIETY JANUARY 5, 1995

Printed in U.S.A.

Geological Society of America
Special Paper 300
1995

Mineralogy, chemistry, and petrography of soils, surface crusts, and soil stones, San Salvador and Eleuthera, Bahamas

Annabelle M. Foos and Roger J. Bain
Geology Department, University of Akron, Akron, Ohio 44325

ABSTRACT

There are three main types of soils in the Bahamas: sandy, organic, and lateritic soils. Sandy soils occur on unconsolidated carbonate sands and consist of unaltered carbonate minerals plus organic matter. Organic soils contain abundant organic material and lack mineral matter. They are most common on flat, rocky lands of the larger Bahamian islands. Lateritic soils are thin and discontinuous, and occur on lithified Pleistocene eolian and beach ridges. They have low SiO_2/Al_2O_3 ratios and contain calcite, aragonite, hematite, goethite, hydroxy-interlayered clay (HIC), boehmite, and quartz. The following petrographic features were observed in pedogenically altered Pleistocene grainstones: rhizoliths, pedotubules, alveolar textures, calcified root hairs, *Microcodium*, laminated micrite, clotted micrite, soil pisoids, circumgranular cracking, horizontal fractures, microbial borings, and iron-rich clay accumulations. The characteristics of Bahamian soils result from a complex interaction of the five major soil-forming factors: climate, topography, vegetation, parent material and time.

INTRODUCTION

Paleosols, ancient soil horizons, have been recognized in Quaternary carbonates and are useful stratigraphic indicators, representing periods of subaerial exposure (Land et al., 1967; Harmon et al., 1983; Garrett and Gould, 1984; Carew and Mylroie, 1985). Paleoenvironmental interpretations, such as paleoclimate, of paleosols in Paleozoic carbonates are common (Goldhammer and Elmore, 1984; Prather, 1985; Ettensohn et al., 1988; Wright, 1988; Goebel et al., 1989). The objective of this chapter is to describe modern Bahamian soils, pedogenically altered limestones, and the present-day conditions of soil formation, so they can be used as modern analogues to ancient paleosols and aid in their stratigraphic and paleoenvironmental interpretations.

In general, Bahamian soils are very thin and discontinuous, with a lack of well-developed horizons. They are alkaline with a high carbonate content. Because they are deficient in potassium and nitrogen, they have a low fertility. The Bahamian Land Resource Survey (Little et al., 1977) defines the following soils that are recognized throughout the Bahamas: sandy soils with humus and leafmould, leafmould soils on rock, muck soils, aluminous lateritic soils, and immature lateritic soils. In addition, there are sandy soils containing concretions of caliche and lime-silt soils, which are restricted to the more arid islands of the southern Bahamas. According to the U.S. *Soil Taxonomy* (Soil Survey Staff, 1975), the sandy soils would be classified as Entisols, the leafmould and muck soils as Histosols, and the aluminous lateritic and immature lateritic soils as Ultisols.

FACTORS OF SOIL FORMATION

Soil scientists define the major factors of soil formation as climate, topography, vegetation, parent material, and time (Jenny, 1941). No one factor plays a dominant role in determining the type of soil formed; rather, the characteristics of a given soil result from a complex interaction of all factors.

Foos, A. M., and Bain, R. J., 1995, Mineralogy, chemistry, and petrography of soils, surface crusts, and soil stones, San Salvador and Eleuthera, Bahamas, *in* Curran, H. A., and White, B., Terrestrial and Shallow Marine Geology of the Bahamas and Bermuda: Boulder, Colorado, Geological Society of America Special Paper 300.

Climate

The Bahamas have a marine tropical climate with warm, moist summers and cool, dry winters. There is a decrease in the annual rainfall from 160 cm on the northern islands (Abaco) to 65 cm on the southeastern islands (Inagua). This trend is accompanied by an increase in the potential evapotranspiration from 125 cm in the north to 190 cm in the southeast. There is a balance between precipitation and evapotranspiration for the northern islands, whereas for the southeastern islands, evapotranspiration exceeds precipitation (Sealey, 1985). The more arid climate of the southeastern islands is reflected in the occurrence of soils containing pedogenic carbonate nodules and evaporite deposits that are absent on the northern islands. San Salvador and Eleuthera are intermediate between these two extremes, with potential evapotranspiration slightly exceeding precipitation. The average annual rainfall on San Salvador and Eleuthera are 115 and 114 cm, and the average temperatures are 26° and 25°C, respectively.

Topography

Relief in the Bahamas is very low, the maximum being 60 m on Cat Island. The highest ridge on San Salvador is 37 m above sea level. The topography is constructional rather than erosional. The major topographic features are the ridges, which represent eolian dunes and beach ridges. Saline to hypersaline lakes occupy the depressions between the ridges.

On the bankward side of larger islands such as Andros, extensive, flat "rock land" is underlain by Pleistocene intertidal and ooid shoal deposits (Sealey, 1985). Lateritic soils occur on Pleistocene ridges, whereas leafmould soil occurs on "rock land." On Pleistocene deposits, karst processes have resulted in the development of a low-relief solution surface and subsurface caverns. Due to the high permeability of the limestone surface, streams are absent. In addition, the irregular solution surface inhibits overland and throughflow of runoff. As a result, transport of soil is minimal or very localized, and typical soil catenas, as described from siliciclastic soils, are not observed. However, barren surfaces may be observed in areas with mature subsurface karst where the soil materials have been transported downward into caverns.

Vegetation

Present-day vegetation ranges from succulent plants on modern dunes to mixed broadleaf coppice in the island interior. The succulent plants of the coastal dunes have shallow root systems that extract pellicular water from the vadose zone. In contrast, plants of the broadleaf coppice have vertical root systems that draw on phreatic ground water. Semeniuk and Meagher (1981) described calcretes formed by these two contrasting vegetation assemblages. Calcretes with large, well-developed rhizomorphs form by plants with vadose root systems. Carew and Mylroie (1985) have observed that this type of calcrete is restricted to regressive Pleistocene dune facies on San Salvador.

Parent material

From a chemical and mineralogic point of view, all the Bahamian soils have a similar parent material—calcium carbonate (calcite and aragonite) plus airborne dust (dominantly illite). However, the porosity and permeability characteristics of the underlying carbonates vary significantly, depending on the original sediment characteristics, degree of diagenetic alteration, and age. The nature of the underlying carbonates plays an important role in soil formation by controlling the moisture regime of the soil. Sandy soils occur on unconsolidated dunes, whereas lateritic soils occur on lithified eolian deposits.

Time

The maximum age of the soils can be estimated by determining the age of the underlying limestones. Eolian dunes and beach ridges were deposited during Pleistocene sea-level highstands (Bain, 1991). The large ridges in the interior of San Salvador are the oldest and have been dated at 220 Ka (Foos and Muhs, 1991). Another suite of ridges is associated with the Pleistocene coral reefs, which have been dated at 125 Ka (Chen et al., 1991; Carew and Mylroie, 1987). A third set of ridges can be related to the Holocene sea-level rise. There is a possible relation between the mineralogy of the lateritic soils and their age. Aluminum-rich laterites containing boehmite occur on 220-Ka-old ridges, whereas immature lateritic soils occur on 125-Ka-old ridges.

METHODS

The location of sampling sites on San Salvador is given in Figure 1. Samples were also collected at Governors Harbour and Hatchet Bay on Eleuthera (see Foos, 1991a, for location map). A total of 60 samples were analyzed. Chemical and mineralogic analyses were performed on samples of the B horizon that were collected at a depth of approximately 10 cm. Petrographic analysis was performed on surface crusts and soil stones associated with lateritic soils developed on Pleistocene grainstones.

The chemical composition of the <2-mm fraction was determined with inductively coupled plasma spectrometry. Due to the presence of organic matter, carbonate, and hydrous phases, the loss on ignition values was high (average, 35%). In order to compare variations in the chemical compositions, the oxides were normalized to 100%. Mineral compositions were determined by X-ray diffraction on a Philips APD 3720 instrument. Samples were X-rayed before and after removal of organic matter, carbonates, and free iron oxides. Organic matter was removed with Na-hypochlorite neutralized to a pH of 9

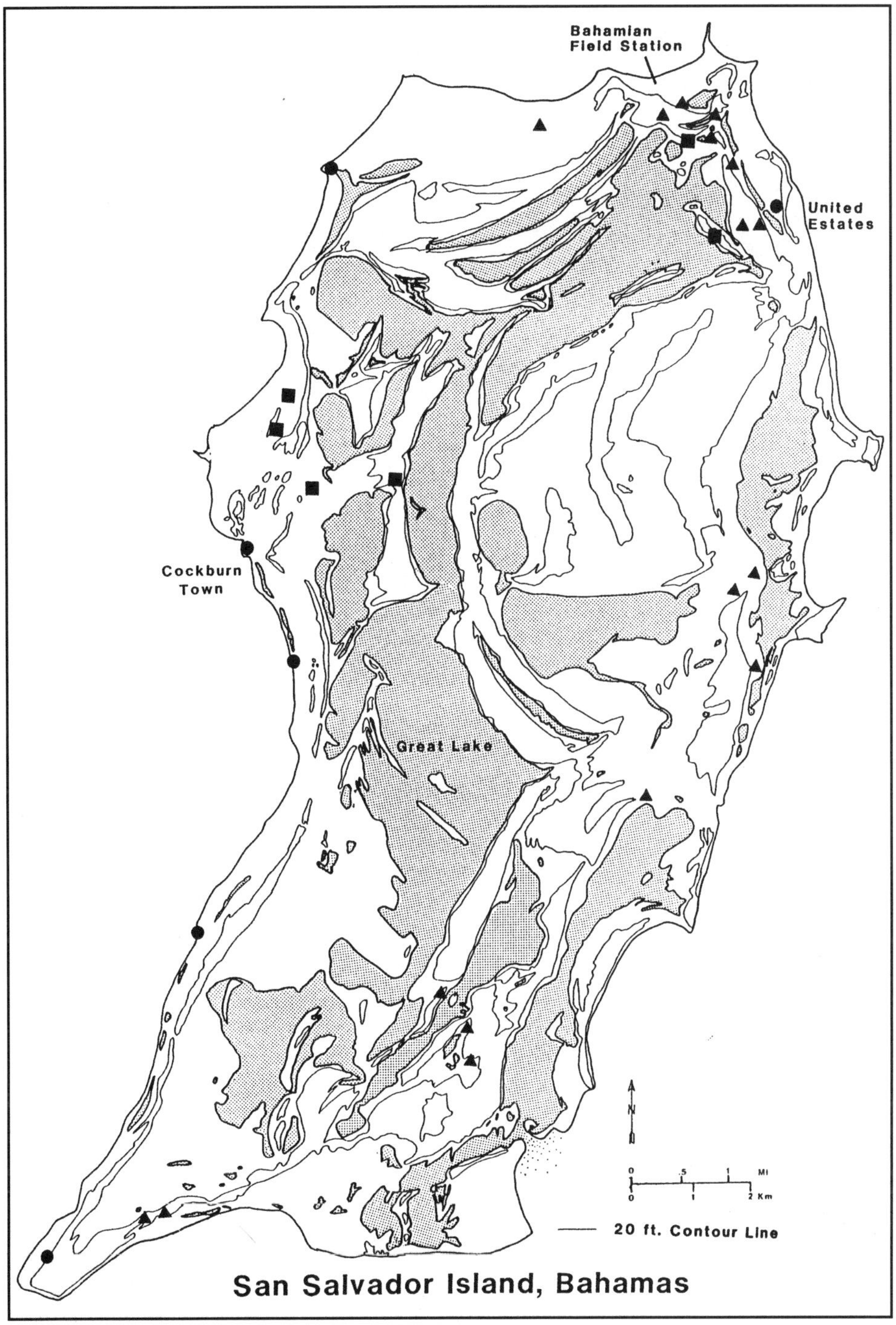

Figure 1. Location of soil sample sites on San Salvador. Black circles indicate sandy soils; black triangles, immature lateritic soils; black squares, aluminous lateritic soils.

(Jackson, 1985). Clay-sized carbonates were removed by rinsing the samples with dilute (5%) hydrochloric acid. Free iron oxides were removed using the sodium citrate–bicarbonate–dithionite method (Jackson, 1985).

Comparisons of treated samples with untreated samples indicates that this treatment did not significantly alter the clay minerals. The clays were saturated with Mg^{+2} and K^{+}, solvated with ethylene glycol, and heated to 400° and 550°C to aid in their identification. Aluminum substitution in hematite and goethite was determined by measuring the peak position of the 300 and 111 reflections, respectively (Norrish and Taylor, 1961; Schwertmann et al., 1979). Total carbonate concentration was calculated from CaO values prior to correction for loss on ignition. The relative concentrations of carbonate minerals in soil

Figure 2. Sandy soil profile developed on unconsolidated eolian sands at Fernandez Bay, San Salvador, showing the upper organic-rich layer grading downward into pure carbonate sand. Also note the root penetration resulting in a lack of primary sedimentary structures. (Scale given in inches.)

stones and surface crusts were estimated by comparison of X-ray diffraction peak areas of the unknowns with pure mineral standards of aragonite, high-Mg calcite, and low-Mg calcite.

DESCRIPTION OF SOILS

Sandy soils

The sandy soils develop on unconsolidated carbonate sands. They consist of an upper gray to grayish brown layer in which organic matter has accumulated; this grades downward into pure carbonate sand (Fig. 2). Penetration of the carbonate sands by plant roots disrupts the primary bedding.

The major oxide in sandy soils is CaO, reflecting their high carbonate content (Table 1). X-ray diffraction indicates that the major carbonate mineral is aragonite with lesser amounts of high-Mg calcite and low-Mg calcite. The mineralogy of these soils is similar to the underlying carbonate sediments, indicating that there has been very little alteration of the carbonate minerals.

Sandy soils have been recognized in Pleistocene eolianites and have been referred to as protosols (Vacher and Hearty, 1989; Carew and Mylroie, 1991). They are massive units that lack primary sedimentary structures within cross-bedded eolianites and display linear voids produced by roots.

Organic soils

Leafmould soil and muck soils contain abundant organic material and lack mineral matter. Leafmould soil occurs over flat, rocky land that is common on the bankward sides of Andros, Abaco, Grand Bahama, and New Providence. It consists of humus overlying less than 15 cm of humic sandy earth that overlies an irregular surface of limestone (Sealey, 1985). The muck soils are thick peaty soils that occupy wide hollows subject to periodic flooding.

The organic-rich soils are poorly represented on San Salvador, and only one muck soils was analyzed for clay mineralogy in this study. This sample contains minor amounts of illite and chlorite. Identification of illite and chlorite was based on the presence of 10 and 14 Å reflections, which did not shift after treatment with ethylene glycol and did not collapse when heated.

Lateritic soils

The aluminous lateritic soils and immature lateritic soils are restricted to older, lithified Pleistocene deposits and commonly

TABLE 1. CHEMICAL ANALYSES OF SELECTED BAHAMIAN SOILS BY WEIGHT PERCENT

Sample	SiO_2	Al_2O_3	Fe_2O_3	CaO	MgO	K_2O	TiO_2	$\frac{SiO_2}{Al_2O_3}$
Sandy Soil								
1	0.2	0.3	0.2	97.3	1.2	0.0	0.0	
Immature Lateritic Soils								
2	48.9	30.6	10.8	1.7	2.8	2.1	2.1	1.60
3	43.0	31.1	13.1	5.7	2.9	2.2	1.7	1.38
4	37.7	32.0	13.3	10.0	3.5	1.0	1.6	1.18
5	26.9	24.2	9.8	34.0	2.4	1.0	1.3	1.11
6	5.4	5.6	2.2	84.8	0.9	0.3	0.3	0.96
Aluminous Lateritic Soils								
7	16.5	17.0	7.1	56.1	1.7	0.4	0.9	0.97
8	25.0	29.0	12.3	27.0	2.34	1.2	1.4	0.86
9	27.1	33.2	13.7	20.6	2.0	1.1	1.7	0.82
10	25.4	47.2	20.2	2.0	1.8	0.6	2.6	0.54

*Weight percent normalized to 100% after subtraction of LOI values.

occur on eolian ridges (Foos, 1991a). First described by Ahmad and Jones in 1969, we know these soils form by the accumulation of insoluble alumina-silicates and iron oxides on hard limestones. The source of the insoluble material is airborne dust, transported by trade winds from North Africa (Muhs et al., 1990). These soils occur in shallow solution depressions. Limestone is exposed at the surface between the localized soil accumulations. Lateritic soils are rarely more than a meter thick and generally are stony (Fig. 3).

The minerals of Bahamian lateritic soils include calcite, aragonite, hematite, goethite, hydroxy-interlayered clay (HIC), boehmite, and quartz. The initial composition of the airborne dust from North Africa consists of illite, quartz, kaolinite, chlorite, and feldspar (Glaccum and Prospero, 1980). With the exception of quartz, all of these minerals have been altered and the noncarbonate mineral assemblage of the Bahamian soils represents a pedogenic assemblage.

The chemical compositions of selected lateritic soils are given in Table 1. The SiO_2/Al_2O_3 ratio of a soil is an indicator of its chemical maturity. As primary minerals undergo chemical weathering, leaching of silica is accompanied with a relative enrichment of Al, Fe, and Ti. All the Bahamian soils analyzed have a very low SiO_2/Al_2O_3 ratio (<2.0) and soils containing boehmite (AlOOH) have SiO_2/Al_2O_3 ratios less than 1.0. There is no relationship between the SiO_2/Al_2O_3 ratio and $CaCO_3$ concentration (Fig. 4), which indicates that the chemical maturity is not a function of carbonate content. Criteria for separating aluminous lateritic soils from immature lateritic soils include the presence of boehmite and a SiO_2/Al_2O_3 ratio of less than 1.0.

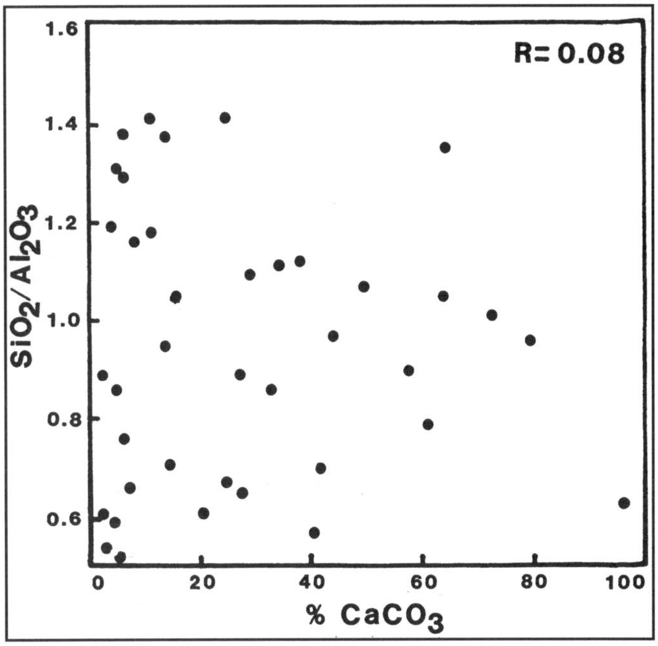

Figure 4. Plot of SiO_2/Al_2O_3 versus $CaCO_3$ of Bahamian soils, indicating a lack of correlation between the two.

Figure 3. Recently cleared field of aluminous lateritic soil near the airport on San Salvador, showing the thin, discontinuous, stony nature of the soils.

The CaO concentration is highly variable, ranging from 2 to 85%, and reflects the carbonate content of the soils, which ranges from 2 to 97%. The major carbonate mineral is low-Mg calcite with minor amounts of aragonite present. The main source of carbonates in the solum is the breakdown of carbonate grainstones. The presence of carbonates, which act as a buffer, results in a relatively high soil pH, ranging from 7.8 to 8.8.

Up to 20% Fe_2O_3 was observed in the lateritic soils. More than 90% of this iron occurs as clay-sized free iron oxides, hematite, and goethite (Foos, 1991a). These minerals play an important role in determining the soil color, which ranges from red (10R 2/2) to yellow-red (10YR 7/6). Bright red colors are due to the presence of hematite; yellow-brown colors are attributed to goethite. The iron minerals hematite and goethite have aluminum substituted in their structure. Goethite consistently has the maximum amount of aluminum allowed in its structure, 28 mol%, whereas the aluminum content of hematite is more variable, ranging from 1 mol% to the maximum amount allowed, 15 mol%. Aluminum-rich goethite is characteristic of highly weathered subtropical and tropical soils such as Oxisols, Ultisols, bauxites, and saprolites (Fitzpatrick and Schwertmann, 1982; Schwertmann, 1985). In the stability diagrams for the Fe_2O_3-Al_2O_3-H_2O system presented by Trolard and Tardy (1987), goethite precipitated in equilibrium with boehmite will have the maximum amount of Al substitution, and hematite Al substitution increases with decreasing H_2O activity.

Quartz, a minor component of the soils, is probably inherited from the original detrital material blown in from North Africa. Quartz occurs as angular to subrounded silt- to sand-sized grains. Surface textures characteristic of large-scale dissolution of quartz grains were not observed, suggesting quartz is relatively stable in this environment (Foos, 1991b). The loss of silica resulting in the low SiO_2/Al_2O_3 ratio is most likely from the breakdown of clay minerals and feldspar rather than dissolution of quartz.

Boehmite (AlOOH) was identified based on the presence of X-ray diffraction reflections at 6.11 and 3.16 Å. The presence of boehmite in soils is rare, gibbsite ($Al(OH)_3$) being the Al phase most commonly observed (Hsü, 1989). Trolard and Tardy (1987) have shown that at 25°C and H_2O activities less than 0.72, boehmite is the stable Al phase. As a soil undergoes drying, low water activities occur within cryptovoids of the unsaturated zone. The thin nature of the lateritic soils, combined with the high porosity and permeability of the underlying limestone, results in low moisture retention allowing the soils to dry between intermittent rainfalls. This repeated drying of the soils results in a low H_2O activity, causing boehmite to be stable over gibbsite.

The major clay mineral in the lateritic soils is hydroxy-interlayered clay. HICs are common constituents of soils. They have been referred to in the literature by other names such as dioctahedral vermiculite, dioctahedral chlorite, chlorite-like clay, and intergradient chlorite-vermiculite. They are 2:1 clays with interlayers of incomplete gibbsite sheets, or "islands" of Al-hydroxy polymers (Barnhisel and Bertsch, 1989). An untreated, air-dried HIC has X-ray diffraction basal reflections at 14.1, 7.1, 4.76, 3.57 Å, which do not shift after treatment with ethylene glycol (Fig. 5). An 060 reflection at 1.50 Å indicates this mineral is dioctahedral. Heating the clays to 550°C after saturation with K^+ causes partial collapse of the 14 Å peak to 12 Å. Some immature lateritic soils contain HIC with very low levels of interlayering where the 14 Å peak collapsed to 10 Å when heated, and the collapse occurred at a lower temperature (300°C) (Fig. 6). HIC formation is favored by frequent wetting and drying cycles (Barnhisel and Bertsch, 1989), caused by the low moisture retention of the lateritic soils, along with the intermittent nature of the rainfall.

DESCRIPTION OF SOIL STONES AND SURFACE CRUSTS

Most of the Pleistocene deposits on San Salvador that are exposed at the surface have undergone pedogenic alteration. An alteration rind covers surface exposures and lines solution cavities. It most likely began to form shortly after the sediments were stabilized by vegetation and continues to form at the present time by a complex array of biogenic and inorganic processes. The alteration rinds may be massive, poorly laminated, or brecciated. They commonly have a brown to reddish color due to the incorporation of iron-rich clay material from the solum. Well-laminated crust may form by accretion of micrite on an impermeable surface. However, the majority of the subaerial alteration on San Salvador consists of *in situ* micritization of the carbonate grainstones.

Shallow depressions on a microkarst surface are the locations of solum accumulation. Moisture retention is high in areas of solum accumulation and low where limestone is exposed at the surface. Two types of samples were investigated in this study: limestone clasts, which were buried within the solum (soil stones), and crusts, which were exposed at the surface.

Mineralogic alteration of soil stones and surface crusts is indicated in Table 2. The relative abundance of aragonite,

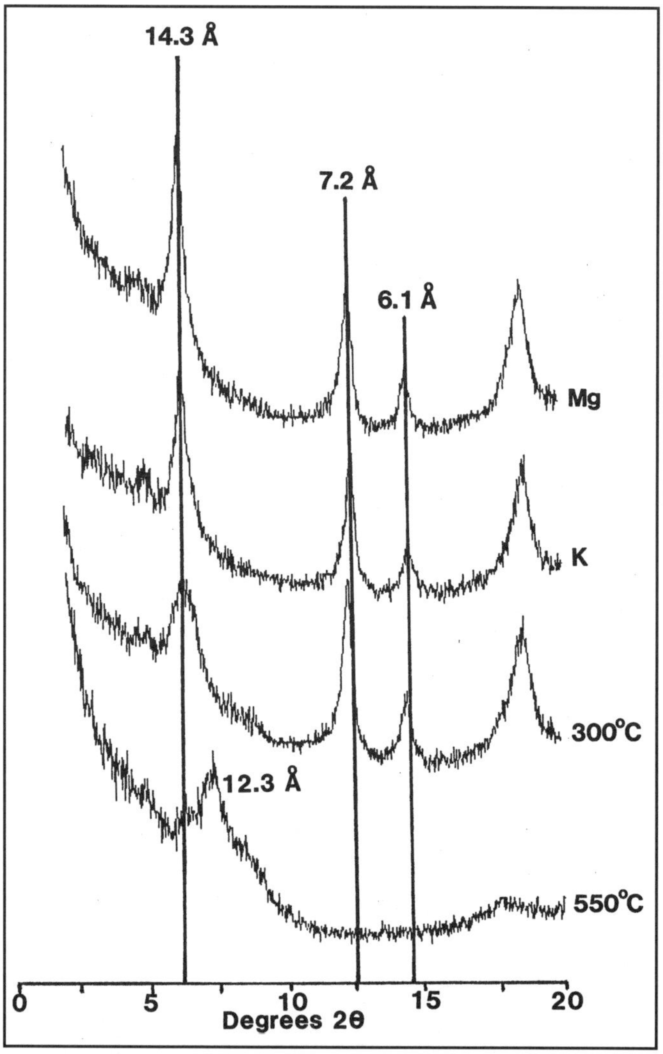

Figure 5. X-ray diffraction patterns of hydroxy interlayered clay (HIC) and boehmite from an aluminous lateritic soil. Mg, saturated with Mg^+; K, saturated with K^+; 300°C, saturated with K^+ and heated to 300°C; 500°C, saturated with K^+ and heated to 500°C; 6.1 Å, boehmite reflection.

high-Mg calcite, and low-Mg calcite in fresh sediments is highly variable and dependent on the ooid/bioclastic content and the abundance of aragonite versus high-Mg calcite species among the bioclasts. High-Mg calcite is a common component of the primary sediments and unaltered Pleistocene grainstones, its concentration being dependent on the source area and the degree of diagenetic alteration (Van Kauwenbergh and Bain, 1985). High-Mg calcite is notably absent in both the surface crusts and the soil stones, indicating that it is highly unstable in a pedogenic environment.

The surface crust commonly consists of a dense micrite, which is reflected in its low aragonite content and high low-Mg calcite content. That the aragonite content of the soil stones is more variable can be attributed to a number of factors such as position (depth within solum), original source, and residence time in the solum. Micritization is favored by frequent periods of wetting and drying. Samples 4, 5, and 6 of Table 2 were collected from one soil profile at depths of 10, 20, and 35 cm, respectively. These samples show an increase in the aragonite content with depth, which may reflect less frequent periods of wetting and drying in deeper parts of the soil profile. Sample 3, which contained no aragonite, was originally a surface crust that was eroded and incorporated in the solum; sample 7, which contains abundant aragonite, could by a recently eroded fragment of unaltered grainstone.

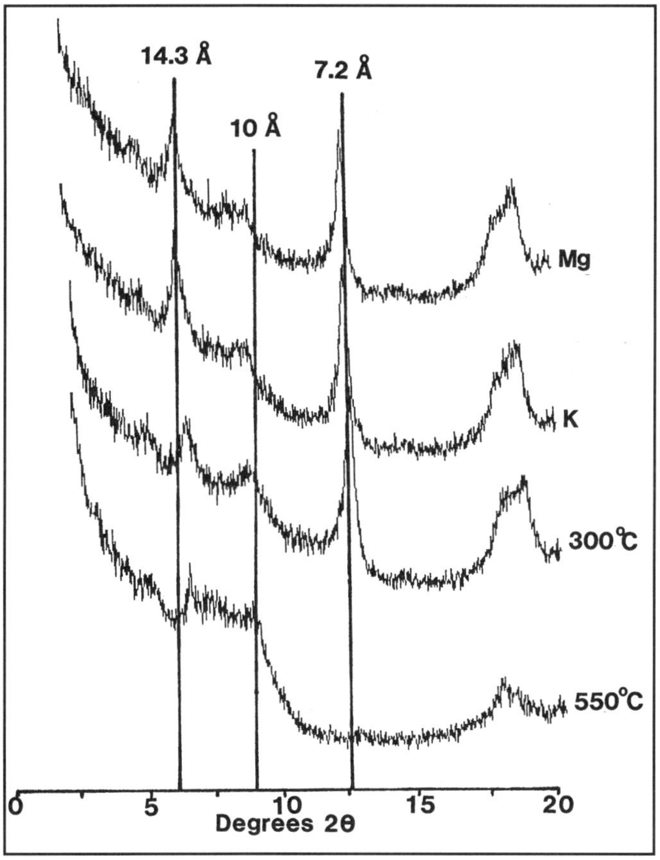

Figure 6. X-ray diffraction patterns of incompletely interlayered HIC from an immature lateritic soil. Mg, saturated with Mg^+; K, saturated with K^+; 300°C, saturated with K^+ and heated to 300°C; 500°C, saturated with K^+ and heated to 500°C.

Petrographic features recognized in San Salvador soil stones and subaerial crusts include rhizoliths, pedotubules, alveolar textures, calcified root hairs, *Microcodium*, laminated micrite, clotted micrite, soil pisoids, circumgranular cracking, horizontal fractures, microbial borings, and iron-rich clay accumulations (Bain and Foos, 1993). Rhizoliths, pedotubules, alveolar textures, calcified root hairs, and *Microcodium* are all root-related features. Rhizoliths are organosedimentary structures produced by roots through accumulation and/or cementation within and around roots, or replacement of roots by mineral matter (Klappa, 1980). They are recognized by the disruption of original depositional texture and fabric, and micritization of adjacent limestone. Micrite apparently replaces original granular textures as well as fills the root channelway as the root decays. The degree of micritization decreases away from the root over a distance of a centimeter or less.

Scanning electron microscope analyses by Jones and Ng (1988) suggest that microflora associated with roots play an important role in micritization of surrounding sediments. Pedotubules are elongate, irregular pores that interconnect over long distances and formed by rootlet penetration with subsequent decay (Braithwaite, 1983). Pedotubules with complex anastomosing patterns are common in Bahamian subaerial crust (Fig. 7A).

Alveolar texture in carbonates was described by Steinen (1974) and interpreted as resulting from rootlets penetrating sediment, causing micritization of adjacent channel walls. The term alveolar texture describes cylindrical to irregular pores, which may be calcite-filled, separated by a network of anastomosing micrite walls. Alveolar texture probably represents millimeter-

TABLE 2. CARBONATE MINERALOGY OF PEDOGENICALLY ALTERED CARBONATES AND UNALTERED SEDIMENTS

Sample	Aragonite (wt%)	Low-Mg Calcite (wt%)	High-Mg Calcite (wt%)
Surface Crust			
1	0	100	n.d.*
2	4	96	n.d.
Soil Stones			
3	0	100	n.d.
4	9	91	n.d.
5	25	75	n.d.
6	55	45	n.d.
7	83	17	n.d.
Unaltered Sediments			
8	86	1	12

*n.d. = not detected.

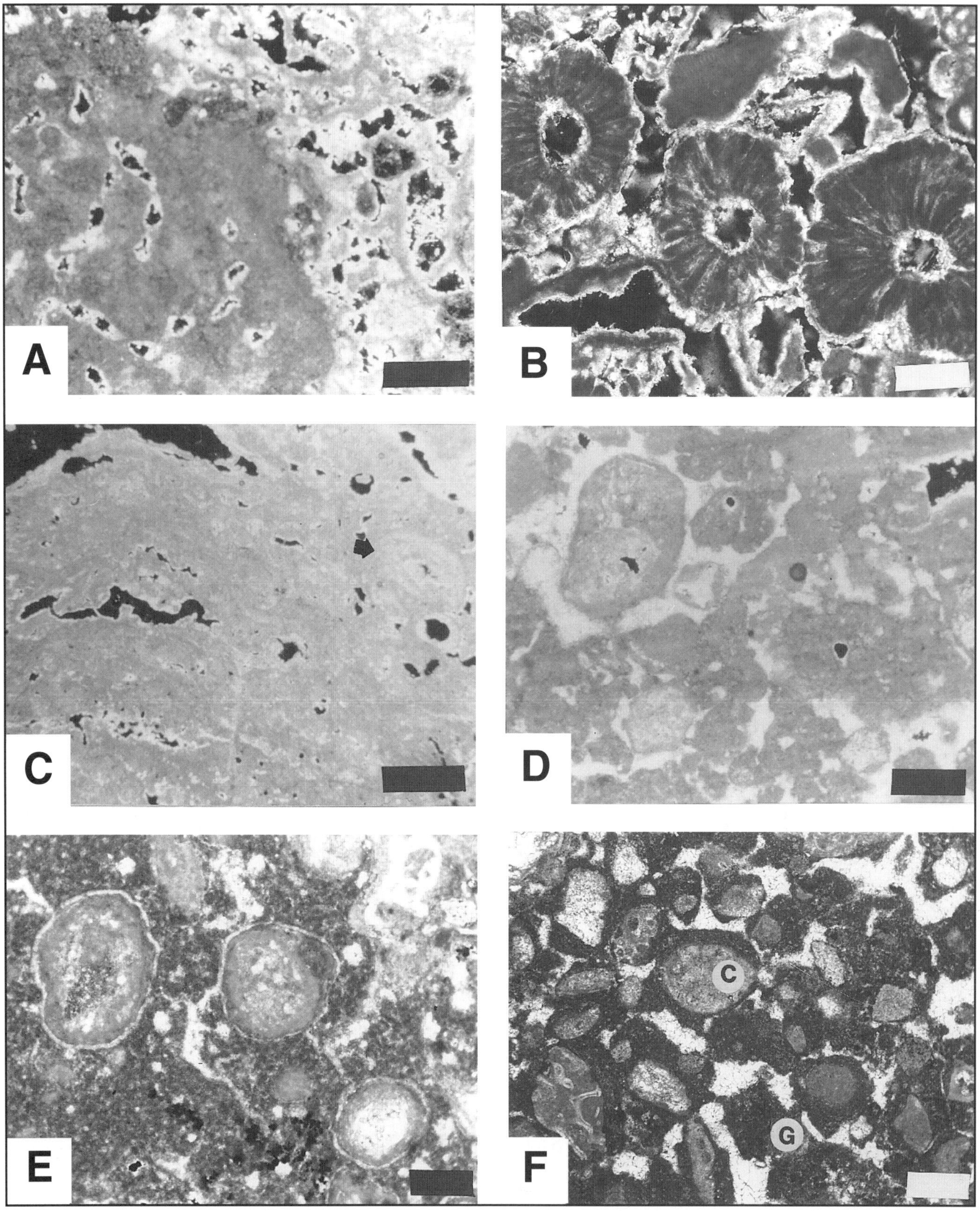
A
B
C
D
E
F
C
G

sized rhizoliths (Esteban and Klappa, 1983). A similar feature consists of calcified root hairs, 20 μ in diameter, which bridge intergranular and root pores. *Microcodium* consists of elongate, radiating, petal-like calcite prisms or ellipsoids, up to 1 mm in length, grouped in clusters (Fig. 7B). *Microcodium* is interpreted to be the result of calcification of a symbiotic association of soil fungi and plant roots (Esteban and Klappa, 1983).

Common in Bahamian surface crust are straight microscopic borings that penetrate both grains and cement. Borings are believed to be the work of fungi because of their size (2 μ) and their straight pattern; however, other microbes could also be responsible. The size of these borings is about 0.25 times the size of irregular, intragranular algal borings that formed while grains were still in their depositional environment. The fact that the fungal borings penetrate both grains and vadose cement indicates boring occurred under subaerial conditions.

Laminated micrite (Fig. 7C) occurs as surface crusts and lines solution cavities, and may form either by accretion on, or alteration of, grainstone. A detailed description of these crusts is given in Multer and Hoffmeister (1968). Clotted micrite consists of irregularly shaped, rounded to angular peloids of micrite, separated by blocky calcite (Fig. 7D) and can be observed in both surface crust and soil stones. It forms by a combination of precipitation of micrite, shrinkage, and microfracturing, and results in a total obliteration of the original texture and fabric. A grainstone composed of well-sorted, fine sand-sized grains can be altered to clotted micrite with peloids ranging in size from 0.5 to 1.0 mm.

Micritic soil pisoids appear as irregularly laminated, concentric micritic grains, equant to elongate, and commonly display circumgranular cracking (Fig. 7E). Soil pisoids are formed by precipitation (chemically or biochemically) of micrite on a nucleus. The circumgranular cracking is believed to be the result of repeated wetting and drying.

Horizontal fracturing results from episodic wetting and drying. It is well developed in soil stones that are only partially buried in the solum and most likely formed by vertical movement of capillary water derived from the underlying solum. There is no evidence of roots associated with the most recently formed fractures, suggesting this is an inorganic process. Micritization occurs preferentially along fractures. Once fractures are established, plant roots follow these planes of weakness, expanding the cracks, which can subsequently be infiltrated with solum.

Figure 7. Petrographic features from recent subaerial crust. A, Anastomosing pedotubules within micrite; also shown is alveolar texture in upper right corner. X-nicols; scale bar = 0.3 mm. B, *Microcodium*. X-nicols; scale bar = 0.3 mm. C, Laminated micrite; also shown are elongate pores and pisoids (arrow). X-nicols; scale bar = 1.2 mm. D, Clotted micrite. X-nicols; scale bar = 0.3 mm. E, Soil pisoids with circumgranular cracking. X-nicols; scale bar = 0.3 mm. F, Iron-rich glaebules (G) and iron-coated carbonate grains (C). X-nicols; scale bar = 0.3 mm.

The mineralogy of insoluble residues in subaerial crusts is similar to the overlying solum suggesting that iron-rich clay material from the overlying solum is incorporated into the pedogenically altered limestone (Foos, 1991a). Large voids produced by solution around roots are infilled with glaebules, coated grains, and pisoids. The glaebules are irregular to rounded, 0.1 to 0.8 mm in diameter, and composed mostly of iron-rich clay material (Fig. 7F). Carbonate grains and intraclasts are coated with iron-rich clay material and grade into iron-rich pisoids. Iron-rich clay material, which was carried by colloidal suspension into the limestone, lines pores that are later filled with equigranular calcite. Alternating laminae of iron-rich clay and micrite are common in the laminar crust.

CONCLUSIONS

Bahamian soils consist of three main types: sandy, organic, and lateritic soils. Sandy soils develop on unconsolidated carbonate sand and show little alteration of the carbonates being composed primarily of aragonite. Organic soils develop on flat, low, rocky surfaces and are depleted in mineral matter. Lateritic soils are very thin and discontinuous and possess pedogenic clays, boehmite, and iron oxide minerals that result from the weathering of airborne dust.

Limestone displays features resulting from pedogenic alteration, which occur as surface crusts on bedrock or within soil stones buried in the solum. The brown to reddish color of altered limestone results from the incorporation of iron-rich clay material. Carbonates are altered mineralogically by a decrease in aragonite and high-Mg calcite and an increase in low-Mg calcite. Original textures and grain structures are destroyed and replaced by micrite. Pedogenic alteration features present include alveolar textures, *Microcodium*, rhizoliths, pedotubules, calcified root hairs, laminated micrite, clotted micrite, horizontal and circumgranular fractures, soil pisoids, and microbial borings.

Soils on San Salvador and Eleuthera formed in response to the following conditions: a warm, tropical climate in which potential evapotranspiration slightly exceeds precipitation; low relief with a constructional rather than erosional topography; vegetation ranging from succulent plants to mixed broad leaf coppice; a parent material of carbonate sands or grainstones plus airborne dust; and a time frame of less than 220,000 yr.

ACKNOWLEDGMENTS

We acknowledge Don Gerace and the staff of the Bahamian Field Station for logistical support of our field work.

REFERENCES CITED

Ahmad, N., and Jones, R. L., 1969, Occurrence of aluminous lateritic soils (bauxites) in the Bahamas and Cayman Islands: Economic Geology, v. 64, p. 804–808.

Bain, R. J., 1991, Distribution of Pleistocene lithofacies in the interior of San Salvador Island, Bahamas, and possible genetic models, *in* Bain, R. J., ed., Proceedings Fifth Symposium on the Geology of the Bahamas: San Salvador, Bahamian Field Station, p. 11–22.

Bain, R. J., and Foos, A. M., 1993, Carbonate microfabrics related to subaerial exposure and paleosol formation, *in* Rezak, R., ed., Carbonate Microfabrics: New York, Springer-Verlag, p. 19–27.

Barnhisel, R. I., and Bertsch, P. M., 1989, Chlorites and hydroxy-interlayered vermiculite and smectite, *in* Dixon, J. B., and Weed, S. B., eds., Minerals in Soil Environments (2nd ed.): Madison, Wisconsin, Soil Science Society of America, p. 729–788.

Braithwaite, C. J. R., 1983, Calcrete and other soils in Quaternary limestones: Structures, processes and applications: Journal Geological Society of London, v. 40, p. 351–363.

Carew, J. L., and Mylroie, J. E., 1985, Pleistocene and Holocene stratigraphy of San Salvador Island Bahamas, with references to marine and terrestrial lithofacies at French Bay, *in* Curran, H. A., ed., Geological Society of America, Orlando Annual Meeting Field Trip Guidebook: Ft. Lauderdale, Florida, CCFL Bahamian Field Station, p. 11–61.

Carew, J. L., and Mylroie, J. E., 1987, A refined geochronology for San Salvador Island, Bahamas, *in* Curran, H. A., ed., Third Symposium on the Geology of the Bahamas: Ft. Lauderdale, Florida, CCFL Bahamian Field Station, p. 35–44.

Carew, J. L., and Mylroie, J. E., 1991, Some pitfalls in paleosol interpretation in carbonate sequences: Carbonates and Evaporites, v. 6, p. 69–74.

Chen, J. H., Curran, H. A., White, B., and Wasserburg, G. J., 1991, Precise chronology of the last interglacial period: ^{234}U-^{230}Th data from fossil coral reefs in the Bahamas: Geological Society of America Bulletin, v. 103, p. 82–97.

Esteban, M., and Klappa, C. F., 1983, Subaerial exposure environment, *in* Scholle, P. A., Bebout, D. G., and Moore, C. H., eds., Carbonate Depositional Environments: American Association of Petroleum Geologists Memoir 33, p. 1–54.

Ettensohn, F. R., Dever, G. R., Jr., and Grow, J. S., 1988, A paleosol interpretation for profiles exhibiting subaerial exposure "crusts" from the Mississippian of the Appalachian Basin, *in* Reinhardt, J., and Sigleo, W. R., eds., Paleosols and Weathering through Geologic Time: Principles and Applications: Geological Society of America Special Paper 216, p. 49–79.

Fitzpatrick, R. W., and Schwertmann, U., 1982, Al-substituted goethite—an indicator of pedogenic and other weathering environments in South Africa: Geoderma, v. 27, p. 335–347.

Foos, A. M., 1991a, Aluminous lateritic soils, Eleuthera, Bahamas: A modern analog to carbonate paleosols: Journal of Sedimentary Petrology, v. 61, p. 340–348.

Foos, A., 1991b, Mineralogy of Bahamian soils, *in* Bain, R. J., ed., Proceedings, Fifth Symposium on the Geology of the Bahamas: San Salvador, Bahamian Field Station, p. 75–80.

Foos, A. M., and Muhs, D. R., 1991, Uranium-series age of an oolitic-peloidal eolianite, San Salvador Island, Bahamas: New evidence for a high stand of sea at 200–225 ka: Geological Society of America Abstracts with Programs, v. 23, no. 1, p. 31.

Garrett, P., and Gould, S. J., 1984, Geology of New Providence Island, Bahamas: Geological Society of America Bulletin, v. 95, p. 209–220.

Glaccum, R. A., and Prospero, J. M., 1980, Saharan aerosols over the tropical North Atlantic—Mineralogy: Marine Geology, v. 37, p. 295–321.

Goebel, K. A., Bettis, E. A., III, and Heckel, P. H., 1989, Upper Pennsylvanian paleosol in Stranger shale and underlying Iatan limestone, southwestern Iowa: Journal of Sedimentary Petrology, v. 59, p. 224–232.

Goldhammer, R. K., and Elmore, R. D., 1984, Paleosols capping regressive carbonate cycles in the Pennsylvanian Black Prince limestone, Arizona: Journal of Sedimentary Petrology, v. 54, p. 1124–1137.

Harmon, R. S., and 8 others, 1983, U-series and amino-acid racemization geochronology of Bermuda: Implications for eustatic sea-level fluctuation over the past 250,000 years: Palaeogeography, Palaeoclimatology, Palaeoecology, v. 44, p. 41–70.

Hsü, P. H., 1989, Aluminum oxides and oxyhydroxides, *in* Dixon, J. B., and Weed, S. B., eds., Minerals in Soil Environments (2nd ed.): Madison, Wisconsin, Soil Science Society of America, p. 331–378.

Jackson, M. L., 1985, Soil chemical analysis—Advanced course (2nd ed.), Madison, Wisconsin, published by author, 895 p.

Jenny, H., 1941, Factors of Soil Formation: New York, McGraw-Hill, 281 p.

Jones, B., and Ng, K. C., 1988, The structure and diagenesis of rhizoliths from Cayman Brac, British West Indies: Journal of Sedimentary Petrology, v. 58, p. 457–467.

Klappa, C. F., 1980, Brecciation textures and tepee structures in Quaternary calcrete (caliche) profiles from eastern Spain; The plant factor in their formation: Geological Journal, v. 15, p. 81–89.

Land, L. S., Mackenzie, F. T., and Gould, S. J., 1967, Pleistocene history of Bermuda: Geological Society of America Bulletin, v. 78, p. 993–1006.

Little, B. G., and 7 others, 1977, Land Resources of the Bahamas: A Summary: Surrey, United Kingdom, Land Resources Division, Ministry of Overseas Development, 133 p.

Muhs, D. R., Bush, C.A., Stewart, K. C., Rowland, T. R., and Crittenden, R. C., 1990, Geochemical evidence of Saharan dust parent material for soils developed on Quaternary limestones of Caribbean and western Atlantic islands: Quaternary Research, v. 33, p. 157–177.

Multer, H. G., and Hoffmeister, J. E., 1968, Subaerial laminated crusts of the Florida Keys: Geological Society of America Bulletin, v. 79, p. 183–192.

Norrish, K., and Taylor, R. M., 1961, The isomorphous replacement of iron by aluminum in soil goethites: Journal of Soil Science, v. 12, p. 294–306.

Prather, B. E., 1985, An upper Pennsylvanian desert paleosol in the D-zone of the Lansing–Kansas City Groups, Hitchcock County, Nebraska: Journal of Sedimentary Petrology, v. 55, p. 213–221.

Schwertmann, U., 1985, The effect of pedogenic environments on iron oxide minerals, *in* Stewart, B. A., ed., Advances in Soil Science, v. 1: New York, Springer-Verlag, p. 171–200.

Schwertmann, U., Fitzpatrick, R. W., Taylor, R. M., and Lewis, D. G., 1979, The influence of aluminum on iron oxides. Pt. II. Preparation and properties of Al-substituted hematites: Clay and Clay Minerals, v. 27, p. 105–112.

Sealey, N. E., 1985, Bahamian Landscapes: An introduction to the Geography of the Bahamas: London, United Kingdom, Collins Caribbean, 96 p.

Semeniuk, V., and Meagher, T. D., 1981, Calcrete in Quaternary coastal dunes in southwestern Australia: A capillary-rise phenomenon associated with plants: Journal of Sedimentary Petrology, v. 51, p. 47–68.

Soil Survey Staff, 1975, Soil Taxonomy: Soil Conservation Service USDA, Agricultural Handbook 436, 485 p.

Steinen, R. P., 1974, Phreatic and vadose diagenetic modification of Pleistocene limestone; petrographic observations from subsurface of Barbados, West Indies: American Association of Petroleum Geologists Bulletin, v. 58, p. 1008–1024.

Trolard, F., and Tardy, Y., 1987, The stabilities of gibbsite, boehmite, aluminous goethite and aluminous hematites in bauxites, ferricretes and laterites as a function of water activity, temperature and particle size: Geochimica et Cosmochimica Acta, v. 51, p. 945–957.

Vacher, H. L., and Hearty, P., 1989, History of stage 5 sea level in Bermuda: Review with new evidence of a brief rise to present sea level during substage 5a: Quaternary Science Reviews, v. 8, p. 159–168.

Van Kauwenbergh, S. J., and Bain, R. J., 1985, Diagenesis of the carbonate rocks of San Salvador Island, Bahamas, *in* Teeter, J. W., ed., Proceedings, Second Symposium on the Geology of the Bahamas: Ft. Lauderdale, CCFL Bahamian Field Station, p. 279–296.

Wright, V. P., 1988, Paleokarsts and paleosols as indicators of paleoclimate and porosity evolution: A case study from the Carboniferous of South Wales, *in* James, N. P., and Choquette, p. W., eds., Paleokarst: New York, Springer-Verlag, p. 329–341.

Manuscript Accepted by the Society January 5, 1995

Printed in U.S.A.

Geological Society of America
Special Paper 300
1995

Roles of organics and water in preneomorphic and early neomorphic alteration of coralline aragonites from San Salvador Island, Bahamas

Susan J. Gaffey
Department of Earth and Environmental Sciences, Rensselaer Polytechnic Institute, Troy, New York 12181
Victor P. Zabielski
George Marshall Engineering Geologists, R.D. 1, Box 45, Averill Park, New York 12018
Charles Bronnimann
Otsuka Electronics, 2555 Midpoint Drive, Fort Collins, Colorado 80525

ABSTRACT

Coral skeletons are three-component systems composed of aragonite and of H_20 and organic matter that occur together in inter- and intracrystalline voids. The earliest dissolution of aragonite in *Diploria* heads occurs not in the meteoric, but in the marine phreatic environment, within the abandoned portions of still-living colonies. Dissolution begins at sclerodermite centers and along growth lines where water and organic matter are concentrated, and subsequent dissolution and neomorphic alteration in the meteoric environment continue at these same H_2O- and organic-rich sites.

Loss of intercrystalline organic matter begins immediately after a portion of the skeleton is vacated by the polyp, and much of the original protein and lipid is lost in a matter of years or decades. Intracrystalline organics break down more slowly and are still present in Pleistocene samples. While intercrystalline organic matter remains, it protects crystal surfaces from contact with pore waters. Because organic matrix materials are known to bind calcium and induce precipitation of carbonate, intact organic matrix may also chemically stabilize the carbonate it helped precipitate. The decrease in the strength of hydrogen bonding and the increase in molecular motion of H_2O in inclusions in dead and Pleistocene *Diploria* samples appear related to the breakdown of organics with which H_2O is associated, and may increase the ability of H_2O in inclusions to serve as a medium for diagenetic alteration. Breakdown of organic compounds may also change the chemistry of waters in inclusions and facilitate the first dissolution and neomorphism of aragonite.

The earliest dissolution and neomorphic alteration in *Diploria* skeletons occurs within diagenetic microenvironments separated by a few millimeters to hundreds of micrometers from similar sites showing different styles of alteration or no alteration at all. Petrographic evidence indicates that many sites of early alteration are isolated both from each other and from the environment exterior to the skeletal element containing them. Trace element analyses of coralline aragonite, calcite cements, and neomorphic calcites, however, indicate that the diagenetic system is partially open, and that reaction sites at which neomorphic alteration occurs are intermittently in contact with formation pore waters.

Gaffey, S. J., Zabielski, V. P., and Bronnimann, C., 1995, Roles of organics and water in preneomorphic and early neomorphic alteration of coralline aragonites from San Salvador Island, Bahamas, *in* Curran, H. A., and White, B., Terrestrial and Shallow Marine Geology of the Bahamas and Bermuda: Boulder, Colorado, Geological Society of America Special Paper 300.

INTRODUCTION

Skeletal carbonates are precipitated within the cells and tissues of carbonate-secreting organisms in physical isolation from the marine environment. Because different organisms exert differing degrees of control over the mineralogy, chemical composition, size, and morphology of the crystals making up their skeletons, the carbonates produced by a given organism can differ significantly both from skeletal materials produced by other types of organisms and from abiogenic carbonates precipitated under the same ambient conditions (e.g., Mann, 1983). Thus, marine carbonate sediments are composed of a much more varied suite of materials with a wider range of diagenetic potentials than is generally recognized.

Corals exert less control over carbonate precipitation than many other types of organisms, and the crystal morphology and the trace element and stable isotopic composition of aragonite in coral skeletons are similar to those of abiogenically precipitated marine aragonites (Constantz, 1986a,b). However, coralline aragonites have a larger number of crystal defects than abiogenic aragonite cements and contain numerous voids filled with water and organic matter (Johnston, 1979, 1980; Green et al., 1980; Bruni and Wenk, 1985; Constantz, 1986a,b; Gaffey, 1985, 1988, 1990). Although the mineral component of coral skeletons has been the subject of intensive study, the water and organics they contain have received much less attention; thus, one goal of this work was to more completely characterize the water and organic contents of coralline aragonites.

Diagenesis of biogenic aragonites begins in the marine environment as soon as the organism's life processes no longer maintain its disequilibrium with the external environment, and early compositional and textural changes affect a coral skeleton's response to later diagenetic environments and processes (Schroeder, 1969, 1984). Considerable work has been done on the preneomorphic and neomorphic diagenetic alteration of the aragonite in coral skeletons, but changes in the organic materials and water they contain, and their possible effects on the aragonite with which they are associated, have received much less attention. Laboratory experiments have shown that the fluid inclusions in coralline aragonites serve as a medium for mineralogic alteration on heating (Gaffey et al., 1991). A second goal of this work was to determine whether these nonmineral components also affected the alteration of coral skeletons in the natural environment.

MATERIALS AND METHODS

Corals used in this study were collected on and around San Salvador Island, Bahamas. Localities from which modern and fossil materials were obtained are shown in Figure 1. Modern samples included those that were alive when collected, and dead material that was attached to still-living colonies or that was obtained from marine bottom sediments and from beaches. Samples collected live, or skeletons of dead colonies from the marine environment that were overgrown with soft algae were cleaned in 50% commercial grade NaOCl before being returned to the laboratory. Sections cut parallel to the direction of growth were examined with a binocular microscope and in thin sections. Broken surfaces of some samples were examined with a JEOL840 scanning electron microscope. Mineralogy of samples was determined by x-ray diffraction (XRD) of powdered samples.

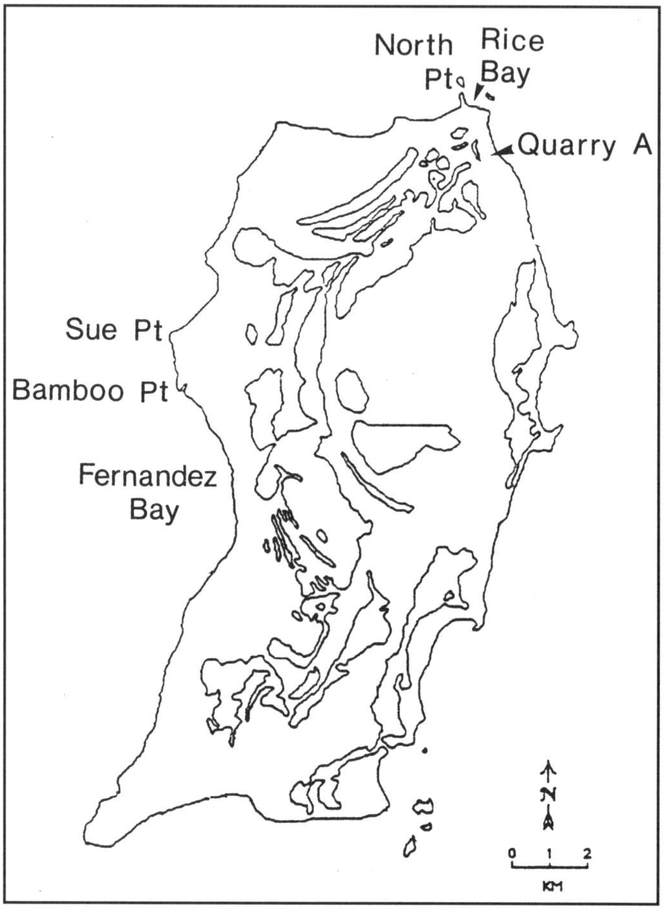

Figure 1. Map of San Salvador Island, Bahamas, showing localities from which samples used in this study were obtained (after Clark et al., 1989.)

High-precision ultraviolet (UV), visible (V), and near-infrared (NIR) (0.3 to 2.7 μm) reflectance spectra with 10-nm resolution were obtained from areas 2 mm in diameter using the spectrophotometer at RELAB at Brown University, which has been described by Pieters (1983). Because water is lost from inclusions fractured by crushing, spectra were obtained from polished slabs 8 to 15 mm thick. Most slabs were bleached in 30% H_2O_2 for at least 1 day, although spectra were also obtained from some unbleached slabs. Although H_2O_2 has since been found to etch carbonate skeletal material (Gaffey and Bronnimann, 1993), reflectance spectroscopy is a volume rather than a surface technique, penetrating at least four crystal diameters into the sample, so that the damaged surfaces would not have materially affected the results of spectral analyses. Unless otherwise specified, spectra were obtained from areas in samples that were free of cements and sediment infill, as determined by examination with hand lens and binocular microscope. Three specimens of *Diploria strigosa* were selected for detailed reflectance spectrographic study because the structure and size of the corallites made it possible to obtain spectral measurements of separate elements within the skeleton and to characterize intrasample variability and heterogeneity. Spectra were obtained from five spots each on the septa and thecae of the live and Pleistocene samples (see discussion of coral structure below) and on four spots each on the septa and thecae of the dead coral head collected from bottom sediment. Other samples of *Diploria* and *Acropora* were analyzed in less detail.

In all spectra, error bars are smaller than the lines used to plot the data. UVVNIR spectral properties of carbonate minerals are discussed in detail in Gaffey (1986, 1987), H_2O- and OH-containing phases in Gaffey (1988), and organic materials occurring in carbonate skeletons in Gaffey (1990). Calculation of relative areas of water absorption bands, which provide a sensitive measure of total water content (Gaffey et al., 1991), is described in Zabielski and Gaffey (1989). Analytical errors in relative band areas, determined by moving band edges to longer or shorter wavelengths by 10 nm, are <1%. Relative intensities of water absorption bands were calculated as described in Gaffey (1988), but the 2.3-µm rather than the 2.5-µm carbonate absorption was used to correct for particle-size effects, as a 2.5-µm H_2O absorption overlaps with the 2.5-µm carbonate band (Zabielski and Gaffey, 1989).

^1H CRAMPS (combined rotation and multiple pulse) and wideline NMR (nuclear magnetic resonance) spectra of bleached and unbleached crushed samples of modern and fossil materials were obtained at the Colorado State University National Center for NMR Applications. Acquisition of spectra is described in Gaffey et al. (1991). Water contents of the 125- to 500-µm fraction of crushed samples prepared as described in Gaffey et al. (1991) were determined from weight loss on heating to 300°C for 24 hr or 400°C for 2 hr, and from wideline NMR spectra, in which integrated intensity of the absorption feature is proportional to the number of H nuclei producing the absorption (Becker, 1980). NMR spectral and heating analyses were performed on splits of the same bulk sample. Larger fragments of coral were not used in heating analyses because they decrepitate violently on heating and material is lost from sample holders, rendering weight determinations inaccurate. Because water is lost from inclusions fractured by crushing, results of these analyses provide a lower limit for water contents of coral samples.

Polished sections of fossil corals from Quarry A cut parallel to the axis of growth were analyzed for Mg, Ca, Sr, and Na using a JEOL733 microprobe with a 15-kV accelerating voltage, a 15-nA current, and counting times of up to 40 sec. Beam diameters of 5 to 20 µm were used, depending on the size of the feature being analyzed. Because of the small size (10 to 20 µm) of many features, beam damage at all spots was examined after each series of analyses. If beam damage was not confined to the feature of interest (e.g., aragonite cement, calcitized sclerodermite centers), the analysis was discarded. Analytical errors for individual analyses for Ca, Mg, Sr, and Na varied with concentration of the element, and in aragonites were in the range of ±0.5, 9 to 20, 2 to 4, and 5 to 10%, respectively. Errors for analyses in calcites were ±0.5, 1 to 4, 8 to 20, and 25%, respectively. Data presented are averages of more than 35 analyses for each petrographic feature.

GEOLOGIC SETTING

San Salvador Island consists of Holocene and Pleistocene carbonate sediments of subtidal, reef, beach and dune facies ranging up to 700 k.y. in age (Carew and Mylroie, 1985). Pleistocene coral samples were obtained from the Sue Point patch reef, described in detail by White (1989), and from Quarry A, where corals formed part of a community growing on shallow subtidal hardgrounds (Bain, 1985). U/Th dating of a *Diploria* specimen from Quarry A yielded an age of 140 k.y. (Carew and Mylroie, 1987), and corals from the reef at Sue Point were 122 to 123 k.y. in age (Chen et al., 1991).

BACKGROUND

Structure of scleractinian coral skeletons

The basic structural unit from which all the elements of coral skeletons are constructed is sclerodermite (Fig. 2A,D), a polycrystalline bundle of fibrous high-Sr aragonite crystals radiating out from a center of calcification composed of equant aragonite crystals 1 µm or less in diameter (Fig. 2B). Each corallite, the skeleton of an individual polyp in a colony, consists of the theca, a wall surrounding the coral polyp, the septa, vertical partitions or supports radiating in from the theca, and dissepiments, thin, approximately horizontal partitions that support the polyp and separate it from the void left by upward growth of the coral (Wells, 1956).The theca in *Diploria* skeletons is dense and solid, and forms a thick partition between the corallites, while the theca in *Acropora* sp. is composed of horizontal spines or bars growing out from and connecting the septa. Corallites in *A. palmata* are connected by the coenosteum, an open grillwork of horizontal and vertical rods (Wells, 1956). More complete illustrations and discussions of coral structure can be found in James (1974), Gvirtzman and Friedman (1977), and Constantz (1986b).

Voids of several types occur within the structural elements of coral skeletons. Void spaces with diameters of 1 µm or less occur between aragonite crystals in the centers of calcification (Fig. 2B) (Wells, 1956; Sorauf, 1972; James, 1974). Radially oriented voids occur between elongated aragonite crystals and are more numerous in the inner portions of sclerodermites than in the outer portions, where the

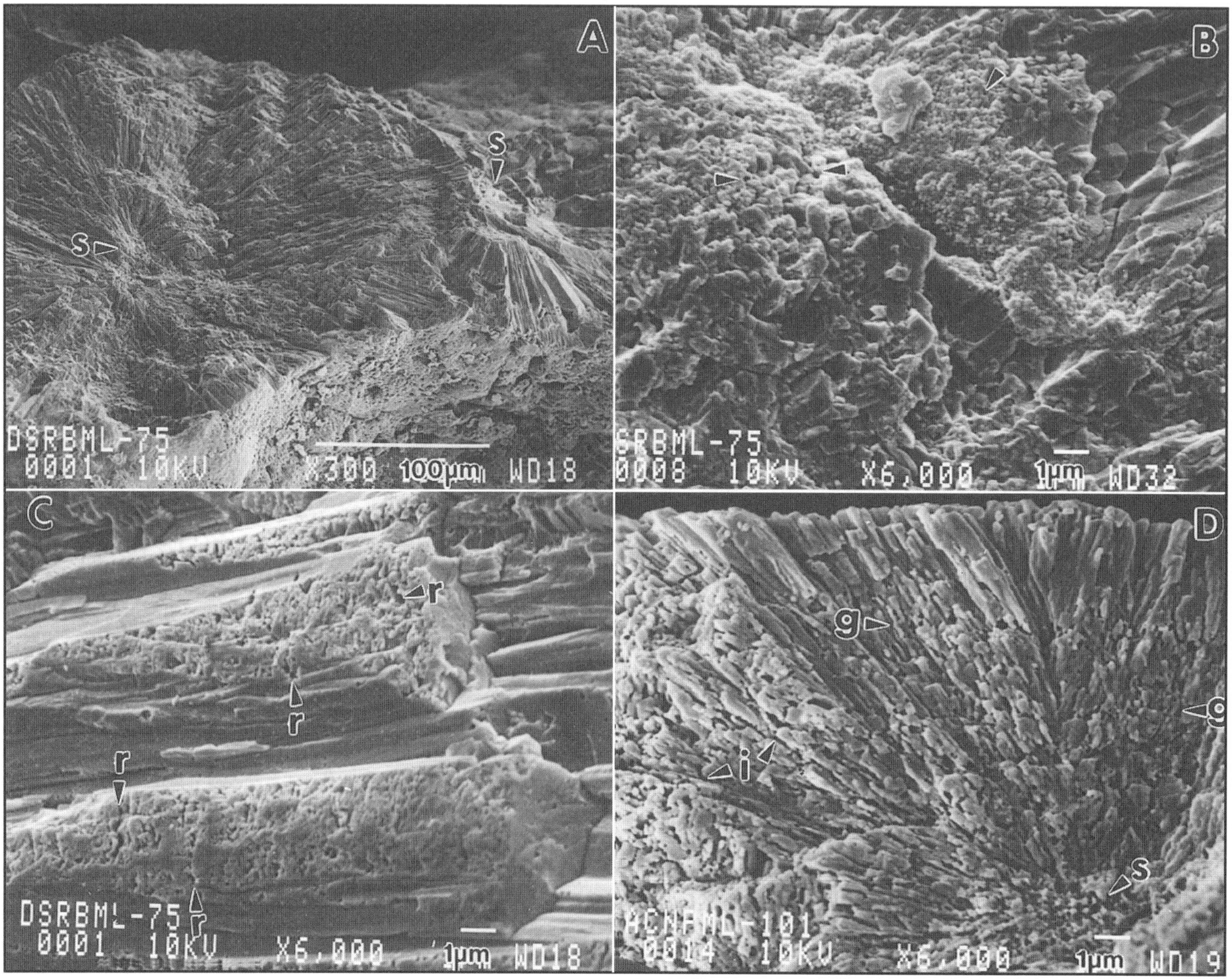

Figure 2. Structure of corals. A, Sclerodermites in *Diploria* skeleton are composed of bundles of fibrous aragonite crystals radiating out from centers of calcification (s). B, The center of calcification in a modern *Diploria* is composed of submicrometer-sized crystals. It also contains minute voids, a few of which are indicated by arrows. Radiating fibrous aragonite crystals can be seen to right and at top of photo. C, Intracrystalline voids (r) are visible in fractured crystals in a *Diploria* skeleton. D, A sclerodermite from *Acropora cervicornis* has prominent lines of voids between aragonite crystals (i). Compare to photomicrographs of *Diploria* (B and C) taken at same magnification.

crystals are coarser and more tightly packed (Fig. 2A,D) (Sorauf, 1972; James, 1974; Gvirtzman and Friedman, 1977; Johnson, 1980; Constantz, 1986b).

Void spaces occur within as well as between crystals of coral skeletons (Fig. 2C). Intracrystalline voids and growth dislocations range in size down to the limits of resolution of the transmission electron microscope (TEM) (Green at al, 1980; Bruni and Wenk, 1985), spanning a continuum of sizes between what would be termed inclusions and what would be classed as crystal defects (Gaffey et al., 1991). Lines or zones of inter- and intracrystalline voids also occur approximately perpendicular to the long axes of the aragonite crystals (Fig. 2D) and probably correspond to the growth lines described by Barnes (1970) and Sorauf (1980) that are particularly prominent in the secondary deposits that thicken the septa and thecae (Wells, 1956).

Aragonite crystals are narrower and less tightly packed in *Acropora cervicornis* and *palmata* than in *Diploria* skeletons (Pittman, 1974; Constantz, 1986b) (Fig. 2C,D). Examination with SEM indicates that in *A. cervicornis* intercrystalline voids are more abundant than intracrystalline ones (Fig. 2D), and that the zones of voids along growth lines are more prominent in *Acropora* than in *Diploria* skeletons.

Water and organics in skeletons

Some workers have stated that the voids in coral skeletons contain seawater (e.g., Walls et al., 1977). However, coral skeletons are precipitated within the tissues of the coral polyp (Wells, 1965; Johnston, 1980) and inclusions are filled with water and organics derived from the animal (e.g., James, 1974; Gvirtzman and Friedman, 1977; Johnston, 1979, 1980; Green et al., 1980; Constantz, 1986a,b; Gaffey, 1986, 1990; Lecuyer and O'Neil, 1994). Organic matter, consisting of proteins, lipids, and in some species, chitin, is thought to make up only ~0.1 wt% of coral skeletons (Johnston, 1980; Lowenstam and Weiner, 1989) (although spectral data indicate this figure may be low; see below). Coralline aragonites also contain up to 3 wt% H_2O (Bruni and Wenk, 1985; Gaffey, 1985, 1988; Zabielski and Gaffey, 1989; Gaffey, 1990; Gaffey et al., 1991; Lecuyer and O'Neil, 1994), and bleaching and heating experiments indicate water is associated with the organic material (Gaffey, 1988; Gaffey et al, 1991). The widths and positions of H_2O absorption features in reflectance spectra and the results of NMR experiments indicate H_2O molecules in inclusions in coral skeletons are more strongly hydrogen bonded than molecules in liquid H_2O. In addition, unlike molecules in liquid H_2O, those in inclusions do not rotate freely but undergo slow wagging, rocking, or twisting motions, apparently due to their association with the organic material (Gaffey et al., 1991).

RESULTS

Petrography of coral samples

Diploria. The early diagenesis of scleractinian corals has been the subject of intensive study for many years, and a number of the petrographic features observed in these samples have been documented by previous authors (see Macintyre, 1984, for summary). Samples of live *Diploria* collected from Rice Bay (sample nos. DSRBML-75 and -91) and Bamboo Point (DSBPML-77) contain borings, many still occupied by microbial endoliths or lined with organic films. Interiors of corals also contain minor amounts of aragonite cement occurring as meshes of needles, coatings on algal filaments, or syntaxial overgrowths on crystals in the coral skeleton.

The oldest parts of *Diploria* heads that were collected live show evidence of dissolution at some sclerodermite centers (Fig. 3A,B), along growth lines, and along some endolith borings. Although these corals were treated with dilute NaOCl, which can cause minor etching or pitting of carbonate (Gaffey and Bronnimann, 1933), the limited and patchy occurrence of the dissolution and its isolated location within the interior of the coral head indicate it is not an artifact of bleaching treatment.

Although the dead coral DSRBMD-92 was collected from bottom sediments in the marine environment and was heavily overgrown with macroalgae, its outer surface shows the blackened, irregular surface typical of carbonates exposed in the shallow subtidal and inter- and supratidal zones (Folk, et al., 1973; James and Choquette, 1984; Strasser, 1984). The sample does show some dissolution of the exterior surfaces of septa and dissepiments, as well as dissolution along endolith borings. Cements include syntaxial overgrowths of aragonite, and micritic cements and abundant calcified algal filaments that x-ray diffraction indicates are high-Mg calcite (8 to 14 mol% $MgCO_3$ determined from the position of the d_{104} reflection using the curves of Milliman, 1974). Sample DSRBMD-92 contains both micro- and macroborings, some of the latter filled with internal sediment composed of ooids and bioclasts. Surfaces of some macroborings are encrusted by the foraminifera *Homotrema rubrum* which still retains its original color. All cements in DSRBMD-92 are typical of those precipitated in marine environments (e.g., James and Choquette, 1983; Tucker and Wright, 1990). Blackening and dissolution of carbonate surfaces, termed "phytokarst" by Folk et al. (1973), can occur very quickly (in as little as 4 yr) (James and Choquette, 1984). The lack of extensive dissolution, low-Mg calcite cements, or neomorphic alteration of aragonite to low-Mg calcite, indicate the sample was only briefly exposed to the intertidal or supratidal environment.

A *D. labyrinthiformis* sample collected from the marine vadose zone at Bamboo Point, well rounded by wave action, also contains endolith borings and minor high-Mg calcite cements.

The three Pleistocene samples of *Diploria* from Quarry A, including the *D. strigosa* sample selected for detailed spectroscopic analyses (DSQAPL-80), show evidence of both carbonate dissolution and precipitation. Degree and style of dissolution vary considerably with individual samples. Although some portions of a given coral head are quite pristine, surfaces of septa, thecae, and dissepiments in other portions of the same coral show evidence of dissolution; in some cases, portions of dissepiments have been completely dissolved away, leaving only a residue of organic material. In general, septa show more evidence of alteration than thecae. Preferential dissolution at sclerodermite centers, resulting in centrosclerodermite voids (Fig. 3C,D) (Gvirtzman and Friedman, 1977), and parallel to the long axes of fibrous aragonite crystals, causing enlargement of radially oriented intercrystalline pores (Fig. 3D), are common phenomena described by James (1974), Gvirtzman and Friedman (1977), Sorauf (1980), Constantz (1986b), and Dullo (1986).

Preferential dissolution has also occurred along growth lines or discontinuities within the sclerodermites (Fig. 3D,E), and within intracrystalline voids, which, in some cases, become connected with intercrystalline pores (Fig. 3F). Dissolution affecting inter- and intracrystalline voids, unlike that affecting outer surfaces of skeletal elements, selectively removes carbonate along edges parallel to c axes of crystals, and along crystal defects (Fig. 3F).

Low-Mg calcite is found within skeletal elements at sites

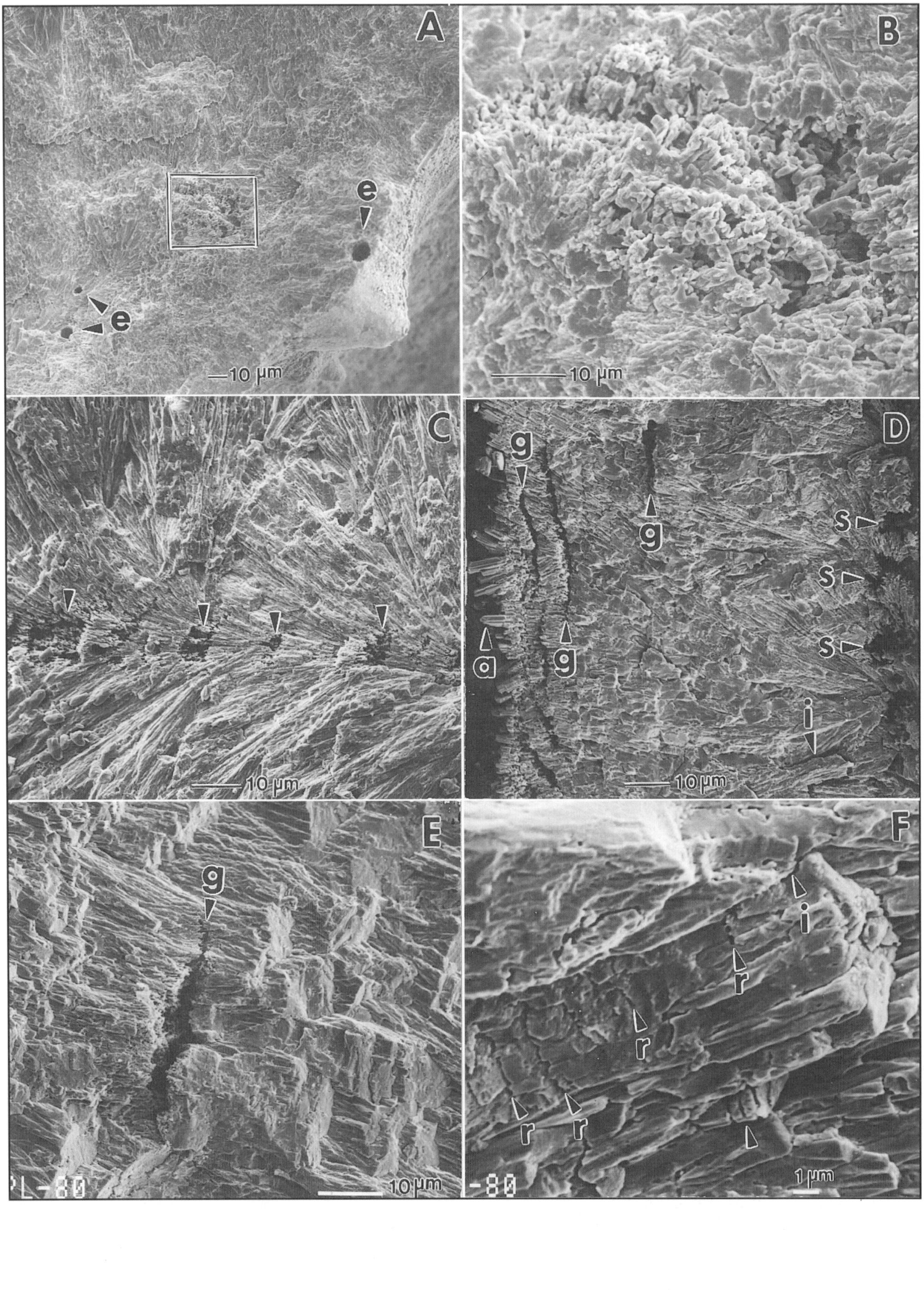
A
e
e
10 μm
B
10 μm
C
10 μm
D
g
g
g
a
s
s
s
i
10 μm
E
g
PL-80
10 μm
F
i
r
r
r
r
-80
1 μm

Figure 3. Dissolution features in modern (A and B) and fossil (C through F) *Diploria* skeletons. A, Dissolution features occur in the base of *Diploria* colony that was alive when it was collected. The box outlines the leached center of a sclerodermite shown in closeup in B. Other dissolution features formed in the marine environment include endolith borings (e). B, A closeup of the area of dissolution at the center of sclerodermite in A shows an increase in void space compared to pristine centers like that in Figure 2B. C, Leaching of centers of sclerodermites in Pleistocene *Diploria*, indicated by arrows. D, Preferential leaching of centers of sclerodermites (s), intercrystalline voids (i), and growth lines (g) in Pleistocene *Diploria*. Note the aragonite cements (a) are relatively unaffected by dissolution in comparison to aragonite of coral skeleton. E, Preferential dissolution along a growth line (g). F, Dissolution results in enlargement of intracrystalline (r) and intercrystalline (i) voids. Compare to Figure 2C.

similar to those at which dissolution has occurred, including centers of calcification (Fig. 4A,B), as observed by James (1974), Gvirtzman and Friedman (1977), Sorauf (1980), Constantz (1986b), and Dullo (1986). Calcite also occurs along growth lines (Fig 4C,D). A void space <1 µm in width between the aragonite and calcite, like those surrounding neomorphic calcites in corals described by Pingitore (1976), occurs at some sites of alteration (Fig. 4B,C) but not at others (Fig. 4A,D). Some calcites contain relict aragonite crystals, typical of neomorphic calcites (Fig 4B) (Sandberg, 1984). Calcites at sclerodermite centers and along growth lines presumably formed by neomorphic alternation of preexisting aragonite, as described

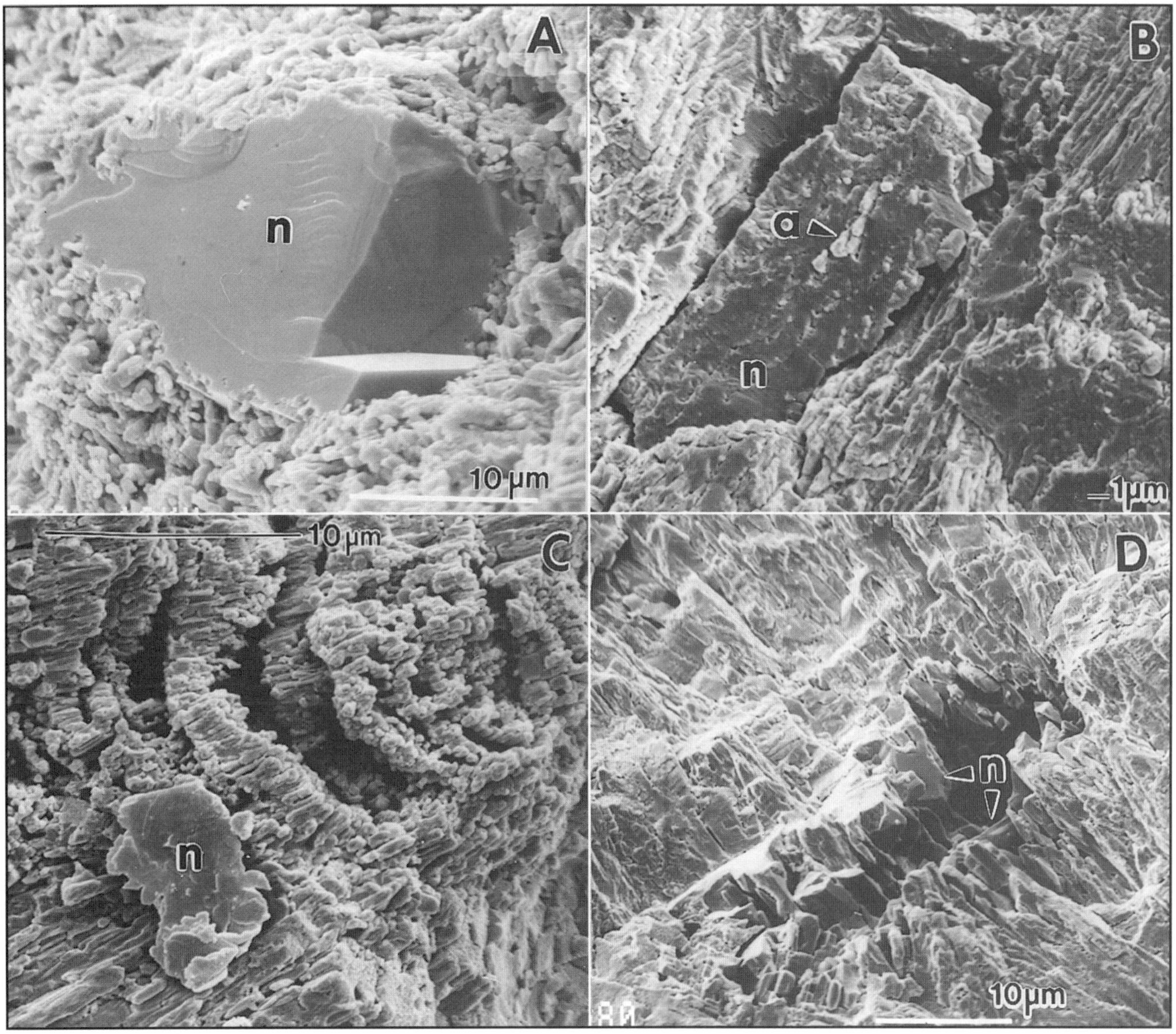

Figure 4. Neomorphism in *Diploria* skeletons. Low-Mg calcite appears smoother and darker than aragonite and does not show the fine structural detail of the original aragonite it replaces. A, Neomorphic calcite (n) replacing center of sclerodermite. B, Neomorphic calcite (n), which, unlike that in A, contains aragonite relicts (a). C, Neomorphic calcite (n) occurring along a growth line. D, Neomorphic calcite (n) formed along growth line (compare to Fig. 3E).

by the above authors. However, Saller (1992) noted that the distinction between cementation and neomorphism is not always clear, and dissolution of aragonite and subsequent precipitation of calcite can occur at distinctly different times and even within different diagenetic environments.

Observations made during microprobe analysis and with SEM indicate that calcitized sclerodermite centers are concentrated in horizons parallel to growth banding. Zones in which neomorphic calcite occurs are separated by unaltered zones, or zones in which sclerodermite centers have been affected by dissolution alone. An attempt to clarify the relationship between neomorphic alteration and growth banding was made by XRD analysis of samples removed with a small drill from different density bands in sample DSQAPL-80. However, results were inconclusive, due to the very small amounts of calcite in the samples.

Cements occur in primary voids between skeletal elements in the Pleistocene *Diploria*. Minor amounts of aragonite cements, which occur as meshes of needles or as syntaxial overgrowths (Fig. 5A), are typical of those known to form in the marine environment (e.g., Tucker and Wright, 1990). Pleistocene *Diploria* also contain two types of low-Mg calcite (as indicated by microprobe analyses; see below) cements. One forms isolated rhombs up to 100 µm in diameter (Fig. 5B), a texture typical of the meteoric phreatic environment (e.g., Budd, 1988). The occurrence of coarse calcite cement is very limited, particularly in DSQAPL-80, the sample selected for detailed spectroscopic study. More commonly, calcite cements occur as thin (10 to 20 µm) discontinuous rims, often overgrowing aragonite cements (Fig. 5A) or as meniscus cements, both characteristic of the vadose meteoric environment (see James and Choquette, 1984, for summary and references).

Meteoric cements occur on outer surfaces of skeletal elements that show dissolution in their interiors (Fig. 5C). Although in places a physical connection can be seen between meteoric cements and calcite filling shallow, near-surface endolith borings (Fig. 5A) and leached growth lines, voids more than a few tens of micrometers from the surface are generally unaffected by calcite cementation (Fig. 5C).

Acropora. Samples of *A. cervicornis* and *A. palmata* collected live are largely free of marine cements and endolith borings. A branch of dead *A. cervicornis* that was still attached to the living colony is encrusted by coralline red algae and serpulids. The interior contains some macro- and microborings

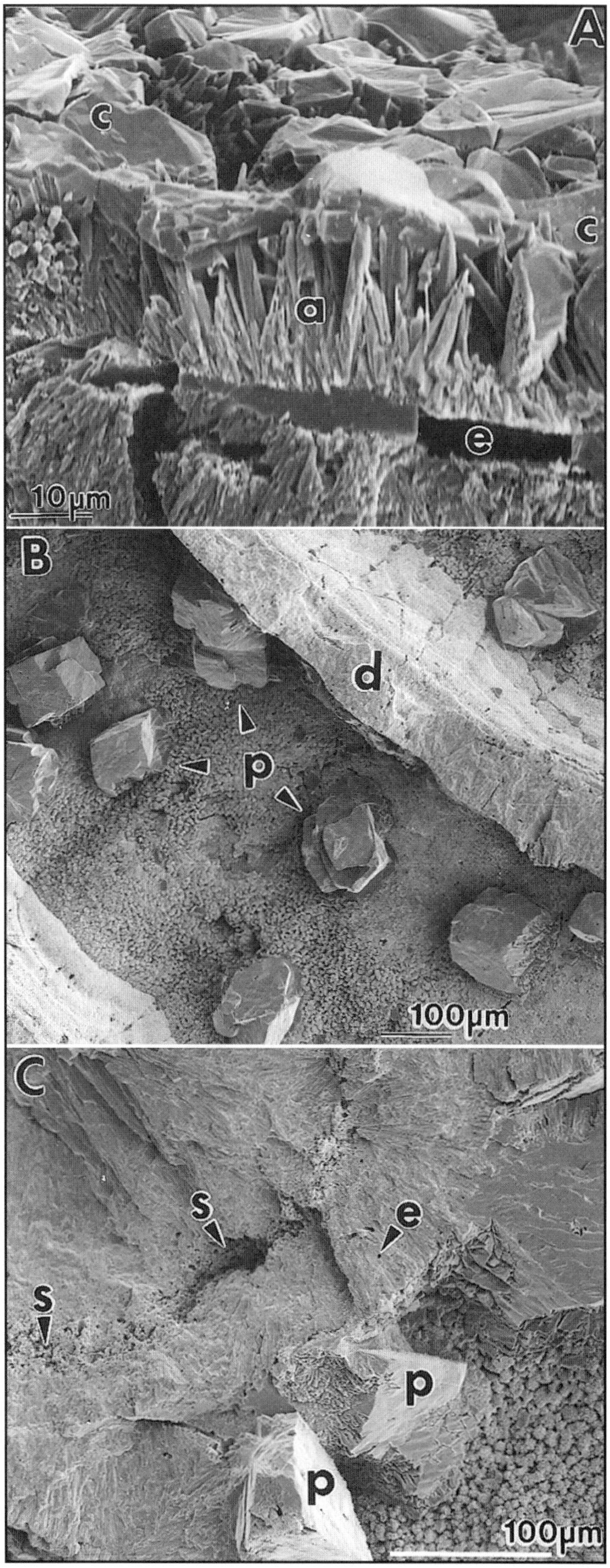

Figure 5. Cementation in Pleistocene *Diploria* skeletons. A, A thin rind of calcite cement (c) has overgrown acicular aragonite cements (a). Low-Mg calcite cements also partially fill near-surface endolith boring (e). B, Coarse, rhombic calcite cements (p) precipitated in voids between dissepiments (d). The texture and distribution of cements indicate pores were filled with water at the time of their precipitation. C, Coarse, rhombic calcite cements (p) occur on the outer surface of septum. Note that leached centers of sclerodermites (s) and endolith borings (e) do not contain cements or neomorphic calcite.

and minor amounts of aragonite cement. A dead branch of *A. palmata,* also still attached to a live colony, contains some high-Mg calcite cements, identified by XRD, and microbial endolith borings. A sample of dead *A. cervicornis* obtained from subtidal sediments is tightly cemented with aragonite. The Pleistocene *A. cervicornis* sample obtained from the fossil reef at Sue Point contains both aragonite cements syntaxially overgrowing aragonite crystals in the skeleton and coarsely crystalline low-Mg calcite crystals filling the remaining void spaces. Dissolution has occurred along growth lines and radially oriented intercrystalline voids. Degree of alteration varies over distances of a few millimeters, ranging from areas showing little alteration to "chalky" zones (James, 1974) to complete removal of coralline aragonite leaving only aragonite cements.

GEOCHEMICAL ANALYSES

Spectroscopic data

Typical VNIR reflectance spectra of the three *Diploria* samples are shown in Figure 6. Average relative areas of the 1.4- and 1.9-μm water bands with variances for spectra of modern and fossil corals are given in Table 1. Relative areas of water bands calculated from spectra for each spot analyzed on the modern and Pleistocene *Diploria* samples are shown in Figure 7. Plots of average relative band areas versus average relative band intensities for the three *Diploria* samples are shown in Figure 8.

^{1}H CRAMPS NMR spectra of modern *A. cervicornis* and of modern and fossil *D. strigosa* are shown in Figure 9. Water contents of the 125- to 500-μm fraction of ground, bleached samples of *Diploria* and *Acropora* determined from wideline NMR spectra, assuming all H was present as H_2O, are 1.95 and 2.23 wt% H_2O, respectively.

Microprobe data

Results of analyses of five different petrographic features in fossil *Diploria* from Quarry A are summarized in Figure 10. The petrographic features and number of individual analyses for each are: coralline aragonite (41), aragonite cements (35), neomorphic calcite (48), narrow calcite rim and meniscus cements (37), and coarse calcite cements (46). Because cements in endolith borings included both aragonite and calcite with a

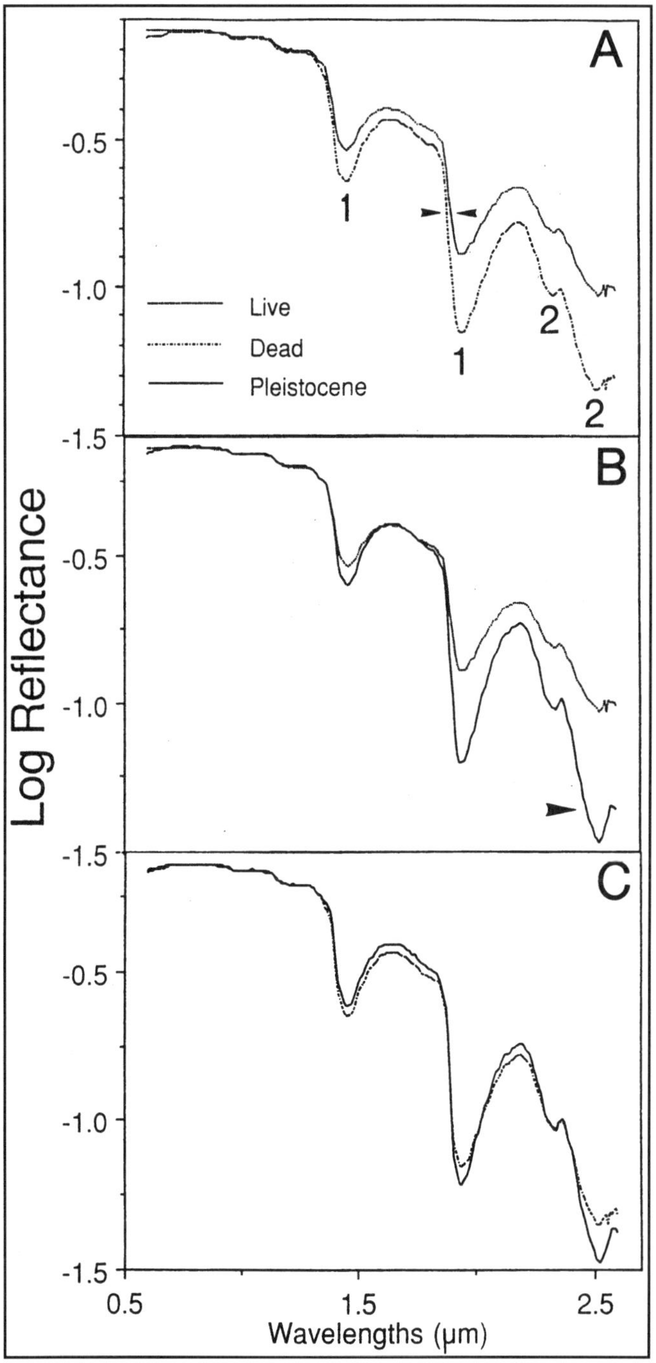

Figure 6. Representative spectra of modern live and dead and of Pleistocene *Diploria strigosa* skeleton. A, Spectra of live and dead *D. strigosa.* Numbers indicate absorption features due to H_2O near 1.4 and 1.9 μm (1) and $CaCO_3$ near 2.3 and 2.5 μm (2). H_2O absorption bands have shifted to shorter wavelengths in spectrum of dead sample (arrows) reflecting a decrease in strength of hydrogen bonding of H_2O. Although relative areas of the 1.4- and 1.9-μm H_2O bands have decreased in spectra of the dead specimen (from 11.09 to 7.26, and from 20.52 to 13.82, respectively) the intensity of the bands has increased, reflecting narrowing of the H_2O bands and a decrease in the range of bonding energies for H_2O in the dead sample. B, Spectra of modern live and Pleistocene samples. Relative areas of 1.4- and 1.9-μm bands in the spectrum of the Pleistocene samples are 5.98 and 12.73, respectively. Increased drop-off at longer wavelengths, indicated by arrow near 2.5 μm, reflects shift of the O-H stretching fundamental, centered near 3 μm, to shorter wavelengths. C, Spectra of dead and Pleistocene samples. Differences between dead and Pleistocene samples are much less marked than those seen in A and B. Some narrowing and a shift to shorter wavelengths of the H_2O absorption features is apparent in the spectrum of the Pleistocene sample.

TABLE 1. AVERAGE RELATIVE BAND AREAS AND VARIANCES FOR THE 1.4- AND 1.9-μm WATER BANDS IN SPECTRA OF *DIPLORIA STRIGOSA* AND *ACROPORA CERVICOMIS*

Sample		1.4-μm Band	Variance	1.9-μm Band	Variance	No. of Analyses
Diploria strigosa						
DSRBML-75	Septa	5.73	1.09	17.76	3.47	5
	Thecae	8.54	3.71	17.13	3.72	5
DSRBMD-92	Septa	4.68	0.48	11.87	0.17	4
	Thecae	5.95	1.38	13.41	1.01	4
DSQAPL-80	Septa	4.04	0.13	10.98	0.63	5
	Thecae	5.74	0.27	12.99	0.17	5
Acropora palmata						
APGMBL-9		7.41	0.50	18.62	1.33	7
Corallite		7.69	0.76	19.43	0.44	4
Coenosteum		7.05	0.01	17.54	0.24	3

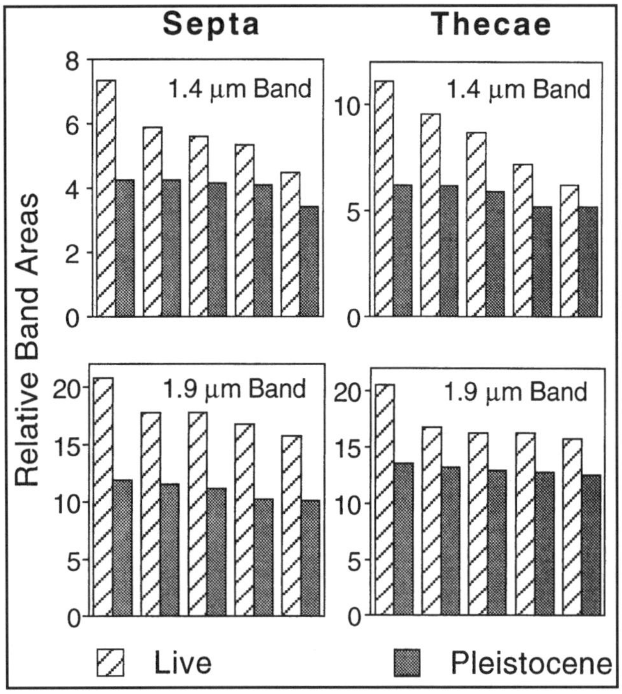

Figure 7. Relative areas of H_2O bands for all spectra obtained from the live and Pleistocene *Diploria* samples. H_2O content is lower and H_2O distribution is less heterogeneous in the Pleistocene sample.

range of Mg contents, and because the number of analyses (11) was limited, these data are not presented here.

Heating data

Mean water contents and standard deviations for coral samples determined by weight loss on heating were 2.48 ± 0.10 wt% (five analyses) for modern *Diploria* (DSRBML-91), 2.53 ± 0.17 wt% (four analyses) for fossil *Diploria* (DSQAPL-80), and 2.75 ± 0.15 wt% (three analyses) for modern *Acropora* (ACNPML-101). Because these samples were ground and bleached before analysis and the organic contents were low, and because the samples were heated under N_2 to prevent oxidation of organic matter, the bulk of the weight change is due to water loss.

DISCUSSION

Amount and distribution of water in coral skeletons

Coral skeletons are compositionally heterogeneous, and total water content varies over distances of millimeters to centimeters within a single skeleton. Data in Figure 7 show relative areas of water bands in spectra of each individual 2-mm spot analyzed on the modern live and Pleistocene *Diploria strigosa* samples described above. The spectrum of each spot is actually an average of a large number of individual spectral measurements, and the error bars are smaller than the lines or dots used to plot the spectrum. Thus differences in areas and intensities of absorption features between spectra obtained from different spots do not reflect analytical errors, but represent real variations in the spectral properties of the skeleton. In the spectra of the *Diploria* head that was collected live, areas of water bands indicate that relative variations in total water content up to 60% can occur within a single skeleton. In general, total water contents are higher and show greater variability in the thecae than in the septa (Fig. 7 and Table 1).

Water is less heterogeneously distributed in the *A. Palmata* skeleton than in *Diploria* (variances in Table 1). The smaller variances in the *Acropora* data are due in part to the smaller size of the corallites and of the individual crystals making up *Acropora* skeletons. Corallites in *Acropora* skeletons are only 1.5 to 2.0 mm in diameter (Constantz, 1984), and the 2-mm spots from which spectra were obtained cover more than one skeletal element, resulting in some averaging of intracorallite variations in water content. However, the differences in variance also reflect differences in the structure of the skeletons produced by the two genera of corals. Measurements show that little difference exists between water contents of the corallite and the coenosteum (the area between corallites jointly constructed by neighboring polyps), in agreement with Wells (1956) observations that structures of the septa and coenosteum are similar.

Figure 8. A, Average band area plotted as a function of average band intensity for the three *D. strigosa* samples. In all cases, except the 1.4-μm feature in the thecae, band intensities are greater (but band areas are smaller) in spectra of dead and fossil samples than in spectra of live samples. This reflects narrowing of the features, seen in the spectra shown in Figure 6. B, Schematic drawings illustrating changes in area, full width at half height, and intensity of H_2O bands with time. If the intensity increases while the area decreases, plane geometry requires that the width of the features decreases to compensate for the increase in intensity. Values given are in arbitrary units.

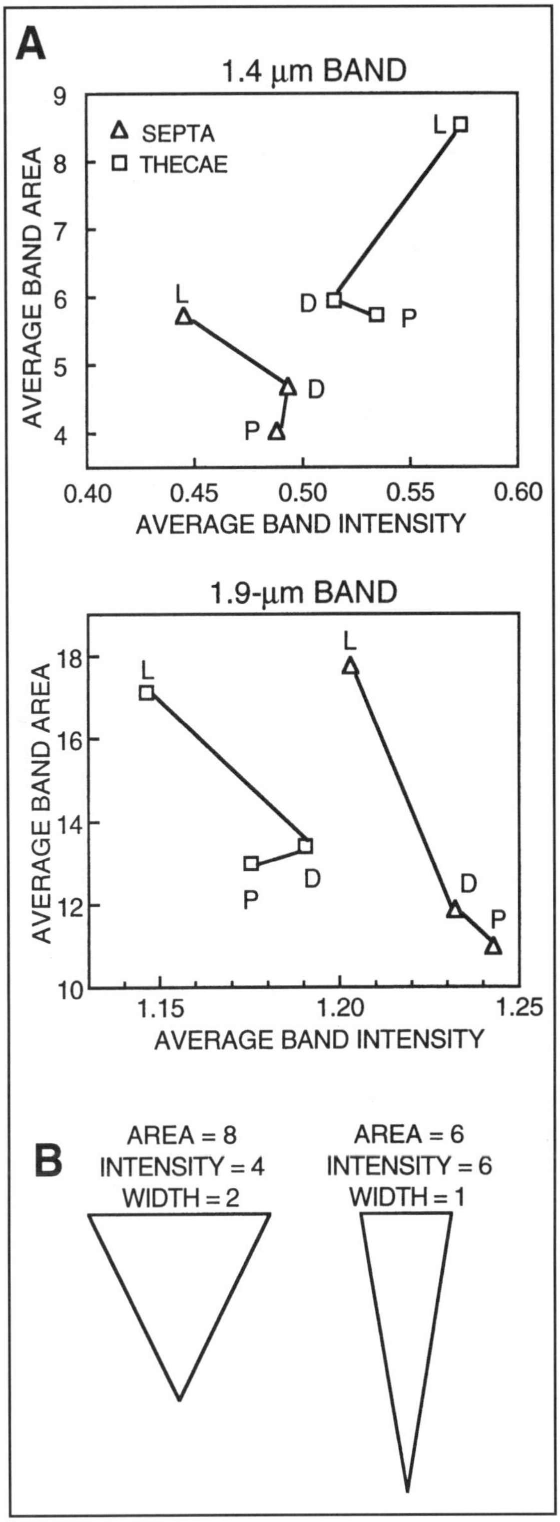

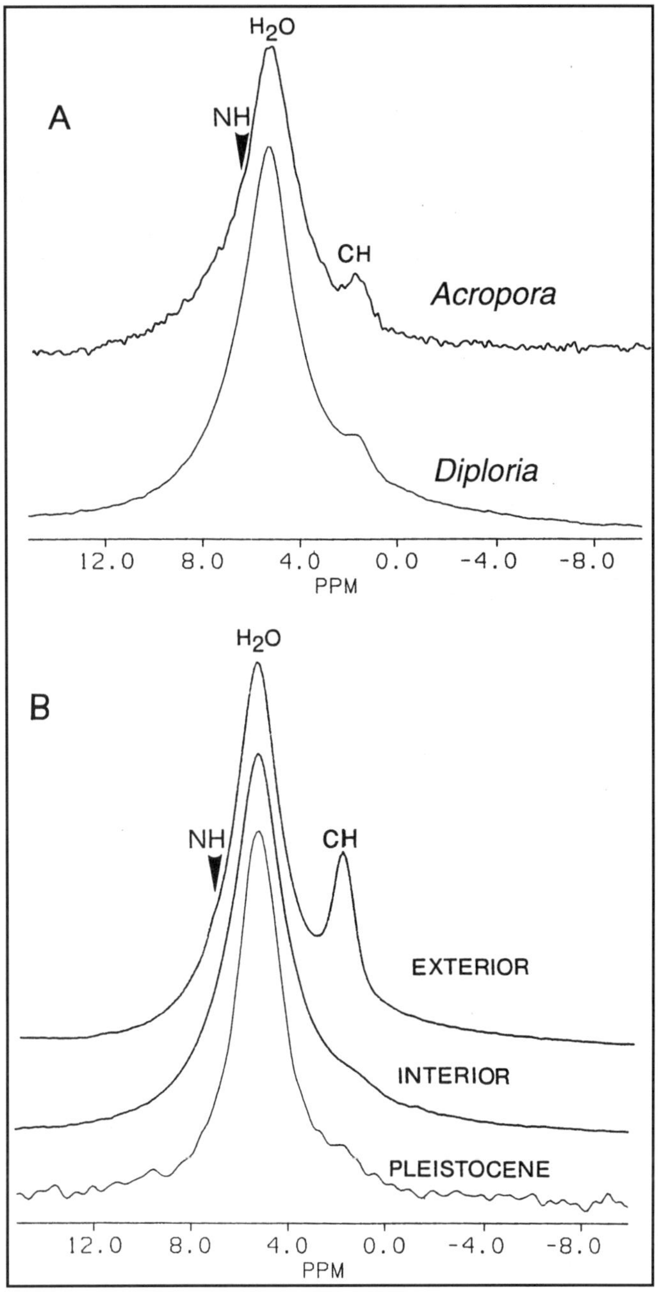

Figure 9. ^{1}H CRAMPS NMR spectra of modern and fossil corals. A, Spectra of *Acropora cervicornis* collected at North Point, and *Diploria strigosa* from Fernandez Bay. H_2O produces the resonance near 5 ppm, CH in lipids that near 1.5 ppm, and NH in proteins produces the resonance near 7 ppm. Both samples were bleached for 3 days in 5% NaOCl. B, Spectra of unbleached modern and fossil *D. strigosa* samples. The decrease in intensity of the 1.5-ppm feature and the masking of the 7-ppm feature indicate loss of lipid and protein from samples. Narrowing of the 5-ppm feature in the spectrum of the fossil sample reflects decrease in range of bonding energies and increased mobility of H_2O in the fossil sample (see Gaffey et al., 1991).

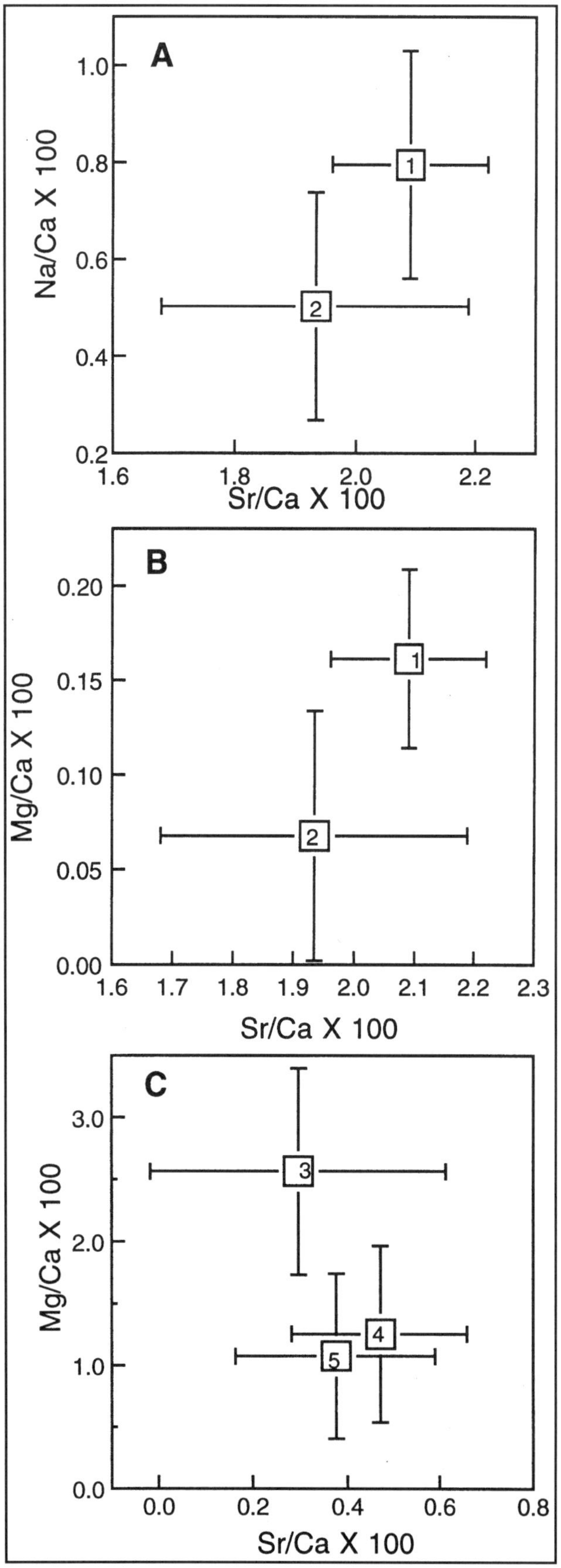

Figure 10. Probe data for aragonites (A and B) and calcites (C) in Pleistocene *Diploria* samples. Error bars are 1σ of the mean. A, Average Na and Sr contents of coralline aragonite (1) and aragonite cements (2). B, Average Mg and Sr contents of coralline aragonite (1) and aragonite cements (2). C, Average Mg and Sr of coarse calcite cements (3), neomorphic calcite (4), and calcite rim cements (5). Na concentrations in calcites were below the detection limits for the operating conditions used in these analyses. The error bars for Sr in the coarse calcite cements are larger than the mean Sr values, and the concentration of Sr in the coarse calcite cements is essentially 0.

Reflectance spectra of the *Acropora* sample that was tightly cemented with acicular aragonite indicate that the water content of aragonite cements is lower than that of coralline aragonites. Relative areas of the 1.4-μm water bands in spectra of the uncemented live and dead *A. cervicornis* branches were 7.41 and 8.82, respectively, while the 1.9-μm areas for the same spectra were 17.62 and 18.52. The relative areas of water bands calculated from spectra of the tightly cemented *A. cervicornis*, on the other hand, were 6.58 ± 0.22 and 15.31 ± 0.16 (three measurements). Spectra of the cemented *Acropora* sampled both coral skeleton and cements, so that water band areas represent an average of water contents from both. The lower water content of cements is in good agreement with previous observations that aragonite cements contain fewer growth dislocations and crystal defects than biogenic aragonites (e.g., James, 1974; Constantz, 1986b).

Organic matter in coral skeletons

Spectral data indicate that proteins and lipids occur in *Diploria* and *Acropora* skeletons, and that total organic contents are significantly higher than the generally accepted value of 0.1 wt%. A ^{1}H CRAMPS spectrum of a sample taken from the outermost part of a *Diploria* colony still occupied by polyps (Fig. 9B, exterior) shows a strong resonance near 5 ppm due in large part to water. Features due to -CHO- and -CHN- in proteins also occur in the 5-ppm region, although the magnitude of their contribution to that resonance is undetermined at present. Asparitic and glutamic acid predominate in proteins in coral skeletons (Young et al., 1971; Mitterer, 1978), and a weaker resonance near 7 ppm is attributable to NH_3 in these amino acids (Wilson, 1987). The CRAMPS spectrum also contains a strong feature near 1.5 ppm. The 1.5-ppm feature in spectra of these same coral samples was originally assigned to mineral hydroxyl (Gaffey et al., 1991). However, reflectance spectra of heated coral samples lack OH features, and the 1.5-ppm feature is weaker in spectra of bleached than of unbleached samples (Fig 9), indicating the material producing it is organic. Thus the 1.5-ppm feature in these coral spectra is best assigned to CH in the lipids corals are known to contain (Young et al., 1971; Johnston, 1980; Wade, 1987; Wilson, 1987).

The intensities of features in ^{1}H CRAMPS NMR spectra are approximately proportional to the number of nuclei causing the absorption (Becker, 1980). Thus, the strengths of the 1.5- and 7.0-ppm resonances relative to that of the 5-ppm resonance, known to be produced by ~2 wt% H_2O, as determined from wideline spectra of bleached powders, indicate that the organic content of newly precipitated coralline aragonites is probably higher than the generally accepted value of 0.1 wt%. The spectra of the bleached samples (Fig. 9A) contain the 1.5-ppm resonance, indicating that both *Diploria* and *Acropora* contain small amounts of lipids in intracrystalline voids. In addition to lipids, proteins are known to occur in powdered, bleached coral samples (Young et al., 1971). Although the 7-ppm NH resonance is masked by the stronger 5-ppm resonance in the *Diploria* spectrum, it is still evident in the spectrum of the bleached *Acropora* sample, indicating that protein also remains in intracrystalline voids.

VNIR reflectance spectra of coral samples also contain absorption features produced by NH in proteins, CH in all types of organics, and by photosynthetic pigments produced by microbial endoliths (Gaffey, 1990).

Diagenetic changes in organics and water

Destruction of organic material. Breakdown of organic compounds is assumed to be one of the earliest diagenetic changes affecting carbonate skeletons, beginning in the marine environment immediately after abandonment by the polyps (e.g., Schroeder, 1969; James, 1974; Gvirtzman and Friedman, 1977). Yet little work has been done to monitor the magnitudes and rates of changes in the organic contents of biogenic carbonates during diagenesis. Johnston (1979, 1980) found that, when first formed, each aragonite crystal is surrounded by an organic sheath. However, the sheath disintegrated with time, and organic materials obtained at depth within a skeleton differed significantly from those present when the skeleton first formed. Both VNIR reflectance (Gaffey, 1990) and ^{1}H CRAMPS NMR (Fig. 9B) spectral show the weakening of NH and CH absorption features with time. The intensity of the 1.5-ppm resonance due to lipids and the 7-ppm resonance attributable to proteins in spectra of the sample obtained from the interior of the modern coral head are similar to those in spectra of bleached samples (Fig. 9A,B), indicating much or all of the intercrystalline organic material in the interior of the coral head was lost while the colony was still alive. The similarity of the intensities of the 1.5-ppm resonances in spectra of bleached and Pleistocene samples could indicate that much of the organic material remaining in the Pleistocene sample is contained within intracrystalline voids. However, loss of organics is a heterogeneous process and does not occur uniformly throughout the coral head, as indicated by organic residues left by dissolution of dissepiments in some parts of fossil coral samples.

Breakdown of organic material can be mediated by the activity of bacteria or can occur by inorganic processes such as oxidation or hydrolysis. Breakdown of intercrystalline organics is probably caused by microbial activity. Interiors of coral heads are infested with microbial endoliths (Lukas, 1974), and bacteria and fungi capable of breaking down organic compounds have been isolated from apparently intact portions of skeletons, as well as from older portions of colonies affected by biologic or physical erosion (Johnston, 1980). Endolith borings in the samples in the present study were commonly observed to run along lines of sclerodermite centers, which could facilitate destruction of the organic matter concentrated there. Organics in intracrystalline inclusions, on the other hand, would presumably be isolated from bacterial activity, and the fact that organic material is still retained in intracrystalline voids in Pleistocene corals may indicate alteration of intracrystalline organics takes place by slower abiotic processes.

Changes in bonding of H_2O. Changes in bonding interactions of H_2O molecules could increase the ability of water in inclusions to serve as a medium for dissolution and reprecipitation of the $CaCO_3$ in the crystals surrounding them. Reflectance and NMR spectral show that H_2O molecules in inclusions in the skeleton of the fossil coral samples display more liquid-like behavior (that is, a decrease in strength of hydrogen bonding and an ability to undergo molecular rotation) than H_2O in inclusions in samples collected live.

A shift to shorter wavelengths of the 1.4- and 1.9-μm absorption bands in the spectra of the dead and Pleistocene samples relative to the positions of the same absorption features in the spectrum of the modern sample (Fig. 6A,B), indicates a decrease in the energy of hydrogen bonding for H_2O (Hamilton and Ibers, 1968). Although the center of the O-H stretching fundamental of H_2O occurs near 3 μm, outside the spectral region covered in this study, a shift in its position is evident in the 2,3- to 2.7-μm region of coral spectra (Fig. 6A,B), where the wings of the strong O-H stretching fundamental cause the marked dropoff of the spectrum toward longer wavelengths (Gaffey, 1988). Although the total water content of the dead and Pleistocene samples is lower than that of the modern sample, and the 3-μm absorption is correspondingly less intense, the dropoff at longer wavelengths is more pronounced in the spectrum of the Pleistocene sample, due to the shift of the 3-μm absorption to shorter wavelengths.

A decrease in the strength of hydrogen bonds for H_2O in coral samples is also indicated by the narrowing of H_2O features in VNIR spectra (Hamilton and Ibers, 1968), seen in Figure 6 and in the plots of average band area versus average band intensity for the three *Diploria* samples in Figure 8. With the exception of the 1.4-μm band in the spectra obtained from the thecae, the H_2O features in the spectra of the dead and Pleistocene samples have smaller relative areas but greater relative intensities than the same features in spectra of live samples. If the H_2O absorption features are approximated as triangles, it can be seen that a decrease in band area, accompanied by an increase in band intensity (or height of the triangle) necessitates a decrease in band width (Fig. 8B). Narrowing of the H_2O features can also indicate a decrease in the range of bonding energies for H_2O.

Broad 5-ppm resonances in NMR spectra of modern coral

samples can be attributed to slow molecular motion for H_2O. The narrowing of the 5-ppm resonance in the spectrum of the fossil coral (Fig. 9B) indicates molecular rotation of H_2O molecules, an interpretation confirmed by dephasing experiments (see Gaffey et al., 1991, for discussion and references).

Laboratory studies indicate that the types of changes in bonding interactions of water molecules in inclusions observed in this study are attributable to the breakdown of the organic material with which the H_2O is associated (Gaffey et al., 1991). Spectral data obtained in the present study indicate that both the breakdown of organic compounds and associated changes in molecular motion and bonding energies of H_2O can occur very rapidly, within a matter of years or decades. The indeterminate age and the evidence for a period of subaerial exposure of the dead *Diploria* skeleton (DSRBMD-92) might indicate that subaerial exposure was necessary to initiate the observed changes in chemical environments for H_2O in coral skeletons. However, slight narrowing of the 1,4- and 1.9-μm features and an increased dropoff toward longer wavelengths, indicating a shift in the 3.0-μm fundamental, are evident even in spectra from the interior portions of the *Diploria* skeleton collected live (DSRBML-75). Growth banding indicates the coral head was ~25 yr old, indicating changes in the chemical environments for H_2O in inclusions begins very early, while skeletons are still in the marine environment. As would be expected, spectral differences between the interiors and outer portions of live coral colony are less pronounced than those between spectra of the live and dead *Diploria* samples seen in Figure 6A. However, differences in positions and widths of water bands between spectra of modern samples collected live and of dead samples obtained from the sediment are much more pronounced than differences between dead and Pleistocene samples for all species studied, again indicating changes in water and organics in coral skeletons occur very early.

Loss of H_2O. Coral skeletons in this study appear to have undergone a progressive loss of H_2O with time. Spectra of the live, dead, and fossil *Diploria strigosa* samples show a progressive decrease in relative band areas of both the 1.4- and 1.9-μm absorption bands (Figs. 7 and 8). Band areas for all *Acropora* and *Diploria*, as well as other genera of corals studied also show a decrease in relative areas of H_2O bands with time. However, a decrease in relative areas of the 1.4- and 1.9-μm bands may result from one or a combination of the following factors: (1) a decrease in the absorption coefficient for H_2O due to a decrease in hydrogen bonding energy (Hamilton and Ibers, 1968); (2) the loss of proteins that produce absorption features that partially overlap with those of H_2O and contribute to total band areas (Gaffey, 1990); or (3) the decay of hygroscopic organic matter and concomitant loss of the water with which it is associated. The fact that H_2O bands show an increase rather than a decrease in intensity, which would result from the occurrence of (1) and/or (2), indicates that (3) is the dominant process.

There is also a decrease in heterogeneity in the distribution of water within the skeleton over time, as indicated by variances in Table 1 and data in Figure 7, and much of the water loss probably occurs from intercrystalline voids. Water content of the bleached 125- to 500-μm fraction of the modern and Pleistocene *Diploria* samples were virtually identical. Since much of the water in these ground, bleached samples is contained in intracrystalline voids (Gaffey et al., 1991), the water contents of the intracrystalline voids are largely unaffected by early diagenetic alteration.

Intraskeletal dissolution and neomorphic alteration of corals

Heterogeneity characterizes the early diagenetic alteration of *Diploria*. Sclerodermite centers or growth lines that have been preferentially altered to low-Mg calcite can occur a few millimeters from centers and growth lines that have been affected by dissolution alone, or that show no evidence of alteration.

Role of organics. Diagenetic alteration of the mineral components of coral skeletons may be linked to diagenesis of the organic matrix (Johnston, 1979). As long as organic matter remains intact, it coats $CaCO_3$ and isolates it from pore waters, preventing dissolution.

Organic matrix materials may also contribute to the chemical stability of skeletal carbonates. Although the role, if any, played by the organic material in the formation of coral skeletons is a subject of debate (e.g., Johnston, 1980; Constantz, 1986a), some components of organic matrices are known to bind calcium and other metal ions, and laboratory studies have shown that decalcified organic material removed from skeletons can promote the growth of $CaCO_3$ crystals that mimic the morphology of the natural skeletal material (e.g., Mitterer, 1975; Wheeler and Sikes, 1984). If, as some workers suggest, the organic material in coral skeletons constitutes an organic matrix that can bind Ca^{2+} or act as a template for crystal nucleation and growth (Johnston, 1979, 1980), it may contribute to the stability of the aragonite it helped precipitate. In addition, hydrogen bonding of H_2O to organic compounds and the restricted motion of H_2O molecules in inclusions may prevent it from acting as a medium for dissolution or alteration. Breakdown of organic matter would remove all these stabilizing factors.

When organic matter begins to break down it may facilitate the alteration of carbonates. Decay of organic matter exposes additional crystal surfaces to pore waters (Schroeder, 1969). Shallow, tropical marine waters are saturated with respect to aragonite. However, breakdown of organic matter can result in undersaturation of marine pore waters with respect to calcium carbonate, resulting in dissolution of carbonate (e.g., Tribble et al., 1990; Walter and Burton, 1990). Gvirtzman and Friedman (1977) concluded from density measurements and mercury porosimetry data that centers of calcification in reef corals collected from the marine environment were enlarged and interconnected by dissolution. Early dissolution observed in the present study occurred at sites within the coral head that have the highest concentrations of organics, i.e., at centers of sclerodermites (Fig. 3A,B), along growth lines, and around endolith borings, within the abandoned portions of skeletons of still-living colonies.

Role of water. H_2O inclusions in coral skeletons can serve as a medium for alteration of aragonite to low-Mg calcite in the laboratory (Gaffey et al., 1991). One goal of the present work was to determine whether fluid inclusions could play a similar role in neomorphic alteration in the natural environment.

Availability of water controls the rate of dissolution and reprecipitation of $CaCO_3$, and hence the rate of neomorphic alteration of coralline aragonites (e.g., Pingitore, 1976). Diffusion of cations through static films of H_2O a micrometer or less in width is generally accepted as the mechanism for neomorphic alteration of corals in the meteoric vadose environment (e.g., Kinsman, 1969; Pingitore, 1976; Constantz, 1986b). All skeletal carbonates contain water in the form of fluid inclusions in quantities sufficient to serve as a medium for diffusion, and infiltration by meteoric waters may not be required to initiate dissolution. A number of authors (James, 1974; Pingitore, 1976; Gvirtzman and Friedman, 1977; Schroeder, 1980; Bruni and Wenk, 1985; Constantz, 1986b; Dullo, 1986) have noted that dissolution and neomorphism of coral skeletons in the meteoric vadose environment often begin not at the outer surfaces of the septa, thecae, or dissepiments, but in their interiors at the centers of calcification where primary inclusions are most abundant.

Petrographic evidence indicates that earliest diagenetic alteration of *Diploria* is heterogeneous, occurring within isolated diagenetic subenvironments within coral heads. Although breakdown of intercrystalline matter and enlargement of centrosclerodermite voids increase the porosity of the skeletons of Faviids, the family of corals to which *Diploria* belongs, it has little or no effect on their permeability (Gvirtzman and Friedman, 1977). Sites of dissolution and neomorphism in the Pleistocene *Diploria* in this study occur within a few millimeters of each other. Cements can be seen to have precipitated on the outer surfaces of some skeletal elements whose interiors show evidence of dissolution alone (Fig. 5C). In other cases, meteoric cements are seen to fill voids near the surfaces of skeletal elements, but not to extend to voids farther within (Fig. 5A,C). Either all cementation precedes dissolution, which is known not to be the case, or dissolution, neomorphism, and cementation occurred within separate, localized diagenetic subenvironments with different pore water chemistries.

The physical isolation of sites of early dissolution and neomorphism, however, may be apparent rather than real and connections among them and with formation pore systems may be present, but not visible in the plane of the sections studied. Trace element chemistry and isotopic composition of neomorphic calcites can be used to determine water/rock ratios and the degree of openness of a diagenetic system. During neomorphic alteration of aragonites in an open system, Sr and Na will diffuse away and Mg will move toward the reaction site. Neomorphic calcite precipitated in an open diagenetic system will have higher Mg concentrations and lower Sr and Na concentrations than the aragonite it replaces. On the other hand, within a completely closed system concentration of Sr and Mg in pore waters will rise as dissolution of aragonite and precipitation of neomorphic calcite proceeds until the calcites precipitated have the same trace element composition as the aragonites from which they were derived (e.g., Kinsman, 1969; Pingitore, 1978; Brand and Veizer, 1980).

Facilities were not available to determine the isotopic composition of the small amounts of neomorphic calcites in these samples, but trace element chemistry was determined using microprobe, and the data are summarized in Figure 10. Alteration of coralline aragonites to calcite by heating in the lab did not cause a change in Sr or Mg concentrations (Gaffey, et al., 1991). Neomorphic calcites in the Pleistocene *Diploria*, however, have lower Sr and higher Mg contents than the aragonites surrounding them (Fig. 10). These trace element analyses could be explained in two ways:

1. There is an intermittently open diagenetic system. The observations in this study may fit Pingitore's (1976) two-water model for vadose meteoric diagenesis, in which dissolution and alteration take place across static, thin films of water that only intermittently come into contact with ground waters passing through the sediments. When pore waters at sites of neomorphic alteration are in communication with formation waters, Sr is transported from, and Mg to, the site of calcite precipitation, resulting in neomorphic calcites with lower Sr and higher Mg concentrations than the precursor aragonites. The comparatively high Sr and low Mg contents of some neomorphic calcites (Fig. 10C) could reflect the presence of aragonite relicts. The similarity of Mg, Sr, and Na contents of the neomorphic calcites and rim and meniscus cements would give added support to existence of a partially open diagenetic system.

2. There is a lack of equilibration of a closed diagenetic system. Because of the differences in partitioning coefficients for trace elements in aragonite and calcite, the first calcites precipitated by neomorphosing fluids will have lower Sr contents than the aragonites from which they are derived (Kinsman, 1969; Pingitore, 1978; Brand and Veizer, 1980). Aragonite dissolution may not have been extensive enough to raise the Sr concentrations in pore fluids to levels required to initiate precipitation of high Sr calcites. Mg concentrations are higher in the neomorphic calcites than in the surrounding aragonite, indicating Mg has been imported to the system (Fig 10B,C). However, several authors (e.g., Cross and Cross, 1983; Bar-Matthews et al., 1993) have found that before neomorphism begins Mg is preferentially lost from coralline aragonites while Sr contents remain the same or even increase. Thus changes in Mg and Sr concentrations in intraskeletal pore waters may not follow the path predicted from idealized models. Additional Mg could be derived from endolithic algae. Infestation by endolithic algae produces the commonly observed green banding in corals (Lukas, 1974). Loss of Mg from algal sheaths and from algal pigments, resulting in the loss of color from photosynthetic pigments as seen in the visible region of reflectance spectra (Baker and Louda, 1986; Gaffey, 1990) can yield Mg that is incorporated in diagenetic carbonates (Gebelein and Hoffman, 1973). Preferential

loss of Mg from aragonites and addition of Mg from endolithic algae could explain the fact that Mg contents of neomorphic calcites are higher than those in surrounding aragonites.

The differences in trace element chemistry between the neomorphic calcites and the coarse calcite cements are not currently understood. Communication between formation waters and sites of alteration in the interiors of skeletal elements would most likely occur when pores were saturated with fluid rather than when vadose conditions prevail. The petrography of the coarse calcite cements indicates pores were filled with fluid during their precipitation, and their comparatively high Mg concentrations and the virtual lack of Sr indicate precipitation in a chemically open system. Although it might be expected that the trace element chemistry of the neomorphic calcites would also reflect the occurrence of this open diagenetic system, the trace element chemistry of the neomorphic calcites more closely resembles that of the vadose cements.

At present it is not possible to determine whether the H_2O which served as a medium for reaction in the neomorphic alteration of the *Diploria* samples was that incorporated during growth of the skeleton, seawater that permeated the coral after disintegration of organics, or meteoric water to which the corals were exposed after the fall of sea level. Additional data are needed to determine the diagenetic role, if any, of water incorporated in skeletal carbonates during growth. Lecuyer and O'Neil (1993) have recently determined that H_2O in fluid inclusions in skeletal carbonates has very high δO^{18} values. Analyses of the isotopic composition of H_2O in inclusions and of neomorphic calcites formed at this very early stage of alteration in fossil corals could shed additional light on the nature of pore fluids in these samples.

CONCLUSIONS

Coralline aragonites are multicomponent systems composed of aragonite and H_2O associated with organic matter in inter- and intracrystalline inclusions. Early dissolution and neomorphic alteration of coral skeletons begins at sites with the highest concentrations of water and organics. Breakdown of organics and some dissolution of the skeleton at centers of sclerodermites, along growth lines, and around endolith borings occur in the marine environment within the abandoned portions of still-living colonies. Dissolution and neomorphism to low-Mg calcite in the meteoric environment occur at the same organic- and water-rich sites.

If, as some workers suggest, the organic matter in coral skeletons comprises an organic matrix that can bind calcium and induce and control precipitation of aragonite, it may serve to stabilize the aragonite it precipitated. Loss of organics exposes mineral surfaces to pore fluids and can change pore water chemistry to facilitate dissolution. Breakdown of organic matter in coral skeletons causes changes in the bonding interactions of the H_2O with which the organic materials are associated in inter- and intracrystalline inclusions, which may allow it to act as a medium for diagenetic alteration. Because water is a primary component of skeletal carbonates incorporated during growth, preservation of ancient aragonites in the geologic record may require isolation from bacteria and oxygen that can break down organic compounds within skeletons, as well as from pore fluids to serve as a medium for alteration.

ACKNOWLEDGMENTS

We thank Carle Pieters of Brown University for the use of her spectrophotometer, and Steve Pratt of RELAB at Brown for his assistance in obtaining reflectance spectra. We thank Don and Kathy Gerace and the staff of the Bahamian Field Station for their logistical support in the field. Thanks also go to Kenneth Towe and an anonymous reviewer whose helpful comments significantly improved the manuscript. Funding for this work was provided by National Science Foundation Grants EAR-8721094 and EAR-9117905, and NMR spectra were obtained at the Colorado State University National Center for NMR Applications funded by NSF Grant CHE-8616437.

REFERENCES CITED

Bain, R. J., 1985, Subtidal-beach-dune sequence, Quarry A, *in* Curran, H. A., ed., Pleistocene and Holocene Carbonate Environments on San Salvador Island, Bahamas: Geological Society of America, Orlando, Annual Meeting Field Trip Guidebook: San Salvador, CCFL Bahamian Field Station, p. 63–72.

Baker, E. W., and Louda, W., 1983, Porphyrins in the geological record, *in* Johns, R. B., ed., Biological Markers in the Sedimentary Record: New York, Elsevier, p. 125–225.

Bar-Matthews, M., Wasserburg, G. J., and Chen, J. H., 1993, Diagenesis of fossil coral skeletons: Correlation between trace elements, textures, and $^{234}U/^{238}U$: Geochimica et Cosmochimica Acta, v. 57, p. 257–276.

Barnes, D. J., 1970, Coral skeletons: An explanation of their growth and structure: Science, V. 170, p. 1305–1308.

Becker, E. D., 1980, High Resolution NMR: New York, Academic Press, 354 p.

Brand, U., and Veizer, J., 1980, Chemical diagenesis of a multicomponent carbonate system—1. Trace elements: Journal of Sedimentary Petrology, v. 50, p. 1219–1236.

Bruni, S. F., and Wenk, H.-R., 1985, Replacement of aragonite by calcite in sediments from the San Cassiano Formation (Italy): Journal of Sedimentary Petrology, v. 50, p. 1219–1236.

Budd, D. A., 1988, Petrographic products of freshwater diagenesis in Holocene ooid sands, Schooner Cays, Bahamas: Carbonates and Evaporites, v. 3, p. 143–164.

Carew, J. L., and Mylroie, J. E., 1985, The Pleistocene and Holocene stratigraphy of San Salvador Island, Bahamas, with reference to marine and terrestrial facies in French Bay, *in* Curran, H. A., ed., Pleistocene and Holocene Carbonate Environments on San Salvador Island, Bahamas: Geological Society of America, Orlando, Annual Meeting Field Trip Guidebook: San Salvador, CCFL Bahamian Field Station, p. 11–61.

Carew, J. L., and Mylroie, J. E., 1987, A refined geochronology for San Salvador Island, Bahamas, *in* Curran, H. A., ed., Proceedings, Third Symposium on the Geology of the Bahamas: San Salvador, CCFL Bahamian Field Station, p. 35–44.

Chen, J. H., Curran, H. A., White, B., and Wasserburg, G. J., 1991, Precise chronology of the last interglacial period: ^{234}U–^{230}Th Data from fossil coral reefs in the Bahamas: Geological Society of America Bulletin,

v. 103, p. 82–97.

Clark, D. D., Mylroie, J. E., and Carew, J. L., 1989, Texture and composition of Holocene beach sediment, San Salvador Island, Bahamas, *in* Mylroie, J., ed., Proceedings, Fourth Symposium of the Geology of the Bahamas: San Salvador, Bahamian Field Station, p. 83–105.

Constantz, B. R., 1984, Functional comparison of the microarchitecture of *Acropora palmata* and *Acropora cervicornis*: Palaeontographica Americana, v. 54, p. 948–952.

Constantz, B. R., 1986a, Coral skeleton construction: a physiochemically dominated process: Palaios, v. 1, 152–157.

Constantz, B. R., 1986b, The primary surface area of corals and variations in their susceptibility to diagenesis, *in* Schroeder, J. H., and Purser, B. H., eds., Reef Diagenesis: Berlin, Springer-Verlag, p. 53–76.

Cross, T. S., and Cross, B. W., 1983, U, Sr, and Mg in Holocene and Pleistocene corals *A. palmata* and *M. annularis*: Journal of Sedimentary Petrology, v. 53, p. 387–594.

Dullo, W.-C., 1986, Variation in diagenetic sequences: An example from Pleistocene coral reefs, Red Sea, Saudia Arabia, *in* Schroeder, J. H., and Purser, B. H., eds., Reef Diagenesis: Berlin, Springer-Verlag, p. 77–90.

Folk, R. L., Roberts, H. H., and Moore, C. H., 1973, Black phytokarst from Hell, Cayman Islands: Geological Society of America Bulletin, v. 87, p. 2351–2360.

Gaffey, S. J., 1985, Reflectance spectroscopy in the visible and near-infrared (0.35–2.55 μm): Applications in carbonate petrology: Geology, v. 13, p. 270–273.

Gaffey, S. J., 1986, Spectral reflectance of carbonate minerals in the visible and near infrared (0.35-2.55 microns): calcite, aragonite, and dolomite: American Mineralogist, v. 71, p. 151–162.

Gaffey, S. J., 1987, Spectral reflectance of carbonate minerals in the visible and near infrared (0.35–2.55 μm): Anhydrous carbonate minerals: Journal of Geophysical Research, v. 92, p. 1429–1440.

Gaffey, S. J., 1988, Water in skeletal carbonates: Journal of Sedimentary Petrology, v. 58, p. 397–414.

Gaffey, S. J., 1990, Skeletal vs. nonbiogenic carbonates—UV–visible–near IR (0.3-2.7-mm) reflectance properties, *in* Coyne, L. M., McKeever, S. W. S., and Blake, D. F., Spectroscopic Characterization of Minerals and Their Surfaces: American Chemical Society Symposium Series, v. 415: Washington D.C., American Chemical Society, p. 94–116.

Gaffey, S. J., and Bronnimann, C. E., 1993, Effects of bleaching on mineral and organic phases in skeletal carbonates: Journal of Sedimentary Petrology, v. 63, p. 752–754.

Gaffey, S. J., Kolak, J. J., and Bronnimann, C. E., 1991, Effects of drying, heating, annealing, and roasting on carbonate skeletal material, with geochemical and diagenetic implications: Geochimica et Cosmochimica Acta, V. 55, p. 1627–1640.

Gebelein, C. D., and Hoffman, P., 1973, Algal origin of dolomite laminations in stromatolitic limestone: Journal of Sedimentary Petrology, v. 43, p. 603–313.

Green, H. W., II, Lipps, J. H., and Showers, W. J., 1980, Test ultrastructure of fusulinid Foraminifera: Nature, v. 283, p. 853–855.

Gvirtzman, G., and Friedman, G. F., 1977, Sequence of progressive diagenesis in coral reefs, *in* Frost, S. H., Weiss, M. P., and Saunders, J. B., eds., Studies in Geology No. 4, Reefs and Related Carbonates—Ecology and Sedimentology: Tulsa, Oklahoma, American Association of Petroleum Geologists, p. 357–380.

Hamilton, W. C., and Ibers, J. A., 1968, Hydrogen Bonding in Solids: New York, W. A. Benjamin, 284 p.

James, N. P., 1974, Diagenesis of Scleractinian corals in the subaerial vadose environment: Journal of Paleontology, v. 48, p. 785–799.

James, N. P., and Choquette, P. W., 1983, Diagenesis 6. Limestones—The sea floor diagenetic environment: Geoscience Canada, V. 10, p. 162–179.

James, N. P., and Choquette, P. W., 1984, Diagenesis 9. Limestones—The meteoric diagenetic environment: Geoscience Canada, V. 11, p. 161–194.

Johnston, I. S., 1979, The organization of a structural organic matrix within the skeleton of a reef-building coral: Scanning Electron Microscopy, v. 2, p. 421–431.

Johnston, I. S., 1980, The ultrastructure of skeletogenesis in hermatypic corals: International Review of Cytology, v. 67, p. 171–214.

Kinsman, D. J., 1969, Interpretation of Sr^{2+} concentrations in carbonate minerals and rocks: Journal of Sedimentary Petrology, V. 39, p. 486–508.

Lecuyer, C., and O'Neil, J. R., 1994, Stable isotope compositions of fluid inclusions in biogenic carbonates: Geochimica et Cosmochimica Acta, v. 58, p. 353–363.

Lowenstam, H. A., and Weiner, S., 1989, On Biomineralization: New York, Oxford University Press, 324 p.

Lukas, K. J., 1974, Two species of the chlorophyte genus *Ostreobium* from skeletons of Atlantic and Caribbean reef corals: Journal of Phycology, v. 10, p. 331–335.

Macintyre, I. G., 1984, Preburial and shallow-subsurface alteration of modern Scleractinian corals: Palaeontographica America, v. 54, p. 229–244.

Mann, S., 1983, Mineralization in biological systems: Structure and Bonding, v. 54, p. 125–174.

Milliman, J. D., 1974, Marine Carbonates: New York, Springer-Verlag, 375 p.

Mitterer, R. M., 1978, Amino acid composition and metal binding capability of the skeletal protein of corals: Bulletin of Marine Science, v. 28, p. 173–180.

Pieters, C. M., 1983, Strength of mineral absorption features in the transmitted component of near-infrared reflected light: First results from RELAB: Journal of Geophysical Research, v. 88, p. 9534–9544.

Pingitore, N. E., Jr., 1976, Vadose and phreatic diagenesis: Processes, products, and their recognition in corals: Journal of Sedimentary Petrology, v. 46, p. 985–1006.

Pingitore, N. E., Jr., 1978, The behavior of Zn^{2+} and Mn^{2+} during carbonate diagenesis: Journal of Sedimentary Petrology, v. 48, p. 799–814.

Pittman, E. D., 1974, Porosity and permeability changes during diagenesis of Pleistocene corals, Barbados, West Indies: Geological Society of America Bulletin, v. 85, p. 1811–1820.

Saller, A. H., 1992, Calcitization of aragonite in Pleistocene limestones of Eniwetok Atoll, Bahamas, and Yucatan—An alternative to thin-film neomorphism: Carbonates and Evaporites, v. 7, p. 56–73.

Sandberg, P. A., 1984, Recognition criteria for calcitized skeletal and nonskeletal aragonites: Palaeontographica Americana, v. 54, p. 272–281.

Schroeder, J. H., 1969, Experimental dissolution of calcium, magnesium, and strontium from Recent biogenic carbonates: A model of diagenesis: Journal of Sedimentary Petrology, v. 39, p. 957–1073.

Schroeder, J. H., 1984, The petrogenetogram of corals: Spatial variations in diagenetic sequences: Palaeontographica Americana, v. 54, p. 261–271.

Sorauf, J. E., 1972, Skeletal microstructure and microarchitecture in Scleractinia (Coelenterata): Paleontology, v. 15, p. 11–23.

Sorauf, J. E., 1980, Biomineralization, structure, and diagenesis of coelenterate skeleton: Acta Palaeontologica Polonica, v. 25, p. 327–341.

Strasser, A., 1984, Black-pebble occurrence and genesis in Holocene carbonate sediments (Florida Keys, Bahamas, and Tunisia); Journal of Sedimentary Petrology, v. 54, p. 1097–1109.

Towe, K. M., and Thompson, G. R., 1972, The structure of some bivalve shell carbonates prepared by ion-beam thinning: Calcareous Tissue Research, v. 10, p. 38–48.

Tribble, G. W., Sansone, F. J., and Smith, S. V., 1990, Stoichiometric modeling of carbon diagenesis within a coral reef framework: Geochimica et Cosmochimica Acta, v. 54, p. 2439–2449.

Tucker, M. E., and Wright, V. P., 1990, Carbonate Sedimentology: Oxford, United Kingdom, Blackwell, 482 p.

Wade, L. G., Jr., 1987, Organic Chemistry: Englewood Cliffs, New Jersey, Prentice–Hall, 1377 p.

Walls, R. A., Ragland, P. C., and Crisp, E. L., 1977, Experimental and natural early diagenetic mobility of Sr and Mg in biogenic carbonate: Geochimica et Cosmochimica Acta, v. 41, p. 1731–1737.

Walter, L. M., and Burton, E. A., 1990, Dissolution of Recent platform carbonate sediments in marine pore fluids: American Journal of Science, v. 290, p. 601–643.

Wells, J. W., 1956, Scleractinia, *in* Moore, R. C., ed., Treatise on Invertebrate Paleontology, Part F: Boulder, Colorado, Geological Society of America, p. 328–443.
Wheeler, A. P., and Sikes, C. S., 1984, Regulation of carbonate calcification by organic matrix: American Zoologist, v. 24, p. 933–944.
White, B., 1989, Field guide to the Sue Point fossil coral reef, San Salvador, Bahamas, *in* Mylroie, J. E., ed., Proceedings, Fourth Symposium on the Geology of the Bahamas: San Salvador, Bahamian Field Station, p. 353–366.
Wilson, M. A., 1987, N.M.R. Techniques and Applications in Geochemistry and Soil Chemistry: New York, Pergamon, 353 p.
Young, S. D., O'Connor, J. D., and Muscatine, L., 1971, Organic material from Scleractinian coral skeletons—II. Incorporation of ^{14}C into protein, chitin, and lipid: Comparative Biochemistry and Physiology, v. 40B, p. 945–958.
Zabielski, V. P., and Gaffey, S. J., 1989, Feasibility of use of VNIR reflectance spectroscopy for analysis of water and organic content in skeletons of scleractinian corals, *in* Mylroie, J. E., ed., Proceedings, Fourth Symposium on the Geology of the Bahamas: San Salvador, Bahamian Field Station, p. 367–381.

Manuscript Accepted by the Society January 5, 1995

Geological Society of America
Special Paper 300
1995

Karst development in the Bahamas and Bermuda

John E. Mylroie
Department of Geosciences, Mississippi State University, Mississippi State, Mississippi 39762
James L. Carew
Department of Geology, University of Charleston, Charleston, South Carolina 29424
H. L. Vacher
Department of Geology, University of South Florida, Tampa, Florida 33620

ABSTRACT

Bahamian caves formed mostly by dissolution resulting from mixing of fresh and saline waters at the margin of an island ground-water lens have been defined as "flank margin caves." The location of these caves provides an indication of sea-level position at the time the caves formed. Dissolution of the caves was rapid compared to the rate of inversion of aragonite to calcite, as indicated by the presence of up to 40% primary aragonite in the wall rock of some Bahamian caves. Bahamian cave chambers with volumes in excess of 14,000 m^3, which formed during a sea-level highstand that lasted no more than 15,000 yr, provide evidence for chamber dissolution rates as high as 1 m^3/yr. Slope and scarp retreat of ridges eventually unroofed flank margin caves, and may have ultimately produced features that appear remarkably similar to abandoned coastal bioerosion notches. Pit caves and banana holes are also common in the Bahamas. The former are vadose pathways dissolved from the surface; the latter are phreatic pockets dissolved at the top of the fresh-water lens.

In contrast, Bermudian caves are mostly collapse caves. The original dissolution that led to these collapse caves occurred during sea-level lowstands and was produced by vadose cave streams perched on the contact between limestone and volcanic rocks. Flank margin caves are small and uncommon on Bermuda; pit caves and banana holes are also rare.

Interior topographic depressions are also dissimilar in Bermuda versus the Bahamas. Bermuda has a positive water budget (precipitation exceeds evapotranspiration) so ground-water flow is either through or outward from interior marshes; this flow transports CO_2-enriched water into the surrounding limestone. Dissolution of the carbonate bedrock is accelerated by this acidic water, and as a result topographic depressions have deepened and expanded over the course of multiple glacioeustatic sea-level changes, Lateral expansion of these inshore basins has erased evidence of earlier flank margin caves. Alternatively, in the southeast Bahamas conditions are more arid, resulting in a negative water budget, and ground water is discharged from adjacent land areas into interior saline lakes. Both waters are saturated with respect to $CaCO_3$, even after mixing, and bed-rock dissolution does not occur. As a result, interior basins have maintained their original depositional morphology.

Mylroie, J. E., Carew, J. L., and Vacher, H. L., 1995, Karst development in the Bahamas and Bermuda, *in* Curran, H. A., and White, B., Terrestrial and Shallow Marine Geology of the Bahamas and Bermuda: Boulder, Colorado, Geological Society of America Special Paper 300.

INTRODUCTION

The islands of Bermuda and the Bahamas are composed entirely of Quaternary carbonates; marine facies crop out at low elevations and eolian units occur at all elevations. The islands have a topography that has been shaped by two interacting agents: constructional processes of carbonate deposition, and destructive processes of karst dissolution and coastal erosion (Bretz, 1960). Glacioeustatic sea-level changes that occurred during the Quaternary have produced repeated discrete episodes of deposition and denudation. Despite the overall similarities of Bermuda and the Bahamas, the differences in climate and subsurface geology have resulted in karst landforms that are quite different.

The Bahamas and Bermuda contain two major types of karst landforms that are readily accessible and persistent through time: caves and interior basins. This chapter reviews these features in terms of processes that have affected their evolution. We omit two other aspects of Bahamian and Bermudian karst because of space considerations—littoral karst (phytokarst or biokarst), which is rarely found as a paleokarst feature (Desrochers and James, 1988); and blue holes and other related submerged caves, which are not readily accessible. For a thorough discussion of these topics, see Viles (1988) for biokarst; and Palmer (1985), Iliffe (1987), Smart et al. (1988), and Smart and Whitaker (1990) for the submerged cave systems.

CAVES AND RELATED FEATURES IN THE BAHAMAS AND BERMUDA

Hydrologic setting

The hydrologic setting of the Bahamas and Bermuda is different from the usual continental setting in two principal ways. First, the islands are completely covered by limestones. Precipitation sinks as diffuse input into the limestone, and surface streams are absent (autogenic input: Mylroie, 1984a). There is no surface water derived from an impervious noncarbonate catchment (allogenic input). As a result, the hydrologic and chemical potential of the meteoric water is dispersed instead of concentrated. Second, nonmarine ground water (fresh or brackish) occurs in a lens that floats on the saline marine water that permeates the islands from below. The size and shape of the fresh-water lens depend on recharge, the size of the island, and the permeability of the carbonate rock (Vacher, 1988; Budd and Vacher, 1991). High permeabilities and low recharge rates produce a thin lens, low permeabilities and high recharge rates produce a thicker lens. In these islands, high permeabilities and thin lenses occur in older rocks with enhanced permeability from secondary dissolutional porosity.

Mixing of the fresh water in the lens with underlying saline ground water generally decreases the state of saturation of the ground water with respect to calcite (Plummer, 1975). The amount of saturation reduction and the resulting degree of renewed chemical aggressiveness depend on the relative amount and initial chemistry of the waters that are mixed, but the process commonly results in water that is undersaturated, even if the fresh and saline waters are each saturated with respect to $CaCO_3$ prior to mixing (Plummer, 1975).

In continental settings, such as the Yucatan Peninsula region of Mexico, this mixing process has been demonstrated to produce geomorphically significant dissolution (Back et al., 1986). Carbonate islands differ from continental settings like the Yucatan in that fresh-water discharge is less because of the smaller island catchment (Mylroie and Carew, 1995). Despite the limited fresh-water flow volumes, significant dissolution does take place in carbonate islands. (Smart et al., 1988; Vogel et al., 1990). Recent work in the Bahamas has demonstrated that oxidation/reduction reactions taking place within the fresh-water lens and/or mixing zone may provide extra dissolutional potential to produce significant voids in a limited time (Bottrell et al., 1993; Mylroie and Balcerzak, 1992).

Flank margin caves

The fresh-water lens thins toward the margin of an island, so the flow in the lens is discharged through an ever-decreasing cross-sectional area. The resulting increase in discharge velocity plays an important role in producing maximum dissolution (Sanford and Konikow, 1989). At the lens margin, superposition of the vadose/phreatic contact at the top of the lens with the fresh-water/saline-water contact at the bottom of the lens places two sites of mixing in close proximity to one another (Mylroie and Carew, 1988, 1990). Density stratification of organics at both mixing horizons may lead to oxidation/reduction reactions that also increase dissolutional potential (Bottrell et al., 1993). The combined result is that cave development by dissolution is maximized at the margin of the fresh-water lens, where the two mixing zones converge. As the fresh-water lens is commonly contained within eolian ridges, the margin of the lens discharges to the sea under the flank of the ridge. The caves produced by dissolution under these conditions have accordingly been labeled flank margin caves (Mylroie, 1988; Mylroie and Carew, 1990; Vogel et al., 1990).

Cave morphology. Examination of flank margin caves, currently in the vadose zone, that formed during past higher stands of sea level demonstrates that these caves have a characteristic morphology (Fig. 1). Typically, these caves are dominated by large globular chambers that are broad in the horizontal plane but vertically restricted (Fig. 2A). The caves are usually entered where erosion of the slope of the ridge containing them has breached the large chamber. Other entries may be provided by vertical vadose shafts from the surface that intersect the caves. At the rear of the chamber there is usually a series of smaller chambers that change into tubular passages as the cave proceeds into the interior of the ridge. Commonly there are many cross-connections between adjacent chambers and passages that give the caves a maze-like character. The passages that penetrate into the ridge end abruptly (Fig. 2D). The chamber and passage walls are often etched into a variety of dissolution pockets and tubes ranging in size from meters to centimeters (Figs. 2B, 2C). In some situations, the chambers are arranged side-by-side along the strike of the hillslope like beads on a string (Vogel et al., 1990). The configuration of these caves occurs at a variety of scales. The general pattern can be seen in caves that have main chambers only a few meters across, and in caves where the chambers are up to 70 m across and contain more than 14,000 m^3 of void (Mylroie et al., 1991). Flow markings, such as ablation scallops (Curl, 1966), are absent.

The morphology of flank margin caves reflects their origin by mixing of fresh water and sea water at the margin of the island. The large main chambers of the caves behave as mixing chambers (Mylroie and Carew, 1990). Through time, the mixing front migrates headward up the routes of the incoming diffuse flow, in a manner analogous to a model (involving different chemistry) proposed by Rhoades and Sinacori (1941). This process adds separate passages and chambers to the main mixing chamber and produces a complex set of interconnections. When sea level falls, dissolution ends, leaving passages that end abruptly at blank walls.

Flank margin caves are found in various stages of formation. All stages have the central main chamber under the flank of the enclosing ridge, but many caves have main chambers with various stages of development of subsidiary tubes and chambers leading into the ridge. The subsidiary tubes and chambers, when present, are always found with an intervening main chamber between them and the flank of the ridge. This relationship indicates that the caves developed inward from the margin of the lens.

The caves developed in the thin margin of the fresh-water lens, so the vertical range of their phreatic development was limited. Flank margin caves, while commonly having horizontal extents of more than 100 m, have vertical development of less than 10 m. The caves must have formed at shallow depths and are not the result of deep phreatic (bathyphreatic) circulation (Mylroie and Carew, 1986).

The morphology of the cave wall includes numerous blind pockets, bed-rock spans, and dead-end tubular passages of a variety of dimensions, which resembles the wall morphology of caves produced in other mixed-water situations. Examples include the caves of the Guadalupe mountains of New Mexico that formed by mixing of H_2S-rich basin water with O_2-rich shallow phreatic water (Hill, 1987), and the caves of the Black Hills of South Dakota that developed by mixing of different thermal waters (Palmer and Palmer, 1989). These types of caves have been termed hypogenic caves (Palmer, 1991). Despite the variety of results of mixing conditions shown by hypogenic caves, they have a dissolutional morphology very similar to that in flank margin caves (Mylroie, 1991).

Flank margin caves should not be interpreted as conduits. These phreatic chambers received and transmitted water as diffuse flow. The caves were mixing chambers that grew as the site of mixing migrated into the diffuse flow contributions discharging from the fresh-water lens.

Relation to sea level. Flank margin caves provide an indication of sea-level position when they formed. Although dissolution caves in an island interior can theoretically develop anywhere within a fresh-water lens that may be up to tens of meters thick, caves at the edge of the island can only have formed in the thinning margin of the lens, which is closely tied to the position of sea level (Mylroie and Carew, 1988). In the Bahamas, and to a lesser extent in Bermuda, there are flank margin caves with dissolutional ceiling elevations up to 6 m above modern sea level that formed during a past interglacial sea-level highstand; most likely the highstand associated with oxygen isotope substage 5e (Mylroie et al., 1991; Carew and Mylroie, 1995). These caves are subaerial today because sea level is now lower, and the fresh-water lens is also lower in elevation than it was when the caves formed.

At times of higher sea level, especially the +6m highstand associated with substage 5e, the Bahama Islands consisted only of individual eolian ridges that projected above broad lagoons. The fresh-water lenses, therefore, were very small; nevertheless they were capable of producing the flank margin caves observed today. Flank margin caves also have been located by submersible on the margin of the San Salvador Island platform at depths of 105 and 125 m (Carew and Mylroie, 1987). These caves, now flooded by marine water and no longer active, represent development a low sea-level positions during the Pleistocene when the Bahama islands consisted of large emergent platforms. Flank margin cave development in the Bahamas at the margin of islands, therefore, occurred under a wide range of island sizes and lens characteristics during the Pleistocene.

The flank margin caves of the Bahamas and Bermuda have developed in rocks that are less than 1,000,000 yr old. Some of these caves in the Bahamas are in rocks as young as 125,000 yr old (Schwabe et al., 1993). Stalagmites from flank margin caves have yielded reliable U/Th ages of up to 70,000 yr, thus indicating that the caves were formed and drained by that time (Carew and Mylroie, 1995). Some Bahamian caves contain evidence of multiple submergence events, in that subaerial cave sediments and speleothems have undergone subsequent phreatic dissolution (Garrett and Gould, 1984; Mylroie et al., 1991).

The elevation of many of the subaerial flank margin caves in the Bahamas is such that only a sea-level highstand event that reached an elevation of at least +6 m could have provided the conditions for their formation. The sea-level highstand associated with oxygen isotope substage 5e seems to provide the only conditions that meet these criteria. Most earlier highstands were not high enough, and the caves produced during those that were high enough would have been lowered below modern sea level as a result of isostatic subsidence of the Bahamas (Mylroie et al., 1991; Carew and Mylroie, 1995). The sea-level highstand of substage 5e was above present sea-level elevation for less than 15,000 yr (Carew and Mylroie, 1995). Individual subaerial dissolution chambers with volumes up to 14,000 m^3 exist

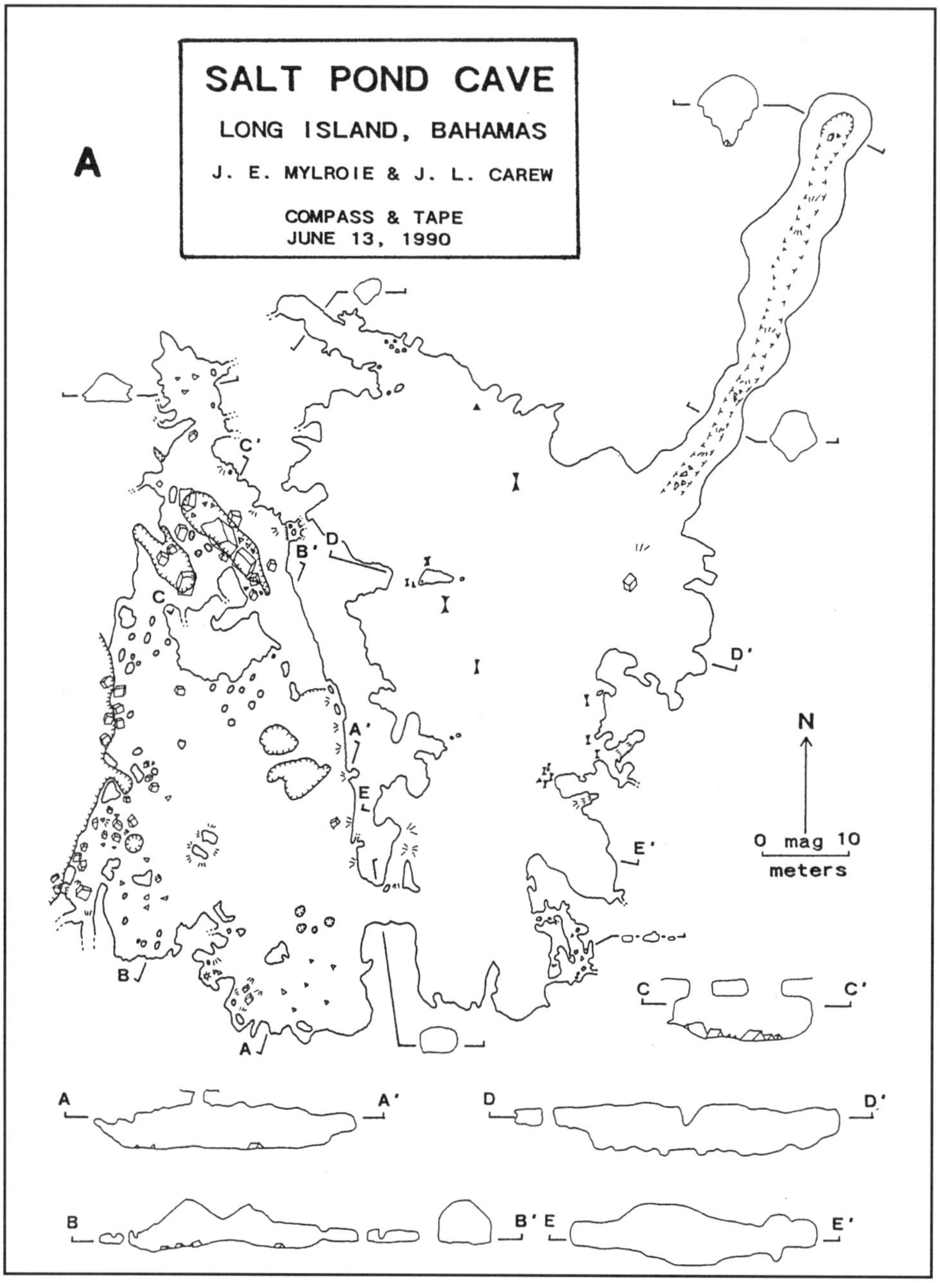

Figure 1 (on this and facing page). Maps with cross sections of Bahamian flank margin caves. A, Salt Pond Cave, Long Island; B, Bahamas West Cave, New Providence Island, Bahamas. The maps show typical flank margin cave development that includes a large central chamber, or chambers, with maze-like passage development toward the interior of the ridge containing the chambers. Surficial erosion and collapse provide entry into the caves. Note the many isolated bedrock pillars and thin rock partitions separating some passages and chambers, and the abrupt termination of some passages that trend into the ridge. Hachured lines indicate connections of the cave to the surface; rectangular blocks indicate areas of collapse debris; solid triangles joined at their apex indicate stalactites and stalagmites; sets of three short diverging lines indicate a slope, downward in the direction of divergence. From Mylroie et al. (1991).

in the Bahamas; therefore, such chambers must have formed at a rate of almost 1 m^3/yr.

Relation to diagenesis. In the Yucatan, the mixing zone at the base of the fresh-water lens has been documented to be a zone of differential dissolution of aragonite and a site of significant diagenesis (Back et al., 1986). The wall rocks of flank margin caves on San Salvador Island, Bahamas, contain a diagenetic record of marine phreatic, fresh-water phreatic, and vadose conditions (Vogel et al., 1990; Schwabe et al., 1993). In about 40% of the examined flank margin caves, dolomite was present in

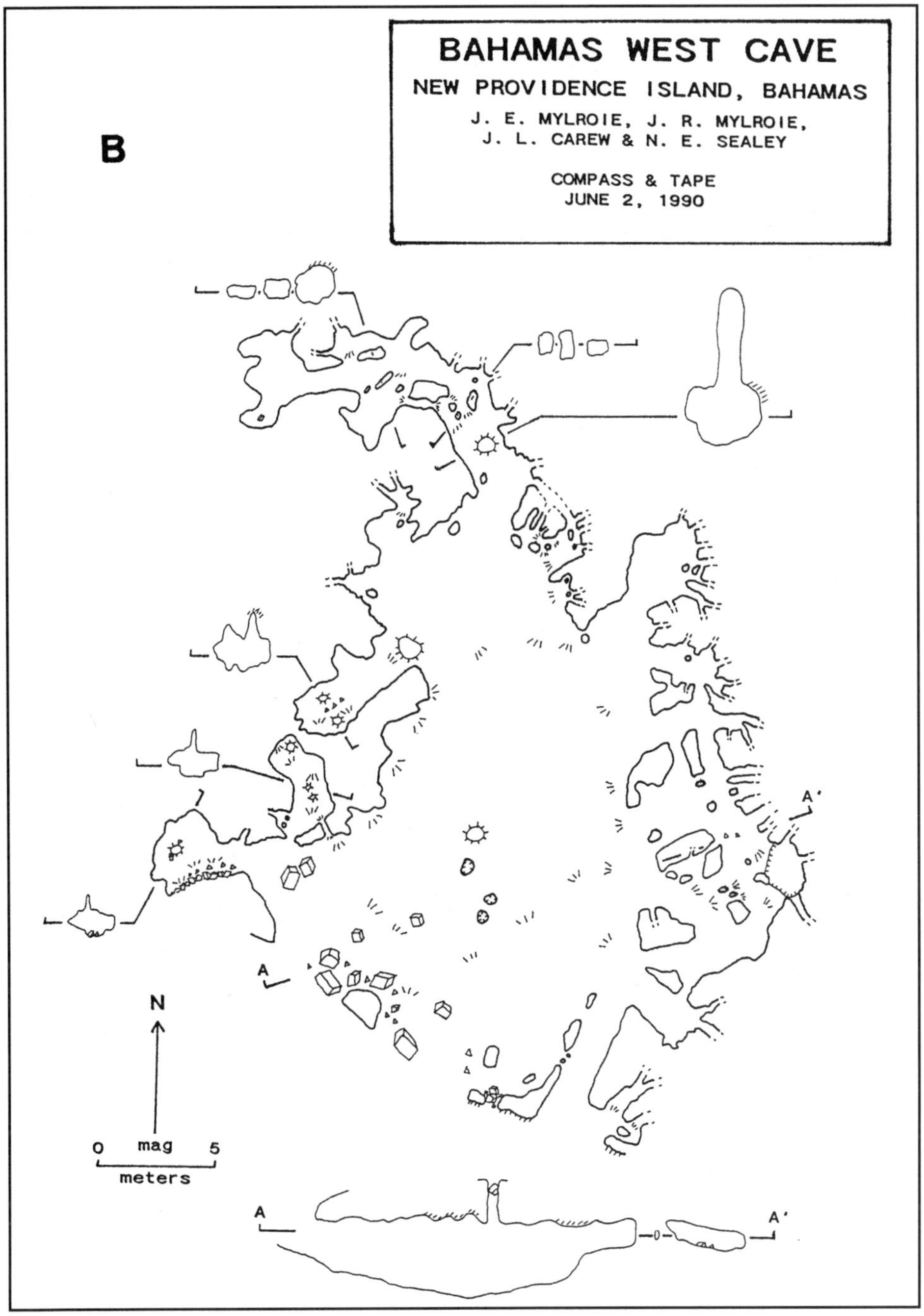

minor amounts (Schwabe et al, 1993). These diagenetic signatures reflect the changing conditions to which the rocks have been exposed as sea level fluctuated during the Late Quaternary.

In some flank margin caves, the wall rock is still up to 40% primary aragonite (Vogel et al., 1990). This indicates that dissolution of the caves is fast compared to the rate of aragonite inversion to calcite. The large size of many of these caves, together with the marine and fresh-water phreatic diagenetic record, had to be produced within a time window of not more than 15,000 yr.

Bioerosion notches versus breached flank-margin caves. Throughout the Bahamas, flank margin caves are found in various stages of degradation by surficial process. In the early stages, this degradation results in the tangential intersection of the main cave chamber by the retreating hillslope, which produces a small entrance to an otherwise intact cave. As surficial erosion continues, the cave chambers are further dissected and exposed (Fig. 3A). In advanced stages of erosion, all that remains are a portion of the roof and a curving back wall (Fig. 3B).

In the terminal stage of subaerial erosion, the remnant of a

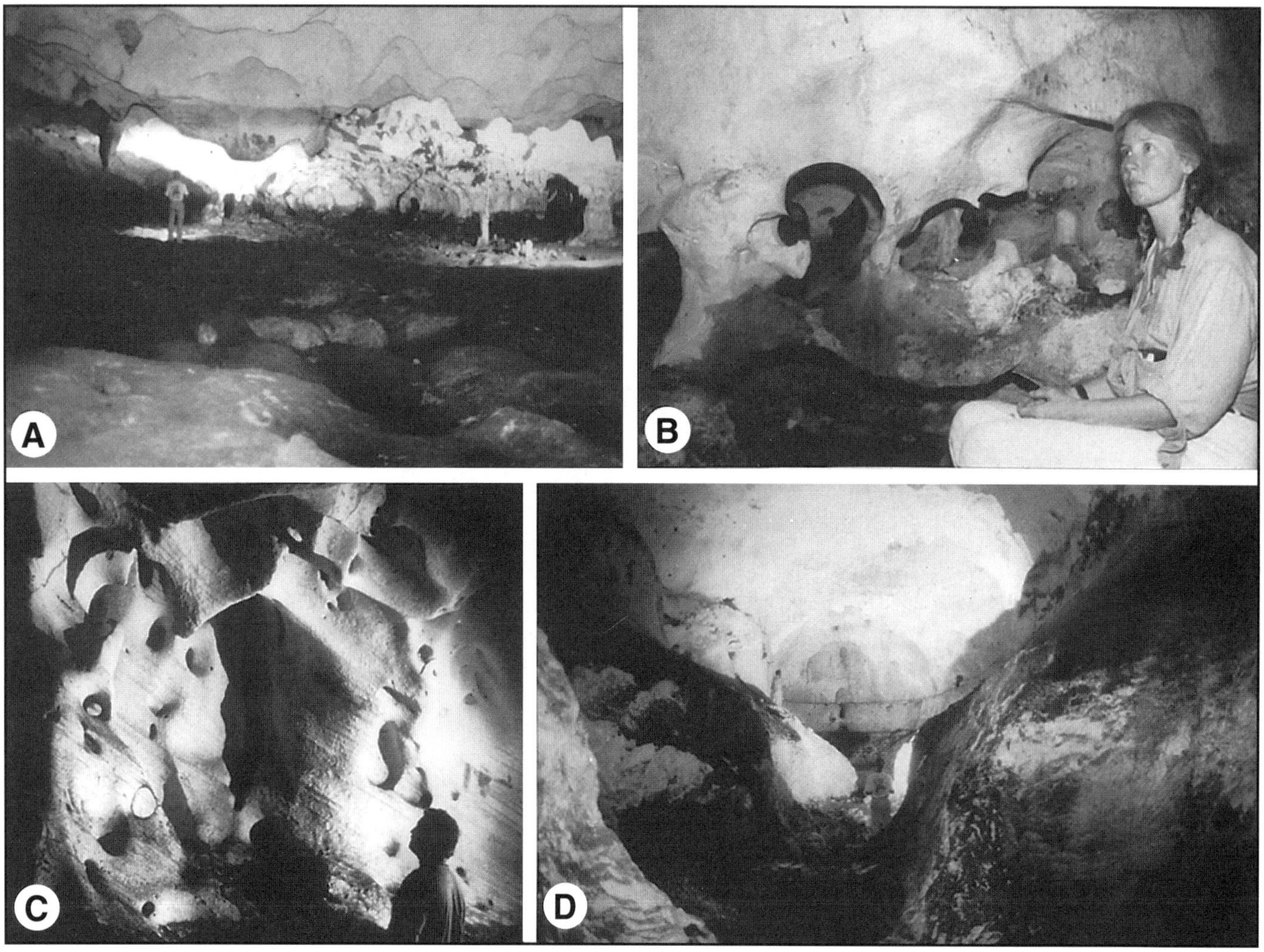

Figure 2. Photographs of the interiors of flank margin caves. A, Main inner chamber in Salt Pond Cave (Fig. 1), Long Island, Bahamas. Note the dissolutional surfaces on the walls, floor, and ceiling; the lack of collapse material; and the limited height of the chamber compared to its horizontal extent. Chamber volume is in excess of 14,000 m^3. B, Small-scale dissolutional fretwork from Harry Oakes Cave, New Providence Island, Bahamas. C, Dissolutional surface in Hamilton's Cave, Long Island, Bahamas. Note the foreset beds of the enclosing eolianite. D, Dead-end passage in Salt Pond Cave, Long Island, Bahamas (Fig. 1). Note the abrupt termination of the passage and the well-developed wall notches, which cut across the foreset beds of the enclosing eolianite.

flank margin cave consists of a reentrant in the hillside that may extend horizontally for tens to hundreds of meters. Such reentrants appear remarkably similar to notches produced by bioerosion along modern rocky carbonate coasts (Mylroie and Carew, 1991). Because both bioerosion notches and flank margin caves form at sea level, the reinterpretation of the origin of many inland scarp reentrants in the Bahamas does not alter their usefulness as indicators of past sea-level position.

Caves of the Bahamas

Flank margin caves are common throughout the Bahama islands; we have mapped them on North Andros, South Andros, New Providence, Eleuthera, San Salvador, Long, and Great Inagua islands. The lateral dimensions of these caves vary from a few meters to 500 m, but their vertical development is limited to less than 10 m (generally +1 to +6, rare examples from –2 to +7 m). The caves are notably free of collapse debris and sediment, which makes the morphology of the dissolved void easy to examine. Maps of representative Bahamian caves are in Mylroie (1988) and Mylroie et al. (1991).

There are also a large number of small, vertical-sided depressions in the Bahamas (Fig. 4). A few lead downward into cave systems. Some vertical shafts, called pit caves, have diameters of a meter or two and depths up to 15 m. Other depressions, called banana holes by Bahamians, are up to 10 m across

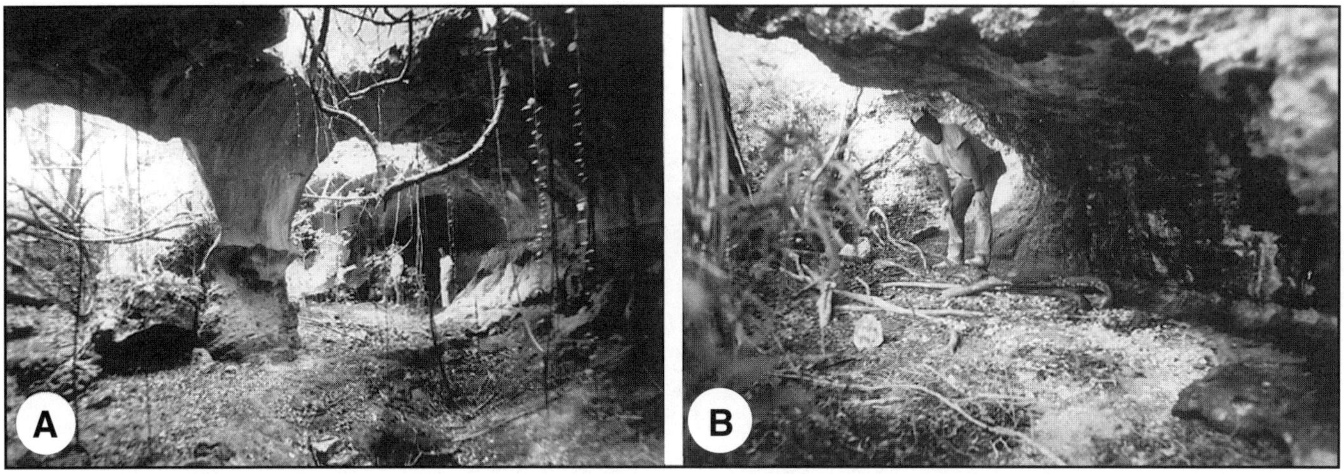

Figure 3. Photographs that illustrate the development of hillside reentrants by collapse of flank margin caves. A, Outer chamber of Harry Oakes Cave, New Providence Island, Bahamas, showing the collapse of the chamber's outer wall and ceiling, which has left a remnant arch. Note the overhung, curved nature of the back (right-hand) wall of the chamber. B, Remnant of an almost completely eroded flank margin cave chamber on San Salvador Island, Bahamas. Note the similarity to a bioerosion notch.

and only a few meters deep. The deep, narrow pit caves are found at all elevations in the Bahamas, including elevations well above any past sea-level highstand (Mylroie, 1988). The banana holes, especially the larger ones, are generally found at elevations of 6 m or less. The banana hole elevations are within the range of the oxygen isotope substage 5e sea-level highstand. In most cases there are obvious morphologic differences between pit caves and banana holes; however, transitional forms and overlap of morphology occur, which can lead to confusion in determining the origin of these karst features. The most difficult of these caves to interpret are those where collapse and surface dissolution have greatly altered the original morphology.

The origin of pit caves has been a subject of much debate (Pace et al., 1993). There seem to be four possible modes of development (Mylroie, 1990): (1) upward stoping from underlying caves of phreatic origin, (2) concentration of surface flow on resistant paleosol surfaces, (3) downwardly corroding accumulations of organic matter (Smart and Whitaker, 1989), and (4) localized vadose flow. Evidence to date indicates that pit caves are the result of localized vadose flow (Pace et al., 1993).

The pit caves at higher elevations and the banana holes and flank margin caves at low elevations are interpreted to be part of a single flow system (Pace et al., 1993) (Fig. 5). Pit caves are conduits that carry vadose water collected from the surface-weathered zone (epikarst) into the limestone. The pits yield their water to diffuse-flow pathways while still in the vadose zone. This vadose seepage enters the phreatic zone of the fresh-water lens, and moves by diffuse flow toward the lens margin, where it enters the mixing chambers of flank margin caves. In rare situations where pit caves have been observed to intersect existing flank margin caves, they pass on through the floor of the caves. This relationship indicates that these particular pit caves represent a vadose phase (most likely younger) unrelated to the flank margin caves they penetrate, as opposed to being hydrologically coupled during cave formation.

Banana holes in some cases are the result of collapse of the ceiling of a flank margin cave, as observed by the position of the banana hole on the lower portion of the flank of a ridge. Most banana holes, however, are found on lowland plains far from the margin of a past fresh-water lens. Many of these banana holes have partial, or almost complete, roofs covering low, wide chambers that have wall morphology typical of that expected from a phreatic origin.

Banana holes are the result of discrete sites of dissolution near the top of a past fresh-water lens, most likely the result of vadose/phreatic mixing and oxidation/reduction reactions. The site of formation of banana holes in what was the upper interior of a past fresh-water lens, coupled with the absence of dolomite in the wall rock of banana holes, seem to indicate that marine waters were not directly involved in banana-hole dissolution (Pace et al., 1993). The banana holes formed in the phreatic, diffuse-flow pathway that connected the vadose pit caves to the flank margin caves, but, like the flank margin caves, they were not conduits but phreatic voids produced by mixing of waters of differing chemistry.

Caves of Bermuda

In general, the caves of Bermuda are different from those found in the Bahamas. Subaerial flank margin caves are known from very few locations; pit caves and banana holes are also extremely rare on Bermuda. Examples of flank margin caves can be seen at +6 m elevation in the Government Quarry area (Mylroie, 1984b) and on the trail to Spittle Pond. The rarity of flank margin caves, and even rarer occurrence of pit caves and banana holes in Bermuda, is problematic. The cause may be cli-

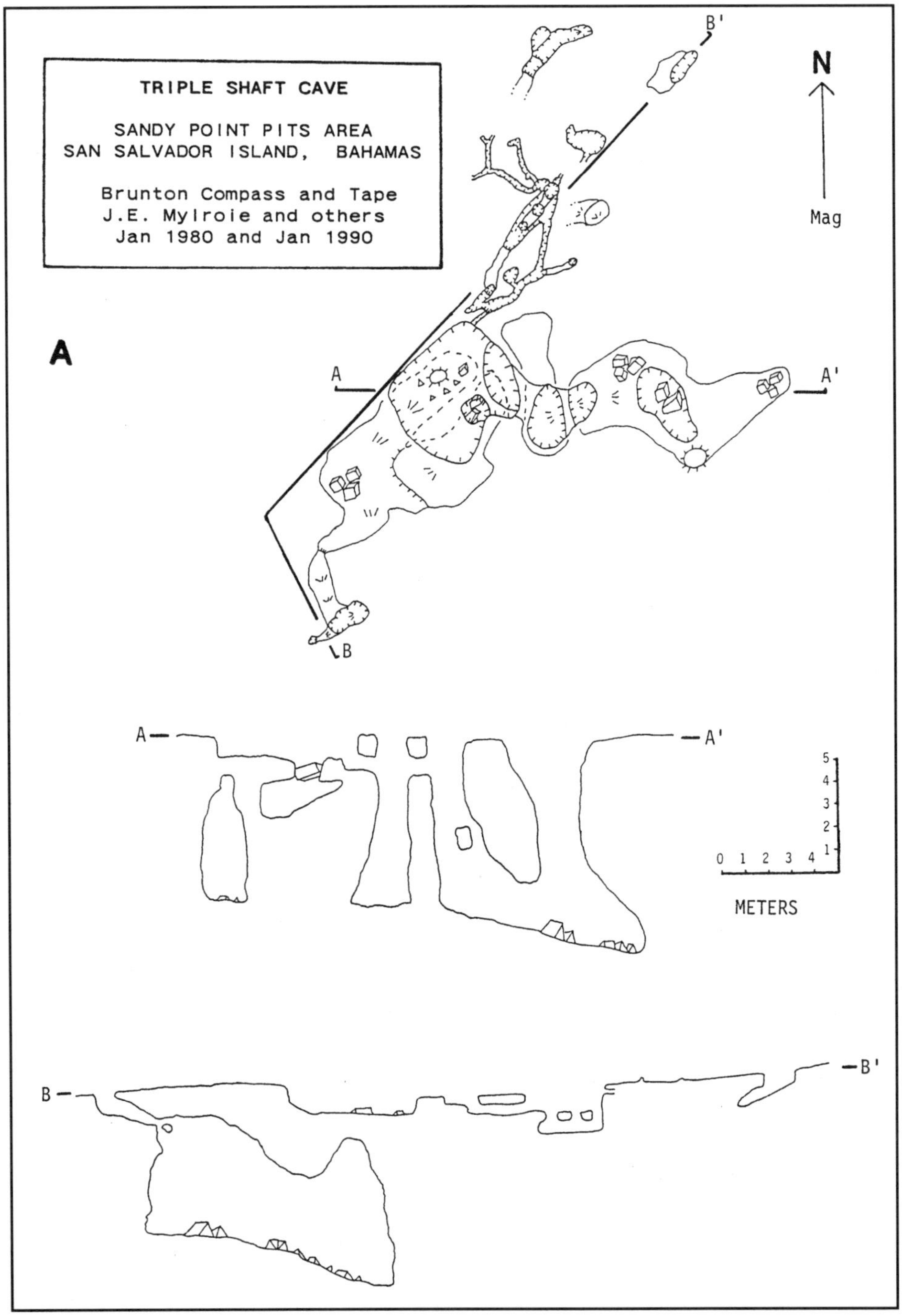

Figure 4 (on this and facing page). A and B, Maps with cross sections of a typical Bahamian pit cave complex and a banana hole. Surface depressions and entrances shown by hachured lines, other symbols as in Figure 1. A, Triple Shaft Cave, San Salvador Island, Bahamas. The entrances are 28 m above sea level, which is within the vadose zone of any Pleistocene sea-level position. The pit caves and associated chambers show vertical markings associated with vadose flow, and they end downward in sand chokes. B, Clifton Banana Hole, New Providence Island, Bahamas. The opening has formed by collapse into an oval void similar to a flank margin cave, but the banana hole is located on a level plain 1 km from the coast at Clifton. The cave entrance is 7 m above sea level. The cave floor is bedrock veneered with collapse debris.

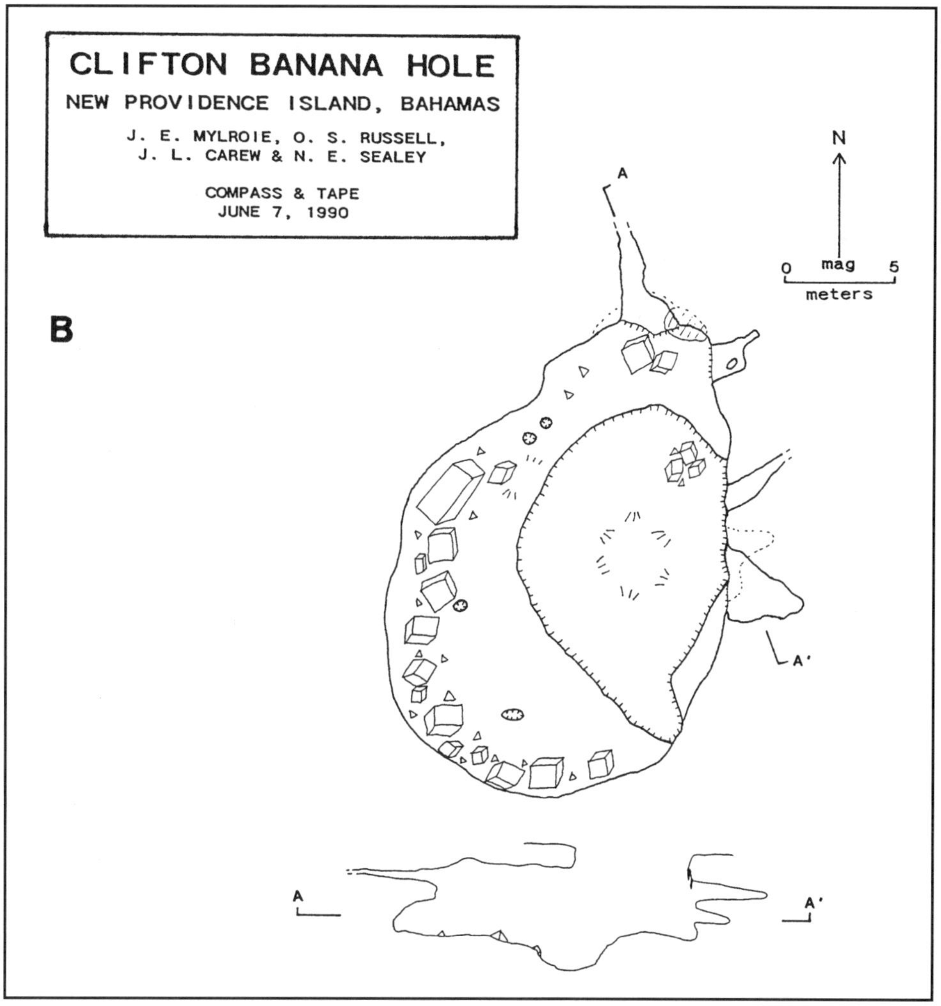

matic; Bermuda's wetter, more humid climate produced denudation rates that removed these near-surface caves. Such climatic variables have implications for larger karst landforms as well, as discussed later for basin evolution. The famous caves of Bermuda, such as Crystal Cave and Leamington Caves, which are operated as tourist attractions, are collapse structures that resulted from stoping from large voids that are now below sea level (Palmer et al., 1977; Mylroie, 1984b).

The collapse of caves of Bermuda have a characteristic morphology (Fig. 6) in which there are no bed-rock floors; instead, the caves are floored by blocks of collapse material, or by sediment and speleothems deposited on top of collapse debris. The entrances are collapse features that provide access at the top or side of a collapse chamber. The caves may continue laterally as a series of overlapping collapse chambers, producing caves of considerable length and complexity. Nearly all of the large caves can be followed downward below sea level (Fig. 7); some collapse chambers extend to more than 24 m below sea level (Iliffe, 1987). Dating of speleothems from these collapse caves has shown that some of them have been in existence for at least 200,000 yr (Harmon et al., 1983). Some of the collapse voids are immense, for example, a chamber in Church Cave (Figs 6 and 7) with dimensions of 100 by 40 m (Mylroie, 1984b).

Cave divers have also found large dissolutional cave systems, such as Green Bay Cave (Iliffe, 1987), at depths up to 18 m. These cave systems developed during a sea-level lowstand, when much of the 775-km^2 area of the Bermuda Bank was subaerially exposed and the fresh-water lens was larger. No correspondingly large dissolution caves have been found above sea level in Bermuda.

The contrasting subsurface geology of Bermuda and the Bahamas also accounts for the difference in cave morphologies (Mylroie, 1984b). The limestones of Bermuda, in contrast to those of the Bahamas, form a relatively thin cap above igneous basement. In most of the Bermuda platform, the volcanic rocks that underlie the limestones are less than 75 m below sea level (Officer et al., 1952), and in some cases are as shallow as –33 m (Peckenham, 1981). During the sea-level lowstands of the Pleistocene, the limestone-basalt contact was within the vadose zone of the emergent platform. During those times, descending vadose water was channeled by the irregularities of the basalt surface, and followed the available downstream pathways at the

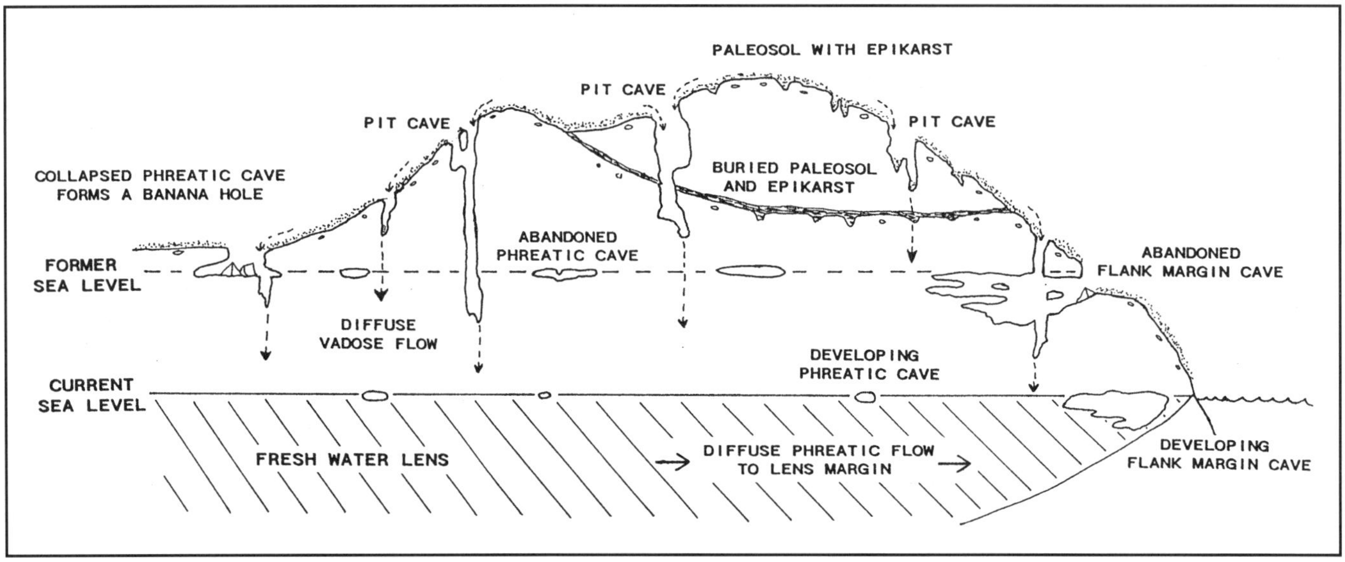

Figure 5. Diagrammatic representation of the hypothesized flow system from pit caves to flank margin caves. Surface water excavates vadose pit caves by downward dissolution, the water eventually continuing by diffuse flow through the remainder of the vadose zone to the fresh-water lens. Diffuse flow in the phreatic zone excavates flank margin caves by mixing dissolution at the margin of the lens. Isolated phreatic voids also dissolve at the top of the lens; on collapse, they form banana holes. The results are dissolutional voids separated by a diffuse flow regime. The location of banana holes and flank margin caves is controlled by sea-level position, but vadose pit caves at high elevations are independent of sea-level position. In some cases, vadose pit caves intersect flank margin caves formed during earlier sea-level highstands.

contact. Large dissolutional cave passages resulted from this concentration of vadose flow. A similar cave origin has been proposed by Jennings (1968) to explain caves at the contact between Quaternary limestone and igneous rocks on Kangaroo Island, Australia. Collapse structures associated with similar contacts have also been shown to control cenote development in Yucatan, Mexico (Pope et al., 1991).

The large collapse caves above sea level today in Bermuda have been produced by collapse of rock into dissolutional voids formed at the limestone-basalt contact. As the collapse proceeded upward, the caves breached to the surface (Mylroie, 1984b). The deep voids that accepted the collapse material must have been larger than the resultant collapse chambers above sea level, unless dissolutional processes removed some of the collapse material.

TOPOGRAPHIC LOWS IN THE BAHAMA ISLANDS AND BERMUDA

Evolution of basins

San Salvador Island, Bahamas, is approximately the same size as Bermuda (Fig. 8), and the rocks of both islands comprise a mixture of marginal-marine facies, eolianites, and paleosols (Bretz, 1960; Carew and Mylroie, 1985, this volume; Vacher et al., 1989). The landscape of both San Salvador and Bermuda has taken shape in the past 1,000,000 yr. Both islands have experienced the same history of glacioeustatic sea-level fluctuations, and both have been constructed by the accumulation and accretion of marginal-marine and terrestrial limestones. However, the physiographies of San Salvador and Bermuda are fundamentally different. Where San Salvador has broad shallow interior lakes, Bermuda is dominated by inshore basins, sounds, and reaches that create a very irregular, curving, and recurving shoreline (Fig. 8). If the geologic history and depositional regime are similar in the two areas, then why should the geomorphology be dominated by marshes and cliffed inshore water basins on one island and shallow inland lakes on the other? The reason for these differences lies in the evolution of interior topographic lows that formed initially as depressions between the Pleistocene dune ridges.

San Salvador. San Salvador Island contains many saline and hypersaline lakes. These lakes occur in interdune depressions that retain most of their original depositional morphology. On San Salvador, cliffed, rocky margins occur only along open coastlines; the restricted lagoons with marine connections, such as Pigeon Creek; and small, fresh-water ponds and blue holes, where the surface area of the water body is small and the rock contains a fresh-water lens. On San Salvador the large interior topographic lows have not significantly expanded, and they are now occupied by inland saline lakes.

Bermuda. In Bermuda, interior topographic depressions have expanded, and many are now sites of inshore reaches and sounds. A conceptual model for this evolution is outlined by

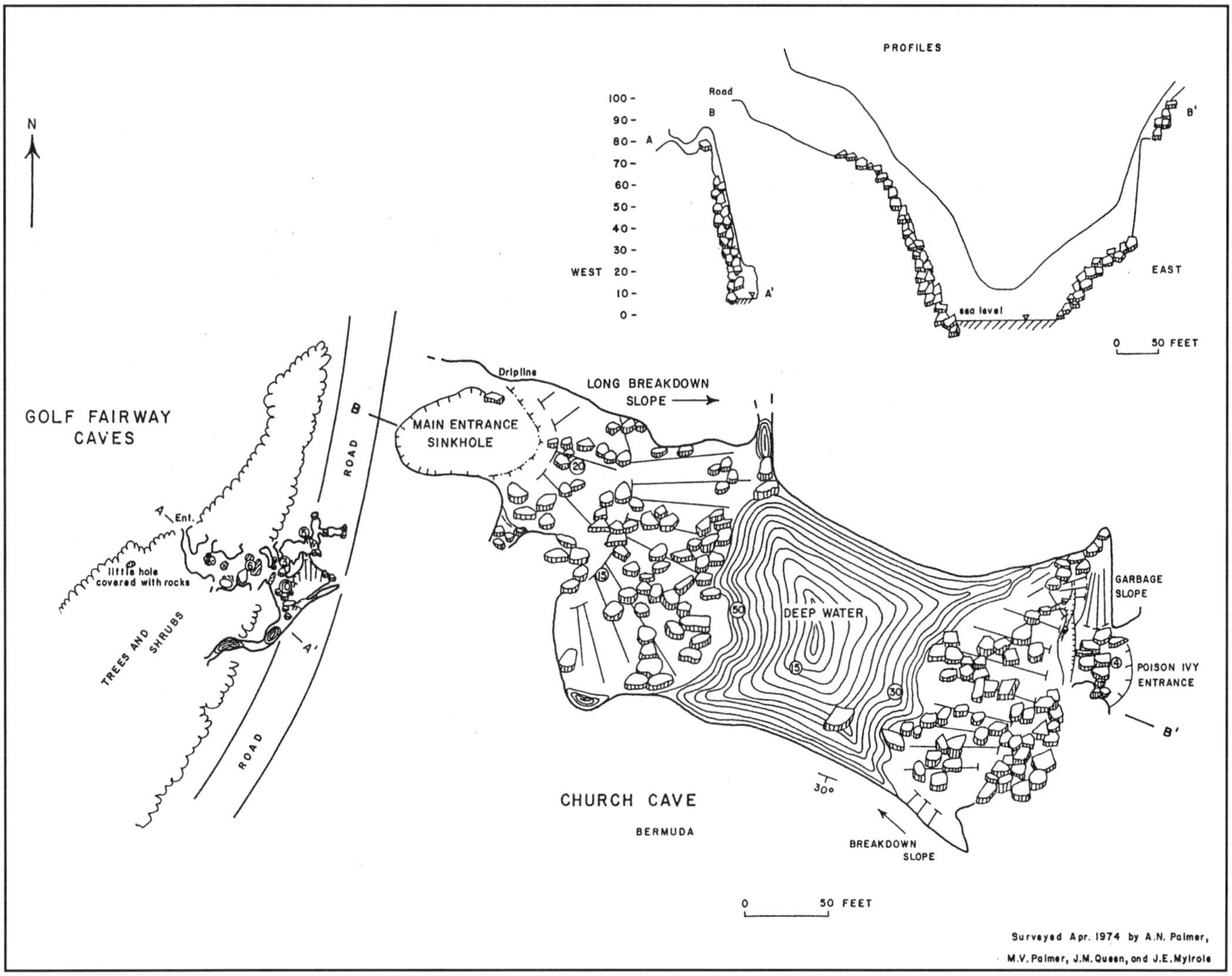

Figure 6. Map with cross sections of Church Cave on Bermuda, a typical collapse cave. Symbols are the same as in Figure 1, with water at sea level shown by line pattern. The large water-filled chamber is an area between two collapses that have prograded to the surface to form the two entrances to the cave. No bedrock floor is visible (adapted from Mylroie, 1984b).

Vacher (1978). The geologic evidence is related to the occurrence of the two major geomorphic terrains of Bermuda. In Younger Bermuda (Sayles, 1931), the eolianites retain their depositional morphology (Bretz, 1960). In Older Bermuda (Sayles, 1931), eolianites are erosionally lowered and segmented (Bretz, 1960) and invaded by inshore water bodies (Bretz, 1960; Vacher, 1978).

Marshes occur landward of the contact between Younger and Older Bermuda, in a major topographic depression between the Rocky Bay Formation and older units, or between the Belmont Formation and older units (Vacher et al., 1989; Hearty et al., 1992). The paleosol associated with the contact has been eroded through and occurs upslope from the bottom of the depression. Thus the depression occupied by the marsh has been deepened by erosion through the paleosols and adjacent eolianites.

The setting of individual depressions of the inshore basins is much like that of the marshes, with three important differences: (1) the depressions are connected to the sea, (2) they occur within the older Town Hill Formation, and (3) they occur landward of paleosols within or below the Town Hill Formation (Vacher et al., 1989; Hearty and Vacher, this volume), in the same way that modern marsh depressions are associated with the paleosols beneath the Rocky Bay and Belmont Formations. Commonly there are small island remnants of older eolianites in the sounds.

The model (Vacher, 1978; Vacher and Mylroie, 1991) for the evolution of the topographic lows on Bermuda has the following steps (Fig. 9):

1. A linear topographic depression originates because a dune ridge forms on the seaward flank of an older ridge. This

Figure 7. Photograph of a view looking west across the lake in Church Cave, Bermuda (Fig. 6); figure floating in foreground, and figure seated in the right background, for scale. Note the large pile of collapse debris that forms the far shore. The collapse pile continues beneath the floating figure to form the floor of the lake (photograph by A. N. Palmer).

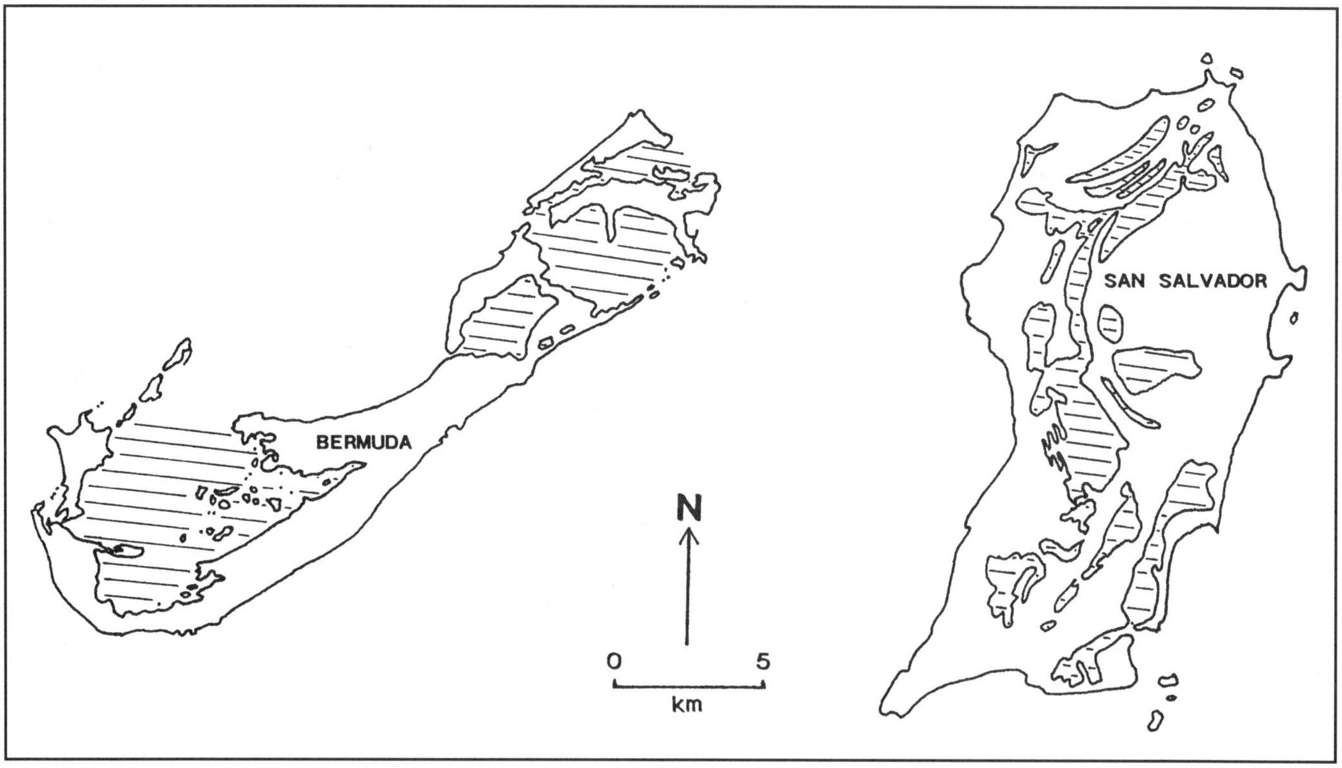

Figure 8. Maps of Bermuda and San Salvador Island, Bahamas, drawn at the same scale. Interior water bodies shown by diagonal lines. Note the complex Bermuda shoreline with many reaches and sounds. San Salvador is dominated by inland lakes which, even at this scale, are unaltered and reveal their origin as drowned interdune swales.

occurs during interglacial stages when the platform is partially flooded and producing carbonate sediment.

2. The undulating landscape is erosionally lowered, generally through dissolution by CO_2-enriched waters passing downward from active soils. This occurs on emergent dunes during both interglacial and glacial stages. Smart and Whitaker (1989) have shown in the Bahamas that biogenic CO_2 production in the soil is an important contributor to landscape dissolution. Using hydrogeochemical data in Plummer et al. (1976), Vacher (1978) estimated that the rate of landscape lowering in Bermuda is about 5 m per 10^5 yr.

3. In a later interglacial stage, the deepened depression is intersected by the water table, which rises with rising sea level, and becomes a marsh. Recharge of the fresh-water lens by CO_2-enriched water form the marsh (Plummer et al., 1976) results in phreatic dissolution of the depression margin, and the enlarged depression fills with peat.

4. With the onset of the next glacial stage, the water table drops below the peat-filled basin, and the peat is removed by oxidation and leaching. During this process, the basin is deepened further by the downward-percolating vadose waters that scavenge CO_2 from the oxidizing peat. Vadose dissolution, collapse of previously formed phreatic dissolution chambers, and scarp retreat all combine to enlarge the depression. The continued dissolutional lowering of the dune surface develops saddles within the eolianite ridges bordering the depressions, so that some depressions are vulnerable to marine flooding during a subsequent interglacial stage and elevated sea level.

5. With the rise in sea level of the subsequent interglacial stage, where climatic conditions permit, the depressions again become fresh-water marshes, with consequent peat deposition and depression enlargement. If sea level rises high enough, or if dissolutional lowering and widening of the depressions are sufficient, the depressions may instead open to the sea and become marine sounds.

6. During the interglacial stages when the basins are inshore sounds, they are expanded laterally and become cliffed because of bioerosion, as documented by Neumann (1965) for Harrington Sound. In addition, enhanced dissolution in the sounds occurs from the mixing of small amounts of sea water with the fresh-water lens. In the geomorphic evolution of Bermuda, there is a complete shift in the role of ground water from dissolution associated with fresh-water recharge near marshes to dissolution associated with brackish water discharge to the sounds.

7. Expansion and deepening of the depressions continue with additional glacial/interglacial cycles. During interglacial stages, the depressions are enlarged to basins by bioerosion, and the cliffs are sapped by mixing-zone–related dissolution. With falling sea levels in transition to glacial lowstands, marshes again develop and peats are deposited in deeper, closed-off portions within the sounds. During the glacial lowstand stages, these peats are leached, and the deep basins are deepened further. Some of the caves made at the basalt-limestone contact during earlier glacial stages, and in the fresh-water lens during interglacial stages, may also collapse and enlarge depressions.

Hydrologic explanation of depression development. Vacher and Wallis (1992) pointed out a fundamental difference between the geometry of the fresh-water lenses in Bermuda and the southern Bahamian islands. Although those authors discussed Great Exuma specifically, their observations apply equally well to other semi-arid Bahamian islands, including San Salvador (Davis and Johnson, 1989).

In Bermuda, the Ghyben-Herzberg fresh-water lens extends from shore to shore. Flow vectors are "centrifugal" in that they diverge from an interior maximum or axis. Variations in the thickness of the lens reflect proximity of the shoreline and across-island variations in hydraulic conductivity (Vacher, 1978). Specifically, the marshes have no significant effect on the general configuration of the lens.

At Great Exuma, on the other hand, separate fresh-water lenses occur in land areas between the interior lakes and ponds (Wallis et al., 1991). The major lens is in Pleistocene eolian limestones (Little et al., 1976). A smaller lens along the eastern shoreline is in a Holocene strandplain. Saline ponds occur between these two lenses. The ground-water flow is centripetal (inward) in the vicinity of the marshes and ponds (Wallis et al., 1991).

Davis and Johnson (1989) showed that on San Salvador the fresh-water lenses occur beneath the eolian ridges and discharge into the lakes that occur between the ridges. As in Great Exuma, and unlike Bermuda, a single lens does not extend across San Salvador; rather, there are a series of discrete lenses, and they underlie topographic highs.

The difference in the nature of fresh-water lenses in Bermuda, versus Great Exuma and San Salvador, is due to differences in climate (Vacher and Wallis, 1992). In Great Exuma, potential evapotranspiration greatly exceeds rainfall (by ~0.5 m/yr), so the interior lakes and ponds are sites of net ground-water loss. The situation is similar on San Salvador. In Bermuda, potential evapotranspiration is a little less than the rainfall; therefore, the marshes do not produce a net extraction from the lens.

The lakes and ponds of Great Exuma and San Salvador have the same effect as a pumped well; that is, they upcone the fresh-water/saline-water interface. Thus in Bermuda, where ground-water flow radiates out from CO_2 sources, the marshes and intereolianite topographic lows evolve through a marsh stage into inshore reaches and sounds. This does not happen in the more arid Bahamas because ground water flows centripetally toward the topographic lows (Vacher and Mylroie, 1991).

CONCLUSIONS

There are significant differences between the karst landforms of the Bahamas and Bermuda. Bermuda has greatly enlarged and modified interdune depressions, whereas the Bahama islands, especially those at the southeastern end of the archipelago, have lakes that occupy interdune depressions that

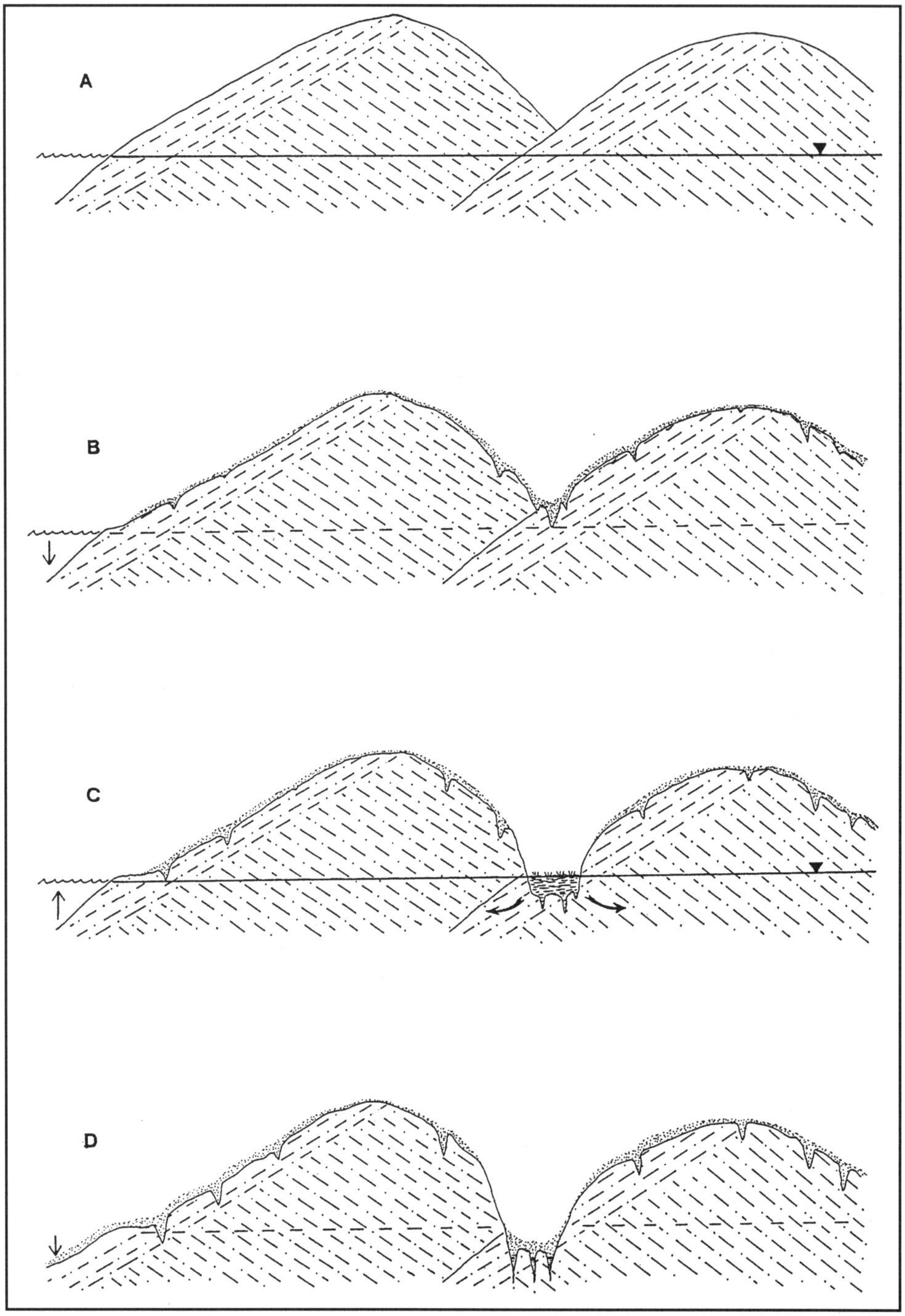

Figure 9 (on this and facing page). Diagrams illustrating the evolution of topographic lows between eolianites under Bermudian climatic conditions. Dunes are diagrammatic; the left-hand dune is seaward of and younger than the right-hand dune. Horizontal dashed line is an arbitrary interglacial sea-level position for reference purposes, with vertical arrow showing the most recent change in sea level. The sequence presented here ignores coastal processes and possible overstepping of the depression by later eolianites. A, Initial conditions following deposition of the younger, seaward dune. B, Landscape lowering by vadose dissolution. C, Marsh development by intersection of the depression by the fresh-water lens during an interglacial sea-level highstand. Arrows in the ground-water lens indicate recharge of the lens by water from the marsh. Note that the contact between the two dunes now lies on the left-hand wall of the depression. D, Accelerated vadose deepening during glacial sea-level lowstand caused in part by peat oxidation. E, Reoccupation of the depression by fresh water during a subsequent interglacial sea-level highstand. F, Consequences of marine invasion of the depression, including flank margin cave development from mixing of fresh and marine waters (arrows in ground-water lens). G, Scarp retreat, vadose processes, and cave collapse enlarge the depression during glacial sea-level lowstand.

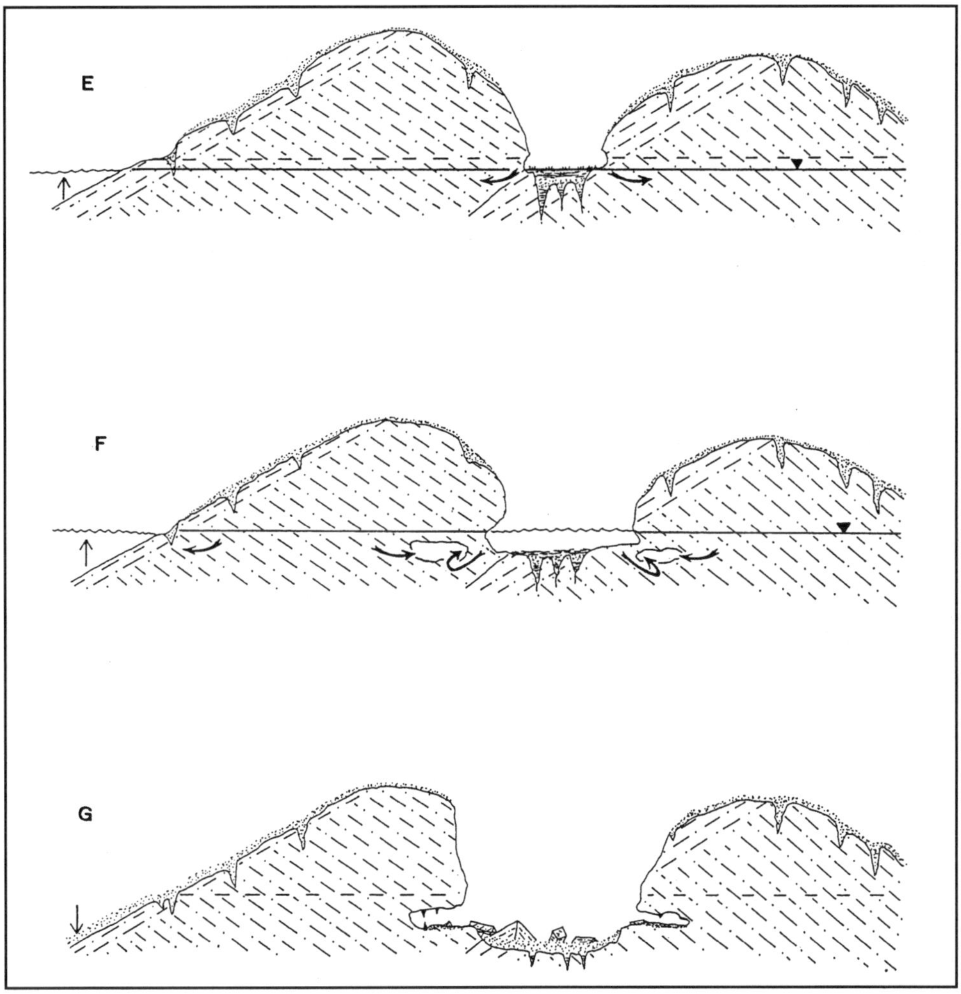

are little changed from their original depositional configuration. This difference is a direct result of the climatic differences between the two regions. The positive water budget of Bermuda promotes interdune enlargement, whereas the negative water budgets of the southeast Bahamas leads to preservation of the original depositional topography.

Caves are also very different in the two island groups. This dissimilarity is the result, in part, of different subsurface geology. On Bermuda, voids excavated by vadose streams near the limestone-basalt contact during sea-level lowstands have stoped upward to yield the present subaerial collapse caves. A comparable contact does not occur at shallow depths in the Bahamas, and collapse caves are uncommon. In the Bahamas, flank margin caves are large and abundant, and pit caves and banana holes are common; in Bermuda, flank margin caves are small and rare, and pit caves and banana holes are nearly nonexistent. The karst landforms and caves found in Bermuda and the Bahamas have differences that relate to different rates of development. In the Bahamas, large chambers can develop within 15,000 yr (up to a rate of 1 m^3/yr), faster than aragonite can completely invert to calcite. On Bermuda, however, surficial dissolutional processes are also rapid. Flank margin caves, by definition, form on the margins of the ridges enclosing them, a site vulnerable to surficial erosion. Flank margin caves on Bermuda are rapidly removed by the processes that enlarge depressions at the expense of hillslopes. In the Bahamas, this process is much slower, and flank margin caves at all stages of degradation are found. Where degradation is advanced, the remnant flank margin caves form cliff reentrants that mimic abandoned bioerosion notches. The same rapid surface denudation that has eliminated most flank margin caves on Bermuda may well have been responsible for the removal or burial of pit caves and banana holes on Bermuda, but the issue is still unresolved.

ACKNOWLEDGMENTS

We thank the many institutions and researchers who have helped to support the research that has led this work. In particular, we thank the Bahamian Field Station, San Salvador Island, Bahamas, Donald T. Gerace, C.E.O, Daniel Suchy, executive director, and the Bermuda Biological Station, Tony Knapp, director. The University of Charleston, Mississippi State University, and the University of South Florida also provided support. Many colleagues shared ideas and criticisms with us about island karst, especially Bill Back, Janet Herman, Conrad Neumann, Mike Pace, Art Palmer, Rob Palmer, Neil Sealey, Stephanie Schwabe, Joe Troester, Pete Vogel, Carol Wicks, and David Wingate.

REFERENCES CITED

Back, W., Hanshaw, B. B., Herman, J. S., and Van Driel, J. N., 1986, Differential dissolution of a Pleistocene reef in the ground-water mixing zone of coastal Yucatan, Mexico: Geology, v. 14, p. 137–140.

Bottrell, S. H., Carew, J. L., and Mylroie, J. E., 1993, Bacterial sulphate reduction in flank margin environments: Evidence from sulphur isotopes, *in* White, B., ed., Proceedings, Sixth Symposium on the Geology of the Bahamas: San Salvador, Bahamian Field Station, p. 17–21.

Bretz, J. H., 1960, Bermuda: A partially drowned, late mature, Pleistocene karst: Geological Society of America Bulletin, v. 71, p. 1729–1754.

Budd, D. A., and Vacher, H. L., 1991, Predicting the thickness of fresh-water lenses in carbonate paleo-islands: Journal of Sedimentary Petrology, v. 61, p. 43–53.

Carew, J. L., and Mylroie, J. E., 1985, The Pleistocene and Holocene stratigraphy of San Salvador Island, Bahamas, with reference to marine and terrestrial lithofacies at French Bay, *in* Curran, H. A., ed., Pleistocene and Holocene carbonate environments on San Salvador Island, Bahamas, Geological Society of America, Orlando Annual Meeting Field Trip Guidebook: Ft. Lauderdale, Florida, CCFL Bahamian Field Station, p. 11–61.

Carew, J. L., and Mylroie, J. E., 1987, Submerged evidence of Pleistocene low sea levels on San Salvador, Bahamas, *in* Cooper, R. A., and Shepard, A. N., eds., National Oceanographic and Atmospheric Administration Undersea Program Symposium Series for Undersea Research, v. 2, p. 167–175.

Carew, J. L., and Mylroie, J. E., 1995, Quaternary tectonic stability of the Bahamian Archipelago: Evidence from fossil coral reefs and flank margin caves: Quaternary Science Reviews, v. 14, no. 2, p 145–153.

Curl, R. L., 1966, Scallops and flutes: Transactions of the Cave Research Group of Great Britain, v. 7, p. 121–160.

Davis, R. L., and Johnson, C. R., Jr., 1989, Karst hydrology of San Salvador, *in* Mylroie, J. E., ed., Proceedings, Fourth Symposium on the Geology of the Bahamas: San Salvador, Bahamian Field Station, p. 118–135.

Desrochers, A., and James, N. P., 1988, Early Paleozoic surface and subsurface paleokarst: Middle Ordovician carbonates, Mingen Islands, Quebec, *in* James, N. P., and Choquette, P. W., eds., Paleokarst: New York, Springer Verlag, p. 183–210.

Garrett, P., and Gould, S. J., 1984, Geology of New Providence Island, Bahamas: Geological Society of America Bulletin, v. 95, p. 209–220.

Harmon, R. S., Mitterer, R. M., Kriausakui, N., Land, L. S., Schwarcz, H. P., Garrett, P., Larson, G. J., Vacher, H. L., and Rowe, M., 1983, U-series and amino-acid racemization geochronology of Bermuda: Implications for eustatic sea-level fluctuation over the past 250,000 years: Palaeogeography, Palaeoclimatology, Palaeoecology, v. 44, p. 41–70.

Hearty, P. J., Vacher, H. L., and Mitterer, R. M., 1992, Aminostratigraphy and ages of Pleistocene limestones of Bermuda: Geological Society of America Bulletin, v. 104, p. 471–480.

Hill, C. A., 1987, Geology of Carlsbad Caverns and other caves in the Guadalupe Mountains, New Mexico: New Mexico Bureau of Mines and Mineral Resources Bulletin 117, 150 p.

Iliffe, T. M., 1987, Observations on the biology and geology of anchialine caves, *in* Curran, H. A., ed., Proceedings, Third Symposium on the Geology of the Bahamas: Ft. Lauderdale, Florida, CCFL Bahamian Field Station, p 73–80.

Jennings, J. N., 1968, Syngenetic karst in Australia, *in* Williams, P. W., and Jennings, J. N., eds., Contributions to the Study of Karst: Canberra, Australian National University Research School of Pacific Studies Publication G5, p. 41–110.

Little, B. G., Cant, R. V., Buckley, D. K., Jeffries, A., Stark, J., and Young, R. N., 1976, Land Resources of the Commonwealth of the Bahamas, Vols. 6A and 6B, Great Exuma, Little Exuma, and Long Island: Suribiton, Surrey, United Kingdom, Ministry of Overseas Development, Land Resources Division, 130 p.

Mylroie, J. E., 1984a, Hydrologic classification of caves and karst, *in* LaFleur, R. G., ed., Groundwater as a Geomorphic Agent: Boston, Allen & Unwin, p. 157–172.

Mylroie, J. E., 1984b, Speleogenetic contrast between the Bermuda and Bahama Islands, *in* Teeter, J. W., ed., Proceedings, Second Symposium on the Geology of the Bahamas: Ft. Lauderdale, Florida, CCFL Bahamian Field Station, p. 113–128.

Mylroie, J. E., 1988, Karst of San Salvador, *in* Mylroie, J. E., ed., Field Guide to the Karst Geology of San Salvador Island, Bahamas: Ft. Lauderdale, Florida, Bahamian Field Station, p. 17–44.

Mylroie, J. E., 1990, Development of karst depressions in the Bahama Islands [abs.]: Geo^2, v. 17, p. 77.

Mylroie, J. E., 1991, Cave development in the glaciated Appalachian karst of New York: Surface-coupled or saline-freshwater mixing hydrology?, *in* Kastning, E. H. ed., Proceedings, Appalachian Karst Symposium: Huntsville, Alabama, National Speleological Society, p. 85–90.

Mylroie, J. E., and Balcerzak, W. J., 1992, Interaction of microbiology and karst processes in Quaternary carbonate island aquifers, *in* Stanford, J. A., and Simons, J. J., eds., Proceedings, First International Conference on Ground Water Ecology: Bethesda, Maryland, American Water Resources Association, p. 37–46.

Mylroie, J. E., and Carew, J. L., 1986, Minimum duration for speleogenesis, *in* Comissió organitzadora del IX Congrés Internacional d'Espeleologia, eds., Proceedings, Ninth International Congress of Speleology: Barcelona, Spain, Union Internacional de Espeleologia, p. 249–251.

Mylroie, J. E., and Carew, J. L., 1988, Solution conduits as indicators of Late Quaternary sea level position: Quaternary Science Reviews, v. 7, p. 55–64.

Mylroie, J. E., and Carew, J. L., 1990, The flank margin model for dissolution cave development in carbonate platforms: Earth Surface Processes and Landforms, v. 15, p. 413–424.

Mylroie, J. E., and Carew, J. L., 1991, Erosional notches in Bahamian carbonates: Bioerosion or groundwater dissolution?, *in* Bain, R. J., ed., Proceedings, Fifth Symposium on the Geology of the Bahamas: San Salvador, Bahamian Field Station, p. 85–90.

Mylroie, J. E., and Carew, J. L., 1995, Karst development in carbonate islands, *in* Budd, D. A., Saller, A., and Harris, P. M., eds., Unconformities in Carbonate Strata: Their Recognition and the Significance of Associated Porosity: American Association of Petroleum Geologists Memoir 63, chap. 3, p. 55–76.

Mylroie, J. E., Carew, J. L., Sealey, N. E., and Mylroie, J. R., 1991, Cave development on New Providence Island and Long Island, Bahamas: Cave Science, v. 18, no. 3, p. 139–151.

Neumann, A. C., 1965, Processes of recent carbonate sedimentation in Harrington Sound, Bermuda: Bulletin of Marine Science, v. 15, p. 987–1035.

Officer, C. B., Ewing, M., and Wuenschel, P. C., 1952, Seismic refraction measurements in the Atlantic Ocean: Pt. IV, Bermuda Rise and the Nares Basin: Geological Society of America Bulletin, v. 63, p. 777–808.

Pace, M. C., Mylroie, J. E., and Carew, J. L., 1993, Petrographic analysis of vertical dissolution features on San Salvador Island, Bahamas, *in* White, B., ed., Proceedings, Sixth Symposium on the Geology of the Bahamas: San Salvador, Bahamian Field Station, p. 109–123.

Palmer, A. N., 1991, Origin and morphology of limestone caves: Geological Society of America Bulletin, v. 103, p. 1–25.

Palmer, A. N., and Palmer, M. V., 1989, Geologic history of the Black Hills, South Dakota: National Speleological Society Bulletin, v. 51, p. 72–99.

Palmer, A. N., Palmer, M. V., and Queen, J. M., 1977, Geology and the origin of caves in Bermuda, *in* Ford, T. D., ed., Proceedings, Seventh International Speleological Congress: Sheffield, United Kingdom, British Cave Research Association, p. 336–338.

Palmer, R. J., 1985, The blue holes of the Bahamas: London, United Kingdom, Johathan Cape, 184 p.

Peckenham, J. M., 1981, On the nature and origin of some Paleogene melilititic pillowed lavas, breccias, and intrusives from Bermuda [M.S. thesis]: Halifax, Nova Scotia, Canada, Dalhousie University, 307 p.

Plummer, L. N., 1975, Mixing of sea water with calcium carbonate ground water, *in* Whitten, E. H. T., ed., Quantitative Studies in the Geologic Sciences: Geological Society of America Memoir 142, p. 219–236.

Plummer, L. N., Vacher, H. L., Mackenzie, F. T., Bricker, O. P., and Land, L. S., 1976, Hydrogeochemistry of Bermuda: A case history of ground-water diagenesis of biocalcarenites: Geological Society of America Bulletin, v. 87, p. 1301–1326.

Pope, K. O., Ocampo, A. C., and Duller, C. E., 1991, Mexican site for K/T impact crater?: Nature, v. 351, p. 105.

Rhoades, R., and Sinacori, M. N., 1941, Pattern of ground-water flow and solution: Journal of Geology, v. 49, p. 785–794.

Sanford, W. E., and Konikow, L. F., 1989, Porosity development in coastal carbonate aquifers: Geology, v. 17, p. 249–252.

Sayles, R. W., 1931, Bermuda during the Ice Age: American Academy of Arts and Sciences, v. 66, p. 381–468.

Schwabe, S. J., Carew, J. L., and Mylroie, J. E., 1993, The petrology of Bahamian Pleistocene eolianites and flank margin caves: Implications for Late Quaternary island development, *in* White, B., ed., Proceedings, Sixth Symposium on the Geology of the Bahamas: San Salvador, Bahamian Field Station, p. 149–164.

Smart, P. L., and Whitaker, F., 1989, Controls on the rate and distribution of carbonate bedrock dissolution in the Bahamas, *in* Mylroie, J. E., ed., Proceedings, Fourth Symposium on the Geology of the Bahamas: San Salvador, Bahamian Field Station, p. 313–321.

Smart, P. L., and Whitaker, F., 1990, Active circulation of saline groundwaters in carbonate platforms: Evidence from the Great Bahama Bank: Geology, v. 18, p. 200–203.

Smart, P. L., Dawans, J. M., and Whitaker, F., 1988, Carbonate dissolution in a modern mixing zone: Nature, v. 335, p. 811–813.

Vacher, H. L., 1978, Hydrogeology of Bermuda—Significance of an across-the-island variation in permeability: Journal of Hydrology, v. 39, p. 207–226.

Vacher, H. L., 1988, Dupuit-Ghyben-Herzberg analysis of strip-island lenses: Geological Society of America Bulletin, v. 100, p. 580–591.

Vacher, H. L., and Mylroie, J. E., 1991, Geomorphic evolution of topographic lows in Bermudian and Bahamian Islands: Effect of climate, *in* Bain, R. J., ed., Proceedings, Fifth Symposium on the Geology of the Bahamas: San Salvador, Bahamian Field Station, p. 221–234.

Vacher, H. L., and Wallis, T. N., 1992, Comparative hydrology of Bermuda and Great Exuma Island, Bahamas: Groundwater, v. 30, p. 15–20.

Vacher, H. L., Rowe, M. P., and Garrett, P., 1989, The geologic map of Bermuda: London, United Kingdom, Oxford Cartographers, and Hamilton, Bermuda, Public Works Department.

Viles, H. A., 1988, Organisms and karst geomorphology, *in* Viles, H. A., ed., Biogeomorphology: New York, Basil Blackwell, p. 319–350.

Vogel, P. N., Mylroie, J. E., and Carew, J. L., 1990, Limestone petrology and cave morphology on San Salvador Island, Bahamas: Cave Science, v. 17, p. 19–30.

Wallis, T. N., Vacher, H. L., and Stewart, M. T., 1991, Hydrogeology of freshwater lens beneath a Holocene strandplain, Great Exuma, Bahamas: Journal of Hydrology, v. 125, p. 93–109.

MANUSCRIPT ACCEPTED BY THE SOCIETY JANUARY 5, 1995

Printed in U.S.A.

Geological Society of America
Special Paper 300
1995

Introduction: Bermuda geology

H. Allen Curran and Brian White
Department of Geology, Smith College, Northampton, Massachusetts 01063

The arcuate cluster of islands that form Bermuda is an emergent part of the isolated Bermuda Platform, located in the western North Atlantic Ocean about 1,000 km east-southeast of Cape Hatteras, North Carolina (see Vacher, Hearty, and Rowe, the lead chapter of this section, and their Fig. 1). Although subtropical in location (32°20′N, 64°45′W), the platform and islands present an all-carbonates system, with the lithofacies of the islands dominated by eolianites. Topographic relief is generally greater than that commonly found in the Bahamas, with some hills reaching elevations of more than 50 m above sea level. The geology of the islands is well exposed along the rocky coastal areas, in roadcuts, and in quarries, and the islands recently have been geologically mapped in their entirety (Vacher et al., 1989).

Our understanding of the stratigraphy of Bermuda is in a mature state of development in comparison with that of the Bahamas. Initial stratigraphic models were developed in classic studies by Verrill (1907) and Sayles (1931), who recognized that the patterns of lithofacies were directly related to changes in Pleistocene sea level. Land et al. (1967) emphasized that the tectonically stable setting of Bermuda makes interpretation of its stratigraphy important in determining Pleistocene eustatic sea-level changes. The geochronology of Bermuda's lithofacies has been tied to sea-level changes that have occurred over the past 250,000 yr (Harmon, et al., 1983).

The lead chapter of this section by Len Vacher, Paul Hearty, and Mark Rowe presents a comprehensive review, further refinement, and modern interpretation of the stratigraphy and geochronology of Bermuda. Used in conjunction with the recent geologic map of the islands, this work undoubtedly will serve as the basis for geologic studies on Bermuda for many years to come. Furthermore, the methods and reasoning for stratigraphic interpretation employed in this study can be used as a standard for others to follow in investigating similar carbonate island terranes elsewhere in the world.

The three remaining chapters of this section present more detailed studies of various aspects of Bermuda geology. Pleistocene beach lithofacies of the South Shore of Bermuda are the subject of a multi-faceted investigation by Dieter Meischner, Rüdiger Vollbrecht, and Dieter Wehmeyer. By combining stratigraphic, sedimentologic, and petrologic analyses, these authors are able to document the position and effects of three mid- to late Pleistocene interglacial sea-level positions. In their geochemical study of solution pipe soils, Stanley Herwitz and Daniel Muhs demonstrate that the parent materials for the soils are wind-transported and must have multiple source areas far from Bermuda, with a possible contribution from the Great Plains area of North America. In the final chapter of this volume, John Hartsock, Donald Woodrow, and Brooks McKinney present an analysis of the fracture systems present in the Pleistocene limestones of northeastern Bermuda. They conclude that the dominant fracture patterns likely result from control by a buried volcanic caldera underlying Castle Harbour. This hypothesis, and many others that might be proposed based on the unique geology of Bermuda, remains open for future investigation.

REFERENCES CITED

Harmon, R. S., and 8 others, 1983, U-series and amino-acid racemization geochronology of Bermuda: Implications for eustatic sea-level fluctuation over the past 250,000 years: Palaeogeography, Palaeoclimatology, Palaeoecology, v. 44, p. 41–70.

Sayles, R. W., 1931, Bermuda during the Ice Age: Proceedings of the American Academy of Arts and Sciences, v. 66, p. 381–468.

Vacher, H. L., Rowe, M. P., and Garrett, P., 1989, The geological map of Bermuda: Hamilton, Bermuda Government, Ministry of Works and Engineering, scale 1:25,000.

Verrill, A. E., 1907, The Bermuda Islands, Pt. IV, Geology and paleontology; and Pt. V, An account of the coral reefs: Transactions of the Connecticut Academy of Arts and Sciences, v. 12, p. 45–348.

Manuscript Accepted by the Society January 5, 1995

Curran, H. A., and White, B., 1995, Introduction: Bermuda geology, *in* Curran, H. A., and White, B., Terrestrial and Shallow Marine Geology of the Bahamas and Bermuda: Boulder, Colorado, Geological Society of America Special Paper 300.

Printed in U.S.A.

Geological Society of America
Special Paper 300
1995

Stratigraphy of Bermuda: Nomenclature, concepts, and status of multiple systems of classification

H. L. Vacher
Department of Geology, University of South Florida, Tampa, Florida 33620
P. J. Hearty
College of the Bahamas, P.O. Box N-4912, Nassau, The Bahamas
M. P. Rowe
Ministry of Works and Engineering, P. O. Box 525, Hamilton HM CX, Bermuda

ABSTRACT

Bermuda, where Pleistocene eolianite was first named and analyzed in terms of glacioeustasy, can be considered the type locality of the carbonate eolianite facies. In Bermuda, this facies consists of bioclastic eolianite and associated beach deposits, weakly developed calcarenitic paleosols, and red to reddish brown paleosols termed "terra rossa" by many authors. The eolianite formed as large retention ridges, which prograded inland into a vegetated landscape; the sediment was derived from the surrounding submerged platform. Eolianite deposition in Bermuda was an interglacial phenomenon. It occurred mainly after the initial submergence, which is recorded by deposits of rocky shorelines and pocket beaches like those of the present-day erosional coastline. The terra-rossa paleosols represent relatively long hiatuses in the carbonate buildup, and therefore intervene between the limestones of successive interglacial stages; however, these paleosols do not equate exactly to glacial stages. The weakly developed calcarenite paleosols represent minor breaks within and between interglacial substages.

Although sea-level history is the ultimate control of the mosaic of facies and the succession of stratigraphic units in Bermuda, attempts to define lithostratigraphic units in terms of sea-level history led to unmappable columns. Lithostratigraphic units now have been mapped successfully throughout Bermuda. These units parallel time-stratigraphic units but, in themselves, do not permit resolution of successive interglacial stages in the lower part of the column. Enhanced resolution is provided by allostratigraphic and aminostratigraphic units. The latter are zones defined by amino acid racemization (AAR) ratios that provide a measure of relative age. By use of interpretive kinetic models, the ratios have been converted to age estimates that allow direct correlation with the deep-sea oxygen isotope stages.

Among the principal findings of stratigraphic mapping and analysis in Bermuda are: that shoreward superposition (lateral accretion) is the key mapping principle; and that the stratigraphic column includes a significant record of oxygen isotope stages 5, 7, 9, and 11, as well as an older history represented by the Walsingham Limestone at the base of the exposed section.

Vacher, H. L., Hearty, P. J., and Rowe, M. P., 1995, Stratigraphy of Bermuda: Nomenclature, concepts, and status of multiple systems of classification, *in* Curran, H. A., and White, B., Terrestrial and Shallow Marine Geology of the Bahamas and Bermuda: Boulder, Colorado, Geological Society of America Special Paper 300.

INTRODUCTION

Since Sayles (1931), it has been recognized that the stratigraphy of Bermuda's carbonate eolianites and associated beach deposits and paleosols is directly related to Pleistocene sea-level changes. Land et al. (1967) argued that the stratigraphy of Bermuda is particularly important because of Bermuda's tectonically stable setting; Harmon et al.(1983, p. 57) likened the record to that of a Pleistocene "tide gauge." The stratigraphy has also provided the temporal framework for using Bermuda as a natural laboratory to determine geologic rates of mineralogic, petrologic, and geochemical changes during meteoric diagenesis of bioclastic grainstones (Morse and Mackenzie, 1990): and patterns of evolution, as exemplified by the endemic land snail *Poecilozonites* (Gould, 1969; Eldredge and Gould, 1972).

Completion of the first detailed geologic map of Bermuda (Vacher et al., 1989) and results from aminostratigraphic analysis (Hearty et al., 1992; Hearty and Vacher, 1995) have led to a refinement of concepts on which stratigraphic classification in Bermuda is based, and a better appreciation of the richness of the Bermuda record. This chapter summarizes this view of Bermuda stratigraphy, including the development of the current nomenclature, the understanding of physical stratigraphic and facies relationships, the conceptual basis of time-stratigraphic interpretation, and the current status of the multiple systems of stratigraphic classification that can be applied.

BERMUDA

Bermuda comprises a group of limestone islands located about 1,000 km east of Cape Hatteras, North Carolina. Five islands account for nearly all of Bermuda's 50 km^2 of land, although there are more than 150 islands and islets in the island group. The islands lie near the southern margin of a 650-km^2 platform (Fig. 1), the geology of which has been detailed by Upchurch (1970) and Vollbrecht (1990). This platform, which is the site of the northernmost coralgal reefs in the North Atlantic, is presently submerged to an average depth of about 20 m. The platform lies atop a completely buried volcanic seamount (Pirsson, 1914; Aumento and Ade-Hall, 1973) that was last active in the Oligocene (Reynolds and Aumento, 1974; Tucholke and Vogt, 1979; Galehouse, 1979).

CARBONATE EOLIANITE FACIES IN BERMUDA

Bermuda can be considered the type locality of the carbonate eolianite facies. Sayles (1931) coined the term "eolianite" for the bioclastic dune rocks that make up more than 90% of Bermuda. Although Sayles (1931, p. 390) intended the word to include "all sedimentary rocks that were deposited by the wind," it generally is now used to mean "eolian sands that have been cemented by calcium carbonate in the subaerial environment" (Gardner, 1983, p. 265), or, more specifically, "calcite-cemented coastal dunes of Quaternary age" (Fairbridge and Johnson, 1978, p. 279). Both definitions apply to Bermuda. Similar deposits tend to occur around the fringes of the world's carbonate belt (Fairbridge and Johnson, 1978). Other examples in the Northern Hemisphere include the Bahamas (Ball, 1967; Garrett and Gould, 1984; Carew and Mylroie, 1985; White and Curran, 1988); coastal Mexico (Ward, 1975); San Clemente and San Nicolas Islands, California (Muhs, 1992); and Mediterranean coastlines (Butzer and Cuerda, 1962; Yaalon and Laronne, 1971; Hearty et al., 1986). Southern Hemisphere localities include western Australia (Fairbridge and Teichert, 1953; Semeniuk and Johnson, 1982) and Lord Howe Island (Gaffney, 1983).

Throughout the range of the carbonate eolianite facies, the eolianites are associated with coastal marine deposits and paleosols. The chief features and terminology of this assemblage of facies in Bermuda are briefly reviewed in this section as a prerequisite for considering the stratigraphic classification and analysis.

Eolian deposits

The eolian deposits (i.e., the rock "eolianite" as the word was used by Sayles, 1931) of Bermuda are bioclastic grainstones that occur in roughly linear, shore-parallel bodies. These lithosomes also have been termed "eolianites" by Vacher (1973) and subsequent authors. The eolianites formed as linear ridges by lateral coalescence of lobate coastal dunes (Bretz, 1960; Mackenzie, 1964a) that stood as much as a few tens of meters above the source beaches. As described by Vacher (1973, p 374), "[the dunes] advanced inland but the advance was a landward progradation through the leeside accretion that accompanied upward growth. They did not advance by the reworking of sediment in the entire dune body, as is the case of classic migratory dune ridges." The occurrence of casts and molds of palmetto trees (Vacher and Harmon, 1987) shows that the dunes advanced into a vegetated landscape. The dunes can be classified as transverse retention ridges (McKee, 1979), and were depositionally similar to so-called "precipitation ridges" (Cooper, 1958) of the Oregon coast.

As shown by Mackenzie (1964b), the orientation of foresets sampled across Bermuda indicates that the formative winds blew from all directions. According to Vacher (1973), who mapped foresets within eolianites and their constituent lobate-dune bodies, geometric relationships of foresets of adjoining lobes and their directional correlation with present-day gale-force winds suggest that strong winds were more important than prevailing winds in shaping the large dune bodies. Gales currently occur on about 36 days a year (Vacher, 1973). The significance of major storms in the development of these dunes is also indicated by the typical occurrence of enormous sets of foresets that remain unbroken or uninterrupted (by soils or bioturbation) for several tens of meters across the eolianites. It is possible that eolianites were deposited mostly during a small number of major storms (with recurrence intervals of tens or

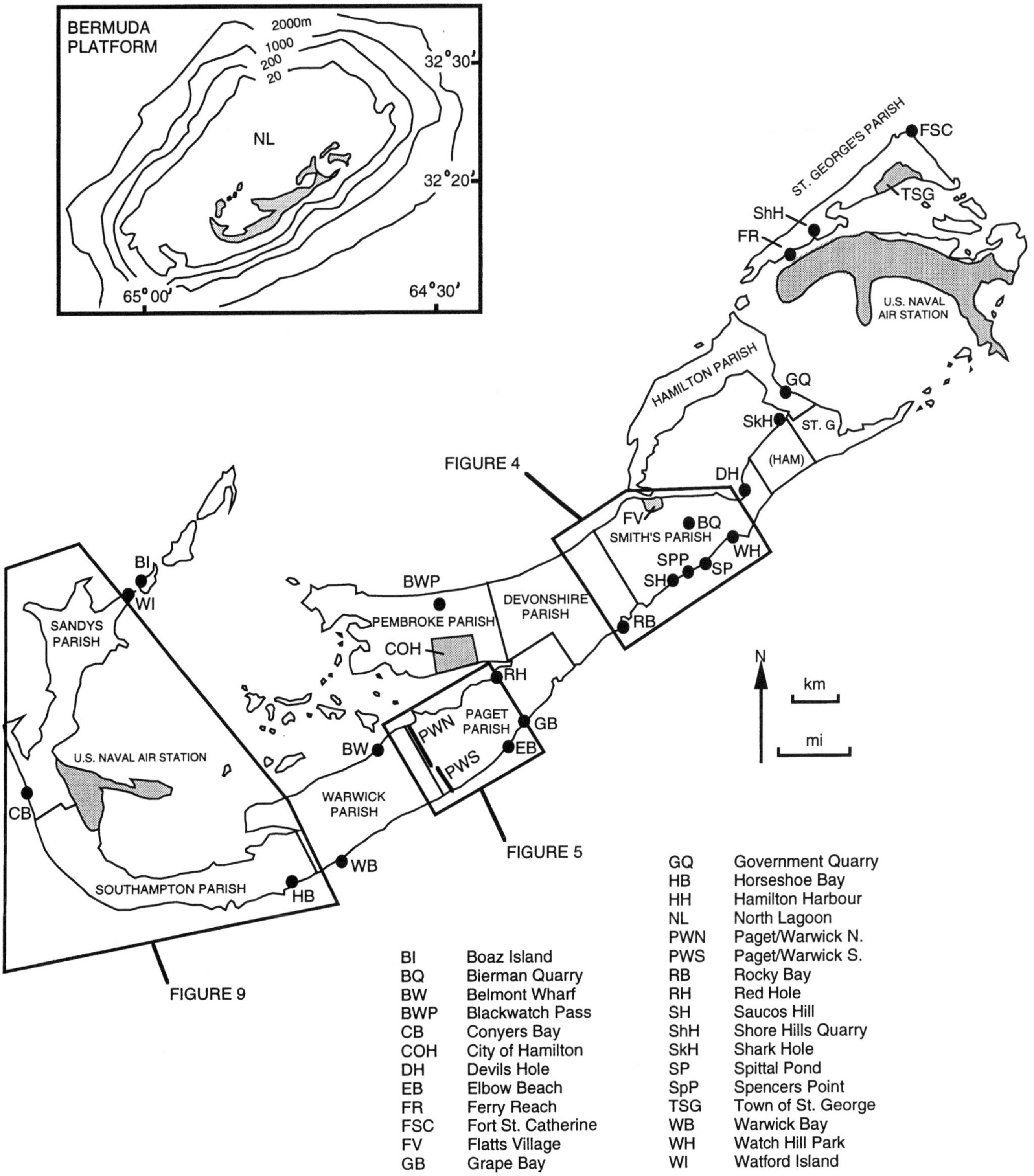

Figure 1. Location of Bermuda, the nine parishes, and localities mentioned in text.

even hundreds of years) when conditions of sediment supply were optimal. Between the storms, sediment mostly accumulated as temporary storage on seaward-prograding beaches.

Marine deposits

The coastal marine rocks are of three types (Vacher, 1971, 1973): erosional coastline, depositional coastline, and protected-coastline deposits.

Erosional coastline deposits are discontinuous lenses and pods of fossiliferous conglomerates and calcarenites deposited along rocky shores and pocket beaches. The deposits occur within notches, on benches, and against paleocliffs. The sediment is derived from erosion of headlands and biogenic sediment production in the pocket bays. The modern and Pleistocene rocky shore deposits of western Australia (Semeniuk and Johnson, 1985) are like these Bermudian deposits. Vollbrecht and Meischner (1993) and Meischner et al. (this volume) have described two good examples (their younger unit, in each case).

Depositional coastline deposits are long, shore-parallel wedges consisting of beach deposits analogous to those described by Inden and Moore (1983). A particularly good example is described by Meischner et al. (this volume). In general, the deposits are bioclastic grainstones that typically contain no whole shells. In some cases, it is difficult to distinguish these shoreline-marine deposits from those of the windward part of eolianites where low-angle conformable cross-beds are common. The depositional coastline deposits formed during times of a positive sediment budget, when sediment was transported to the shore from offshore sediment sources (reefs and lagoon: Upchurch, 1970; Vacher, 1973). These beach deposits typically grade laterally or commonly upward into an eolianite (Bretz, 1960; Meischner et al., this volume).

Protected-coastline deposits, typically onlapping the leeward foresets of older eolianites, are formed along the shoreline of an inshore water body. In terms of whole and well-preserved shells, these deposits are the most fossiliferous in Bermuda. The deposits are rare, probably because of erosion accompanying lateral expansion of the inshore water bodies (Neumann, 1965; Vacher, 1978; Vacher and Mylroie, 1991; Mylroie et al., this volume).

Paleosols

Paleosols in Bermuda are of two types (Sayles, 1931): red to reddish brown paleosols and weakly developed calcarenite paleosols.

The first, red to reddish brown paleosols, stand out in contrast to the cream-colored limestones. The description of the physical character of the red soils of Eleuthera by Foos (1991) fits these Bermudian paleosols. The Bermudian red paleosols are thickest and best developed in topographic lows (Sayles, 1931; Ruhe et al., 1961; Vacher, 1973). They are associated with downward protruding soil pipes, which have been termed "palmetto stumps" by Sayles (1931), "roots" by Bretz (1960), and "solution pipes" by Land et al. (1967); the origin of these structures has been most recently analyzed by Herwitz (1993), who gives evidence of a compromise—tree-guided dissolution. Thin but locally prominent calcretes occur within and at the base of these red paleosols (Mauritsen, 1983).

Termed "soils of weathering" by Sayles (19310, the red paleosols were first thought to be the insoluble residue of large amounts of eolianite (Verrill, 1907; Sayles, 1931). Bretz (1960) and Ruhe et al. (1961) first used the term terra rossa for these Bermudian paleosols and specifically meant to imply only "red soils overlying limestone, but not necessarily derived from it" (Ruhe and others, 1961, p. 1138). It is now recognized that the noncarbonate fraction of these paleosols was derived from atmospheric circulation (Bricker and Mackenzie, 1970), perhaps partly, but probably not entirely, from the Sahara (Herwitz and Muhs, this volume). The name terra rossa is now firmly entrenched for these Bermudian paleosols. They are probably equivalent to Alfisols and Ultisols of the U.S. Soil Taxonomy (D. R. Muhs, personal communication, 1994).

The second type, weakly developed calcarenite paleosols, was described as "embryonic" by Bretz (1960) and termed "protosol" by Vacher and Hearty (1989) and Hearty and others (1992). These weakly developed paleosols are probably equivalent to Entisols, Inceptisols, and minimally developed Alfisols in the U.S. Soil Taxonomy (D. R. Muhs, personal communication, 1994). According to Ruhe et al. (1961), these protosols consist 96% or more of carbonate sand. Decay of organics and pedogenic processes have provided some color—usually buff, tan, or brown—but as quantified by Ruhe et al (1961), this is usually less than two units of chroma.

Sayles (1931) and Bretz (1960), emphasizing process, called the protosols "accretionary soils," indicating "eolian sand that accumulated so slowly that, under concomitant weathering, . . . [they attained] only a faint color . . . from included plant carbon" (Bretz, 1960, p. 1737). We picture a process where accumulation and pedogenesis were virtually synchronous, as sand was trapped by the coastal grasses. In some other cases, we suspect, bioturbation and pedogenesis overprinted an already wholly accumulated eolian or beach deposit.

STRATIGRAPHIC NOMENCLATURE AND ITS EVOLUTION

The history of stratigraphic nomenclature in Bermuda (Table 1) can be divided into two periods. In the first period, the stratigraphic column was based on correlation of isolated, mainly coastal, localities; the correlation was from interpreted, relatively detailed sea-level history (Sayles, 1931; Bretz, 1960; Land et al., 1967), and later, U-series ages (Land et al., 1967; Harmon et al., 1981, 1983). These stratigraphic columns were functionally a set of time-stratigraphic units with a lithostratigraphic nomenclature (Vacher, 1973).

During the second period, the field effort was focused on a

TABLE 1. HISTORY OF STRATIGRAPHIC NOMENCLATURE IN BERMUDA

Verrill, 1907	Sayles, 1931	Land et al., 1967	Vacher, 1973, 1974	Vacher et al., 1989
Paget Fm (ls and soils)	Southampton Fm (eol)	Southampton Fm	Paget Fm, Upper mbr	Southampton Fm
Devonshire Fm	McGalls Soil	St. Georges Soil	Paget Fm, Lower mbr	Rocky Bay Fm
Washington Fm	Somerset Fm (eol)	Spencers Point Fm	Shore Hills Soil	Devonshire mbr
(ls and soils)	Signal Hill Soil	Pembroke Fm	Belmont Fm	Shore Hills Geosol
	Warwick Fm (eol)	Harrington Soil	Unnamed soil	Belmont Fm
	St. Georges Soil	Devonshire Fm	Walsingham Fm	Ord Road Geosol
	Pembroke Fm (eol)	Shore Hills Soil		Town Hill Fm, Upper mbr
	Harrington Soil	Belmont Fm		Harbour Road Geosol
	Devonshire Fm (mar)	Soil?		Town Hill Fm, Lower mbr
	Shore Hills Soil	Walsingham Fm		Castle Harbour Geosol
	Belmont Fm (mar)			Walsingham Fm
	Walsingham Fm (eol)			

stratigraphic column that could be mapped in the field (Vacher, 1973; Vacher et al., 1989; Rowe, 1990). Later it was coordinated with an independent system of aminostratigraphic zones (Hearty and Vacher, 1992, 1995). Also during the second period, the mapping effort became part of a ground-water exploration program, which was most concerned with Bermuda's interior regions where rock exposures are more limited. It was from study of these interior regions that the presence of a significant middle Pleistocene succession was recognized (Vacher and Harmon, 1987; Hearty et al., 1992; Hearty and Vacher, 1995).

Before geologic mapping

The first stratigraphic column (Verrill, 1907) was a threefold subdivision (Table 1) consisting of a marine unit (Devonshire Formation) separating an underlying sequence of indurated eolianites (Walsingham Formation) and an overlying sequence of "normal" eolianites and undifferentiated paleosols (Paget Formation). Noting that foresets of the eolianites extend below sea level, Verrill (1907) reasoned they formed when the platform was tectonically elevated relative to its present position.

Sayles (1931) was the first to recognize the connection between Pleistocene glacial-interglacial cycles, sea-level fluctuations, and the superposition of marine limestones, eolianites, and paleosols in Bermuda. To Sayles, the submerged eolianites indicated that sand seas spread across the platform during sea-level lowstands of glacial stages; conversely, marine limestones and paleosols, which alternate with eolianites, were seen to represent interglacials. Sayles's (1931) column, which was designed to reveal the cyclicity of the record, consisted of two marine limestones (Belmont and Devonshire Formations), five eolian units, and five paleosols (Table 1). Accordingly, Sayles correlated these units with five glacial advances and four recessions as recognized in the North American mid-continent.

Although Bretz (1960) found Sayles's (1931) column to be inadequate, he did not formally propose a new one. His principal contribution to the evolution of stratigraphic thinking in Bermuda was the reinterpretation of the timing of eolianite deposition. His key observation was that beach deposits intergrade with eolianite. Therefore, according to Bretz (1960, p. 1730), the eolianites formed as coastal dunes that "did not migrate far from their feeding grounds, the shore lines." They must have formed during interglacial epochs. This interpretation has been adopted by all subsequent workers and is expanded upon in this chapter.

Stratigraphic nomenclature was incidental to the main effort of Land et al. (1967) which was to determine Bermuda's geologic history as a framework for understanding diagenetic pathways, Pleistocene sea levels, and *Poecilozonites* evolution. Land et al. (1967), however, did agree with the criticisms by Bretz (1960) and revised Sayles's (1931) column accordingly. Thus Land et al. (1967) combined the upper three eolian units and two paleosols into the Southampton Formation and inserted a post-Pembroke marine unit, the Spencer's Point Formation (Table 1). Those authors also redefined type localities for the post-Walsingham units; their justification was that Sayles's column was unworkable because Sayles's (1931) type localities were scattered throughout Bermuda in such a way that superposition of the various units could not be demonstrated. The type localities defined by Land et al. (1967) are in a single area, a short stretch of the south shore from Rocky Bay to Spencer's Point (see Fig. 1 for localities mentioned in text).

Formulation of the column of Land et al. (1967) was tightly interwoven with interpretation of sea-level history. According to Land et al., (1967, p. 1005), ". . . [the stratigraphic units] cannot be differentiated on the basis of lithologic or paleontologic criteria alone. . . . The geometric and temporal limits of Bermudian formations, i.e., their boundaries, are dictated by the conceptual scheme into which they are placed—the rise and fall of the Pleistocene sea".

In practice, both the column and the sea-level curve of Land et al. (1967) were derived from correlation of four principal local sections, as shown in the diagrammatic mosaic of Fig-

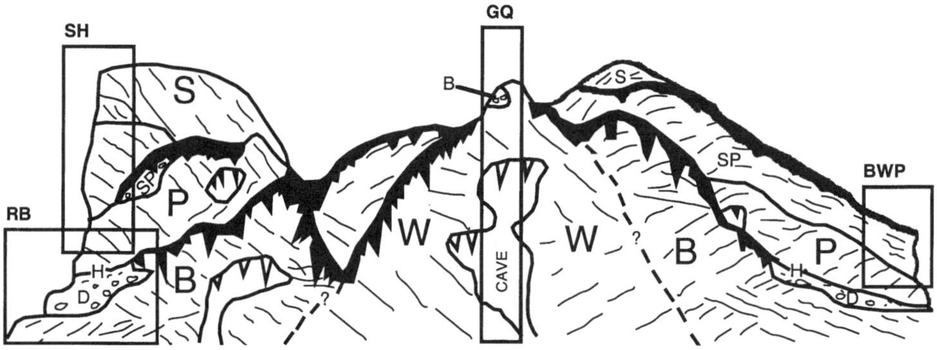

Figure 2. Schematic mosaic of lithostratigraphic units in Bermuda as portrayed by Land et al. (1967). Stratigraphic units: S = Southampton; SP = Spencer's Point; P = Pembroke; H = Harrington; D = Devonshire; B = Belmont; W = Walsingham. Rectangles refer to four classic localities: SH = Saucos Hill (and continuation to Spencer's Point); RB = Rocky Bay (Devonshire Bay of earlier papers); GQ = Government Quarry; BWP = Blackwatch Pass.

ure 2. Rocky Bay (RB of Fig. 2) and Saucos Hill (SH) were the type localities of their post-Walsingham units. The high elevation of interpreted beach deposits along the north shore at Blackwatch Pass (BWP) was taken as verification of a post-Devonshire highstand (Spencer's Point). Government Quarry (GQ) was both the type section for the Walsingham (following Sayles, 1931) and the location of a marine conglomerate at +22 m that Land et al. (1967) classified as Belmont.

Development of a map-based stratigraphy

Vacher (1973) found that the units of earlier columns were not mappable. He concluded that the mappability requirement of formations would be best accomplished by grouping the units of the column of Land et al. (1967) into multi-facies limestones bracketed by terra-rossa paleosols that represent island-wide unconformities (Land et al, 1967). Accordingly, Vacher (1973) defined the Paget Formation to include all the units above the Shore Hills Soil of Land et al. (1967). The St. George's Soil (Land et al., 1967) was not used as a formation boundary because it was known that Lower Paget and Upper Paget eolianites (Pembroke and Southampton, respectively, of Land et al., 1967) occur in succession at some places with no terra rossa between them (Vacher, 1973).

Completion of the geologic mapping (Vacher et al., 1989; Rowe, 1990) resulted in the present column, which is shown in Figure 3 and Tables 1 and 2. It consists of five formations and four geosols (paleosols used as named stratigraphic units; North American Commission of Stratigraphic Nomenclature, 1983). The Paget Formation of Vacher (1973) is elevated to Paget Group (Fig. 3), consisting of Southampton and Rocky Bay Formations (Table 1), which do not have a geosol between them. The large, complex interval between the Rocky Bay and the Walsingham Formations is subdivided (Table 1) into two formations, the Belmont (sensu Land et al., 1967) and the Town Hill Formations, which are each bracketed by geosols. The Town Hill Formation, in turn, is subdivided into upper and lower members, which are separated by a geosol. These two members are not recognized as formations because they cannot be distinguished in areas of poor exposure. Additional terra-rossa paleosols occur in the lower part of the column (in the Lower Town Hill and Walsingham) but are not used to separate lithostratigraphic units nor used as geosols.

As shown in Figure 3, the stratigraphic column of Vacher et al. (1989) can be considered as a composite of partial sections at several localities. These sections occur in two principal areas, from which the stratigraphic names were derived. The first is southern Devonshire and Smith's Parishes (Fig. 4), which include the following important localities: (1) Rocky Bay (RB of Fig. 3), which is the type locality of the Shore Hills and Harrington Soils and Belmont, Devonshire, and Pembroke Formations of Land et al. (1967) and the Rocky Bay Formation of Vacher et al. (1989); (2) the shoreline at Saucos Hill (SH), McGall's Hill, and Spencer's Point, all of which are in the type area of the St. Georges Soil and Spencer's Point and Southampton Formations of Land et al. (1967); (3) Spittal Pond (SP), where a succession similar to the Devonshire, Harrington, and Pembroke of the Rocky Bay occurs within the Belmont Formation (Vacher and Harmon, 1987); (4) Watch Hill Park (WH), where the Belmont marine deposits lie against a sea cliff cut in upper Town Hill eolianite which, inland and at a higher elevation (along South Road), is overlain by the Ord Road Geosol and Belmont eolianite; and (5) Bierman Quarry (BQ; Rocky Heights Quarry on the geologic map [Vacher et al, 1989], and more recently, DeSilva Quarry), the type locality of the Town Hill Formation of Vacher et al. (1989), where Upper Town Hill, Lower Town Hill, and Walsingham eolianites occur in direct superposition with intervening terra-rossa paleosols (Hearty and Vacher, 1995); Belmont eolianite and the Ord Road Geosol were also present earlier in the quarry operations (Vacher and Harmon, 1987).

The second area (Fig. 5) lies between the south shore and

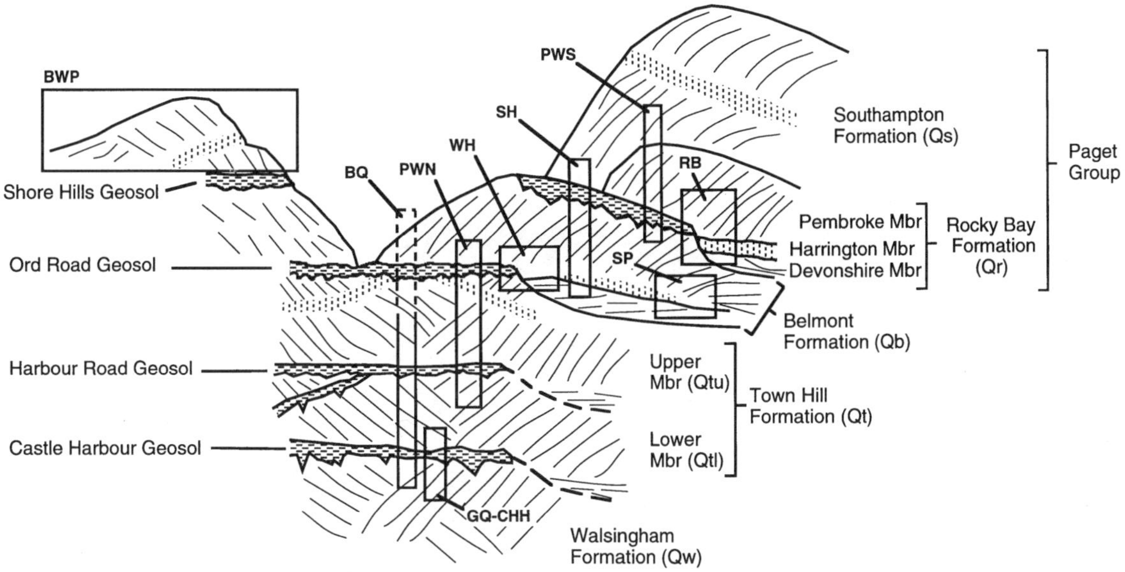

Figure 3. Schematic mosaic of lithostratigraphic units in Bermuda according to Vacher et al. (1989) and abbreviations of stratigraphic units (e.g., Qs, Qr) used in other figures. Localities not included in Figure 2: BQ = Bierman Quarry (Fig. 4); GQ-CHH = Government Quarry area, including Castle Harbour Hotel; PWN = Paget/Warwick North (Fig. 5); WH = Watch Hill Park (Fig. 4); PWS = Paget/Warwick South (Fig. 5); SP = Spittal Pond (Fig. 4).

TABLE 2. PARALLEL SYSTEMS OF STRATIGRAPHIC CLASSIFICATION

Litho-		Soil-	Allo-	Amino-	Oxygen Isotope Stage	Age* (ka)
Paget Group	Southampton Fm		S3	C	5a	85 (c)
			S2			
			S1			
	Rocky Bay Fm		R2	E	5a	125 (c)
			R1	E/F		170 (a:P)†
		Shore Hills Geosol				
	Belmont Fm		B2	F	7	200 (c)
			B1			250 (a:wr)
		Ord Road Geosol				
	Town Hill Fm: Upper member		UT2	G	9	350 (a:P,wr)
			UT1			
		Harbour Road Geosol				
	Town Hill Fm: Lower member		LT3	H	11	450 (a:P,wr)
			LT2			
			LT1	J		>700 (a:wr)
		Castle Harbour Geosol				
	Walsingham Fm		Undiff	K		>880 (a:wr)

*c = From U-series dates on corals (Harmon et al., 1983; used as calibration for AAR estimates; a = estimates from AAR (Hearty et al., 1992; Hearty and Vacher, 1995); P = *Poecilozonites* samples; wr = whole-rock samples. Age estimates are within plus or minus about 15% (Hearty et al., 1992).
†Probably an overestimate.

Hamilton Harbour along the boundary of Paget and Warwick Parishes (Cobb's Hill Road). Along this transect, which was first described by Sayles (1931), the entire section from Lower Town Hill to Southampton is laid out in a pattern of successive onlap. Particularly important is a section along Harvey Road with three critical exposures: the contact between Southampton and Rocky Bay eolianites (a weakly developed paleosol) at South Road just east of Harvey Road; the contact between Rocky Bay and Belmont eolianites (Shore Hills Geosol) in the quarried recesses along Harvey Road; and the contact between Belmont and Upper Town Hill (type Ord Road Geosol) along both Ord Road and Harvey Roads near their intersection. The

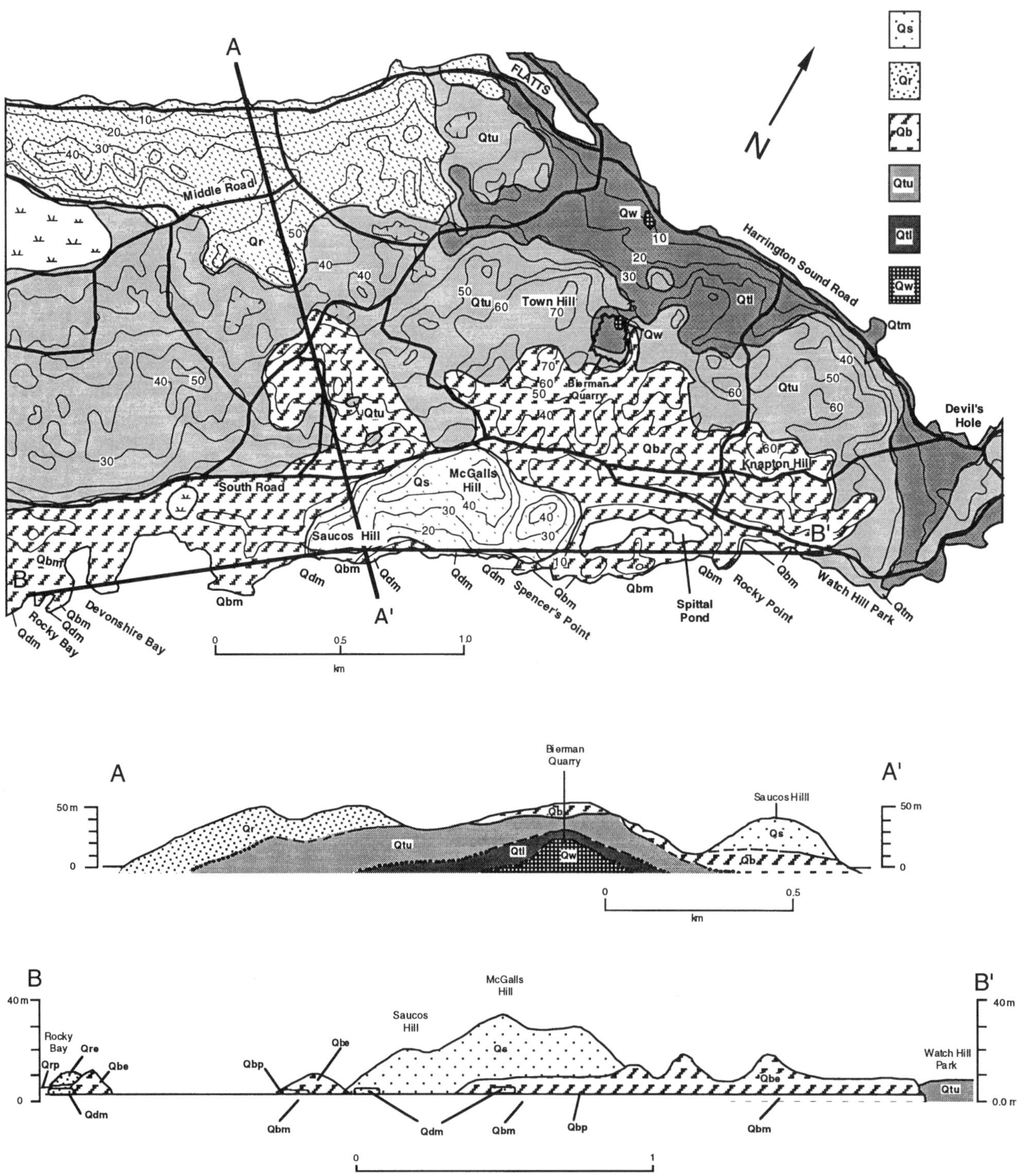

Figure 4. Geologic map and cross sections of Smith's and eastern Devonshire Parishes. See Figure 1 for location and Figure 3 for stratigraphic abbreviations. Additional abbreviations: Qdm, Qrp, and Qre are marine (Devonshire Member), protosol, and eolianite facies of Qr, respectively; Qbm, Qbp, and Qbe are marine, protosol, and eolianite facies of Qb, respectively. Topographic contours in meters. Adapted from Vacher et al. (1989).

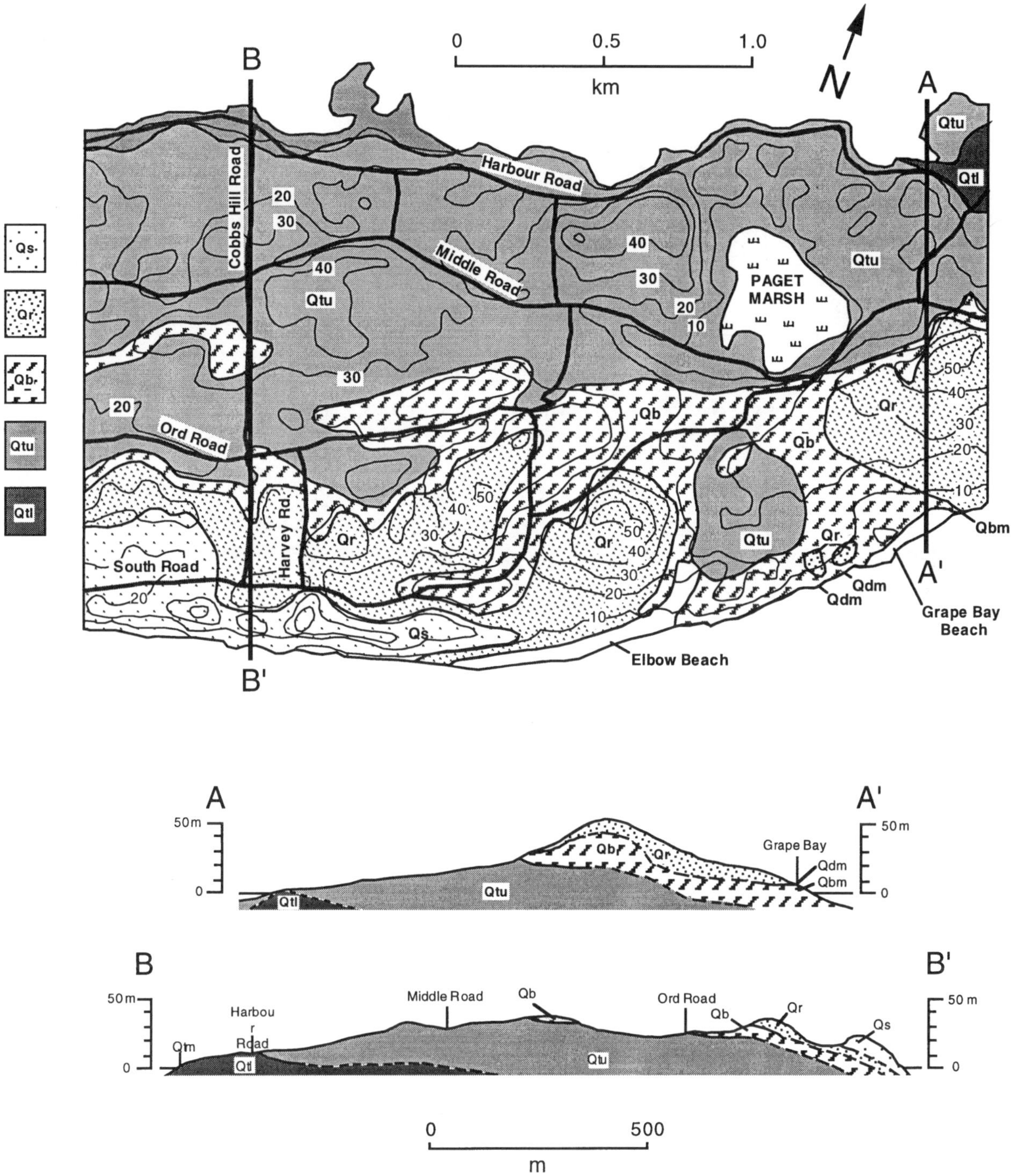

Figure 5. Geologic map and cross sections of the area around the Paget/Warwick boundary. See Figure 1 for location. Topographic contours in meters. Adapted from Vacher et al. (1989).

type Harbour Road Geosol occurs farther north (down section) at the intersection of Cobb's Hill and Harbour Roads.

Vacher et al. (1989) retain the Devonshire, Harrington, and Pembroke of Land et al. (1967) as successive members within the Rocky Bay Formation where the latter consists of a succession of marine limestone, protosol, and eolianite. This succession is widespread along the coastline. There are, however, several places where the Rocky Bay Formation consists of some other combination of facies; one example is the north shore area of Pembroke and Devonshire Parishes (e.g., Blackwatch Pass, Fig. 6; BWP of Figs. 2 and 3). In this area, the members are not recognized; specifically, the notion of Devonshire and

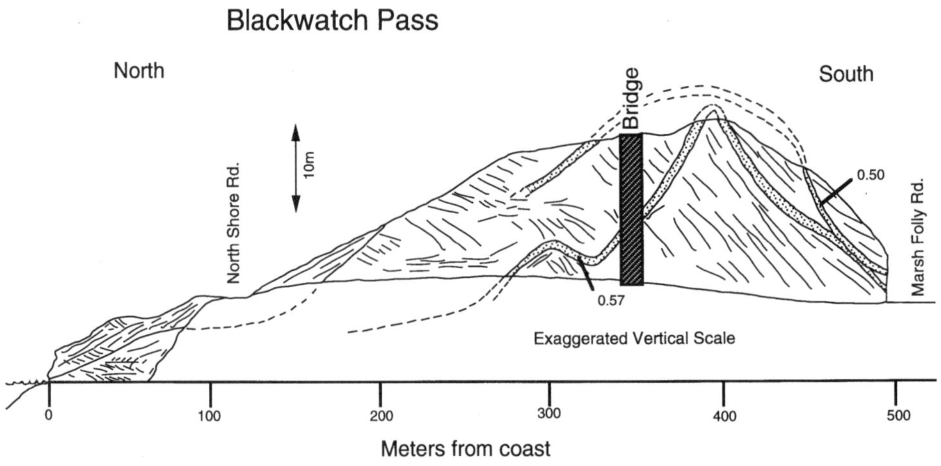

Figure 6. Cross section showing field relations at Blackwatch Pass, including adjacent north shore. Numbers refer to amino-acid racemization ratios on *Poecilozonites* from the two protosols exposed in the roadcut.

Harrington eolianites (Bretz, 1960; Land and Mackenzie, 1970) is discouraged because it represents use of members as time-stratigraphic units. Also, a succession of marine limestone, protosol, and eolianite occurs repeatedly through the column, not only in the Rocky Bay Formation. In particular, such a succession occurs in the Belmont Formation in the classic south shore section of Smith's Parish (Fig. 4, Section B-B′ including SH and SP of Fig. 3). Recognition that the protosol and eolianite of this succession lie within the Belmont (thus do not correlate with the succession at Rocky Bay, RB) eliminates the need for the Spencer's Point Formation of Land et al. (1967) (Vacher and Hearty, 1989).

In light of the results of geologic mapping (Vacher et al., 1989), it is now evident that the meaning of several of Sayles's (1931) units were in fact changed by Land et al. (1967) when they redefined type localities. For example, Sayles's (1931) type locality for the Belmont was Shore Hills Quarry (now adjacent to the Bermuda Biological Station), and he took the name from Belmont Wharf on nearly the other side of Bermuda; the relevant deposits at these two places are mapped as Walsingham and Lower Town Hill, respectively, by Vacher et al. (1989). Similarly, Sayles's (1931) type Pembroke (in the city of Hamilton) is mapped as Upper Town Hill, and his Shore Hills (at Shore Hills Quarry), Harrington (at Shark Hole), and St. Georges (town) are mapped as Castle Harbour–Harbour Road composite, the Harbour Road Geosol, and a paleosol (Shore Hills) separating Belmont and Southampton, respectively. Thus, as suspected by Land et al. (1967), some of Sayles's (1931) type sections are not in stratigraphic order. Vacher et al. (1989) retained the nomenclature sensu of Land et al. (1967) because of its clearer meaning and usefulness.

LATERAL ACCRETION: THE FORM OF SUPERPOSITION IN BERMUDA

Sayles (1931, p. 446) identified the cardinal feature of Bermudian stratigraphy: "The present Bermuda has therefore evolved from a much smaller, older Bermuda by a process of accretion." As illustrated by Land et al. (1967) in their diagrammatic mosaic (Fig. 2), successive units are not tabular island-wide layers; instead, they are discontinuous, irregularly shaped sediment bodies with considerable depositional relief. Because this relief is much larger than the difference in relative sea level between successive highstands, coastal dune complexes of later interglacials generally accumulated on the outside margin of the deposits of earlier interglacials. Therefore, stratigraphic units are arranged in lateral as well as vertical succession. The geologic map (Fig. 7) documents the relation in detail; in general, the section gets younger toward the external shorelines. Thus, the Walsingham and Town Hill Formations occur in the interior of the island next to the inshore water bodies, and the Belmont, Rocky Bay, and Southampton Formations successively, and roughly radially, offlap this core.

Although lateral accretion is the general rule, there are numerous exceptions where younger eolian units overstep older ones. This overstepping typically occurs where the eolianites of successive interglacials are markedly different in size. For example, the very large Southampton and Rocky Bay eolianites of western Bermuda mostly, and in places completely, bury the small Belmont eolianites of that area. Even more striking is the upper member of the Town Hill Formation, which covers the Lower Town Hill nearly throughout Older Bermuda (Vacher et al., 1989; Hearty and Vacher, 1995). Clearly, more eolian sediment was deposited in some interglacials than in others.

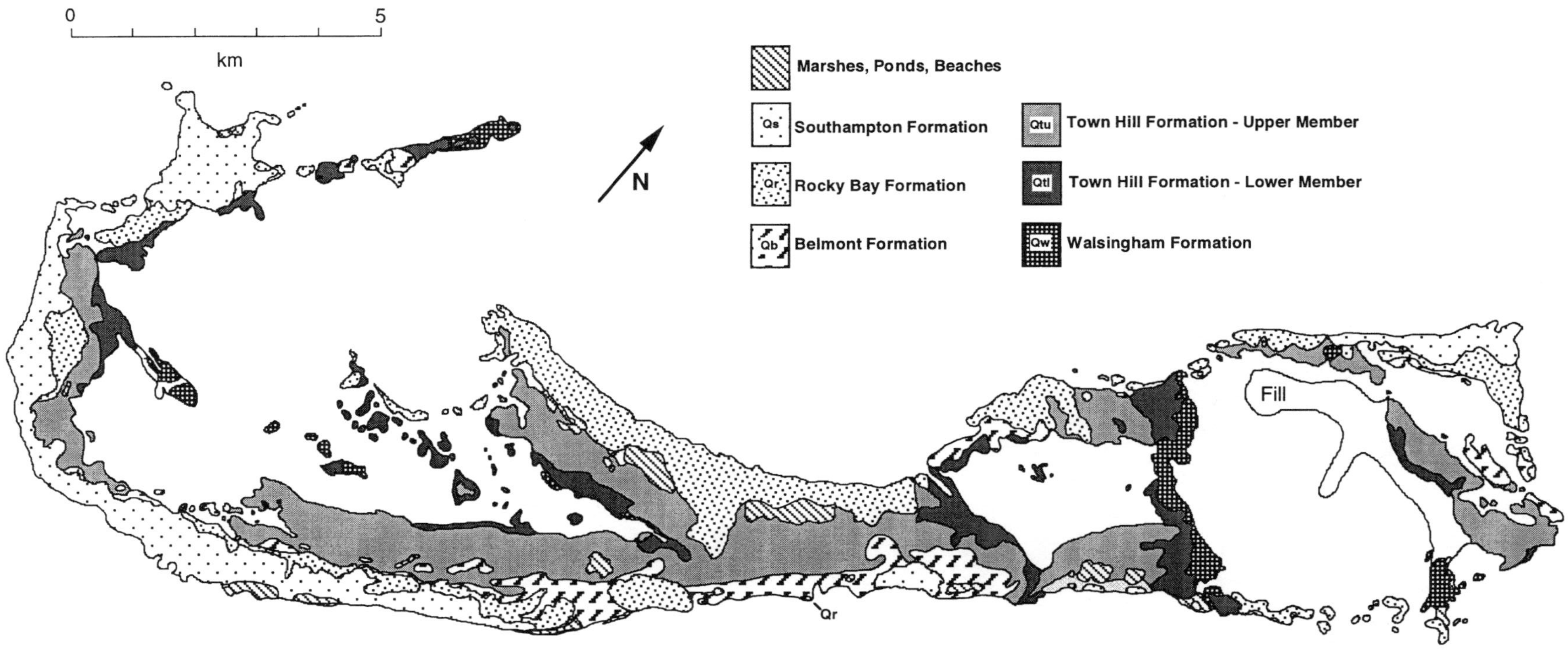

Figure 7. Geologic map of Bermuda generalized from Vacher et al. (1989).

According to Hearty and Vacher (1995), oxygen isotope stages 5 and 9 (Shackleton and Opdyke, 1973) were the most important in terms of volumes of sediment accumulated.

CONCEPT OF TIME-STRATIGRAPHIC INTERPRETATION OF EOLIANITES AND PALEOSOLS

Beach dune transitions and the timing of eolianite deposition

The earliest interpretations of Bermuda eolianites (Verrill, 1907; Sayles, 1931) were that they formed at times of lower relative sea level. This inference was prompted by the widespread and striking occurrence of foresets at the present water line; it was also consistent with the absence in the Holocene of Bermuda of large dune ridges comparable to those of the Pleistocene. The current interpretation, that the large eolianites formed during interglacials comparable to that of the present (Bretz, 1960, and later authors), was based on recognition of beach-dune transitions above present sea level and the identification of the eolianites as coastal dunes. Indeed, mapping of eolianites in relation to coeval beach deposits (Vacher, 1973; Vacher et al., 1989) has shown that the dunes did not advance more than a few hundred meters from the shoreline; therefore, at the time the eolianites on the present island were deposited, the shoreline (hence sea level) must have been nearby, even in the case of the eolianites with the partially submerged foresets. From examination of the facies relationships along Bermuda's entire shoreline, especially along the inshore water bodies, it is possible to make the following refinement: where the timing can be worked out in detail, it generally can be shown that deposition of the large eolianites followed the initial transgression and submergence of the platform (see also Meischner et al., this volume).

Exposed beach-dune transitions typically occur in the type of facies mosaic shown in Figure 8A. In the proximal (seaward) part of the complex, depositional coastline marine deposits (unit 5 of Fig. 8A) grade upward and without break into gently inclined eolian cross-beds (6) representing a beach ridge that nucleated the growth of the main, eolian part (8) of the complex; this is best seen in the Belmont Formation at Spittal Pond (Vacher and Harmon, 1987). In the distal (landward) part of the complex, there is a vertical succession of erosional coastline deposits (4), back-beach protosol (7), and eolian foresets (8). The coastal erosion surface (3) associated with the distal marine deposits truncates the terra rossa (2) formed on the limestone substrate (1) of earlier interglacials. The foresets overstep the coastal erosion surface so that, at the landward extremity of the complex, foresets overlie terra rossa and earlier limestones. Erosion of the complex back to the position of the protosol and erosional coastline facies results in exposure of the classic three-part succession that, in the Rocky Bay Formation, has been subdivided into the Devonshire, Harrington, and Pembroke Members (Fig. 3). At Saucos Hill, the Belmont consists of a similar three-part succession due to erosion of the complex back to the position of the protosol and depositional coastline beach facies (Vacher and Harmon, 1987). A comparable succession is common in the Town Hill Formation of interior shorelines (Belmont Wharf, Red Hole, Devil's Hole, Ferry Reach), although identification of the beach facies (as distinct from near-the-beach eolian) is less certain.

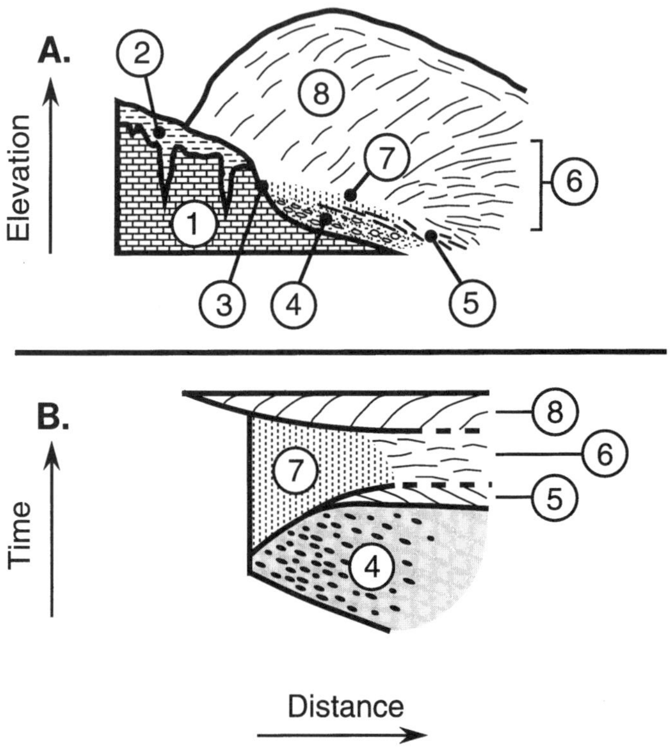

Figure 8. A, Physical mosaic, and B, Wheeler diagram of shallowing-upward sequence deposited during sea-level highstand. Key to units are: 1, limestone of previous interglacial; 2, terra-rossa paleosol; 3, erosion surface; 4, erosional coastline, conglomeratic marine deposit; 5, deposits of an extensive beach derived from offshore sediment sources; 6, beach ridge which nucleates later eolianite; 7, back-beach protosol; 8, eolianite. Explanation in text.

Time-stratigraphic interpretation of the components of the facies mosaic of Figure 8A is shown in the time-length diagram of Figure 8B, in which the material ordinate axis of the length-length diagram (regular cross section) is replaced by a time axis. Sloss (1984) has named such time-versus-geography plots "Wheeler diagrams" from their use in the classic paper (Wheeler, 1958) that drew the distinction between lacuna (the time represented by an unconformity), hiatus (the constituent interval of nondeposition), and erosional vacuity (the time represented by the eroded rocks).

Although those distinctions do not apply in the problem being discussed here, the Wheeler diagram (Fig. 8b) also facilitates analysis of the sequence of depositional events producing a facies mosaic (Fig. 8A). Thus, as shown in Figure 8B, the first deposits represented by the mosaic are those (unit 4) of erosional coastlines. As sediment was delivered to the shoreline, the pocket beaches prograded seaward; the back part of the beach developed as grassed-over supratidal accumulation of sand (the protosol, 7) washed and blown in from the beach. As delivery of offshore sediment increased, the beach sediment budget became more positive, and long, seaward-progradational beaches (5) developed. Beach ridges (6), and finally, major landward-prograding dune ridges (8) grew with the continued delivery of offshore sediment. The dunes became the major sink for offshore sediments delivered to the shoreline. As shown in Figure 8B, the protosol represents the time needed to develop a positive sediment budget in which offshore-derived sediment overflows to the dunes.

In many cases, the evolution of the shoreline to that of a positive sediment budget was accompanied by a drop in sea level. This is particularly evident in coastal exposures of the Rocky Bay Formation. For example, at many localities, the Harrington protosol and overlying Pembroke eolianite extend down to the present water level and clearly indicate that sea level had dropped several meters from its maximum position (up to 5 m above sea level and marked by erosional coastline deposits of the Devonshire in these cases), before deposition of the eolianites. At other places (e.g., the "Devonshire eolianites" of Bretz, 1960, at Watford Island), the implied drop in sea level is considerably less.

Although detailed relationships within beach-dune transitions generally indicate that eolianite deposition occurred when sea level was falling, we do not believe that a drop in sea level caused the eolianite deposition. Indeed, Belmont beach-dune transitions along the south shore of Devonshire Parish indicate that some deposition of eolianite occurred during the rise and at the peak of the relevant sea-level highstand. Probably, passage of time is the critical factor: with sufficient time, offshore sediment sources build up and transport routes to the shoreline develop. If a few thousand years were required before sufficiently large volumes of sediment were delivered to the shoreline, then sea level could well have been dropping, coincidentally, in accord with its 20,000-yr cyclicity (Broecker and van Donk, 1970; Hays et al., 1976; Martinson et al., 1987).

Recognition that deposition of the large eolianites lags the initial transgression is a refinement of Bretz's (1960) conclusion that eolianites are formed during interglacials and is consistent with the observation that eolianite volumes like those of the Pleistocene have not been deposited during the Holocene. Indeed, the present south shore is a good analogue for the shoreline facing the platform margin during deposition of the Devonshire marine deposits (substage 5e). All of Bermuda's famous beaches and small dunes (Vacher, 1973) occur in short segments totaling less than 15% of this 19-km-long margin-facing shoreline. Clearly, the ca. 4,000 yr that sea level has been at about its present position in Bermuda (Neuman, 1969; Ellison, 1993) has not been sufficient time for eolianite production to be established at Pleistocene volumes. According to Garrett and Scoffin (1977), the main sediment production off the south shore is on a prominent terrace between the line of reefs and the platform edge, and this sediment is mostly backed up against the step on the shoreward side of the terrace. The sediment has overtopped this step in a few places; and those are the places where the beaches and dunes occur (e.g., Elbow Beach, Warwick Bay, Horseshoe Bay). This accumulation represents a start, and, with time, more sediment will be produced, and a greater fraction of that produced should reach shore (P. Garrett, personal communication, 1982). Delivery would be enhanced, of course, if the shoreline receded toward the sediment source with a drop in sea level.

Although most of the exposed beach-dune transitions indicate that sea level was dropping during deposition of the eolianite, there are many eolianites for which there are no data, or equivocal data, on the amount or sign of sea-level change at the time of deposition. Obvious examples include eolianites deposited during sea-level highstands that peaked below the present position of sea level. Another example is the Rocky Bay eolianites at Blackwatch Pass (Fig. 6), a locality that has been debated many times (Bretz, 1960; Land et al., 1967; Vacher, 1973; Harmon et al., 1981, 1983), and that faces the North Lagoon (Fig. 1), not the platform margin. It is now thought (Harmon et al., 1981, 1983; Hearty et al., 1992) that none of the low-angle beds exposed in the roadway (Fig. 6) are marine, that the marine deposits (if any are present) are limited to the lowest few meters exposed in the cliffs, and that the beach (?) and eolian deposits along the north shore correlate in time with the Devonshire Member. AAR ratios (Hearty et al., 1992) indicate that *Poecilozonites* in the upper of the two protosols exposed in the roadcut (Fig. 6) correlate in time with the Harrington member of the south shore, and that *Poecilozonites* fossils in the lower (and most prominent) protosol are significantly older. These data admit to a variety of interpretations regarding sea-level history and its relation to eolianite deposition. It is entirely possible that the timing of deposition of eolian sediment derived from the heart of the North Lagoon is different from that derived from the platform margin (Vacher, 1973). The facies mosaic of Figure 8 is inferred largely from exposures along the margin-facing shoreline.

The timing of eolianite deposition on Bermuda is different than that documented in the Bahamas, where eolianite deposition was also an interglacial phenomenon (Garrett and Gould, 1984; White et al., 1984; Carew and Mylroie, 1985, this volume; Hearty and Kindler, 1993a,b; White and Curran, this volume). In the Bahamas, there are also large transgressive eolianites. The most obvious examples are the Holocene eolianites of the Rice Bay Formation of San Salvador (Carew and Mylroie, 1985, this volume; Hearty and Kindler, 1993a) and correlative units at Lee Stocking Island (Kindler, this volume) and indeed throughout the Bahamian archipelago (Hearty and Kindler, 1993b). Comparable eolianites do not occur in Bermuda, probably because of less sediment production.

Alternation of limestones and terra rossas and the time-stratigraphic meaning of terra rossas

Although Bermuda's eolianites and marine deposits formed during interglacial stages, it is an oversimplification to equate Bermuda's terra rossas, which alternate with the limestones, to glacial stages. Terra rossas simply represent unconformities (Land et al., 1967). This is illustrated by the schematic Wheeler diagram of Figure 9, which shows the kind of time-stratigraphic relations that occur along strike of the eolianites comprising the Belmont, Rocky Bay, and Southampton Formations in western Bermuda. For the sake of illustration and argument, the effects of erosion, which are second-order in this case, are ignored; that is, the discontinuous nature of the eolianites in this area is considered to be due to nondeposition, and the amount of degradation of eolianite due to chemical erosion is small relative to its original thickness; therefore, erosional vacuities (Wheeler, 1958) do not complicate the diagram. The point to be made is that disjunct glacial stages plot as "time-distance *layers*" within the terra-rossa field of the Wheeler diagram, but the terra-rossa field is not limited to layers. Using the definitions of Wheeler (1958), the total terra-rossa field is a hiatal holosome with respect to the limestone, not a stack of soil holostromes. This is so because the time-area field of terra rossa includes interglacial time owing to the lensing out and overstepping of the various eolian bodies and the time that eolianite deposition lags behind the initial sea-level rise. A similar conclusion was drawn by Carew and Mylroie (1991) in their discussion of Bahamian paleosols.

What is the minimum length of time required for a time-parallel tongue of the hiatal holosome (of a Wheeler diagram such as Fig. 9) to be physically manifested by a terra-rossa paleosol? The best-constrained answer can be derived from the classic south shore area of Devonshire and Paget Parishes, where the Shore Hills Geosol is bracketed between Devonshire marine deposits with U-series coral ages of 125 ka (Harmon et al., 1983) and Belmont marine deposits correlated with coral-bearing deposits dated at 204 ka (Harmon et al., 1983). The hiatus, 80,000 yr for this moderately developed terra rossa, seems a reasonable time to accumulate a Bermudian soil by dust fallout. Taking values of 3% for the amount of insolubles (Herwitz and Muhs, this volume), 30 cm for the thickness of the Shore Hills between these Belmont and Rocky Bay deposits, and 1.7 g/cc for a bulk density (D. R. Muhs, personal communication, 1994), the implied accumulation rate is about 0.02 g/cm^2/1,000 yr. This rate is an order of magnitude less than the rates (0.1 to 0.5 g/cm^2/1,000 yr) of fallout documented for Barbados and Miami (Muhs et al., 1990, and references therein), which are more favorably located to receive Saharan dust.

Time stratigraphic meaning of protosols

The weakly developed calcarenite paleosols occur throughout the column and in three types of settings:

(1) Between an underlying marine deposit and an overlying eolianite (Fig. 8). As already discussed, these protosols mark the transition from erosional to depositional coastlines within a single sea-level highstand and represent no more than a few thousand years.

(2) Lenses between eolian deposits that comprise a single morphologic unit. These protosols may represent short, local pauses in the accumulation of eolian sand—perhaps no more than the recurrence interval of major storms. Such protosols are especially common in the Southampton Formation and the upper part of the upper member of the Town Hill Formation. At places in Sandy's and Southampton Parish as many as five protosols can be seen in succession within the Southampton Formation in a small area. ARR ratios on *Poecilozonites* indicate that several protosols that occur in vertical succession in the Southampton have nearly equivalent age (Hearty et al, 1992). At one particular residential exposure near Conyer's Bay, three protosols are interlayered with high-angle foresets, conformably and at about 1-m intervals; these protosols apparently mark reactivation surfaces. Such intermittent deposition of eolian foresets on the distal accumulation margin of the largest and highest dunes suggests that the packets of foresets represent rare events (large storms) and that the protosols represent the times between those events (see also Carew and Mylroie, 1991).

(3) Extensive layers between eolian deposits that comprise different morphologic units. These protosols probably have more stratigraphic significance, such as a marker for island-wide discontinuities resulting from the precession-related sea-level lowstands within an interglacial stage (Broecker and van Donk, 1970; Hays et al., 1976; Martinson et al., 1987). As yet, petrologic/geochemical criteria have not been developed to distinguish such stratigraphically important protosols from the local and temporally insignificant ones.

STATUS OF MULTIPLE SYSTEMS OF STRATIGRAPHIC CLASSIFICATION

The Pleistocene section of Bermuda can be subdivided stratigraphically by a variety of criteria. Each criterion gives rise to a different system of stratigraphic classification (North American Commission of Stratigraphic Nomenclature, 1983). Five such systems—consisting of litho-, soil-, allo-, amino-, and chronostratigraphic units—are shown in Table 2. The status of these and three others—bio-, morpho-, and hydrostratigraphic units—are revealed in this section.

Lithostratigraphy

The five formations of Figure 3 and Table 2 represent an attempt to subdivide a lithologic gradation into distinguishable units. The formations are all similar in that they consist of the same material: bioclastic grainstone deposited in coastal beaches and dunes. They are different only because of the different amounts of time that they have experienced meteoric dia-

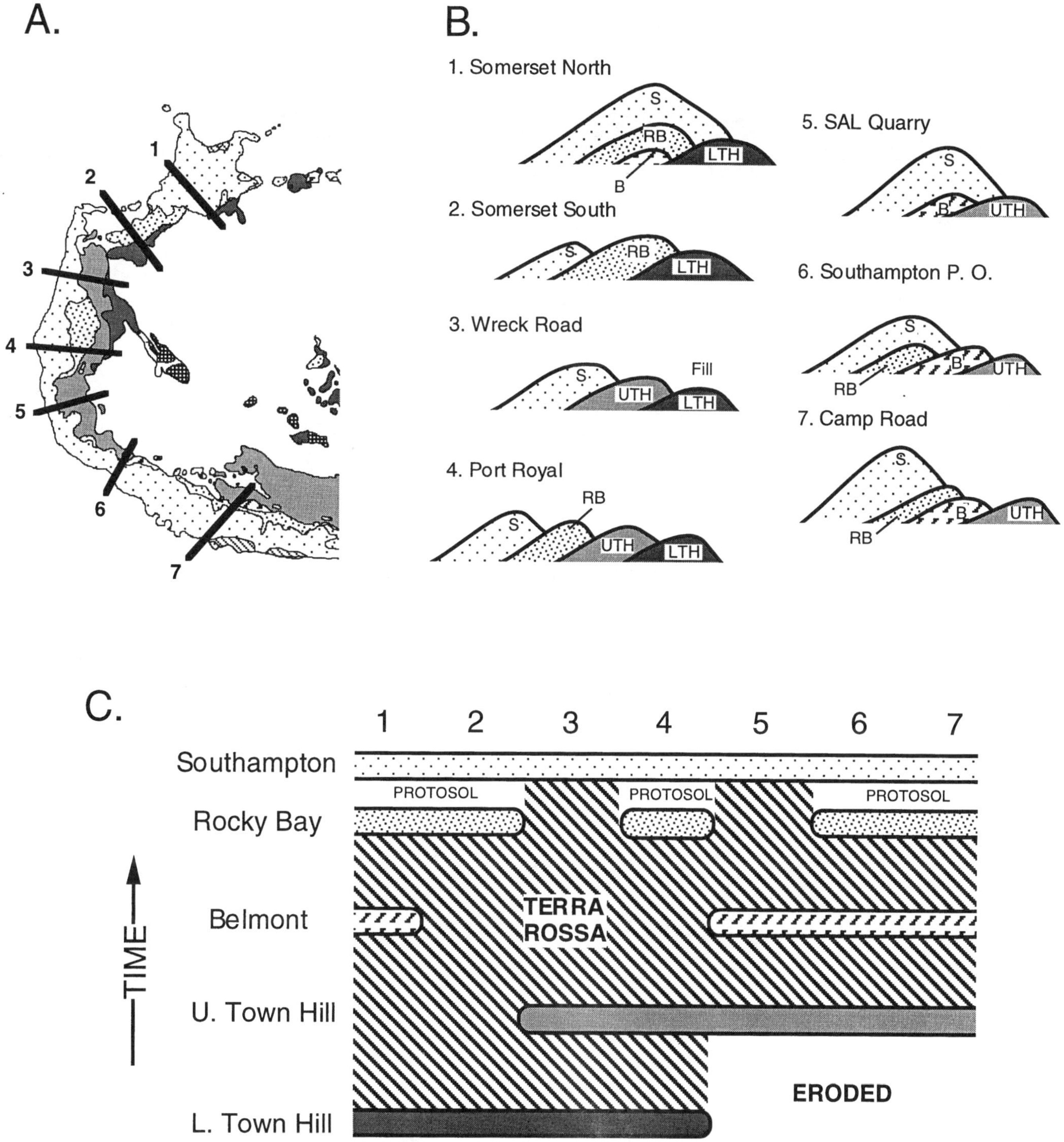

Figure 9. A, Location, B, schematic cross sections, and C, schematic Wheeler diagram of seven transects in western Bermuda. The transects extend, from left to right, from the external shoreline to an interior shoreline (east of north; see Fig. 7). Note that time represented by terra-rossa soil varies from place to place because eolianites do not extend indefinitely along strike. Effects of erosion are ignored in the time-length holosome (Wheeler, 1958) of the terra-rossa soil.

genesis. If it were not for lithologic differences resulting from differences in cumulative diagenesis (Vacher, 1973), albeit clouded by different intensities of diagenetic overprints in different diagenetic environments (Land et al., 1967; Land, 1970; Schroeder, 1973; Vollbrecht, 1990; Vollbrecht and Meischner, 1993), then there would be no way of dividing the section into lithostratigraphic units: there would be only one formation, presumably the Bermuda Limestone, with many constituent allostratigraphic units representing various sea-level cycles.

The formations were mapped (Vacher et al., 1989) in the field by tracing out individual depositional units and terra-rossa paleosols, and from lithologic differences as they affect the appearance of large exposures of eolianite in, for example, cliffs, roadcuts, commercial quarries, and ubiquitous backyard recesses cut into the eolian hills. These large exposures allow visual sampling of representative parcels of what is volumetrically the most important combination of depositional and diagenetic facies in Bermuda: eolianite altered in the intermediate aerated zone.

The two extremes in the lithologic gradation of vadose-altered eolianite are: eolianite consisting of loosely cemented, Mg-calcite and aragonite grains (typical of the Southampton Formation), and "ringing-hard" calcitic limestone (typical of the Walsingham Formation). The Rocky Bay, Belmont, and Town Hill Formations occur within the transition between these two extremes. Differences in appearance of the field exposures of the rock relate to cementation and grain alteration. Consolidation of the Rocky Bay, Belmont, and Upper Town Hill, for example, is such that these units are commonly quarried for building stone by literally sawing into the hillside; the Southampton is too friable, and the Lower Town Hill and Walsingham are too indurated. Individual grains in the Southampton, Rocky Bay, and Belmont eolianites, in contrast to the Town Hill, appear "fresh," and, depending on the constituent, still colored (e.g., *Homotrema*). Vertical sections of Rocky Bay eolianite typically show meter-scale uncemented zones conforming to the foresets; the Belmont is more uniformly cemented. Friable zones are common also in the Town Hill Formation, especially in the upper member; these zones have a chalky friability (reflecting dissolution of aragonite bioclasts) that contrasts with the sandy, uncemented zones higher in the section. Cross-bedding in weathered roadcuts in the Town Hill, especially the lower member, is less conspicuous than in the Rocky Bay and, especially, the Belmont because of more thorough cementation. Additionally, there is commonly a differential cementation of the fine grains at the surfaces of graded cross-beds in the Rocky Bay and Belmont. In Town Hill eolianites, especially in the lower member, there are meter-scale bodies of tightly cemented limestone like that typical of the Walsingham Formation.

There is a large overlap in the lithology of the successive formations, even if attention is restricted to large exposures of vadose-altered eolianite. The nature of the overlap—as well as the strategy of defining lithostratigraphic units relative to the known underpinning of glacioeustatic periodicity—is shown by the conceptual model illustrated in Figure 10. According to this conceptual model, the Bermuda lithology ("diagenetic properties" of vadose-altered eolianite) evolves from a low-variability initial condition (y_0) to a final condition (y_∞), also of low variability; further, between those extremes, the diagenetic properties approach y_∞ at an ever-decreasing rate. The simplest curve following from these assumptions is $y = 1 - e^{-at}$, where a is a rate constant, t is "diagenetic time," y is a measure of the diagenetic property, and y_0 and y_∞ are assumed to be 0 and 1, respectively (Fig. 10). If the rate constant, a, is a random variable rather than a constant (Lafon and Vacher, 1975), there is an envelope rather than a single curve representing the variation of the diagenetic property with diagenetic time. This envelope describes how the lithologic variability at a given diagenetic time varies through the transitional period. As shown in Figure 10, the width of the envelope increases while the slope of the envelope is large during the early part of the transitional period; during the later part of the transitional period, the width of the envelope decreases while the slope of the envelope is small. Although we have given an equation (representing first-order growth to a limit) and many other mathematical expressions could be argued from the original basic assumptions, all of the critical features of the conceptual model—the presence of an envelope, the general shape of the envelope, and the general relation between the width and slope of the envelope—would be unaffected.

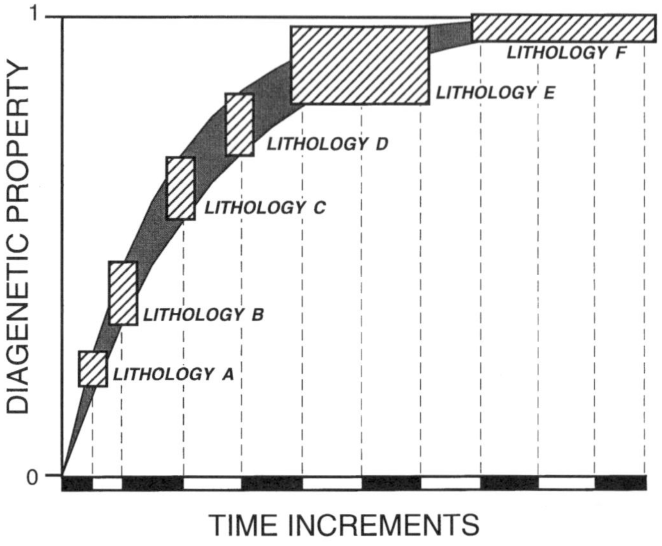

Figure 10. Conceptual model of temporal variation of a lithologic parameter that approaches a limit (final condition) with a progressively decreasing rate. The envelope of variation is sampled at periodic time increments. The rectangular boxes indicate groups of incremental samples that are combined to minimize lithologic overlap of the groups. The model illustrates, conceptually, why lithostratigraphic units in the lower part of the section in Bermuda span multiple interglacial-glacial cycles.

The problem posed by formulating a set of lithostratigraphic units from the diagenetic transition in Bermuda is conceptually like sampling the envelope of Figure 10 at periodic increments—where the time is stratigraphic age rather than diagenetic time, and the width of the envelope represents "lithologic variability" of a stratigraphic unit (sample) representing a particular age. Ideally, one wants the successive *lithostratigraphic* units to be disjunct lithologically; as shown in Figure 10, the success of attaining this goal is determined by the combination of slope and width of the diagenetic envelope. There is little overlap between lithologies B, C, and D—which represent samples of equal-time increments—because of the large slope despite the large width of the envelope. In the older part of the curve, the samples need to be grouped (E and F) in order to define units (the groups) that might be differentiable. In contrast, in the very early part of the transition, units of a finer time increment are lithologically disjunct (A and B). Thus, returning to the Bermuda lithostratigraphy, the Belmont represents an incremental sample (packet of interglacial limestone bracketed by terra rossa) in the middle of the transition; the Town Hill Formation, in the later part of the transition, spans at least three increments; and the Southampton and Rocky Bay Formations in the early part of the transition correspond to finer time increments.

It is worth repeating that "mappability" of Bermuda formations by lithology (Vacher et al., 1989)—and hence use of our lithostratigraphic units—follows from focusing on a single depositional/diagenetic environment (vadose-altered eolianite). Giving equal attention to the other diagenetic facies (e.g., phreatic zones, perched water tables, preferred passageways in the vadose zone, upper aerated zone beneath paleosols) results in the kind of variability found by Land et al. (1967, p. 1002): "Bermuda formations are *not* (their emphasis) diagenetic units." Even with the restrictions noted here, however, our units would not be mappable were it not also the case that terra-rossa paleosols call attention to lithostratigraphic boundaries, and the eolianites of successive formations are commonly topographically differentiated. Despite these aids, the lithologic overlap is still such that the uncertainty for the identification of a given formation at a particular outcrop is probably plus or minus one unit (e.g., Belmont can be confused with some Town Hill or Rocky Bay, but not with Southampton or Walsingham). Closer identification is obtained, then, by mapping the exposed unit into areas where superposition with others can be established.

Soil stratigraphy

The four geosols (Table 2) are all terra rossas. Unlike earlier columns (Sayles, 1931; Land et al., 1967), protosols are not used as soil stratigraphic units.

The four geosols are classified on the basis of the overlying lithostratigraphic units (i.e., the Shore Hills Geosol is the sub-Paget soil; the Harbour Road Geosol is overlain by the Upper Town Hill). They are not classified according to the time interval (or glacial stage) that they are thought to represent; indeed, the time represented by a given geosol varies from place to place because of the lenticularity of overlying and underlying units. (Fig. 9). Because of this variation, the physical character of the paleosols varies from place to place. Also, the physical character varies according to paleotopographic position: thin, weakly developed carbonate-rich terra rossas on the highs grade downslope to thick, clayey terra rossas with dissolution pits in the lows (Vacher, 1973). The various geosols, therefore, are not mappable on the basis of appearance in the field; rather, they are identified by stratigraphic position. After completion of the mapping, however, it does appear that the Ord Road and Castle Harbor Geosols are generally thicker and more clay-rich than the others.

Biostratigraphy

The endemic land snail *Poecilozonites* occurs in the terra rossas and eolianites and is abundant in the protosols. Like land snails of other isolated oceanic islands, *Poecilozonites* has undergone a dramatic evolutionary radiation (Gould, 1969). A total of 10 species classified into three subgenera have been recognized in the fossil record (Gould, 1969). The phylogeny of one of the subgenera, *P. (Poecilozonites)*, has been detailed in the classic work of Gould (1969), who based his stratigraphy on the column of Land et al. (1967). The correspondence of that phylogeny and column is shown in Figure 11A, which was prepared by Warren Allmon (personal communication, 1991) from data and discussion in Gould (1969). There is obviously a large overlap in the ranges. As noted by Gould (Land et al., 1967), however, although there is no snail guide fossil for the eolianites, the apparent relatively early extinction of *P. cupula* and the persistence of *P. bermudensis* could be of some utility in recognizing stratigraphic position.

Many of the localities that were sampled by Gould (1969) are older than previously thought. Reclassification of these localities according to the stratigraphic column of Vacher et al. (1989), and assuming the morphologic relations discussed by Gould (1969), results in the ranges shown in Figure 11B (W. Allmon, personal communication, 1989). The effect of the stratigraphic changes is to stretch out the chart, and, apparently, to move back the extinction of *P. cupula*. It is possible that the occurrence of *P. cupula* may indicate a pre-stage 5 stratigraphic position.

Hydrostratigraphy

Fresh ground-water lenses in Bermuda are strongly controlled by the lateral variation in hydraulic conductivity in the upper part of the saturated zone (Vacher, 1978). Vacher (1974) mapped this variation in terms of three hydrostratigraphic units, and, unfortunately, used lithostratigraphic names for them—Paget, Belmont, and Walsingham, in order of increasing hydraulic conductivity. Paget and Belmont rocks, although very permeable on an absolute scale (hydraulic conductivities on the order of 100 and 1,000 m/day, respectively), differ from the cavernous Walsingham in that they are not too permeable for fresh-water

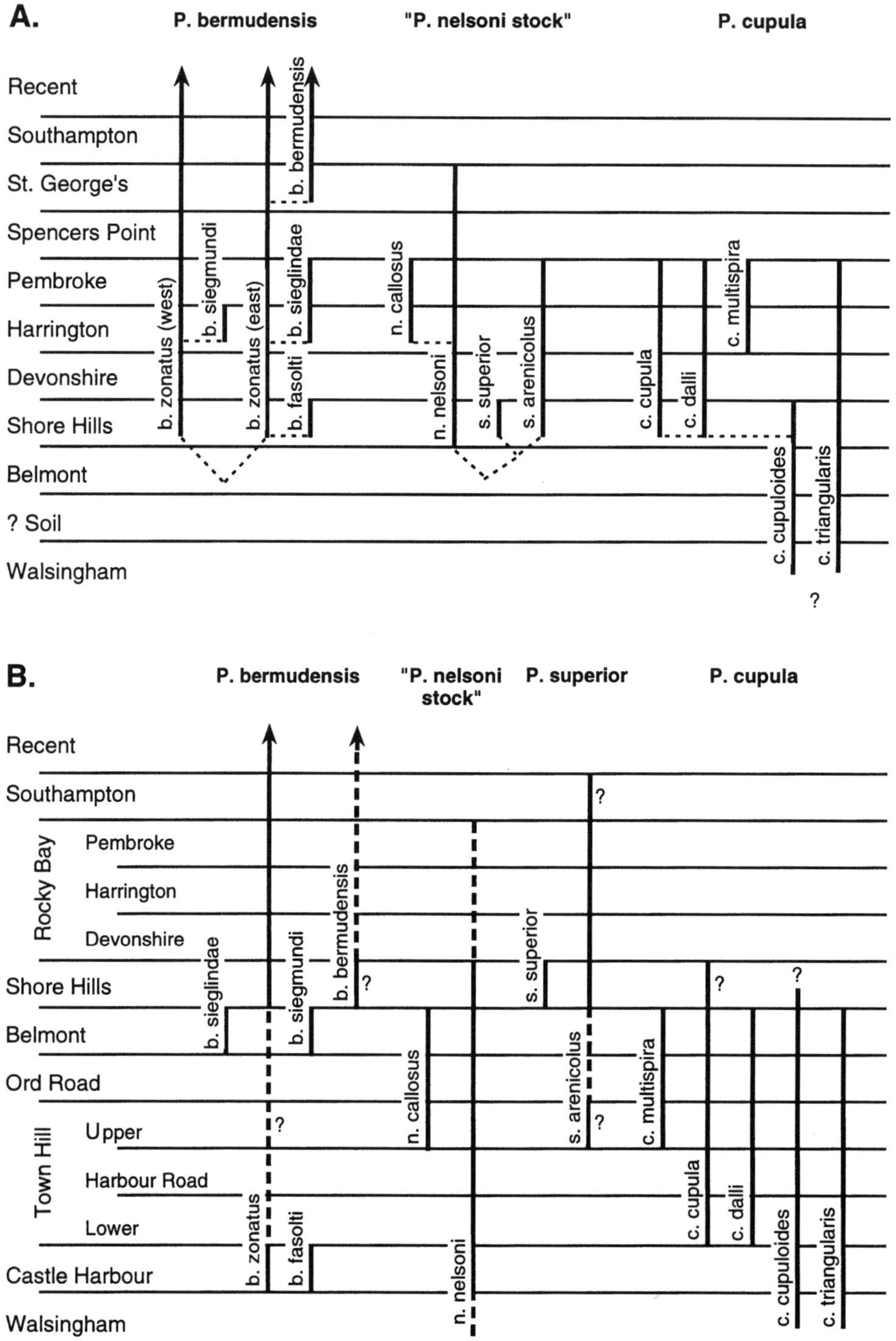

Figure 11. A, Stratophenetic evolutionary trees for the three "species groups" in the subgenus *P. (Poecilozonites)*, based on the conclusions of Gould (1969). B, Revised stratigraphic ranges for the nominal subspecies within the four species of *P. (Poecilozonites)*, recognized by Gould (1969), employing the stratigraphy of Vacher et al., (1989) and the locality data of Gould (1969). No branching order of subspecies within species is implied. (W. Allmon, personal communication, 1991.)

lenses to develop; the lens is thinner and more mixed with sea water in Belmont rocks than in Paget rocks. From this association of lithostratigraphic names with hydro-stratigraphic features, the three units came to be labeled the Paget, Belmont, and Walsingham aquifers in Bermuda's ground-water literature (e.g., Vacher, 1974, 1978; Plummer et al., 1976; Rowe, 1984; Vacher et al., 1990). There are no confining beds recognized in this hydrostratigraphy.

Aside from the poor practice of using the same name for two types of stratigraphic units, the previously used hydro-stratigraphic names are inappropriate for two reasons. First, the Belmont of Vacher (1974) was intended to represent all the limestones between the Walsingham and Paget; these rocks are now classified as the Belmont and Town Hill Formations (Tables 1 and 2). Second, subsequent mapping has shown that much of the Belmont Formation (sensu Table 2) is hydraulically like the Paget "aquifer"; therefore, nearly all of the Belmont "aquifer" of earlier papers actually consists of Town Hill formation. The more recent practice of the Bermuda Government (Rowe, 1991) is to use Langton aquifer (from Langton Hill) for the lower-permeability unit (corresponding to the Paget aquifer of earlier papers) and Brighton aquifer (from Brighton Hill) for the higher-permeability unit (corresponding to the Belmont aquifer of earlier papers). According to this classification, the Langton aquifer consists of the Paget Group and Belmont Formation, and the Brighton aquifer consists of the Town Hill Formation.

Morphostratigraphy

Two scales of morphostratigtaphic units have been used in Bermuda. The largest units are the two geomorphic provinces, Younger and Older Bermuda, noted by Sayles (1931). In Younger Bermuda, the eolianites retain their depositional morphology. In Older Bermuda, the eolianites have been degraded by chemical erosion and their area invaded the expanding inshore water bodies (Bretz, 1960; Vacher, 1978; Vacher and Mylroie, 1991; Mylroie et al., this volume).

Vacher (1973) tried mapping topographic ridges that were thought to represent single depositional units as smaller scale morphostratigraphic units. This was not totally successful, even within Younger Bermuda, because of the complexities due to overstepping of successive eolianites. From this experience, it was evident that the stratigraphic analysis based on morphology alone, or even relying heavily on it, would lead to misinterpretation in Bermuda. To give one example, failure to realize that the major quarried ridge in southern Paget and Warwick Parishes (Fig. 5) is underlain by two terra rossas (now identified as the Shore Hills and Ord Road Geosols), instead of just one (Shore Hills), delayed recognition by a number of years that a major middle Pleistocene section (the Town Hill Formation) is present. Morphology, on the other hand, has been useful in tracing successive units defined by other criteria (lithostratigraphic or allostratigraphic), by guiding the search for contacts (most notably paleosols).

Allostratigraphy

Allostratigraphic units, which represent depositional cycles (North American Commission on Stratigraphic Nomenclature, 1983), are generally bounded above and below by terra rossas or extensive, presumably regional, protosols. There are significantly more allostratigraphic units than there are formations. Successive allostratigraphic units are not necessarily lithologically different. Presumably, if each interglacial stage consisted of a sea-level rise to and fall from a single maximum, and if the deposits of successive interglacial stages were lithologically different, then there would be the same number of allostratigraphic units and formations.

As used here (Table 2), the allostratigraphic units are informal and are designated alphanumerically. The letter, or letters, of the designation indicates the umbrella lithostratigraphic unit. The number is a count upward from the base.

There are at least three allostratigraphic units in the Southampton Formation. The youngest, S3, is discussed in detail by Vacher and Hearty (1989). It is a small and localized group of marine and eolian deposits that onlaps the seaward margin of the main eolianites of the Southampton. Both S1 and S2 are composed of only eolianites. The two can be traced along the south shore of Paget and Warwick Parishes. S1 is limited to low elevations along the shoreline and appears to be only the top part of a nearly totally submerged eolianite ridge. The larger S2 oversteps S1 and produces the high dune-shaped hills along the shoreline.

The R1 and R2 units occur along the lagoon-facing north shore. These units are the two major complexes exposed at Blackwatch Pass (Fig. 6), and they are separated by an extensive protosol. The cycle represented by the Devonshire, Harrington, and Pembroke Members correlates time-stratigraphically with R2. Only one Rocky Bay allostratigraphic unit, R2, is recognized along margin-facing shorelines.

B1 and B2 occur in southern Smith's, Devonshire, and Paget Parishes. Eolianites of the two are separated by a protosol, and they are also morphologically distinct (Hearty and Vacher, 1995). B2 is the unit that occurs along the shoreline and includes the type Belmont at Rocky Bay. Two depositional cycles are also recognized in the marine facies at Grape Bay (Meischner et al., this volume).

In most areas where it occurs, the eolian facies of the Upper Town Hill can be divided into two allostratigraphic units, UTH1 and UTH2. These units are separated by a regionally mapped, relatively clay-rich protosol (Vacher et al., 1989). They generally occur as different ridges, at least in part.

At least three depositional cycles occur within the Lower Town Hill at Bierman's Quarry. The contact between the lowest two is a thin terra rossa. A similar terra rossa occurs within the

Lower Town Hill at the Naval Air Station, St. George's Parish (Vacher et al., 1989).

Aminostratigraphy

The extent of postmortem epimerization of the amino acid L-isoleucine to D-alloisoleucine in fossils can be used to assess the relative age of the fossils, and, with some assumptions and calibration, make an estimate of their numerical age (Mitterer, 1975). The first application of the amino acid racemization (AAR) technique in Bermuda (Mitterer and Kriausakul, *in* Harmon et al., 1983) was on *Poecilozonites* and showed great promise for subdividing fossiliferous deposits that could be neither dated by U-series methods (owing to an absence of corals) nor differentiated biostratigraphically. This promise was confirmed when it was shown that AAR ratios (D-alloisoleucine/L-isoleucine, or simply A/I) on marine molluscs from the younger, coastal deposits fall into natural clusters that parallel the lithostratigraphy (Hearty and Hollin, 1986; Vacher and Hearty, 1989). The grouping of ratios allowed definition of "aminozones" (Nelson, 1982; Hearty and Hollin, 1986) and thus a system of aminostratigraphic classification.

The aminostratigraphic classification has been extended throughout the Bermuda column (Hearty et al, 1992; Hearty and Vacher, 1995). A summary of the current status is shown in Table 3. Because epimerization rate is a function of taxon as well as time and temperature, different numerical values for the zoned ratios apply to different sample materials. The values for marine shells in Table 3 represent a composite from five different molluscs, for which the ratios have been converted to those of *Glycymeris*, the one that was most often sampled (conversions are in Vacher and Hearty, 1989). The whole-rock ratios are from samples of bioclastic grainstone (generally eolianite) and, therefore, represent a natural mix of taxa that racemize at a variety of rates. The fact that the succession of whole-rock ratios parallels the others attests to the minimally changing environmental provenance of Bermuda's grainstones.

The parallelism of the amino- and lithostratigraphy is shown in Table 2 and is discussed elsewhere in more detail (Hearty et al., 1992; Hearty and Vacher, 1995). It should be noted here, however, that the two classifications were derived independently. The island was mapped, and the map was in press, before the collaborative sampling by Hearty, Vacher, and Rowe. The fact that there is agreement between the AAR ratios and the lithostratigraphy in all but four individual samples (or in 97% of the 257 shells) supports the utility of the method, including its application to whole-rock samples in Bermuda. The agreement also supports the lithostratigraphic classification.

TABLE 3. AMINOZONES AND CORRESPONDING AMINO ACID RACEMIZATION RATIOS

Amino Zone	Mean A/I* Marine Shells	Mean A/I* *Poecilozonites*	Mean A/I* Whole Rock
C	0.42	0.40	0.23
E	0.57	0.49	0.27
E/F		0.57	
F (F2)	0.69	0.61	0.39
(F1)			0.49
G		0.78	0.56
H		0.91	0.69
J			0.92
K			1.11

*A/I = D-alloisoleucine/L-isoleucine. From Hearty et al., 1992. Uncertainties (SD) range from 0.01 to 0.05 and are mostly 0.02 to 0.04.

Time stratigraphy and geochronology

Correlation of Bermudian stratigraphic units with the Pleistocene glacial and interglacial stages defined from the deep-sea oxygen isotope record (Shackleton and Opdyke, 1973) must be done by geochronologic means. Two such methods have been applied in Bermuda: alpha-spectrometric U-series ages of corals (Harmon et al., 1978, 1981, 1983) and interpretation of AAR ratios (Hearty and Hollin, 1986; Vacher and Hearty, 1989; Hearty et al, 1992; Hearty and Vacher, 1995).

U-series ages. Corals are rare in Bermuda, so only a small fraction of the deposits can be dated radiometrically, and those that can are not necessarily located where stratigraphic relations permit definitive classification of the coral-bearing deposits. The vast majority of the U-series ages are about 125 ka from the Devonshire Member and clearly indicate its correlation with substage 5e of the deep-sea oxygen isotope record and with a major highstand of sea level recorded by the classic reefs of Barbados and New Guinea (Mesolella et al., 1969; Bloom et al., 1974). Two ages indicate correlation of isolated marine deposits with substage 5a and a highstand of sea level at about 80 ka recorded by the uplifted reefs of Barbados and New Guinea (Mesolella et al., 1969; Bloom et al., 1974) and a terrace in California (Muhs, 1992). One of these 80-ka deposits in Bermuda is included in an S3 unit at Fort St. Catherine, which was mapped as Southampton (Upper Paget of Vacher, 1973) before the U-series dating was done; the age has subsequently been substantiated by AAR ratios (Hearty et al., 1992) and mass spectrometric U-series analyses (Ludwig et al., 1994).

Additional stage 5 ages have been obtained from the patchy, scattered, and mostly removed marine conglomerates along the shore at Saucos Hill, McGall's Hill, and Spencer's Point. There are five ages (Harmon et al., 1981, 1983): one at 121 ka, a typical Devonshire result; three ranging from 97 to 115 ka, suggesting correlation with substage 5c and one of the uplifted reefs of the Barbados and New Guinea sections; and, curiously, one at less that 2 ka. Although Harmon et al. (1981, 1983) make a case that some of these marine conglomerates correlate with the substage 5c highstand, we are not convinced (Vacher and Hearty, 1989) that the ages require that these marine deposits be younger than the Devonshire of substage 5e. The four Saucos Hill ages (Harmon et al., 1983), ranging from 121 to 99 ka (with laboratory uncertainties ±6 ka), are reminiscent of the 15 alpha-spectrometric U-series ages obtained from corals of the substage 5e Rendezvous Hill (Barbados III) Terrace in an extensive geochronologic study by Ku et al. (1990): The Rendezvous Hill values range from 134 to 99 ka; have laboratory uncertainties of ±3 ka, and include four values less than 110 ka. Allowing the uncertainty to reflect the various results from different samples rather than laboratory precision, the Saucos Hill result is 108 ± 9 ka and the Rendezvous Hill result is 117 ± 10 ka.

In addition to the stage 5 ages, a few results indicate correlation of isolated marine deposits with stage 7. For each of these deposits, the stratigraphic relations at the site are consistent with, but not definitive of, their classification as Belmont. The stage 7 correlation of one of these deposits (Boaz Island) has been supported by quantitative geochronologic interpretation of AAR ratios (see below). Thus the interpretation (Vacher and Hearty, 1989) of available U-series ages in Bermuda is that the S3 unit of the Southampton, the R2 of the Rocky Bay Formation, and the Belmont Formation correlate with substage 5a (ca. 80 ka), substage 5e (ca. 125 ka), and stage 7 (ca. 200 ka), respectively.

AAR age estimates. Because of greater availability of sample material and more definitive stratigraphic relations at the sample localities, the correlation of lithostratigraphy and time stratigraphy has significantly more tie-points through aminostratigraphy calibrated with U-series ages than through the U-series ages alone. Hearty et al. (1992) estimated ages from AAR ratios in Bermuda in two ways. In the first method, they assumed the ages of Aminozones C, E, and F from the U-series dates of the same unit in Bermuda, and, using log-log plots for each sample material, they interpolated and extrapolated the ages of the other aminozones. In the second method, they assumed the age of a single calibration unit, Aminozone E at 125 ka, and calculated ages of the other aminozones using the model of apparent parabolic kinetics (Mitterer and Kriausalak, 1989). The results support the correlations from the U-series ages: the upper part of the Southampton and the Belmont correlate with substage 5a and stage 7, respectively. The AAR age estimates also indicate that the Upper Town Hill and the upper units of the Lower Town Hill correlate with stages 9 and 11, respectively.

With these geochronologic interpretations of AAR ratios, the allostratigraphic units line up easily with the deep-sea isotope stages with one exception, as shown in Table 2. The AAR age estimate from the *Poecilozonites* in the protosol above R1 appears to indicate that R1 is too old for substage 5e, although there is no terra rossa between it and R2. We suspect that R1 is, in fact, early substage 5e and that the calculated value is an over estimate. The calculation assumes a single parabolic racemization curve, which would be consistent with a single temperature. Taking account of bimodal (interglacial, glacial) temperatures would lead, conceptually, to a more jagged curve. If the *Poecilozonites* were from the early part of a fast-racemization sector, the age calculated from the single, "average parabola" would be too large.

CONCLUSIONS

Bermuda is exceptional in the number of roadcuts, commercial and household quarries, and coastal cliffs that permit examination of the complex three-dimensional mosaic of eolianites, marine deposits and paleosols that comprise this bank of marginal marine Pleistocene carbonates. The stratigraphy has been examined and reexamined over several decades. Now there is a geologic map that documents the richness of the section. What has been learned? Following are some of the concepts and lessons.

(1) The arrangement of stratigraphic units reflects lateral accretion.

(2) Although the limestones formed during interglaciations, and glaciations were times of soil formation in Bermuda, it cannot be said that the terra rossas formed (exclusively) during glaciations. The paleosols simply represent hiatuses in the buildup of the limestone bank. The times they represent vary geographically according to the lensing out of eolianites.

(3) Although deposition of Bermuda's eolianites was exclusively an interglacial phenomenon, it can also be said that deposition of the large eolianites that characterize Bermuda, in general, lagged behind the initial transgression and platform submergence. Some time is required for the development of offshore sediment sources and shoreward transport routes. This concept, derived from the mosaic of Pleistocene beach-dune transitions, is consistent with the absence of Holocene eolianites in Bermuda.

(4) Although the mosaic of facies and the succession of major limestone packages is ultimately controlled by sea-level history, such history cannot be used to define lithostratigraphic units, because units thus defined are not mappable.

(5) Basing lithostratigraphic units on lithologic features observable in the field means that lithostratigraphic units are identified from properties related to extent of diagenesis in conjunction with stratigraphic position (including lateral accretion). In assessing lithologies it is important to use large representative exposures and compare bodies of rock from the same combination of depositional and diagenetic environments. Although such lithostratigraphic units parallel time-stratigraphic subdivisions, the resolution of time increments will inevitably decrease downsection.

(6) Within the time frame of Bermuda's exposed stratigraphy, AAR ratios provide useful time-stratigraphic correlations. With land snails in the paleosols and marine molluscs in many of the marine deposits, considerably more units are suitable for AAR analysis than U-series analysis of corals. With whole-rock ratios, now even eolianites can be used. In practice, AAR ratios provide an independent system of stratigraphic classification with time-parallel aminozones providing relative age. Correlation of these zones with the deep-sea oxygen isotope stages follows from calculation of numerical age estimates using a kinetic model calibrated to local U-series ages. Mitterer and Kriausakul's (1989) model of apparent parabolic kinetics has proved successful for this purpose.

(7) It would be procedurally incorrect to define or identify formations on the basis of AAR ratios in the same way that it is incorrect to base lithostratigraphic units on fossil content, depositional cycles, inferred sea-level history, or geologic time. We advocate using multiple systems of classifications, each with a single parameter (lithology, soils, snails, AAR ratios, cycles, and finally time).

(8) Although recitation of all the implications to Pleistocene sea-level history is beyond the scope of this chapter, a few essentials should be noted. Given that the shoreline must have been nearby to leave eolianites on the present island and that marine deposits occur at low elevations, there is no question that sea level was close to its present position, at least once, during each of the last four major interglacials (stages 5, 7, 9, and 11). By "close," we wish to imply that the preferred position of sea level during interglacials was within, say, ±6 m of its present position; thus we agree with the characterization by Meischner et al. (this volume) of the Bermuda sea-level record as that of a yo-yo. In terms of the volume of sediment produced (or at least preserved) and the complexity of history recorded by allostratigraphy, however, the interglacials were not equal: stages 5 and 9 were the most important. The last very high sea-level stand (+22 m at Government Quarry) occurred early in the history recorded by the exposed stratigraphy (Hearty and Vacher, 1995).

ACKNOWLEDGMENTS

We thank Warren Allmon for his help with *Poecilozonites* and Dan Muhs for his advice about paleosols. We appreciate the helpful reviews and comments by Dan Muhs, Jim Carew, John Mylroie, F. T. Mackenzie, and S. J. Gould on an earlier version of this manuscript. Finally, H. L. Vacher gratefully acknowledges his stratigraphy professors, Harry Wheeler (1964, Washington) and Larry Sloss (1965, Northwestern), whose classes stimulated a lasting fascination with the subject. Bermuda Biological Station for Research Contribution 1372.

REFERENCES CITED

Aumento, F., and Ade-Hall, J. M., 1973, Deep Drill 1972: Petrology of the Bermuda drill core [abs]: EOS (Transactions, American Geophysical Union), v. 54, p. 485.

Ball, M. M., 1967, Carbonate sand bodies of Florida and the Bahamas: Journal of Sedimentary Petrology, v. 37., p. 556–591.

Bloom, A. L., Broecker, W. S., Chappell, J. M. A., Matthews, R. K., and Mesolella, K. J., 1974, Quaternary sea level fluctuations on a tectonic coast: New ^{230}TH/^{234} U dates from the Huon Peninsula, New Guinea: Quaternary Research, v. 4, p. 185–205.

Bretz, J H., 1960, Bermuda: a partially drowned late nature Pleistocene karst: Geological Society of America Bulletin, v. 71, p. 1729–1754.

Bricker, O. P., and Mackenzie, F. T., 1970, Limestones and red soils of Bermuda, Discussion: Geological Society of America Bulletin, v. 81, p. 2523–2524.

Broecker, W. S., and van Donk, J., 1970, Insolation changes, ice volumes and the O^{18} record in deep sea cores: Reviews of Geophysics and Space Physics, v. 8, p. 169–198.

Butzer, K. W., and Cuerda, J., 1962, Coastal stratigraphy of southern Mallorca and its implications for the Pleistocene chronology of the Mediterranean Sea: Journal of Geology, v. 70, p. 398–416.

Carew, J. L., and Mylroie, J. E., 1985, The Pleistocene and Holocene stratigraphy of San Salvador Island, Bahamas, with reference to marine and terrestrial lithofacies at French Bay, *in* Curran, H. A., ed., Pleistocene and Holocene Carbonate Environments on San Salvador Island, Bahamas: Geological Society of America, Orlando Annual Meeting, Field Trip Guidebook: Ft. Lauderdale, Florida, CCFL Bahamian Field Station, p. 73–93.

Carew, J. L., and Mylroie, J. E., 1991, Some pitfalls in paleosol interpretation in carbonate sequences: Carbonates and Evaporites, v. 6, p. 69–74.

Cooper, W. S., 1958, Coastal sand dunes of Oregon and Washington: Geological Society of America Memoir 72, 169 p.

Eldredge, N., and Gould, S. J., 1972, Punctuated equilibria: An alternative to phyletic gradualism, *in* Schopf, T. J. M., ed., Models in Paleobiology: San Francisco, W. H. Freeman, p. 82–115.

Ellison, J. C., 1993, Mangrove retreat with rising sea-level, Bermuda: Estuarine, Coastal and Shelf Science, v. 37, p. 75–87.

Fairbridge, R. W., and Johnson, D. L., 1978, Eolianite, *in* Fairbridge, R. W., and Bourgeois, J., eds. The Encyclopedia of Sedimentology: Stroudsburg, Pennsylvania, Dowden, Hutchinson and Ross, p. 279–282.

Fairbridge, R. W., and Teichert, C., 1953, Soil horizons and marine bands in the coastal limestone of western Australia: Journal and Proceedings of the Royal Society of New South Wales, v. 86, p. 68–86.

Foos, A. M., 1991, Aluminous, lateritic soils, Eleuthera, Bahamas: A modern analog to carbonate paleosols: Journal of Sedimentary Petrology, v. 61, p. 340–348.

Gaffney, E. S., 1983, The cranial morphology of the extinct horned turtle, *Meiolania platyceps,* from the Pleistocene of Lord Howe Island, Australia: Bulletin of the American Museum of Natural History, v. 175, p. 361–480.

Galehouse, J. S., 1979, Heavy mineralogy and provenance of volcaniclastic turbidites of Site 386, *in* Tucholke, B. E., Vogt, P. R., and others, eds., Initial Reports of the Deep Sea Drilling Project, v. 43, p. 407–410.

Gardner, R. A. M., 1983, Aeolianite, *in* Goudie, A. S., and Pye, K., eds., Chemical Sediments and Geomorphology: London, Academic Press, p. 265–300.

Garrett, P., and Gould, S. J., 1984, Geology of New Providence Islands, Bahamas: Geological Society of America Bulletin, v. 95, p. 209–220.

Garrett, P., and Scoffin, T. P., 1977, Sedimentation on Bermuda's atoll rim: Proceedings, Third International Coral Reef Symposium, Miami, Florida, p. 87–95.

Gould, S. J., 1969, An evolutionary microcosm: Pleistocene and Recent history of the land snail *P.* (*Poecilozonites*) in Bermuda: Bulletin of the Museum of Comparative Zoology, v. 138, p. 407–532.

Harmon, R. S., Schwarcz, H. P., and Ford, D. C., 1978, Late Pleistocene sea level history of Bermuda: Quaternary Research, v. 9, p. 205–218.

Harmon, R. S., Land, L. S., Mitterer, R. M., Garrettt, P., Schwarcz, H. P., and Larson, G. J., 1981, Bermuda sea level during the last interglacial: Nature, v. 289, p. 481–483.

Harmon, R. S., and 8 others, 1983, U-series and amino-acid racemization geochronology of Bermuda: Implications for eustatic sea-level fluctuation over the past 250,00 years: Palaeogeography, Palaeoclimatology, Palaeoecology, v. 44, p. 41–70.

Hays, J. D., Imbrie, J., and Shackleton, N. J., 1976, Variations in the Earth's orbit, Pacemaker of the Ice Ages: Science, V. 194, p. 1121–1132.

Hearty, P. J., and Hollin, J. T., 1986, Aminostratigraphy of Quaternary shorelines in Bermuda: Geological Society of America Abstracts with Programs, v. 18, p. 633.

Hearty, P. J., and Kindler, P., 1993a, New perspectives on Bahamian geology: San Salvador Island, Bahamas: Journal of Coastal Research, v. 9, p. 577–594.

Hearty, P. J., and Kindler, P., 1993b, An illustrated stratigraphy of the Bahama Islands: In search of a common origin: Bahamas Journal of Science,

v. 1, p. 28–45.

Hearty, P. J., and Vacher, H. L., 1995, Quaternary stratigraphy of Bermuda: A high-resolution pre-Sangamonian rock record: Quaternary Science Reviews, v. 13, p. 685–697.

Hearty, P. J., Miller, G. F., Stearns, C. E., and Szabo, B. J., 1986, Aminostratigraphy of Quaternary shorelines in the Mediterranean basin: Geological Society of America Bulletin, v. 97, p. 850–858.

Hearty, P. J., Vacher, H. L., and Mitterer, R. M., 1992, Aminostratigraphy and ages of Pleistocene limestones of Bermuda: Geological Society of America Bulletin, v. 104, p. 471–480.

Herwitz, S. R., 1993, Stemflow influences on the formation of solution pipes in Bermuda eolianite: Geomorphology, v. 6, p. 253–271.

Inden, R. F., and Moore, C. H., 1983, Beach environment, *in* Scholle, P. A., Bebout, D. G., and Moore, C. H., eds., Carbonate Depositional Environments: American Association of Petroleum Geologists Memoir 33, p. 211–265.

Ku, T.-L., Ivanovich, M., and Luo, S., 1990, U-series dating of last interglacial high sea stands: Barbados revisited: Quaternary Research, v. 33, p. 129–147.

Lafon, G. M., and Vacher, H. L., 1975, Diagenetic reactions as stochastic processes: Application to the Bermudian eolianites, *in* Whitten, E. H. T., ed., Quantitative Studies in the Geological Science: Geological Society of America Memoir 142, p. 187–204.

Land, L. S., 1970, Phreatic versus vadose meteoric diagenesis of limestones: Evidence from a fossil water table: Sedimentology, v. 14, p. 175–185.

Land, L. S., and Mackenzie, F. T., 1970, Field guide to Bermuda geology: Bermuda Biological Station Special Publication 4, 14 p.

Land, L. S., Mackenzie, F. T., and Gould, S. J., 1967, The Pleistocene history of Bermuda: Geological Society of America Bulletin, v. 78, p. 993–1006.

Ludwig, K. R., Muhs, D. R., Halley, R. B., and Shinn, E. A., 1994, Sea level records at 80,000 BP from tectonically stable platforms: The Florida Keys and Bermuda: American Quaternary Association, Program and Abstracts, 13th Biennial Meeting, p. 225.

Mackenzie, F. T., 1964a, Geometry of Bermuda calcareous dune cross-bedding: Science, v. 144, p. 1449–1450.

Mackenzie, F. T., 1964b, Bermuda Pleistocene eolianites and paleowinds: Sedimentology, v. 3, p. 51–64.

Martinson, D. G., Pisias, N. G., Hays, J. D., Imbrie, J., Moore, T. C., Jr., and Shackleton, N. J., 1987, Age dating and the orbital theory of the ice ages: Development of a high-resolution 0 to 300,000-year chronostratigraphy: Quaternary Research, v. 27, p. 1–29.

Mauritsen, M. V., 1983, Studies of diagenesis of Bermuda limestones: 1. The calcretes; 2. Modern marine cement in a Pleistocene eolianite [M.S. thesis]: Pullman, Washington State University, 104 p.

McKee, E. D., 1979, Sedimentary structures in dunes, *in* McKee, E. D., ed., A study of global sand seas: U.S. Geological Survey Professional Paper 1052, p. 83–134.

Mesolella, K. J., Matthews, R. K., Broecker, W. S., and Thurber, D. L., 1969, The astronomical theory of climatic change: Barbados data: Journal of Geology, v. 77, p. 250–274.

Mitterer, R. M., 1975, Ages and diagenetic temperatures of Pleistocene deposits of Florida based upon isoleucine epimerization in *Mercenaria*: Earth and Planetary Science Letters, v. 28, p. 275–282.

Mitterer, R. M., and Kriausakul, M., 1989, Calculation of amino acid racemization ages based on apparent parabolic kinetics: Quaternary Science Reviews, v. 8, p. 353–357.

Morse, J. W., and Mackenzie, F. T., 1990, Geochemistry of Sedimentary Carbonates: New York, Elsevier, 707 p.

Muhs, D. R., 1992, The last interglacial-glacial transition in North America: Evidence from uranium-series dating of coastal deposits, *in* Clark, P. U., and Lea, P. D., eds., The Last Interglacial-Glacial Transition in North America: Geological Society of America Special Paper 270, p. 31–51.

Muhs, D. R., Bush, C. A., Stewart, K. C., Rowland, T. R., and Crittenden, R. C., 1990, Geochemical evidence of Saharan dust parent material for soils developed on Quaternary limestones of Caribbean and western Atlantic island: Quaternary Research, V. 33, p. 157–177.

Nelson, A. R., 1982, Aminostratigraphy of Quaternary marine and glaciomarine sediments, Qivitu Peninsula, Baffin Island: Canadian Journal of Earth Science, v. 19, p. 945–961.

Neumann, A. C., 1965, Processes of recent carbonate sedimentation in Harrington Sound, Bermuda: Bulletin of Marine Science, v. 15, p. 987–1035.

Neumann, A. C., 1969, Quaternary sea-level data from Bermuda: Resumes des Communications, VIIIe Congres INQUA, Paris, p. 228–229. [Also, Quaternaria, 1971, v. 14, p. 41–43.]

North American Commission on Stratigraphic Nomenclature, 1983, North American Stratigraphic Code: American Association of Petroleum Geologists Bulletin, v. 67, p. 841–875.

Pirsson, L. V., 1914, Geology of Bermuda Island: The igneous platform: American Journal of Science, v. 38, p. 189–206, 331–334.

Plummer, L. N., Vacher, H. L., Mackenzie, F. T., Bicker, O. P., and Land, L. S., 1976, Hydrogeochemistry of Bermuda: A case history of ground-water diagenesis of biocalcarenites: Geological Society of America Bulletin, v. 87, p. 1301–1316.

Reynolds, P. R., and Aumento, F. A., 1974, Deep Drill 1972: Potassium-argon dating of the Bermuda drill core: Canadian Journal of Earth Sciences, v. 11, p. 1269–1273.

Rowe, M. P., 1984, The freshwater "Central Lens" of Bermuda: Journal of Hydrology, v. 73, p. 165–176.

Rowe, M. P., 1990, An explanation of the geology of Bermuda, with reference to the Geological Map of Bermuda (1989): Hamilton, Bermuda Ministry of Works and Engineering, 28 p.

Rowe, M. P., 1991, Bermuda, *in* Falkland, A., ed., Hydrology and Water Resources of Small Islands: A Practical Guide: Paris, United Nations Environmental, Scientific, and Cultural Organization, p. 333–338.

Ruhe, R. V., Cady, J. G., and Gomez, R. S., 1961, Paleosols of Bermuda: Geological Society of America Bulletin, v. 72, p. 1121–1142.

Sayles, R. W., 1931, Bermuda during the Ice Age: American Academy of Arts and Sciences, v. 66, p. 381–468.

Schroeder, J. H., 1973, Submarine and vadose cements in Pleistocene Bermuda reef rock: Sedimentary Geology, v. 10, p. 179–204.

Semeniuk, V., and Johnson, D. P., 1982, Recent and Pleistocene beach/dune sequences, Western Australia: Sedimentary Geology, v. 32, p. 301–328.

Semeniuk, V., and Johnson, D. P., 1985, Modern and Pleistocene rocky shore sequences along carbonate sequences along carbonate coastlines, southwestern Australia: Sedimentary Geology, v. 44, p. 225–261.

Shackleton, N. J., and Opdyke, N. D., 1973, Oxygen isotope and paleomagnetic stratigraphy of equatorial Pacific core V28–238: Oxygen isotope temperatures on a 10^5 and 10^6 year time scale: Quaternary Research, v. 3, p. 39–55.

Sloss, L. L., 1984, Comparative anatomy of cratonic unconformities, *in* Schlee, J. S., ed., Interregional Unconformities and Hydrocarbon Accumulation: American Association of Petroleum Geologists Memoir 36, p. 1–6.

Tucholke, B. E., and Vogt, P. R., 1979, Western North Atlantic: Sedimentary evolution and aspects of tectonic history, *in* Tucholke, B. E., Vogt, P. R., and others, eds., Initial Reports of the Deep Sea Drilling Project, v. 43, p. 791–825.

Upchurch, S. B., 1970, Sedimentation on the Bermuda Platform [Ph.D. thesis]: Evanston, Illinois, Northwestern University, 145 p. (Also, U.S. Lake Survey, Research Report 2-2, 172 p.).

Vacher, H. L., 1971, Late Pleistocene sea-level history: Bermuda evidence [Ph.D thesis]: Evanston, Illinois, Northwestern University, 147 p.

Vacher, H. L., 1973, Coastal dunes of Younger Bermuda, *in* Coates, D. R., ed., Coastal Geomorphology: Binghamton, State University of New York, p. 355–391.

Vacher, H. L., 1974, Ground water hydrology of Bermuda: Hamilton, Bermuda Public Works Department, 85 p.

Vacher, H. L., 1978, Hydrogeology of Bermuda—Significance of an across-the-island variation in permeability: Journal of Hydrology, v. 39, p. 207–226.

Vacher, H. L., and Harmon, R. S., 1987, Geological Society of America Penrose Conference Field Guide to Bermuda Geology: Hamilton, Bermuda Biological Station for Research, 48 p.

Vacher, H. L., and Hearty, P. J., 1989, History of Stage 5 sea level in Bermuda: Review with new evidence of a brief rise to present sea level during Substage 5a: Quaternary Science Reviews, v. 8, p. 159–168.

Vacher, H. L., and Mylroie, J. E., 1991, Geomorphic evolution of topographic lows in Bermudian and Bahamian islands: Effect of climate, *in* Bain, R. J., ed., Proceedings, Fifth Symposium on the Geology of the Bahamas: San Salvador, Bahamian Field Station, p. 221–234.

Vacher, H. L., Rowe, M. P., and Garrett, P., 1989, The Geologic Map of Bermuda: London, Oxford Cartographers; Hamilton, Bermuda Ministry of Works and Engineering.

Vacher, H. L., H. L., Bengtsson, T. O., and Plummer, L. N., 1990, Hydrology of meteoric diagenesis: Residence time of meteoric ground water in island fresh-water lenses with application to aragonite-calcite stabilization rate in Bermuda: Geological Society of America Bulletin, v. 102, p. 223–232.

Verrill, A. E., 1907, The Bermuda Islands, P. IV, Geology and paleontology, and P. V, An account of the coral reefs: Connecticut Academy of Arts and Sciences Transactions, v. 12, p. 45–348.

Vollbrecht, R., 1990, Marine and meteoric diagenesis of submarine Pleistocene carbonates from the Bermuda carbonate platform: Carbonates and Evaporites, v. 5, p. 13–95.

Vollbrecht, R., and Meischner, D., 1993, Sea level and diagenesis—A case study of Pleistocene beaches, Whalebone Bay, Bermuda: Geologische Rundschau, v. 82, p. 148–162.

Ward, W. C., 1975, Petrology and diagenesis of carbonate eolianites of northeastern Yucatan Peninsula, Mexico, *in* Wantland, K. F., and Pusey, W. C., III, eds., Belize Shelf—Carbonate sediments, clastic sediments and ecology: American Association of Petroleum Geology Studies in Geology 2, p. 500–571.

Wheeler, H. E., 1958, Time stratigraphy: American Association of Petroleum Geologists Bulletin, V. 42, p. 1047–1063.

White, B., and Curran, H. A., 1988, Mesoscale physical sedimentary structures and trace fossils in Holocene carbonate eolianites from San Salvador Island, Bahamas: Sedimentary Geology, v. 55, p. 163–184.

White, B., Kurkjy, K. A., and Curran, H. A., 1984, A shallowing-upward sequence in a Pleistocene coral reef and associated facies, San Salvador, Bahamas, *in* Teeter, J. W., ed., Proceedings, Second Symposium on the Geology of the Bahamas: San Salvador, CCFL Bahamian Field Station, p. 53–70.

Yaalon, D. H., and Laronne, J., 1971, Internal structures in aeolianites and palaeowinds, Mediterranean coast: Israel: Journal of Sedimentary Petrology, v. 41, p. 1059–1064.

Manuscript Accepted by the Society January 5, 1995

Printed in U.S.A.

Geological Society of America
Special Paper 300
1995

Pleistocene sea-level yo-yo recorded in stacked beaches, Bermuda South Shore

Dieter Meischner, Rüdiger Vollbrecht, and Dieter Wehmeyer
Abteilung Sediment-Geologie, Institut für Geologie und Paläontologie, Goldschmidt-Strasse 3, D 37077 Göttingen, Germany

ABSTRACT

On the tectonically stable Bermuda Carbonate Platform (Atlantic Ocean, 64°50'W; 32°20'N), sea-level fluctuations during the Pleistocene left a pile of marine and eolian limestones (interglacial) intercalated with red soils (glacial). Each marine highstand deposited a new, basically similar sequence of sediments. Along Bermuda South Shore, as many as three Pleistocene beach sequences occur, stacked one above another. In this study, we describe the principal sequence of onlapping/offlapping beach deposits, and assign actual sections in Grape Bay to individual highstands that correlate to isotopic stages 5e (Rocky Bay), and tentatively, 7ac and 7e (Upper and Lower Belmont).

The diagenetic imprint is studied to link phases of marine and meteoric cementation with times of transgression, highstand, and subaerial exposure. Even in the older beaches (Belmont), primary high-Mg calcite and aragonite grains are little affected by diagenetic alteration. Cements occur as repeated high-Mg calcite rinds around grains, separated by abrasion and weak meteoric alteration. The younger beach (Rocky Bay) was cemented in the meteoric environment. No evidence was found for an extended Belmont-time fresh-water lens and meteoric alteration in the Belmont postulated by Land (1970) to explain the striking bench-like appearance of the indurated Belmont beach deposits.

Based on observations from Grape Bay, other outcrops along South and North Shores, and along inshore cliffs, we conclude that each of the interglacial highstands as recorded here reached a position within 8 m of the Recent sea level.

INTRODUCTION

The Bermuda Carbonate Platform caps a Cretaceous basalt volcano that received its last supply of intrusive magma in the Oligocene period (Reynolds and Aumento, 1974) and since then has undergone leveling by erosion and thermal subsidence. Carbonate production started in the Miocene and continues to the Present. The average thickness of the limestone cap is approximately 75 m, most of it Pleistocene age (Pirsson, 1914; Officer et al., 1952; Schenk, 1973).

Global, eustatic changes of sea level with an amplitude of more than 120 m have affected the tectonically stable Bermuda Platform during the Quaternary (Sayles, 1931; Bretz, 1960; Land et al., 1967; Harmon et al., 1983; Vacher et al., this volume). During warm, interglacial periods, at high sea level, carbonates formed throughout the platform, covering preexisting coral reef and lagoon bottoms. During times of low, glacial sea level, the platform was emergent and experienced fresh-water diagenesis and karstification. With each sea-level retreat, large areas of freshly formed carbonate sands became available for eolian transport and deposition as coastal dunes. They comprise by far the largest part of the present-day Bermuda Islands (Fig. 1) and reach to more than 70 m above present datum, i.e., well beyond any interglacial sea level (Land et al., 1967).

Meischner, D., Vollbrecht, R., and Wehmeyer, D., 1995, Pleistocene sea-level yo-yo recorded in stacked beaches, Bermuda South Shore, *in* Curran, H. A., and White, B., Terrestrial and Shallow Marine Geology of the Bahamas and Bermuda: Boulder, Colorado, Geological Society of America Special Paper 300.

The spectacular eolianites are intercalated with red soils that form during times of continuous subaerial exposure by solution of limestone and accumulation of windblown dust (Land et al., 1967; Bricker and Mackenzie, 1970; Vacher et al., this volume). High, interglacial sea levels have left erosional features, infills of rubble and shell, and in rare cases, autochthonous accretionary growth, mainly of calcareous algae and vermetids on the older eolianites; these formed steep cliffs, especially along the southern and western shores of Bermuda, and around inshore waters such as Harrington Sound. In a number of places, marine coastal sands interfinger with the sequence of eolianites and soils (Bretz, 1960; Land et al., 1967). Depositional structures in these sediments can be attributed to sedimentation in the upper subtidal, the actual beach, and the transition into the coastal dune. This chapter focuses on a particularly fine example from Grape Bay (Fig. 2), where extended remnants of three Pleistocene beaches are exposed in the cliffs and on the flat faces of the Recent shore.

BERMUDA STRATIGRAPHY

Pleistocene stratigraphy in Bermuda is based on the repeated succession of eolianites and red soils (Verrill, 1907; Sayles, 1931; Land et al., 1967; Vacher, 1971, 1973; Vacher and Harmon, 1987; Vacher et al., 1989, this volume). Earlier authors integrated marine strata to the stratigraphic column as local formations (Sayles, 1931; Land et al., 1967). Recently, attempts have been made to correlate Bermuda Pleistocene formations and marine intercalations with the well-established global deep-sea isotopic record that reflects changes between glacial and interglacial periods, and hence, global sea-level fluctuations (Harmon et al., 1981, 1983; Vacher and Hearty, 1989; Vollbrecht, 1990; Hearty et al., 1992, Vacher et al., this volume). The present state of correlation is summarized in Table 1.

There exists some uncertainty with regard to the older formations. No uranium-series ages are available for the correlation of formations older than isotopic stage 7. However, amino acid racemisation (AAR) analyses allow a relative ranking and hence support age estimates (Hearty et al., 1992).

PLEISTOCENE BEACHES ON BERMUDA SOUTH SHORE

Nature of the record

All along Bermuda South Shore, from Trott's Bay to Elbow Beach, Belmont and Rocky Bay deposits occur plastered against, or overstepping, older eolianites. Two extreme situations can be observed:

1. On steep erosional shores such as Watch Hill Park (Fig. 3), marine Rocky Bay is preserved as rubble and encrustations in caves and crevices in the older rocks (both Belmont and Town Hill) and in a rare case, as an accretional lip of algal-vermetid growth with exactly the same geometry and composition as the Recent lip that grows around low tide (Oertel, 1970). Belmont occurs as a thin veneer of well-sorted marine and eolian calcarenite with changing seaward and landward dip. The highest such transition occurs at 7.5 m above present datum.

2. On low, sedimentary shores such as Grape Bay (Fig. 4), a multitude of planar and cross-bedded calcarenites occur that are subdivided by erosional interfaces. On first sight, the geometry of the beds is confusing, but as a rule, planar beach and cross-bedded sublittoral beds occur on the seaward edges of the Recent beach. In more backshore positions, landward-dipping eolianites predominate; toward land, they become divided from each other by rhizoturbated protosols that, within tens of meters, give way to red soils.

The most complete outcrop is in Grape Bay, a locality already intensely surveyed by Sayles (1931), Land et al. (1967), and Vacher et al. (1989). On closer inspection, three sets of beach and backshore beds can be discerned. Each starts from an erosional base (on the lowermost set, this is only locally seen). The bed immediately above the base forms a coarse drape over preexistent relief and may develop into coarse conglomerate with marine shell and rubble in erosional notches and pockets of the older rocks. Each of the sedimentary sets consists of a predominantly convex landward-dipping lower unit and a planar seaward-dipping upper unit. Intensely cross-stratified beds interpreted as sublittoral occur between the two units of the upper two sediment sets. In their landward sections, all three sets grade into eolianites.

The three sets of coastal sediments described above are interpreted as products of cycles of transgression, beach development, regression, and coastal dune superimposition. The individual sets are separated from one another by erosional interfaces that can be mapped over the whole of Grape Bay and beyond. They are interpreted as representing phases of longer lasting sea-level lowstands, and hence are sequence boundaries. The upper sequence is assigned to the Rocky Bay Formation in the sense of Vacher et al. (1989). The protosol at the base of the eolian foresets ends at the Shore Hills geosol.

The lower two sequences are identified as Belmont. They rest on older eolianites probably of the Town Hill Formations. This relation has been clearly established by mapping (Vacher et al., 1989). There is no doubt on this age relation among workers on Bermudan geology. Some local problems, however, may arise from the interplay of the complex geometry of the coastal sediment bodies and the complicated shoreline that experienced strong erosion and collapse during Holocene time.

Of the three sedimentary sequences described above, each is most complete where the two others are thin or absent. Along South Shore the predominance of the individual sequences changes rapidly, sometimes within a few meters. This points to the deposition of each sequence on a previous rugged shoreline that had formed by erosion of indurated earlier rocks. A sedimentary sequence is best developed where it fills a preexistent

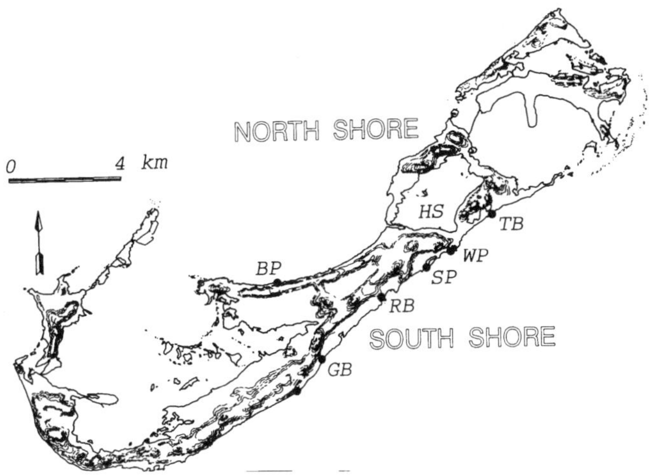

Figure 1. Map of the Bermuda Islands. South Shore faces the open ocean, whereas North Shore opens onto the lagoon, and is protected from oceanic swell, although it receives impact from heavy winterly gales. Localities: BP = Blackwatch Pass, TB = Trott's Bay, WP = Watch Hill Park, SP = Spittal Pond, RB = Rocky Bay, GB = Grape Bay, HS = Harrington Sound.

Figure 2. Location map of Grape Bay. Figures indicate positions of outcrops and refer to figures in the text. A = access from White Sands via public right-of-way. (After Bermuda Government, Bda 111, Edition 2—Bda 1974, original scale 1:2500).

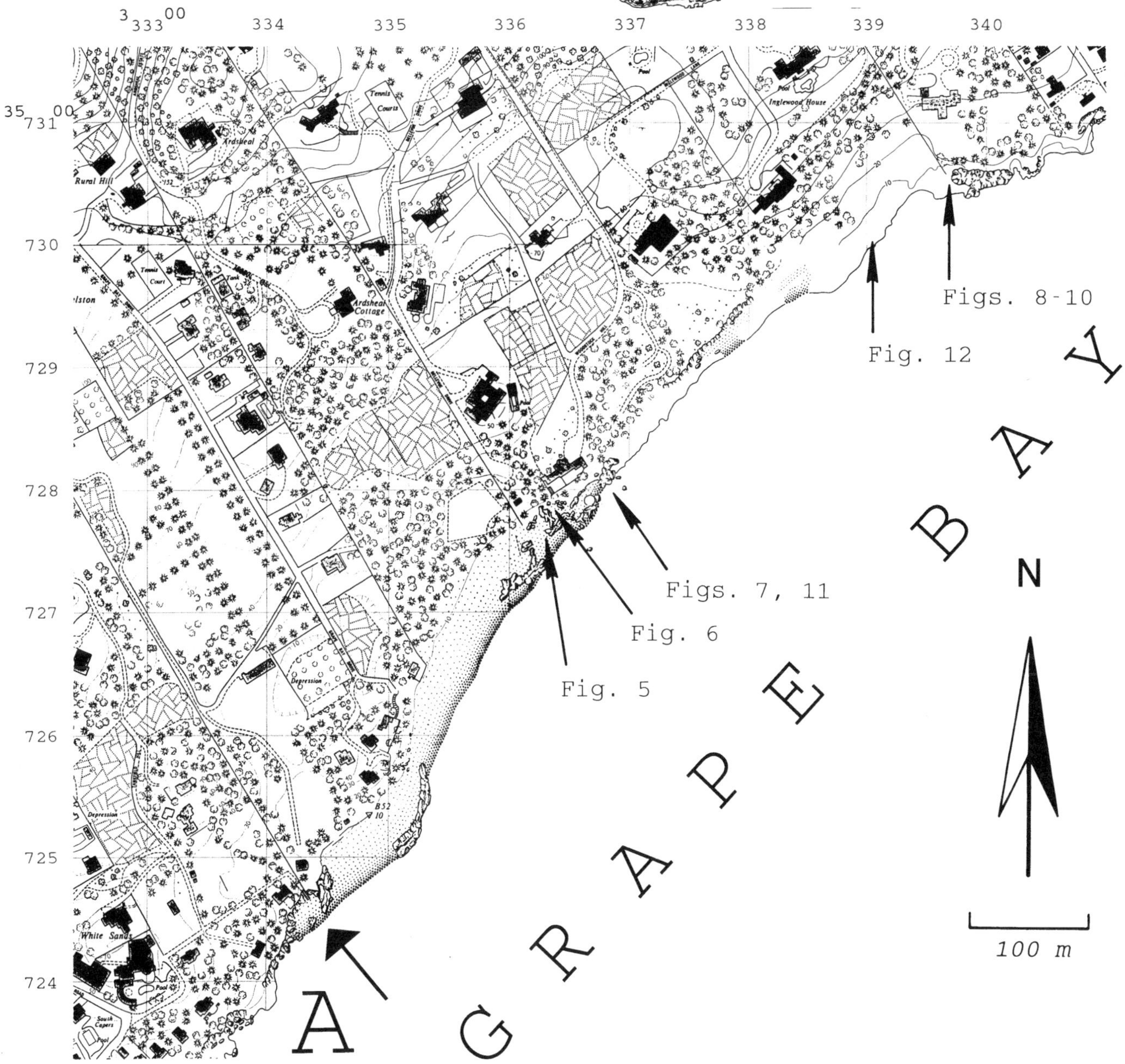

TABLE 1. PLEISTOCENE STRATIGRAPHY OF BERMUDA*

Stratigraphic Unit (this chapter)	Vacher et al. (this volume)	Supposed $\delta^{18}O$-isotope Stage
PAGET GROUP	PAGET GROUP	5†,§
Southampton Fm.	Southampton Fm. (S3, S2, S1)	5a§
Rocky Bay Fm. (RB)	Rocky Bay Fm. (R2, R1)	5e†
Shore Hills Geosol		6
BELMONT GROUP	BELTMONT FM. (B2, B1)	7†
Upper Belmont Fm. (B2)		7ac**
Lower Belmont Fm. (B1)		7e**
Ord Road Geosol		8
TOWN HILL GROUP	TOWN HILL FM.	
Upper Town Hill Fm.	Upper member (UT2, UT1)	9‡
Harbour Road Geosol		10
Lower Town Hill Fm.	Lower member (LT3, LT2)	11‡
UNNAMED UNITS (in Harrington Sound)	Lower member (LT1)§§	>11‡
Castle Harbour Geosol		
WALSINGHAM FM.	WALSINGHAM FM.	

*After Vacher et al., 1989, slightly modified with respect to the ranking of units.
†Harmon et al., 1983.
§Vacher and Hearty, 1989.
**Vollbrecht and Meischner, 1993, this chapter.
‡Hearty et al., 1992; Vacher et al, this volume.
§§Named "Unmapped Unit" by Hearty et al., 1992.

depression or pocket bay, and where it has been preserved from erosion during subsequent sea-level highstands.

A complete description of the three sequences must therefore take into account those features that are visible in more distant outcrops but that may be poorly developed or absent in Grape Bay, in addition to the variation of structures in Grape Bay itself. The following description proceeds from southwest to northeast and stratigraphically from Rocky Bay down to Lower Belmont.

Actual sections in Grape Bay

On entering Grape Bay shore platform (Figs. 2 and 4) from White Sands via the public right-of-way, one stands on planar beds of Lower Belmont (which are discussed below). The bay extends to the northeast for about 800 m. Most of it is bare rock with small sandy pockets and a few more spectacular erosional remnants of bedded calarenites.

Rocky Bay sequence. The first major section is in the center of the bay where a vertical wall projects across the shore platform (Fig. 5) opposite Bellevue Drive. The wall consists of the Rocky Bay (RB) sequence. Sedimentation is initiated on an erosional face with coarse drape: conglomeratic beds with a few *Cittarium pica* (Linnaeus) and other shells and corals. Where bedding is visible, it tends to be parallel to the base of the section. The following onlapping beds have convex landward-dipping foresets, but are subdivided by long, continuous seaward-dipping erosional interfaces. Bedding turns to planar, gently seaward dipping in the upper part of the section. This is interpreted as a regressive accretionary beach succession.

The Rocky Bay sequence is completed by an outcrop (Fig. 6) just 15 m behind the wall in Figure 5. A protosol rests with an irregular lower face on beach beds, and is in turn overstepped by a dune with landward-dipping foresets. More detail of the lower part of the sequence is seen in a cliff (Fig. 7) approximately 20 m north-northwest of the outcrop of Figure 5. Large pebble- to cobble-size slabs of older rocks occur in the lower part of Rocky Bay. The landward dip of the transgressive sand is partly oversteepened due to seaward sliding of the Rocky Bay on the inclined Belmont/Rocky Bay interface.

The Rocky Bay sequence is best developed at the northeastern end of Grape Bay in a large wall and between deeply eroded blocks (Figs. 8–10). A more complex association of landward- and seaward-dipping cross beds and lenticular sand bodies occurs between the basal drape and the protosol, the remnants of which top individual rock pillars and disappear landward underneath the dune foresets. Figure 10 shows the most complete section, with approximately 1 m of conglomerate containing numerous *Cittarium pica* and other marine shells and corals at the base, variegated cross-stratified sublittoral, planar beach, and finally rhizoturbated protosol. Note that the Belmont is deeply eroded at this locality, giving way to a thicker development of coarse basal Rocky Bay.

Upper Belmont sequence. The Upper Belmont sequence (B2) is only decimeters thick where the Rocky Bay is more complete, similar to that in the localities shown in Figures 5 and 8 through 10. Very regular karren relief is locally developed on top of B2, as well as on B1 (Fig. 11). In the center of Grape Bay, B2 is reduced to a thin bank, and the two karren surfaces follow closely above one other. Exactly the same relief can be seen in the Recent supralittoral splash zone of South Shore, e.g., at Watch Hill Park.

The surface of the Lower Belmont, like that of the Upper Belmont in places, is settled by randomly scattered, solitary fixisessile vermetids in most places in the bay (Fig. 6). This growth can be followed underneath the Upper Belmont that is still in situ. It also covers older rock (Town Hill?) where it protrudes through the Lower Belmont. This makes the horizon with vermetids consistent over the whole extent of Grape Bay, and constantly intervening between Lower and Upper Belmont, or where the Lower Belmont was eroded before deposition of the Upper Belmont, between pre-Belmont and Upper Belmont. Thus the vermetids became attached after deposition and lithification of Lower Belmont, but before the sedimentation of the basal drape of Upper Belmont.

The Upper Belmont at the northeastern end of Grape Bay

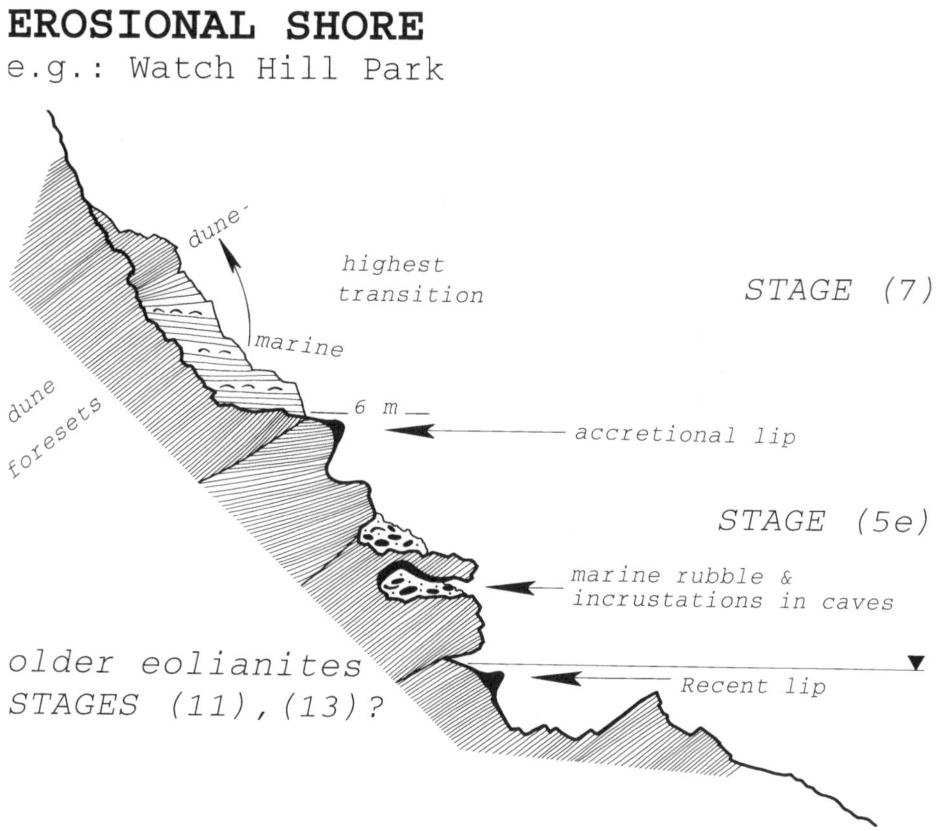

Figure 3. Erosional shore, composite section of Watch Hill Park. Remnants of high sea-level products occur plastered against preexistent, heavily eroded eolianites. Isotopic stage 5e is represented by an algal-vermetid lip of the same kind as the Recent one—but 5.8 m above present sea level—and by submarine incrustations and marine rubble (shells and pebbles) in caves and crevices. Marine stage 7 deposits have been partly eroded before the deposition of stage 5e material. The transition from beach to dune bedding in stage 7 is preserved approximately 7.5 m above present sea level.

(Fig. 9) exhibits, over a smooth surface of Lower Belmont with vermetids, a coarse drape of some 10 cm overstepped by onlapping, convex-stratified sublittoral sand. The rest of the sequence is eroded at this locality, but is more complete just 60 m southwest in a low cliff (Fig. 12). An approximately 2-m-thick set of vividly cross-bedded strata over coarse drape with rich marine fauna, turns upward into planar, gently seaward-dipping beach foresets. Individual sand sheets steepen along a hinge line into a ramp. The hinge moves seaward with accretion of the beach, which is the feature of a seaward-prograding, downlapping beach.

The cross-bedded section is exposed along a low cliff some 80 m long parallel to the shore. In plan view, the sets of cross beds form long, continuous ridges and troughs roughly parallel to the shore. The strike of the structures is 50°. In longitudinal sections, the individual sets show current foresets that dip predominantly to 230°, but in single layers can be reversed to 50°. Hence, the cross bedding was produced by elongated sublittoral ripples that trended parallel, or at low angle, to the Upper Belmont beach, and migrated following the longitudinal component of onshore waves. Today, similar ripples can be seen in light weather just below the low water line of beaches. The cross-bedded section of the Upper Belmont sequence is therefore interpreted as upper sublittoral, an interpretation that is in agreement with Land and Mackenzie (1970).

As shown in Figure 12, the Upper Belmont foreshore is directly overlain by an erosional remnant of Rocky Bay regressive beach. In other places, the Upper Belmont sequence grades landward into back beach and coastal dune in much the same way as the Rocky Bay. A more complete outcrop of this transition was described from the type section in Devonshire Bay by Land and Mackenzie (1970).

Lower Belmont sequence. Extended, flat, gently seaward-dipping Belmont beach beds form the major part of the modern rocky foreshore of Grape Bay. From the deeply eroded cliffs in the center of the bay, hinge lines run northeast at a low angle to the Recent shore. Seaward of the hinges, Lower Belmont beach beds (B1) dip more steeply (Fig. 7). A typical section shows planar beach-face beds dip 6° to 8°, steepening along a ramp to 22° to 24°. This is a consistent feature in the upper, regressive sections in both Belmont sequences. The B1

DEPOSITIONAL SHORE

Grape Bay, composite section

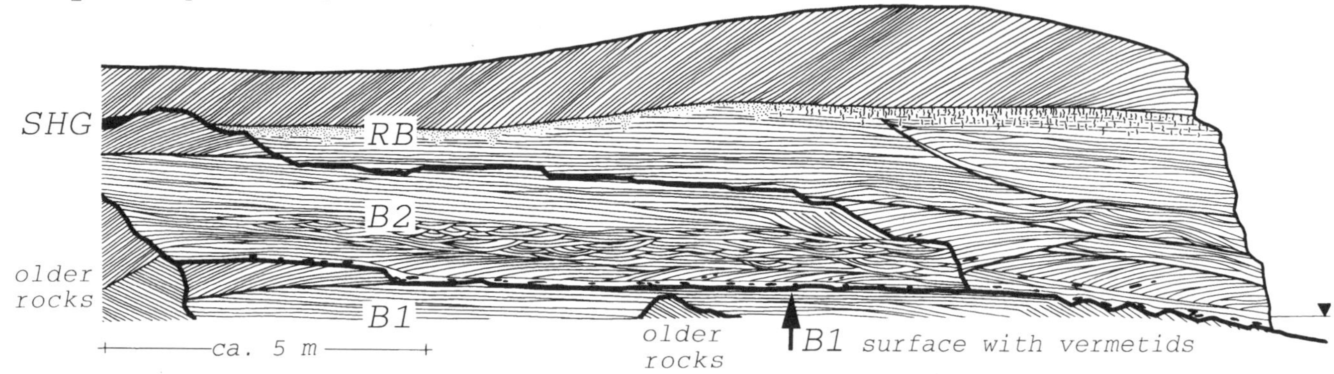

Figure 4. Depositional shore, composite section of Grape Bay. Remnants of three Pleistocene beaches are exposed in cliffs and on the Recent rocky shore face. The figure is a compromise in that each of the Pleistocene sets is best developed only in the absence of the others, i.e., when they are emplaced where older sets had been eroded before, and when they have not undergone later erosion. Lower Belmont (B1) deposits are dominated by shore face and regressional features. Upper Belmont (B2) sections are more complete; they overstep Lower Belmont with coarse drape, followed by landward-dipping transgressive sand and cross-stratified sublittoral sands. The following regressive beach section resembles that of the Lower Belmont, but occurs approximately 2 m higher above Recent sea level. Rocky Bay (RB) has a more pronounced transgressive base with coarse rubble and numerous shells mainly of *Cittarium pica*. Bedding in this part is convex landward, and seaward turns sigmoidal. Structures in the marine beds are more irregular and more complex than in the Belmont. The marine strata are overlain by a protosol with strong rhizoturbation that abuts landward against the lateritic Shore Hills geosol (SHG).

ramp occurs seaward and approximately 2 m lower than its B2 counterpart.

The same ramps can be found in Devonshire Bay, Rocky Bay, in the extended outcrops along the shore opposite Spittal Pond, and at the foot of Saucos Hill. There is not much more of Lower Belmont exposed in Grape Bay. At the northeastern end of the bay, erosion reaches little deeper in the section and exposes landward-dipping beds.

The surface of the Lower Belmont, like the Upper, is highly indurated. Regular jointing occurs approximately parallel and at right angle (140°) to the shore, the latter direction being more straight and more pronounced. Locals call this pattern "chequer board formation." The 140° joints are parallel to a straight textural fabric of the sandstone that may have been produced by orientation of the grains by onwashing surf. The otherwise smooth surface is locally scarred by slightly curved to straight burrows of *Echinometra lucunter* (Linnaeus) that have formed during the Upper Belmont transgression.

Petrography

As a result of changing sea level, the Pleistocene beaches at Grape Bay were exposed to various marine and meteoric diagenetic environments. The amount of diagenetic alteration is surprisingly small, considering the extended time span, from deposition to present, of meteoric influence on the Belmont rocks.

Constituent grains. Molluscan and red algal grains, benthic foraminifera (mainly Soritidae), and intraclast-like, multispecific fragments of biogenic crusts (red algae, foraminifera) comprise the bulk of both Paget and Belmont beaches. Corals, *Halimeda*, echinoderms, sponges, and lithoclasts are minor components. This suite of skeletal grains suggests that, during Younger Pleistocene interglacials, reefal environments persisted along Bermuda's southern shoreline as they do today. A high degree of reworking, both onshore and offshore, is indicated by the great number of multispecific grains and by lithoclasts that range from sand-size grains to slabs of rock about 1 m long. Grainstones and fine rudstones constitute the dominant sediment type of any beach studied. Conglomerates and coarse rudstones are confined to the basal part of each marine unit.

Diagenetic alteration. Differences in mineralogic composition of marine Rocky Bay and Belmont strata (Fig. 13) result from variable primary composition and from distinctly different paths of diagenesis. Marine diagenesis is evident in both marine Belmont units but is negligible in the marine Rocky Bay. In fact, the two marine Belmont units are almost indistinguishable on petrographic evidence alone.

Marine diagenesis in the Belmont beaches is expressed by locally altered high-Mg calcite beachrock cements (Fig. 14) that precipitated around the constituent grains. Subsequent meteoric alteration, often prior to the precipitation of the next high-Mg calcite cement layer, partly transformed these marine high-Mg calcite crusts into low-Mg calcite with elevated Mg content. Microprobe data point to an $MgCO_3$-content of 9 to 10 mol% within the cement crusts. This is probably an average over Mg-rich and Mg-depleted zones too small to be resolved by the electron beam (approximately 30 μm diameter). Alteration remained incomplete in that it usually did not affect the high-Mg calcitic skeletal grains. None of the studied rock samples displays extensive neomorphism or calcitization.

Preservation of skeletal high-Mg calcite, addition of small rinds of marine high-Mg calcite cement, and missing pervasive meteoric alteration are responsible for the high portion of high-Mg calcite in Belmont rocks. In contrast, Rocky Bay sands are devoid of any significant interparticle marine cement. In striking similarity to the Rocky Bay ("Pembroke") eolianite as depicted by Friedman (1964), lithification of the marine Rocky Bay is due to meteoric, blocky low-Mg calcite cement, which locally displays meniscus outlines. Solution structures and low-Mg calcite tangential needle cements indicate proximity to a subaerial, possibly soil-covered exposure surface. Similar meteoric-vadose features are sometimes also found in Belmont rocks below the Belmont–Rocky Bay interface (Figs. 15 and 16).

Discussion

Origin of sequences. In spite of their differences, the sets of beach deposits show striking similarities in their overall geometry and in the sequential order of structures. The interfaces between Town Hill and Lower Belmont, Lower and Upper Belmont, Upper Belmont and Rocky Bay sequences, all have the same karren relief, are settled by vermetids, and may be bored by sea urchins, as are the Recent rocky littoral and upper sublittoral surfaces.

Considering the persistency and the lateral continuity of both the sedimentary sequences and the interfaces, we conclude that the Pleistocene beach sediments in Grape Bay document three sequences of transgression, formation of a beach and sandy sublittoral, and of regression and downlapping of a planar beach, followed by the development of a weak soil on the dried sand, and by a landward migrating coastal dune. The sequences are asymmetrical; the transgressive or onlapping part is underrepresented by sediment, but comprises the major part of the time involved.

Typically, the transgression proceeds over the barren surface of the indurated beach that remains from the preceding marine highstand. During this phase, little sediment is available. The indurated substrate can be settled by sessile carbonate producers such as vermetids and *Chama*, and burrowed by hard-bottom dwellers such as *Echinometra*. Sedimentation starts with rubble from the contemporaneous erosion of older rocks, with marine shells (especially *Cittarium pica* and *Linga* sp.), and with an admixture of coarse sand. This material forms a structureless coarse drape some 10 cm thick, but can accumulate to more than 1 m in erosion holes and crevices of the rocky shore. The coarse drape is followed by landward-dipping, convex cross-bedded sublittoral sand near the peak of the sea-level highstand. The highest sublittoral beds may be used for the recon-

Figure 5. Actual section in central Grape Bay opposite Bellevue Drive. Lower Belmont (Belmont 1) beach dips gently seaward, and steepens in the right part of the section. Upper Belmont (Belmont 2) is confined to a few decimeters and is overlain, with erosional contact, by coarse, partly conglomeratic Rocky Bay. Bedding is convex landward in the lower and more landward part of the section, but is interrupted by seaward-dipping erosional faces. Transition from landward to seaward dip is well exposed in the landward part of the section. Planar beach bedding is in the uppermost layers.

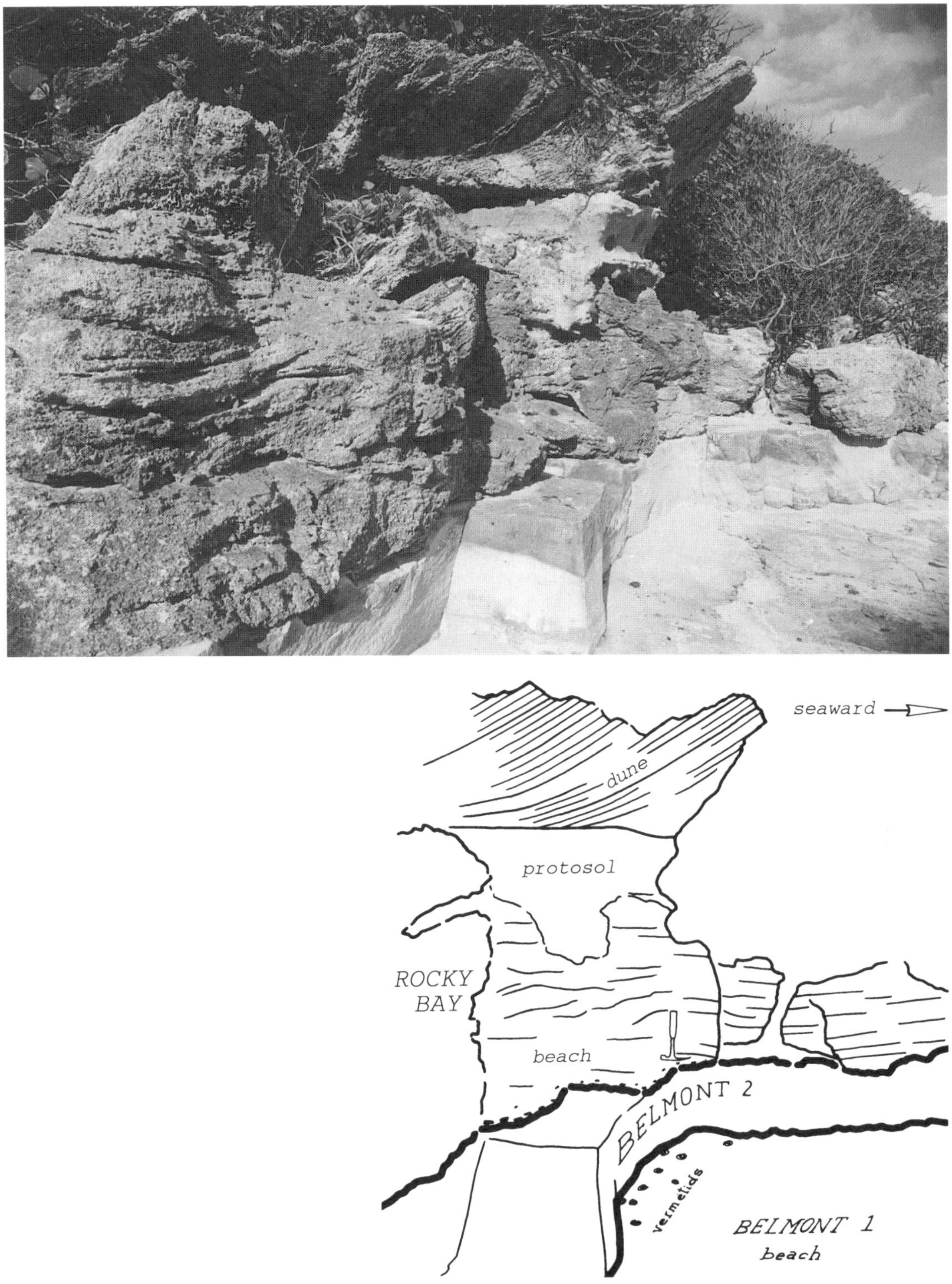

Figure 6. Outcrop approximately 15 m behind the person photographed against the wall in Figure 5. A plaster of fossil vermetids occurs between Belmont 1 and 2. They became attached at the transgression of the younger Belmont sea level over an already lithified Belmont 1 rock. Marine Rocky Bay is capped by a protosol, in turn overstepped by steep dune foresets. Hammer for scale is around the cliff edge.

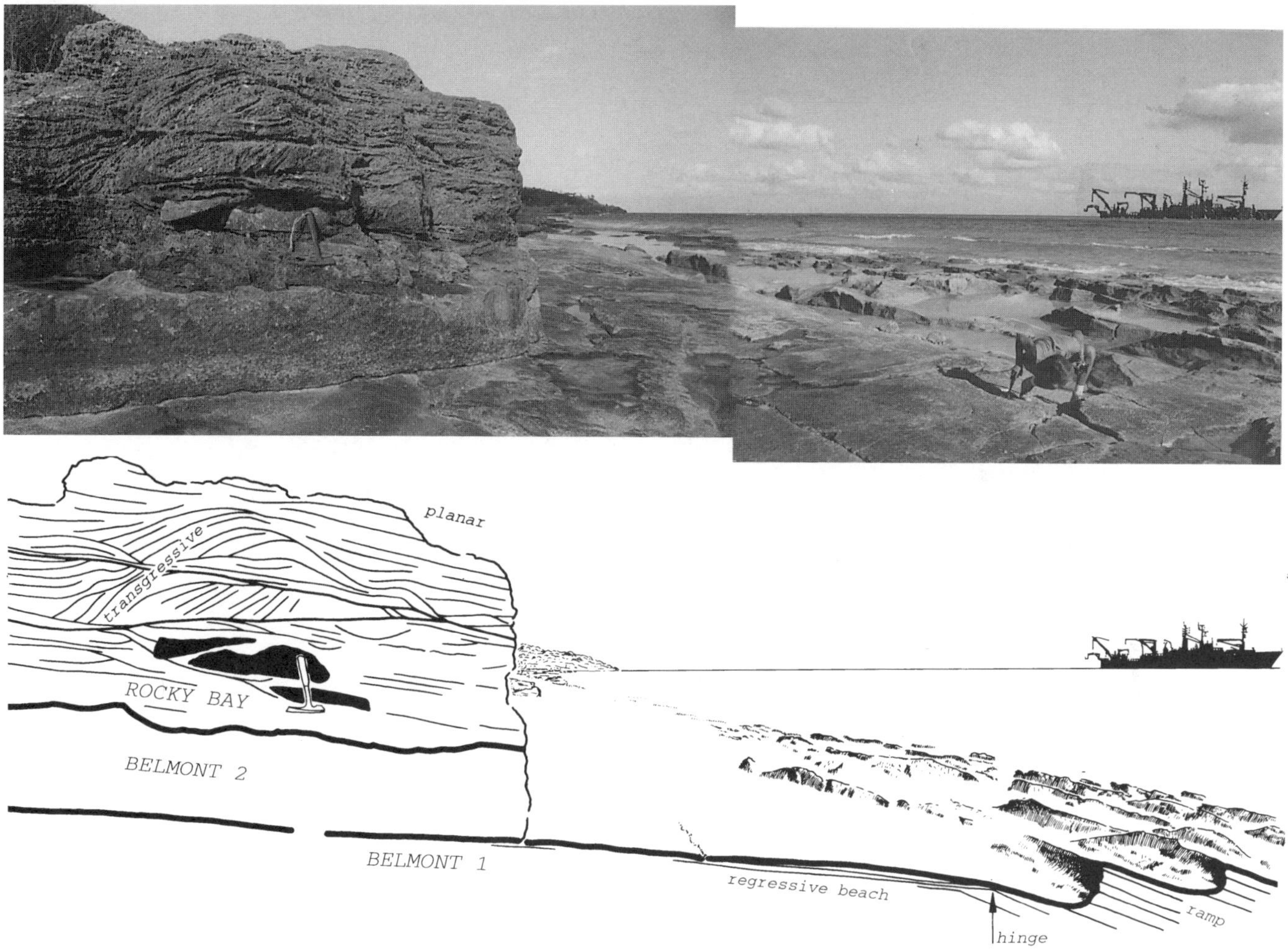

Figure 7. Approximately 80 m northeast of the outcrop in Figure 5. The Belmont 1 regressive beach steepens along a hinge line that runs at low angle to the Recent shoreline. Note boulders in Rocky Bay. Landward cross beds in Rocky Bay are oversteepened due to slumping. Picture combined from two wide-angle photographs.

struction of the minimum sea-level elevation. Mean sea level should be approximately 1 m above this datum, according to the modern beach profile along the same coast. At this time, a planar beach develops along the actual shoreline, and sand is blown out to form incipient back beach dunes. No clear remnants of the actual beach berm were found in the sections.

Regression starts with the progradation of the planar foreshore over the sublittoral sediment. Near the former low water line, the dip of the individual sand sheets steepens suddenly to some 22° to 24° along a ramp. The seaward lowering of this ramp may be taken as an indication of a drop in sea level. It appears that the dried sand was colonized by backshore vegetation, and a protosol with rhizoturbation developed; finally a coastal dune was blown over the section at already lowered sea level. Curran (1992, Fig. 8, p. 114) described a fine example from Doe Bay where the regressive planar beach—in our interpretation—is capped by heavily bioturbated coarse calcarenite from which backshore *Psilonichnus upsilon* burrows penetrate the foreshore beds. This puts the highest marine record of the Lower Belmont at 1.5 m in this place. No such burrows have been found in Grape Bay.

Differences between individual sequences. Rocky Bay and Belmont Beach sequences differ markedly. Upper and Lower Belmont are very similar in structures and grain sizes, so that the existence of two separate sea-level highstands has been overlooked until now. The shell fauna at the base of the Upper Belmont is diverse and contains endobenthonic species together with hard-bottom dwellers. Bedding is regular; even the cross-bedded sublittoral sediments are arranged in well-organized ridge and runnel parallel to the beach.

By contrast, the Rocky Bay is coarse, poorly sorted sand and gravel. Conglomeratic layers at the base contain numerous shells of the littoral welk *Cittarium pica* and rock-dwelling *Chama* sp., together with other shell and coral debris. Cross

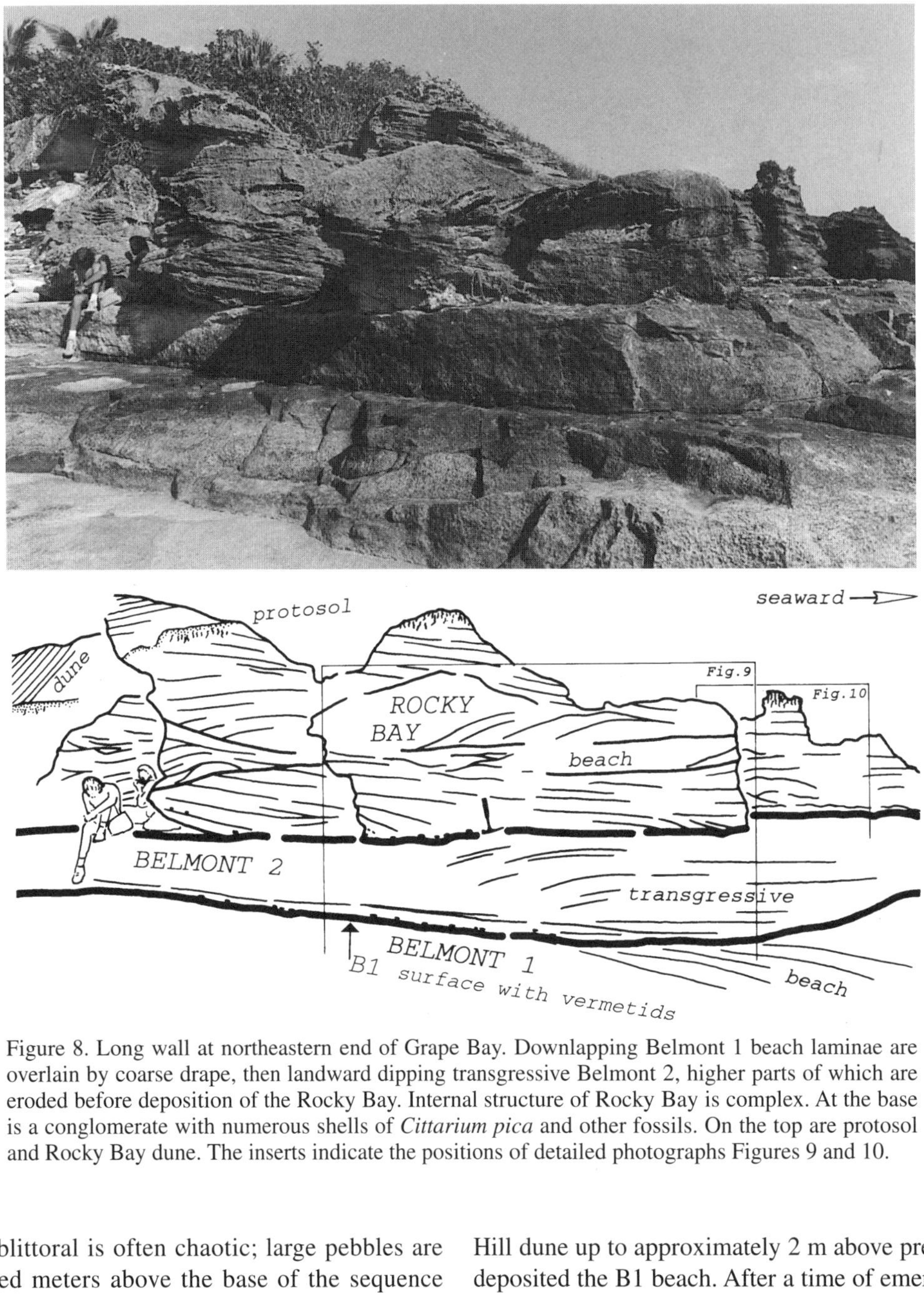

Figure 8. Long wall at northeastern end of Grape Bay. Downlapping Belmont 1 beach laminae are overlain by coarse drape, then landward dipping transgressive Belmont 2, higher parts of which are eroded before deposition of the Rocky Bay. Internal structure of Rocky Bay is complex. At the base is a conglomerate with numerous shells of *Cittarium pica* and other fossils. On the top are protosol and Rocky Bay dune. The inserts indicate the positions of detailed photographs Figures 9 and 10.

bedding in the sublittoral is often chaotic; large pebbles are sometimes admixed meters above the base of the sequence (Fig. 7).

This contrast can be explained in terms of successive sea-level highstands. The rocky substrate that underlies the Belmont and the Rocky Bay, respectively, was formed during the Town Hill sequences of sea-level highs. The Lower Town Hill, with marine shell, rests on Walsingham in Harrington Sound. The corresponding sea level may have been 2 m above present datum. This may correspond to the older beach in Whalebone Bay as described by Vollbrecht and Meischner (1993). Upper Town Hill sea level has been lower, probably just below the present level, as may be judged from extended backbeach and incipient dune stratification throughout the extent of this formation.

The Lower Belmont transgression overstepped the Town Hill dune up to approximately 2 m above present sea level and deposited the B1 beach. After a time of emergence, the freshly indurated beach was again flooded by the Upper Belmont transgression which reached at least 7.5 m above present datum. This level is clearly indicated by the position of the highest B2 sublittoral as a minimum mark in Grape Bay and by exceptionally fine marine/dune transitions at Spanish Rock and at Watch Hill Park (Fig. 3). Both of the Belmont transgressions (supposedly 7ac, 7e) reached higher than the foregoing sea-level highs. The Belmont record is mainly sedimentary and consists of newly formed skeletal carbonate that was well sorted in the sublittoral zone and on slightly inclined, extended beaches.

By contrast, the Rocky Bay transgression lagged behind the Upper Belmont at 5.8 m and remained there for a long time as determined from the accretional lip in Watch Hill Park

Figure 9. Detail of Figure 8. Upper Belmont (B2) with sharp, erosional contact on B1, coarse drape at base of B2 overstepped by onlapping, convex landward-dipping strata. Fossil vermetids on indurated surface of B1.

Figure 10. Detail of Figure 8. Rocky Bay, thick basal conglomerate overlain by predominantly landward-dipping, convex cross beds, capped by even beach stratification and rhizoturbated protosol. Note Belmont deeply eroded in the right lower corner.

(Fig. 3) and from clear-cut water tables in Harrington Sound. New evidence presented by Chen et al. (1991) puts the duration of the stage 5e sea-level high at approximately 10,000 yr. Belmont beach deposits were deeply eroded during this time, and the new, Rocky Bay, sublittoral, and beach sediments filled into recesses and notches of the erosional shore. Much of the material is reworked older carbonate that was deposited on narrow, discontinuous benches to form unstable packages of cross-bedded sediments that would tend to slide and slump seaward or become intermittently eroded.

The Recent Holocene sequence is even more erosional than the Rocky Bay. Sandy beach is restricted to a number of protected bays, whereas most of the shore exhibits heavily washed rocky faces even where sand was deposited during

Figure 11. Belmont 2 (B2) rests with erosional contact on Belmont 1 (B1), and is in turn overlain by marine Rocky Bay (RB). The surfaces of the Belmont 1 and 2 show karren relief, perfect analog to the relief in the Recent splash zone along the rocky South Shore. Vermetid growth occurs on the surfaces of both B1 and B2.

foregoing highstands. This is best demonstrated by Grape Bay itself, where a few fortnightly sandy patches do not coalesce into a beach.

In our opinion, the arbitrary succession of different sea-level elevations accounts for the differences among the products of individual sea-level highstands (see above).

Ages of the sequences. From work by Harmon et al. (1978, 1981, 1983) and Vacher and Hearty (1989), it has become clear that Rocky Bay corresponds to isotopic stage 5e (Table 1). The corresponding sea level of 5.8 m is in good accord with the level reported from the Bahamas by Neumann and Moore (1975), and from other places worldwide. The age of the Belmont is fixed by a uranium/thorium age of 200,000 yr on an in situ coral head in an isolated deposit on Boaz Island (Harmon et al., 1983). This age corresponds to stage 7 and is also supported by amino acid racemization (AAR) dating (Hearty et al., 1992). Because the two Belmont beach units are so similar in every respect, including constituent grains, structures, and diagenetic imprint, they have been treated as a single entity prior to our survey. It is very likely that the erosional interface between Lower and Upper Belmont represents a retreat of sea level. Because the global deep-sea record of the stage 7 highstand tends to be bimodal (7ac/7e or 7.1,7.3/7.5, Prell et al., 1986), we correlate the Upper and Lower Belmont of this chapter with isotope stages 7ac and 7e, respectively.

This dating does not conflict with the well-established stratigraphic column by Hearty et al. (1992), who have assigned the Town Hill Formation to stages 9 and 11 (Upper and Lower Town Hill). In terms of absolute ages, the central part of Bermuda South Shore has remained at about its present position through repeated accretion, cementation, and marine erosion of littoral sediments for approximately 400,000 yr.

Diagenetic imprint. Diagenesis of the Rocky Bay strata is rather straightforward: aside from thin marine cement fringes within some of the skeletal grains, marine precipitates are missing. Lithification is by meteoric cements with typical vadose features, including fine crystal size, fibrous habit, patchy distribution, or meniscus outlines. This style of lithification is not uncommon in Rocky Bay deposits elsewhere on the platform, i.e., in beach sediments (Vollbrecht and Meischner, 1993) as well as in submerged lagoonal sediments considered to be stage 5e (Vollbrecht, 1990). Where interparticle marine cementation actually occurs, it usually affects the underlying substrate rather than the Rocky Bay sediments themselves, thus indicating precipitation during transgression, prior to any substantial sedimentation.

Conditions were somewhat different with the Belmont transgressions. In the Belmont beaches, grains with multicyclic cement rims are abundant, indicating a much closer temporal inferfingering of cementation and sedimentation compared with the Rocky Bay beach. Given that these grains were produced during Belmont time and are not altogether lithoclasts of much older age, they must have been subjected repeatedly to initial beachrock cementation followed by weak meteoric alteration, reworking, and grain remobilization.

From diagenetic evidence, we cannot draw conclusions on the duration of subaerial exposure of the surface of the Lower Belmont where the vermetids had settled. But considering the

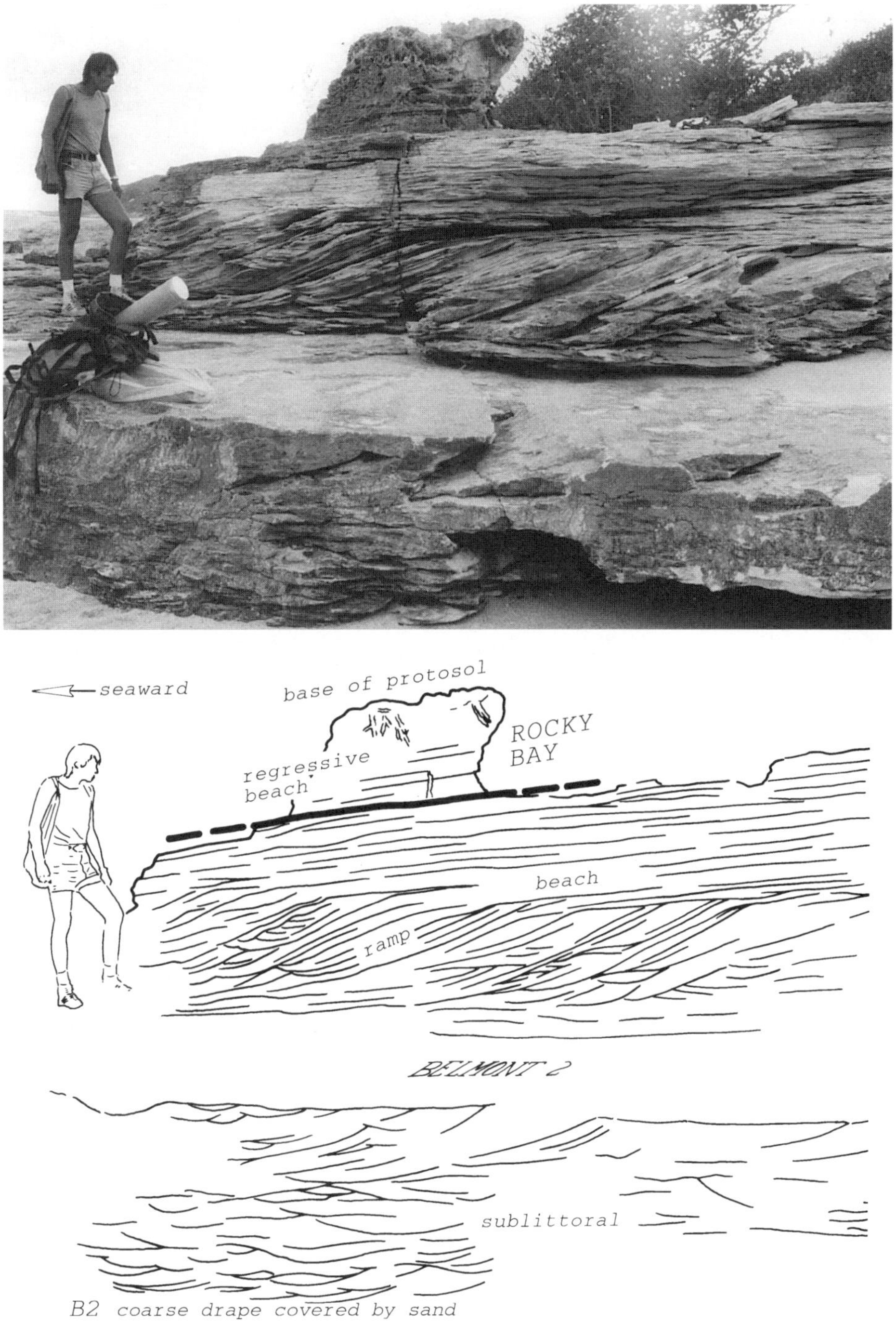

Figure 12. Complete section of marine Upper Belmont 60 m before northeastern end of Grape Bay. Rich marine fauna in B2 coarse drape (not visible in the photograph). Cross-bedded sublittoral reaches several meters above present sea level and pinches out landward.

overall persistency of depositional structures as depicted in Figure 4 and the development of a karren relief (e.g., Fig. 11), we believe that the vermetids colonized an indurated substrate freshly flooded by the rising B2 sea level.

No extensive neomorphism or pervasive calcitization is seen in Belmont samples from Grape Bay. This precludes any active Belmont-time freshwater lens in Grape Bay, such as the one visualized by Land (1970) for Belmont deposits in Devonshire Bay. Instead, diagenetic features suggest inefficient, meteoric-vadose or stagnant meteoric-phreatic alteration during and after Belmont time.

Significance. Bermuda is among the few places where bias

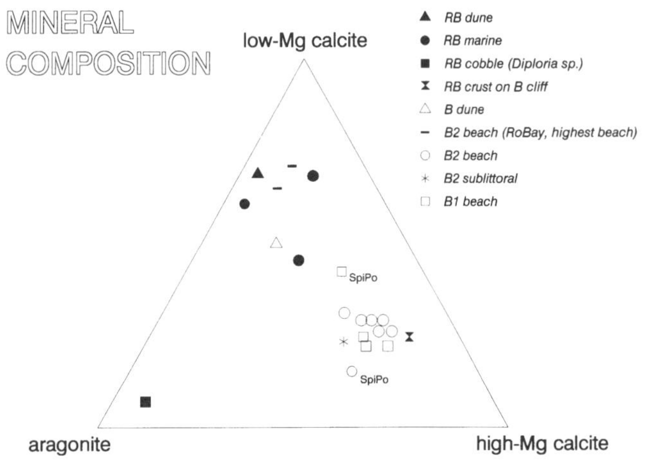

Figure 13. Mineralogic composition of Pleistocene sediments sampled at Grape Bay, Rocky Bay, and Spittal Pond. RB = Rocky Bay, B2 = Upper Belmont, B1 = Lower Belmont, SpiPo = Spittal Pond.

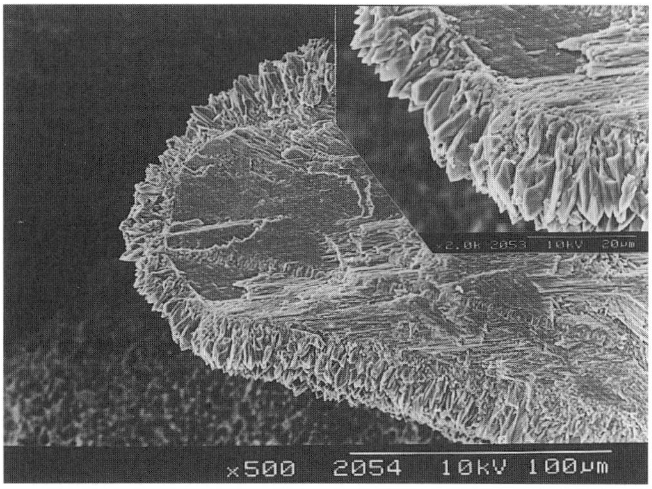

Figure 14. Scanning electron micrograph of marine isopachous cement crust composed of calcite, with an average of 9 to 10 mol% $MgCO_3$ (see text). Note elongated growth form of cement crystals and unaltered crossed lamellar structure within molluscan grain. Lower Belmont.

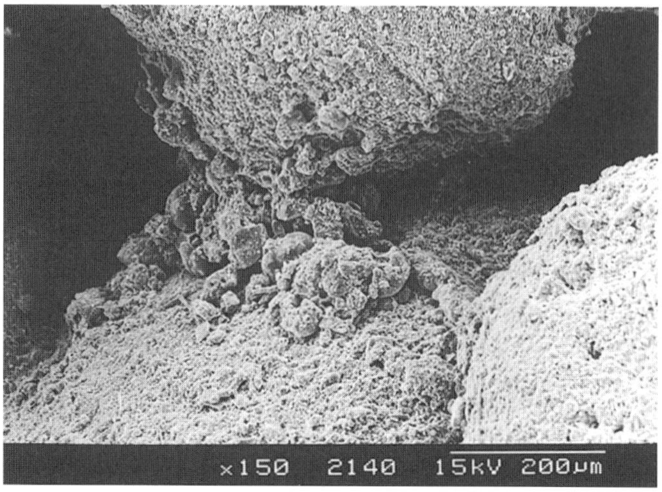

Figure 15. Scanning electron micrograph of cementation by microcrystalline to very finely crystalline meteoric low-Mg calcite. Note bimodal grain-size distribution, with fines concentrated along contacts of larger grains. Upper Belmont, coarse drape.

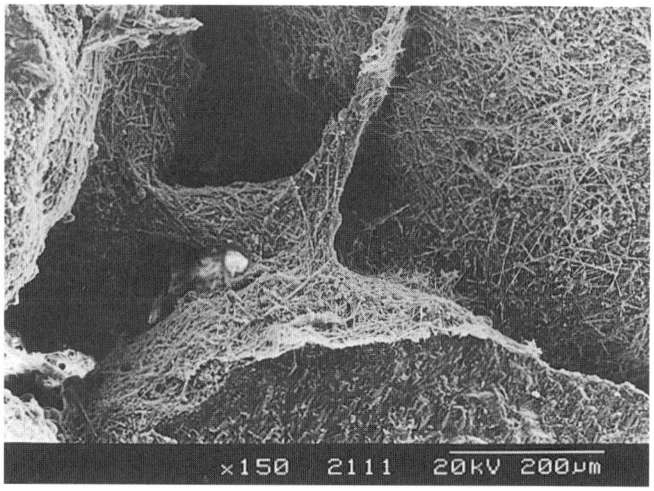

Figure 16. Scanning electron micrograph of alveolar structure composed of low-Mg calcitic tangential fibers, indicating meteoric-vadose alteration close to a subaerial exposure surface. Upper Belmont, immediately below Belmont-Rocky Bay interface.

on fossil sea-level readings is minimal. The seamount is firmly anchored in the oceanic crust and tectonically stable. Subsidence is restricted to the rate that derives from the relative drift away from the Mid-Atlantic Ridge; this is below 1 m per 100,000 yr in the Quaternary. The top of the leveled basalt is about 75 m below present sea level (Officer et al., 1952). Elastic crustal bulging following loading and unloading of polar ice is minimal around the latitude of Bermuda. Effects of water burden should be minimal; at least there are no differences expected between sea-level highstands of similar elevation. Pleistocene sea levels recorded on Bermuda should therefore be taken as indicators of global sea-level changes.

All of the interglacial sea levels involved in the formation of Bermuda South Shore range within 10 m vertically, i.e., from –2 to +8 m. Corrected for thermal subsidence, the range remains the same, and all levels plot above the Holocene sea level. This narrow range indicates a striking regularity in the extent of deglaciation during warm phases of the Pleistocene that may correspond to the stability of the atmospheric and marine heat budget. The Holocene sea level is the lowest of the major highstand levels since at least 400,000 yr.

Submarine dives on the slopes of the Bermuda pedestal (Fricke and Meischner, 1985) yielded similarly constant glacial sea levels between –114 m (Late Wisconsin) and approximately –130 m. This has caused us to think of Pleistocene sea-level changes as of a yo-yo; reflecting the findings of Broecker et al.

(1985), who proposed the question of a bimodal stability of the ocean/atmosphere system based on deep-sea sediments. In this context, interglacial beach deposits should be considered a dependable source of information on the elevation of interglacial sea levels. The relative height of the peak levels recorded in Bermuda South Shore beaches, e.g., by Vacher and Hearty (1989), Hearty et al. (1992), and ourselves, differs from the levels one would expect from $^{18}O/^{16}O$ peaks in oceanic foraminifera (Prell et al., 1986; Chappell and Shackleton, 1986).

SUMMARY AND CONCLUSIONS

1. Pleistocene beach and sublittoral deposits on Bermuda South Shore consist of remnants of three interglacial sea levels that correspond to isotopic stages 5e and two highstands within stage 7, here tentatively assigned to 7ac and 7e. They rest on eolian deposits of the foregoing stages 9 and 11; the elevation of their corresponding sea levels can be derived precisely from a number of localities.

2. All of the recorded sea levels range within 10 m. When corrected for thermal subsidence of the Bermuda seamount, all the interglacial sea levels have reached above present datum.

3. The younger formations are arranged as sequences of transgressive, onlapping and regressive, downlapping, sublittoral and beach sands.

4. Differences between individual sequences in terms of sedimentary structures, grain sizes, and apparent sorting are explained by different elevations of successive sea levels, and therefore, changes from more sedimentary (Belmont) to erosive (Rocky Bay) coastal environments.

5. The South Shore of Bermuda has remained near its present position for at least 400,000 yr.

6. Multiple marine and meteoric cementation events are preserved in the Belmont sequences, whereas lithification of the Rocky Bay is solely meteoric. In general, diagenetic alteration, even in the Belmont beaches, is small; primary aragonite and high-Mg calcite grains are little affected.

ACKNOWLEDGMENTS

We thank our friend Max Schwanitz, who helped us during field work and acted as a scale to Figure 5. We are indebted to Hartmut Scholz and Gerald Hartmann for their careful and patient guidance during scanning electron microscope and microprobe studies, respectively, and to Gabriela Meyer, who typed the first draft of the manuscript. We are also grateful to the Bermuda Biological Station for providing support during the field studies. Special thanks go to Len Vacher and P. J. Hearty for their very valuable, meticulous reviews.

This work was supported by Deutsche Forschungsgemeinschaft (German Science Foundation) Grant Me 267/25 (to D.M.). This is contribution No. 1350 of the Bermuda Biological Station for Research.

APPENDIX: METHODS

Samples were impregnated with dyed resin and thin-sectioned following standard procedures. High-Mg calcite was stained according to Choquette and Trusell (1978). X-ray analyses were run on a Philips PW 1800 diffractometer (45 kV, 40 mA, Cua1/2), and the data evaluated following the peak-area method of Milliman (1974). MgCO3 contents of calcites were determined assuming a linear peak shift between calcite, d = 3.036 Å, and dolomite, d = 2.886 Å (Scholle, 1978; Richter, 1984). Microprobe data were collected at the Institut für Geochemie, Göttingen, on an ARL-SEMQ Electron Microprobe (acceleration voltage 15 keV, emission current 100 µA, sample current 20 nA on dolomite standard (Voigt, 1984), beam diameter 30 µm, TAP spectrometer, count times: peak 20 s, background 10 s; error 2% of standard.

REFERENCES CITED

Bretz, J. H., 1960, Bermuda: A partially drowned, late mature Pleistocene karst: Geological Society of America Bulletin, v. 71, p. 1729–1754.

Bricker, O. P., and Mackenzie, F. T., 1970, Limestones and red soils of Bermuda: Geological Society of America Bulletin, v. 81, p. 2523–2524.

Broecker, W. S., Peteet, D. M., and Rind, D., 1985, Does the ocean-atmosphere system have more than one stable mode of operation?: Nature, v. 315, p. 21–25.

Chappell, J., and Shackleton, N. J., 1986, Oxygene isotopes and sea level: Nature, v. 324, p. 137–140.

Chen, J. H., Curran, H. A., White, B., and Wasserburg, G. J., 1991, Precise chronology of the last interglacial period: ^{234}U-^{230}Th data from fossil coral reefs in the Bahamas: Geological Society of America Bulletin, v. 103, p. 82–97.

Choquette, P. W., and Trusell, F. C., 1978, A procedure for making the Titan-yellow stain for Mg-calcite permanent: Journal of Sedimentary Petrology, v. 48, p. 639–641.

Curran, H. A., 1992, Trace fossils in Quaternary, Bahamian-style carbonate environments: The modern to fossil transition, *in* Maples, C. G., and West, R. R., eds., Trace Fossils: Short Courses in Paleontology, no. 5, p. 105–120.

Fricke, H., and Meischner, D., 1985, Depth limit of Bermudan scleractinian corals: A submersible survey: Marine Biology, v. 88, p. 175–187.

Friedman, G. M., 1964, Early diagenesis and lithification in carbonate sediments: Journal of Sedimentary Petrology, v. 45, p. 777–813.

Harmon, R. S., Schwarcz, H. P., and Derek, C. F., 1978, Late Pleistocene sea level history of Bermuda: Quaternary Research, v. 9, p. 205–218.

Harmon, R. S., Land, L. S., Mitterer, R. M., Garrett, P., Schwarcz, H. P., and Larson, G. L., 1981, Bermuda sea level during the last interglacial: Nature, v. 289, p. 481–483.

Harmon, R. S., and 8 others, 1983, U-series and amino-acid racemization geochronology of Bermuda: Implications for eustatic sea-level fluctuation over the past 250,000 years: Palaeogeography, Palaeoclimatology, Palaeoecology, v. 44, p. 41–70.

Hearty, P. J., Vacher, H. L., and Mitterer, R. M., 1992, Aminostratigraphy and ages of Pleistocene limestones of Bermuda: Geological Society of America Bulletin, v. 104, p. 471–480.

Land, L. S., 1970, Phreatic versus vadose meteoric diagenesis of limestones: Evidence from a fossil water table: Sedimentology, v. 14, p. 175–185.

Land, L. S., and Mackenzie, F. T., 1970, Field guide to Bermuda geology: Bermuda Biological Station Special Publications, no. 4, p. 1–14.

Land, L. S., Mackenzie, F. T., and Gould, S. J., 1967, Pleistocene history of Bermuda: Geological Society of America Bulletin, v. 78, p. 993–1006.

Milliman, J. D., 1974, Recent sedimentary carbonates. Pt. 1: Marine carbonates: Berlin, Springer, 375 p.

Neumann, A. C., and Moore, W. S., 1975, Sea level events and Pleistocene coral

ages in the Northern Bahamas: Quaternary Research, v. 5, p. 215–224.
Oertel, G. F., 1970, Preliminary investigation of intertidal bioconstructional features along the South Shore of Bermuda: Bermuda Biological Station Special Publications, v. 6, p. 99–108.
Officer, C. B., Ewing, M. and Wuenschel, P. C., 1952, Seismic refraction measurements in the North Atlantic Ocean. Pt. IV: Bermuda, Bermuda Rise and Nares Basin: Geological Society of America Bulletin, v. 63, p. 777–808.
Pirsson, L. V., 1914, Geology of Bermuda Island; the igneous platform: American Journal of Science, ser. 4, v. 38, p. 189–206.
Prell, W. L., Imbrie, J., Martinson, D. G., Morley, J. J., Pisias, N. G., Shackleton, N. J., and Streeter, H. F., 1986, Graphic correlation of oxygene isotope stratigraphy application to the Late Quaternary: Paleoceanography, v. 1, no. 2, p. 137–162.
Reynolds, P. R., and Aumento, F. A., 1974, Deep drill 1972: Potassium-argon dating of the Bermuda drill core: Canadian Journal Earth Science, v. 11, p. 1269–1273.
Richter, D. K., 1984, Zur Zusammensetzung und Diagenese natürlicher Mg-Calcite: Bochumer Geologische und Geotechnische Arbeiten, v. 15, p. 1–310.
Sayles, R. W., 1931, Bermuda during the ice age: Proceedings, American Academy of Arts and Science, v. 66, p. 381–468.
Schenk, P. E., 1973, Deep drill 1972: Pleistocene stratigraphy: Eos Transactions (American Geophysical Union), v. 54, p. 486.
Scholle, P. A., 1978, A color illustrated guide to carbonate rock constituents, textures, cements, and porosities: American Association of Petroleum Geologists Memoir, v. 27, p. 1–241.
Vacher, H. L., 1971, Late Pleistocene sea-level history: Bermuda evidence [Ph.D. thesis]: Evanston, Illinois, Northwestern University, 153 p.
Vacher, H. L., 1973, Coastal dunes of Younger Bermuda, *in* Coates, D. R., ed., Coastal Geomorphology: Binghamton, State University of New York, Publications on Geomorphology, p. 355–391.
Vacher, H. L., and Harmon, R. S., 1987, Geological Society of America Penrose Conference Field Guide to Bermuda Geology (April 1987): Bermuda, Penrose Conference, Bermuda Biological Station, 48 p.
Vacher, H. L., and Hearty, P., 1989, History of stage 5 sea level in Bermuda: Review with new evidence of a brief rise to present sea level during substage 5a: Quaternary Science Reviews, v. 8, p. 159–168.
Vacher, H. L., Rowe, M. P., and Garrett, P., 1989, The geological map of Bermuda: Hamilton, Bermuda Government, Ministry of Works and Engineering, scale 1:25,000.
Verrill, A. E., 1907, The Bermuda Islands. Pt. IV, Geology and Paleontology; Pt. V, An Account of the Coral Reefs: Connecticut Academy Arts and Sciences Transactions, v. 12, p. 45–348.
Voigt, I., 1984, Geochemische und $^{87}Sr/^{86}Sr$-Isotopenbestimmungen an Gangarten und am Nebengestein der Erzlagerstätte Grund (Oberharz) [Doctorate thesis]: Göttingen, Georg-August-Universität, 170 p.
Vollbrecht, R., 1990, Marine and meteoric diagenesis of submarine Pleistocene carbonates from the Bermuda Carbonate Platform: Carbonates and Evaporites, v. 5, p. 13–95.
Vollbrecht, R., and Meischner, D., 1993, Sea level and diagenesis—A case study on Pleistocene beaches, Whalebone Bay, Bermuda: Geologische Rundschau, v. 82, no. 2, p. 248–262.

Manuscript Accepted by the Society January 5, 1995

Printed in U.S.A.

Geological Society of America
Special Paper 300
1995

Bermuda solution pipe soils: A geochemical evaluation of eolian parent materials

Stanley R. Herwitz
Graduate School of Geography, Clark University, Worcester, Massachusetts 01610
Daniel R. Muhs
U.S. Geological Survey, MS 424, Box 25046, Denver Federal Center, Denver, Colorado 80225

ABSTRACT

Solution pipes found in the Quaternary eolian and marine carbonates of Bermuda are filled with reddish to reddish-brown soil material. The bulk of the soil is composed of clay and silt-sized quartz and aluminosilicate clay minerals. The carbonates are of high purity and, therefore, are not likely to have been the parent material. Previous workers have hypothesized that Saharan dust may have been the soil parent material. The fine-grained component of loess from the Mississippi River Valley of North America also could have contributed. Paleoclimate models indicate that both North Africa and North America could have been important source areas during both glacial and interglacial periods. Immobile element concentrations in Bermuda soil samples collected from the interiors of solution pipes were determined for the purpose of geochemical fingerprinting and comparisons with the hypothesized parent materials. Immobile element ratios using Al, Ti, Zr, Y, and Th suggest that neither Saharan dust nor lower Mississippi River Valley loess were the sole contributors to Bermuda soils. Eolian dust from at least one other source area such as the Great Plains may have contributed parent material to the soils of Bermuda.

INTRODUCTION

Calcareous eolianites of Quaternary age are common on many tropical and subtropical coastlines and islands where substantial supplies of carbonate sand have been mobilized by the wind. There have been many studies of the structural and diagenetic features of Quaternary eolianites (e.g., Mackenzie, 1964a; Gardner, 1983; Esteban and Klappa, 1983; McKee and Ward, 1983; Pye and Tsoar, 1990). Such features include bounding surfaces, horizontal laminar calcretes, rhizoliths, and vertically oriented solution pipes. Solution pipes are characteristically lined with secondary calcite and commonly filled with clay-rich sediments derived from overlying soil B horizons (Ruhe et al., 1961; Herwitz, 1993).

On Bermuda (Fig. 1), which consists mainly of calcareous eolianite formations and some localized marine facies (Vacher et al., 1989), soil-filled solution pipes are commonly exposed as erosional remnants in the form of pedestals above the surrounding host eolianite or marine facies (Fig. 2A). The solution pipes also are exposed in vertical sections at road cuts and in quarries, and as cylindrical pillars in areas of active shoreline erosion (Fig. 2B). The interiors of the pipes are filled with reddish to reddish-brown (2.5YR 3/4 to 5YR 4/4) soil materials that previously supported terrestrial vegetation, as evidenced by the presence of carbonate rhizoliths (Fig. 2C). The pipes originally were referred to as "palmetto stumps" because it was thought that they were fossil casts of the trunks or taproots of the endemic palm tree species *Sabal bermudana* (Verrill, 1902). Livingston (1944) suggested that they may be fossil casts of the endemic cedar *Juniperus bermudiana*.

Solution pipes in eolianites have been reported along the east coast of the Mediterranean (Day, 1928), in southeastern Africa (Coetzee, 1975), and Western Australia (Fairbridge, 1950; Blackburn et al., 1965). We have observed them in Qua-

Herwitz, S. R., and Muhs, D. R., 1995, Bermuda solution pipe soils: A geochemical evaluation of eolian parent materials, *in* Curran, H. A., and White, B., Terrestrial and Shallow Marine Geology of the Bahamas and Bermuda: Boulder, Colorado, Geological Society of America Special Paper 300.

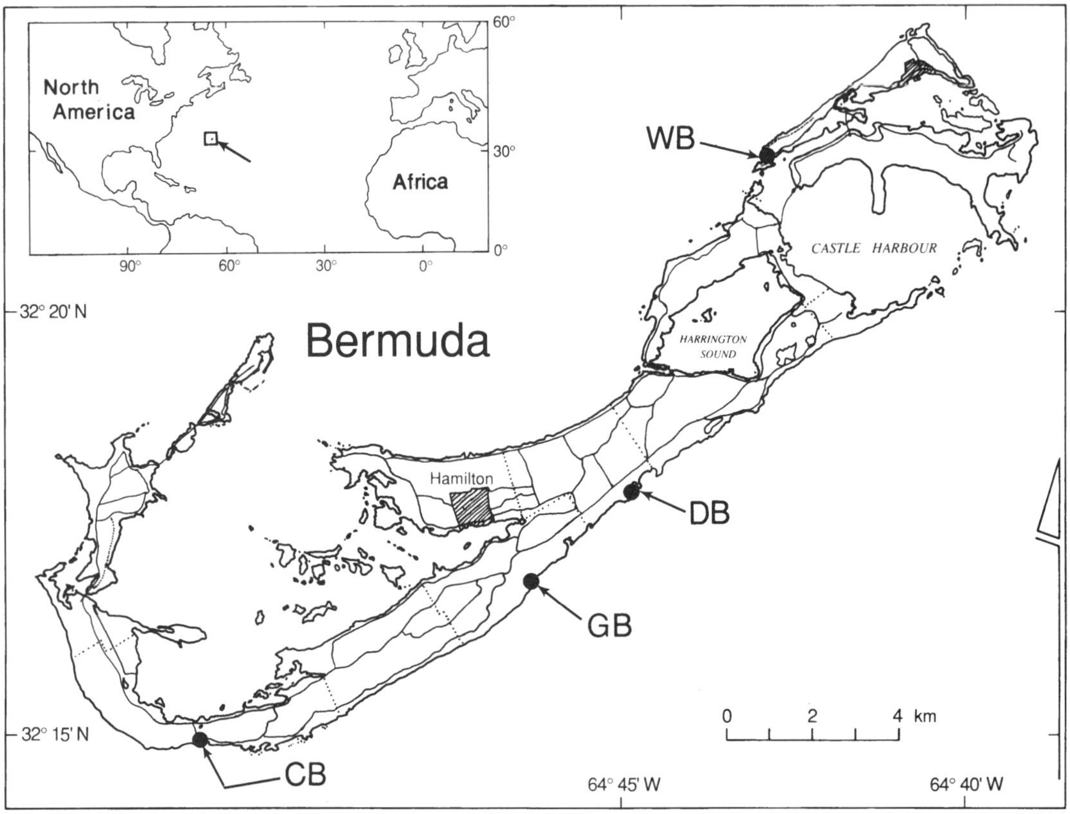

Figure 1. Location of soil-filled solution pipe sample localities on Bermuda: CB = Church Bay; DB = Devonshire Bay; GB = Grape Bay; WB = Whalebone Bay.

Figure 2. Soil-filled solution pipes (see arrows) in contrasting stages of exposure. A, Remains of lower sections of solution pipes in the form of raised soil-filled pedestals projecting above surrounding host carbonate at Whalebone Bay. B, Soil-filled solution pipes recently exposed by shoreline erosion of the surrounding host eolianite at Church Bay. C, Close-up view of the interior of one of the Church Bay solution pipes showing soil- and rhizolith-filled interior.

ternary eolianites on San Nicolas and San Miguel Islands off the coast of southern California. Some workers have assumed that the soil-filled pipes of Bermuda are simply the product of nonbiogenic solutional processes (Sayles, 1931; Ruhe et al., 1961; Plummer et al., 1976). Reference to Bermudan palmetto stumps, nevertheless, has persisted (Bretz, 1960; Land et al., 1967; Land and Mackenzie, 1969; Vacher and Harmon, 1987; Rowe, 1990), perhaps because the nonbiogenic hypothesis does not explain their clustered distribution pattern, their tapering cylindrical shape, and their smooth inner calcite casing. Herwitz (1993) recently conducted a field study of stemflow drainage on Bermuda, and hypothesized that the pipes may be the product of concentrated acidic stemflow inputs from long-lived trees. The present study is aimed at understanding the origin of the soil materials that fill the interiors of these pipes.

Sayles (1931) noted "heavy red clay fillings" in the interiors of solution pipes, and reported the occurrence of typically continental minerals such as quartz (comprising as much as 43% of the insoluble fraction), orthoclase, pink garnet, hornblende, and muscovite. Sayles proposed the following possible explanations for the occurrence of these minerals: (1) ingestion and transport of these mineral grains by birds migrating between Nova Scotia and the West Indies; (2) hurricane transport of sand from the Atlantic coastal plain; and (3) attachment of the minerals to marine algae and jellyfish, and subsequent transport by sea currents.

The first detailed study of the soils of Bermuda was by Ruhe et al. (1961), who documented the presence of kaolinite and vermiculite in both the horizontal soils and the soil-filled solution pipes, but these workers did not discuss the possible parent materials. Blackburn and Taylor (1969, 1970) suggested that Bermuda soils are residual, derived from the noncarbonate impurities in the eolianites. Bricker and Prospero (1969) and Bricker and Mackenzie (1970) argued that the soils were derived from wind-transported continental parent material based on similarities in the mineralogy of the soils and airborne dust collected on Bermuda. Bricker and Prospero (1969) suggested that the dust on Bermuda was derived from Africa. The formation of the soils on Bermuda from atmospheric dust now appears to be the generally accepted interpretation (e.g., Hearty et al., 1992), but there are no published geochemical data of which we are aware that document the source area(s) of the eolian parent materials.

Muhs et al. (1987, 1990) used immobile element ratios as geochemical "fingerprints" and concluded that red, reddish-brown, and dark brown clay-rich soils on relatively pure Quaternary carbonates on Barbados, Jamaica, the Bahamas, and the Florida Keys are probably derived from Saharan dust (Fig. 3) transported across the Atlantic Ocean by the northeast trade winds (Fig. 4). In the present study, we also adopt the geochemical fingerprinting approach to evaluate possible parent materials for the soil in the solution pipes of Bermuda. We use present-day atmospheric circulation patterns and global paleoclimate models to determine possible source regions for dust transport to Bermuda.

ATMOSPHERIC TRANSPORT OF DUST TO BERMUDA

Much research is being conducted on the transport of continental aerosols to the Atlantic Ocean. The main concern is to determine the eastward transport of pollutants from North

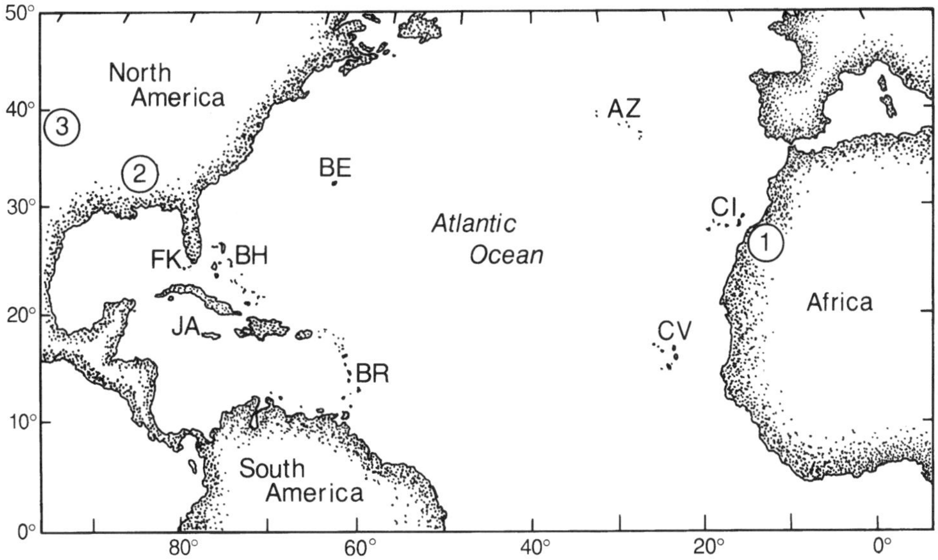

Figure 3. Location of Bermuda (BE) shown in relation to North America, the Florida Keys (FK), the Bahamas (BH), Jamaica (JA), Barbados (BR), South America, the Azores (AZ), Cape Verde Islands (CV), the Canary Islands (CI), Africa, and three continental source areas of atmospheric dust: (1) the western Sahara, (2) the lower Mississippi River Valley, and (3) the Great Plains.

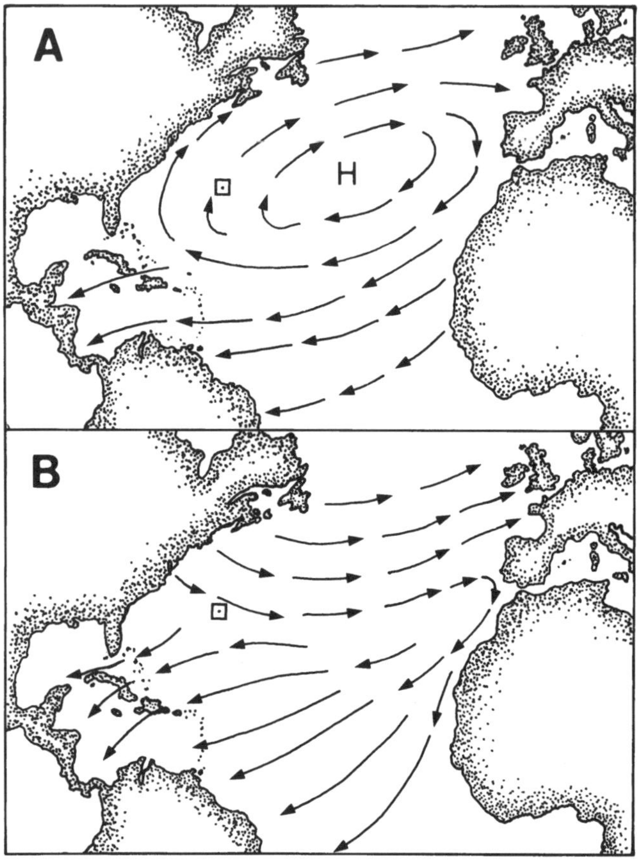

Figure 4. Circulation pattern of northeast trade winds and prevailing westerlies shown in relation to Bermuda (square/dot symbol) and Bermuda high-pressure cell (H). A, Summer season pattern; (B), winter season pattern.

America (e.g., Jickells et al., 1982; Galloway et al., 1988). Chen and Duce (1983) quantified the present-day contribution of mineral aerosols to Bermuda, and they reported that both North Africa and North America are contributing sources, with North Africa contributing more on an annual basis. Saharan dust is transported westward during the summer months by the northeast trade winds in a latitudinal belt mainly between 10°N and 25°N (Fig. 4A) at altitudes of 1.5 to 3.5 km (Rydell and Prospero, 1972; Glaccum and Prospero, 1980; Prospero, 1981; Prospero et al., 1981; Talbot et al., 1986). At 32°N, Bermuda is at the northernmost limit of the northeast trade winds. Saharan dust reaching Bermuda would most likely involve trade winds that circulate around the southern portion of the Bermuda high pressure cell (Fig. 4A). During winter, when the airflow pattern is displaced toward the equator (Prospero and Nees, 1977; Savoie and Prospero, 1977; Prospero et al., 1981), there is no apparent windflow pattern that could transport Saharan dust to Bermuda (Fig. 4B). It is during the winter months that Bermuda receives dust from North America via the prevailing westerlies.

During full-glacial times, sea-surface temperatures and atmospheric circulation patterns were not the same as they are in the present-day interglacial (Keffer et al., 1988). General circulation model (GCM) experiments have shown the probable atmospheric circulation that existed during full-glacial time at ~18 Ka (Kutzbach and Guetter, 1986; Kutzbach, 1987; Webb et al., 1987). The Northern Hemisphere storm tracks were displaced considerably south of their present positions in both summer and winter. However, these GCM experiments also suggest that the northeast trade winds may have been stronger than present during full-glacial summers. GCM maps show easterly surface winds reaching the vicinity of Bermuda during full-glacial summers (Fig. 5A) (Kutzbach and Guetter, 1986; Kutzbach, 1987; Webb et al., 1987). Sediment data from several deep-sea cores from the eastern Atlantic indicate that northwest Africa generated significantly more dust during full-glacials than during interglacials (Bowles, 1975; Kolla et al., 1979; Pokras and Mix, 1985; Grousset et al., 1989). Hence, even under full-glacial conditions, North Africa may have contributed to the formation of the Bermuda soils.

Westerly winds from North America would have reached Bermuda during full-glacial winters (Fig. 5B). A likely source of dust carried to Bermuda by westerly winds at these times would be the fine-grained (<10 μ in diameter) component of

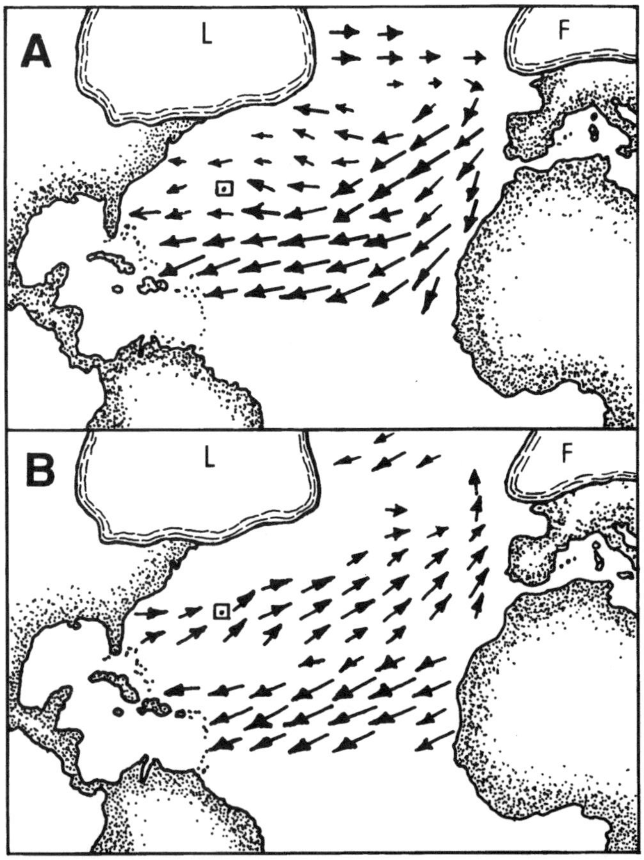

Figure 5. Surface wind circulation pattern during full glacial time at ~18 Ka based on a general circulation model (GCM). A, Summer season pattern; B, winter season pattern. Location of Bermuda shown by square/dot symbol. Also shown are the Laurentide (L), and Fennoscandian (F) ice sheets. Modified from Kutzbach and Guetter (1986).

glacial loess. Based on studies of contemporary eolian transport of silts from glacial outwash in Alaska, loess transport from continental glacial ice fronts in North America would be expected to have occurred mainly during summers, when outwash sedimentation was at a maximum (Flint, 1971). Farther south, however, along drainages distant from the North American continental ice fronts, it is possible that loess transport and deposition also took place during glacial winters. Given the latitude of Bermuda, we hypothesize that the most likely loess sources in North America would be in the lower Mississippi River Valley (Fig. 3) in the states of Arkansas, Mississippi, and Louisiana (Fig. 6).

QUATERNARY STRATIGRAPHY OF BERMUDA

Most of the surficial geology of Bermuda is dominated by Quaternary eolianites with some occurrences of marine deposits that grade into or are interstratified with the eolianites (Hearty et al., 1992; Vacher et al., 1989, this volume). The dunes, which coalesced to form irregularly defined, transverse dune ridges, were lobate-shaped sand bodies (Mackenzie, 1964a,b). Stratigraphic units on Bermuda are defined as packages of eolianite and beach/marine facies separated from older and younger sediment packages by well-developed soils (Vacher et al., 1989). These soils indicate periods of nondeposition of either eolianite or beach/marine deposits. From youngest to oldest, the major stratigraphic units (Vacher et al., this volume) are: (1) the Southampton Formation (~85 Ka); (2) the Rocky Bay Formation (~125 Ka); (3) the Belmont Formation (200 to 225 Ka); (4) the Town Hill Formation, which has both upper and lower members (upper, 325 to 350 Ka; lower, 430 to 475); and (5) the Walsingham Formation (>880 Ka). The ages of the Southampton, Rocky Bay, and Belmont Formations are derived from U-series dating of corals from the beach or marine facies of the units (Harmon et al., 1983); the age estimates of the older units are based on amino acid ratios of whole-rock eolianite samples and amino acid ratios of land snails from the eolianites, calibrated by the U-series–dated units (Hearty et al., 1992).

MATERIALS AND METHODS

Geochemical analyses were conducted on soil samples collected from the interiors of subaerially exposed solution pipes and on samples of the surrounding host eolianite or marine facies at four localities on Bermuda (Fig. 1). The stratigraphic units represented were solution pipes in the Southampton, Belmont, and upper Town Hill Formations (Figs. 7 and 8). Because soil horizons are not well defined within the solution pipes and there is no significant change in soil texture as a function of depth, the soil samples collected at different depths (>3 cm) from each solution pipe were combined to form composite samples. At least two solution pipes were sampled at each of the four localities.

The soil and host carbonate samples were passed through a 2-mm sieve, and then ground to <200 mesh. The concentrations of Al, Ti, Zr, Y, Nb, and Th were of particular interest because these are elements that have high (>3) ionic potentials and are not normally mobilized during chemical weathering and soil formation. Major element concentrations were determined by wavelength-dispersive x-ray fluorescence (Taggart et al., 1987), and all trace elements except Th were measured by energy-dispersive x-ray fluorescence (Johnson and King, 1987). Concentrations of Th were determined by sealed-can gamma-ray spectrometry (Bunker and Bush, 1966).

For Saharan dust, we used Al, Ti, Zr, Y, and Th data reported by Glaccum (1978) and Rydell and Prospero (1972), based on airborne dust samples collected from Sal Island (Cape Verde Islands), Barbados, and Miami, Florida. Saharan dust reaching the western Atlantic Ocean is dominated by quartz and mica, with lesser amounts of kaolinite, chlorite, calcite, and feldspars (Glaccum and Prospero, 1980). For lower Mississippi River Valley loess, we used geochemical data for the relatively unweathered portions of loess (i.e., soil C horizons) of various ages in Mississippi and Louisiana reported by Miller et al. (1986) and Pye and Johnson (1988). The loess samples (Fig. 6) were from units of several ages (Peoria loess, Crowley's Ridge loess, Sicily Island loess, and units 1 through 4 of Pye and Johnson [1988]). These loess units range in age from the late

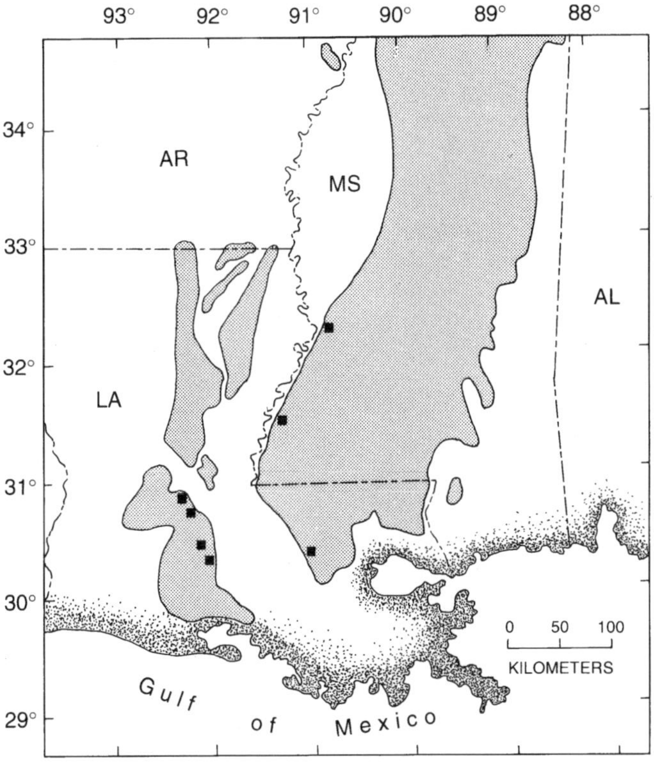

Figure 6. Distribution of loess (shown by shading) in the lower Mississippi River Valley and sample localities (solid squares) of Miller et al. (1986) and Pye and Johnson (1988) used in this study. Distribution of loess based on Thorp and Smith (1952) for Arkansas and Mississippi, and Miller et al. (1986) for Louisiana.

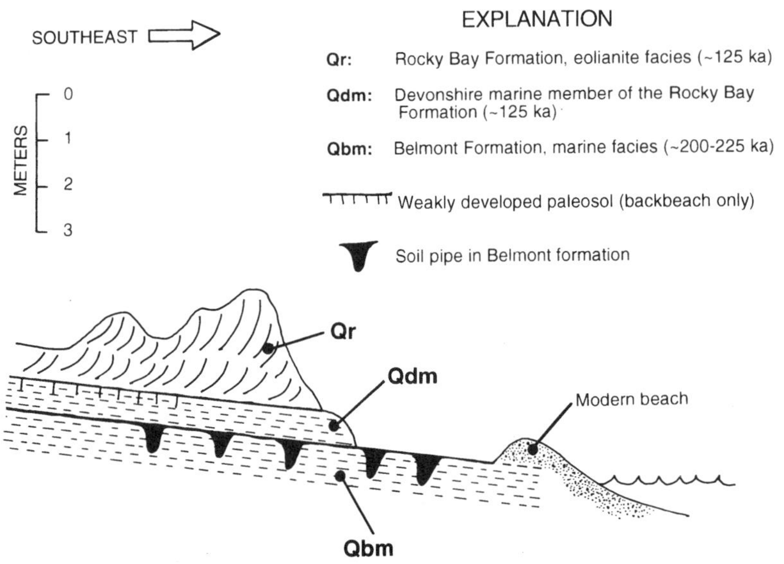

Figure 7. Coastal exposure in the vicinity of Grape Bay, Bermuda, showing relation of soil pipes to eolianite and marine units. Vertical scale is approximate.

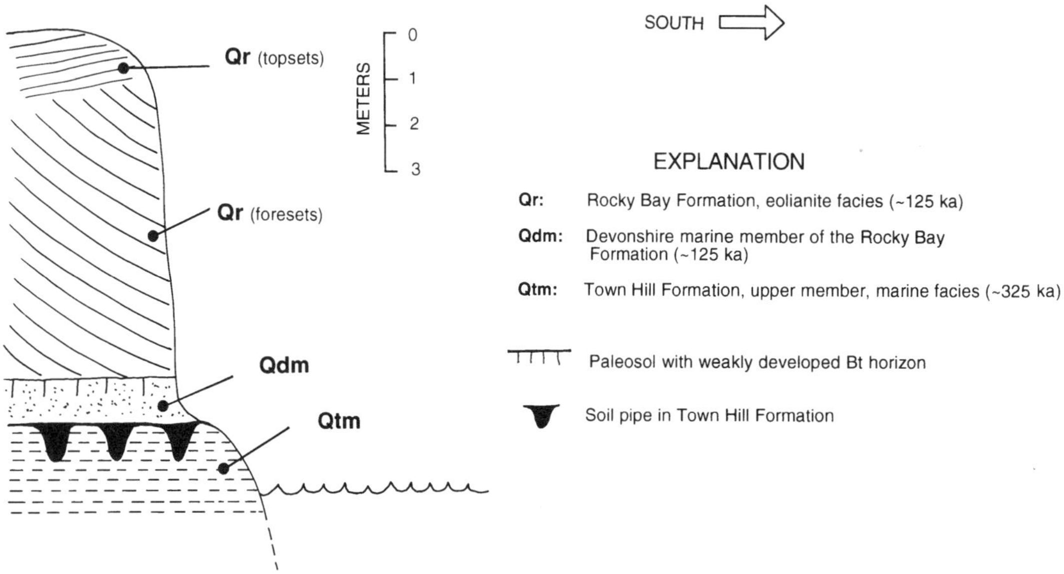

Figure 8. Exposure on the north side of Whalebone Bay, Bermuda, showing relation of soil pipes to eolianite and marine units.

Wisconsin to at least the mid-Pleistocene, based on radiocarbon dating, thermoluminescence dating, and aminostratigraphy (Miller et al., 1986; Pye and Johnson, 1988; Clark et al., 1989). Bulk lower Mississippi River Valley loess has mostly quartz, dolomite, and feldspars; clay fractions of the loess are dominated by smectite with lesser amounts of kaolinite and mica (Ruhe, 1984; Miller et al., 1986; Pye and Johnson, 1988).

It is important to note that the lower Mississippi River Valley loess data are based on analyses of bulk loess samples, which includes all particle sizes found in the deposit. Certain element ratios in the <10 μ fraction may differ from the same ratios in bulk loess; for example, Al_2O_3/TiO_3 may be higher in the <10 μ fraction, although Ti/Y and Ti/Zr may remain relatively constant. Miller et al. (1986) reported that Ti/Zr values in the clay-rich B horizons of Mississippi River Valley loess are not significantly different from the Ti/Zr values in the clay-poor C horizons.

RESULTS AND DISCUSSION

Geochemistry of host carbonate rocks

The major element analyses indicate that the eolianites and marine facies of Bermuda are extremely pure carbonates (Table 1), which is in agreement with the observations of

Bricker and Mackenzie (1970). The concentrations of SiO_2 and Al_2O_3, the major components of clays, are <0.5% in all carbonate samples we analyzed. Many major elements are not present in detectable amounts. Assuming that all of the measured CaO was present in $CaCO_3$, and that the loss on ignition (LOI) is due to CO_2 loss, then CaO plus LOI is a reasonable approximation of carbonate content. These calculations show that Bermuda carbonates are 97 to 100% $CaCO_3$. The presence of heavy mineral grain impurities has been noted in the area of Castle Harbour (Fig. 1); it was attributed to localized subaerial exposure of the submerged volcanic platform early in Bermuda's geologic history but never again reexposed (Land et al., 1967). Volcanic detritus is most evident in the marine facies of the upper member of the Town Hill Formation at Whalebone Bay; however, Bricker and Mackenzie (1970) rejected the hypothesis that the quartz and aluminosilicate clays of Bermuda soils are derived from the basaltic platform. We observed minute amounts of black volcanic grains in hand specimens of eolianites at other localities, but these impurities constitute <1% of the samples we examined.

If Bermuda soils were derived solely from the host carbonates and their limited noncarbonate fraction, we would expect to see soil morphology dominated by secondary carbonates such as laminar, platy, or nodular calcretes (Esteban and Klappa, 1983). In the source of our field work, we saw little or no calcrete development in soils exposed in natural outcrops and in road cuts. Variable concentrations of carbonate rhizoliths were found in the soil-filled interiors of most of the solution pipes; however, there was no evidence of a carbonate-derived soil profile. The reddish, clay-rich soils, whether horizontal or in solution pipes, usually overlie or cut directly across primary sedimentary structures such as topset and foreset beds with little intervening secondary carbonate. Our findings, therefore, are in agreement with the view that the Bermuda soils originate from an external, eolian parent material (Bricker and Mackenzie, 1970). Although dust from both North America and the Sahara reaches Bermuda today (Chen and Duce, 1983), we did not observe a modern "blanket" of dust covering the Bermudan landscape. Whatever dust has recently fallen onto Bermuda has been fully incorporated into the surface soils, similar to what Muhs et al. (1990) reported for dust-derived soils on Caribbean islands.

Immobile element ratios: Bermuda soils and hypothesized parent materials

Employing the geochemical fingerprinting approach, Al_2O_3/TiO_2, Ti/Y, Ti/Th, and Ti/Zr values were determined for the soils in Bermuda solution pipes and the two hypothesized parent materials. Not all of these immobile element ratios can be used to discriminate between the two parent materials. For example, the mean Al_2O_3/TiO_2 value for Saharan dust collected from Sal Island, Barbados, and Miami, Florida is 16 ± 2 (mean ± 1 standard deviation; n = 30), which is not significantly different from Louisiana and Mississippi loess, which is 13 ± 2 (n = 18). Similarly, the mean Ti/Y value for Saharan dust is 159 ± 12, which is not significantly different from the loess value of 137 ± 28. Thorium data are not yet available for loess, so it is not possible to use Ti/Th values for comparisons with Saharan dust and the Bermuda soils. However, Ti/Zr for Saharan dust is 37 ± 6, whereas loess is 9 ± 2. Despite the fact that only one immobile element ratio clearly distinguishes the two hypothesized parent materials, useful comparisons with the Bermuda soils can be made using the other ratios.

Two soil samples (W3 and W4) have anomalously high Ti, Zr, and Nb concentrations (Tables 2 and 3). Repeat analyses were performed using a different x-ray fluorescence spectrometer, and the same values were obtained within analytical uncertainty. Because these two samples are from the Whalebone Bay

TABLE 1. MAJOR ELEMENT CONCENTRATIONS (WT. %) IN BERMUDA CARBONATE ROCKS

Strat. Unit*	Age† (Ka)	Locality§	Sample No.	SiO_2	Al_2O_3	Fe_2O_3	MgO	CaO	Na_2O	TiO_2	P_2O_5	LOI**	$CaCo_3$‡
Qs	85	CB	C14	0.12	....	....	2.33	53.0	....	....	0.07	44.9	97.9
Qs	85	CB	C15	....	....	....	2.69	52.1	....	....	0.06	44.9	97.0
Qs	85	CB	C16	0.12	0.11	....	2.84	52.3	....	....	0.06	44.8	97.1
Qb	200	DB	D7	0.18	0.20	....	0.85	55.7	....	....	0.08	44.3	100.0
Qb	200	DB	D8	....	0.12	....	0.89	55.6	....	....	0.06	44.3	99.9
Qb	200	GB	G8	0.12	0.12	....	0.73	53.8	....	....	0.07	45.6	99.4
Qb	200	GB	G9	0.17	0.11	....	1.87	53.4	....	....	0.06	44.7	98.1
Qtu	325	WB	W7	0.20	0.21	0.08	1.03	55.5	....	....	0.12	44.0	99.5
Qtu	325	WB	W8	0.10	0.10	....	2.61	52.4	0.19	0.04	0.05	44.7	97.1

K_2O and MnO not detectable in any samples.
*From Vacher et al., 1989; Qs = Southampton; Qb = Belmont; Qtu = upper Town Hill Formations.
†Estimated ages from Harmon et al., 1983; Vacher and Hearty, 1989; and Hearty et al., 1992.
§Locality abbreviations: CB = Church Bay; DB = Devonshire Bay; GB = Grape Bay; WB = Whalebone Bay.
**LOI = Loss on ignition at 900°C.
‡$CaCO_3$ content estimated from sum of CaO and LOI.

area, the anomalous Ti, Zr, and Nb concentrations may be due to the volcanic heavy mineral grains present locally in the marine facies of the host upper Town Hill Formation. We collected modern Whalebone Bay beach sand that contains high concentrations of heavy mineral grains which appear to be derived from shoreline erosion of the Town Hill marine facies. These beach sediments, even without complete carbonate removal, have Ti contents of ~29%, Zr contents of >800 ppm, and Nb contents of >4,000 ppm. It is interesting to note, however, that the concentrations of Ti, Zr, and Nb in the Whalebone Bay beach sediments and soils are all significantly higher than the concentrations of these elements reported for most basic vol-

TABLE 2. MAJOR ELEMENT CONCENTRATIONS (WT. %) IN BERMUDA SOLUTION PIPE SOILS

Host Unit*	Locality†	Sample No.	Max. Age§ (Ka)	SiO_2	Al_2O_3	Fe_2O_3	MgO	CaO	Na_2O	K_2O	TiO_2	P_2O_5	MnO	LOI**
Qs	CB	C2	85	7.52	10.2	4.10	1.67	36.2	0.33	0.21	0.29	1.20	0.04	37.3
Qs	CB	C3	85	7.61	9.93	4.00	1.76	36.4	0.42	0.12	0.31	1.12	0.04	37.3
Qs	CB	C4	85	22.8	23.2	10.0	2.85	6.81	1.43	0.48	1.20	3.41	0.09	25.1
Qs	CB	C5	85	22.0	24.3	10.0	2.92	7.58	1.09	0.55	1.20	2.12	0.09	27.2
Qs	CB	C7	85	17.1	19.5	7.91	2.41	17.1	1.25	0.49	0.80	2.20	0.09	29.4
Qb	DB	D1	200	16.4	21.1	6.97	2.81	19.3	0.75	0.28	0.83	0.51	....	29.8
Qb	DB	D2	200	16.7	22.3	7.20	2.86	18.0	0.80	0.30	0.86	0.52	....	29.4
Qb	DB	D3	200	17.1	14.6	6.29	2.27	25.3	0.74	0.53	0.76	0.42	0.10	31.4
Qb	DB	D5	200	18.0	22.5	8.38	2.61	15.0	0.97	0.35	1.06	0.61	0.02	28.4
Qb	GB	G6	200	4.32	3.49	1.49	2.81	45.2	0.27	0.05	0.16	0.17	....	42.2
Qb	GB	G7	200	6.30	5.27	2.26	2.25	42.3	0.33	0.06	0.24	0.23	....	40.7
Qtu	WB	W1	325	8.39	14.3	7.21	1.81	31.9	0.27	0.32	1.34	0.66	0.04	33.4
Qtu	WB	W2	325	9.25	15.2	7.42	1.69	30.6	0.36	0.37	0.86	0.56	....	33.3
Qtu	WB	W3	325	5.41	11.0	7.28	1.44	33.8	0.39	0.11	7.97	0.64	0.05	30.7
Qtu	WB	W4	325	8.30	13.0	9.22	3.03	26.2	1.07	0.09	7.34	0.37	0.14	29.2

*From Vacher et al., 1989; Qs = Southampton; Qb = Belmont; Qtu = upper Town Hill Formations.
†Locality abbreviations: CB = Church Bay; DB = Devonshire Bay; GB = Grape Bay; WB = Whalebone Bay.
§Maximum age estimates based on Harmon et al., 1983; Vacher and Hearty, 1989; and Hearty et al., 1992.
**LOI = Loss on ignition at 900°C.

TABLE 3. TRACE ELEMENT CONCENTRATIONS (PPM) IN BERMUDA SOLUTION PIPE SOILS

Host Unit*	Locality†	Sample No.	Max. Age§ (Ka)	Rb	Sr	Y	Zr	Nb	Th
Qs	CB	C2	85	94	5160	478	64	94	10
Qs	CB	C3	85	69	4940	468	44	...	7
Qs	CB	C4	85	51	6720	476	140	...	20
Qs	CB	C5	85	10	4410	515	122	...	24
Qs	CB	C7	85	51	5200	691	89	...	20
Qb	DB	D1	200	7	1000	336	147	...	19
Qb	DB	D2	200	4	919	368	145	...	19
Qb	DB	D3	200	19	881	428	131	...	13
Qb	DB	D5	200	4	1180	594	163	...	...
Qb	GB	G6	200	8	1160	104	28	...	4
Qb	GB	G7	200	9	1060	231	34	...	6
Qtu	WB	W1	325	19	1160	226	135	59	20
Qtu	WB	W2	325	29	877	322	111	22	16
Qtu	WB	W3	325	4	787	669	269	591	74
Qtu	WB	W4	325	2	855	108	349	509	...

*From Vacher et al., 1989; Qs = Southampton; Qb = Belmont; Qtu = upper Town Hill Formations.
†Locality abbreviations: CB = Church Bay; DB = Devonshire Bay; GB = Grape Bay; WB = Whalebone Bay.
§Maximum age estimates based on Harmon et al., 1983; Vacher and Hearty, 1989; and Hearty et al., 1992.

canic rocks (Pearce and Cann, 1973). Because of the anomalously high concentrations of Ti, Zr, and Nb in soil samples W3 and W4 due to local contamination by heavy mineral grains, we have not included them in our overall calculations of immobile element ratios.

Comparison of immobile element ratios in the Bermuda soils with the hypothesized parent materials indicates that the soils probably have been influenced by more than one parent material, and probably include source areas not yet identified. The Ti/Zr value, which distinguishes Saharan dust from Mississippi River Valley loess, shows considerable scatter for the Bermuda soils (Fig. 9), far greater than that observed by Muhs et al. (1990) for soils on Caribbean islands and other western Atlantic islands at lower latitudes. The Ti/Zr values for the soils in solution pipes in the Belmont Formation (~200 Ka) plot closely within range of Ti/Zr values for Saharan dust. Most of the Ti/Zr values for soils from solution pipes in the Southampton (~80 Ka) and upper Town Hill (~325 Ka) Formations plot above the Saharan range (Fig. 9). Because of these relatively high Ti/Zr values and because Mississippi River Valley loess has lower Ti/Zr values than Saharan dust, it does not appear that Bermuda soils are a simple mix of Saharan dust and Mississippi River Valley loess.

A complex parent material assemblage for Bermuda soils is also indicated by the Al_2O_3/TiO_2 values. The Al_2O_3/TiO_2 values of soils from solution pipes in the Southampton and Belmont Formations plot above the ranges for both Mississippi River Valley loess and Saharan dust, strongly suggesting a different parent material (Fig. 10). In the case of the Ti/Y values, soils from all three formations plot well below the range of values for loess and Saharan dust, again implying at least a third parent material having a distinct composition (Fig. 11). A few soils, one from each formation, have Ti/Th values that plot within the range of Saharan dust, but most plot below this range (Fig. 12).

Collectively, the four immobile element ratios suggest the following interpretations: (1) the soils of Bermuda are not derived exclusively from either lower Mississippi River Valley loess or Saharan dust; (2) the soils are not derived from simple mixing of the two hypothesized parent materials; and (3) at

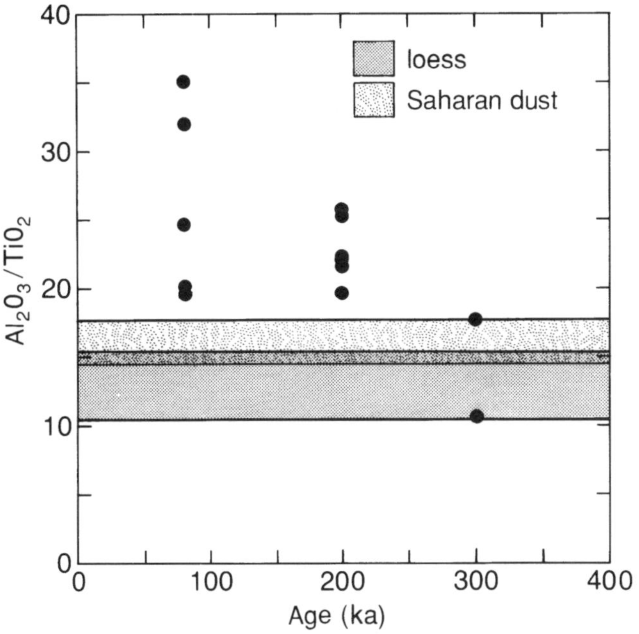

Figure 10. Al_2O_3/TiO_2 values of Bermuda soils from solution pipes formed in carbonates of different ages compared to the range (mean ± 1 standard deviation) of Al_2O_3/TiO_2 values characterizing Saharan dust and Mississippi River Valley loess.

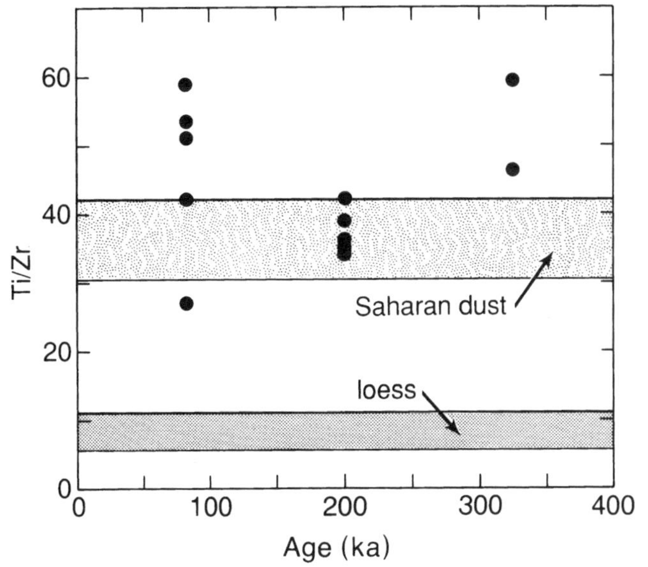

Figure 9. Ti/Zr values of Bermuda soils from solution pipes formed in carbonates of different ages compared to the range (mean ± 1 standard deviation) of Ti/Zr values characterizing Saharan dust and Mississippi River Valley loess.

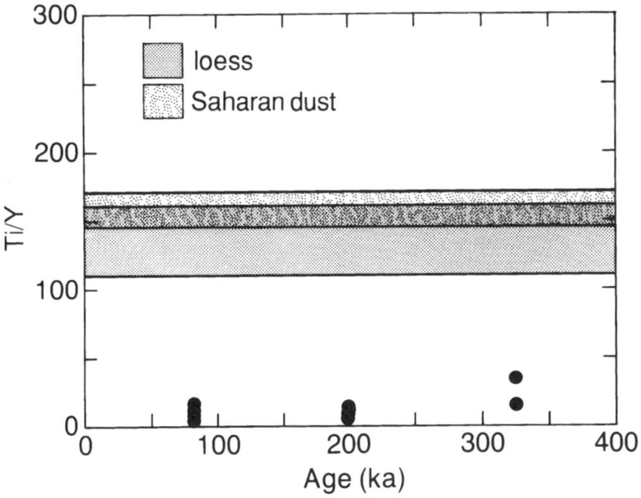

Figure 11. Ti/Y values of Bermuda soils from solution pipes formed in carbonates of different ages compared to the range (mean ± 1 standard deviation) of Ti/Y values characterizing Saharan dust and Mississippi River Valley loess.

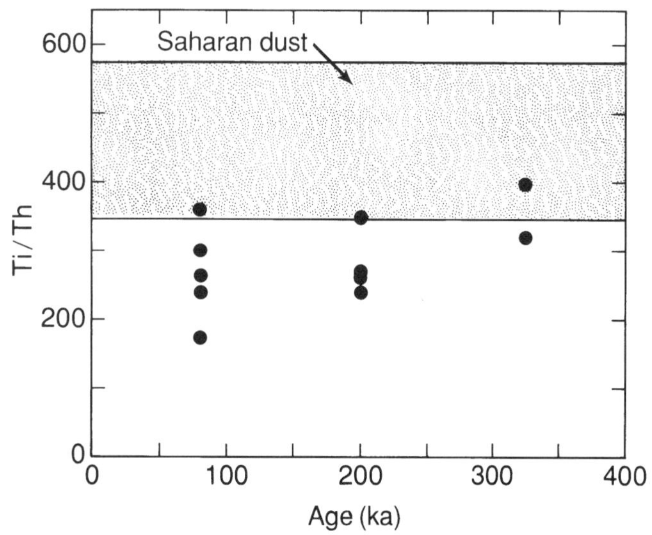

Figure 12. Ti/Th values of Bermuda soils from solution pipes formed in carbonates of different ages compared to the range (mean ± 1 standard deviation) of Ti/Th values characterizing Saharan dust.

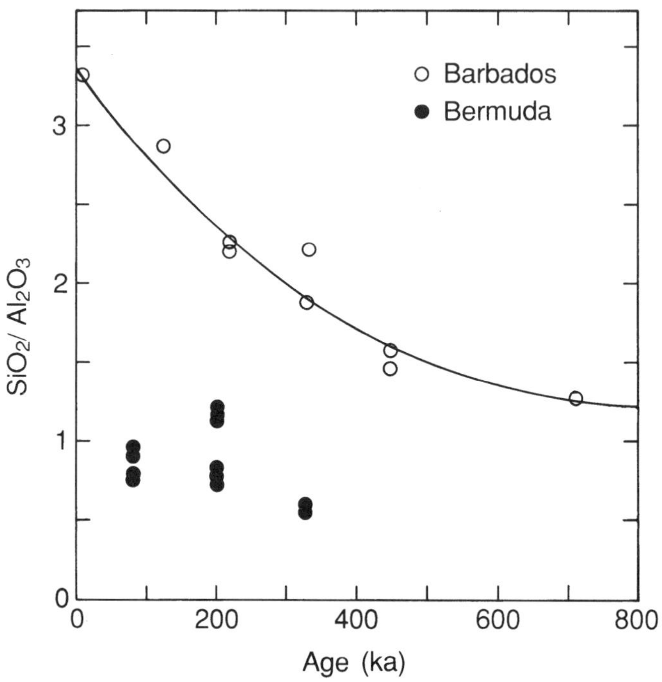

Figure 13. SiO_2/Al_2O_3 values of dust-derived soils in Barbados plotted as a function of terrace age (Muhs et al., 1990) compared to SiO_2/Al_2O_3 values of Bermuda soils from solution pipes formed in carbonates of different ages.

least one other source area must be contributing to the soils and may in fact be the most important contributor.

Degree of chemical weathering of soils in Bermuda solution pipes

A comparison of the degree of weathering of the Bermuda solution pipe soils with soils on Barbados reef terraces reinforces the interpretation that Saharan dust is not the only parent material of Bermuda soils. A useful measure of chemical weathering in soils is the SiO_2/Al_2O_3 value. Because Si is lost through mineral alteration by hydrolysis, while Al generally is retained, the SiO_2/Al_2O_3 value usually decreases as weathering proceeds. Muhs et al. (1990) found that soils (using weighted profile averages of all horizons) derived primarily from Saharan dust on uplifted reef terraces of Barbados showed a systematic decrease in their SiO_2/Al_2O_3 values as a function of terrace age (Fig. 13). This trend is explained by the fact that the clay mineralogy of the younger Barbados soils is dominated by mixed-layer kaolinite-smectite, whereas soils on the oldest terraces have decreasing amounts of smectite and the oldest soil has only kaolinite. On Eleuthera Island in the Bahamas, Foos (1991) reported SiO_2/Al_2O_3 values of 0.8 for soils developed from Saharan dust on carbonate eolianites. She interpreted these relatively low values to mean that soils on Eleuthera Island have experienced considerable chemical weathering. This interpretation is supported by the soil mineral assemblage, which consists of boehmite, hydroxy-interlayered clay, hematite, and goethite.

The SiO_2/Al_2O_3 values in the Bermuda pipe soils range from 0.59 to 1.24, with a mean value of 0.88 (Fig. 13). All of these values, when plotted as a function of carbonate age, fall below the Barbados time-trend line and are much closer to the values reported for Eleuthera soils by Foos (1991) (Fig. 13). In their study of samples of horizontal soils and soils from the interiors of solution pipes, Ruhe et al. (1961) reported that the soils of Bermuda are characterized by amorphous material, kaolinite, vermiculite, and gibbsite (in some cases), but no mica or smectite. This mineralogy explains the low SiO_2/Al_2O_3 values, suggesting either that Bermuda soils have experienced a significantly greater degree of chemical weathering than Barbados soils of comparable age or that the parent material arrived in a "preweathered" condition (i.e., arrived on Bermuda with an initially low SiO_2/Al_2O_3 value). If Bermuda soils have experienced a greater degree of chemical weathering than Barbados soils of comparable age and the two islands had the same soil parent material, a difference in climate is a possible explanation for the different SiO_2/Al_2O_3 values. Localities with higher temperatures and a higher ratio of precipitation (P) to potential evaporation (PE) should experience more chemical weathering (Birkeland, 1984).

On Barbados, both P and PE vary as a function of elevation; soils on the higher terraces receive greater mean annual P totals and the rates of PE are lower than on the lower terraces (Rouse, 1966). For example, on the lower 125- and 190-Ka terraces of Barbados, P is 1,295 to 1,448 mm yr^1 and PE is 1,676 to 1,763 mm yr^1; whereas on the highest terrace (estimated to be about 700 Ka by Muhs et al., 1990), P is 1,981 to 2,007 mm yr^1 and PE is 1,168 to 1,245 mm yr^1 (Rouse, 1966). The P/PE values on Barbados are 0.74 to 0.86 for the 125-Ka and 190-Ka terraces, 0.89 to 1.03 for the 220-Ka and 320-Ka terraces, 0.95 to 1.10 for the 460-Ka terrace, and 1.59 to 1.72 for the 700-Ka terrace.

On Bermuda, P is 1,463 mm yr[1] and PE is about 1,366 mm yr[1] (Macky, 1957). The P/PE value of 1.07 for Bermuda is significantly higher than the P/PE values for the 125-Ka and 190-Ka terraces on Barbados, but not significantly different from the P/PE values for the older (>220 Ka) terraces. Thus, the amount of moisture available for chemical weathering might explain some of the differences in SiO_2/Al_2O_3 values for the lower terrace soils on Barbados and Bermuda soils of comparable age. The amount of available moisture, however, cannot explain the differences in SiO_2/Al_2O_3 values for the older soils. Assuming that the present climatic differences between Barbados and Bermuda are representative of past climatic gradients, mean annual temperature differences as well as the P/PE values argue against a more rapid rate of chemical weathering on Bermuda compared to Barbados. Present mean annual air temperature at Bridgetown, Barbados is 26°C, while on Bermuda it is 21°C. Due to the higher temperatures, slightly higher (not lower) chemical weathering rates would be expected on Barbados.

One could argue that on Bermuda, the rate of dust influx might be lower than on Barbados, and therefore the rate of chemical weathering exceeds the rate of parent material accumulation. However, Bricker and Prospero (1969) reported that during a comparable time period of measurement, dust flux on Bermuda was a factor of six greater than that on Barbados. We conclude that neither differences in rates of chemical weathering nor rates of dust fall can explain the differences between the SiO_2/Al_2O_3 values of Bermuda and Barbados soils. The differences in the SiO_2/Al_2O_3 values, therefore, appear to be related to differences in the initial composition of the parent materials.

A Great Plains dust source for Bermuda soils?

Examination of satellite imagery suggests a third possible source area for an eolian soil parent material on Bermuda. During February 1977, a major dust storm originating in eastern Colorado and the Texas–New Mexico border area of the Texas panhandle passed over the southern United States over a 4-day period (McCauley et al., 1981). Satellite imagery from this period shows that dust reached Georgia, South Carolina, North Carolina, and much of the western Atlantic Ocean, possibly including Bermuda. On a coastal pier near Savannah, Georgia, Windom and Chamberlain (1978) collected dust-fall from this storm and found that the material was about 90% clay-sized particles, with a mineralogy dominated by mica, but also included quartz, feldspars, chlorite, and kaolinite. These observations suggest that sources in the central Great Plains and southern High Plains of the United States (Holliday, 1991) could provide dust to Bermuda. The next step in our ongoing study of soil genesis on Bermuda will involve geochemical analyses of possible dust source areas in the Great Plains.

CONCLUSIONS

Based on our initial studies of soil materials in the carbonate solution pipes of Bermuda, we conclude the following:

1. Bermuda eolianites and their associated marine facies are extremely pure carbonates based on major element analyses and, therefore, are not the parent material of the soils. We agree with Bricker and Mackenzie (1970) that eolian dust derived from outside the island is the soil parent material.

2. Compared to soils on carbonates of comparable age in lower latitudes in the Caribbean and western Atlantic, the immobile element composition of soils in Bermuda solution pipes is more variable. The results of these analyses suggest that more than one parent material may have contributed to Bermuda soils.

3. Most values of Al_2O_3/TiO_2, Ti/Zr, Ti/Y, and Ti/Th in Bermuda soils are significantly different from the values that characterize Saharan dust and lower Mississippi River Valley loess. An important finding is that the values for the Bermuda soils are not intermediate between the range of values for the two hypothesized parent materials. We conclude, therefore, that Bermuda soils are not a simple mixture of the two possible sources, and that at least one additional source must be contributing eolian dust to Bermuda.

4. SiO_2/Al_2O_3 values in Bermuda soils are significantly lower than SiO_2/Al_2O_3 values in Barbados soils that are of comparable age and are thought to be derived from Saharan dust. Climate differences cannot explain the differences in SiO_2/Al_2O_3 values between the two islands. This interpretation reinforces the conclusion, based on the immobile element ratios, that Saharan dust is not the only parent material of the soils in Bermuda solution pipes. Other possible North American source areas outside of the lower Mississippi River Valley, such as the central and southern Great Plains, must be examined in more detail.

ACKNOWLEDGMENTS

This study was supported in part by Clark University and in part by the Global Change and Climate History Program of the U.S. Geological Survey. We thank D. Lines, R. Heminway, S. Kerns, D. Lajoie, I. Phillips, and S. Kaplan for assistance with the field sampling. J. Taggart, A. Bartel, C. Bush, D. Siems, J. Evans, P. Maat, and S. Mahan generated the chemical data. We appreciate helpful discussions on the geology of Bermuda with L. Vacher and M. Rowe, and on paleoclimatology with J. Kutzbach. L. Vacher and H. Markewich read an earlier version of this work and made many helpful suggestions for its improvement.

REFERENCES CITED

Birkeland, P. W., 1984, Soils and Geomorphology: New York, Oxford University Press, 372 p.

Blackburn, G., and Taylor, R. M., 1969, Limestones and red soils of Bermuda:

Geological Society of America Bulletin, v. 80, p. 1595–1597.

Blackburn, G., and Taylor, R. M., 1970, Limestones and red soils of Bermuda; Reply: Geological Society of America Bulletin, v. 81, p. 2525–2526.

Blackburn, G., Bond, R. D., and Clarke, A. R. P., 1965, Soil development associated with stranded beach ridges in south-east south Australia: CSIRO Australian Soil Publication, v. 22, p. 1–65.

Bowles, F. A., 1975, Paleoclimatic significance of quartz/illite variations in cores from eastern equatorial North Atlantic: Quaternary Research, v. 5, p. 225–235.

Bretz, J. H., 1960, Bermuda; A partially drowned, late mature, Pleistocene karst: Geological Society of America Bulletin, v. 71, p. 1729–1754.

Bricker, O., and Mackenzie, F. T., 1970, Limestones and red soils of Bermuda; Discussion: Geological Society of America Bulletin, v. 81, p. 2523–2524.

Bricker, O., and Prospero, J. M., 1969, Airborne dust on the Bermuda Islands and Barbados [abs.]: Eos (Transactions, American Geophysical Union), v. 50, p. 176.

Bunker, C. M., and Bush, C. A., 1966, Uranium, thorium, and radium analyses by gamma-ray spectrometry (0.184–0.352 million electron volts), *in* Geological Survey Research 1966: Geological Survey Professional Paper 575-B, p. B164–B181.

Chen, L., and Duce, R. A., 1983, The sources of sulfate, vanadium and mineral matter in aerosol particles over Bermuda: Atmospheric Environment, v. 17, p. 2055–2064.

Clark, P. U., Nelson, A. R., McCoy, W. D., Miller, B. B., and Barnes, D. K., 1989, Quaternary aminostratigraphy of Mississippi Valley loess: Geological Society of America Bulletin, v. 101, p. 918–926.

Coetzee, F., 1975, Solution pipes in coastal aeolianites of Zululand and Moçambique: Transactions of the Geological Society of South Africa, v. 78, p. 323–333.

Day, A. E., 1928, Pipes in the coast sandstone of Syria: Geological Magazine, v. 65, p. 412–415.

Esteban, M., and Klappa, C. F., 1983, Subaerial exposure, *in* Scholle, P. A., Bebout, D. G., and Moore, C. D., eds., Carbonate depositional environments: American Association of Petroleum Geologists Memoir 33, p. 1–54.

Fairbridge, R. W., 1950, The geology and geomorphology of Point Peron, Western Australia: Journal of the Royal Society of Western Australia, v. 34, p. 35–72.

Flint, R. F., 1971, Glacial and Quaternary Geology: New York, John Wiley, 892 p.

Foos, A. M., 1991, Aluminous lateritic soils, Eleuthera, Bahamas: A modern analog to carbonate paleosols: Journal of Sedimentary Petrology, v. 61, p. 340–348.

Galloway, J. N., Artz, R. S., and Pueschel, R. F., 1988, WATOX-85; An aircraft and ground sampling program to determine the transport of material between the United States and Bermuda: Atmospheric Environment, v. 22, p. 2345–2360.

Gardener, R. A. M., 1983, Aeolianite, *in* Goudie, A. S., and Pye, K., eds., Chemical Sediments and Geomorphology: London, United Kingdom, Academic Press, p. 265–300.

Glaccum, R. A., 1978, The mineralogy and elemental composition of mineral aerosols over the tropical North Atlantic; The influence of Saharan dust [M.S. thesis]: Miami, Florida, University of Miami, 161 p.

Glaccum, R. A., and Prospero, J. M., 1980, Saharan aerosols over the tropical North Atlantic—Mineralogy: Marine Geology, v. 37, p. 295–321.

Grousset, F., Buat-Menard, P., Boust, D., Ru-Cheng, T., Baudel, S., Pujol, C., and Vergnaud-Grazzini, C., 1989, Temporal changes of aeolian Saharan input in the Cape Verde abyssal plain since the last glacial period: Oceanologica Acta, v. 12, p. 177–185.

Harmon, R. S., and 8 others, 1983, Uranium-series and amino-acid racemization geochronology of Bermuda; Implications for eustatic sea-level fluctuations over the past 250,000 years: Palaeogeography, Palaeoclimatology, Palaeoecology, v. 44, p. 41–70.

Hearty, P. J., Vacher, H. L., and Mitterer, R. M., 1992, Aminostratigraphy and ages of Pleistocene limestones of Bermuda: Geological Society of America Bulletin, v. 104, p. 471–480.

Herwitz, S. R., 1993, Stemflow influences on the formation of solution pipes in Bermuda eolianite: Geomorphology, v. 6, p. 253–271.

Holliday, V. T., 1991, The geologic record of wind erosion, eolian deposition, and aridity on the southern High Plains: Great Plains Research, v. 1, p. 6–25.

Jickells, T. D., Knap, A. H., Church, T. M., Galloway, J. N., and Miller, J. M., 1982, Acid rain on Bermuda: Nature, v. 297, p. 55–57.

Johnson, R. G., and King, B., 1987, Energy-dispersive x-ray fluorescence spectrometry, *in* Baedecker, P. A., ed., Methods for Geochemical Analysis: U.S. Geological Survey Bulletin 1770, p. F1–F5.

Keffer, T., Martinson, D. G., and Corliss, B. H., 1988, The position of the Gulf Stream during Quaternary glaciations: Science, v. 241, p. 440–442.

Kolla, V., Biscaye, P. E., and Hanley, A. F., 1979, Distribution of quartz in late Quaternary Atlantic sediments in relation to climate: Quaternary Research, v. 11, p. 261–277.

Kutzbach, J. E., 1987, Model simulations of the climatic patterns during the deglaciation of North America, *in* Ruddiman, W. F., and Wright, H. E., Jr., eds., North America and adjacent oceans during the last deglaciation; The Geology of North America: Boulder, Colorado, Geological Society of America, v. K-3, p. 425–446.

Kutzbach, J. E., and Guetter, P. J., 1986, The influence of changing orbital parameters and surface boundary conditions on climate simulations for the past 18,000 years: Journal of the Atmospheric Sciences, v. 43, p. 1726–1759.

Land, L. S., and Mackenzie, F. T., 1969, Field Guide to Bermuda Geology: 1969, Penrose Conference, St. George's Island, Bermuda, Bermuda Biological Station, 36 p.

Land, L. S., Mackenzie, F. T., and Gould, S. J., 1967, Pleistocene history of Bermuda: Geological Society of America Bulletin, v. 78, p. 993–1006.

Livingston, W., 1944, Observations on the structure of Bermuda: Geographical Journal, v. 104, p. 40–48.

Mackenzie, F. T., 1964a, Geometry of Bermuda calcareous dune cross-bedding: Science, v. 144, p. 1449–1450.

Mackenzie, F. T., 1964b, Bermuda Pleistocene eolianites and paleowinds: Sedimentology, v. 3, p. 52–64.

Macky, W. A., 1957, The rainfall of Bermuda: Bermuda Meteorological Office Technical Note 8, 58 p.

McCauley, J. F., Breed, C. S., Grolier, M. J., and Mackinnon, D. J., 1981, The U.S. dust storm of February 1977: Geological Society of America Special Paper 186, p. 123–147.

McKee, E. D., and Ward, W. C., 1983, Eolian environment, *in* Scholle, P. A., Bebout, D. G., and Moore, C. D., eds., Carbonate depositional environments: American Association of Petroleum Geologists Memoir 33, p. 131–170.

Miller, B. J., Day, W. J., and Schumacher, B. A., 1986, Loesses and loess-derived soils in the lower Mississippi Valley, *in* Guidebook for soils-geomorphology tour: New Orleans, Louisiana, American Society of Agronomy, 144 p.

Muhs, D. R., Crittendon, R. C., Rosholt, J. N., Bush, C. A., and Stewart, K. C., 1987, Genesis of marine terrace soils, Barbados, West Indies; Evidence from mineralogy and geochemistry: Earth Surface Processes and Landforms, v. 12, p. 605–618.

Muhs, D. R., Bush, C. A., Stewart, K. C., Rowland, T. R., and Crittendon, R. C., 1990, Geochemical evidence for Saharan dust parent material for soils developed on Quaternary limestones of Caribbean and western Atlantic islands: Quaternary Research, v. 33, p. 157–177.

Pearce, J. A., and Cann, J. R., 1973, Tectonic setting of basic volcanic rocks determined using trace element analyses: Earth and Planetary Science Letters, v. 19, p. 290–300.

Plummer, L. M., Vacher, H. L., Mackenzie, F. T., Bricker, O. P., and Land, L. S., 1976, Hydrogeochemistry of Bermuda; A case history of ground-water diagenesis of biocalcarenites: Geological Society of America Bulletin, v. 87, p. 1301–1316.

Pokras, E. M., and Mix, A. C., 1985, Eolian evidence for spatial variability of late Quaternary climates in tropical Africa: Quaternary Research, v. 24,

p. 137–149.

Prospero, J. M., 1981, Arid regions as sources of mineral aerosols in the marine atmosphere: Geological Society of America Special Paper 186, p. 71–86.

Prospero, J. M., and Nees, R. T., 1977, Dust concentration in the atmosphere of the equatorial North Atlantic; Possible relationships to the Sahelian drought: Science, v. 196, p. 1196–1198.

Prospero, J. M., Glaccum, R. A., and Nees, R. T., 1981, Atmospheric transport of soil dust from Africa to South America: Nature, v. 289, p. 570–572.

Pye, K., and Johnson, R., 1988, Stratigraphy, geochemistry, and thermoluminescence ages of lower Mississippi Valley loess: Earth Surface Processes and Landforms, v. 13, p. 103–124.

Pye, K., and Tsoar, H., 1990, Aeolian Sand and Sand Dunes: London, United Kingdom, Unwin Hyman, 396 p.

Rouse, W. R., 1966, The moisture balance of Barbados and its influences on sugar cane yield: McGill University Climatological Research Series 1, p. 1–54.

Rowe, M. P., 1990, An explanation of the geology of Bermuda: Hamilton, Bermuda, Ministry of Works and Engineering, 28 p.

Ruhe, R. V., Cady, J. G., and Gomez, R. S., 1961, Paleosols of Bermuda: Geological Society of America Bulletin, v. 72, p. 1121–1142.

Rydell, H. S., and Prospero, J. M., 1972, Uranium and thorium concentrations in wind-borne Saharan dust over the western equatorial north Atlantic Ocean: Earth and Planetary Science Letters, v. 14, p. 397–402.

Savoie, D. L., and Prospero, J. M., 1977, Aerosol concentration statistics for the northern tropical Atlantic: Journal of Geophysical Research, v. 82, p. 5954–5964.

Sayles, R. W., 1931, Bermuda during the ice age: Proceedings of the American Academy of Arts and Sciences, v. 66, p. 381–467.

Taggart, J. E., Jr., Lindsay, J. R., Scott, B. A., Vivit, D. V., Bartel, A. J., and Stewart, K. C., 1987, Analysis of geologic materials by wavelength-dispersive x-ray fluorescence spectrometry, *in* Baedecker, P. A., ed., Methods for geochemical analysis: U.S. Geological Survey Bulletin 1770, p. E1–E9.

Talbot, R. W., Harriss, R. C., Browell, E. V., Gregory, G. L., Sebacher, D. I., and Beck, S. M., 1986, Distribution and geochemistry of aerosols in the tropical North Atlantic troposphere; Relationship to Saharan dust: Journal of Geophysical Research, v. 91, p. 35173–35182.

Vacher, H. L., and Harmon, R. S., 1987, Field Guide to Bermuda Geology: 1987, Penrose Conference, St. George's Island, Bermuda, Bermuda Biological Station, 49 p.

Vacher, H. L., and Hearty, P., 1989, History of stage 5 sea level in Bermuda; Review with new evidence of a brief rise to present sea level during substage 5a: Quaternary Science Reviews, v. 8, p. 159–168.

Vacher, H. L., Rowe, M., and Garrett, P., 1989, Geological Map of Bermuda: Hamilton, Bermuda, Public Works Department, scale 1:25,000.

Verrill, A. E., 1902, The Bermuda Islands; Their scenery, climate, productions, natural history, and geology: Transactions of the Connecticut Academy of Arts and Sciences, v. 11, p. 413–956.

Webb, T., III, Bartlein, P. J., and Kutzbach, J. E., 1987, Climatic change in eastern North America during the past 18,000 years; Comparisons of pollen data with model results, *in* Ruddiman, W. F., and Wright, H. E., Jr., eds., North America and adjacent oceans during the last deglaciation; The Geology of North America: Boulder, Colorado, Geological Society of America, v. K-3, p. 447–462.

Windom, H. L., and Chamberlain, C. F., 1978, Dust-storm transport of sediments to the north Atlantic Ocean: Journal of Sedimentary Petrology, v. 48, p. 385–388.

Manuscript Accepted by the Society January 5, 1995

Geological Society of America
Special Paper 300
1995

Fracture systems in northeastern Bermuda

John K. Hartsock*
Department of Geology, University of Maryland, College Park, Maryland 20740
Donald L. Woodrow and D. Brooks McKinney
Department of Geoscience, Hobart and William Smith Colleges, Geneva, New York 14456-3397

ABSTRACT

Geologic investigations in northeastern Bermuda have revealed the presence of unique systems of fractures in the Pleistocene limestones capping the volcanic basement. The fractures are particularly well exposed at the southwestern end of St. George's Island in zones of anastomosing, steeply dipping to vertical open fractures, with a dominant set striking generally northeastward and a subordinate set striking for the most part at right angles to the dominant set.

Control of coastal erosion by the dominant set of fractures is evident along the entire northwestern coast of St. George's Island, while erosion of the northeastern coast appears to be controlled by the subordinate set. These intersecting sets are believed to be responsible for the distinctive angularity, or corner-like shape, of the northeastern end of the island. The relationship of this "corner" to the quasi-circular water body of Castle Harbour, which is underlain by a buried volcanic caldera, together with a slight shift in strike of the dominant set of fractures from north-northeast to northeast, reflecting the curvature of the northwestern quarter of Castle Harbour, points toward a probable connection between the configuration of the caldera and the patterns of the fracture systems.

The fractures may have been formed by collapse of limestone strata into previously existing deep caverns at or near the limestone-basalt interface. The occurrence of periodic earthquakes in Bermuda would have aided this process. Other possible causes may be the occurrence of minor volcanic activity or minor fault displacement in the volcanic base during Pleistocene or Recent times.

INTRODUCTION

Field observations and photogeologic studies from 1984 to the present have revealed an unusual set of fracture systems in the carbonate bedrock of northeastern Bermuda. These fracture systems, which parallel the linear trends of the coastlines and which seem to bear a geometrical relationship to a buried volcanic caldera, have not been found to be prevalent in southwestern Bermuda. Comparison with fracture systems on other limestone islands in various parts of the world indicates an origin singularly unique to Bermuda's geologic history.

PRESENT INVESTIGATIONS

In 1984 and 1986, Woodrow and McKinney, with field parties of geology students from Hobart and William Smith Colleges, Geneva, New York, mapped details of the fracture pattern and prepared a statistical study of fracture orientation in and around Whalebone Bay, near the southwestern end of St. George's Island. In 1987, J. K. Hartsock measured the strikes of fractures in the Ferry Point–Whalebone Bay area on St. George's Island and on the point of land 1 km south of Ferry Point, near Coney Island. In 1990 and 1991 Hartsock made a

*Present address: 1002 Gadsden Avenue, Silver Spring, Maryland 20905

Hartsock, J. K., Woodrow, D. L., and McKinney, D. B., 1995, Fracture systems in northeastern Bermuda, *in* Curran, H. A., and White, B., Terrestrial and Shallow Marine Geology of the Bahamas and Bermuda: Boulder, Colorado, Geological Society of America Special Paper 300.

photogeologic study of the fractures in the same areas and extended the study northeastward 2 km along St. George's Island (Hartsock, 1991). This chapter presents a summation of the above investigations, as well as previous relevant investigations in this area, and an analysis of possible modes of origin of the fracture system.

GENERAL GEOLOGIC FRAMEWORK OF BERMUDA

The Bermuda group of islands occupies the southern rim of the Bermuda Pedestal, the largest of three volcanic seamounts on the Bermuda Rise, an upswelling of the deep ocean floor approximately 1,000 km east of Cape Hatteras (Fig. 1). The islands and islets, numbering over 120, make up a close-grouped archipelago some 24 km long from northeast to southwest and 2 to 5 km from northwest to southeast. They are composed of eolian limestone and shallow-water marine limestone, ranging from more than 200 m thick in the southwestern part of Bermuda to less than 100 m thick in the northeastern part (James and Schenk, 1983, p. 3). These limestones, together with a layer of carbonate sediments and coral reefs extending over 10 km northward beneath the shallow waters of North Lagoon, form a relatively thin veneer capping the volcanic base. The islands and their northward extension are often referred to as the Bermuda Platform (Fig. 2).

The limestones making up the islands are Pleistocene in age, deposited during higher stands of sea level. The volcanic rocks underlying the limestone range in age from middle Cretaceous beneath the western part of Bermuda to Oligocene beneath the eastern part. This Oligocene phase is characterized by sill intrusion into the original Cretaceous volcanic beds, resulting in the volcanic surface here being over 100 m higher than it is beneath the western part of Bermuda (James and Schenk, 1983, p. 5)

Varying dips in the Pleistocene limestone strata reflect depositional modes rather than structural deformation. The shallow-water marine limestones dip gently toward the sea, away from their original strandlines. The eolianites, which are much more prevalent than the marine limestones, represent subaerial carbonate dune deposits. Their most prominent feature is high-angle landward-dipping foreset beds, while their less distinguishable topset beds dip gently seaward with numerous scour-and-fill structures. Fossil soil horizons occur at intervals within the limestone formations, marking periods when the land surface was exposed to subaerial weathering.

DESCRIPTION OF COASTLINES

A comparison of the southwestern and northeastern ends of Bermuda reveals a marked contrast in the coastal outlines. As shown in Figure 2, the southwestern end is characterized by a curvilinear outline suggestive of sand hook formation by wave and current action. The northeastern end, however, demonstrates definite angular and linear trends. From Coney Island to Whalebone Bay (Fig. 3), the general trend of the coast is north-northeastward. From a point slightly northeast of Whalebone Bay to just before St. Catherine Point, at the extreme northeastern end of St. George's Island, the coast—except for minor irregularities—is remarkable for its consistent northeasterly trend, as well as for its nearly continuous line of sea cliffs (Public Works Department, Bermuda, 1984). Just south of St. Catherine Point the cliffed coastline changes its trend to the southeast, down to its end at Town Cut (see Fig. 6). Paget Island continues this southeasterly trend, although the sea cliffs are not in evidence here. The effect of these intersecting coastline trends is to give a distinctive angularity, or corner-like appearance, to the northeastern end of Bermuda in contrast to the curvilinear appearance of the southwestern end.

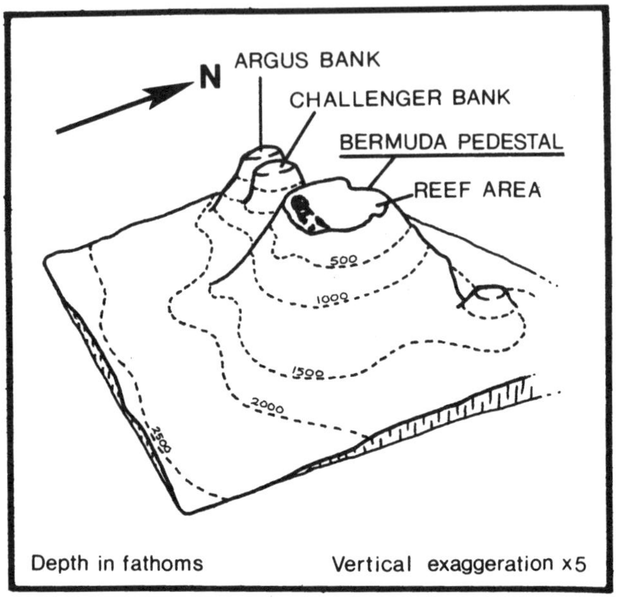

Figure 1. Bermuda Pedestal and adjacent sea floor. (After James and Schenk, 1983.)

FRACTURE SYSTEMS

Most of the fractures observed had no discernible displacement. The general term "fracture" as used in this chapter includes both joints and the relatively few observed faults.

Coney Island to Whalebone Bay

Personal inspection of the fractures was made by J. K. Hartsock along the northwest coastal area from the point of land just south of Coney Island to a short distance beyond Whalebone Bay on St. George's Island. The area surrounding Whalebone Bay was studied briefly by Hartsock, and in much greater detail by Woodrow and McKinney. Aerial photographs (Canadian Forces Liaison Officer, Bermuda, 1986) were used by Hartsock to supplement his observations and tentatively extend them northeastward along St. George's Island to a point

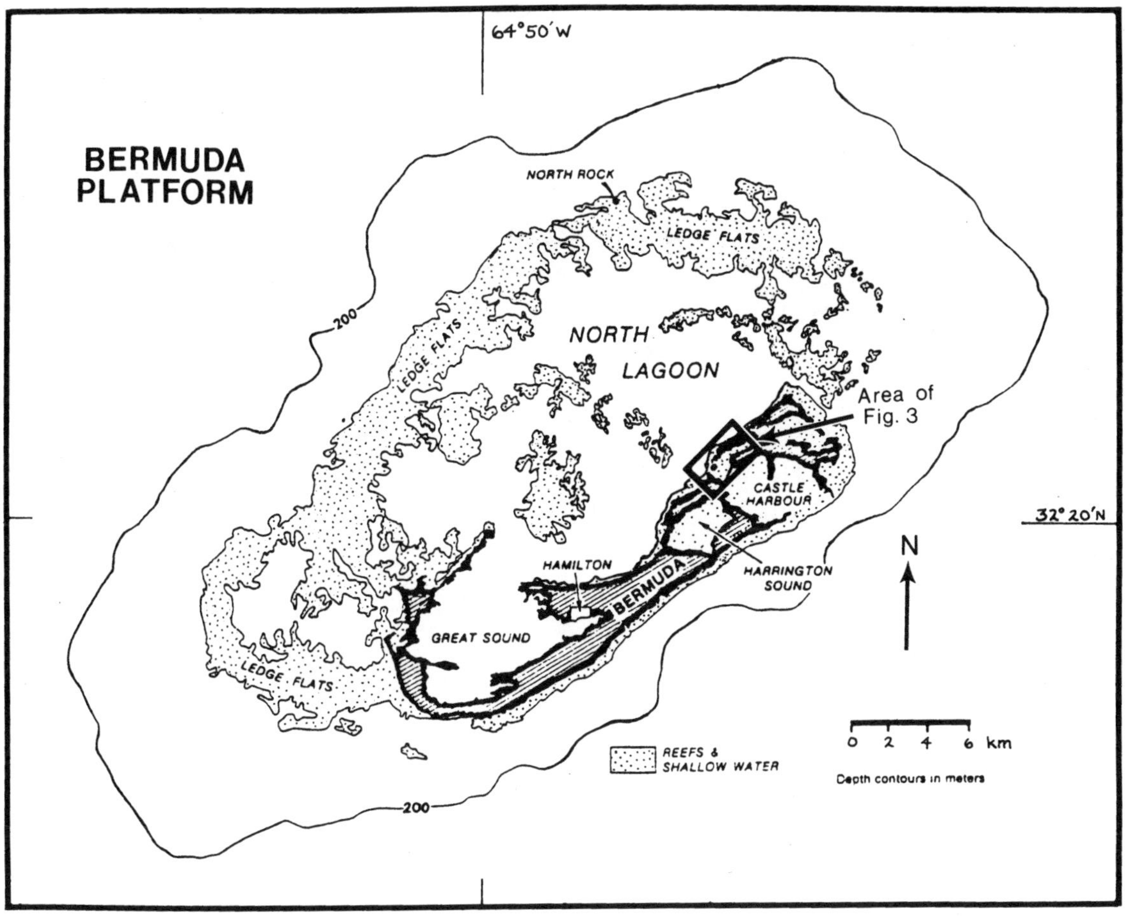

Figure 2. Bermuda Platform capping the Bermuda Pedestal. (After James and Schenk, 1983.)

opposite the Bermuda Biological Station. The following description summarizes Hartsock's measurements of the orientations of 10 fractures with Brunton compass and 26 fractures from aerial photographs. Details of these features are shown in Figures 3 and 4.

On the point of land south of Coney Island, a dominant set of well-exposed fractures occurs in the compact limestone forming the point. They have a general strike slightly east of north, the same as on Coney Island.

This same strike occurs in the dominant set of fractures on Ferry Point and can be traced northeastward past Whalebone Bay about 200 m. Here, the strike shifts abruptly from north-northeast to northeast. The coastline likewise shifts its trend and maintains a northeasterly trend nearly to the northeastern corner of St. George's Island, just short of St. Catherine Point.

The bottom vegetation zones offshore also demonstrate this same shift in trend from north-northeast to northeast, probably reflecting the control of fractures in the bedrock of the seabottom here.

A subordinate set of fractures, striking for the most part at right angles to the dominant set, can be observed on Coney Island, on Ferry Point, around Whalebone Bay, and to the northeast of Whalebone Bay.

The fractures on Ferry Point are particularly well exposed. They occur as zones of anastomosing, vertical to steeply dipping open joints, frequently bordered by prominent belts of redeposited calcite (Fig. 4).

Whalebone Bay

In 1984 and 1986, Woodrow and McKinney measured the orientations of 101 fractures around Whalebone bay and presented the results graphically (Fig. 5). The purpose of these investigations was to provide data to test the hypothesis that the configuration of the bay is structurally controlled. The test was made by visual comparison of joint data and map geometry of the bay.

The fractures occur throughout the rock column here as vertical to steeply dipping joints widened by solution and then partly refilled with calcite. On weathered rock surfaces, both at the bay shore and away from it, fractures usually are expressed geomorphically as narrow, sharp ridges of resistant calcite, with central open fissures, rising 1 to 10 cm above the bare rock. Where soil is developed, the calcite ridges may jut through the grass. Individual ridges can be traced as distinct features for tens or even a few hundreds of meters. Fracture traces also are

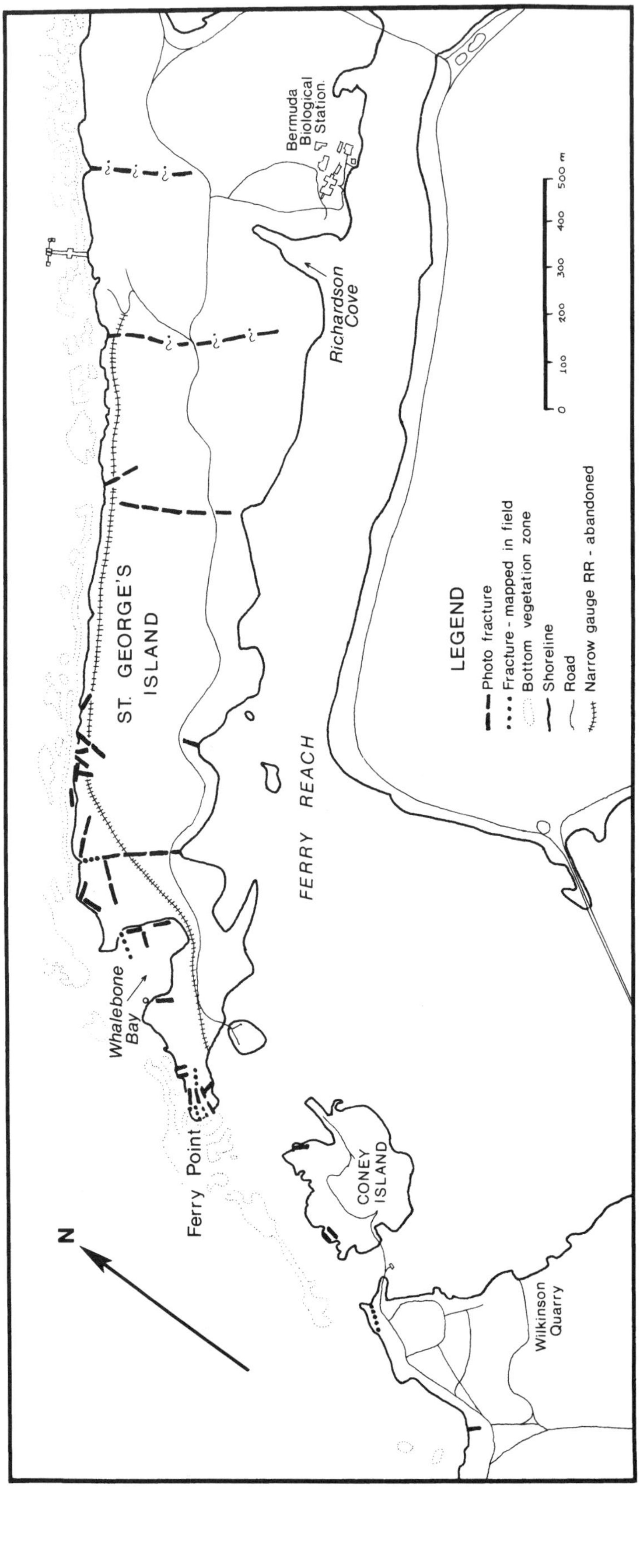

Figure 3. Fractures on Coney Island and southwestern end of St. George's Island. (Shorelines and cultural features adapted from Canadian Air Force aerial photographs.)

Figure 4. Open joints on Ferry Point. Photograph shows central opening and prominent zones of secondary calcite deposited along walls of dominant joint. Smaller subordinate joint crosses the dominant joint at approximately right angles. Pencil (lower part of photograph) = 13 cm long.

visible on the widely scattered patches of exposed rock of the bay floor as indistinct grooves a centimeter or two wide, extending for a few meters. Most fractures are not straight but curve gently, forming arcuate features convex to the northwest.

Orientations were measured with a Silva compass for all the fractures intersected on a traverse line that circled the bay, set back from its shore by about 10 m. The west side of the bay is marked by a wide rock expanse, and all fractures seen on it were also measured. Orientations of fractures on the bay floor were measured with a diver's wrist compass.

In Figure 5 the small rose diagrams A through D illustrate the wide variation in orientations from the general trend accounted for by the arcuate curvature of the fractures. The large rose diagram E incorporates all the data from A through D, and demonstrates the existence of a primary, dominant set of fractures oriented in a generally northeast-southwest direction, with secondary, subordinate sets at large angles to this.

Remainder of St. George's Island and Paget Island

No fractures were mapped by the authors northeast of the Bermuda Biological Station. However, the consistent linear trend of the northwest coastline from this point to Tobacco Bay, just southwest of St. Catherine Point (see Fig. 6), is interpreted as evidence of fracture control of coastal erosion here.

Likewise, while no fractures were mapped along the coast southeast of St. Catherine Point, the linear trend to the southeast of the cliffed coastline here (Public Works Department, Hamilton, Bermuda, 1984, Sheets 5/3, 5/7, and 5/11) is interpreted as control of erosion of the coast by northwest-southeast–striking fractures. These would be parallel to the northwest-southeast–striking, or subordinate, set of fractures mapped on the western half of St. George's Island.

The general southeasterly trend of the northeast coastline of Paget Island may also be due to control of coastal erosion by the same set of fractures, although the sea cliffs are not in evidence here.

RELATIONSHIP TO CASTLE HARBOUR CALDERA

The angular corner formed by the northwestern and northeastern coastlines of St. George's Island partially encloses St. George's Harbor and, south of that, the quasi-circular water body of Castle Harbor (Fig. 6). The slight shift in coastal trend from north-northeast to northeast near Whalebone Bay parallels the general curvature of Castle Harbor in this sector.

A continuous air-gun seismic reflection survey of Castle Harbour (Gees and Medioli, 1970, Pt. I) revealed the presence of a volcanic rock surface, varying from about 100 ft (about 30 m) below sea level around the edges of Castle Harbour to about 300 ft (about 95 m) below sea level near the center. This volcanic subbottom is overlain by a sequence of limestones varying from several tens of feet (5 to 10 m) in thickness along the edge of Castle Harbour to over 200 ft (over 60 m) in the center (Fig. 7). This bowl-like configuration of volcanic subbottom was interpreted by Gees and Medioli to be the remains of an ancient volcanic caldera that predated the deposition of the sedimentary strata.

MODES OF FRACTURE ORIGIN

The geologic literature has few references to fracture systems on limestone island groups such as Bermuda. A comparison of Bermuda's fracture systems with those in similar island settings elsewhere sheds little light on the origin of Bermuda's fractures.

In the Marshall Islands of the Western Pacific, Emery et al. (1954, p 162) described cracks in the reef flat of Bikini atoll that "parallel the trend of the reef for distances of several hundred feet," and showed a photograph of one such crack. These were considered to be tension cracks resulting from the compaction of reef debris. In none of the other Marshall Islands described by Emery were cracks or rock fractures of any sort mentioned. However, similar features were referred to on Funafuti, an atoll in the Ellice Islands, some distance southeast of the Marshall Islands.

Tracey et al. (1964) described may well-developed systems of fractures in the extensive limestone areas of Guam, in the Mariana Islands. These were directly related to faulting, which is consistent with Guam's active tectonic history. This is true also for Tinian and Saipan, likewise located in the Marianas arc (D. B. Doan, 1992, personal communication).

Neither the examples from Bikini nor those from Guam, Tinian, and Saipan shed light on the origin of Bermuda's fracture systems. Since the cracks described for Bikini all occur on an active reef flat, the situation is not comparable to that of Bermuda. In the cases of Guam, Tinian, and Saipan, the rock

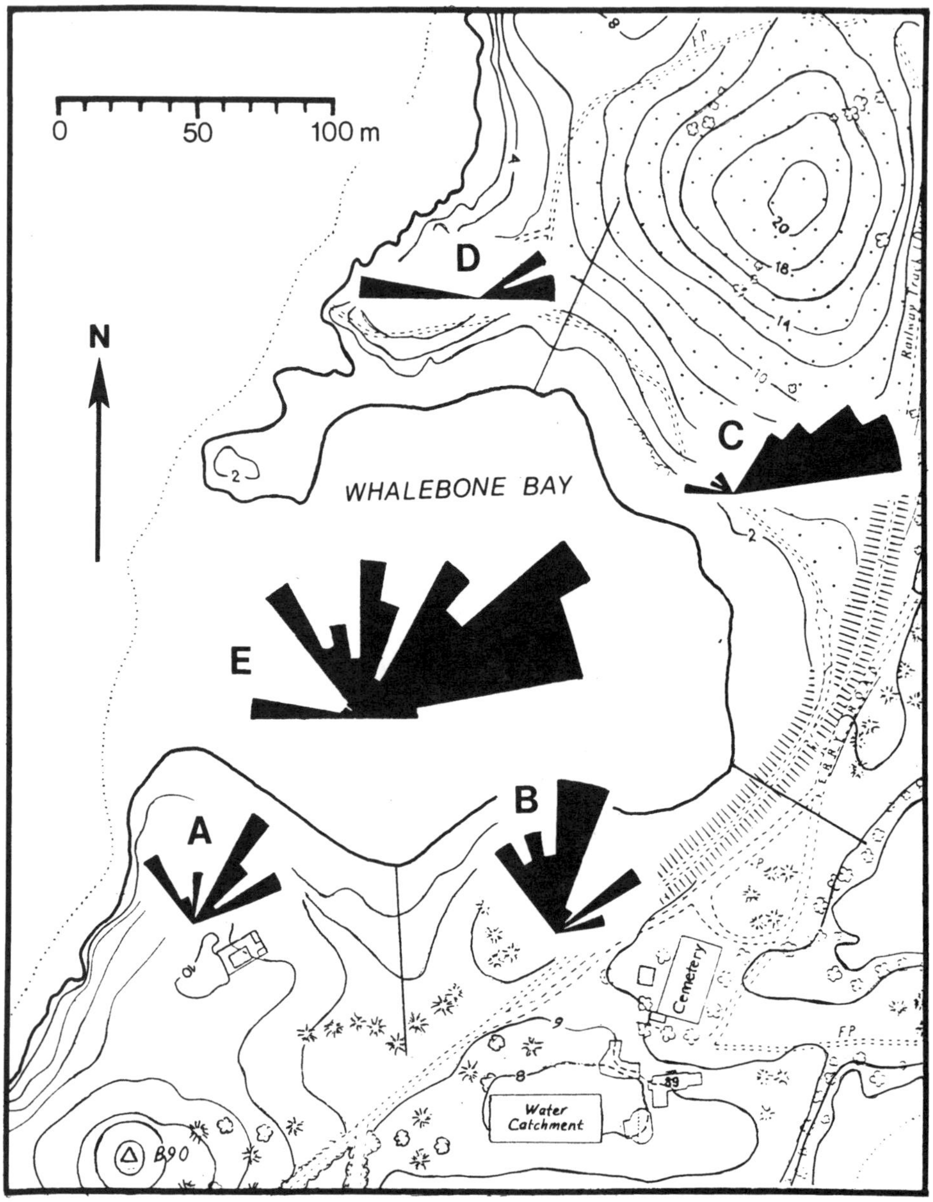

Figure 5. Rose diagrams from fracture orientation data around Whalebone Bay gathered by Woodrow and McKinney field parties. Small diagram A is based on measurements of 20 fractures; diagram B, 34 fractures; diagram C, 31 fractures; and diagram D, 16 fractures. The large diagram, E, incorporates all 101 measurements from diagrams A, B, C, and D.

fractures are directly related to tectonic activity of an intensity far too great to permit comparison with Bermuda.

In the Bahama Islands, Aby et al. (1992) conducted an analysis of over 800 fracture orientations on Lee Stocking Island, located near the edge of the bank bordering Exuma Sound, and related their development primarily to sea-level lowering during the last glacial stage. They postulated that tensile stress, initiated by release of hydrostatic confining pressures when steep, subaerial cliffs on the bank margin were exposed and undermined by wave action from the lowered sea, caused fracturing subparallel to the bank margin.

This factor is not believed to have played a significant role in causing the fractures in northeastern Bermuda since the comparable edge of the bank there, with oceanic depths beyond, does not lie near the coastline as at Lee Stocking Island. Instead, it lies beyond the ledge flats bounding North Lagoon, over 10 km north of the coastline, as shown in Figure 2.

Smart and Whitaker (1988) mentioned cavernous porosity, which was preferentially developed along major fractures paralleling the bank margin on South Andros and Grand Bahama, Bahamas. They did not, however, postulate how the fractures may have been formed.

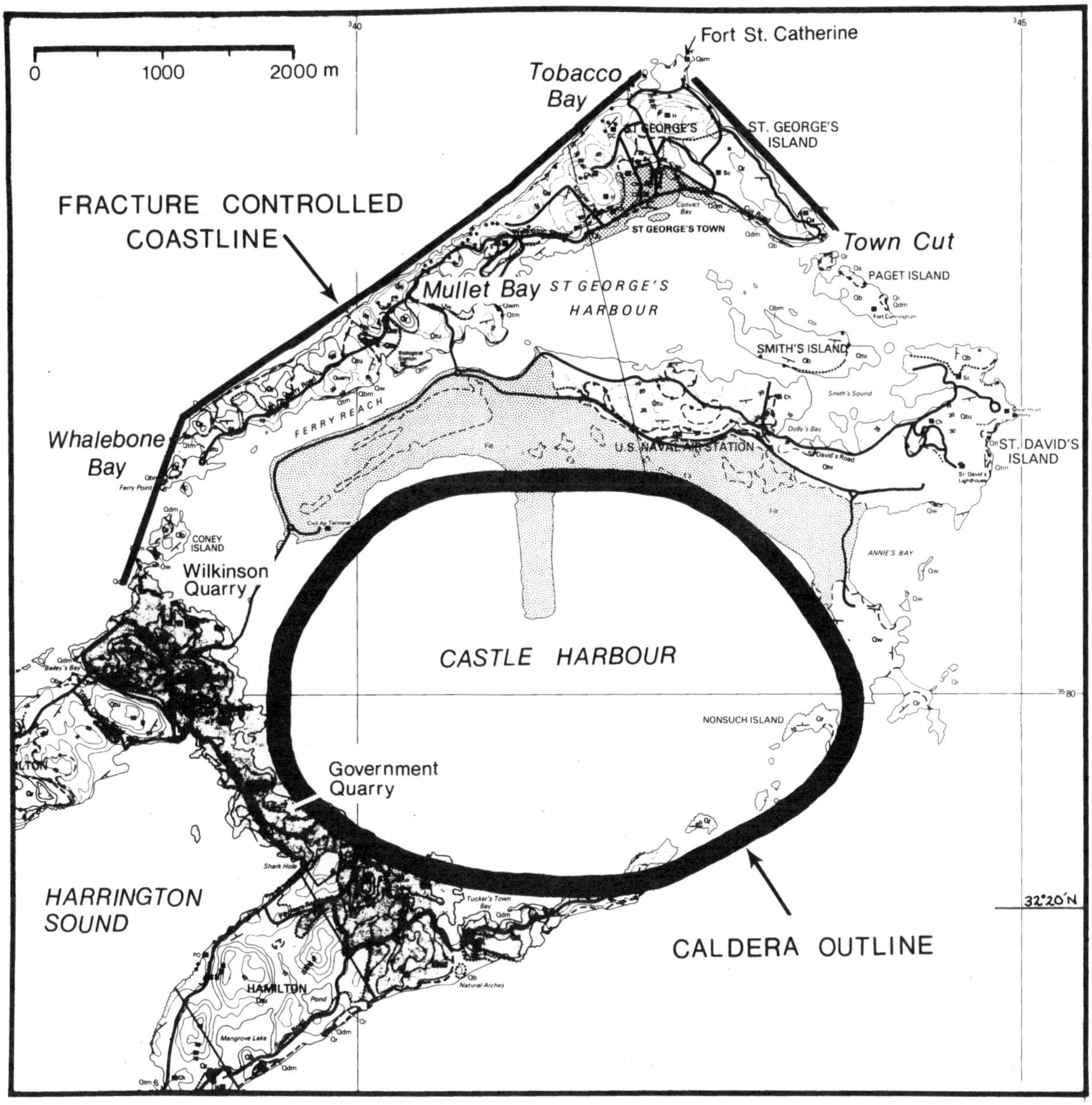

Figure 6. Relationship of fracture-controlled coastlines on Prince George's Island to Castle Harbour Caldera. Approximate outline of caldera at about 30 m below sea level is shown. (Map is adapted from Vacher et al., 1989.)

Scheidegger (1976) made a statistical study of joint orientation in Bermuda. He interpreted the joint trends as reflecting the geotectonic stress field and stated that the main stress orientations in Bermuda are a north-south compression and an east-west stress relief, with the joint orientations running at approximately 45° to these directions. These are related to "movement of the North Atlantic tectonic plate from the Mid-Atlantic ridge towards North America. The joints would . . . correspond to faults . . . caused by the stresses present in the moving plate" (Scheidegger, 1976, p. 7).

Scheidegger's statement applies to the general orientation of joints over most of Bermuda and explains the general prevalence of northeast- and northwest-striking joints. It does not, however, account for the intensity of fracturing in the Ferry Point–Whalebone Bay area nor for the conformance of fracture strikes to part of the curvature of Castle Harbour.

Cavern collapse

Bretz (1960) described 10 caverns in line along the Castle Harbour–Harrington Sound isthmus and its northward extension (Fig. 6). Most of these showed some indication of fracture openings leading from the surface to solutional caverns at depth.

M. P. Rowe, (1991, written communication) described

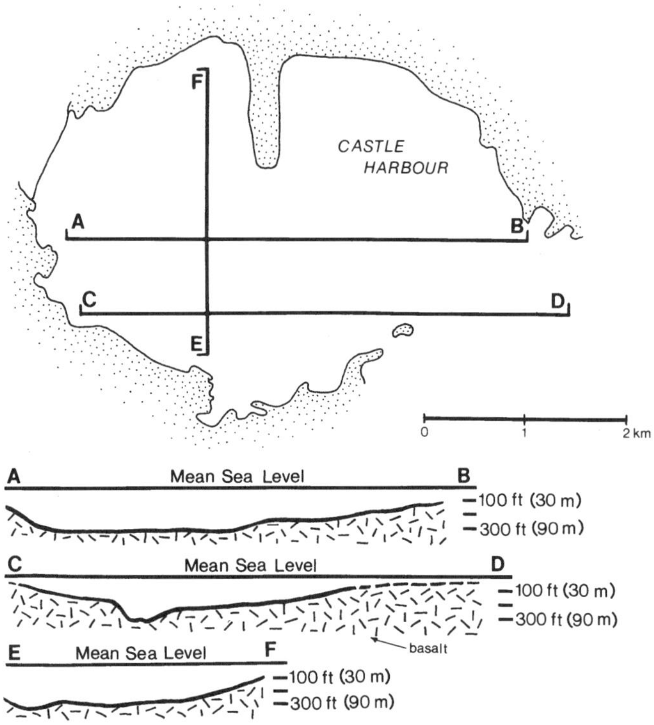

Figure 7. Simplified cross sections of basalt-sediments interface beneath Castle Harbour, with sediments removed. (After Gees and Medioli, 1970, Pt. 1.)

north-striking reverse faults at Government Quarry (Fig. 6), on the isthmus, associated with a large dome-shaped block of limestone separated at the top and on both sides from the remainder of the outcrop. This block of limestone was believed to have dropped as a result of cavern collapse at a deeper level, possibly near the limestone-basalt contact. Additionally, the eastern section of Harrington Sound itself was believed to be a product of cavern collapse (Gees and Medioli, 1970, Pt. II; Rowe, 1990).

Again, Bretz (1960, p. 1745) described cave walls showing fracture patterns at Wilkinson Quarry, a short distance south of Coney Island (Figs. 3 and 6). In addition, the quarry showed "three very marked fissures parallel to each other and to the near-by shore line. . . . Each is steeply inclined shoreward." He also described a fault, parallel to the above fissures, with the downthrown block dropped about 10 ft (about 3 m) on the shoreward side of the quarry: "Rock in the downdropped block is considerably fractured and constitutes a founder breccia."

Rowe (1991, written communication) also described a reverse fault at Wilkinson Quarry with one slip surface that has opened up, indicating association with cavern collapse. The displacement is about 0.75 m and the strike of the fault is approximately NE-SW. The dip of this fault is over 70°. In a later written communication (1992), Rowe also mentioned that there are a number of caverns in the vicinity of Wilkinson Quarry, both to the southeast and to the northwest, adjacent to the Coney Island area.

The apparent cause for all these fracture systems was collapse of the overlying limestone strata into previously existing large, deep caverns believed to have been formed (Rowe, 1990) by concentrated ground water flow at depth. It is probable that this ground water flow was guided by the sloping outer flanks of the Castle Harbour caldera at or near the limestone-basalt interface. The lowered sea-level datum during the last glacial stage, with resultant lowered water table and improved drainage gradient, would have enhanced this process.

Invoking this explanation of deep cavern collapse also could provide a rationale for the formation of the fracture systems from Coney Island past Whalebone Bay and along the entire length of St. George's Island. With a family of such caverns at roughly the same base level partially encircling the caldera on the west, north, and northeast, the configuration of the caldera should be a potent factor in controlling the loci of cavern collapses and patterns of resulting fractures.

Thus far no such caverns have been found at depth beneath Coney Island or St. George's Island. However, the generally cavernous nature of the oldest rock unit (Walsingham Formation) exposed on Bermuda suggests that caverns could be expected on Coney Island and St. George's Island wherever this rock unit is developed.

A scrutiny of the Public Works Department topographic sheets reveals numerous enclosed depressions and conical hills on both islands, indicating either development of karst topography resulting from cavern collapse or possibly development of dune and swale topography resulting from eolian processes.

The geometry of the shoreline of Whalebone Bay suggests that, at least in part, the configuration of the bay is controlled by fracture patterns, with the fractures themselves being produced by solution-induced subsidence in the underlying limestones.

Bretz (1960, p. 1744) interpreted Richardson Cove on Ferry Reach (Fig. 3) as "another shore-line indentation caused by cave failure. The evidence consists of a surviving small cave opening on its slopes, a number of excellent showings of cave-onyx deposits, and some stalactites."

Mullet Bay, the largest shoreline indentation into the south shore of St. George's Island (Fig. 6), may likewise be interpreted to have resulted from cavern collapse.

Finally, M. P. Rowe (1991, written communication) described a small reverse fault on the coastline north of Whalebone Bay, striking to the northeast with displacement of about 20 cm and showing drag features in a narrow zone of fault breccia. By inference and comparison with the faults at Government Quarry and Wilkinson Quarry to the west of Castle Harbour, this fault may also be interpreted as the result of cavern collapse.

The divergence of the northwestern coastline of St. George's Island from the curving outline of Castle Harbour (Fig. 6) could indicate a greatly decreased angle of dip on the buried volcanic surface in this quarter. This would result in the loci of deep cavern formation, collapse, and fracture formation being situated farther out from the caldera rim than in the case of the deep caverns to the west, where proximity to Castle Harbour implies a much more steeply dipping western flank to the caldera.

Earthquakes

In the summer of 1988 two earthquakes, magnitudes 5.1 and 5.3 on the Richter scale, were experienced in Bermuda (J. Filson, 1991, personal communication). Their epicenters were located on the sea floor, on the Bermuda Rise near the base of the Bermuda Pedestal. In 1965, an earthquake of comparable intensity originated from the sea floor north of Bermuda. It can be assumed that other shocks have occurred in the vicinity throughout the past. Such earthquakes could be expected to contribute to the process of cavern collapse and resulting fracture formation, especially during a period of lowered sea level with resultant water-table lowering and caverns thereby emptied of water. Origin of the earthquakes may be attributed to crustal stress engendered by movements of the North American tectonic plate westward from the Mid-Atlantic Ridge, as inferred by Scheidegger (1976).

Quaternary volcanic activity

The Castle Harbour caldera presumably could have originated from roof collapse of the volcano during the episode of Oligocene sill intrusion. Although original formation of the caldera predated the deposition of Pleistocene sediments, the possibility of later minor volcanic activity should be considered.

No direct evidence of Pleistocene or Recent volcanic activity, such as dikes or sills intruded into the limestone, has yet been recognized on Bermuda. However, abundant indirect indications of this possibility exist in the sedimentary strata. Volcanic cobbles have been found in the basal limestones quarried in Government Quarry. Grains of volcanic material (obsidian, magnetite, apatite, garnet, goethite) are incorporated in the somewhat younger limestones outcropping on Whalebone Bay, as well as in the present-day beach sands (James and Schenk, 1983).

All of the above indications admittedly could be the result of sedimentary processes. Wave erosion of the volcanic surface by the advancing Pleistocene sea would indeed have produced volcanic gravel and sand that could become incorporated into the basal limestone sediments. These in turn could then be eroded and redeposited into progressively younger sediments, up to the present-day beach sands.

This does not, however, remove the possibility of as yet undiscovered small dikes or sills having been intruded into the limestone strata of the islands. The disruptive effects to the sedimentary beds of such minor intrusions should therefore be considered as one possible factor playing a part in the origin of the rock fractures under discussion. Further investigation, however, would be necessary to explore this possibility.

Basement faulting

The presence of a submerged wave-eroded terrace around the entire Bermuda Platform at a depth of about 35 fathoms (64 m), developed in an interval of stillstand during the post-Pleistocene sea-level rise, would seem to preclude the likelihood of any major fault cutting through the volcanic pedestal and displacing the platform after the terrace was eroded (Stanley and Swift, 1968). This does not, however, eliminate the possibility of such fault displacement of the platform before the end of the Pleistocene Epoch or the possibility of minor fault displacement, then or later, in the volcanic basement within the confines of the platform and circumscribing terrace. Further investigations, however, would be required to explore these possibilities and their effects in producing fracture systems in the overlying sedimentary strata.

CONCLUSIONS

The major cause of the unique fracture systems in northeastern Bermuda is collapse of the limestone strata into previously existing large, deep caverns. These caverns are believed to have been formed by concentrated ground water flow at or near the limestone-basalt interface, guided by the sloping outer flanks of the Castle Harbour caldera on the west, north, and northeast sides. Lowered sea-level datum during the last glacial stage, with resultant improved drainage gradient, could be expected to enhance this process. The occurrence of occasional earthquakes in Bermuda also would contribute to cavern collapse and resulting fracture formation. With a family of such deep caverns at roughly the same base level partially encircling the caldera, the configuration of the quasi-circular caldera would thus be a potent factor in controlling the loci of cavern collapse and the patterns of resulting fracture systems.

Other possible causes for fracture formation could be intrusion of small volcanic dikes and/or sills into the limestone strata and fault displacement in the volcanic basement beneath Bermuda. Further investigations, however, would be required to properly evaluate these possibilities.

ACKNOWLEDGMENTS

The authors are much indebted to Mark P. Rowe, hydrogeologist, Ministry of Works and Engineering, Hamilton, Bermuda, for his invaluable assistance in pointing out many of the geologic features discussed here, and in supplying information concerning general geologic details in Bermuda. Thanks are also due to David B. Doan, Division of International Minerals, U.S. Bureau of Mines, Washington, D.C., for his information on the geology of Tinian and Saipan, in the Marianas island arc, and to John Filson, Geologic Division, Branch of Global Seismology and Geomagnetism, U.S. Geological Survey, Reston, Virginia, for his information concerning earthquakes in the vicinity of Bermuda. Thanks are likewise due to Antonio V. Segovia, Department of Geology, University of Maryland, College Park, Maryland, and to Jan Kutina, Department of Chemistry, American University, Washington, D.C., who reviewed the manuscript and offered many helpful comments and suggestions. The Canadian Air Force provided vertical aerial photographs of the western end of St. George's Island and Coney Island, while the Bermuda Department of Tourism supplied numerous oblique aerial photographic slides that were useful in supplementing the vertical photographs. The authors are also indebted to the staff of the Bermuda Biological Station for Research for their logistical support.

REFERENCES CITED

Aby, S. B., Aalto, K. R., and Dill, R. F., 1992, Origin and significance of filled fractures along a carbonate bank margin, Lee Stocking Island, Exumas, Bahamas: Geological Society of America Abstracts with Programs, Cordilleran Section, p. 1.

Bretz, J. H., 1960, Bermuda: a partially drowned, late mature, Pleistocene karst: Geological Society of America Bulletin, v. 71, p. 1729–1754.

Canadian Forces Liaison Officer, Bermuda, and 415 (MP) Squadron, CFB Greenwood, Nova Scotia, Canada, 1986: Black and white vertical aerial photographs 1234C–1266C, approximate scale 1:8,000.

Emery, K. O., Tracey, J. I., and Ladd, H. S., 1954, Geology of Bikini and nearby atolls: Pt. 1. Geology: U.S. Geological Survey Professional Paper 260-A, p. 1–265.

Gees, R. A., and Medioli, F., 1970, A continuous seismic survey of the Bermuda Platform, Pt. I: Castle Harbour: Maritime Sediments, v. 6, p. 21–25. Pt. II: Harrington Sound: Maritime Sediments, v. 6, p. 118–120.

Hartsock, J. K., 1991, Fracture systems in Bermuda: Geological Society of America Abstracts with Programs, Northeastern and Southeastern Sections, v. 23, p. 42.

James, N. P., and Schenk, P. E., 1983, Field guide to Pleistocene and modern carbonates of Bermuda: Bermuda Biological Station for Research Special Publication 25, 72 p.

Public Works Department, Hamilton, Bermuda, 1984, Topographic sheets 5/2, 5/3, 5/6, 5/7, 5/9, 5/10, 5/11; Series E8111 (BDA 111), scale 1:2,500.

Rowe, M. P., 1990, An explanation of the geology of Bermuda (written with reference to the Geological Map of Bermuda, 1989): Hamilton, Bermuda, Ministry of Works and Engineering, 28 p.

Scheidegger, A. E., 1976, Joints on Bermuda: Rivista Italiana Di Geofisica e Scienze Affini, v. III, p. 101–105.

Smart, P. L., and Whitaker, F. F., 1988, Controls on the rate and distribution of carbonate bedrock dissolution in the Bahamas: *in* Mylroie, J., ed., Proceedings, Fourth Symposium on the Geology of the Bahamas: San Salvador, Bahamian Field Station, p. 313–321.

Stanley, D. J., and Swift, D. J. P., 1968, Bermuda's reef-front platform: Bathymetry and significance: Marine Geology, v. 6, p. 429–500.

Tracey, J. I., Jr., Schlanger, S. O., Stark, J. T., Doan, D. B., and May, H. G., 1964, General geology of Guam: U.S. Geological Survey Professional Paper 403-A, p. A1–A104.

Vacher, H. L, Rowe, M. P., and Garrett, P., compilers, 1989, Geological map of Bermuda: Hamilton, Bermuda, Ministry of Works and Engineering, scale 1:25,000.

MANUSCRIPT ACCEPTED BY THE SOCIETY JANUARY 5, 1995

Printed in U.S.A.

Index

[Italic page numbers indicate major references]

H

I

J

K

L

T

U

Typeset and printed in U.S.A. by Johnson Printing, Boulder, Colorado, U.S.A.